Urban Operations

Plus

Military Mountaineering

Department of Defense

Pentagon
Publishing

Field Manual
No. 3-06

Headquarters
Department of the Army
Washington, DC, 26 October 2006

Urban Operations

Contents

Distribution Restriction: Approved for public release; distribution is unlimited.

***This publication supersedes FM 3-06, 1 June 2003.**

Figures

Historical Vignettes

Preface

Doctrine provides a military organization with a common philosophy, a language, a purpose, and unity of effort. Rather than establishing a set of hard and fast rules, the objective of doctrine is to foster initiative and creative thinking. To this end, FM 3-06 discusses major Army operations in an urban environment. This environment, consisting of complex terrain, a concentrated population, and an infrastructure of systems, is an operational environment in which Army forces will operate. In the future, it may be the predominant operational environment. Each urban operation is unique and will differ because of the multitude of combinations presented by the threat, the urban area itself, the major operation of which it may be part (or the focus), and the fluidity of societal and geopolitical considerations. Therefore, there will always exist an innate tension between Army doctrine, the actual context of the urban operation, and future realities. Commanders must strike the proper balance between maintaining the capability to respond to current threats and preparing for future challenges.

PURPOSE

This manual provides the analytical tools for evaluating an urban operation to determine if the operation is necessary for overall mission success. It also provides the means to understanding and determining the impacts of the urban environment on military operations and provides information on managing, taking advantage of, and mitigating the effects of those impacts as appropriate. As such, this manual demonstrates how to apply the doctrinal principles in FM 3-0 to this unique environment.

SCOPE

Chapter 1 introduces theoretical and historical perspectives of urban operations that serve as the underlying basis for the rest of the manual. While this manual has incorporated lessons learned from recent and ongoing operations, it has deliberately taken a broad and varied historical perspective in order to remain relevant to future threats and circumstance differing from those that the Army currently faces. Chapter 2 discusses the characteristics of urban centers and populations as well as their impact on operations. It is unlikely that Army forces will ever operate in a benign urban environment; therefore, Chapter 3 discusses the varied nature of potential urban threats. Chapter 4 describes the effects of the urban environment on warfighting functions and tactics. An understanding of the complexities of the urban environment, the nature of the enemy as an adaptive, learning opponent, and the effects of the environment on warfighting capabilities and skills is essential to sound decision making. Chapters 5 and 6 discuss the potential costs of urban operations and risk reduction measures that the commander and his staff must consider early in their planning. These chapters also outline an urban operational framework and specific urban considerations that create the foundations necessary for successfully applying operational doctrine to an urban environment.

The second part of the manual (Chapters 7 – 10) discusses how urban operations are conducted and resourced. Urban operations include major offensive and defensive operations in urban environments as well as stability or civil support operations ranging from peace operations and counterterrorism to disaster relief and humanitarian assistance. For the different types of operations—offense, defense, and stability or civil support—the purpose, characteristics, organization, and considerations are discussed. However, commanders consider that most urban operations will normally involve the simultaneous execution of offense and defense with stability or civil support (although proportional emphasis will shift over time) and plan accordingly.

APPLICABILITY

This manual is intended for Army commanders and their staffs at the brigade level and above. It addresses full spectrum operations that Army units will execute in urban settings. However, users should also consult JP 3-06 for specific joint information. Additionally, users should be familiar with FM 3-06.1, FM 3-06.11, TC 90-1, and urban operations chapters, appendices, or sections found in other infantry, armor, combined arms, and proponent field manuals for the tactics, techniques, and procedures (TTP) and appropriate proponent information necessary to conduct tactical urban operations at the brigade level and below. This publication applies to the Active Army, the Army National Guard (ARNG)/Army National Guard of the United States (ARNGUS), and United States Army Reserve unless otherwise stated.

ADMINISTRATIVE INSTRUCTIONS

When this FM 3-06 was approved, doctrine was incomplete for incorporating stability operations and civil support operations in place of stability operations and support operations. This manual incorporates stability and reconstruction and civil support operations doctrine found in FM 1 and FMI 5-0.1; it will be revised to reflect the future updates of FM 3-0 and FM 3-07 and, when published, FM 3-28. (Note: Stability and reconstruction operations has been redesignated stability operations to comply with Department of Defense Directive 3000.05.)

Chapter 2 defines "city" according to a population size. However, in historical vignettes and accounts, the term "city" may be applied in its common usage without specific regard to size to maintain conformity with most other historical reports.

In this manual, the term "threat" is applied broadly to include an enemy force (conventional or unconventional), an armed belligerent in a peace operation, antagonistic or unfriendly elements of the civilian population, or some other hazardous condition in the urban environment that negatively influences mission accomplishment.

The term military operations on urban terrain (MOUT) is replaced by urban operations (UO). MOUT is an acronym from FM 90-10, which is superseded by this manual.

Otherwise, the glossary lists most terms used in FM 3-06 that have joint or Army definitions. Where Army and joint definitions are different, (Army) follows the term. Definitions for which FM 3-06 is the proponent manual (the authority) are marked with an asterisk (*). The proponent or amplifying manual for other terms is listed in parentheses after the definition.

The manual attempts to incorporate a broad range of historical vignettes into each chapter where the account supports the doctrinal line of reasoning. Two historical vignettes, however, were included as appendices (A and C) because of their longer lengths.

Unless this publication states otherwise, masculine nouns or pronouns do not refer exclusively to men.

The proponent for this publication is the United States Army Training and Doctrine Command. Send written comments and recommended changes on DA Form 2028 (Recommended Changes to Publications and Blank Forms) directly to Commander, U.S. Army Combined Arms Center and Fort Leavenworth, Combined Arms Doctrine Directorate, ATTN: ATZL-CD, (FM 3-06), 201 Reynolds Avenue, Fort Leavenworth, KS 66027-2337. Send comments and recommendations by e-mail to webb-cadd@leavenworth.army.mil. Follow the DA Form 2028 format or submit an electronic DA Form 2028.

This page intentionally left blank.

This page intentionally left blank.

Chapter 1
The Urban Outlook

Today's security environment demands more from Army leaders than ever before. Army leaders must not only be able to lead Soldiers but also influence other people. They must be able to work with members of other Services and governmental agencies. They must win the willing cooperation of multinational partners, both military and civilian. But ultimately, the Army demands self-aware and adaptive leaders who can compel enemies to surrender in war and master the circumstances facing them in peace. Victory and success depend on the effectiveness of these leaders' organizations. Developing effective organizations requires hard, realistic, and relevant training.

FM 1

Given the prevalence of large cities throughout the world, Army forces will likely be required to conduct operations in, around, and over large urban areas. These operations will normally be in support of a joint force commander (JFC) conducting military operations pursuant to U.S. national security policy. This manual is designed to facilitate the planning and conduct of the full range and spectrum of land operations in a complex urban environment. Each urban environment and urban operation is unique; prescribing a specific doctrinal "solution" for each situation is impossible. Instead, this manual provides a framework to commanders and their staffs for understanding the urban environment, for analyzing and deciding whether or not to initiate urban operations (UO), and for applying operational doctrine to this complex environment. It also provides a broad base of historical vignettes and examples to help develop a refined analytical perspective and stimulate thought. The manual also includes some planning points and tactics and techniques to assist in preparing for and conducting UO. This information provides a foundation for approaching major UO. Combined with other joint and Army doctrine, this information also will help commanders and their staffs learn to adapt and succeed in this challenging environment.

THE PROSPECT OF URBAN OPERATIONS

1-1. The world is undergoing massive urbanization. Although exceptions exist, an overall trend of migration from rural to urban areas is occurring throughout the globe. (Australia, one of the world's most urbanized countries, is actually becoming less urbanized.) This trend is especially evident in developing nations. Combined with the exponential growth of the global population in the last quarter century, this migration has created massive urban areas that hold the centers of population, government, and economics in their respective regions. In many cases, rapid urbanization has overburdened already weak infrastructures, scarce resources, and a fragile economic base. As urbanization has changed the demographic landscape, potential enemies recognize the inherent danger and complexity of this environment to the attacker, and may view it as their best chance to negate the technological and firepower advantages of modernized opponents. Given the global population trends and the likely strategies and tactics of future threats, Army forces will likely conduct operations in, around, and over urban areas—not as a matter of fate, but as a deliberate choice linked to national security objectives and strategy, and at a time, place, and method of the commander's choosing.

Army Urban Operations

Army forces conduct UO either as one component of a larger operation or as a single operation focused totally on a specific urban environment. Major Army UO are often part of a joint and multinational effort requiring rigorous interagency and civil-military coordination that typically includes the full spectrum of military operations. Commanders of Army major operations must determine if UO are essential to mission accomplishment. If so, commanders must carefully integrate the operations into campaign planning to support the operational objectives of the JFC.

Army leaders conducting UO must—

- Understand the urban environment to determine decisive points.
- Shape the operation to set the conditions for success.
- Precisely mass the effects of combat power to thoroughly engage the decisive points that lead to centers of gravity.
- Continually consolidate gains essential to the retention of the initiative.
- Transition the urban area to the control of another force or agency or, ultimately, back to legitimate and functioning civilian control.

AN URBAN PERSPECTIVE

1-2. As a subset of all Army operations, UO are operations focused on an urban environment. UO include full spectrum operations—offensive, defensive, and stability or civil support—that may be executed, either sequentially or (more likely) simultaneously, during the conduct of a single urban operation (see figure 1-1). UO may be the sole mission of the commander or one of several tasks nested in a larger operation. Regardless of the types of operations conducted or whether the urban area is the single focus of the operation or only one component of a larger operation or campaign, the complex urban environment significantly affects the overall conduct of the mission.

1-3. When conceptualizing urban operations, commanders must understand two important terms: urban area and urban environment. The first is a subset of the second. An ***urban area*** **is a topographical complex where man-made construction or high population density is the dominant feature.** Focusing on urban areas means concentrating on the physical aspects of the area and their effects on weapons, equipment, line-of-sight, and tactics, techniques, and procedures. The ***urban environment*** **includes the physical aspects of the urban area as well as the complex and dynamic interaction and relationships between its key components—the terrain (natural and man-made), the society, and the supporting infrastructure—as an overlapping and interdependent system of systems.**

1-4. Importantly, commanders must also understand and consider that critical elements of the infrastructure may lie far beyond the area's physical confines. For example, the generating source providing power to the urban energy system is part of that system but may be located well outside of the urban area. Similarly, effects of the interaction between components of the infrastructure, located both inside and outside the urban area, extend well into smaller, neighboring urban areas and surrounding rural areas and often form their political, economic, and cultural focus. Understanding the total urban environment is essential to planning and conducting full spectrum urban operations.

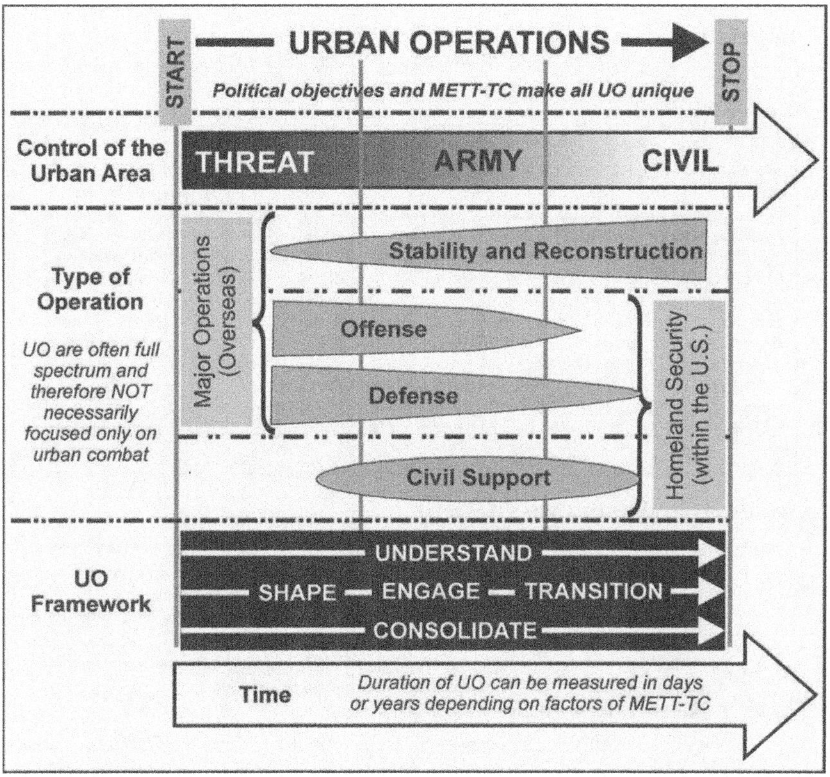

Figure 1-1. Full spectrum urban operations

HISTORICAL SIGNIFICANCE OF URBAN AREAS IN WARFARE

1-5. Urban areas always have been central to, or have significantly influenced, military operations. One of the first urban-centered battles was the siege of Troy at the beginning of Greek history. Moreover, much of the history of early Greece revolved around wars between its city-states or with Persia and centered on the conquest, siege, or blockade of cities. Five hundred years later, the Roman Empire replaced Greece as the dominant world power; although, urban areas remained central to the Roman method of warfare. Even Rome's history can be viewed as a microcosm of urban warfare over the past two thousand years. Though military operations within the physical confines of many of these historic urban areas were not the norm, the focus of these operations was their conquest or control.

1-6. Although Rome last saw combat in 1944, urban areas have been no less prominent in warfare since that time. Seoul in Korea, Beirut in Lebanon, Panama City in Panama, Grozny in Chechnya, Sarajevo in Bosnia-Herzegovina, Kabul in Afghanistan, and Baghdad in Iraq have been centers of conflict in the last 50 years. Urban areas, now more pervasive than ever before, will continue to be essential to successful operational and strategic warfighting. Today, armies cannot expect to execute major military operations without the influence of the urban environments within their area of operations.

Rome: A Microcosm of Urban Warfare

During two millennia, Rome has been the center of at least 12 battles. The Gauls lay siege to Rome first in 387 BC. That first siege lasted six months and ended after the barbarians burnt much of the city. The surviving patrician families paid a ransom for the withdrawal of Brennus' army. From 408 to 410 AD, the Goth leader, Alaric, successfully besieged Rome no less than three times. The Byzantine General Belisarius captured Rome twice from the Goths and withstood siege inside the city once between 536 and 549. Five hundred years later in 1084, Norman adventurer Robert Guiscard captured medieval Rome and sacked the city during a dispute between the Pope and the Holy Roman Empire. Forces of the Holy Roman Empire again stormed and captured the city to punish the Pope in 1527. During the Italian Revolution in 1849, a French army supporting the Pope captured the city from the Italian revolutionary army under Garibaldi. In 1944, the last military action took place in and around Rome when the U.S. Fifth Army captured the city from the retreating German army. Rome's turbulent history—fought over ethnic and religious differences, prestige, and military necessity—demonstrates the importance of urban areas in warfare and the various causes and combatants within this complex environment.

Strategic Importance of Urban Areas

1-7. Several reasons have attracted (and continue to attract) armies to combat in urban areas:

- A military force chooses to position itself in an urban area to capitalize on the perceived defensive advantages offered by the environment. In contrast, an opposing force, by analyzing the factors of the situation, determines that it must enter the urban area to attack and destroy its enemy (or devote essential combat power to their isolation).
- The urban environment's people (their allegiance and support), infrastructure, capabilities, or other resources have or can be of significant operational or strategic value.
- The urban area has significant symbolic importance.
- The urban area's geographical location dominates a region or avenue of approach.

1-8. Russia's 1994 experience in Chechnya illustrates an increasingly important motivation for conducting urban operations. The Chechen rebels, after failing to engage Russian forces outside the city, chose to turn Grozny into the main battlefield. Leaders of the defeated Chechen conventional forces recognized that fighting in the urban area provided them their best chance for success. The complexities of urban combat and the perceived advantages of defending an urban area mitigated their numerical and technological inferiority. The urban area provided the Chechens protection from fires, resources, interior lines, and covered and concealed positions and movement. Given such advantages offered by the environment, smaller or less-sophisticated military forces have similarly chosen to fight in urban areas either as a deliberate strategy or to escape certain destruction in open terrain.

1-9. Such advantages of operating in an urban environment also prompt forces to conduct an urban operation to facilitate a larger campaign plan and decisive battle in another location. The urban operation can focus the enemy on the urban area and allow other forces to conduct operations elsewhere. From a defensive perspective, an urban defense may gain time and space to reorganize forces in new defensive positions, to divert enemy forces from other critical tasks, or to prepare to conduct offensive operations. To some extent, these reasons motivated Soviet forces defending Leningrad and Stalingrad from the Germans in World War II. The stubborn defense permitted the Soviets to reorganize for later offensive operations. From an offensive perspective, an attack on an urban area may be a shaping operation used to divert enemy resources from the decisive operation that will follow.

1-10. Armies also fight in an urban area to obtain some critical feature or resource in the area, such as a port facility. The desire to control an important seaport and access to the Persian Gulf largely motivated the Iranian and Iraqi struggle for Basra in the 1980s. Earlier, in 1944, British forces fought German units in Arnhem for control of the Rhine River Bridge. Other key infrastructure of the urban environment may have operational or strategic significance and can compel military forces to attack or defend the area. As urban areas account for an increasing share of a country's national income, often generating over 50 percent of gross national product, the strategic implications for their control or influence become even greater.

1-11. Urban areas are often located on terrain that dominates a region or an avenue of approach. In these cases, offensive armies capture these areas to proceed with security to another objective. Conversely, defensive forces commonly defend the area to deny the area of operations. To illustrate, Cassino, Italy during World War II stood astride the critical highway approach up the Liri valley to Rome. The allies had to attack and capture the monastery to facilitate the allied offensive north. Cassino's location made bypassing virtually impossible. Likewise, in the early 1980's, Israeli army urban operations in Beirut were a result of its strategic location near the Israeli security zone; various Arab insurgent and terrorist groups used Beirut as a base for attacks against Israel. Beirut evolved as the major base of the Palestine Liberation Organization, a major opponent of Israel. Beirut's location made it a security threat to Israel and thus compelled several major Israeli operations in the urban area (see Appendix A).

1-12. Another reason for engaging in urban operations is the symbolic—historical, cultural, political, and even economic—importance of many urban areas. Often, capital cities—such as Rome, Paris, Seoul, Berlin, and Baghdad—are identified as the strategic centers of gravity of their respective nations. Possessing or threatening these urban areas may impact directly on the outcome of a conflict. The objective of Germany's wars with France in 1870 and 1914 was ultimately Paris. Napoleon's 1812 campaign had as its objective Moscow, as did Hitler's 1941 offensive into Russia. The objective of the Soviet 1945 offensive was Berlin, and the North Vietnamese 1975 offensive had as its objective the South's capital of Saigon. Still, history also reminds us that commanders must assess the sustainability and decisiveness of operations directed toward these "prestige" objectives. For example, in 1812, Napoleon captured Moscow but had to evacuate it within 30 days. He lacked supplies and shelter, failed to destroy the Russian Army, and failed to defeat the political will of the Czar and the people. Similarly, the North Korean occupation of Seoul during the Korean War was equally indecisive.

U.S. Army's Experience in Urban Operations

1-13. The U.S. Army has a varied history of conducting urban operations. The American Revolution saw the Army conduct several urban operations. These operations included the unsuccessful defense of New York, the successful attack on Trenton, and the decisive siege and attack on British forces at Yorktown. The Mexican War also had a successful assault on the fortified city of Monterey and the decisive siege of Mexico City. During the American Civil War, the armies, in the tradition of Napoleonic maneuver warfare, avoided urban areas and fought in the open. However, the opposing armies frequently made urban areas their objective because of their importance as railheads. Success in the siege of several key urban areas—Vicksburg, Atlanta, and Petersburg—contributed to the Northern victory.

1-14. Following the Civil War, the U.S. Army faced no large-scale urban combat for several generations. The Indian Wars, the Spanish-American War, the Philippine Insurrection, and even World War I did not require the Army to fight in large urban areas. Between the Civil War and World War II, the U.S. Army fought in several urban areas worldwide supporting U.S. commitments. These limited urban combat operations were small but essential parts of what are currently called urban stability operations. From 1900 to 1901, the Army provided public security for a sector of Peking, China of around 50,000 inhabitants. The Army conducted UO and, in the course of the operation, the 9th U.S. Infantry suffered 20-percent casualties while fighting in Tientsin. Punitive expeditions to places such as Siberia, Cuba, Philippines, Central America, and Mexico put the Army in various urban situations that required using military power, notably, the occupation and security of Vera Cruz, Mexico in 1914. In the context of these smaller-scale contingencies, UO became a staple of U.S. Army employment.

1-15. World War II forced the Army to grapple with the issues of large-scale urban combat almost immediately. In his 1941 defense of the Philippines, General MacArthur examined how to defend Manila.

Manila represented a large, modern, friendly urban area, which was the capital city of a close U.S. ally. Defending the urban area posed numerous challenges. Ultimately General MacArthur determined that he could best conduct its defense outside the city by defeating the enemy forces in combat on the invasion beaches or shortly after they landed. When Japanese forces defeated MacArthur's Philippine Army in a series of engagements, MacArthur had to decide how best to protect the friendly populace of Manila. He had two choices: abandoning the city or waging a costly defense that would likely result in the city's destruction, thousands of noncombatant casualties, and no operational advantage. He had little choice but to declare Manila an open city and move his forces to Bataan to wage an operational defense in the vain hope that a counteroffensive could relieve his isolated force. On 2 January 1942, Japanese forces entered Manila unopposed.

1-16. Had General MacArthur decided to defend Manila, his forces would have found scant doctrine in the Army regarding how to fight in an urban area. Doctrine for urban operations did not appear until early 1944, when faced with the possibility of fighting through the larger urban areas of Western Europe. At this time the U.S. Army published FM 31-50. This manual had the first formal discussion of how the Army viewed urban combat. It was based on the Army's limited experiences in the Mediterranean theater and the study of German and Soviet experiences on the Eastern front.

1-17. FM 31-50 emphasized a deliberate pace, individual and small unit initiative, the liberal use of direct and indirect firepower, and decentralized command and execution. It focused on the urban area (as opposed to the environment); however, it did include policies towards the noncombatants. The manual was also focused at the regimental combat team level. Complementing the doctrine of FM 31-50 was the 1944 operations manual, FM 100-5. This latter manual rightly emphasized the importance of combined arms actions and the need for extensive reconnaissance of prepared and defended cities. The Army successfully implemented this doctrine in several major instances of urban combat, most notably the capture of the first German city, Aachen, and hundreds of small-scale urban assaults on cities, towns, and villages across France, the Benelux, and Germany. Army forces also successfully employed this urban combat doctrine during the liberation of Manila in 1945.

1-18. The legacy of this era of Army operations was an effective tactical solution to urban offensive combat: isolate the urban area, seize a foothold, and expand the foothold block by block until occupying the entire urban area and destroying the enemy. The doctrine's emphasis on firepower kept friendly casualties to a minimum. Unfortunately, when enemy forces stoutly defended the urban area, the emphasis on firepower resulted in the area's virtual destruction and high casualties among noncombatants.

1-19. The doctrinal approach honed in World War II remained the accepted Army approach to urban combat to the century's end. The last successful implementation occurred during the Korean War and the final liberation of Seoul. The Vietnam conflict did not offer the Army opportunities or the requirement to practice urban combat or test and refine doctrine on a large scale. The largest urban battle, Hue, was a chaotic tactical battle that validated most of the historical lessons of urban combat without generating any new doctrinal insights for large-scale urban warfare.

1-20. From the mid-1950s through the 1990s, the Army conducted UO in the U.S. in support of civil authorities during civil unrest and anti-Vietnam protests. Some operations involved numerous active and Reserve Component forces engaged in restoring public order. The Detroit riots of 1967 and the Los Angeles riots of 1992 required the commitments of active and National Guard units. In 1968, the Army deployed over 35,000 troops to Washington D.C., Chicago, and Baltimore following the death of Dr. Martin Luther King, Jr.

1-21. In the 1970s and 1980s, Army doctrine predominantly focused on urban areas and successfully fighting a conventional ground war against Soviet and Warsaw Pact forces in Central Europe. FM 90-10 (1979) described how to conduct urban operations against Soviet forces in Germany. Its concepts were never tested other than in simulation, and its approach to urban combat was not substantially different from that practiced by the Army since World War II. Despite previous doctrine's admonition to avoid cities, the Army has had to fight in them in diverse circumstances.

1-22. In the 1990s and early 21st century, commanders and planners began to more carefully consider the impact of urban environments on the overall conduct of campaigns and major operations. Maneuver warfare characterized Operation Desert Storm in 1991; however, urban environments became prominent during subsequent operations in Somalia, the Balkans, and again in Iraq. (During Operation Iraqi Freedom in 2003, for example, Baghdad was considered as a strategic center of gravity for planning.) This was a result of adversaries recognizing the asymmetric advantages to be gained by fighting superior U.S. forces in an urban environment as well as a general worldwide trend of increasing urbanization. The evolving nature of Army operations in the urban environment was recognized in the Army doctrine of full spectrum operations. No longer could combat operations be considered exclusively as a separate phase of the major operation. Instead, full spectrum operations—combat and stability or civil support—would be conducted simultaneously. It also became clear to Army commanders that while one type of operation can often dominate for a period, simultaneous, full-spectrum urban operations were now the norm and an in-depth understanding of complex urban societies essential to overall success.

MODERN ARMY URBAN OPERATIONS

1-23. Modern urban operations will be full spectrum operations executed jointly and will often contain a multinational and interagency component. They will span the entire range of possible applications of military power. At higher echelons, these separations are often viewed as levels or scales of intensity. For the tactical units conducting urban operations, these divisions appear indistinct, as the intensity is often high despite where the operation falls within the range of military operations.

JOINT, INTERAGENCY, AND MULTINATIONAL

1-24. The urban environment is too multifaceted for a single-service, single-agency, or single-dimensional solution. Generating desired effects and avoiding unintended negative consequences in this complex environment requires careful integration of joint (and often multinational) forces and interagency capabilities throughout all phases of the operation. Effective interagency collaboration will help plan effects, supporting actions, and measures of effectiveness to ensure that military actions complement diplomatic, economic, and informational activities.

1-25. Joint urban operations (JUOs) in which Army forces are a major component will be primarily land operations. These operations may take place within the context of a joint campaign conducted by a joint force land component commander or a joint task force (JTF) commander. Or they may be an Army operation under an ARFOR commander who himself operates for a JFC, depending on the organization of the theater's joint command structure. In the later case, the JFC will manage joint issues in the urban area.

1-26. The JFC conducting JUOs will focus on effectively organizing his forces for UO and tasking them in accordance with their service capabilities. His guide for the conduct of the JUO will be the joint operational tasks described in JP 3-0. JP 3 06 will provide the JTF commander specific guidance regarding the conduct of joint operational tasks in the urban environment. Army commanders will execute tasks assigned by the JFC and advise him on using Army forces and capabilities. Army commanders will also ensure that Army UO are nested within the JFC's concept of operations. Also, the ARFOR commander will request support through the JFC from other service and functional commanders who have urban capabilities critical to the success of Army UO. See Appendix D for more information on joint capabilities in an urban environment.

1-27. The military conflict will not be an end to itself. It is inevitably a means of transition from a perceived unsatisfactory state of affairs to an improved end. Therefore, urban operations will require the careful orchestration of military and civilian capabilities to achieve success. An early identification of potential requirements will allow for the proper allocation of responsibilities between military and civilian agencies and a clear understanding of what is possible for each to achieve. Such understanding will preclude the establishment of false perceptions from all concerned, but particularly from the urban population and the public at large.

1-28. Interagency cooperation will include the Army as part of the Department of Defense, elements of the Department of State, and other various governmental and nongovernmental organizations. Urban operations demand the expertise and abilities of various organizations and a synergistic unity of effort if strategic goals aimed at bringing about a more satisfactory political, social, diplomatic, economic, and military situation are to be achieved. However, there will often be a gap in time before bringing many governmental capabilities to bear that the Army and other military forces—to include coalition partners—will have to fill. Routine interagency training, planning, and coordination will help to decrease that gap in time to its minimum. As it is likely that civilian organization leaders will be in charge during some phases of the overall operation, they should lead planning at those times. A careful survey of urban requirements and rapid communication to participating civilian organizations will also help speed civilian capabilities to where they are needed most.

Winning the Peace in Iraq:
The Requirement for Full Spectrum Operations

We found that if we concentrated solely on establishing a large security force and targeted counterinsurgent combat operations—and only after that was accomplished, worked toward establishing a sustainable infrastructure supported by a strong government developing a free-market system—we would have waited too long. The outcome of a sequential plan allowed insurgent leaders to gain a competitive advantage through solidifying the psychological and structural support of the populace.

Further, those who viewed the attainment of security solely as a function of military action alone were mistaken. A gun on every street corner, although visually appealing, provides only a short-term solution and does not equate to long-term security grounded in democratic process. Our observation was born not from idealism, but because it creates the essence of true security, protecting not only our soldiers, but Iraq, the region, and, consequently, our homeland.

Major General Peter Chiarelli
Commander, 1st Cavalry Division

FULL SPECTRUM OPERATIONS

1-29. Army forces will conduct full spectrum operations within urban areas. Army commanders conduct full spectrum operations abroad by executing offensive, defensive, and stability urban operations as part of an the joint, interagency, and multinational effort described above. Army forces within the United States and its territories conduct full spectrum operations by combining offensive, defensive, and civil support operations to support homeland security. The situation will mandate that one type of operation—offense, defense, stability, or civil support—dominates the urban operation. Depending on whether the operation is overseas or within the United States, however, commanders will often find themselves executing offensive and defensive operations and stability or civil support operations simultaneously. In fact, waiting until all combat operations are concluded before beginning stability or civil support operations often results in lost, sometimes irretrievable, opportunities. Unsurprisingly, the mission determines the dominant type of operation, with the other types of Army operations conducted to shape the AO for mission success. The dominant type of operation will vary between different urban areas even in the same campaign.

Offense

1-30. Against a large conventional enemy in a major urban area with a large civil population present, offensive operations will likely require a greater commitment of Army resources than in other environments. Urban offensive operations will also incur the greatest risks to Army forces and noncombatants. Within defensive and stability or civil support operations, forces may conduct tactical offensive UO, such as counterattacks to maintain the initiative or raids to eliminate elements disrupting the stability or civil support operation.

Defense

1-31. Defensive UO are generally conducted as a shaping operation within a larger major operation. These temporary operations often set conditions for successful offensive operations, stability operations, or civil support operations. Commanders often conduct defensive UO within other types of operations to protect essential facilities in the urban area, protect flanks against counterattack, prevent the breakout of isolated enemies, or protect valuable supply bases or vulnerable convoy routes. Army forces conducting defensive UO must creatively use the environment to enhance their combat power.

1-32. In UO, essential facilities will likely include urban ports and airfields required by sister services to support ground operations. Therefore, Army commanders will need to carefully plan integrated air and base defense operations with air and naval component commanders. Protection requirements increase dramatically as air and naval assets are most vulnerable when aircraft are on the ground and ships are docked at port. Without a carefully synchronized and integrated base defense plan, Army commanders risk sporadic or total loss of support by air and naval forces—forces essential to successful UO.

Stability

1-33. Stability operations in an urban environment require offensive and defensive operations, combined with other tasks unique to each stability operation. Army forces conduct urban stability operations for various reasons, including noncombatant evacuation operations, peace operations, or support to insurgencies (see Chapter 9). Urban stability operations will require an offensive capability to destroy any military capability that overtly threatens its objectives before that military threat can adversely affect the operation. Army forces employ defensive capabilities to safeguard themselves as well as secure critical places, populations, or infrastructure in the urban area. Commanders may also employ defensive capabilities to separate and protect one faction from another. Various stability operations will also require the distribution of food or aid and the protection or assistance of agencies conducting economic or humanitarian activities.

Civil Support

1-34. Army civil support operations in an urban environment aid other U.S. agencies and organizations in mitigating the consequences of natural and man-made disasters. In response to the disaster, civil support operations require the equipment, personnel, or organizational abilities of Army forces rather than the Army's combat capabilities. In a civil support mission, these capabilities often involve Army transportation, medical, quartermaster, or engineer forces. Unless conducted in conjunction with homeland defense, urban civil support operations will seldom require combat; however, commanders must determine if force protection threats exist that could hamper Army civil support operations. During homeland security operations, defensive and offensive capabilities will be required to defeat hostile armed forces.

PREPARING FOR FUTURE URBAN OPERATIONS

1-35. To operate successfully in a complex urban environment requires a thorough understanding of the urban environment and rigorous, realistic UO training. Training should be conducted by the complete combined arms team and cover full spectrum operations to include appropriate tactics, techniques, and procedures (see FM 3-06.11 and TC 90-1). It should also replicate—

- The psychological impact of intense, close combat against a well-trained, relentless, and adaptive enemy.
- The effects of noncombatants—including governmental and nongovernmental organizations and agencies—in close proximity to Army forces. This necessitates—
  - An in-depth understanding of culture and its effects on perceptions.
  - An understanding of civil administration and governance.
  - The ability to mediate and negotiate with civilians including the ability to effectively communicate through an interpreter.
  - The development and use of flexible, effective, and understandable rules of engagement.
- A complex intelligence environment requiring lower-echelon units to collect and forward essential information to higher echelons for rapid synthesis into timely and useable intelligence for all levels of command. Understanding the multifaceted urban environment necessitates a bottom-fed approach to developing intelligence (instead of a top-fed approach more common and efficient for open terrain and conventional threats). It also emphasizes the need for intelligence reach and a truly collaborative approach to the development and sharing of intelligence.
- The communications challenges imposed by the environment as well as the need to transmit large volumes of information and data.
- The medical and logistic problems associated with operations in an urban area including constant threat interdiction against lines of communications and sustainment bases.

Developing Joint, Interagency, and Multinational Operability

1-36. UO training extends from the individual Soldier to the joint level including the integration of conventional and special operations forces. Preparedness also includes enhancing interoperability in regards to urban multinational and interagency operations. Training, as well as campaign and contingency plan creation, should include significant and sustained participation by civilian agencies. Fruitful multinational and interagency relationships must be cultivated before the onset of operations, that is, before Soldiers and their coalition counterparts are making decisions—many with strategic implications and often when they are under fire. Joint, interagency, and multinational collaboration will help design effects, supporting actions, and measures of effectiveness necessary to ensure that military actions in urban environments complement the diplomatic, informational, and economic activities necessary to achieve strategic objectives.

Conducting Live, Virtual, and Constructive Training

1-37. Force preparedness mandates integrating the actual use of urban terrain, exercises at urban training sites, simulations, or any combination into tactical- and operational-level intra- and interservice training. This type of multi-faceted training will help commanders develop a better understanding of the complexity of the urban environment and enable them to execute missions across full spectrum operations. Careful use of these facilities will also allow Army forces to fully integrate urban operations within self-development, institutional, and operational training.

Ensuring Every Soldier is an Urban Warrior

1-38. In a complex urban environment, every Soldier—regardless of branch or military occupational specialty—must be committed and prepared to close with and kill or capture threat forces in an urban environment. Every Soldier must also be prepared to effectively interact with the urban area's noncombatant population and assist in his unit's intelligence collection efforts. Each urban operation will be unique and commanders—given an opportunity to surge and conduct mission-specific post-mobilization and in-theater training—will need to conduct an analysis of the tasks requiring emphasis or modification to fit the conditions of the operational environment. In UO, every soldier will likely be required to—

- Perform advanced rifle marksmanship to include advanced firing positions, short-range marksmanship, and night firing techniques (unassisted and with the use of optics).
- Operate their unit's crew-served weapons.
- Conduct urban reconnaissance and combat patrolling.
- Enter and clear buildings and rooms as part of an urban attack or cordon and search operation.
- Defend an urban area.
- Act as a member of a combat convoy (including specific drivers training).
- Control civil disturbances.
- Navigate in an urban area.

1-39. While not all-inclusive and necessarily urban-specific, other critical individual and collective UO tasks (often modified for the urban environment) might include—

- Conduct troop-leading procedures.
- React to contact, ambush, snipers, indirect fire, and improvised explosive devices.
- Establish an observation point, personnel or vehicle checkpoint, or roadblock.
- Secure a disabled vehicle or downed aircraft.
- Call for indirect fire and close air support.
- Create and employ explosive charges.
- Handle detainees and enemy prisoners of war.
- Treat and evacuate casualties.
- Accurately report information.
- Understand the society and culture specific to the area of operations.
- Use basic commands and phrases in the region's dominant language.
- Conduct tactical questioning.
- Interact with the media.
- Conduct thorough after-action reviews.

1-40. Commanders must understand that all Soldiers will require urban-specific equipment to conduct many urban-specific tasks. Critically, commanders cannot expect supporting forces to fight alongside and support ground maneuver units in an urban environment without the appropriate equipment and training.

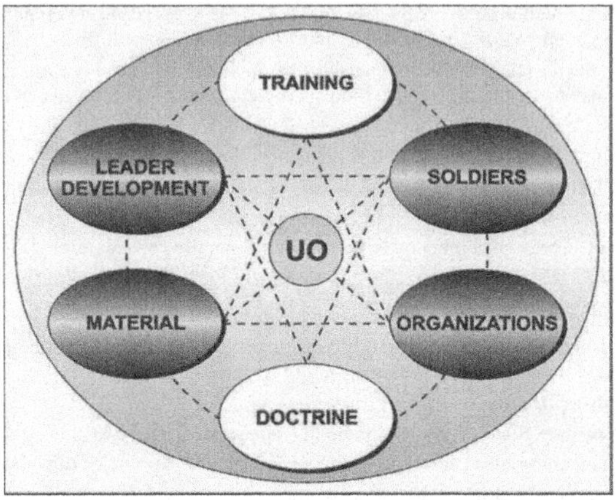

Figure 1-2. UO and DOTMLPF

Continual Adaptation and Innovation

1-41. Realistic UO training (as well as the conduct of real world operations) has the added benefit of identifying operational requirements and resultant changes necessary in our doctrine, organizations, training, materiel, leadership and education, personnel, and facilities (DOTMLPF) (see figure 1-2 and FM 100-11). While technology (materiel) and organizational changes are important, confident Soldiers (personnel) remain the decisive means for success. Technology and organizational changes will be a critical enabler to attain better understanding of the urban environment, enhance command and control on the noncontiguous battlefield, and achieve the agile, simultaneous, and precise effects required in UO. In the future, technology may lead to a radically new operational concept and approach to UO. Still, competent leaders and well-trained, disciplined, and adaptive Soldiers will remain the decisive means for the Army to succeed in this complex and multidimensional urban environment.

Chapter 2

Understanding the Urban Environment

From a planning perspective, commanders view cities not just as a topographic feature but as dynamic entities that include hostile forces, local population, and infrastructure. Planning for urban operations requires careful IPB, with particular emphasis on the three-dimensional nature of the topography and the intricate social structure of the population.

FM 3-0

Of all the environments in which to conduct operations, the urban environment confronts Army commanders with a combination of difficulties rarely found elsewhere. Its distinct characteristics result from an intricate topography and high population density. The topography's complexity stems from the man-made features and supporting infrastructure superimposed on the natural terrain. Hundreds, thousands, or millions of civilians may be near or intermingled with soldiers—friendly and enemy. This second factor, and the human dimension it represents, is potentially the most important and perplexing for commanders and their staffs to understand and evaluate. To this end, this chapter provides information essential to understanding the urban environment and conducting an effective intelligence preparation of the battlefield (see FM 2-01.3, FMI 2-91.4, and Appendix B).

Although urban areas possess general similarities, each environment is distinct and will react to and affect the presence and operations of Army forces differently. A tactical technique effective in one area may not be effective in another area due to physical differences, such as street patterns or the type of building construction. An Army policy popular with one urban group may cause resentment and hostility in another due to diverse cultural differences. All difficulties potentially exist, and they increase the complexity for Army forces operating in urban areas. These difficulties range from conventional military forces to disease and starvation (see Chapter 3) to a pervasive media—often acutely present in intricate combinations. Thus, commanders at all levels must make extraordinary efforts to assess and understand their particular urban environment to plan, prepare for, and execute effective urban operations (UO).

A COMPLEX ENVIRONMENT

2-1. Urban areas vary depending on their history, the cultures of their inhabitants, their economic development, the local climate, available building materials, and many other factors. This variety exists not only among urban areas but also within any particular area. The ever-changing mix of natural and man-made features in urban areas present commanders with some of the most difficult terrain in which to conduct military operations.

2-2. Although urban areas possess similar characteristics, no two are identical. The sprawl of Los Angeles, for example, bears little physical resemblance to New Delhi. Societal characteristics most significantly affect each area's uniqueness and complexity. While complex, information about the terrain, its potential effects on operations, and how it changes over time may be determined with some degree of certainty. However, the human dimension is much more difficult to understand and assess, particularly its effects on military operations. Like any environment, the side that can best understand and exploit the positive and mitigate the negative effects of the urban environment has the best chance of success.

2-3. Whether a large metropolis or a small village, each urban environment has identifiable components that constantly change and interact. This "system of systems" consists of the terrain, the society, and the infrastructure that links the two (see figure 2-1).

2-4. These systems are not separate and distinct categories, but rather overlapping and interdependent systems, acting dynamically with each other. Thoroughly analyzing these elements, along with the other factors of mission, enemy, weather, troops and support available, and time—

- Contributes to the accuracy of the commanders' situational understanding.
- Potentially lessens the number and cost of close combat engagements.
- Allows commanders to develop courses of action that apply appropriate resources against decisive points.

2-5. In stability operations and civil support operations, this understanding allows commanders to engage and dominate the decisive points critical to maintaining peace or restoring normalcy to the urban environment. Although each system is categorized into subordinate components or subsystems, commanders must often "step back" and visualize each system, the complex urban environment, and their area of operations (AO) as a unified whole. This "systems thinking" aids commanders in uncovering key relationships and intersections that can help reveal centers of gravity (COGs) and decisive points.

2-6. To comprehend the urban environment and its components to the fullest extent possible, commanders must carefully integrate and employ tactical reconnaissance forces, special operations forces (SOF)—to include psychological operations (PSYOP) and civil affairs units—and a myriad of other human intelligence (HUMINT) assets and regional, language, and cultural experts. The societal aspects and integrating infrastructure will challenge commanders' assessment and understanding. These aspects will also require greater dependence on nonmilitary and nongovernmental organizations (NGOs) and host-nation agencies for their information, knowledge, and expertise. This last consideration requires commanders to develop effective techniques and procedures for coordinating, interacting, and, to the greatest extent possible, synchronizing activities with these agencies.

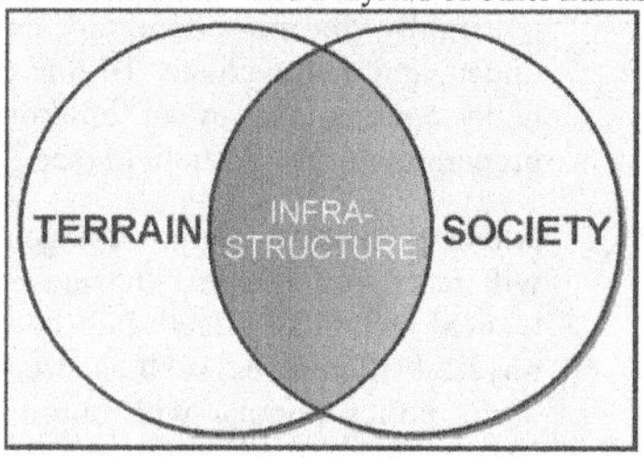

Figure 2-1. Keys to understanding the urban environment

URBAN TERRAIN

2-7. Although complex and difficult to penetrate with many intelligence, surveillance, and reconnaissance (ISR) assets, the terrain is the most recognizable aspect of an urban area. Truly understanding it, however, requires comprehending its multidimensional nature. The terrain consists of natural and man-made features, with man-made features dominating; an analysis considers both. Buildings, streets, and other infrastructure have varied patterns, forms, and sizes. The infinite ways in which these factors can intertwine make it difficult to describe a "typical" urban area. However, these various factors provide a framework for understanding the complex terrain in an urban area. Furthermore, man-made features significantly affect military systems and Soldiers, and thus tactics and operations. General effects on urban operations are discussed in this chapter. Specific effects on warfighting functions (see Chapters 4 and 10) and the spectrum of operations (see Chapters 7, 8, and 9) are interwoven throughout the manual.

MULTIDIMENSIONAL BATTLEFIELD

2-8. Urban areas present an extraordinary blend of horizontal, vertical, interior, exterior, and subterranean forms superimposed on the natural relief, drainage, and vegetation. An urban area may appear dwarfed on a map by the surrounding countryside. In fact, the size and extent of the urban area of operations is many times that of a similarly sized portion of undeveloped natural terrain. A multi-storied building may take up the same surface area as a small field, but each story or floor contains approximately an equal area as the ground upon which it sits. In effect, a ten-story building can have eleven times more defensible area than "bare" ground—ten floors and the roof. It is the sheer volume and density created by this urban geometry that makes UO resource intensive in time, manpower, and materiel.

2-9. Like natural disasters, UO can radically alter the physical characteristics of the urban terrain in ways not experienced in other environments. They may cause (either intentionally or not) uncontrollable fires or the loss of electricity. A power outage can cause flooding (especially in subsurface areas) by shutting down pumping stations. Entire buildings may be destroyed, eliminating reference points, leaving large piles of rubble, altering fields of fire, and making movement and transportation extremely difficult. Additionally, buildings and other urban structures, damaged but not destroyed, can become (or remain) effective obstacles and possible booby traps. Even without enemy exploitation, their weakened construction and unstable structure increase the risk of injury to Soldiers and civilians moving within them. (Engineer expertise will often be needed to determine whether the buildings can support occupation by Army forces or civilians.) Yet, even the total collapse of a building may not eliminate its defenders. Of additional concern, the likely presence of toxic industrial material (TIM) can create additional obstacles and health hazards.

2-10. Commanders in other environments normally address the depth, breadth, and height of their AO in terms of two areas: airspace and surface. In an urban environment, they broaden their scope to include supersurface and subsurface areas (see figure 2-2) that voluminously extend the commanders area of operations. Although spatially separated, each area may be used as an avenue of approach or mobility corridor, line of communications (LOC), or engagement area.

2-11. Supersurface and subsurface areas magnify the complexity of the urban physical environment. Commanders must consider activities that occur outside buildings and subterranean areas (the external space) as well as the activities happening unseen in buildings and subterranean systems (the internal space). This internal volume further challenges command, control, and intelligence collection activities and increases the combat power required to conduct UO. Commanders must develop methods and techniques to help themselves, their staffs, and their subordinate commanders and staffs to represent, visualize, and reference these multiple dimensions. Increasing the difficulty, such dimensions can change rapidly simply due to continued urban growth or, as described earlier, the effects of nature and UO themselves.

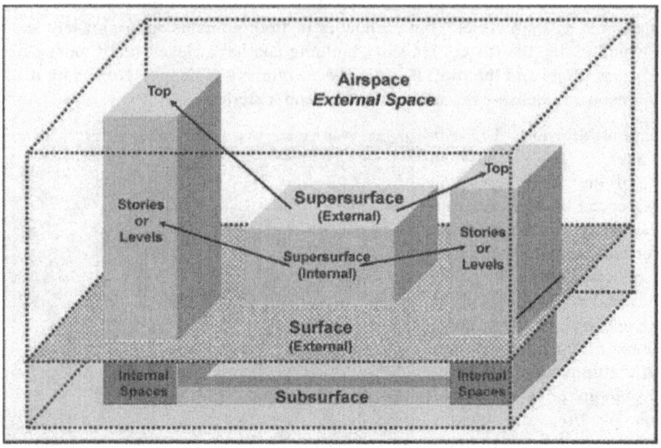

Figure 2-2. The multidimensional urban battlefield

Airspace

2-12. As in all other environments, aircraft and aerial munitions use the airspace as rapid avenues of approach in urbanized areas. Forces can use aviation assets for observation and reconnaissance, aerial attack, or high-speed insertion and extraction of Soldiers, supplies, and equipment. Some surface obstacles in an urban area, such as rubble, do not affect flight (though they may prevent the take-off and landing of aircraft). Buildings of varying height and the increased density of towers, signs, power lines, and other urban constructions, however, create obstacles to flight and the trajectory of many munitions (masking). Similarly, these obstacles can restrict a pilot's line of sight as well as physically limit low-altitude maneuverability in the urban airspace. Excellent cover and concealment afforded enemy gunners in an urban area increases aviation vulnerability to small arms and man-portable air defense systems (MANPADS), particularly when supporting ground forces. The potential for a high volume of air traffic (military and civilian) over and within urban airspace (including fixed-wing, rotary-wing, and unmanned aircraft systems) may become another significant hazard and necessitate increased airspace command and control measures.

Surface

2-13. Surface areas apply to exterior ground-level areas, such as parking lots, airfields, highways, streets, sidewalks, fields, and parks. They often provide primary avenues of approach and the means for rapid advance. However, buildings and other structures often canalize forces moving along them. As such, obstacles on urban surface areas usually have more effect than those in open terrain since bypass often requires entering and transiting buildings or radical changes to selected routes. Where urban areas border the ocean or sea, large lakes, and major rivers, the surface of these bodies of water may provide key friendly and threat avenues of approach or essential LOCs—a significant consideration for Army commanders. As such, amphibious, river-crossing, and river operations may be integral parts of the overall urban operation (see FM 3-05.212 and FM 3-90.12).

2-14. Larger open areas—such as stadiums, sports fields, school playgrounds, and parking lots—are often critical areas during urban operations. They can provide locations for displaced civilians, interrogation centers, and prisoner of war holding facilities. These areas also can afford suitable aircraft landing and pickup zones and artillery firing locations. They can provide logistic support areas and aerial resupply possibilities because they are often centrally located. Finally, large open areas (and immense or unusually shaped structures) within urban areas are often easier to see—especially from the air—and can serve as excellent target reference points from which to shift or control fires.

Supersurface

2-15. These areas include the internal floors or levels (intrasurface areas) and external roofs or tops of buildings, stadiums, towers, or other vertical structures. They can provide cover and concealment; limit or enhance observation and fields of fire; and restrict, canalize, or block movement. However, forces can move within and between supersurface areas creating additional, though normally secondary, avenues of approach. Rooftops may offer ideal locations for landing helicopters for small-scale air assaults and aerial resupply. First, however, engineers must analyze buildings for their structural integrity and obstacles. Such obstacles include electrical wires, antennas, and enemy-emplaced mines (although personnel may be inserted by jumping, rappelling, or fast roping from a hovering helicopter and extracted by hoist mechanisms). Some rooftops are designed as helipads. Roofs and other supersurface areas may also provide excellent locations for snipers; lightweight, handheld antitank weapons; MANPADS; and communications retransmission sites. They enable top-down attacks against the weakest points of armored vehicles and unsuspecting aircraft. Overall, elevated firing positions reduce the value of any cover in surrounding open areas and permit engagement at close range with less risk of immediate close assault. This area (and the subsurface area) requires commanders to think, plan, and execute ground operations vertically as well as horizontally. In this latter regard, UO share strong similarities with mountain operations (see FM 3-97.6).

Subsurface

2-16. Subsurface areas are below the surface level. They may serve as secondary and, in fewer instances, primary avenues of approach at lower tactical levels. When thoroughly reconnoitered and controlled, they offer excellent covered and concealed LOCs for moving supplies and evacuating casualties. They may also provide sites for caching and stockpiling supplies. Subsurface areas include subterranean areas such as subways, mines, tunnels, sewers, drainage systems, cellars, civil defense shelters, and other various underground utility systems. In older cities, they may include ancient hand-dug tunnels and catacombs. Both attacker and defender can use subsurface areas to gain surprise and maneuver against the rear and flanks of a threat and to conduct ambushes. However, these areas are often the most restrictive and easiest to defend or block. Their effectiveness depends on superior knowledge of their existence and overall design. Army commanders may also need to consider potential avenues of approach afforded by the subsurface areas of rivers and major bodies of water that border urban areas. This particularly applies when operating as part of a joint task force (JTF) task organized with SOF or when opposing a threat with similar capabilities.

Major Urban Patterns

2-17. Four major urban patterns (satellite, network, linear, and segment) can influence UO (see figure 2-3). Central to two of the patterns (satellite and network) is the hub or dominant urban area or pattern around which outlying urban areas or patterns radiate. (A segmented urban area, because it tends to be a larger urban area, can often be a hub.) In offensive and defensive operations, the hub serves as a pivot or strong point; as such, it can become a major obstacle to an attacker. If the attacker chooses to bypass the urban area (hub) located along his axis of advance without first isolating the area, he may expose his flank or LOC to attack from the hub as well as dependent urban areas or subordinate satellite patterns. Because the focus of stability and civil support operations is normally on people, commanders should understand the value and influence of the hub to the economic, political, or cultural well being of the surrounding area. Generally the larger the hub, the greater influence it has on satellite urban areas and surrounding rural areas. Commanders must remember that urban areas are not islands; all are connected to the surrounding rural (and other urban) areas through fluid and permeable boundaries and LOCs.

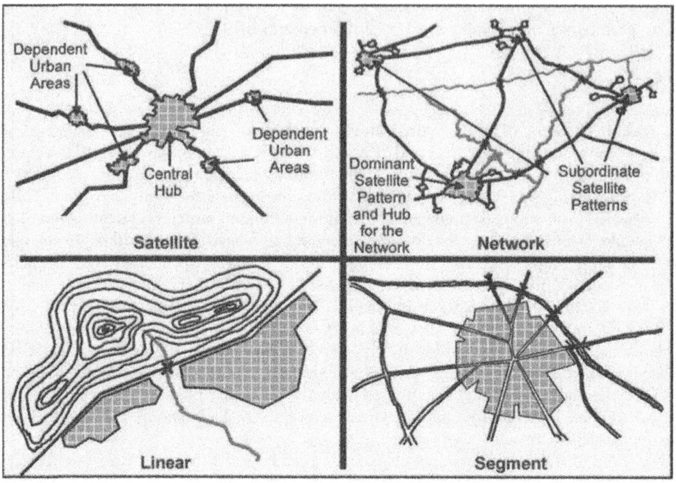

Figure 2-3. Major urban patterns

Satellite Pattern

2-18. This common pattern consists of a central hub surrounded by smaller, dependent urban areas. LOCs tend to converge on the hub. Outlying areas often support the principal urban area at the hub with means of reinforcement, resupply, and evacuation. In some instances, they may serve as mutually supporting battle positions. Commanders should consider the effects of the outlying urban areas on operations within the hub, and, conversely, the effects of operations within the hub on outlying urban areas. Information operations (IO), for example, targeted primarily at key leaders and other civilians located within the hub of a satellite pattern may subsequently influence civilians in outlying urban areas and achieve necessary effects without having to commit specific resources to these outlying areas.

Network Pattern

2-19. The network pattern represents the interlocking of the primary hubs of subordinate satellite patterns. Its elements are more self-sufficient and less supportive of each other, although a dominant hub may exist. Major LOCs in a network extend more than in a satellite pattern and take more of a rectangular rather than a convergent form. Its natural terrain may vary more than in a single satellite array. Operations in one area may or may not easily influence, or be influenced by, other urban areas in the pattern.

Linear Pattern

2-20. Potentially a subelement of the previous two patterns, the linear pattern may form one ray of the satellite pattern or be found along connecting links between the hubs of a network. Most frequently, this pattern results from the stringing of minor urban areas along a confined natural terrain corridor, such as an elongated valley, a body of water, or a man-made communications route. In offensive and defensive operations, this latter form of the linear pattern facilitates developing a series of strong defensive positions in depth, effectively blocking or delaying an attacking force moving along the canalized terrain.

Segment Pattern

2-21. When dominant natural terrain, such as a river or man-made features (canals, major highways, or railways), divides an urban area, it creates a segmented pattern. This pattern often makes it easier for commanders to assign areas of operations to subordinate commanders. However, this pattern may fragment operations and increase risk to an operation requiring mutual support between subordinate units. Still, the segmented urban areas may allow commanders to isolate threats more easily in these areas and focus operations within segments that contain their decisive points. Although an integral part of the whole (the urban area), each segment may develop distinct social, economic, cultural, and political characteristics. This social segmenting may benefit commanders faced with limited assets to influence or control the urban populace. After thoroughly analyzing the society, they may be able to focus IO and populace and resources control measures against only specific segments that affect decisive operations. Commanders may need only to isolate other segments or may need to just monitor for any significant changes in the attitudes, beliefs, or actions of the civilians located there.

LESSER STREET PATTERNS

2-22. Lesser patterns in the urban area result from the layout of the streets, roads, highways, and other thoroughfares. They evolve from influences of natural terrain, the original designer's personal prejudices, and the changing needs of the inhabitants. Street patterns (and widths) influence all warfighting functions; however, they greatly affect movement and maneuver, command and control, and sustainment. (In some portions of older Middle Eastern urban areas, the labyrinths of streets were designed only to allow two loaded donkeys to pass each other; tanks are too wide.) Urban areas can display any of three basic patterns and their combinations: radial, grid, and irregular (see figure 2-4).

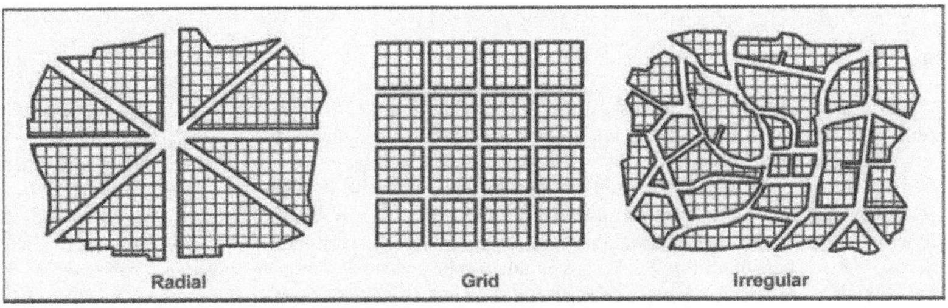

Figure 2-4. Basic internal street patterns

Radial

2-23. Societies of highly concentrated religious or secular power often construct urban areas with a radial design: all primary thoroughfares radiating out from the center of power. Urban areas with this design may signal an important historical aspect in the overall analysis of the urban society. Terrain permitting, these streets may extend outward in a complete circle or may form a semicircle or arc when a focal point abuts a natural barrier, such as a coastline or mountain. To increase mobility and traffic flow, societies often add concentric loops or rings to larger radial patterns. Unless commanders carefully plan boundaries, routes, and axes of advance, their subordinate units' movement or maneuver may be inadvertently funneled toward the center of urban areas with this pattern resulting in congestion, loss of momentum, and an increased potential for ambush or fratricide.

Grid

2-24. The most adaptable and universal form for urban areas is the grid or rectangular pattern: lines of streets at right angles to one another forming blocks similar to the pattern of a chessboard. A grid pattern can fill in and eventually take over an original radial pattern. Grid patterns often appear to ease the assignment of boundaries for subordinates units. However, commanders also consider how the natural terrain influences operations and the establishment of graphic control measures. They also consider the influence of the buildings and other structures lining these streets, such as their height and construction, before assigning boundaries and developing other control measures. Commanders should also consider the following when developing urban graphic control measures:

- Describing boundaries, phase lines, checkpoints, and other graphic control measures by easily recognizable features is as important in urban areas as elsewhere. While easily identifiable urban structures such as unusually tall or oddly-shaped buildings, cemeteries, stadiums, or prominent rail or highway interchanges can be useful references, available natural features are a better descriptor than man-made features that may be altered or unrecognizable. As an aid to air-to-ground coordination, commanders should select features that can be identified by both ground and air forces. Those that help in controlling ground forces may not be easily visible from the air and vice versa.
- Commanders should also consider whether a boundary along an easily-recognizable terrain feature, such as a river, will also be easy to identify by the threat who may seek to "find the seam" and exploit the likely control and coordination difficulties associated with boundaries, especially between higher-level units. This often requires commanders to carefully position a control measure away from the key feature to provide a designated subordinate force with the terrain and space necessary to control the feature. On the other hand, commanders working closely with local authorities during stability and civil support operations may not need to thoroughly understand the physical effect of street patterns on the assignment of boundaries as they might for combat UO. Instead, commanders may choose to assign boundaries overlaid on existing geopolitical boundaries used by local agencies to increase interoperability and aid in unity of effort.

IRREGULAR

2-25. In most urban areas, regardless of the original intent, plan, or vision, existing street patterns emerge from successive plans overlaid one on another. Some are well planned to fit with previous plans while others a haphazard response to explosive urban growth. The result may mix patterns. Urban engineers and planners may specifically design irregular patterns for aesthetic reasons (as in many suburban housing developments) or to conform to marked terrain relief. Irregular street patterns may alert commanders and analysts that the underlying natural terrain may exert greater influence over operations than in other portions of the urban area. Finally, irregular street patterns make the movement and maneuver of forces less predictable. However, a labyrinth of irregular or "twisting" street patterns may increase the possibility of fratricide particularly for units that are trained or accustomed only to grid patterns.

AN URBAN MODEL

2-26. Throughout the world, urban areas have similar form and function. In form, urban areas contain like characteristics, readily divisible into distinct sections or areas. Functionally, they tend to be the centers of population, finance, politics, transportation, industry, and culture. While urban areas may be modeled by several different means, figure 2-5 illustrates the general forms and internal functions. Some forms and functions may overlap. For example, high-rise buildings are located in core areas as well as in outlying areas and may be used for residential purposes. With the rapid urbanization associated with developing nations, the areas displayed in this urban model often manifest themselves less clearly there than in developed nations.

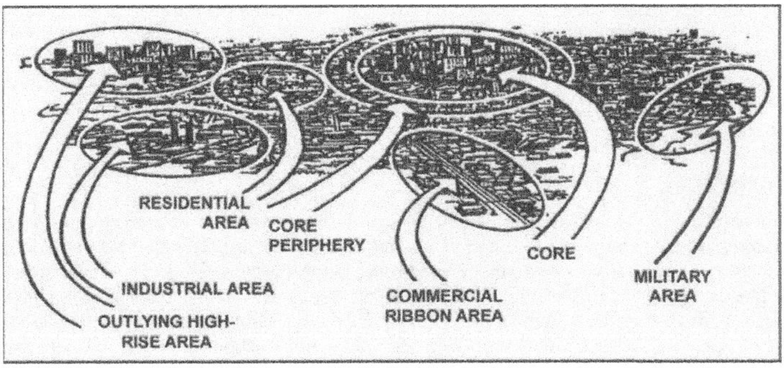

Figure 2-5. Urban functional zones

2-27. This analysis helps to determine, in general terms, potential advantages and disadvantages each portion of the urban area may have toward accomplishing the urban operation. However, construction materials and methods can vary drastically. Commanders and their staff will often need to identify specific building types and construction and understand weapons effects on them. If a commander desires precise effects, the chosen munitions or weapons system must be sufficiently accurate, capable of penetrating the target structure (without exiting the other side), and achieve effects within. Often noncombatants, critical infrastructure, or protected targets are in the vicinity. Commanders may need to determine if the surrounding walls or structures will sufficiently absorb or negate the blast or thermal effects of the weapon. Regardless, understanding the structure and composition of buildings and other structures in the urban AO may be necessary to allow commanders to determine the best means to accomplish the mission.

Core

2-28. The core is the heart of the urban area, the downtown or central business district. Relatively small and compact, it contains a large percentage of the urban area's shops, offices, and public institutions. Often, it houses the headquarters for commercial and financial activities and contains important cultural, historical, and governmental buildings. These activities prefer the core because of its accessibility. As the focal point of the transportation network, residents find the core the easiest part of the urban area to reach. It normally has the densest concentration of multistory buildings and subterranean features (underground parking garages, underground shopping centers, and basements).

2-29. High-rise buildings, varying greatly in height (possibly 50 stories above ground and four stories below ground), make up the cores of many of today's urban areas. Buildings routinely abut one another, with little or no setback from the sidewalks. Building height and density (except in outlying high-rise areas) often decreases from the core to the edge of the residential areas, while the amount of open areas frequently increases. Modern urban planning allows for more open spaces between buildings than found in the cores of older urban areas. Most core areas have undergone constant redevelopment resulting in various types of construction. Commonly, brick buildings abound in the oldest part of the core; framed, heavy-clad structures in the next oldest part; and a concentration of framed, light-clad buildings in the newest part. The outer edge of the core, the core periphery, has ordinarily undergone less change than the core resulting in buildings of uniform height (commonly two to three stories in towns and five to ten stories in larger urban areas).

2-30. Generally, offensive operations focused in core areas (even when effectively isolated) will require greater resources—particularly manpower, time, and information—than in many other parts of the urban area. Mounted maneuver often proves more difficult in core areas because of fewer open areas, buildings closer to the streets, and an increased density of civilian vehicles. Razed buildings in central core areas (especially high-rise buildings) become greater obstacles to mobility as they can collapse on and easily block thoroughfares. Rubble piles provide excellent covered and concealed positions for dismounted threat forces. Consequently, commanders often use more dismounted forces as part of their combined arms operations. Conversely, the core may be critical to urban defensive operations, particularly older areas of heavier construction that afford greater protection. Despite potential difficulties, the core area may be key to accomplishing many stability or civil support missions since it houses much of the human activity that occurs in the urban area.

Industrial Area

2-31. Industrial areas often develop on the outskirts of the urban areas where commercial transportation is easiest (along airfields and major sea, river, rail, and highway routes). The road networks in and around industrial areas are generally more developed and suitable for transportation assets. These areas will likely displace farther from the core and residential areas as urban planners recognize the potential threat of TIM. The dispersed pattern of the buildings provides sufficient space for large cargoes, trucks, and materiel handling equipment. These areas may provide ideal sites for sustainment bases and maintenance sites. While older heavier-clad structures may be found, new construction generally consists of low, large, flat-roofed factory and warehouse buildings with large parking areas and work yards. These structures generally have steel frame and lightweight exterior walls. Multistory structures usually have reinforced concrete floors and ceilings.

Toxic Industrial Chemical	Industrial/Commercial Uses
Ammonia	Commercial Refrigerant, Fertilizer and Food Production, Petroleum, Explosives, Other Chemicals
Arsine	Semiconductor Industry
Boron Trichloride	Organic Catalyst, Soldering Magnesium
Boron Trifluoride	Chemical Catalyst, Aluminum Refining
Carbon Disulfide	Industrial Solvent, Dry Cleaning, Agriculture, Petroleum, Electroplating
Chlorine	Potable Water, Disinfectants, Metal Treatment, Plastics & Rubber
Diborane	Plastics and Rubber
Ethylene Oxide	Industrial Alcohols, Fumigant, Industrial Sterilant
Fluorine	Uranium Processing, Rocket Fuel
Formaldehyde	Plastics, Fertilizers, Preservative/Corrosion Inhibitor, Fungicide and Germicide, Pesticide, Pharmaceuticals
Fuming Nitric Acid	Fertilizers, Explosives, Metal Processing, Pesticides, Rocket Fuel
Hydrogen Bromide	Chemical Industry, Pharmaceuticals
Hydrogen Chloride	Fabrics, Semiconductors
Hydrogen Cyanide	Pesticides, Other Chemicals, Pharmaceuticals, Electroplating
Hydrogen Fluoride	Glass Production, Chemical Catalyst
Hydrogen Sulfide	Metallurgy, Agricultural Disinfectant
Phosgene	Dyes, Pharmaceuticals, Herbicides & Insecticides
Phosphorus Trichloride	Metallurgy, Pesticides and Germicides, Gasoline Additive
Sulfur Dioxide	Paper, Food Processing, Ice Production, Disinfectant, Leather Processing
Sulfuric Acid	Fertilizers, Petroleum, Iron and Steel Production, Battery Electrolyte
Tungsten	Electronics, Other Chemicals

Figure 2-6. Some toxic industrial chemicals and their industrial or commercial uses

2-32. Toxic industrial chemicals and other TIM may be transported through an urban area (by rail, barge, truck, or pipeline) or found stored throughout. However, larger concentrations will exist in industrial areas, and their presence should concern Army forces operating near them. Some TIM are heavier than air and tend to settle in low-lying and subsurface areas.

2-33. Each year, over 70,000 different chemicals are produced, processed, or consumed globally. An estimated 25,000 commercial facilities around the world produce, process, or store chemicals that have a legitimate industrial use yet are also classified as chemical warfare agents. Many other chemicals (not classified as weapons) may still be sufficiently hazardous to pose a considerable threat to Army forces and civilians in urban areas as choking agents or asphyxiates, flammables or incendiaries, water contaminants, low-grade blister or nerve agents, or debilitating irritants. These chemicals can be released either accidentally or deliberately. For example, on 2 December 1984, nearly 40 tons of methylisocyanate used to produce pesticides leaked from a storage tank at Bhopal, India, killing thousands and injuring hundreds of thousands. Figure 2-6 contains a small sampling of other toxic industrial chemicals along with their industrial or commercial usage that commanders may encounter in an urban area. The most common chemicals that may pose an immediate risk to Army forces are highly toxic irritant gases such as ammonia, chlorine, hydrogen chloride, and sulfur dioxide.

2-34. Standard chemical defense equipment may not protect against (and chemical detection devices may fail to detect) many toxic industrial chemicals. Therefore, the risk to Soldiers operating near the chemicals may increase. Commanders must vigilantly identify these potential hazards, carefully consider them as part of their overall vulnerability analysis, factor the analysis into their risk assessment, and execute necessary contamination avoidance measures. (Local urban fire fighters may be a critical source of information for determining the likely locations of TIM.) Any assessment includes the chance that toxic industrial chemicals may be deliberately released by a threat to gain advantage or accidentally released by friendly actions (see FM 3-11.9, FM 3-11.14, FM 3-11.21, and the U.S. Department of Transportation's current version of the Emergency Response Guidebook).

Outlying High-Rise Area

2-35. High-rise areas consist of multistoried apartments, commercial offices, and businesses separated by large open areas, such as parking lots, parks, and individual one-story buildings. High-rise buildings are generally of framed, light-clad construction with thin walls of brick, lightweight concrete, or glass. The automobile, mass transit systems, and improved road networks encourage these areas to grow and function further from the urban core.

2-36. Similar to the urban core, units given the mission to clear these areas, or even portions therein, will need more resources—most notably personnel and time—to accomplish their mission. Commanders should consider courses of action that isolate these entire areas, multiple sections within these areas, or even individual buildings before assigning tasks. Without careful consideration and analysis, some tasks in these areas could unintentionally—but rapidly—drain a unit's resources or unhinge other portions of the major operation. When defending, commanders who can integrate these areas in the defense will present the attacker with similar resource problems and may be appropriate in a defense to delay. However, defending commanders must ensure that the defense is arranged so that this portion cannot be easily isolated and bypassed. Defensive positions in structures may require extensive reinforcement due to light-clad construction.

Residential Area

2-37. Residential areas can be found dispersed throughout the urban area; however, large suburban areas (or sprawl) normally form on the outskirts. Residential areas often consist of row houses or single-family dwellings set in a grid or ringed pattern in a planned development project. Yards, gardens, trees, and fences may separate the buildings in a residential area. In some areas of the world, residential areas may be located in high walled compounds with houses built right up to the edge of the street. Modern residential construction is more often of light-clad, framed wood construction, or brick, however; residential homes formed by poured or pre-cast concrete can be found throughout many parts of the world. The combined population of surrounding suburban areas often far outnumbers that of the urban area proper. Specific suburbs tend toward homogeneity based on ethnicity, religion, economics, or some other social aspect. Commanders must locate and analyze these areas to determine their impact on operations—often the most critical importance is the people located there (see the subsequent discussion in this chapter on the urban society).

2-38. In offensive and defensive operations, commanders should determine whether operations pose an unacceptable physical risk to civilians. If so, they may have to relocate civilians to a safer area, perhaps another residential area. If not, commanders may implement a "stay-put" policy for that area and attempt to isolate the effects of the operation from them. During civil support and stability operations, residential locations may be the initial focal point for operations since most of the permanent population is located there.

2-39. This area also contains an urban phenomenon known as shantytowns. These areas are commonly on unoccupied, low-value land in and around many urban areas in underdeveloped countries. Shantytowns may contain over 50-percent of the total urban population. They usually lack streets and public utilities. The lean-to structures tend to be irregularly laid out, connected by walking paths, and made of any scrap material available: lumber, brick, sheet metal, cardboard, cloth, or vegetation. The random arrangement of structures, the absence of formal street naming and numbering, and often the lack of easily identifiable buildings and terrain create challenges. These challenges include navigation, coordination, and the transmission of accurate information and intelligence. Depending on the operation, the temporary nature of the structures can also mean that mobility can be either more or less restricted than other sections of the urban area. A military force may easily knock down and traverse structures without affecting mobility at all. However, their destruction may cause unacceptable civilian casualties, in which case mobility becomes more restrictive as the narrow paths often do not accommodate vehicular traffic. Similarly, the makeshift materials inhibit weapons effects less than many other parts of the urban area built more solidly. A tank round, for example, may go much farther and injure many more noncombatants than in an area where the primary building material is stone. Regardless, commanders should consider the effects of their operations in this area, to include vehicles and weapons, as the weak structures increase the risk of fratricide, civilian casualties, and large, rapidly spreading fires.

Commercial Ribbon Area

2-40. Commercial ribbon areas are rows of stores, shops, and restaurants built along both sides of major streets that run through and between urban areas. These same types of areas often develop along the roads that connect one urban area to another (strip areas). The buildings uniformly stand two to three stories tall (about one story taller than the dwellings on the streets behind them).

Military Area

2-41. Fortifications and military installations may be found in or near urban areas throughout the world. Historically, they may have been the "seed" responsible for initiating the growth of the present-day urban area. Many countries possess long coastlines and borders with potentially hostile neighbors. To meet their defensive needs, they developed coastal and border defense works that include extensive subsurface facilities, many contiguous to urban areas. North Korea, for example, has built numerous hardened artillery, missile, and command and control facilities along both its coasts and the demilitarized zone which separates them from South Korea. While some fortifications may be inactive (particularly ancient fortifications), they can be rapidly activated and modified by threats (or Army forces) to accomplish their original role. Even if not activated, they may still serve as choke points and major obstacles to movement and maneuver.

2-42. Permanent-type fortifications can be made of earth, wood, rock, brick, concrete, steel-reinforced concrete, or any combination of the above. Some contemporary constructions have been built subsurface and employ heavy armor, major caliber weapons, internal communications, service facilities, and chemical, biological, radiological, and nuclear overpressure systems. Because they have been built specifically for military purposes, commanders and planners should carefully consider the effects of these military constructions on the conduct of UO.

URBAN SOCIETY

2-43. Although intricate, understanding the urban terrain is relatively straightforward in comparison to comprehending the multifaceted nature of urban society. UO often require Army forces to operate in close proximity to a high density of civilians. Even evacuated areas can have a stay-behind population in the tens of thousands. This population's presence, attitudes, actions, communications with the media, and needs may affect the conduct of operations. Homogeneity decreases drastically as the size of the urban area increases. Commanders must take into account the characteristics of a population whose beliefs and interests vary based on factors (see figure 2-7). Analysis and understanding of these societal factors is critical to a successful IO campaign and, thus, the entire operation.

2-44. Civilian populations continually influence, to varying degrees, operations conducted in an urban area. Thoroughly understanding these societal aspects and avoiding "mirror-imaging"—overlaying one's own values and thought processes on top of the person or group one is trying to understand—will help to accurately anticipate civilian actions and response. For example, while bribery and nepotism are not accepted norms in U.S. society, they may be customary practices in the foreign urban environment in which Army forces are operating. If commanders attempt to understand incidences of bribery and nepotism using their own societal norms and values as a basis, they are likely to misinterpret the significance of the events and implement an inappropriate course of action. Critically, however, the development of an in-depth understanding of the urban society does not mean that commanders and their Soldiers should adopt the culture of the foreign urban society. While understanding and respect of the urban area's culture is crucial to successful UO, Soldiers function best in dealing with the urban society when they act in accordance with American culture and values.

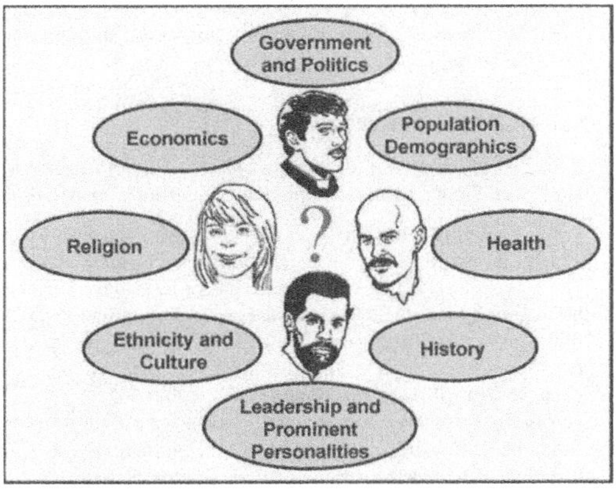

Figure 2-7. Key aspects of the urban society

POTENTIAL CENTER OF GRAVITY

2-45. A COG during an urban operation, particularly in stability and civil support operations, may be the civilian inhabitants themselves—specifically their behavior. However, supportive behavior is generally an advantage in any type of operation. Correspondingly, neutral behavior toward friendly forces is an advantage over hostile behavior. (However, a neutral population normally will not readily provide information to Army forces—often critical to understanding the urban environment and threat. In this light, neutrality can be seen as a disadvantage.) To influence or control their behavior, commanders must first understand the society's complex nature and character. Second, they must understand and accept that every military action (or inaction) may influence the relationship between the urban population and Army forces, and, by extension, mission success or failure. Lastly, they must understand that Army forces may play only a supporting (but essential) role as part of an integrated and synchronized multiagency effort focusing all aspects of national power—diplomatic, informational, military, and economic. With this awareness, commanders can take one or more actions:

- Coordinate and plan operations.
- Implement effective civil-military programs.

- Take the immediate action necessary to maintain support of a friendly populace, neutralize or gain the support of hostile or neutral factions, or do any combination of these activities to achieve precise effects and accomplish the mission.

Without this understanding, commanders increase the risk that their actions, particularly concerning the urban population, may not have the intended and necessary effects.

2-46. Although the factor of civil considerations takes on added significance in UO, it is just one that commanders evaluate. Sometimes, as in most urban counterinsurgency operations, it may be the most important factor to consider as a COG. At other times it may be the least important as to be almost negligible. Its importance is not constant; it changes over time (like all factors). At the beginning of the operation, civil considerations may not be essential to mission accomplishment, but as the operation progresses this factor's importance to success may increase. In other circumstances, the opposite may be true. Overall, commanders should consider three objectives regarding the civilians of the urban area:

- Minimize their interference with urban operations. In offensive and defensive operations this means moving them away from combat operations or establishing measures to shield them from its effects. In all operations, it often requires centralizing them in one or more locations—which may mean keeping them where they are.
- Maximize their support of Army, joint, and multinational forces and government agencies.
- Observe the necessary legal, moral, and humanitarian obligations.

General Population Size

2-47. Urban areas are commonly classified according to the general size of their population instead of landmass. Figure 2-8 lists one method of defining and categorizing urban areas by various population sizes.

2-48. These categories are useful to establish commonality and standardize terms that shape ideas, discussion, and concepts. Smaller populations usually suggest homogeneity among the inhabitants. Homogeneity can make consensus or compromise easier to achieve because fewer opposing viewpoints exist. Given this homogeneity, effects of change are more certain and often easier to determine. However, homogenous does not mean identical. If major social divisions do exist (either physical or ideological), commanders can more easily determine those divisions and their fundamental causes with smaller populations. Treating an urban population as a completely homogenous entity can lead to false assumptions, cultural misunderstandings, and poor situational understanding.

Category	Population
Village	3,000 or less.
Town	Over 3,000 to 100,000.
City	Over 100,000 to 1 million.
Metropolis	Over 1 million to 10 million.
Megalopolis	Over 10 million.

Figure 2-8. Urban areas by population size

2-49. As urban areas expand, the urban patterns begin to blur and the social complexity increases. For example, as satellite patterns continue to grow, the LOCs between a central hub and outlying urban areas may develop and begin to assume a linear urban pattern. Simultaneously, a hub and outlying urban areas may continue to expand until they merge into a single, large metropolis. On a larger scale, a network pattern can grow and unite as a single megalopolis. This growth physically unites smaller urban areas but cannot force conformity of needs and beliefs. It also increases the physical and social complexity of an urban area.

Group Size, Location, and Composition

2-50. Understanding how specific elements of the urban society affect operations (and vice versa), begins with analyzing their size, location, and composition (see figure 2-9). Because commanders must minimize civilian casualties, size and location (without regard to composition) are important initial demographic considerations. After determining the presence and numbers of civilians relative to decisive points, commanders can then decide whether civilian proximity and density represent a significant risk to their mission—refugees clogging LOCs, for example. If civilians are the primary focus of the operation, as in many stability and civil support operations, this same analysis may help to determine decisive points. In this analysis, commanders should consider that urban areas, on many levels, are in constant motion. The densities of circulating people and other traffic often vary according to cultural events or the time of day, such as religious holidays or sporting events and rush hours or market times. Therefore in planning urban operations, commanders must consider the timing, rhythms, or patterns of the population and their vehicular movements in the urban area. Identifying and understanding trends and patterns of activity (and disruptions to them) may provide critical information to commanders.

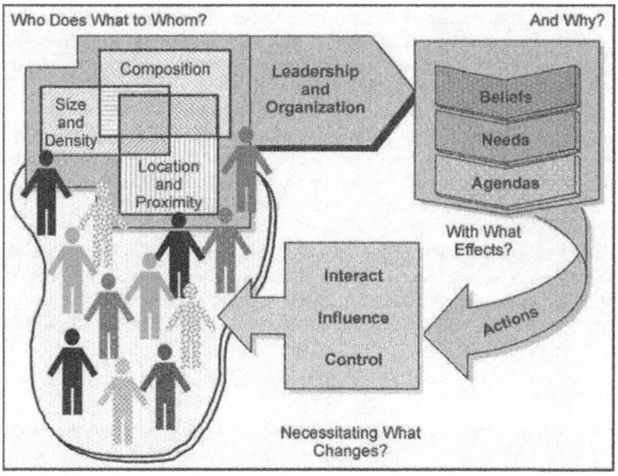

Figure 2-9. Simplified analysis of urban society

2-51. Commanders normally determine the composition of, or the identifiable groups or organizations within, the civilian urban population. Groups may be categorized by race, religion, national origin, tribe, clan, economic or social class, party affiliation, education level, union memberships, age, gender, occupation, or any other significant social demographic. Physical and ideological overlaps (and divisions) often exist between groups. Overlaps may provide early focus for analysis and suggest ways to affect more than one group simultaneously. In some cases, groups may have radically different ideologies but are (or can be) united by a single characteristic. Commanders must understand the intricacies of "who does what to whom." Such understanding furthers identifying the urban society's sources of power, influence (both formal and informal), and decisive points that hold the keys to controlling or protecting this potential COG. (See also the discussion of competing power structures in Chapter 3.) Commanders should have expert, detailed, and current knowledge and information to avoid developing simple formulas of social interaction that may actively mislead and add to a flawed course of action.

Leadership and Organization

2-52. Commanders must also understand how authority and responsibility is held or shared within and between each of the identified groups; they must understand leadership and the social hierarchy. For groups to exert meaningful influence, leadership provides vision, direction, and organized coherence. This leadership can be a function of personality as well as organization. Some groups depend on a charismatic leader to provide cohesion; although in some cultures, the spokesman is not the leader. Others de-emphasize individual leadership and provide redundancy and replacement in decision making. Others combine elements of both these types of leadership and organization. Based solely on personality, a leader may centralize power or, while still being in ultimate control, decentralize decision making and execution to subordinates. In contrast, a single person may head a group while a ruling council actually makes and executes policy. Groups centered on one leader (which may or may not be the officially designated leader) can often produce decisions and initiate actions rapidly but are vulnerable to disruptions if key personalities are removed or co-opted. Groups with shared or redundant leadership take longer to make decisions yet are more resistant to change and outside influence.

2-53. In UO, particularly stability operations, Army commanders and leaders at all levels will devote considerable effort to identifying and cultivating relationships with civilian leaders in their AO. This civilian leadership will include political, religious, tribal or clan, ethnic, and economic leaders. Commanders should consider that their attention toward and discussion with identified leaders may increase (or, some instance, decrease) the targeted leaders' prestige and power. While this may be intentional, commanders must often ensure that the leaders that they chose to deal with are the legitimate and accepted in the eyes of the urban population. Otherwise, they may further imbalance an already weak power structure and exacerbate the situation. In unique circumstances, commanders may need to identify and interact with the leadership of criminal organizations. These civilian leaders will be the means to affect change and allow the urban populace to understand and accept the purpose behind Army operations. As importantly, they will be the conduit for understanding the urban society's sentiments, perceptions, and reactions to Army operations. As such, reliable and trustworthy linguists that can not only interpret the language but can serve as cultural advisors to Army forces will be in great demand. Keeping the local population objectively informed not only of current Army operations but also of the intent and desired end state will often be a key task for commanders. A poor communications effort and a lack of understanding of intent may alienate a commander and cause significant problems in future negotiations. It is imperative that communications be clear and effective and that all Army leaders are aware of its implications. See FM 4-05.401 and FM 3-07 for the effective use of interpreters and the conduct of productive meetings and negotiations.

Interests and Actions

Me and Somalia against the world, me and my clan against Somalia, me and my family against the clan, me and my brother against my family, me against my brother.

Somali Proverb

2-54. Identifying and analyzing groups also helps commanders focus on specific segments of the urban society to determine their beliefs, needs, and agendas. It also helps commanders determine how those interests motivate groups to future action (or inaction)—previous patterns of activity are critical in this regard. This analysis seeks to determine why groups (and their leaders) act as they do. Commanders should consider political, economic, cultural, and religious factors in this analysis. These factors affect all groups to some extent and often provide the basis for their beliefs, needs (actual or perceived), and subsequent behavior. Size and location considerations also apply to each group to help determine to what extent its beliefs or ideologies, needs, and actions may impact the urban operation. However, size and proximity may not accurately indicate actual or potential capabilities. Individuals, small groups, and groups located some distance from the actual conduct of the urban operation may be able to influence large portions of the population. These individuals or groups may have a capability disproportionate to their size and proximity—especially against objectives that are not terrain oriented (as in the case of many stability operations).

Interaction, Influence, or Control

2-55. Commanders must cultivate an understanding of a group's—

- Size, location (and proximity to operations), and composition (to include leadership and organization).
- Interests.
- Capabilities.
- Potential actions (intent) and their effects—if any—on operations.

2-56. Then they can develop or modify courses of action as appropriate. Certain courses of action may be needed to improve the interaction between Army forces and civilians (and between other agencies) to accomplish common goals. Others may be needed to influence favorable support, stabilize neutral groups, or neutralize hostile groups. Still others may require more forceful means to control and protect—never to punish—civilians. Courses of action may include—

- Establishing buffer zones and restricted areas.
- Setting up checkpoints and roadblocks with other travel restrictions on people and goods.
- Screening civilians.
- Conducting negotiations (directly or as a mediator).
- Providing or protecting rations, water, and other critical resources.
- Restoring or improving specific, key infrastructure.
- Enforcing curfews.
- Inspecting facilities.
- Directing amnesty programs.
- Conducting internment and resettlement operations.
- Maintaining a "stay-put" policy.

2-57. Commanders must remember that many measures will require significant resources that may initially be beyond the capabilities of the Army force to impose and enforce. (Where possible, commanders should attempt to use host-nation security forces and local urban law enforcement to control activities.) Other elements of the environment, terrain and infrastructure, may fragment efforts and make it difficult to impose controls throughout the area. A careful assessment and understanding of the urban society's interests (beliefs, needs, and agendas) is essential before implementing any populace and resources control measures. Otherwise inappropriate controls, particularly if they are improperly perceived as punishment, may only aggravate the situation. Finally, an appropriate course of action may require no specific action towards the urban society. In most cases, training and discipline, grounded in cultural understanding and sensitivity, will help mitigate many potential adverse effects resulting from military-civilian interaction. (Commanders should review FM 3-19.15 and FM 3-19.40 for additional civilian control measures and considerations.)

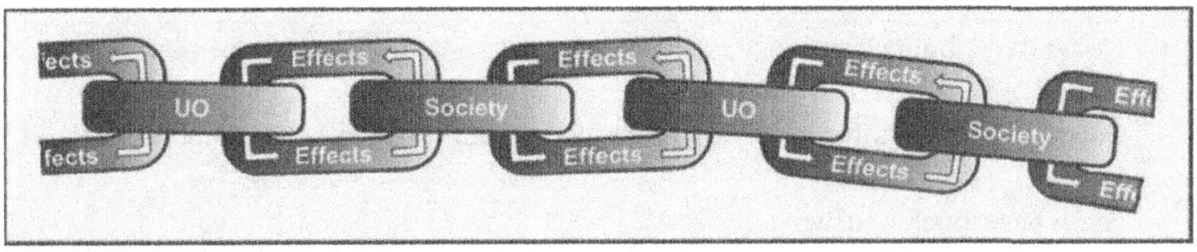

Figure 2-10. UO society chain of effects

A Chain of Effects

2-58. Since urban society is so dynamic and the relationship between various elements so complex, commanders must continually assess how their operations will affect the society and vice versa (see figure 2-10). Specifically, they assess how effectively their measures improve interaction with, influence of, and control over civilians. There is always a difference between intended and actual effects of a specific course of action. Nowhere is this more prominent than dealing with the urban society. This chain of effects frustrates understanding during UO. Therefore, commanders must continuously monitor these effects to make decisions and modifications while planning, preparing, executing, sustaining, and transitioning UO. (Commanders may conduct informal and formal polling activities to assess and understand changing civilian perceptions.) Initially certain aspects of the society, such as religion, may not affect the operation. However, if the threat successfully shapes the perceptions of the urban populace that Army forces are biased against their religion or religious beliefs (or at least critical segments are affected by propaganda), this element may become extremely important. In this instance, the urban commander may need to adjust his IO (to include PSYOP), public affairs (PA) activities, and CMO to counter this propaganda while diverting other combat power to control the populace. Overall, commanders must understand and account for second, third, and higher order effects of their actions and decisions.

Understanding the Effects of Unit and Soldier Actions in Iraq

During OPERATION IRAQI FREEDOM, an armor task force commander described his methodology as that of plotting and measuring everything. After a positive or negative event, he would have his staff evaluate all actions they had conducted before, during, and after the event. This would allow him to correlate activities with outcomes and develop tactics, techniques, and procedures for future success. While not all actions easily correlated to events, this methodological approach helped establish a base line of comparative success for his task force.

Some units assessed the Iraqi culture and developed practical guidelines or rules of interaction for their soldiers to follow as they conducted their missions. These guidelines encompassed a list of actions that their Soldiers should and should not do when dealing with the Iraqi populace. The guiding principle behind these actions was not only to "win the hearts and minds" of the Iraqi people but, more accurately, to establish legitimacy and credibility through appropriate Soldier actions that reflected a combined understanding of the mission and the Iraqi culture.

DO:

- Separate the men from the women and children during cordon and searches.
- Talk to the oldest or most senior male when civilian cooperation is needed.
- Bring something (for example, water, food, or replacement locks) to give residents to compensate for damages when the search ends up being a "dry hole."
- Learn Arabic phrases; say hello and greet people on the street.
- Use female Soldiers to search and talk to Iraqi females.
- Cordon and knock whenever possible instead of kicking in doors needlessly.
- Understand that in Iraq, women do not have the same social status as men.

DO NOT:

- Force a detainee's head to the ground or use your feet to hold someone to the ground.
- Stare at or touch Iraqi women.
- Show obvious disrespect to suspects in front of family members.
- Manhandle the Koran or other religious objects.
- Commit to anything; instead of agreeing to (or denying) an Iraqi's request, it is good practice to say "In'sha Allah" (If God wills it).

URBAN INFRASTRUCTURE

2-59. Urban infrastructures are those systems that support urban inhabitants and their economy. They form the essential link between the physical terrain and the urban society. During urban stability and civil support operations, restoration or repair of urban infrastructure will often be decisive to mission accomplishment. During urban combat operations, destroying, controlling, or protecting vital parts of the infrastructure may be a necessary shaping operation that can isolate a threat from potential sources of support. A threat force operating in an urban area may rely on the area's water, electricity, and sources of bulk fuel to support his forces. This is true particularly when his bases or facilities are physically located in or near the area. Isolating this threat from these sources may require him to generate his own electricity and transport his own water and fuel from outside the urban area. To transport supplies, the threat may rely on roads, airfields, sea or river lanes, and rail lines. Controlling these critical transportation nodes may prevent the threat from resupplying his forces. The control of key radio, television, and newspaper facilities may isolate him from the urban populace (another potential source of support).

INTERDEPENDENCE

2-60. Commanders must understand that destroying or disrupting any portion of the urban infrastructure can have a cascading effect (either intentional or unintentional) on the other elements of the infrastructure. Yet, they may be able to gain an operational advantage while minimizing unwanted and unintended effects. Commanders can control, seize, or secure an essential facility or structure by using precision munitions, electronic disruption of communications, or SOF and conventional ground forces. To gain this advantage, commanders will rely more on the expertise of Army engineer and civil affairs units; local urban engineers, planners, and public works employees; and others with infrastructure-specific expertise. After understanding the technical aspects of the area's systems and subsystems, commanders can then develop the best course of action.

A SYSTEM OF SYSTEMS

2-61. Hundreds of systems may exist. Each system has a critical role in the smooth functioning of the urban area. Simple or complex, all systems fit into six broad categories (see figure 2-11). Commanders should analyze key facilities in each category and determine their role and importance throughout all phases of the urban operation. As there is much overlap between infrastructure systems, this analysis considers each system individually and in relation to others to determine an appropriate course of action toward it.

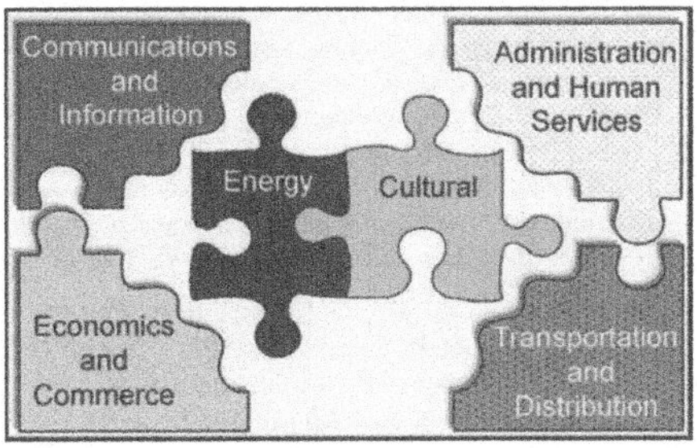

Figure 2-11. Urban infrastructure

A COMBINATION OF STRUCTURES AND PEOPLE

2-62. As depicted earlier in figure 2-1, each element of the infrastructure consists of both a physical (terrain) and human component. For example, the physical component of the electrical segment of the energy infrastructure consists of power stations, substations, a distribution network of lines and wires, and necessary vehicles and repair supplies and equipment. The human component of this same segment consists of the supervisors, engineers, linemen, electricians, and others who operate the system. Commanders must understand and recognize the physical and human components in their assessments.

POTENTIAL IMPACT ON FUTURE OPERATIONS

2-63. Destroying or incapacitating of any of these elements may impact future operations and inhabitants of the urban area. Destroying urban infrastructure during initial phases of an operation may require commanders to assume responsibility for repair, maintenance and clean up, and operation of those same facilities later. Although exceptions will exist, commanders cannot destroy or significantly damage the infrastructure of a foreign urban center during operations and expect the population to remain friendly to U.S. or allied forces. On the other hand, early repair or restoration of critical or essential infrastructure may improve civil-military relations, speed transition back to competent civilian authorities, and, overall, aid in successful mission accomplishment. Still, support from the urban society (albeit of increased importance in UO) is only one factor that commanders weigh while developing appropriate courses of action.

RESOURCE INTENSIVE

2-64. Requirements to protect, restore, or maintain critical infrastructure may divert substantial amounts of resources and manpower needed elsewhere and place additional constraints on subordinate commanders. Civilian infrastructure is often more difficult to secure and defend than military infrastructure. The potentially large and sprawling nature of many systems (such as water, power, transportation, communications, and government), make their protection a challenge. Yet, the infrastructure of an urban area may provide commanders with essential logistics and support. Therefore, the initial expenditure of time and other resources may be necessary to support concurrent or future operations. Legal considerations, however, may affect using the infrastructure and acquiring the urban area's goods and services. Commanders, their staffs, and subordinates (often down to the individual soldier) must know their limits concerning Army authority to commandeer civilian supplies or equipment to facilitate mission accomplishment (see the legal support discussion in Chapter 9). In stability and civil support operations, the safeguard or restoration of critical urban infrastructure for military or civilian use may be a decisive point in the overall operation.

2-65. Keys to understanding the magnitude of the resources and manpower required to restore the infrastructure are an initial infrastructure assessment and, as soon as practical afterward, a detailed infrastructure survey. An initial assessment provides the commander immediate feedback concerning the status of basic services needed to meet the urgent needs of the urban population. The systems assessed are based on mission, enemy, terrain and weather, troops and support available, time available, civil considerations, and the commander's vision of the overall end state. The infrastructure assessment, while typically performed by engineers, may be accomplished by, or in conjunction with, others with sufficient expertise to provide the type and quality of information required. These others may include civil affairs, medical, and chemical personnel. Those tasked with this assessment should routinely consult other Army and coalition forces and governmental and nongovernmental agencies currently operating in the urban area as well as the urban civilian leadership for their informed input.

2-66. While an infrastructure assessment functions to support the resolution of immediate challenges to urban reconstruction and restoration, it also provides the initial basis for determining the conditions for successful transition. However, commanders and planners must continually expand and refine their understanding. As a necessary follow-on, commanders initiate a detailed infrastructure survey. This survey is normally conducted by U.S. Army Corps of Engineers personnel assigned to forward engineer support teams. As with the assessment, the commander should incorporate other technical specialty personnel in the survey team to enhance the quality and accuracy of the product (see FM 3-34.250).

COMMUNICATIONS AND INFORMATION

2-67. This system is comprised of the facilities and the formal and informal means to transmit information and data from place to place. Understanding communication and information infrastructure of an urban area is important because it ultimately controls the flow of information to the population and the enemy. It includes—

- Telecommunications, such as telephone (to include wireless), telegraph, radio, television, and computer systems.
- Police, fire, and rescue communications systems.
- Public address, loudspeaker, and emergency alert systems.
- The postal system.
- Newspapers, magazines, billboards and posters, banners, graffiti, and other forms of print media.
- The informal human interaction that conveys information such as messengers, open-air speeches and protests, and everyday conversations.
- Other inventive informal means such as burning tires and honking horns.

2-68. Perhaps more than any other element of the infrastructure, communications and information link all the other elements in an interdependent "system of systems." It is a critical enabler that helps coordinate, organize, and manage urban activities and influence and control the urban society. Army commanders are acutely aware of the impact that a loss or degradation in communications has on their own operations. The urban environment experiences similar impacts to communication failures; however, urban governments and administrations are generally less prepared to deal with a collapsed communications and information infrastructure than are trained Army forces.

2-69. Militarily, a functioning urban communications and information system can serve as an alternate for both friendly and threat forces and can be easily secured with civilian, off-the-shelf technologies. Threats may make use of commercial systems intertwined with legitimate civilian users, making it unpalatable to prevent use of these assets. Forces can also use these systems to influence public opinion, gain intelligence information, support deception efforts, or otherwise support IO.

Increasing Impact of Computers

2-70. In many urban areas, computers link other elements of the urban infrastructure. They link functions and systems in the urban area and connect the area to other parts of the world. This latter aspect creates important implications for commanders of a major operation. Operations involving this cybernetic function may produce undesirable effects on a greater scale than initially intended. For example, commanders may be able to close or obstruct an urban area's banking system; however, this system may impact the international monetary exchange with unwanted or even unknown effects. The authority to conduct these types of IO will often be retained at the strategic level.

Pervasive Media

2-71. The media is central to the communications and information infrastructure and a critical operational concern. Compared to other operational environments (jungles, deserts, mountains, and cold weather areas), the media has more access to urban operations. This is due largely to airports, sea and river ports, and major road networks; ready access to power sources and telecommunications facilities; as well as access to existing local media structures. Hence, media presence may be pervasive and IO even more critical to success in UO than operations in many other environments.

A Complex Relationship

2-72. A complex relationship exists among information, the public, and policy formulation. Although the degree and manner in which public opinion shapes government policy are difficult to accurately determine, negative visual images of military operations presented by the media can change political objectives and, subsequently, military objectives. As important, media reporting can influence civilian activity in an urban AO to either the advantage or disadvantage of the commander.

Induce Cooperation Through Credibility

2-73. Commanders do not control the media; however, they monitor the flow of information that the news media receives and subsequently reports. Consequently, commanders should plan and execute PA operations that will induce cooperation between the media and Army forces. Successful relations between urban Army forces and the news media evolve from regular interaction based on credibility and trust. More information is usually better than less, except when the release of such information may jeopardize security and the success of the operations and threaten the safety of Soldiers. However, commanders cannot simply withhold information to protect the command from embarrassment. They consider media interests as part of the normal planning process and work to ensure that information presented to the news media is accurate, timely, and consistent with operations security. Since the media will likely arrive in the urban area before the conduct of operations, early deployment of PA assets may be critical. Commanders should synchronize PA activities with CMO and PSYOP. Such action eliminates duplicated effort and ensures a unity of purpose consistent with the IO concept of support (see Chapter 4).

2-74. Failure to provide sufficient information can hamper a commander's ability to conduct the mission. Commanders cannot refuse to deal with particular news media because they consistently report a negative image of Army forces and operations. Poor relationships with any media can result in inaccurate and even biased reporting. Such reporting can cause a public reaction that influences the ability to achieve operational objectives. During the Russian 1994-95 battle against Chechen separatists in Grozny, for example, the Russian military refused to communicate with reporters. The media reported primarily from the perspective of the Chechen rebels. This encouraged both local and international support for the rebels. It also allowed the Chechens, who lacked sophisticated information systems, to use the media to broadcast operational guidance to their forces. (During their second Chechnya campaign of 1999-2000, Russia learned this lesson well and the Russian view of the war dominated domestic public opinion.) On the other hand, successfully engaging the media can serve as a force multiplier. The Army's open and responsive interaction with the media during peacekeeping operations in Bosnian urban areas helped to explain the challenges and successes of Army forces in the Balkans to the public. This helped maintain domestic, international, and local political support for NATO operations and, with a successful command information program, helped maintain Soldiers' morale.

TRANSPORTATION AND DISTRIBUTION

2-75. This element of the infrastructure consists of—

- Networked highways and railways to include bridges, subways and tunnels, underpasses and overpasses, ferries, and fords.
- Ports, harbors, and inland waterways.
- Airports, seaplane stations, and heliports.
- Mass transit.
- Cableways and tramways.
- Transport companies and delivery services that facilitate the movement of supplies, equipment, and people.

Similar to communications and information, this facet provides the physical link to all other elements of the infrastructure.

2-76. Army forces deploying into a theater of operations depend on ports and airfields; seizure and protection of these critical transportation nodes may impact the projection of combat power. Once in theater, transportation and distribution systems in the urban area can contribute greatly to the movement of forces, maneuver, and logistic operations throughout the entire AO. Control of decisive points in this infrastructure may be important to the military operation and to the normal functioning of the urban area (and surrounding rural areas). Supplies traveling through the transportation and distribution system may be military-specific supplies (such as ammunition and repair parts) and supplies for both the military and urban population (such as food, medicine, oil, and gas). The system may also support the movement of military forces and the urban area's population (for which it was designed). Therefore, commanders of a major operation may have to develop innovative methods that limit the transit of threat supplies and reinforcements while facilitating the movement of their own resources and those of civilians. This last consideration attempts to minimize hardship and promote normalcy in the urban area and will increase in significance as the need for legitimacy increases.

2-77. Most urban areas (particularly in developing countries) have two forms of transportation and distribution systems that exist simultaneously: a formal system and an informal or paratransit system. Large organizations, bureaucracy, imported technology, scheduled services, and fixed fares or rates characterize formal systems. Low barriers to entry; family and individual entrepreneur organizations; adapted technology; flexible routes, destinations, and times of service; and negotiated prices characterize the informal system. The informal system is more decentralized and covers a much greater portion of the urban area than the formal system. The informal transportation and distribution system often includes a waterborne element, is more likely to function through turbulence and conflict, and can extend hundreds of kilometers beyond the urban area. Accordingly, commanders should understand both systems to establish effective movement control.

ENERGY

2-78. The energy system provides the power to run the urban area. It consists of the industries and facilities that produce, store, and distribute electricity, coal, oil, wood, and natural gas. This area also encompasses alternate energy sources, such as nuclear, solar, hydroelectric, and geothermal power. Energy is needed for industrial production and is therefore vital to economics and commerce. Among many other things, this system also provides the fuels to heat, cool, and light homes and hospitals, cook and preserve food, power communications, and run the transportation necessary to move people and their supplies throughout the urban area. Loss of an important energy source such as electricity or gasoline, especially for those accustomed to having it, will become an immense area of discontent that the commander of a major urban operation will need to quickly address. Therefore some threats, particularly terrorists and insurgents, may actively target this element of the urban area's infrastructure to erode support for civilian authorities and Army forces.

2-79. Sources of energy may be tens or hundreds of miles away from the urban area itself. Therefore, commanders may exert control without applying combat power directly to the urban area itself by controlling or destroying the source (power generation or refinement plant) or the method of distribution (pipelines or power lines). With electrical energy that cannot be stored in any sizable amount, the latter may be the best means as most major urban areas receive this energy from more than one source in a network of power grids. However, control may be as simple as securing a power station or plant and turning off switches or removing a vital component that could later be restored. On the other hand, lengthy pipelines and power lines may compound security and protection of this element of the infrastructure.

2-80. The number of nations that have invested in nuclear power and nuclear research is increasing. With this increase, the potential for Army forces to operate in urban areas that include (or are near) these facilities also increases. Damage to one of these facilities and potential radiation hazards will present special challenges to commanders of a major operation. To safeguard friendly forces and civilians, commanders will need to employ a blend of peacetime and tactical nuclear contamination avoidance principles (see FM 3-11.14).

ECONOMICS AND COMMERCE

2-81. This system encompasses—

- Business and financial centers to include stores, shops, restaurants, hotels, marketplaces, banks, trading centers, and business offices.
- Recreational facilities such as amusement parks, golf courses, and stadiums.
- Outlying industrial, mineral, and agricultural features to include strip malls, farms, food processing and storage centers, manufacturing plants, mines, and mills.

2-82. An essential aspect of this area during operations may be the political sensitivity of U.S. or allied industries investing and operating in a foreign country, particularly during stability operations. An enemy or a disgruntled civilian population may attack or disrupt commercial activities as a political statement against the United States or our allies. Food production facilities also may assist commanders in Army food services and may be an essential concern during relief operations. During long-term stability operations, visible, material, and tangible economic progress consisting of the creation or restoration (and protection) of businesses, agriculture, and overall jobs will often be critical to—

- Generating or maintaining the urban population's support to Army forces and operations.
- Reducing support to threat forces and operations to include eliminating civilians as a potential manpower pool for insurgent or terrorist organizations and activities.
- Lowering other hostile civilian activities such as protests and riots.
- Transitioning the urban area back to legitimate civilian responsibility and control.

2-83. This element of the infrastructure also consists of the production and storage of toxic industrial chemicals used in agriculture (insecticides, herbicides, and fertilizers), manufacturing, cleaning, and research (to include biological agents). Fertilizer plants may be of specific concern as they contribute to providing a key material in terrorist and insurgent bomb-making activities. A thorough analysis of this element of the infrastructure may also be essential to understanding how urban insurgencies are funded and supported. This helps commanders to understand the true organization of the insurgency as well as to suggest methods to isolate insurgents from their economic or financial support. In their overall assessment of this area of the infrastructure, commanders should also consider the activities and influence of criminal organizations or elements.

ADMINISTRATION AND HUMAN SERVICES

2-84. This wide-ranging system covers urban administrative organizations and service functions concerned with an urban area's public governance, health, safety, and welfare. Together, it encompasses—

- Governmental services that include embassies and diplomatic organizations.
- Activities that manage vital records, such as birth certificates and deeds.
- The judicial system.
- Hospitals and other medical services and facilities.
- Public housing and shelter.
- Water supply systems.
- Waste and hazardous material storage and processing facilities.
- Emergency and first-responder services such as police, fire, and rescue.
- Prisons.
- Welfare and social service systems.

CULTURAL

2-85. This system encompasses many organizations and structures that provide the urban populace with its social identity and reflect its culture. (This infrastructure system overlaps with many recreational facilities included under the economics and commerce infrastructure. For example, an urban society may radically follow soccer matches and teams. Hence, soccer stadiums relate to the society's cultural infrastructure.) Some of these facilities, particularly religious structures, will be protected targets and others may require security and law enforcement protection from looting and pilferage. However, commanders will need to quickly educate, inform, and continually remind the urban populace (and the media) that cultural infrastructure may lose its protected status when used by threats for military purposes. Cultural infrastructure may include—

- Religious organizations, places of worship, and shrines.
- Schools and universities.
- Museums and archeological sites.
- Historic monuments.
- Libraries.
- Theaters.

Restoring and Protecting Essential Services

2-86. Losing the support of essential elements of the infrastructure will have an immediate, destabilizing, and life-threatening impact on the inhabitants of the urban area. In stability and civil support operations, numerous parts of the administrative and human services and energy infrastructure often rise to critical importance before all other elements. Again, however, complete restoration of these essential services is often a lengthy, resource-intensive civil-military operation. (Following the end of major combat operations of OPERATION IRAQI FREEDOM, most units developed one or more of their logical lines of operations oriented along the restoration or improvement of urban infrastructure and essential services. An acronym used by many units to focus and track critical activities within this line of operation was sewer, water, electricity, and trash (SWET). Later, other units modified this acronym to SWEAT and then SWEAT-MS to include concerns for restoring academics [or schools], revitalizing medical facilities, and establishing security [police and host-nation security forces].) Of critical importance will be a simultaneous IO campaign that includes efforts to help ensure that the urban population develops realistic expectations about Army abilities to restore their essential urban infrastructure.

Understanding the Urban Environment: Paris – 1944

The summer of 1944 confronted German General Dietrich von Choltitz with a dilemma. As military commander of greater Paris, he was to eliminate French Resistance internal to the city while defending against approaching Allied units, missions for which he had insufficient forces. (Notably, General Eisenhower, the commander of the Allied forces, wanted to bypass Paris to sustain the offensive. Seizing the French capital would task his forces with the support of tens of thousands of civilians. Eisenhower was nevertheless ordered to capture the city. A political decision resulted in civilian assistance taking precedence over combat operations.)

Choltitz's situation was further complicated by Hitler's demand that he destroy the city, an action the general saw as needlessly destructive (and infeasible given his scant resources). Choltitz's seniors directed the preparation, and later the destruction, of Paris's 45 Seine River bridges. They were the only remaining crossing points over that waterway given Allied bombing of others outside the French capital. Premature destruction would trap German forces defending to their north, a second-order effect that Choltitz used to justify his disobedience of orders demanding the bridges' demolition.

The German general also recognized that some mission-critical elements were part of Paris's social rather than physical infrastructure: the leadership of the various resistance groups and the relationships between them. Choltitz understood that he lacked resources to defeat the many separate factions; he therefore chose the unorthodox (asymmetric) approach of accepting an intermediary's offer of a truce with these groups. Such an agreement provided some measure of the stability needed while Choltitz awaited promised reinforcements. Further, he realized that the resistance factions were by no means united in their goals. Communist elements sought a much different end than those looking toward a de Gaulle-led postwar government. A truce thus set the French Communists (who sought an uprising so as to legitimize their claims to power) against others trying to buy time until Allied forces arrived, forces that included Free French units supportive of de Gaulle.

Although his defense of the capital failed, Choltitz succeeded in harboring his available resources, reducing the effectiveness of the resistance organizations fighting his soldiers, and maintaining withdrawal routes for units north of the Seine. The German commander's analysis in support of these efforts was effective in part because of his insightful (1) identification of critical points that included elements of terrain, citizenry, and infrastructure; (2) understanding of the relationships between these parts; and (3) use of an asymmetric approach to address his lack of sufficient force to otherwise handle the urban densities that challenged him.

This page intentionally left blank.

Chapter 3
Understanding the Urban Threat

... [T]he United States could be forced to intervene in unexpected crises against opponents with a wide range of capabilities. Moreover, these interventions may take place in distant regions where urban environments, other complex terrain, and varied climatic conditions present major operational challenges.

Quadrennial Defense Review Report, 30 September 2001

As the strategic environment has become less stable, more uncertain, and more dangerous, Army forces must be trained and ready to address persistent and evolving urban threats. These threats range from regional conventional military forces, paramilitary forces, guerrillas, and insurgents to terrorists, criminal groups, and angry crowds. These threats can hide in plain sight and become indistinguishable from the noncombatant urban population that may help shield, protect, and sustain them. Although uncertain about events, Army forces can be clear about trends. Increasingly, the Army will face threats that severely differ in doctrine, organization, and equipment, are skilled at developing and adapting techniques to counter Army tactics, techniques, and procedures (TTP), and can fully interact with the three other components of the urban battlefield—terrain, society, and infrastructure. In urban operations, commanders must broaden their concept of the threat to include natural disasters, hunger and starvation, and rampant disease. Further, commanders must plan to contend with many passive urban threats, such as psychological illnesses and toxic industrial materials (TIM). These threats may be found in isolation, but most likely commanders will encounter them in various combinations. Moreover, each new threat will pose a different combination and likely have new capabilities that previous opponents lacked.

ASYMMETRICAL AND ADAPTIVE

3-1. An emphasis on asymmetry to offset U.S. military capabilities has emerged as a significant trend among potential threats and become an integral part of threat objectives and tactics (discussed below). Though, this is not to imply that all future urban threats will fight and operate asymmetrically. Some may be asymmetrical; others may not. Future threats may be a symmetric conventional military; others may be paramilitary, insurgent, or even a nonmilitary threat such as a criminal organization. Asymmetry results when one opponent has dissimilar capabilities—values, organization, training, or equipment—that the other cannot immediately counter. It is not a new concept. It naturally evolves from a sound mission, enemy, terrain and weather, troops and support available, time available, civil considerations (METT TC) analysis by an intelligent, freethinking, and adaptive threat. These asymmetric approaches will include the most advanced, commercially-available technology innovatively applied and mixed with crude, simple, and unsophisticated weapons and TTP. The success of these approaches often depends upon the predictability of the forces on which they are employed. Thus, Army commanders must consider whether their operations and activities lend themselves to pattern analysis by threat forces and, if so, they may need to vary their battle rhythm and take other steps necessary to become less predictable.

WEAPONS OF MASS DESTRUCTION

3-2. A chief asymmetric means of engaging the national power of the United States is to employ weapons of mass destruction (WMD) against the United States or its allies. These weapons can be used against military forces and include high-yield explosives as well as nuclear, biological, and chemical weapons. Operations in urban areas may require concentrating forces and may create a lucrative target for a threat that possesses fewer numbers and less equipment.

3-3. A threat's WMD use will adversely affect the Army's abilities to conduct urban operations (UO) to various degrees. For example, the intervening structures and the effects of urban microclimates complicate the ability to detect and identify radiological, chemical, or biological attacks from a standoff distance. Also, the individual Soldier's ability to recognize his leaders, understand oral and visual commands, and operate increasingly sophisticated equipment is difficult when wearing protective clothing and equipment—particularly if his training proficiency is low. Despite the increased challenges and complexity, Army forces have the best (or better) training and equipment necessary to respond to such an attack compared with most armies around the world or with any civilian organization.

3-4. Although initial casualties could be high, the public can accept military casualties before those of civilians. Therefore, threats may gain an initial tactical advantage but would achieve less asymmetric benefit by directly attacking Army forces. Instead, they may attempt to achieve a strategic advantage by employing WMD against U.S. or allied civilian populations. In doing so, threats hope to use political sensitivity to high civilian casualties to reduce popular support for the United States or its allies. The chance of these attacks occurring in an urban area increases because—

- The area facilitates weapon effects and camouflages delivery means.
- The dense civilian population ensures a high casualty rate.
- The attack (or even the threat of attack) often will receive more publicity and public attention.
- The urban infrastructure is especially vulnerable to WMD, (particularly the systems of the economics and commerce infrastructure located in large urban areas) and may have far-reaching national and global effects.

THREAT OBJECTIVES

3-5. The threat will seek to achieve several key objectives when opposing Army forces operating in an urban environment (see figure 3-1). These objectives focus more on how a threat might fight in an urban area rather than specifically who the threat might be or in what region of the world the conflict might occur. These objectives are more easily achieved in an urban environment due to—

- The high costs in time, material, and manpower involved in UO.
- The limiting effects of urban areas on many technological advantages.
- The proximity of airfields and ports to urban areas.
- The potential moral dilemmas created by exposing numerous civilians to harm or injury.
- The ability of the media to observe and report the threat's version of events.

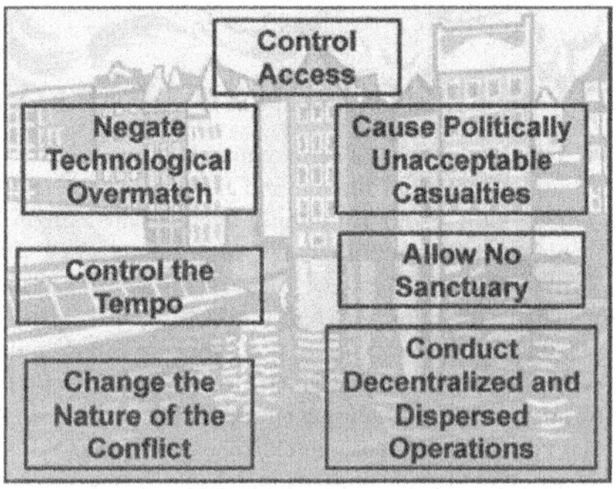

Figure 3-1. Threat objectives

3-6. These objectives complement and overlap each other; however, at their core is the threat need to defeat an enemy of superior numbers, technology, or both.

Control Access

> *... Iraq made no direct effort to impede the buildup in the theater. Planners should consider what might have happened if Iraq had attempted a strategy based on denying access to the region. Planners might also wonder what the outcome would have been if Iraq had attacked US forces in Kuwait before they were ready for the running start. The point is that the conditions in [US Central Command] in 2003 are unlikely to be replicated elsewhere.*
>
> *On Point: The United States Army in Operation IRAQI FREEDOM*

3-7. The Army may not be located where future conflicts are fought. Thus, the Army maintains the ability to rapidly project and sustain combat power over long distances and time spans. This capability demands that Army forces quickly gain and maintain control of seaports or aerial ports of embarkation or debarkation, particularly where the density of U.S. basing and en route infrastructure is low. Commanders gain control of these ports by unopposed (assisted or unassisted) or forcible entry operations. In either case, these phased-entry operations may present potential vulnerabilities, particularly—

- Unsuitable composition of initial or early entry forces lacking necessary combat power for immediate decisive operations.
- Initial command and control difficulties and an immature situational understanding.
- Weak initial port or base defenses without the necessary integration of naval and air component forces.
- Lack of developed logistic support.

3-8. Consequently, threats may attack during initial force projection operations to oppose, delay, disrupt, or otherwise manipulate the build-up of essential combat power into a theater of operations. These attacks may occur anywhere deploying Army forces are located, at overseas bases, at home stations, and even in military communities. Increasingly, deployment facilities such as airfields and ports exist as integral components of urban areas. Threats will invariably use the complex and concealing nature of these urban areas, coupled with the vulnerabilities, to create favorable conditions for their attacks.

Negate Technological Overmatch

3-9. Threats will always strive to force engagements at a time and place most advantageous to them. They may locate military forces and vital military capabilities in urban areas to achieve sanctuary from the effects of Army capabilities and make Army forces and systems more vulnerable to less-sophisticated weapons.

3-10. The clutter of the physical structures, electromagnetic radiation, and population diminishes Army capabilities. This clutter makes it difficult for Army forces to acquire and effectively engage targets at long ranges. In urban areas, the terrain often allows a threat to operate in closer proximity to friendly forces. Therefore, the threat may "hug" friendly forces to avoid the effects of high-firepower standoff weapon systems and degrade their ability to gain or maintain a thorough common operational picture. Additionally, this threat tactic attempts to inhibit friendly commanders from employing some weapon systems and munitions for fear of fratricide. Threats will also seek to hide and operate in culturally sensitive areas and sites, such as places of worship and schools, to establish temporary safe havens and refuge from Army forces operating within the urban environment.

Control the Tempo

3-11. Threats will try to achieve a decisive advantage by setting and controlling the tempo. To prevent the Army's entry into theater, threats may try to create a high operational tempo to take advantage of the inherent weaknesses in power projection operations outlined earlier. As other efforts deny entry, threats may seize the initiative, achieve surprise, and exploit the tempo differential by attacking with heavy conventional forces potentially possessing greater firepower and more rapid ground mobility than the Army's initial-entry forces.

3-12. If they cannot deny entry or end the conflict quickly, threats may use any preparations made in the initial high-tempo period to prolong the event, aiming to degrade U.S. or allied commitment. (Some threats may conduct operationally planning in terms of decades rather than years.) The complex nature of the urban environment slows operations conducted in, around, and over these areas. Threats may maximize this characteristic by fighting (conventionally or unconventionally) for key urban complexes and infrastructure, forcing friendly forces to operate within these areas. If Army operations focus on one or more urban areas, the overall campaign slows. However, even when UO make up only one component of a much larger campaign, they may consume valuable resources needed for other operations and delay the entire campaign. (See also the discussion of Tempo in Chapter 7.)

Tempo

The battle for Aachen, Germany, in the fall of 1944, developed during the U.S. First Army's offensive to breach the Westwall fortifications. Aachen, the ancient capital of Charlemagne, had symbolic political and psychological significance for the Germans and Americans. Furthermore, it was the first city on German soil to face an assault by the Allies. Consequently, the symbolic importance of this first major battle in Germany ensured bitter resistance against American attackers. The Germans surrendered only after the city was destroyed. Expected to take a few days, instead, the battle took weeks. Although the Army had achieved a clear tactical victory, the German defense of Aachen cost the First Army valuable time and resources, and delayed the planned attack to the Rhine River.

CHANGE THE NATURE OF THE CONFLICT

3-13. Threats may attempt to change the fundamental nature of the urban conflict to exploit ambiguous or tenuous political-military objectives. Many nations must gain and maintain domestic popular support to use their armies for political objectives. The threat may attempt to change the nature of the conflict by modifying its strategy and tactics, the environment, or any combination, ultimately hoping to reduce friendly popular support. For example, introducing an urban terrorist threat to U.S. civilians or Soldiers not directly engaged in operations changes the nature of the conflict. This type of threat may not have been an initial consideration, and this change may reverse public support for the operation. Another example, growing U.S. coalition combat power may cause the threat to switch from open maneuver warfare to urban insurgency operations to avoid decisive combat with superior forces and achieve a stalemate. Originally expecting a quick solution or victory, the political leadership may now envision a longer deployment with less chance of lasting success.

CAUSE POLITICALLY UNACCEPTABLE CASUALTIES

3-14. Threat forces may gain an advantage against superior friendly forces by capitalizing on a perceived weakness of many Western nations: the inability to endure continuous losses or casualties for other than vital national interests or losses for which they are psychologically unprepared. A secondary U.S. interest may equate to national survival on the part of a threat. Therefore, the threat (particularly with fanatical leadership) may willingly sacrifice excessive amounts of money, equipment, and people (soldiers and civilians) to achieve victory. Threats may attempt to weaken U.S. resolve and national will to sustain the deployment or conflict by inflicting highly visible, embarrassing, and if possible, large losses on Army forces, even at the cost of similar losses to themselves. Many threat forces will use UO to inflict mass casualties and destroy premier Army weapon and information systems. The physical characteristics of the urban environment support these ambush techniques. Light infantry or insurgents with readily obtainable, hand-held antiarmor weapons can effectively attack armored vehicles and helicopters, no matter how sophisticated, in an urban area.

ALLOW NO SANCTUARY

3-15. Threats will attempt to deny Army forces safe haven anytime and anywhere. Terrorism and insurgency may be some of the tactics used to deny sanctuary to Army forces. They will attack Army forces anywhere, particularly while operating in urban areas where the fear from being attacked from any quarter is often greater. Threats may be or employ state-sponsored or independent terrorists, well equipped and motivated to accomplish their assigned missions.

3-16. Military buildings, facilities, and installations in urban areas are particularly vulnerable to high-yield explosive munitions as well as other clever means to create large explosions. The close-in nature of urban areas, large populations, and high volume of vehicle traffic provide a good environment for target reconnaissance, explosives positioning (conventional or high-yield), and cover for an attack. These attacks will likely be preceded by extensive, careful reconnaissance, necessitating a solid friendly counterterrorism, counterinsurgency, and counterintelligence effort.

CONDUCT DECENTRALIZED AND DISPERSED OPERATIONS

3-17. To a certain extent, decentralized and dispersed operations are an integral part of all threat objectives. However, this concept warrants separate emphasis as an objective since threat forces will likely place great significance on it on future urban battlefields. Both dispersed and decentralized approaches seek to reduce threat vulnerabilities to air power and precision-guided munitions (PGM) while increasing their agility, flexibility, and overall maneuverability in an urban environment.

3-18. Urban terrain tends to fragment and separate forces that operate in it. Threat forces recognize this characteristic, accept it, and make it work to their advantage. They conduct operations from dispersed urban locations to reduce their vulnerability to friendly decisive operations and massed firepower. Although separated, threat forces will attempt to retain the ability to assemble and mass quickly so to strike as opportunities present themselves. Once threat forces complete the operation, they will return to separate locations to avoid potential counterattack. The fluidity and seemingly disjointed appearance of these UO threats will challenge friendly efforts to conduct templating and pattern analysis. Ambushes (air and ground) will be used to deny friendly ground and air reconnaissance of their dispersed locations.

3-19. Dispersed operations normally depend on good command and control to achieve synchronization and massed effects. Threat forces also understand the debilitating effects of the urban terrain on communications and the execution of operations. When they cannot mass their forces or effects, they will depend on decentralized operations to achieve their objectives. They will operate autonomously, guided only by a higher authority's purpose and intent. These operations make them even less vulnerable to massed attacks and PGM as smaller threat forces do not present an objective or target that will allow friendly decisive operations. Again, pattern analysis and templating, although essential, will be extremely difficult. This objective may prolongs the conflict but is central to achieving the other threat objectives.

3-20. When the threat employs decentralized operations, is effectively isolated, or both, Army and coalition forces are much more likely to be able to share lessons learned and emergent TTP than a dispersed threat. Threat isolation, either imposed by Army forces or as a result of the threat's desire to remain concealed from Army intelligence, surveillance, and reconnaissance (ISR) capabilities, reduces the threat's ability to freely and easily exchange information and quickly adapt. As learning organizations, this allows Army forces to develop dominance over a threat's weapons and tactics for longer, albeit temporary, periods of time. This ability to rapidly train and learn is an advantage the Army and coalition forces must maintain while denying the same to future threats.

URBAN THREAT TACTICS

3-21. Urban areas provide a potentially casualty-producing and stress-inducing environment ideally suited for using specific urban tactics. Moreover, urban areas can provide threats with an unmatched degree of cover and concealment from friendly ISR and firepower systems. Understanding how these tactics will be used often requires a concurrent understanding of the threat's culture, customs, language, and philosophical ideals. While active urban threats may vary widely, many techniques will be common to all.

Figure 3-2 outlines a set of tactics available to potential threats opposing mission accomplishment in urban areas. Army forces may use many of these tactics, except those that violate the law, ethics, and morals, to defeat urban threats. Moreover, using asymmetric means is not the sole domain of the threat. Army commanders can also leverage capabilities, create conditions, and plan operations to develop their own asymmetric advantages to accomplish the mission.

USE THE POPULATION TO ADVANTAGE

3-22. Many urban areas may be too large to evacuate completely (if at all). Even if desirable, a military force may have no place to safeguard and secure the inhabitants. Therefore, future UO may see large segments of the populace remain. Offensive and defensive operations may be constrained not only by the terrain, but also by the presence of many civilians. Army forces involved in urban stability and civil support operations will certainly conduct missions in and amongst the residents. These residents may restrict operations and, when gathered in large numbers, may (even without initial hostile intent) present a critical force protection issue for the commander.

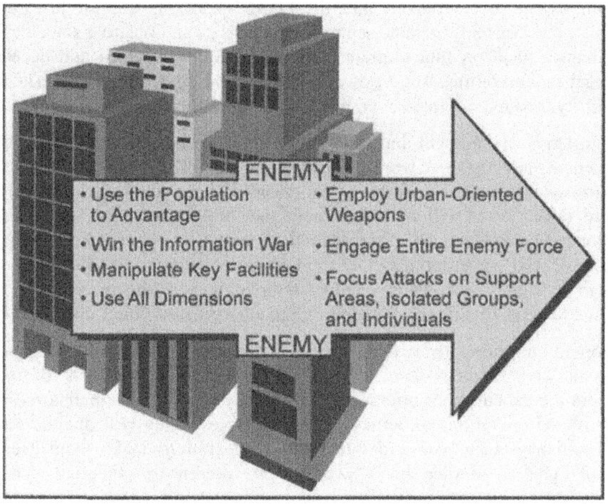

Figure 3-2. Urban threat tactics

Use as Key Terrain and Concealment

3-23. From the threat standpoint, the populace is similar to key terrain: the side that manages it best has an advantage. Threat forces may gain this advantage by using civilians as camouflage, concealment, and a means of deception. Guerrilla and terrorist elements may look no different from any other member of the community. Many foreign conventional and paramilitary troops—in addition to terrorists and insurgents—often have a "civilian" look. Western military forces originally adopted the clean-shaven and close-cut hair standards to combat disease and infection, but future opponents may not adhere to those standards. Instead, they may adopt grooming standards, civilian-looking clothing, and other "nonmilitary" characteristics to make themselves indistinguishable from the civilians. To avoid defeat or capture, threats will often throw down their weapons and attempt to blend into the urban population. Therefore, commanders must adopt effective populace and resource control measures to limit, identify, and track threat movements through the urban area of operation.

Identifying Threats from Noncombatants

During Russia's 1994-95 conflict with Chechnya, Russian forces had difficulty identifying Chechen guerrilla forces from Grozny's noncombatant population. Because their dress and ethnic appearance was identical to that of the urban populace, Chechen soldiers could freely walk around the city, suddenly disappear, and then abruptly reappear firing their weapons from basements, windows, or dark alleyways. To distinguish fighters from peaceful city dwellers, Russian forces began looking at men's shoulders to see if they were bruised (from firing weapons) and their forearms to see if there was burned hair or flesh (from the extraction of cartridges). They closely examined their clothing and smelled for gunpowder residue. To identify a Chechen artilleryman, Russian soldiers checked for glossy spots left by artillery and mortar rounds on the bends and cuffs of sleeves. They also turned pockets inside out to check for a shiny, silvery-leaden hue indicating the former presence of small arms ammunition. Russian forces also recognized a grenade launcher operator or mortar man from fibers and crumpled pieces of gun cotton on their clothing. U.S. Army commanders may need to develop similar, imaginative techniques and rapidly integrate technological means to identify and track threat forces such as—

- Military working dogs, ion spray, and electronic vapor tracers to locate explosives (or their former presence).
- Biometrics—measurable biological characteristics—such as fingerprints, voiceprints, facial scans, and retina scans.
- Electronic and chemical tagging devices.
- Monitoring hospitals for injuries that indicate participation in combat activities. For example, improper use of rocket-propelled grenade launchers may result in significant burns to the lower part of the body due to the back blast.

Gain Cover, Protection, and Increased Mobility

3-24. Threat forces may attempt to gain cover by using the urban inhabitants as human shields. With this increase in protection, they simultaneously increase their mobility. They recognize the Army's focus on developing and applying rules of engagement (ROE). They will take advantage of the restraining effects of international law and the Army ethical values to enhance their mobility in proximity to friendly positions. Knowing the Army's reluctance to cause noncombatant casualties and collateral damage, threats may operate in areas containing civilians and essential facilities to restrict the Army's use of massed or nonprecision firepower. For example, they may use school buses and ambulances to transport fighters and equipment about the urban battlefield. They may also employ "rent-a-crowds"—civilians paid or incited to demonstrate against military forces—armed only with sticks, stones, and Molotov cocktails (a potential asymmetric challenge). Threats can also hide military equipment within sensitive infrastructure—for example, caching weapons in places of worship, schools, or medical facilities. As Army and coalition forces potentially respond by targeting and firing upon these locations, threats will make use of these reactions for their possible propaganda value.

Make Moral Responsibilities a Weakness

3-25. Depending on their successes, threats may use these tactics and skillful information operations that attack national will and coalition sensitivities in an attempt to force the Army to establish more restrictive ROE. For example, a teenaged fighter coerced or enticed to fire a rifle or rocket-propelled grenade launcher can be quickly stripped of his weapon once dead and displayed at the local hospital as a child victim of coalition forces. Threat forces may also take advantage of the Army's moral responsibilities. By herding refugees into friendly controlled areas, threat forces try to make the civilians a burden on the Army's logistic and security resources. Similarly, some threats may attack to kill NGOs and force the remainder of them out of the area to achieve the same burden effect. At the same time, these threats hope to cause the population to lose confidence in the Army's (and potentially the government's) ability to protect and provide them essential services.

3-26. Threat forces, on the other hand, may not abide by international agreements, such as the Geneva conventions. Threats may not take prisoners unless they can be ransomed or made part of a local prisoner exchange. They may even execute friendly prisoners in front of the media to show their "strength" and, more importantly, to cause friendly forces to overreact and lose their legitimacy. Threat forces can then use such an overreaction to unite others with their cause. Pretending to be civilian noncombatants, they may also feign injury or pretend to need assistance (such as a disabled taxi) with the primary motive of luring Army forces into a deliberate ambush. While the threat may gain a short-term advantage by using ethical responsibilities as a weapon, the Army commander's strict adherence to moral principles will, in the long run, contribute greatly to threat defeat—particularly during long-term urban stability operations.

Acquire Intelligence and Logistic Support

3-27. Indigenous threat forces can normally use the local population for intelligence and logistic support far more effectively than can an alien army. Threat forces may manipulate local hires serving among U.S. Soldiers, such as those contracted by the Army for base operation purposes or translator duties. In addition, refugees moving through friendly controlled sectors may provide the threat with information on friendly dispositions, readiness, and intent. (Women and children may be specifically used for this purpose since U.S. Soldiers' cultural bias can create an incorrect perception that these noncombatants pose no threat to security.) Even friendly residents may become unwitting or unwilling informants, providing an enemy or a hostile with vital information on friendly activities, dispositions, and capabilities. However, a threat employing particularly cruel, abusive, or repressive measures may easily turn certain groups in the urban area against them, even when they share a common history, culture, and ethnicity with the civilians. This is more likely in those areas with higher population densities.

3-28. Threat forces may also seek to use some nongovernmental organizations (NGOs). They may try to obtain relief supplies either through the organizations' legitimate relief operations or as a target for theft. Some organizations may even be fronts for weapons, food, ammunition, money, and fighters. For example, during Russia's second conflict in Chechnya (1999-2000), documents purportedly found in Grozny by the Russians listed nations such as Sudan, Nigeria, Niger, and Ivory Coast as sending fighters to Chechnya under the guise of the International Islamic Relief Organization. (Chechen fighters also disguised themselves as Red Cross workers.) This deception increases the need for strict security and force protection measures, close coordination with NGOs operating in urban areas, and closer monitoring of suspect organizations' activities by civil affairs personnel.

Win the Information War

3-29. Threat forces will try to win the information war as much as they will directly oppose Army UO. Threat urban campaigns need not be tactical military successes. They need only to weaken legitimacy and make the opposition's campaign appear unpalatable to domestic, U.S., and world support. As a critical part of a threat's overall information operations, they will use the ever-present media to tell their story. Portable video cameras, commercial radios, and cellular telephones, available and easily concealed, will be as important to many threat actors as weapons and ammunition. Internet access, already firmly established in many urban areas, provides the means to easily disseminate threat propaganda, misinformation, and disinformation through web sites and electronic mail. The internet also allows threats to conduct their own distance learning and education. Threats will make good use of the instantaneity and global reach of television and radio broadcasts, particularly local and international media, to obtain information of immediate tactical value and broadcast their own operational and tactical guidance. Hackers, covered and concealed in the interior spaces of the urban area, may gain access to U.S. sites to manipulate and obtain information to the threat's advantage.

> **Information and the Media**
>
> The media coverage of the urban battle for Hue, South Vietnam, although only one of hundreds of different attacks of the Tet Offensive, affected the will of both the American people and their political leadership. On January 31, 1968, two North Vietnamese Army (NVA)/Vietcong (VC) regiments and two sapper battalions, moving rapidly and with the element of surprise, attacked and seized part of the walled city (Citadel) of Hue. It was the third largest city in South Vietnam, the former capital of a united Vietnam, the capital of Thua Thien province, and a spiritual and cultural center. Initially intending to hold the city for seven days, the NVA/VC retained portions of the city for approximately three weeks against determined U.S. and South Vietnamese attempts to retake it.
>
> Hue marked a revolution in the coverage of war by modern media. It was the first time Americans could sit at home and watch an ongoing battle on the evening news. One of the most intense and savage battles of the Vietnam conflict, it was televised every evening for almost a month. Although the battle for Hue was a tactical victory for the United States, the North Vietnamese clearly achieved strategic success by searing the American consciousness with the high costs of urban warfare. Had U.S. leaders made winning the information war a central part of the overall campaign plan—for example, exposing the American people to the NVA's brutality by publicizing the civilian executions in Hue—civilian support for the war may have been bolstered and a different outcome achieved. See Chapter 7 for a more detailed account of the battle for Hue.

Manipulate Key Facilities

3-30. Threat forces will attempt to identify and quickly seize control of critical components of the urban area to help shape the operational environment. Urban telephone exchanges can provide threats with simple and reliable communications that can be easily secured with off-the-shelf technologies. Existing public address and loudspeaker systems can be used to incite riots or openly command and control dispersed threat forces. (Local language may provide threats with a measure of communications security.) Sewage treatment plants and flood control machinery can be used to implement WMD strategies or to make sections of the urban area uninhabitable. Media stations significantly improve the information operations abilities of the controlling force. Power generation and transmission sites provide means to control significant aspects of civilian society over a large area. Insurgents will attempt to destroy key facilities and critical infrastructure in order to shape the populace's perception that the existing government is incapable of providing for their safety and meeting their basic needs.

Use All Dimensions

3-31. Threats will think and operate throughout the depth, breadth, and height (including supersurface and subsurface areas) of the urban environment. However, threats will often choose to fight from within interior spaces rather than out in the streets and other open areas. Conventional lateral boundaries will often not apply as threat forces control some stories of the same building while friendly forces control others. The use of all three dimensions of an urban area makes identification, reporting, and targeting of enemy locations more difficult for forces accustomed and trained to acquire and engage targets in primarily two dimensions. Because of these target identification and acquisition difficulties, threats will likely use decoys in the urban area to cause erroneous assessments of their combat capabilities, strength, and disposition. They will also attempt to use these decoys to absorb expensive and limited precision-guided munitions as well as cause misallocation of other critical Army resources.

3-32. Roofs and other supersurface areas provide urban threats with excellent observation points and battle positions above the maximum elevation of many weapons. Shots from upper floors strike armored vehicles in vulnerable points. Basements and other subsurface areas also provide firing points below many weapons' minimum depressions and strike at another weakness in most armor. Sewers and subways may provide covered and concealed access throughout the area of operations.

EMPLOY URBAN-ORIENTED WEAPONS

3-33. Whether purpose-built or adapted, many weapons are more useful in an urban environment while others may have significant disadvantages. Urban threat weapons are much like the nature of urbanization and the urban environment: inventive and varied. Many threats will integrate widely available off-the-shelf technologies into their weapon systems and armed forces. However, sniper rifles and small, man-portable, fire-and-forget weapons and demolitions and other improvised explosive devices (IEDs), to include suicide and car bombs, will likely dominate the urban environment. Figure 3-3 lists examples of threat weapons favored in UO.

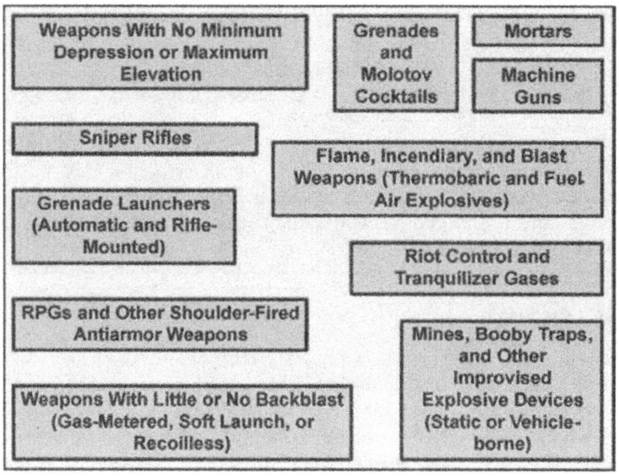

Figure 3-3. Favored threat weapons

ENGAGE ENTIRE ENEMY FORCE

3-34. Threats may attempt to keep all or significant portions of Army forces engaged in continuous operations to increase their susceptibility to stress induced illnesses. UO, by their nature, produce an inordinate number of combat-stress casualties. Continuous operations exacerbate this problem. Threat forces that employ this tactic will often maintain a large reserve or achieve respite by hiding among the civilian population to minimize the psychological impacts on their own forces.

3-35. To accomplish this key tactic, threat UO will likely involve decentralized maneuver, precision fires, and simultaneous operations involving unconventional and special purpose forces and guerilla, insurgent, and terrorist tactics. Threats may also attempt to bait Army forces by ambush or other means with a primary focus of attacking response or reaction forces. Overall, threat forces will take advantage of any exposed weakness in weapons, equipment, and TTP and they will cultivate the flexibility to engage in battles as favorable opportunities present themselves.

Focus Attacks on Support Areas, Isolated Groups, and Individuals

3-36. To supplement the previous tactic, threat forces will seek to target support areas, small groups, leaders and their headquarters, and unaware Soldiers. Their focus on resupply and convoy operations, casualty evacuation, and other sustainment activities, coupled with the compartmented terrain, navigational challenges, and multiple three-dimensional avenues of approach often makes these locations and Soldiers more susceptible to surprise raids and ambushes. Attacks on these areas and groups are conducted to erode the Army's ability to sustain UO, to inflict maximum casualties, and to induce psychological stress. These attacks can be mitigated by careful, regular evaluation of choke points and other restrictive terrain, regular awareness training for units and individuals operating in or transiting through potential incident-prone areas, well-planned and protected convoys, and thorough after-action analysis of incidents.

NEGATIVE EFFECTS OF URBANIZATION

3-37. Many urban areas are the engines for increased industrialism and economic growth as an expanding population provides the labor for manufacturing and service needs. However, rapid and inadequately planned growth can result in undesirable consequences. Uncontrolled urbanization may result in an infrastructure and economic base unable to support the growing population. A large transient, ill-housed, and idle population in a close geographic space may produce strife. Classes, cultures, ethnic groups, and races that might otherwise peacefully coexist can clash under the stress of survival. Uncontrolled urban growth has resulted in the negative effects listed in figure 3-4. In many urban stability and civil support operations, these may be the primary "threats" to mission accomplishment.

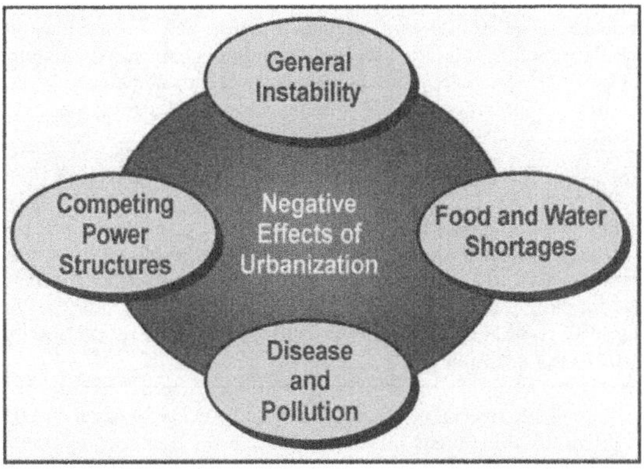

Figure 3-4. Negative effects of urbanization

3-38. Not all urban areas prevail as inherently unstable or hotbeds for unrest. Urban growth due to migration may remove sources of conflict, or it may provide the catalyst for violence. Commanders should recognize the possible effects of uncontrolled urbanization. During their intelligence preparation of the battlefield (IPB), they should determine if these conditions exist. Throughout mission analysis and the development of courses of action, commanders should consider the impact (if any) on their operations. At the same time, they must recognize that UO may create similar problems that may affect the current operation as well as the overall campaign.

General Instability

3-39. Urbanization can enhance stability by generating industrialization and economic growth resulting in more jobs, a higher overall standard of living, and an educated, relatively satisfied populace. However, the population dynamics associated with urbanization can also have an opposite, destabilizing effect. Radical population growth may create overcrowding and generate or aggravate resource and quality of life issues. Intense and destructive competition for employment, housing, and social status may develop in this climate of economic deprivation. The inability of some governments to handle these problems—

- Makes their urban areas potential sources of unrest.
- Increases the likelihood of the Army's involvement in stability operations.
- Complicates all operations conducted in such an urban environment.

Weak civil administrations—perhaps weakened by the effects of previous UO—have difficulty controlling their society, safeguarding their military armaments, and preventing their urban areas from serving as sanctuaries to terrorists, insurgents, and criminal organizations.

3-40. Urbanization in developing countries warrants more concern. Intense migration and growth, coupled with the forced closeness of people once separated by the rural countryside, may stress already struggling institutions, hasten conflict, and lead to overall instability. Commanders must understand that UO, depending on the operation, may either cause massive population movement out of or into urban areas.

3-41. Urban areas with a large youth population may also help to generate conditions for instability. Rural-to-urban migrants tend to be relatively young. In 1999, Cairo, for example, had more than 40 percent of its population younger than 15 years. Young urban populations generate enormous demands for social resources, primarily education and jobs. Even a strong urban economy may fold under the economic expectations of a tremendous influx of young migrants. Disorder and violence may result as hostiles (many nonstate actors) easily mobilize and manipulate the idle young to act politically and criminally. Urbanization and population growth are more dangerous when they combine to produce a cohort of young urban dwellers separated from traditional social controls, such as village elders and clan leaders.

Cultural and Religious Instability

The 1992 bombing of the Babri Masid Mosque in Ayodya, India, enflamed an already intense cultural and religious rivalry between Hindus and Muslims and led to rioting throughout many Indian urban areas. Of the 1,500 who died in conflicts and riots, almost 95 percent died in urban areas. The violence struck Ahmedabab and Bombay most seriously, with related acts of murder, gang rapes, and arson occurring months after the destruction of the mosque.

3-42. Ethnic, religious, and other social issues may become the vents for anger and frustration produced by the high tension of urban life. Major acts of violence and destruction, such as occurred in 1992 in India, can directly threaten a nation's security. Army forces may have to conduct large-scale, stability operations to promote peace and protect national interests. In these cases, all levels of command will be particularly concerned with maintaining impartiality and perceived legitimacy.

Food and Water Shortages

3-43. Rapid urbanization, primarily in developing nations, may lead to severe food shortages that could influence Army forces (or lead to their use). Such shortages may cause instability, massive migration, revolts, or increased support of armed opposition groups. Armed factions may target NGOs that supply aid as a means of furthering dissatisfaction among the populace. In effect, food may become a weapon. Deployed troops may need to provide or support humanitarian food aid networks to keep the humanitarian situation from escalating.

3-44. However, planners must be wary of the unintended consequences of well-intentioned humanitarian assistance operations. For example, providing free, safe food may alleviate starvation, but could also undercut the local agricultural system by reducing demand in the market. If the food is distributed through urban centers, urbanization could increase, further reducing the food supply and adding to the existing strains on the infrastructure. Unless they are trying to use them as a means to separate combatants from noncombatants, commanders should normally use centralized feeding centers as a last resort. Instead, Army forces should bring the food closest to the population to encourage civilians to stay in their homes. If safe areas or camps are created, they should be designed for use over as short a time as is feasible. The general rule should be to return the urban population to their homes as soon as possible. Army forces conducting domestic support or foreign humanitarian assistance operations that cannot maintain the safe food supplies may find the frustrations and hostility of the local population focused on them.

3-45. Water shortages (and quality) are becoming a serious problem in many regions. Sewage, industrial waste, other forms of pollution and deliberate contamination pose threats to existing water supplies. Commanders operating in an urban environment need to know the water supply origins and its treatment, purification, distribution, and vulnerabilities. Before beginning operations, commanders must know if they are providing water for the noncombatants as well as their own forces. Across the spectrum of operations, controlling and protecting a limited water supply is, or may become, an essential operational consideration during UO.

Food and Water Shortages

Countries as varied as Indonesia and Algeria exported their food surpluses only two generations ago but now import up to two-thirds of their basic staples. This cycle has resulted in many countries, which once exported agricultural products, facing the growing cost of imports to feed their urban populations. Estimates predict that in the 2010 timeframe, at least 65 countries (including 30 of Africa's 51 countries) may depend completely on food imports. For some countries, the outlook is even worse. The Democratic Republic of Congo (Zaire), once a net food exporter, faces severe malnutrition and utter starvation for a large portion of its population.

Other estimates predict that by 2025, 2.7 to 3.5 billion people will live in water-deficient countries. Even now in urban areas such as Jakarta, Indonesia, for example, urban authorities cannot provide the necessary water to support its growing population. Less than half of Jakarta's population is supplied by the urban area's water infrastructure; the remainder of the area's residents rely on water from wells and other groundwater supplies that lie only a few feet below the surface. These alternate sources need to be sufficiently boiled or residents risk increased disease. In the northern part of Jakarta, land is actually sinking as the urban area's groundwater supply is overused by its residents. Houses on low-lying areas have to be rebuilt every few years to keep them above sea level. Meanwhile, saltwater from the Java Sea is seeping into the land polluting the remaining groundwater and further exacerbating the situation.

Disease and Pollution

3-46. Urban areas frequently spawn epidemics; therefore, widespread disease may pose a significant threat to Army forces that operate there. In many developing nations, rapid urbanization has occurred without a corresponding upgrade, expansion, or even development of adequate sewage and water systems. Some urban areas have only one toilet for every 750 people or more. In these areas, hundreds of thousands live much as they would in poor villages, yet so confined as to ensure high transmission rates for airborne, waterborne, sexually transmitted, and contact-transmitted microbes.

3-47. In urban areas lacking adequate trash and waste management infrastructure, insect-spread diseases proliferate. Mosquitoes that breed in polluted water, open water tanks, and irrigated urban gardens carry malaria and dengue fever—the leading causes of sickness and death from infectious disease in Latin America and Africa. The problem compounds with growing numbers of bacteria resistant to various antibiotics, a shortage of trained medical personnel, inadequate or insufficient medical facilities and supplies, and unclean agricultural and food-processing practices.

3-48. Pollution also creates critical health problems in developing areas and a potential health risk for intervening Army forces. For example, urban areas in China have recorded five to ten times the levels of sulfur dioxide found in the air of urban areas in the developed world. In parts of Poland, toxic waste has so polluted the land and water that ten percent of the babies have birth defects. Pollution may cause immediate health problems but more often, the insidious effects appear months or years after exposure. As discussed earlier, UO may contribute, either intentionally or unintentionally, to an increase in pollution. Destruction of industrial complexes that use, produce, and store hazardous material may produce toxic gas and smoke pollutants and create significant health concerns to exposed Soldiers and civilians. Soldiers forced to fight within the confines of subsurface areas such as sewers and drainage systems further increases their exposure to contaminants.

3-49. Commanders should initiate force health protection planning early, including analysis of the medical threat and other critical medical information requirements during the IPB process. A medical surveillance system should monitor the daily status of Army personnel throughout the operation. In preparation, all personnel must receive a predeployment medical examination. This exam establishes an accurate baseline health status of the force and ensures that Army forces do not introduce new diseases to an urban area, possibly exacerbating the situation. Conversely, Soldiers not immune to native viruses or possessing a weakened immune system due to continuous operations and the stress associated with UO may put Army forces at a significant disadvantage. An outbreak of plague during an operation would have an effect similar to a chemical or biological attack. The closer that Army forces operate to civilians (the humanitarian assistance operations conducted in Port-au-Prince, Haiti, and Mogadishu, Somalia, for example), the more probable that these situations may occur. (See Chapter 9.)

Competing Power Structures

3-50. Many groups can exist that become strong enough to rival the power of the governing officials and eventually turn the urban area into a system of divergent and competing power structures. These groups can consist of insurgent forces, a merchant class or an economic elite, criminal organizations, or some other significant source of power such as religious organizations (see Chapter 10), clans, or tribes (figure 3-5). In the absence of a legitimate authority, armed factions headed by "warlords" may vie to fill the power void. Sometimes these groups or organizations, normally at odds with each other, may form alliances to achieve specific goals. Commanders must recognize, identify, and understand these alternate urban power bases and, if necessary, develop engagement strategies to neutralize or harness them to accomplish the Army mission. Four of these groups, not previously covered or covered in a subsequent chapter, are discussed in greater detail below.

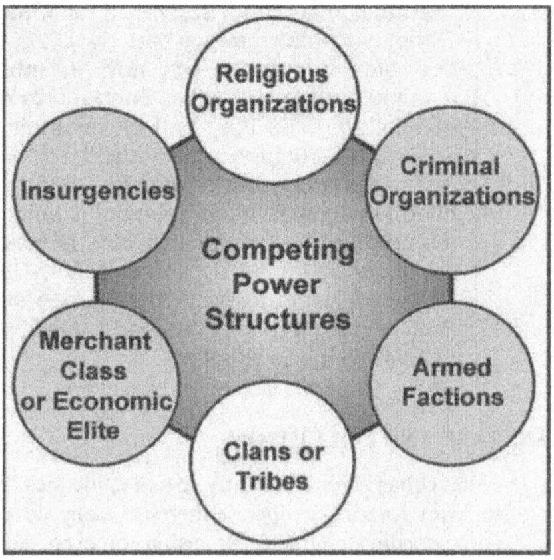

Figure 3-5. Competing power structures

Urban Insurgencies

3-51. As urban migration increases in the developing world, rural guerrillas appear to follow. This transition of insurgencies from rural to urban areas occurs because urban areas offer a rich field of targets for insurgent attacks. People immediately notice any disruption of urban infrastructure, thus having great propaganda value. A concentrated urban population whose allegiance and support are critical to the success of an insurgency is often more susceptible to propaganda, political organization, and terrorism. Insurgents can easily arrange mass demonstrations using available communications facilities, both overt and covert. Travel is effortless and large urban populations provide cover and concealment. On the whole, urban areas, particularly those suffering the effects of general instability outlined earlier, may provide a fertile environment for guerrillas to apply formerly rural insurgent strategies. However, even with a primarily rural-based insurgency, operations in urban areas will often be essential to the insurgents' strategy as they offer them distinct opportunities to disrupt, discredit, and demoralize the government (see FM 3-05.20 and FMI 3-07.22).

[illegible]

Merchant Class or Economic Elite

3-52. Urban areas normally possess a merchant class or economic elite as part of their social structure. In some urban areas, they may carry more power than the local or central state government. They may isolate themselves physically and socially from the sprawling poor yet wield enormous power over the country's political and economic activities. The degree of economic separation between the merchant class and the poor may be small but still socially or politically significant.

3-53. In a vastly impoverished area where the economy of the urban area is severely disrupted, the merchant class will often continue to operate and function and, as a result, achieve a measure of influence. To continue to operate under acute economic turmoil, they may form alliances in criminal organizations and secure loyalties within the government. Outside resources introduced into a crisis area (such as food, water, fuel, and pharmaceuticals) take on increased value, may replace currency as the medium for exchange, and often become the means to amass and hold wealth. One primary way to obtain wealth is to steal it.

3-54. In some turbulent situations that lead to the need for stability operations, commanders may harness the power of the merchant class as a force for peace and stability instead of one that uses crime to achieve economic goals. For example, in a relief situation, instead of competing with the merchant class by distributing food directly to the needy and possibly creating an environment of looting and black marketeering, it may be possible to monetize food. Food assistance from donor governments could be sold to merchants at an attractive price so they have a reliable source of supply. This could, in turn, create a healthy economic system and separate merchants from criminals and gangs.

Criminal Organizations

3-55. Organized criminal groups have grown common in urban areas; have also become an important part of the urban social structure (gangs for example); and can exert considerable influence on governments, people, and military forces conducting UO. These organizations can threaten the successful completion of urban operations both directly and indirectly. Criminals and criminal organizations may directly target Army forces, stealing supplies and extorting money or contracts. Criminals may also violently confront Army forces during mission execution. During many UO, particularly during or following combat, civil disturbances, or large natural disasters, looting (organized or unorganized) may become of critical concern.

For a portion of the population, crime may be the only reliable source of income and resources to support their families. Therefore, UO will often require a combined law enforcement and military response.

3-56. Some large criminal organizations relying on international connections often have better resources and equipment than their insurgent counterparts. (A prime example is drug traffickers.) Their large financial resources, long-reaching connections, and ruthlessness provide them the means to corrupt or intimidate local officials and government institutions. Their tactics parallel those of insurgents. They have developed an intuitive cultural understanding of slum neighborhoods and the ability to lure civilians into criminal activities. They have also mastered the management of mobs. They recruit teenagers and young adults in their efforts against rivals and authorities, just as insurgents muster armies from the youth of rural villages. In many developing nations, there exists an alliance between insurgents and organized criminal groups. In these alliances, the insurgents defend the criminals and the criminals fund the insurgents.

3-57. Excessive criminal activity can undermine the commander's efforts to establish a sense of security among the remainder of the urban populace. Commanders will rely on the ability of their military police (MP) units to assess and understand the potential influence that crime and criminal organizations may have on UO. MPs—in coordination with local police—will need to quickly identify law enforcement gaps to include dysfunctional police organizations and stations and inadequate prison systems. Urban police records may be an important intelligence database to assist in this assessment and may be particularly useful in counterterrorism and counterinsurgency operations. Success in ensuring that law and order is maintained or reestablishing law and order will directly affect the population's perceptions and support of Army operations. Also, MP intelligence operations will contribute immeasurably to the overall IPB effort and aid in understanding of the complexities of the urban society (see FM 3-19.50).

Crime and Criminal Organizations

Crime and poverty plague urban areas such as Rio de Janeiro, Brazil's second largest urban area, and would affect military operations conducted in their limits. Rio has some of the nation's highest negative urban indicators: the largest number of slum dwellers (1 million), the highest murder rate (1 of 700 residents per year), and the highest kidnapping rate (4 per week). In 1989, the homicide rate of the urban area was three times higher than New York City's, and the rate of urban violence has continued to rise. Therefore, law enforcement management may be a critical issue for Army forces operating in urban environments similar to that of Rio de Janeiro.

However, criminal elements or organizations may not always work against Army commanders. They can be co-opted or influenced to serve friendly objectives. For example, during World War II the U.S. Navy worked covertly with the Mafia in New York City to secure the New York harbor from German U-boats believed to be torpedoing ships there. The Mafia controlled most dock activities in New York harbor and was perfectly positioned to monitor other subversive waterfront activity. This capability provided needed information to the Navy for its counterintelligence and security tasks. New York civil authorities, therefore, agreed to permit a Navy-Mafia alliance to operate at the port for the greater good of the country. Although the Mafia was not the preferred ally of the Navy, it had the capability to protect U.S. ships and the interest (patriotism) to help in the war effort. In those circumstances, the temporary alliance worked (see also the civilian threat discussion in Appendix B).

Armed Factions

3-58. A characteristic of some stability operations has been the deterioration or complete collapse of political authority in the country or urban area in crisis. In some cases, "warlords" or the leaders of various armed factions have attempted to fill the power vacuum (see Appendix C for an example). These individuals often have no particular claim to legitimacy. Their power issues from their weapons, not necessarily from their political skills, human services provided, or popular consent (although they have some popular support to remain in their relative position of authority). Others, however, may already have a substantial base of support (for example, a religious base) and then seek to achieve their goals through violent means. In dealing with the leaders of these armed factions during stability operations, it may appear that there are three general courses of action. The first is to aggressively locate and target the leaders and their forces militarily. The second is to ignore them completely, and the third is to meet and work with them diplomatically. Depending on METT-TC, commanders may reduce some of the risks involved in these first three courses of action by combining the last two to create a fourth approach. Nevertheless, the technique chosen must clearly support political and military objectives and lead to the desired end state.

3-59. In some cases, the leaders of armed factions pose a direct threat to Army forces, the urban populace, and the overall accomplishment of the mission. Therefore, it may become necessary to target them and their forces for capture or destruction. Yet, commanders should generally avoid framing the success of a major operation on the death or capture of one or more individuals. Even with increased ISR capabilities, commanders may find an individual manhunt difficult or impossible to achieve. Instead, it may be better to seek to isolate warlords and their forces militarily, while simultaneously seeking to marginalize their influence through a combination of political, economic, and informational means. This does not necessarily mean that commanders must abandon the search if the death or capture the faction leader will contribute to achieving the mission. However, commanders must maintain the capability to strike with speed and precision if the threat leader's location is discovered in the clutter of the complex urban environment.

3-60. In some situations, commanders may attempt to marginalize warlords by ignoring them. However, refusal to acknowledge warlords may increase the threat to Army forces and NGOs. Their militias may attack Army forces to achieve recognition or simply due to misunderstanding or inherent friction between armed forces. On the other hand, dealing with them may provide legitimacy to the exclusion of other elements of the urban population such as professional groups (for example, doctors or teachers), religious leaders, and traditional clan or tribal chiefs—which may have a greater claim to legitimacy and better form the foundation for a reconstituted urban society.

3-61. A compromise between the last two options may offer the best chance for success. In this fourth course of action, commanders can generally recognize these warlords and work with them diplomatically to decrease the risk of armed conflict and civilian casualties; however, this recognition can be kept at staff levels to avoid bestowing any perceived legitimacy on them. Instead, commanders themselves may visibly meet the other elements of society that have a more legitimate claim to political, social, or economic leadership. Inevitably, commanders may need to meet with warlords. In those circumstances, clan or tribal elders, and others who represent traditional authority, should attend the meetings. Moreover, commanders should ask for and give deliberate consideration to their opinions above those of the leaders of the armed factions.

This page intentionally left blank.

Chapter 4

Understanding the Urban Environment's Effects on Warfighting Functions and Tactics

War is, above all things, an art, employing science in all its branches as its servant, but depending first and chiefly upon the skill of the artisan. It has its own rules, but not one of them is rigid and invariable. As new implements are devised, new methods result in its mechanical execution; but over and above all its mechanical appliances, it rests upon the complex factors of human nature, which cannot be reduced to formulas and rules.

Captain Francis V. Greene, 1883

Commanders of major operations should understand the potential effects that the urban environment may have on warfighting functions. They should also understand the possible effects that the urban environment may have on lower-level tactics to properly plan, prepare, and execute major operations that may include UO. Otherwise, commanders may ask their subordinates to achieve effects, accomplish objectives, or adhere to a timetable that is unsupportable due to the constraints imposed by the urban environment. Commanders and their staffs must do more than simply understand the impossible; rather, they must apply the art and science of warfighting to the urban environment and determine what it will take to make it possible.

WARFIGHTING FUNCTIONS

4-1. Understanding the potential effects of the urban environment on warfighting functions permits the urban commander to better visualize his operational environment. With this appreciation, he can conduct a more thorough assessment and thereby determine the most efficient and effective means of employing Army forces. The staff should be intimately familiar with effects in their area of expertise and use that knowledge to understand the problem and develop creative and innovative solutions to achieve their commander's intent.

WARFIGHTING FUNCTIONS

- Intelligence
- Movement and maneuver
- Fire support
- Protection
- Sustainment
- Command and control

INTELLIGENCE

4-2. The intelligence function facilitates understanding of the threat and the environment. The urban environment affects this critical function in many ways. Impacts of the environment on the intelligence function include degraded reconnaissance capability; more difficult IPB process; and an increased importance of credible HUMINT (including the contribution of local civilian liaisons), and an established intelligence reach capability. The Army forces' response to these effects can result in timely, accurate, and actionable intelligence that permits the effective application of other warfighting functions to the mission within the urban environment (see FM 2-0).

Degraded Reconnaissance and Surveillance Capability

4-3. The physical environment creates a major challenge to the intelligence function. The man-made construction in the urban areas provides nearly complete cover and concealment for threats. Although improving many sensor capabilities cannot penetrate the subsurface facilities and much of the space within supersurface areas. The mass of buildings can also defuse electronic signatures. Tall buildings shield movement within urban canyons from aerial observation except from directly overhead. Urban threats may be less technology dependent and may thwart some signals intelligence efforts simply by turning off their radios and using messengers. Threat forces will likely use elements of the civilian telecommunications infrastructure for C2. These systems may include traditional landline phones, cellular telephones, and computer-to-computer or Internet data communications. Most urban telecommunications systems use buried fiber or cables or employ modern digital signaling technology. Such systems are difficult to intercept and exploit at the tactical level.

4-4. From the above, it is evident that these characteristics make it more difficult for the intelligence function to use electronic means to determine threat dispositions and, in offensive and defensive UO, identify decisive points leading to centers of gravity. While the environment limits some typical collection methods, all enemy electronic and human activity creates some form of observable signature and exposes the enemy to potential collection. Seeking ways to take advantage of these vulnerabilities will provide the commander an information advantage over his opponent.

Challenging IPB Process

4-5. The sheer complexity of the environment also challenges the intelligence function. The intelligence function applies the IPB process to the urban environment in accordance with Army doctrine (see Appendix B). With more data points for the IPB process to identify, evaluate, and monitor, this application becomes more demanding. The human and societal aspects of the environment and the physical complexity primarily cause this difference. Relationships between aspects of the environment, built on an immense infrastructure of formal and informal systems connecting the population to the urban area, are usually less familiar to analysts. Thus, the urban environment often requires more specifically-focused intelligence resources to plan, prepare for, execute, and assess operations than in other environments.

4-6. Compounding the challenges is the relative incongruity of all urban environments. No two urban areas are alike physically, in population, or in infrastructure. Thus, experience in one urban area with a particular population and pattern of infrastructure does not readily transfer to another urban area. Any experience in UO is valuable and normally serves as a starting point for analysis, but the intelligence function cannot assume (and treat as fact) that patterns of behavior and the relationships in one urban area mirror another urban area. The opposite is as likely to hold true. The intelligence function will have to study each urban area individually to determine how it works and understand its complex relationships.

4-7. Each characteristic of the urban environment—terrain, society, and infrastructure—is dynamic and can change radically in response to UO or external influences. Civilian populations pose a special challenge to commanders conducting UO. Civilians react to, interact with, and influence to varying degrees Army forces. Commanders must know and account for the potential influence these populations may have on their operations. Intelligence analysts must revisit or continuously monitor the critical points looking for changes, relationships, and patterns.

4-8. The actions of Army forces will affect, positively or negatively, their relationship with the urban population and, hence, mission success. NGOs may deliberately or inadvertently influence civilians. The intelligence function can monitor and predict the reactions of the civil population. However, accurate predictive analysis of a large population requires specific training and extensive cultural and regional expertise.

Increased Importance of Human Intelligence

> ***HUMINT*** is the collection of foreign information—by a trained HUMINT collector—from people and multimedia to identify elements, intentions, composition, strength, dispositions, tactics, equipment, personnel, and capabilities. It uses human sources as a tool, and a variety of collection methods, both passively and actively, to collect information. FM 2-0

4-9. The intelligence function adjusts to the degradation of its technical intelligence gathering systems by increasing emphasis on HUMINT in UO. HUMINT operations may be the primary and most productive intelligence source in UO. In urban offensive and defensive operations, HUMINT gathers information from refugees, immigrants and former citizens (especially previous civil administrators), civilian contractors, and military personnel who have operated in the area. Credible intelligence of this type can help meet requirements, provide more detail, and alleviate some of the need to physically penetrate the urban area with reconnaissance forces. In many urban operations where HUMINT is the primary source of intelligence, acting on single-source reporting is a constant pitfall. Yet, situations may arise where commanders must weigh the consequences of inaction against any potential negative consequences resulting from acting on uncorroborated, single-source information. (See also the Human Capabilities discussion under the Urban ISR portion of Chapter 5.)

4-10. In urban stability operations, HUMINT identifies threats and monitors the intentions and attitudes of the population. A chief source of information contributing to the development of accurate HUMINT, particularly at the tactical level, is reconnaissance forces—especially small-unit dismounted patrols. Urban reconnaissance forces and patrols should be thoroughly and routinely debriefed by unit intelligence personnel to obtain information that aids in developing a clearer picture of the threat and the urban environment. Reliable and trustworthy HUMINT is particularly important in foreign internal defense, counterterrorism, and support to counterdrug operations. Leaders must organize intelligence resources appropriately, and learn and apply valuable techniques, such as pattern and link analysis (see FM 34-3). Additionally, Soldiers, as part of reconnaissance and patrolling training, should be taught to handle captured documents, weapons, material, and equipment as legal evidence much like military and civilian police. Proper "evidence" handling is often a critical intelligence concern in counterterrorism and counterinsurgency operations.

Developing Local Liaisons

4-11. Whenever Soldiers encounter the urban populace, the resulting interaction may become an important source of information the commander can use to answer questions about the threat and the urban environment. While military intelligence units are the primary collectors and processors of HUMINT, commanders are not likely to have enough trained HUMINT Soldiers to satisfy their requirements—particularly in a larger urban environment and during longer-term stability operations. Therefore, commanders may need to cultivate and establish local civilian associations to provide relevant information for decision making and to support the overall HUMINT effort.

4-12. Urban liaisons can be developed through positive civil-military interaction with the urban populace. Critical information may be acquired through interface with the urban leadership (both formal and informal), administration officials, business owners, host-nation support workers, inhabitants along a unit's patrol route, pedestrians at a checkpoint, civilian detainees, or any other human source willing to volunteer information to Army forces or who respond positively to tactical questioning. (Noncombatants are never coerced to provide information.) Commanders may also direct unit leaders to conduct liaisons with specific local leaders and key members of the community to obtain command directed information. Critical information may also come from other U.S. and coalition forces and intelligence organizations operating near or within the commander's AO. To this end, commanders should ensure that collectors operating in an urban area coordinate and deconflict activities and, if possible, outbrief subordinate, geographically-responsible commanders with any relevant information that may affect their current operations. Any relevant information obtained incident to civilian liaison activities should be routinely provided to intelligence staffs not only to gain assistance in verifying the credibility of the information but to share the information with all affected echelons and units.

4-13. NGOs operating in urban areas can also be especially beneficial resources for credible and relevant information about the urban environment. (However, they are generally not a good source for information about the threat since providing such information can violate their neutrality thereby making it difficult for them to achieve their humanitarian aid objectives.) During the 1999 fighting in Kosovo, for example, the Red Cross provided the most accurate figures regarding the number of Kosovar refugees, helping U.S. and other coalition forces to estimate the appropriate level of support required to handle their needs. In addition to a developed understanding of the current needs of the local urban populace, NGOs may also have—

- A network of influential associations.
- Historical archives.
- Extensive understanding of the urban infrastructure.
- Key knowledge of political and economic influences.
- A keen awareness of significant changes in the urban environment.
- Insight into the current security situation.
- Up-to-date web sites and maps.

4-14. While productive civilian associations may become long term, they should not be confused with HUMINT source operations. Only trained HUMINT personnel can recruit and task sources to seek out threat information. Information obtained from these societal connections is normally incidental to other civil-military relationships. For example, as part of infrastructure repair in an urban stability operation, a commander may be instrumental in obtaining a generator for a local hospital. Within the context of this relationship, the commander may develop a rapport with one or more of the hospital's administrators or health practitioners. These civilians may be inclined to provide valuable information about the threat and the urban environment—often on a continuing basis. In any civil-military relationship, however, commanders ensure that the information provided is not tied to promises of assistance or that such assistance is in any way perceived as a means to purchase civilian loyalty.

4-15. Commanders also understand that repeated interaction with any one individual may put that individual and his family in danger from threat forces. Before this potential danger becomes a reality, they should refer their civilian connections to trained HUMINT personnel for protection and continued exploitation. In addition to civilian protection considerations, commanders may also deem it necessary to turn their civilian associations over to trained HUMINT collectors anytime during the relationship if they consider the information that the contact is providing (or may provide) is credible, relevant, and—

- Provides essential threat information on a repetitive basis.
- Helps answer higher-level CCIR.
- Affects operations in another AO.
- Requires interrogation or monetary compensation to obtain.

However, turning a liaison over to trained HUMINT teams does not necessarily preclude maintaining a continued, albeit a more guarded, relationship with the individual.

4-16. In developing these civilian liaisons essential to understanding the urban environment, commanders must avoid the distinct possibility of conducting unofficial source operations by non-HUMINT Soldiers. While prohibited by regulatory guidance (see Defense Intelligence Agency Manual 58-11 for requirements and restrictions for source operations), such actions also run the additional risks of—

- Obtaining unevaluated information that cannot be crosschecked with and verified by other sources of information.
- Creating inequities that result from illegally rewarding contracts which can undermine HUMINT Soldiers who are constrained by intelligence contingency fund regulations.
- Disrupting ongoing HUMINT operations when different sources are seen to be treated differently by non-HUMINT Soldiers vice HUMINT Soldiers.
- Providing non-HUMINT and HUMINT Soldiers with the same information potentially leading to a false confirmation of information.

- Increasing the likelihood that untrained Soldiers may fall victim to a threat deception and misinformation.

Established Intelligence Reach

4-17. Understanding the complex urban environment, particularly the infrastructure and the society, will require more sources of information beyond a unit's organic intelligence capabilities. Therefore, commanders will have to make extensive use of intelligence reach to access information and conduct collaboration and information sharing with other units, organizations, and individual subject matter experts. Before deployment (and throughout the operation), units should establish a comprehensive directory of intelligence reach resources. These resources may include national, joint, Army, foreign, commercial, and university research programs. (Prior to deployment for OPERATION IRAQI FREEDOM, some units established contacts within the local community outside their bases such as police, fire department, and government officials that expanded their reach once in theater—particularly for information regarding civilian infrastructure and urban administration.) Once deployed, intelligence reach includes effective information sharing and collaboration among adjacent units, sister services, coalition partners, and other governmental and nongovernmental agencies operating in the area. Effective information sharing and collaboration requires common network analysis software and databases to be used among all Army forces and, if possible, other governmental agencies.

MOVEMENT AND MANEUVER

4-18. Army maneuver forces—infantry, armor, cavalry, and attack aviation—move to achieve a position of advantage. Entire urban areas or specific threat forces located within may be isolated from affecting other operations and then bypassed. However, when the situation requires entering the urban area to accomplish the mission—such as when bypassed urban threat forces interdict critical lines of communications (LOCs), the environment will significantly affect the Army's ability to move and maneuver (see figure 4-1). These negative effects include canalization, compartmentalization, and increased vulnerability. However, tactics and techniques equip Army forces to overcome these challenges and maneuver successfully. One tactic, effective combined arms task organization, includes an increased dismounted maneuver capability combined with armor and combat engineers, continuous operations, and technological enhancements.

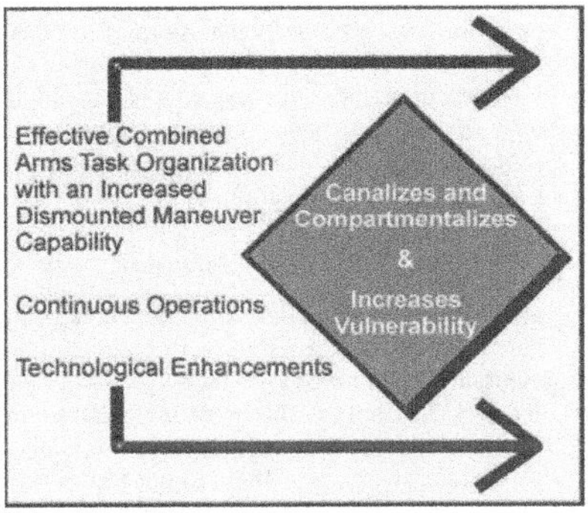

Figure 4-1. Urban maneuver challenges and means to overcome them

4-19. In all UO, mobility operations may allow civilian traffic and commerce to resume, letting the urban area return to some semblance of normalcy (often a critical objective). In stability operations, mobility often focuses on keeping lines of communications open and reducing the threat of explosive hazards to Soldiers and civilians. In civil support operations, mobility may focus on removing storm debris or reducing obstacles caused by destroyed property.

Canalization and Compartmentalization

4-20. The urban terrain will often canalize and compartment forces—and their fires—moving and maneuvering through it. Buildings pose obstacles to both mounted and dismounted movement, forcing units to be canalized along streets. The buildings also block movement between streets, thus compartmenting units. Fires are canalized into open and unmasked areas where vision is unobstructed, producing concentrated fire zones and areas, especially at road intersections and in front of defended positions. Hence, changing directions, repositioning committed forces, reinforcing forces in contact, bypassing threats, and maneuvering to the threat flank become extremely difficult. Units often breach obstacles to help solve this problem. Using helicopters to quickly move forces, both forward into contact and to rear areas as part of repositioning, also permits Army forces to overcome some terrain constraints.

4-21. The canalized and compartmented effects can slow movement and maneuver considerably. However, slowed movement also results from the physically demanding soldier tasks required in an urban environment. Soldiers operate dismounted across rubble and hard surfaces. Operating in three dimensions, they constantly move up the supersurface areas of building interiors and down into basements, cellars, and other subsurface areas. They breach many obstacles and use upper-body strength, ropes, and ladders to scale heights. The inability to see into the next room, floor, or building magnifies stress. The resulting fatigue slows the overall rate of Army force movement and maneuver.

Increased Vulnerability

4-22. The urban environment increases the vulnerability of Army forces executing movement and maneuver in offensive, defensive, and stability or civil support operations. Both the physical terrain and the urban population provide threat cover and concealment. Air movement and maneuver is vulnerable for many of the same reasons. In offensive or defensive operations, enemy forces can remain undetected in buildings and in position to ambush Army forces. Consistent with METT-TC, Army forces should clear buildings along maneuver routes prior to mounted movement along those axes. Failure to clear routes (and effectively mark cleared portions) may expose mounted movement to ambush at close range. Movement back across streets and obstacles may be difficult particularly if the element of surprise was essential in the initial crossing or breach. The same buildings also provide cover and concealment to enemy air defense capabilities—particularly man-portable air defense systems that can be fired from multiple positions hidden amongst the clutter of fires, lights, smoke and dust, and easily concealed and transported in civilian vehicles throughout the urban area. In all operations, but especially stability operations, civilians can conceal threat elements. The threat can then initiate offensive operations against Army forces from close range and where ROE will hamper applying combat power. Thus, maneuver through a dense population can be a high-risk operation.

Combined Arms Task Organization

4-23. Effective combined arms task organization ensures that forces are task organized with infantry—the essential building block for all organizations conducting UO. Infantry protects mounted elements as the combined arms unit moves and maneuvers through the urban area. (In some urban situations, mechanized infantry may not be able to provide dismounted support beyond support to its own vehicles—tanks may require the support of additional light infantry). The infantry destroys the enemy in buildings, bunkers, and subsurface areas where they cannot be defeated by mounted forces and prevents infiltration of threat forces back into hard-won urban terrain. Field artillery aids in dismounted and mounted (to include air) maneuver by suppressing known and suspected enemy positions with precision fires. Attack aviation make best use of their standoff capabilities (see later fire support discussion) and aircraft speed to conduct running and diving fires; in UO, hovering fire is generally avoided. Armored elements protect Soldiers from small arms fire and destroy or suppress enemy positions with precise, direct fire. Carefully protected artillery may also be used in this direct fire role. Armored forces and attack helicopters also can facilitate maneuver through shock action that can have a psychological effect, particularly against less well-trained threats and, in discrete instances, hostile crowds. (Although, commanders consider that the "intimidation" value of any method erodes quickly with its repetitive use.)

4-24. Combined arms also ensure that combat engineers support dismounted maneuver by assisting in covered and concealed maneuver through buildings and off exposed streets. In addition to combat engineers, explosive ordnance disposal teams, military police, chemical personnel, and other with essential expertise to conduct mobility missions significantly reduces mobility and maneuver challenges (see FM 3-34.2). Urban buildings are often obstacles to movement and mobility. Combat engineers, trained and equipped for UO, can turn these obstacles into an advantage by breaching them with "mouse holes" made by explosives, sledgehammers, bulldozers or armored vehicles, or high-strength (diamond or carbide-tipped) cutting devices. These breaches permit dismounted movement through buildings under both cover and concealment.

4-25. Combat engineers must also be trained and equipped to facilitate mounted mobility in the urban environment. Generally, buildings restrict mounted movement to the compartmented and canalized streets. Threats can block streets with roadblocks ranging from sophisticated log and concrete cribs reinforced with antitank and antipersonnel mines to the expedient use of cars, buses, and trucks to create obstacles. Combat engineers must be capable of breaching these obstacles to maintain the coherence of the combined arms team (mounted and dismounted). Combat engineers should be forward, often task organized down to platoon level, and have the expertise and equipment to rapidly reduce point obstacles. It even may be necessary that every armored vehicle (or section of two vehicles) be task organized with an associated engineer squad and vehicle. Because of the increased density and hardness of many urban building and construction materials, heavy engineer equipment (such as the D9 bulldozer) will be in great demand to accomplish mobility (and countermobility, and survivability) functions in an urban environment. However, commanders will need to consider increased protection requirements and the availability of equipment transport to move these slower-moving engineer assets around the urban battlefield.

4-26. A major difference of UO combined arms is in proportion and organization. Although based an accurate METT-TC assessment, UO will often require an increased proportion of dismounted infantry and engineer capabilities; armor may not be required in the same high numbers. As significant, genuine combined arms urban operations are required at lower tactical levels where small, well trained and led units will dominate. Commonly, company level will require true combined arms capability and may include combat engineers, military intelligence, reconnaissance, and artillery. Combined arms teams can then form at platoon and squad levels. Because of this, larger units will need more CA, military intelligence, and combat engineers than habitually attached for combat in more open or less restrictive terrain.

4-27. In determining the appropriate task organization, commanders consider a suitable span of control for subordinate commanders. They also consider the potential of dissipating a unit's combat power, capabilities, and synergy by breaking a unit up into smaller units in an attempt to ensure subordinate maneuver units have a complete combined arms capability. For example, an additional engineer battalion may be task organized to a brigade combat team. In turn, the brigade combat team may task organize this battalion into engineer companies under the control of their subordinate maneuver battalions. If this type of organization continues, maneuver companies may end up with an engineer platoon and, in the end; maneuver platoons each having an engineer squad. Ultimately, a combined arms capability may have been established at lower tactical levels but the parent maneuver unit (in this example, the brigade combat team) may have lost the ability to conduct larger engineer operations without having to re-task organize and potentially disrupt current operations and established relationships. As a guide, urban commanders may consider task organizing to create combined arms organizations at lower tactical levels when operations are predominately offensive or defensive, and bringing those assets back under their own control when the operation transitions to predominately stability or civil support operations.

Continuous Operations and Technology Enhancements

4-28. Two other means to improve Army forces' ability to move and maneuver in urban terrain is through continuous operations and the leveraging of technology, such as the Army's night operations capability. Historically, urban battles have been fought primarily during daylight because of technological limitations and fatigue. By utilizing night vision technologies, accurate situational understanding, a common operational picture (COP), training, and rotated units, Army forces can defeat threats who use the same soldiers in day and night operations and who are less well-equipped and adept at night operations. Night operations are also a means of mitigating the air defense threat against air maneuver. Continuous operations through night maneuver with fresh forces are challenging, but it can overcome many advantages that a stationary force has against maneuver in the urban environment. However, commanders should also consider that streetlights, fires, background illumination (as well as dark building interiors without ambient light), the increased heat absorption of many urban structures, and the skillful use of searchlights by threat forces may limit the effectiveness of night vision devices and make thermal imagery identification difficult.

Countermobility

4-29. Countermobility capabilities in urban terrain are also an essential consideration in all UO. In defensive operations, commanders use countermobility capability to control where the enemy moves in the urban area. Repositioning defensive forces in the urban area can be difficult and obstacles are essential to limiting the enemy's maneuver options. During offensive operations, countermobility protects exposed flanks and air assaulting forces from counterattack. In stability and civil support operations, countermobility operations may take the form of constructing barriers to assist in populace and resources control at critical urban locations.

FIRE SUPPORT

4-30. The fire support function includes the collective and coordinated use of several means to attack targets in the urban area (see Appendix D for joint capabilities). These means include target acquisition data, indirect fire weapons, rotary- and fixed-wing aircraft, offensive information operations (IO), and other lethal and nonlethal means. The urban environment affects these components of the fires function and their employment.

Target Acquisition

4-31. Target acquisition in an urban environment faces several challenges. First, forces have difficulty penetrating the urban environment's increased cover and concealment using sensors and reconnaissance. Acquiring targeting information and tracking targets throughout the depth of the urban area may prove challenging. Moving personnel or vehicular targets are normally easiest to acquire. However, the cover and concealment provided by urban terrain gives moving targets short exposure times requiring firing systems to act rapidly on targeting data. Targeting of opposing indirect fire units by acquisition radar may work more effectively in urban terrain because of the necessary high angles of indirect fire. The urban environment presents similar difficulties for battle damage assessment.

4-32. Targeting challenges are met by innovatively integrating reconnaissance capabilities. These capabilities include SOF, cavalry, unmanned aircraft systems, and aerial observers as well as the standard reconnaissance assets. More artillery systems may need to be used to ensure the responsiveness (rather than the weight) of fires. Positioning numerous artillery systems reduces the dead space (as discussed below) and permits units to establish more direct sensor-to-shooter links.

The Targeting Process

4-33. Heightened concerns for collateral damage will require that commanders pay particular attention to their targeting process. This process ensures that all available combat power, both lethal and nonlethal (including offensive IO), is effectively integrated and synchronized to accomplish the mission. Commanders ensure that techniques and procedures are in place, rehearsed, and understood by all members of their staffs. Additionally, the C2 system must be responsive and agile; otherwise, an elusive and adaptable threat will likely disappear before units can employ the appropriate weapon systems. In an urban area, even 10-digit grid coordinates may not be sufficient to accurately identify targets as buildings may be connected to each other—often throughout the entire block. Target locations, in addition to grid coordinates, may need to routinely include the street address, number of stories, shape, color, or any other distinguishing characteristic essential for ground and air forces to achieve targeting precision. A common urban reference system with graphics, reference points, and other control measures adequate for both ground and air forces may also help facilitate identification of targets and facilitate the rapid clearance of fires.

4-34. Greater concerns exist for the safety and health (environmental matters) of the urban populace and the protection of critical infrastructure and cultural structures. Hence, CA and staff judge advocates (see Chapters 5 and 10 respectively) will play a greater role for the expert advice they can provide regarding these elements of the urban environment. Nonetheless, all members of the staff ensure that operations minimize collateral damage. That responsibility does not end with identifying potential collateral damage; the goal, as always, is successful mission accomplishment. Again, staffs are guided by the commander's intent and work to develop courses of action that incorporate collateral damage concerns (short- and long-term) yet accomplish the mission. This requires a keen understanding of the legal issues and both friendly and enemy weapon systems' effects in an urban environment.

Urban Effects on Fire Support Systems

4-35. Both the physical and human components of the urban area affect how units use fire support weapon systems. The physical aspects of the urban environment, such as the heights and concentration of buildings, may cause significant masking and dead space (see figure 4-2). Intervening buildings that stand three or more stories tall hinder close indirect fire support. Tall buildings can potentially mask several blocks of area along the gun-target line of artillery. (For low-angle artillery fire, dead space is about five times the height of the building behind which the target sits.) The potential for collateral damage to adjacent buildings may also prevent engagement with artillery. Such damage might cause noncombatant and friendly troop casualties and unintentional (and unwanted) rubbling. Commanders can offset these effects by carefully placing artillery positions, repositioning artillery as targets change, and using mortars. Mortars have a steep angle of fall and short minimum ranges as a high-angle alternative to field artillery fire. (In comparison to artillery, dead space for mortar fire is only about one-half the height of the building.) For fixed-winged aircraft, precision munitions and weapons with low explosive yields and near-vertical impact angles resulting in bomb burial can significantly reduce collateral damage. Collateral damage concerns may also cause commanders to—

- Maintain approval authority for some sensitive or protected targets (churches or mosques, for example) at higher echelons of command.
- Restrict attacks to certain times of day.
- Give warning prior to an attack so that noncombatants can evacuate the area.
- Incorporate indigenous forces into the operation.
- Abort an attack unless the required level of precision effects can be achieved.
- Prepare specific branches and sequels to the IO plan to inform the populace why the collateral damage was justified. These plans may include filming or otherwise documenting the operation to thwart threat propaganda and claims of excessive collateral damage.
- Develop and rehearse detailed staff battle drills that address clearance of fires in an urban environment.

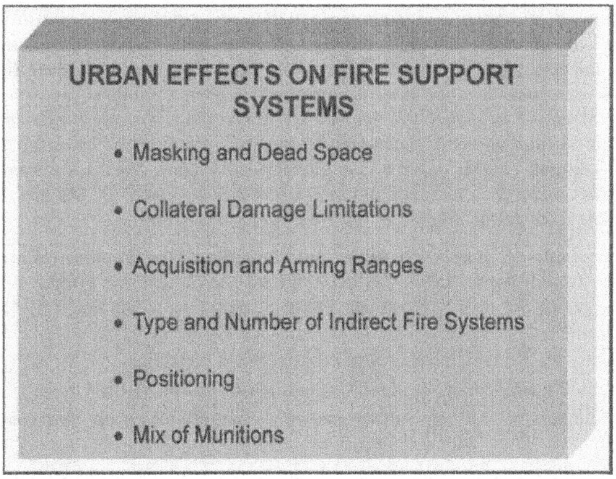

Figure 4-2. Urban effects on fire support systems

4-36. Vertical structures interrupt line of sight (LOS) and create corridors of visibility along street axes. The result is thereby shortened acquisition and arming ranges for supporting fires from attack helicopters and subsequently affected engagement techniques and delivery options. Pilots maintain LOS long enough to acquire targets, achieve weapons delivery solutions, and fly to those parameters. For example, tube-launched, optically tracked, wire-guided heavy antitank missile systems require 65 meters to arm. Similarly, the Hellfire missile requires at least 500 meters to reliably arm and stabilize on the intended target. Thus, attack helicopters firing from longer ranges actually improve the probability of a hit. Poor weather and heavy smoke and dust rising from urban fires and explosions may hinder target identification, laser designation, and guidance for rotary- and fixed-winged aircraft. Poor air-to-ground communications may also hinder effective use of airpower. The close proximity of friendly units and noncombatants requires units to agree on, thoroughly disseminate, and rehearse clear techniques and procedures for marking target and friendly locations. The ability for ground units to "talk-on" aircraft using a common reference system described earlier helps expedite aerial target acquisition (and helps mitigate potential fratricide). FM 3-06.1 details other aviation TTP in an urban environment.

4-37. The urban environment also affects the type and number of indirect fire weapon systems employed. Commanders may prefer high-angle fire because of its ability to fire in close proximity to friendly occupied buildings. Tactically, commanders may consider reinforcing units in UO with mortar platoons from reserve units. This will increase the number of systems available to support maneuver units. Unguided Multiple Launch Rocket Systems (MLRSs) may be of limited use in urban areas due to their exceptional destructive capabilities and the potential for collateral damage. However, commanders may use unguided MLRSs to effectively isolate the urban area from outside influence. Commanders may also employ field artillery systems as independent sections, particularly self-propelled systems, in the direct-fire role; decreasing volume and increasing precision of artillery fire helps minimize collateral damage. While discretely applying the effects of high-explosive and concrete-piercing munitions, these self-propelled systems take advantage of the mobility and limited protection of their armored vehicles.

4-38. The urban area may also affect the positioning of artillery. Sufficient space may not exist to place battery or platoon positions with the proper unmasked gun line. This may mandate moving and positioning artillery in sections while still massing fires on specific targets. Commanders must protect artillery systems, particularly when organized into small sections. Threats to artillery include raids and snipers. Therefore, maneuver and firing units will have to place increased emphasis on securing their positions and other appropriate force protection measures.

4-39. The mix of munitions used by indirect fire systems will change somewhat in urban areas. Units will likely request more precision-guided munitions (PGM) for artillery systems to target small enemy positions, such as snipers or machine guns, while limiting collateral damage. Currently, only conventional tube artillery, not mortars, has this capability. However, large expanses of polished, flat reflective surfaces common in urban areas may degrade laser designation for these munitions (as well as attack helicopter PGM). The vertical nature amplifies the geometrical constraints of many precision munitions. Remote designators need to be close enough to accurately designate but far enough away not to be acquired by the PGM during its flight path. PGMs based on the global positioning system (for instance, guided MLRS or the Air Force's joint direct attack munitions) or other optically guided PGMs may be more effective if urban terrain hinders laser designation.

4-40. The urban environment greatly affects the use of nonprecision munitions. Building height may cause variable time fuses to arm prematurely. Tall buildings may also mask the effects of illumination rounds. Units may choose not to use dual-purpose conventional munitions if (similar considerations apply to Air Force cluster bombs)—

- The enemy has several building floors for overhead protection.
- Dismounted friendly units need rapid access to the area being fired on.
- Large numbers of civilians will operate in the target areas soon after combat operations have ceased.

4-41. Depending on the building construction, commanders may prohibit or limit illumination, smoke, and other munitions because of fire hazards. In particular instances, they may specifically use them for that effect. Structure fires in an urban area are difficult to control and may affect friendly units. Conventional high-explosive munitions may work best against concrete, steel, stone, and other reinforced structures. When not used in the direct-fire role, a greater mass of indirect fire is often required to achieve desired effects. Commanders balance firepower and collateral damage since the rubbling caused by massive indirect fires may adversely affect a unit's ability to maneuver and provide a threat with additional cover and concealment.

4-42. Nonlethal weapons, munitions, and devices can help commanders maintain the desired balance of force protection, mission accomplishment, and safety of noncombatants by expanding the number of options available when deadly force may be problematic. As additional nonlethal capabilities are developed, they are routinely considered for their applicability to UO. In determining their use and employment, commanders, in addition to any previous experience at using these weapons, munitions, and devices, consider—

- **Risk**. The use of nonlethal weapons in situations where lethal force is more appropriate may drastically increase the risk to Army forces.
- **Threat Perspective**. A threat may interpret the use of nonlethal weapons as a reluctance to use force and embolden him to adopt courses of action that he would not otherwise use.
- **Legal Concerns**. Laws or international agreements may restrict or prohibit their use (see Chapter 10).
- **Environmental Concerns**. Environmental interests may also limit their use.
- **Public Opinion**. The apparent suffering caused by nonlethal weapons, especially when there are no combat casualties with which to contrast it, may arouse adverse public opinion.

PROTECTION

4-43. The protection function includes those tasks and systems that preserve the force so that commanders can apply maximum combat power. Preserving the force includes enhancing survivability and properly planned and executed air and missile defense as well as defensive IO (see IO discussion in Chapter 5) and chemical, biological, radiological, nuclear, and high-yield explosive (CBRNE) counterproliferation and consequence management activities (see CBRNE discussion in Chapter 9).

Survivability

4-44. Survivability in the urban environment is a significant force multiplier. Properly positioned Army forces can take advantage of the increased survivability afforded by the physical terrain. Even a limited engineer effort can significantly enhance the combat power of small Army forces. In stability operations, properly planned and constructed survivability positions can enable small groups of Soldiers to withstand the assaults of large mobs, sniping, and indirect fire. Well-protected support bases are often critically essential to minimizing casualties during long-term stability operations and can become a key engineer task.

4-45. While executing major combat operations or campaigns, in particular defensive operations, well planned and resourced engineer efforts can enhance the survivability characteristics of the urban area. These efforts, though still requiring significant time and materiel, can establish defensive strong points more quickly and with greater protection than can be done in more open terrain. Skillfully integrating the strong point into the urban defense greatly increases the overall effectiveness of the defense disproportionately to the number of forces actually occupying the strong point (see Chapter 8).

4-46. Commanders increase survivability by ensuring that all Soldiers have necessary protective equipment and are trained and disciplined in their use. In addition to standard equipment such as helmets, gloves, boots, and chemical protective overgarments, commanders should ensure, as necessary, availability of other protective equipment and material such as—

- Body armor.
- Goggles or ballistic eye protection.
- Knee and elbow protectors.
- Riot control equipment such as batons, face masks, and shields.

- Barrier material such as preformed concrete barriers, wire, sandbags, and fencing material.
- Up-armored or hardened vehicles.
- Fire extinguishers and other fire-fighting equipment.
- Immunizations.

4-47. The Army's urban survivability operations can become complex if the Army is tasked to support survivability operations for civilians. Such operations can range from constructing civil defense shelters or evacuating the population to assisting the population in preparing for or reacting to the use of weapons of mass destruction. However, Army forces are not organized or equipped to support a major urban area's requirements as well as its own mission needs. Normally, Army forces can render this type of support only as a focused mission using a unique, specially equipped task organization.

Air and Missile Defense

4-48. The air and missile defense protects the force from air surveillance and air and missile attack. This system uses—

- The careful massing of air and missile defense combat power at points critical to the urban operation.
- The proper mix of air defense weapon and sensor systems.
- Matched (or greater) mobility to the supported force.
- The integration of the air defense plan into the overall urban operation.
- The integration of Army systems with those of joint and multinational forces.

4-49. Properly planned and executed air and missile defense prevents air threats from interdicting friendly forces and frees the commander to synchronize maneuver and other elements of firepower. Even in a major combat operation or campaign, the enemy will likely have limited air and missile capabilities and so seek to achieve the greatest payoff for the use of these systems. Attacking Army forces and facilities promises the greatest likelihood of achieving results, making urban areas the most likely targets for air and missile attack.

Rotary- and Fixed-Winged Aircraft

4-50. Enemy rotary-wing aircraft can be used in various roles to include air assault, fire support, and combat service support. Some threats may use unmanned aircraft systems to obtain intelligence and target acquisition data on friendly forces. Increased air mobility limitations and targeting difficulties may cause enemy fixed-wing aircraft to target key logistics, C2 nodes, and troop concentrations outside the urban area, simultaneously attacking key infrastructure both in and out of the urban area.

Increased Missile Threat

4-51. The intermediate range missile capability of potential threats has increased to be the most likely air threat to an urban area. Urban areas, particularly friendly or allied, make the most attractive targets because of the sometimes-limited accuracy of these systems. By firing missiles at an urban area, a threat seeks three possible objectives:

- Inflict casualties and materiel damage on military forces.
- Inflict casualties and materiel damage on the urban population.
- Undermine the confidence or trust of the civil population (particularly if allied) in the ability of Army forces to protect them.

4-52. If facing a missile threat, commanders conducting UO work closely with civil authorities (as well as joint and multinational forces) to integrate the Army warning system with civil defense mechanisms. Similarly, Army forces may support urban agencies reacting to a missile attack with medical and medical evacuation support, survivor recovery and assistance in damaged areas, and crowd control augmentation of local police forces. Before such an attack, Army engineers might assist and advise urban officials on how to construct shelters.

Increased Security of Assets

4-53. When defending against an air or missile threat in a neutral or hostile urban environment, air defense assets are concerned with security. Separating air defense locations from high population and traffic centers, as well as augmenting these positions with defending forces, can prevent or defeat threat efforts to neutralize them. Additionally, increased density of UO means increased concentration of all friendly and enemy systems engaged in air and counter-air operations. This density may increase friend and foe identification challenges, air space management challenges, and the overall risk in the conduct of air operations. Finally, limited air defense assets, difficulties in providing mutual support between systems, potential mobility limitations, and other effects of the urban environment increase the need for (and effectiveness of) a combined arms approach to air defense (see FM 44-8).

SUSTAINMENT

4-54. The sustainment function incorporates support activities and technical service specialties, to include maximizing available urban infrastructure and contracted logistics support. It provides the physical means with which forces operate. Properly conducted, the sustainment function ensures freedom of action, extends operational reach, and prolongs endurance. Commanders conducting sustainment to support full spectrum operations must understand the diverse logistic requirements of units conducting UO. They must also understand how the environment (to include the population) can impact sustainment support. These requirements range from minimal to extensive, requiring Army forces to potentially provide or coordinate all life support essentials to a large urban population.

4-55. Commanders and staffs consider and plan for Army sustainment operations that are based in a major urban area. These operations are located in major urban areas to exploit air- and seaports, maintenance and storage facilities, transportation networks, host-nation contracting opportunities, and labor support. These operations are also UO. The commander gains additional factors to consider from basing the sustainment operation in an urban environment. See Chapter 10 for a detailed discussion of urban sustainment considerations.

COMMAND AND CONTROL

Fighting in a city is much more involved than fighting in the field. Here the "big chiefs" have practically no influence on the officers and squad leaders commanding the units and subunits.

Soviet General Vasili Chuikov
during the 1942-43 Battle for Stalingrad

4-56. The command and control function is the related tasks and systems that support the commander in exercising authority and direction. The urban environment influences both the commander and his C2 system (which includes INFOSYS). The leader's ability to physically see the battlefield, his interaction with the human component of the environment, and his intellectual flexibility in the face of change all impact the mission. The C2 system faces difficulties placed on the tactical Internet and system hardware by the urban environment, by the increased volume of information, and by requirements to support the dynamic decision making necessary to execute successful UO.

Unity of Command

4-57. Although severely challenged, the principle of unity of command remains essential to UO. However, the number of tasks and the size of the urban area often require that Army forces operate noncontiguously. Noncontiguous operations stress the C2 system and challenge the commander's ability to unify the actions of his subordinates, apply the full force of his combat power, and achieve success. To apply this crucial principle in an urban environment requires centralized planning, mission orders, and highly decentralized execution. The method of C2 that best supports UO is mission command (see FM 6-0). Mission command permits subordinates to be innovative and operate independently according to clear orders and intent as well as clearly articulated ROE. These orders and ROE guide subordinates to make the right decision when facing—

- A determined, resolute, and adaptive threat.
- A complex, multidimensional battlefield.
- Intermittent or complete loss of communications.
- Numerous potentially hostile civilians close to military operations.
- The constant critique of the media and military pundits.

4-58. Decentralized execution allows commanders to focus on the overall situation—a situation that requires constant assessment and coordination with other forces and agencies—instead of the numerous details of lower-level tactical situations. Fundamentally, this concept of C2 requires commanders who can accept risk and trust in the initiative, judgment, and tactical and technical competence of their subordinate leaders. Many times, it requires commanders to exercise a degree of patience as subordinate commanders and leaders apply mental agility to novel situations.

Political and Media Impact

4-59. Commanders of a major operation consider how the need to maintain a heightened awareness of the political situation may affect their exercise of C2. A magnified political awareness and media sensitivity may create a desire to micromanage and rely solely on detailed command. Reliance on this method may create tactical leaders afraid to act decisively and with speed and determination—waiting instead for expected guidance from a higher-level commander. Threats may capitalize on this hesitation by conducting operations faster than Army forces can react. Mission orders that express the overarching political objectives and the impact of inappropriate actions, combined with training and trust, will decrease the need for detailed command. Leaders must reduce a complex political concept to its simplest form, particularly at the small-unit level. Even a basic understanding will help curtail potentially damaging political actions and enable subordinates to make the often instantaneous decisions required in UO—decisions that support military and political objectives.

Commander's Visualization

> *I heard small-arms fire and RPG explosions and felt shrapnel hit the vehicle.... Land navigation at this time was impossible; every time I tried to look out, I was thrown in a different direction.... At this time, I was totally disoriented and had not realized we were on our own.*
>
> Captain Mark Hollis
> "Platoon Under Fire"

4-60. Leaders at all levels need to see the battlefield to lead Soldiers, make effective decisions, and give direction. Sensors and other surveillance and reconnaissance assets alone cannot provide all the information regarding the urban environment that commanders will need. The focus of lead elements narrows rapidly once in contact with a hostile force limiting their assessment to the local area. Therefore, tactical commanders will not be able to observe operations from long, stand-off ranges. Their personal observation remains as critical in urban areas as elsewhere and helps to preclude commanders from demanding their subordinates accomplish a task or advance at a rate inconsistent with the immediate situation. In urban offensive and defensive operations, seeing the battlefield requires that commanders move themselves and their command posts forward to positions that may be more exposed to risk. Thus, commanders modify their C2 system capabilities to make them smaller, reduce their signature, and increase their mobility. Because of the greater threat to C2, security efforts may be more intense.

4-61. In stability operations, commanders often intervene personally to reassure the urban population and community and faction leaders about the intentions of Army forces. In these type operations, threats may attack leaders to gain the greatest payoff with the least expenditure of resources. Commanders carefully evaluate risk and potential benefits of such exposure. These risks however, cannot stop them from seeing the battlefield, personally intervening in situations as appropriate, and leading their Soldiers.

4-62. Commander's visualization also requires having an accurate understanding of friendly and enemy locations, detailed maps, other appropriate intelligence products, and INFOSYS that accurately depict the urban environment and help establish a COP. The reliability of these items is as important to planning major operations as it is to tactical-level operations. The commander of the major operation ensure that subordinate tactical-level commanders have the necessary products to achieve accurate situational understanding and dominate the urban environment as subordinate commands often lack the personnel or assets to develop these products. Frequently, satellite or aerial imagery is requested to compensate for the drastic changes that can occur due to UO, natural disasters, and outdated or imprecise maps. (Even maps developed and maintained by urban area's administrative activities may not be up-to-date. Extensive and continually expanding shantytowns, for example, may not be mapped at all. Maps may have even been purposefully distorted or critical detail intentionally omitted. The systems used to transliterate some languages such as Arabic and Chinese to Anglicized alphabets often result in the same location being spelled several—and frequently considerably different—ways. Maps may also assign names to features that are completely different than those used by locals to refer to them.)

4-63. Other critical intelligence products needed in the COP include overlays or gridded reference graphics. (Whenever possible, gridded reference graphics should conform to standard military grid reference system formats to reduce the probability of error when entering target coordinates into targeting systems that use global positioning systems.) These products should be developed and distributed to all participants prior to the UO. Overall, their focus should be on ease of reference and usefulness for all forces—ground and air (see Appendix B). Overlays and graphics can also portray important societal information or urban infrastructure, such as—

- Religious, ethnic, racial, or other significant and identifiable social divisions.
- Locations of police, fire, and emergency medical services and their boundaries or zones of coverage.
- Protected structures such as places of worship, hospitals, or other historical and culturally significant buildings or locations.
- Underground subway, tunnel, sewer, or water systems.
- Bridges, elevated roadways, and rail lines.
- Electrical generation (to include nuclear) and gas storage and production facilities and their distribution lines.
- Water and sewage treatment facilities.
- Telephone exchanges and television and radio stations.
- Toxic industrial material locations.

Mental Flexibility

4-64. Commanders conducting UO must remain mentally flexible. Situations can change rapidly because of the complexity of the human dimension. Typical of the change is a stability operation that suddenly requires the use of force. Commanders must be capable of quickly adjusting their mental focus from a noncombat to combat situation. Equally important is dealing with populations during combat operations. Consequently, commanders must also be capable of rapidly adjusting plans and orders for sudden stability or civil support tasks that emerge during or soon after a combat mission. In developing their vision, commanders must consider the second- and third-order effects of UO.

Information Systems

4-65. The urban environment also challenges INFOSYS that support the commander, especially communications. Urban structures, materials, densities, and configurations (such as urban canyons) and power constraints associated with man-portable radios significantly degrade frequency modulation (FM) communications. This causes problems at brigade-level and below where commanders rely heavily on constant FM radio contact with subordinates. Tactical communication problems might also cause an inability to maintain a COP, to give orders and guidance, to request support, or to coordinate and synchronize elements of the combined arms team. Communication problems in urban areas can prevent the achievement of information superiority and contribute directly to mission failure. In UO, allocating critical or high-value communication assets will be significant and essential to the main effort.

4-66. In an urban environment, units and staffs properly prepare for and mitigate the communication problems in urban areas (see figure 4-3). Adequate communications, in most cases, are ensured by—

- Training in and use of retransmission and relay sites and equipment, which may include unmanned aircraft systems (UAS).
- Airborne command posts, satellite communications, high-frequency radios, and other redundant communication platforms and systems.
- Careful positioning of commanders, command posts, and antennas to take advantage of urban terrain characteristics.
- Detailed communications analysis for movement from one AO to another due to the likely density of units operating in the urban environment.

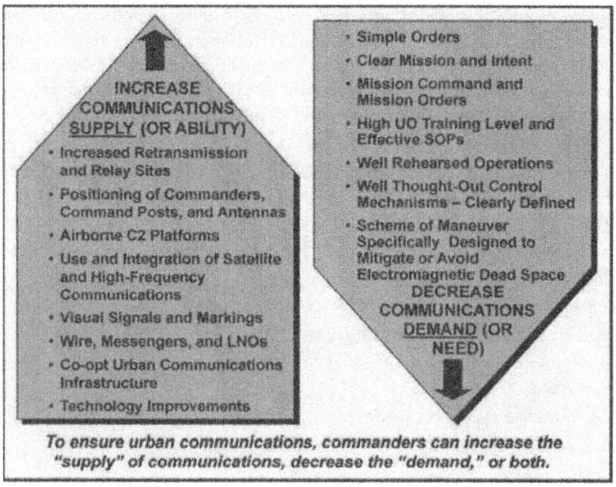

Figure 4-3. Methods to overcome urban communications challenges

4-67. Standing operating procedures (SOPs) for visual markings (both day and night) may assist in command and control. These SOPs indicate unit locations and other essential information. They coordinate with units across common boundaries. Given adequate consideration to limitations on multinational capabilities, these SOPs may assist in command and control and preclude fratricide incidents resulting from loss of FM communications. However, visual signals, including pyrotechnics, are less effective in buildings and enclosed spaces.

4-68. In defensive, stability, or civil support operations, positions often do not change as frequently as in offensive operations. Urban commanders then rely more on military wire (properly camouflaged amongst the civilian communications infrastructure), commercial communications, and messengers. Even in combat, some if not all of the urban area's organic communications structure remains intact for Army use. For example, every building may have one or more telephone distribution boxes that can control hundreds of individual telephone lines. Setting up wire communications using these points is relatively simple but, like all wire communications, is susceptible to wire-tapping. Cellular telephones can usually work well in urban areas; however, locating and destroying the repeater stations or other land-based elements of the cellular telephone system (or the effects of natural disasters) can easily disable them. (Cell phones may be a critical and only means to rapidly communicate with key civilian organizations and important community leaders.) Consequently, the C2 system may use these alternatives to FM communications but with proper operations and physical security procedures in place.

Example of Simple Communications Innovation: Israel's Six-Day War – 1967

In the 1967 battle for Jerusalem, the Israeli Defense Force placed flags on top of cleared buildings so that aircraft providing close air support could monitor the Israeli forward line of troops. They also used a spotlight during the night to mark specific buildings as close air support targets.

4-69. Units will likely use multiple means to communicate throughout the urban area. Hence, commanders emphasize proper operations security procedures (OPSEC) despite the level of security provided by the communications system. This emphasis helps to lessen the probability that Soldiers will inadvertently compromise essential information as they switch from one mode of communications to another (for example, from secure FM radio to unsecured cellular telephones or from classified to unclassified Internet domains).

4-70. Command posts above brigade-level ensure that they can communicate in an urban area without significant disruption. In stability and civil support operations, immediate and reliable communications between tactical and strategic levels may be necessary. Higher commanders anticipate that although the urban area does not significantly challenge their INFOSYS, the area may severely challenge systems at the lower tactical levels. For this reason, information flow from lower to higher may take longer. If the situation is not acceptable, the higher headquarters takes steps to mitigate it, such as increasing the number of liaison officers operating with units engaged in decisive operations. In some instances, the scheme of maneuver may be specifically designed to account for communications interference, propagation characteristics, and electromagnetic dead space. (However, this will require more time, resources, and a detailed urban communications IPB.)

4-71. Finally, urban areas can overload the INFOSYS with information. UO across the spectrum of operations can generate large volumes of information when crises threaten. This sheer volume can easily overwhelm UO commanders and command posts, and the information conduit connecting the two. Training prepares command posts to handle this volume of information and to filter the critical from the merely informative. Staffs must create products (visual or textual) that help their commanders understand the urban environment, not just present them information to know.

KEY TACTICAL CONSIDERATIONS

4-72. Commanders and planners of major UO must thoroughly understand the tactical urban battle. Especially, they must understand the effects of the environment on men, equipment, and systems. The complexity of urban environment changes and often compresses many tactical factors typically considered in the planning process. Figure 4-4 shows some of these compressed factors, however, commanders and their staffs should carefully review FM 3-06.11 for doctrine to support tactical urban operations.

Figure 4-4. Compressed tactical factors

Time

4-73. The time available to think and act is compressed in urban combat operations. The tactical engagements that comprise battles and major UO are often quick and decisive; therefore, higher-level commanders require the ability to conduct battle command on the move so that their decision making remains correspondingly fast. The impact of decisions (or lack of) and the outcome of battle can occur in mere minutes. Often the amount of information and the number of decisions can overwhelm the overall ability of INFOSYS to respond. Commanders have little time to influence tactical actions with resources kept in reserve. Reserves and fire support assets are close to the point of decision so that they can respond in time to make a difference. The terrain causes C2 challenges that further inhibit commanders from responding quickly to changes in the situation. Small unit leaders receive training that emphasizes understanding the commander's intent so that they can recognize tactical opportunities and can act quickly to take advantage of them.

Distances and Density

4-74. Distances in UO are compressed to correspond to the density of threat forces and noncombatants. In open terrain, squads, platoons, and companies may be able to control or influence thousands of meters of space. In UO, large buildings can absorb the efforts of several companies or battalions. Crowds of thousands can assemble in areas of a few hundred meters requiring correspondingly large forces for control. Maximum engagement ranges, as influenced by the urban terrain, are usually closer. Units may require field artillery for direct fire at targets ranging fewer than a hundred meters. Commanders and staffs understand the telescoping nature of the battlefield, the density of threat forces, and the density of noncombatants. In addition to the actual conduct of urban tactical operations, these factors will directly affect training, planning, force deployment, and strength.

4-75. Time-distance considerations are especially important throughout planning cycles. Though distances may be short, the physical nature of the environment can drastically change the planning factors for unit movements. The advance of a battalion may be measured in hundreds of meters per day. Thus, all time and distance calculations that relate to sequencing of forces, synchronizing combat power and other capacities, and making decisions require reevaluation based on the urban conditions.

4-76. Although it may appear counterintuitive, airspace above the urban area may be also be severely compressed as multiple fixed-wing, rotary-wing, and indirect fires (including multinational assets) compete for the same space. Increasing this density is the use of airspace by civilian aircraft and the proliferation of tactical UAS. Due to the potential for a high volume of air traffic, commanders and planers pay close attention to the integration and deconfliction of airspace over urban areas. Commanders should consider specific techniques and procedures for UAS which may include—

- Defining airspace control areas.
- Regulating flight times.
- Creating altitude restrictions.
- Including UAS on the air tasking order.

Combat Power

4-77. The urban terrain can also compress combat power. This terrain increases the utility and effects of some weapons and systems, increasing overall combat power. One system that dramatically demonstrates this effect is the sniper. In open terrain, snipers slightly influence operations. In UO, snipers—well concealed, positioned, and protected—can take on significance disproportionate to their combat capability in other situations (see FM 23-10).

4-78. The density of ground combat power in a given size area is also increased because of the effect of the terrain on ranges. The complex terrain precludes standoff engagement from extended ranges by dispersed forces. Commanders often position weapon systems closer together and at shorter ranges to mass effects on the same target. Thus, commanders may position armored vehicles, which typically position themselves Hundreds of meters from friendly troops and other vehicles, within a few meters of each other to provide mutual support. Targets, which in open terrain are engaged at thousands of meters, are engaged in tens of meters on the urban battlefield.

4-79. The dense clutter of the urban environment also affects target acquisition. Systems, such as radar optimized for open terrain, will not be able to acquire targets as effectively. Decreased acquisition capability equates to diminished combat power. It may also require increasing the density of acquisition systems to compensate for reduced capability.

4-80. Finally, the density of combat power may also increase the vulnerability of Army forces. Many Army systems are protected from enemy systems at longer ranges. The number of enemy systems that can threaten Army forces at a short range increases dramatically. Lack of dispersal will make it more likely that multiple Army systems can be targeted by a single enemy threat.

LEVELS OF WAR

4-81. The levels of war are also compressed in the urban area. The tactical actions of individuals and small units can directly influence operational and even national and strategic objectives. Conversely, the decisions of the President can directly affect the conduct of tactical operations. UO have short cause and effect links between the tactical, operational, and strategic levels of operations. Because of the close media scrutiny of UO, the President can sometimes observe the actions of platoons in real time. For example, the media may film a platoon applying nonlethal force for crowd control. The President can view that film on the nightly news before the platoon even disengages from the action, much less reports formally through the various levels of command. If appropriate, the President can decide and direct the strategic and operational commanders to adjust ROE before the platoon has reported. Therefore, commanders at all levels must understand the urban environment's potential compressive effects on the levels of war. A major impact of these effects can be a lower tolerance for tactical errors and a greater need for detailed planning and precision in execution and weapons' effects (lethal and nonlethal).

DECISION MAKING

4-82. The nature of the urban environment compresses the time available to make decisions and increases the number of decisions to make. This is particularly true at the lower tactical levels. Units observing an urban AO face more potential unknowns than in other situations. A large structure presents many more potential firing positions that are observed than simpler terrain. Movement in one of those windows forces the soldier or unit to quickly make a decision regarding the nature of the target—deciding whether it is a threat or a noncombatant. Incorporating combatant and noncombatant discriminatory considerations into all live-fire training will improve Soldiers' ability to make these critical judgment decisions. Overall, lack of understanding regarding the urban environment requires commanders to rely more on analytic decision making, while greater understanding facilitates adaptability and allows commanders to make rapid and intuitive decisions while still accounting for second, third, and higher-order effects (see Figure 4-5 and FM 6-0). Realistically, however, all decisions made during urban operations will likely require a combination of analytic and intuitive decision-making abilities.

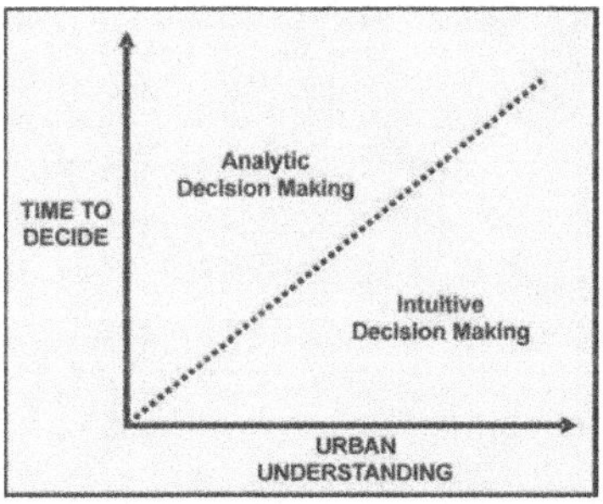

Figure 4-5. Urban understanding and decision making

This page intentionally left blank.

Chapter 5
Contemplating Urban Operations

We based all our further calculations on the most unfavorable assumptions: the inevitability of heavy and prolonged fighting in the streets of Berlin, the possibility of German counter-attacks from outside the ring of encirclement from the west and south-west, restoration of the enemy's defence to the west of Berlin and the consequent need to continue the offensive.

General of the Army, S. M. Shtemenko
describing operational level planning for taking Berlin
The Soviet General Staff at War

In any potential situation and in any area, Army commanders will likely need to assess and understand the relevance and impact of one or more urban areas on their operations. They will also need to determine whether full spectrum urban operations (UO) will be essential to mission accomplishment. UO may be the commander's sole focus or only one of several tasks nested in an even larger operation. Although UO potentially can be conducted as a single battle, engagement, or strike, they will more often be conducted as a major operation requiring joint resources. Such actions result from the increasing sizes of urban areas. Army commanders of a major urban operation must then ensure that UO clearly support the operational objectives of the joint force commander (JFC), requesting and appropriately integrating critical joint resources. Whether the urban operation is the major operation itself or one of many tasks in a larger operation, Army commanders must understand and thoroughly shape the conditions so subordinate tactical commanders can engage and dominate in this complex environment.

A ***major operation*** is a series of tactical actions (battles, engagements, strikes) conducted by various combat forces of a single or several services, coordinated in time and place, to accomplish operational, and sometimes strategic objectives in an operational area.

DETERMINING THE NECESSITY OF URBAN OPERATIONS

5-1. Early in planning, commanders of a major operation must address the necessity and feasibility of conducting operations in urban areas located throughout their areas of operations (AOs). Chapter 1 discussed strategic and operational considerations that compel forces to operate in urban areas. These reasons include the location and intent of the threat force; critical infrastructure or capabilities that are operationally or strategically valuable; the geographic location of an urban area; and the area's political, economic, or cultural significance. Additionally, humanitarian concerns may require the control of an urban area or necessitate operations within it. Several considerations exist, however, that may make UO unnecessary, unwarranted, or prohibitively inefficient. When determining whether to operate in an urban environment, commanders must consider the operational (and accidental) risks and balance them with mission benefits. The factors shown in Figure 5-1 highlight some measures to evaluate the risks associated with UO.

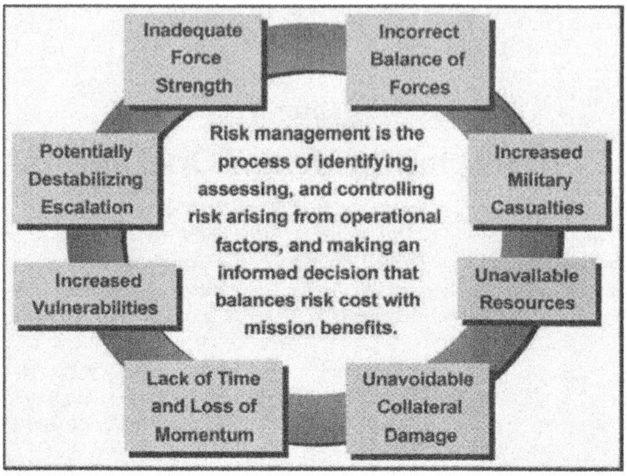

Figure 5-1. Risk management and the associated risks with urban operations

INADEQUATE FORCE STRENGTH

5-2. When facing prospective UO, commanders must consider if they have the necessary troops available to conduct the operation properly and within acceptable risk. Under normal circumstances, large urban areas require many forces merely to establish control. The New York City police department has over forty thousand officers and hundreds of support staff simply to conduct peacetime law enforcement. Major UO, particularly those that are opposed, will often require a significant number of forces. Offensive missions, for example, can require three to five times greater troop density than for similar missions in open terrain. If commanders lack sufficient force to conduct effective operations, they may postpone or consider not initiating those operations until they have the necessary strength. Commanders should also add to their analysis the requirements for troop strength elsewhere in the AO. Additionally, commanders must consider the number (and type) of forces required to make the transition from urban offensive and defensive to stability or civil support operations when determining overall force requirements—not just the number of forces required to realize objectives for major combat operations.

INCORRECT BALANCE OF FORCES

5-3. Along with force strength, commanders must consider the type and balance of forces available. This consideration includes an assessment of their level of training in urban operations. Generally, UO put a premium on well-trained, dismounted infantry units. The operational or tactical necessity to clear threat forces from the dense urban environment, hold hard-won terrain, and interact with the urban population greatly increase dismounted requirements. Therefore, Army forces conducting UO are often force tailored to include a larger infantry component. In addition, special operations forces (SOF) are invaluable in UO. SOF include psychological operations (PSYOP) and civil affairs (CA) forces. They should always be considered as part of the task organization.

5-4. UO must include combined arms to ensure tactical success in combat. Although masses of heavy forces may not be required, successful UO require the complementary effects of all Army forces. Even if an urban operation is unlikely to involve offensive and defensive operations, armor, combat engineers, and field artillery may be essential to mobility and force protection. In urban stability and civil support operations, successful mission accomplishment requires more robust CA organizations. They are also valuable in urban offensive and defensive operations. While commanders may have sufficient combat and combat support forces, they may lack enough sustainment forces to provide the support necessary to maintain the tempo. Again, commanders without balanced types of forces, to include their proficiency in operating in urban environments, should consider alternatives to UO or delaying UO until proper force types are trained and available in sufficient numbers.

INCREASED MILITARY CASUALTIES

5-5. Casualties in UO are more likely than in operations in other environments. In urban offense and defense, friendly and threat forces often engage at close range with little space to maneuver. The urban terrain provides numerous advantages to the urban defender; higher casualties occur among troops on the offensive, where frontal assaults may be the only tactical option. Conversely, defenders with limited ability to withdraw can also suffer high casualties when isolated and attacked. Casualties can be more difficult to prevent in urban stability operations because of the dense complex terrain, the close proximity of the urban population, and the possible difficulty in distinguishing friend from foe. The potential for high casualties and the subsequent need for casualty evacuation under difficult circumstances make the positioning and availability of adequate medical resources another important consideration. Additionally, high-intensity urban combat and the potential for increased stress casualties may require additional units to allow for adequate unit rotations so that Soldiers receive the rest they require.

5-6. Though casualties occur in all operations, commanders should recognize the likelihood of more casualties during large-scale or high-intensity UO. During the battle for Hue in 1968, for example, many company-size units suffered more than 60 percent casualties in only a few days of offensive operations. Commanders conducting urban stability operations must know the casualty risk and how it relates to national and strategic objectives. While a lower risk normally exists in stability operations than in offensive and defensive operations, just one casualty may adversely impact the success of the stability mission. A realistic understanding of the risk and the nature of casualties resulting from UO critically affect the decision-making process. If commanders assess the casualty risk as high, they must ensure that their higher headquarters understands their assessment and that the objectives sought within the urban area are commensurate with the anticipated risk.

UNAVAILABLE RESOURCES

5-7. Offensive and defensive operations in an urban environment put a premium on certain types of munitions and equipment. Forces may want to use vast amounts of precision munitions in the urban environment. At the tactical level, they will likely use more munitions than during operations in other environments. These munitions include—

- Grenades (fragmentation, concussion, stun, riot control, and smoke).
- Mortar ammunition (due to its rate of fire, responsiveness, and high-angle fire characteristic).
- Explosives.
- Small arms.

5-8. Soldiers need access to special equipment necessary to execute small-unit tactics effectively. In urban stability and civil support operations, this equipment may include antiriot gear, such as batons, protective clothing, and other nonlethal crowd-control devices. In urban offensive and defensive operations, special equipment can include sniper rifles, scaling ladders, knee and elbow pads, and breaching equipment such as door busters, bolt cutters, and sledgehammers. Soldiers can conduct UO with standard clothing and military equipment. However, failure to equip them with the right types and quantities of munitions and special equipment will make mission success more difficult and costly. When commanders consider whether to conduct UO, they must evaluate the ability of logistics to provide the necessary resources (see Chapter 10). Considerations must include the ability to supply all Soldiers regardless of branch or military occupational specialty with urban-specific equipment.

UNAVOIDABLE COLLATERAL DAMAGE

5-9. UO require an expanded view of risk assessment. When considering risk to Army, joint, and multinational forces, commanders must also analyze the risk to the area's population and infrastructure. This comprehensive analysis includes the second- and third-order effects of significant civil casualties and infrastructure damage. Collateral damage can influence world and domestic opinion of military operations and thus directly affect ongoing operations. It also influences the postconflict physical environment and attitudes of the population. Negative impressions of the civilian population caused by collateral damage can take generations to overcome. Destroying an urban area to save it is not a viable course of action for Army commanders. The density of civilian populations in urban areas and the multidimensional nature of the environment make it more likely that even accurate attacks with precision weapons will injure noncombatants. Unavoidable collateral damage of sufficient magnitude may justify avoiding UO, which, though it may be tactically successful, may run counter to national and strategic objectives.

Lack of Time and Loss of Momentum

5-10. Commanders conducting major operations must analyze the time required to conduct UO successfully. UO can be time consuming and can require larger quantities of resources. The density of the environment, the need for additional time to conduct a thorough reconnaissance, the additional stress and physical exertion imposed on Army forces operating in urban areas, and the potential requirements to care for the needs of the urban population consume time and slow momentum. Commanders cannot permit UO conducted as a shaping operation to divert resources from the decisive operation. Nor can they allow UO to interrupt critical time lines, unnecessarily slow tempo, or delay the overall operation. Threat forces may conduct UO with the primary purpose of causing these effects. Commanders must recognize that time generally works against political and military objectives and, hence, they must develop plans and operations to avoid or minimize UO that might unacceptably delay or disrupt the larger, decisive operation. Once commanders achieve major combat objectives, however, they will likely need to shift resources and focus on the urban areas that they previously isolated and bypassed.

Increased Vulnerabilities

5-11. Commanders must weigh the potential for increased vulnerabilities when executing UO. The density of the environment makes protection (safety, field discipline, force protection, and especially fratricide avoidance) much more difficult. Forces operating in a large urban area increase their risk of isolation and defeat in detail. Joint capabilities, such as air power, may work less effectively to support a close urban battle than in some other environments. Thus, responding to unexpected situations or augmenting disadvantageous force ratios when applying joint capabilities is significantly more difficult. Although organized, trained, and equipped for success in any environment, the Army vulnerability to weapons of mass destruction (WMD) increases when forces concentrate to conduct UO. Commanders may consider not committing forces or limiting the size of a force committed to an urban area because of increased vulnerability to (and likelihood of) attack by WMD.

5-12. Fratricide avoidance is a matter of concern for commanders in all operations. The complex urban terrain and density of participating forces coupled with typical battlefield effects—smoke, dust, burning fires—and weather effects—fog, snow, rain, and clouds—immensely increase the potential for urban fratricide. Additionally, safety requirements make it difficult to fully replicate training conditions that allow leaders to become more aware of the conditions that contribute to fratricide. The effects of fratricide can be devastating to UO and spread deeply within the Army force. Critical effects include—

- Needless loss of combat power.
- Decreased confidence in leadership, weapons, and equipment. These lead to a loss in initiative and aggressiveness, failure to use supporting combat systems, and hesitation to conduct limited visibility operations.
- Disrupted operations and decreased tempo.
- General degradation of cohesion and morale.

5-13. Therefore, commanders should increase fratricide awareness and emphasis on prevention measures during UO. Causes can be procedural, technical, or a combination of the two and include—

- Combat identification failures due to poor situational understanding, lack of communication, failure to effectively coordinate, and short engagement ranges coupled with the need for quick reaction.

- Location errors involving either the target or enemy forces due to poor situational understanding.
- Inappropriate command and control and fire support coordinating measures; a failure to receive, understand, or adhere to these measures.
- Imprecise weapons and munitions effects such as, an antitank round that penetrates several walls before exploding near friendly forces.

Exacerbating these difficulties will be the likelihood of Army forces conducting operations with (or within proximity of) SOF, coalition forces, and indigenous security forces including local police. (During OPERATION IRAQI FREEDOM, the head of the Australian Defence Forces SOF in Iraq made it a point to carefully coordinate with U.S. Army forces—down to the company level—whenever his personnel would be in an American AO. He felt the danger of fratricide was a much greater risk than the loss of operational security.)

Potentially Destabilizing Escalation

5-14. In the urban environment, Army forces cannot avoid close contact with enemy forces and civilians that may potentially become hostiles. In urban stability and civil support operations, commanders should consider the chance of this contact escalating into confrontation and violence, which may become destabilizing. This consideration may delay, limit, or altogether preclude UO using Army forces.

Consider Alternatives and Risk Reduction Measures

5-15. Since UO are often high risk, commanders should consider courses of action that provide alternatives. When the objective of an urban operation is a facility, commanders should consider replicating that facility outside of the urban area. For example, a critical requirement for an airfield to sustain operations may lead commanders to consider UO to seize or secure one located in an urban area. However, if adequate resources exist (especially time and adequate general engineering support), Army forces may build an airfield outside of the urban area and eliminate the need to conduct the urban operation. Similarly, logistics over-the-shore operations may be an alternative to seizing a port facility. In some situations, the objective of UO may be to protect a political organization such as a government. Relocating the government, its institutions, and its personnel to a safer area may be possible. Commanders can also design an operation to avoid an urban area. For example, if an urban area dominates a particular avenue of approach, use a different avenue of approach. Using a different avenue of approach differs from isolating and bypassing because the entire operation specifically makes the urban area irrelevant.

5-16. If commanders execute UO, they must assess potential hazards, and then they develop controls to either eliminate or reduce the risks to Army forces. The first means to offset risk is always to ensure a thorough understanding of the urban environment and its effects on operations by all members of the force. Other measures to mitigate risk may include—

- Detailed planning to include thorough intelligence preparation of the battlefield and the development of appropriate branches and sequels.
- Integrated, accurate, and timely intelligence, surveillance, and reconnaissance (ISR).
- Clear missions and intent, which includes a well-articulated end state that looks beyond the cessation of combat operations.
- Sufficient reserves and rotation of forces.
- Vigilant physical security precautions to include increased use of barriers and other defenses, particularly when urban areas are used as support areas.
- Operative communications and other information systems (INFOSYS).
- Effective populace and resources control measures.
- Comprehensive and flexible rules of engagement (ROE) continuously reviewed to ensure they remain adequate for the situation.
- Sufficient control measures (which often include a common urban reference system) and standard marking and identification techniques that adequately consider limited visibility concerns for both air and ground forces. Measures should allow commanders to satisfactorily

control UO and minimize fratricide without unreasonably restricting subordinate commanders' ability to accomplish assigned missions. Commanders must ensure that all subordinate units thoroughly disseminate any approved nonstandard reference systems.

- Proper targeting procedures (including effective fire support coordinating measures and a streamlined legal review of targets), positive identification of targets, and controlled clearance of fires. The goal is achievement of precise (yet rapid) effects with both lethal and nonlethal means. In close air support, positive air-to-ground communications are essential to coordinate and authenticate markings.
- Well-synchronized information operations (IO) that begin before introducing Army forces into the urban environment and well through transition. Commanders should emphasize vigilant operations security (OPSEC) particularly when operating closely with the media, nongovernmental organizations (NGOs), and elements of the civilian population.
- Active and effective integrating, synchronizing, and coordinating among all forces, agencies, and organizations involved in the operation. Commanders should allow adequate planning and rehearsal time for subordinates.
- Responsive, sustainable, and flexible urban sustainment.
- Forces well trained in joint, interagency, multinational, and combined arms UO.
- The creation of adaptable, learning organizations. This requires thorough after-action analyses conducted during actual operations as well as after training exercises. In addition to official Army sites (such as the Center for Army Lessons Learned and the Battle Command Knowledge System), commanders must create a unit-level system to allow hard-won, lessons learned and tactics developed to be immediately passed on to other units and soldiers—even in the midst of an operation. This system may be technology-based, procedural, or both. (During OPERATION IRAQI FREEDOM, the 1st Cavalry Division developed an effective web-based knowledge network—called CAVNET—that allowed them to actively capture and share lessons learned among subordinate units.)

INTEGRATION INTO LAND OPERATIONS

5-17. The commander of the major operation, after determining that urban operations are required, will then integrate the urban operation into his overall operation. He does this by articulating his intent and concept for the urban operation to his subordinates. The commander of the major operation must also set the conditions for successful tactical urban operations by his subordinates. He should define ROE, focus ISR efforts, task organize his capabilities, ensure information superiority, design the operational framework, and initiate and sustain effective coordination with other agencies and organizations (see FM 6-0).

CONCEPT OF THE OPERATION

5-18. The commander's concept of the operation should address all operationally important urban areas in his AO. It should also articulate his vision of the urban operation through directions to his staff and subordinates. Subordinate commanders should address urban areas that the higher commander does not specifically address. The commander's concept should discuss each urban area in terms of task and purpose (see FM 5-0). The commander should also describes his vision of the situation's end state in terms of—

- The threat.
- The urban environment (terrain, society, and infrastructure).
- Friendly forces.
- The conditions necessary to transition control of urban areas within his AO to another agency or back to legitimate civilian control.

Rules of Engagement

5-19. National- or joint-level command authorities may develop urban-specific ROE. If not, Army commanders, as part of their assessment, must determine if urban-specific ROE are required for their situation and provide supplemental ROE. However, commanders must forward any conflicts or incongruities to their higher headquarters for immediate resolution.

5-20. Developing effective ROE relies on thoroughly understanding the national and strategic environment and objectives. It also relies on understanding how to conduct urban operations at the tactical level including weapons effects. For example, broad ROE may result in significant collateral damage and civilian casualties. Even in a major operation or campaign, significant collateral damage caused during UO can make postcombat operations difficult. Such damage may even change national and international public opinion or threaten the achievement of national and strategic objectives. In contrast, restrictive ROE can hamper tactical operations causing mission failure, higher friendly casualties, or both. ROE are often part of essential elements of friendly information (EEFI), protected to reduce the potential for threat exploitation. Even in a limited urban operation, ROE will frequently need to change as circumstances warrant. Therefore, commanders should plan ROE "branches" for anticipated changes in the operational environment.

5-21. In urban operations, ROE are flexible, detailed, and understandable. They should preclude the indiscriminate use of deadly force while allowing soldiers latitude to finish the mission and defend themselves. ROE should recognize that the urban area is not homogenous and may vary according to the key elements of the threat and environment: terrain, society, and infrastructure. To be effective, ROE are consistent throughout the force (an increased challenge in multinational urban operations), and soldiers are thoroughly trained and familiar with them.

Enemy Considerations

5-22. The nature of an urban enemy affects ROE as well. Commanders must consider the type of enemy weapon systems, the degree of defensive preparation, the ability to target enemy vulnerabilities with precision systems, and the ability to distinguish combatant from noncombatant.

Terrain Considerations

5-23. ROE may vary according to the terrain or physical attributes of an urban area. Physical factors may drive the ROE to preclude certain types of munitions. For example, if the construction of a portion of the area is sensitive to fire, then ROE may preclude using incendiary munitions in that area. The ROE may lift this prohibition when units move into areas of masonry construction. Toxic industrial chemicals or radiological contaminants in an industrial area may also affect ROE.

Societal Considerations

5-24. The societal or human dimension of the urban environment will often affect ROE the most. Commanders must base the ROE development on a thorough understanding of the civilian population and threat. They evaluate the loyalty of the population, its dynamic involvement in activities that affects the operation, and its size and physical location. A population that is present and supports Army forces will likely elicit more restrictive ROE than a hostile population actively supporting forces opposing the Army forces. A neutral population, not actively involved in behavior affecting Army forces, supports consideration of more restrictive ROE. In all cases, ROE conforms to the law of war. However, ROE may be much more restrictive than the law of war requires.

5-25. The location of the population also affects ROE. The evacuation or consolidation of noncombatants into controlled, safe areas may result in less restrictive ROE. A U.S. or allied population that remains in the urban area conducting routine business in and amongst Army forces during noncombat UO will normally require the most stringent ROE.

Infrastructure Considerations

5-26. Commanders must consider the urban infrastructure when developing ROE. Urban infrastructure vital to current or future Army operations may dictate that commanders adjust ROE to ensure that critical elements of the infrastructure remain intact during the conduct of operations. If Army forces conduct an urban operation to capture port facilities, for example, the ROE should address damage to the key facilities that are the objective of the operation.

RESOURCE ALLOCATION

5-27. Commanders of a major operation must ensure that subordinate tactical commanders have the resources necessary to conduct UO effectively. They assign appropriate forces to subordinate commanders tasked to conduct UO; support them with Army forces at the operational level; and request and coordinate their support by joint resources.

Task Organization

5-28. Task organizing subordinate units for urban operations will depend largely on the nature of the operation. Some units, however, are always part of the task organization to ensure the success of UO. Infantry, CA, aviation, military police, PSYOP, military intelligence, and engineers are units required for all urban operations across full spectrum operations. Other type forces—such as armor, artillery, and chemical—have essential roles in specific types of urban operations but may be less applicable across the spectrum of operations. Commanders and staffs of a major operation must understand their mission, the particular urban environment in which they operate, and the general effects of the environment across the warfighting functions to allocate the appropriate forces to their tactical commanders. See Chapter 4 for details.

Operational-Level Support

5-29. Commanders of a major operation should also support the tactical commander with forces remaining under their direct control. These forces can include Army SOF, such as CA, PSYOP, and Special Forces, ground and air cavalry, aviation, logistics, engineers, and communications support. These forces may not be under operational control of the supported command, but their efforts are synchronized and coordinated.

Coordinating and Requesting Joint Support

5-30. Commanders of a major operation will often provide forces to the JFC as well as receive assets. They will also coordinate for and integrate joint assets to support the tactical battle. These assets will usually include air support, such as close air support, tactical airlift, and aerial reconnaissance and surveillance. Intelligence support comes in the form of reach to strategic and national intelligence capabilities and to space-based systems. This reach to space assets provides reliable, robust long-range communications, environmental monitoring, ISR, positioning and navigation, and warning of enemy missile launch. Special operations capabilities can assist the tactical mission with special reconnaissance and direct action (including special operations aviation) against high-payoff targets. Joint resources also provide the Army forces augmentation by Marine ground forces. In coastal areas, Navy forces and Coast Guard elements assist Army forces with security, sealift, and fire support. Additionally, the potential agility, reach, and power projection provided by Navy forces and platforms, combined with their self-sustaining capability, can play a vital role in supporting UO. This is especially true when UO are conducted in the littorals, require forcible entry operations, and when host nation infrastructure is severely taxed or destroyed. Commanders of a major operation will coordinate with the JFC regarding available joint resources and their allocation. They then ensure that their efforts coordinate with and complement those of tactical Army forces in the urban area. Appendix D discusses the potential contribution of joint capabilities to Army UO.

URBAN ISR

5-31. Commanders at all levels require accurate and timely information to conduct assessments for successful urban operations. This is critical to planning and execution. Senior commanders have a large role in coordinating the urban ISR effort. National strategic sources (as well as open sources) provide most of the information that commanders and staffs require on the characteristics of the human dimension, the physical properties of the terrain, and the infrastructure. The general characteristics of these aspects of the urban environment do not change drastically over time, with one exception. Military operations or natural disasters can change physical characteristics radically. Analysts can obtain crucial information through diligent research of intelligence databases and open sources. However, the disposition and composition of the urban threat is time sensitive and not likely to be discovered through this type of investigation. Due to the effects of the urban environment, deceptive efforts may influence the threat more easily. The urban population is dynamic and critical aspects of this dimension must be updated or confirmed as a prelude to urban operations. Surveillance and reconnaissance (including geospatial assets) provide relevant information regarding threat dispositions, composition and the state of the population, and the specifics of the urban terrain. Successful urban operations depend on the successful conduct of urban reconnaissance. (See also the earlier discussion of effects on the intelligence and command and control warfighting functions in Chapter 4. While counterintelligence (CI) is an essential ISR component, it will be covered later in this chapter under IO.)

Challenges

5-32. The most significant challenge to urban ISR is physical. The physical organization and complexity of the urban terrain, both man-made and natural, challenges national strategic, operational, and tactical ISR capabilities. Commanders should understand the challenges when planning and allocating time and resources to their ISR efforts. They acknowledge that subordinate commanders will face similar challenges. Therefore, commanders consider subordinate capabilities, limitations, and needs when planning, requesting, allocating, and prioritizing ISR assets and capabilities.

Imagery Capabilities

5-33. A significant national, strategic, and operational ISR capability is imagery. Even at the tactical level, units employ unmanned aircraft systems (UAS) and vehicle and dismounted electro-optics, commercial cameras (still and video), and co-opt existing civilian surveillance systems to gain situational understanding. However, the structures of the urban area significantly degrade the information that imagery acquires and may make it susceptible to physical deception measures.

5-34. Imagery, however, is an excellent source regarding the arrangement and nature of many other physical aspects. It can be used to create updated mapping products and provide significant detail of major portions of the infrastructure. Imagery can also reveal what may be happening in structures through detailed study of patterns and other exterior indicators. Yet, the bulk of a skillful threat's forces, well positioned and concealed inside or underneath structures in the urban area, are largely immune from rapid detection by overhead imaging systems. The volume of civilian movement in an urban area will itself provide a degree of camouflage and increase the difficulty of employing pattern analysis. The success in 1999 of the Yugoslavian army concealing heavy forces when confronting NATO indicates the limits of these assets to penetrate an urban area. This leaves a requirement to supplement overhead imagery in the urban environment with a reliable vehicle-mounted or handheld imaging system. For example, thermal sights on a vehicle patrolling an urban street during limited visibility may note the hot engine of a vehicle on the side of the road, possibly indicating suspicious activity. Simply providing patrols with digital cameras or video cameras can also help provide needed information and fill in information gaps.

5-35. The requirement for high altitude and low oblique imagery in more open terrain has been supplemented by a growing need for detailed, close-up, and nearly vertical overhead shots of specific portions of the urban area. This forces collection managers and system operators to move expensive and scarce imagery collection systems closer to the target or to employ flight profiles that place them at greater risk from threat air defense systems. While gun-camera images from aircraft can provide a stand-off reconnaissance platform, UAS imagery may be one of the fastest, least risky methods by which commanders can conduct reconnaissance to update and verify current maps of the urban area and determine routes, obstacles such as damaged and destroyed buildings, and intact and destroyed bridges and other significant infrastructure. Increased UAS proliferation will require greater attention to airspace command and control procedures as well as the movement and protection of ground control stations.

5-36. As the media will likely be present in the urban environment, commanders may consider devoting intelligence resources (to include linguists) to monitoring the various local, national and international sources, particularly the medium of television. The media coverage of threat forces may provide important intelligence and relevant information of immediate tactical value. Commanders, however, should always be cognizant that threats may recognize that Army forces will be using this as a source of intelligence and, feeling no compulsion to be truthful, may use it for their own deception means.

Electronic Capabilities

5-37. The physical attributes of the urban area also diminish the effectiveness of electronic ISR capabilities. Buildings and other structures significantly disrupt radio communications in an urban area. Buildings not only make tactical radio communications difficult for the user, they also make them difficult to locate, intercept, and jam. The range and clarity of frequency modulation (FM) signals significantly diminish when antennas are located inside buildings or when buildings block line of sight between the source and receiving station. To mitigate this effect on signals intelligence means, detection capabilities often move closer to the transmission source. Thus, the threat's vulnerability to compromise by means of his FM and other wireless communications in an urban environment is much less than in many other environments.

5-38. Without losing tactical surprise and increasing risk, units cannot effectively use many electronic detection and surveillance capabilities until urban combat is imminent or perhaps already begun. Ground surveillance radars will have limited uses in urban areas because of the lack of wide-open spaces in which they best operate, but can be used along the periphery of urban areas to detect infiltration and exfiltration of threat forces. However, remote systems such as the remotely monitored battlefield sensor system and the platoon early warning device may have great utility in monitoring many of the numerous avenues of approach that cannot be kept under human surveillance due to manpower constraints. Among many urban applications, remote sensors and early warning devices can be used to monitor subsurface passageways, the entrances and exits to buildings, fire escapes, the perimeters of sustainment bases, and covert foot trails between urban and rural areas.

Human Capabilities

5-39. The limits on imagery and electronic ISR capabilities place a premium on human-based visual reconnaissance. Commanders have three types of human reconnaissance assets to augment electronic reconnaissance resources: special reconnaissance, conventional combat reconnaissance, and human intelligence (HUMINT) gathered by military intelligence from individuals and multimedia. The urban environment poses several challenges to these capabilities.

5-40. The urban area challenges special reconnaissance in several ways. First is the access to the urban area. Although avenues of approach may be numerous, concealed avenues of approach into a defended urban area may be limited and thoroughly covered. Air access is also more difficult because aircraft are detected more easily, airspace is smaller, drop and landing zones are limited or not secure, and more air defense systems probably exist. Still, special reconnaissance efforts to penetrate the urban area can be successful using unconventional techniques including high-altitude low-opening parachutes or underwater penetration.

5-41. Special reconnaissance then faces a second challenge: moving in and identifying targets in the urban area. Stealth movement in an occupied urban area is exceptionally difficult. Repositioning to new or alternate positions is also dangerous. The soldiers' ability to conceal themselves among the civil population can mitigate some of these challenges but includes inherent risks of a different nature. Also difficult is establishing observation positions that provide a field of view of several targets.

5-42. Lastly, special reconnaissance may face navigational and reporting challenges. Special reconnaissance's ability to locate themselves and communicate critical locations and routes are challenged by—

- Differences in language and numbering systems. Road signs, if they exist at all, may be removed or deliberately altered to confuse.
- Irregular street patterns.
- Outdated maps.
- Intervening structures that impede communications and global positioning systems.
- Changes to the landscape due to the effects of UO or natural disasters.
- Featureless shantytowns.

5-43. Conventional reconnaissance faces many of the same challenges as special reconnaissance. Conventional reconnaissance also may lack the advantage of surprise and the special equipment and training that provides special reconnaissance stealth capability. Conventional reconnaissance is not likely to operate undetected by the civilian population. Given the constraints discussed above on other sources, conventional reconnaissance units will likely begin their mission with much less information than they would have on threat dispositions in a less complex environment. Commanders may choose to have their reconnaissance elements fight for information in the urban area. While this high-risk option is more favorable under fluid conditions, it can be used at any time. It requires careful planning, rehearsal, and formulation of information requirements.

5-44. As previously indicated, human intelligence will be one of the most valuable sources for information regarding the situation inside an urban area. HUMINT may take advantage of the proximity and large numbers of potential informants to gather information about threat activities and capabilities. It is especially valuable because it can address all elements of the environment. HUMINT sources can describe political and religious nuances that may be significant to commanders. Such information is useful for insights regarding the human dimension but extremely difficult to obtain from other means. This intelligence also can describe the infrastructure uncovering essential details of how the infrastructure functions. Obtaining good HUMINT requires skilled interrogators and linguists—often in large numbers. Commanders must know and account for some of the possible shortcomings of HUMINT:

- It is susceptible to the influence of the threat; the threat can threaten and influence the source as well as local interpreters. Commander may need to consider that people may not want to talk through a local interpreter because they are providing information they might not want others to know; anonymity may be essential. On the other hand, a local interpreter may provide community access. Yet, those trusted by the locals may not be truthful with Army forces. Conversely, those truthful with Army forces may not be trusted by the indigenous population particularly if these interpreters come from another country, region, or demographic group. In such cases, commanders may need more than one interpreter.
- It is limited by the accuracy of the source and interpreter's perceptions. A local interpreter may have difficulty understanding and translating military topics or the intricacies of urban infrastructure. Some local translators may want to please both participants, so they may shade the conversation, telling each side what they want to hear. Finally, accuracy may be affected by the collector's own perceptions if not firmly grounded in an understanding of social, cultural, and other local conditions.
- Multiple echelons collecting from a given area may inadvertently collect from the same sources and thereby give the false impression of independent sources confirming the same data. To help mitigate this possibility, commanders should ensure that all HUMINT sources are properly recorded on intelligence source registers.
- It may not be timely. The process of identifying and cultivating a source (particularly in an environment where most civilians support threat forces), gathering information, analyzing the information, and providing the intelligence to commanders can be extremely time consuming.
- Some informants may come from unscrupulous or sordid elements of the urban society and may also have their own agenda. They may attempt to use protection afforded them by their relationship with Army forces to conduct activities (even atrocities) that will compromise

political and military objectives. In some cases, a local interpreter's own agenda may influence his translation. (For example, a local schoolteacher used as a translator may report to the collector that the source believes that repairing schools in their urban area is critical, when actually the source may have said that potable water is most needed.)

Conducting Urban ISR

5-45. To be successful, ISR efforts (national to tactical level) are exceptionally comprehensive and synchronized. Moreover, success necessitates the integration of all ISR sources into operational and tactical planning. This requires that ISR assets deploy and execute early, diversify, properly focus, and integrate under a comprehensive ISR plan. Successful ISR also requires flexibility to adapt to the operational and tactical needs of the commander (see Figure 5-2). Commanders must ensure that the appropriate intelligence architecture (including robust links to joint and multinational elements) is established and tested prior to execution of ISR effort.

Early Deployment

5-46. One of the first requirements for effective urban ISR is the early deployment and employment of assets. The complex urban terrain presents a significant challenge. Commanders should consider that ISR assets will normally take longer to gather data amid the complexity.

5-47. Limited national, strategic, and operational imagery intelligence (IMINT) and signals intelligence (SIGINT) capabilities should be considered and quickly requested. If they are approved, they are tasked and deployed or repositioned to begin urban ISR operations. This takes time. Fortunately, spacing the ISR effort over time permits the analysis of the information or data as it is received. Such time also permits subsequently refining the ISR effort before all assets are committed.

5-48. SOF or conventional units will require significantly more time to execute reconnaissance missions and maintain an acceptable survivability rate. Urban reconnaissance operations require additional time for stealthy insertion into the urban area. IMINT and SIGINT capabilities are used to identify possible locations of high-value targets and corresponding observation positions; this helps minimize time-consuming and high-risk repositioning in the urban area. Again, reconnaissance units may require extensive time to observe from observation positions for indicators of threat activity and disposition and identify patterns.

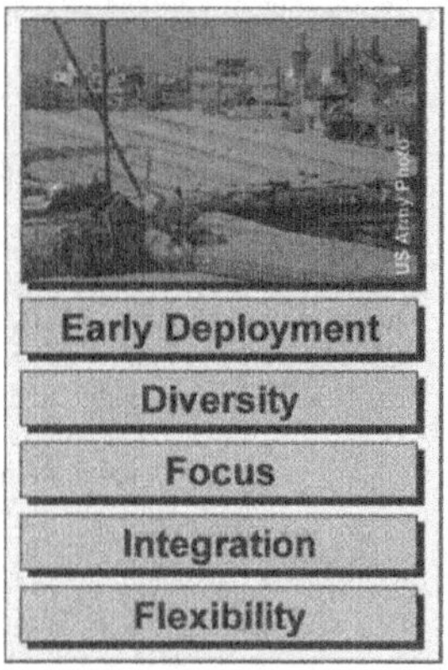

Figure 5-2. Urban ISR considerations

5-49. As conventional combat forces prepare to commit to the urban area, conventional reconnaissance precedes their actions. Conventional reconnaissance will often be a slow and methodical effort. Such forces may need time to reconnoiter the interior of structures for snipers and other small threat teams. They may also need time to deploy and destroy snipers and small delaying elements and to breach harassing obstacles. If necessary, they may need time to mass the combat power necessary to fight through enemy security and continue the reconnaissance.

Diversity

5-50. No single ISR capability can solve the riddle of the urban defense. In order to gain an accurate common operational picture of the complex urban terrain necessary to focus combat power on decisive points, commanders must employ diverse ISR capabilities. These capabilities will each contribute pieces of relevant information to permit identifying operational objectives and leveraging tactical combat power to achieve those objectives quickly. Higher-level commanders should understand that tactical reconnaissance capabilities alone often cannot provide all the tactical information required for success at lower echelons.

5-51. Using diverse capabilities challenges the threat's ability to conduct deception and otherwise defeat the friendly ISR effort. A threat that focuses on minimizing his vulnerability to overhead imagery (for example, satellites) may increase his reliance on communications and thus his vulnerability to SIGINT. At the same time, he may decrease his ability to detect the actions of ground reconnaissance units. Conversely, a threat that actively campaigns to detect ground reconnaissance may make himself more vulnerable to SIGINT and IMINT. Moreover, as Army forces enter urban areas, threat activity and susceptibility to SIGNINT often increases.

5-52. Diverse capabilities also facilitate the tactical ISR effort. Tactical reconnaissance units often consist of small dismounted teams and small combined arms teams with a dismounted element and an armor-protected mounted element. Combat engineers with their technical expertise and breaching capability are essential to the combined arms reconnaissance effort. The teams' movements are synchronized and coordinated with other assets, such as UAS and air cavalry reconnaissance. These teams use several movement techniques including infiltration, with the primary objective of conducting zone reconnaissance along key axes that support brigade and battalion actions against decisive points. To accomplish this mission, reconnaissance reconnoiters the proposed routes and alternate approaches. This supports deception and contingency planning. (Reconnaissance of alternate routes and objectives also applies to aerial reconnaissance; helicopters and UAS are not invisible. They can alert a threat to impending operations.) Infiltration of dismounted reconnaissance is made easier when a threat focuses on combined arms reconnaissance teams. Aerial reconnaissance, such as air cavalry and UAS, provides early warning of threat elements to ground reconnaissance, identifies obstacles and ambush sites, and helps select the routes for ground reconnaissance. Air elements may also reduce the mobility of counterreconnaissance forces.

Focus

5-53. Another key to successful ISR is the ability to focus the assets on commander's critical information requirements (CCIR). This focus begins with mission analysis and the commander's initial planning guidance. It is incrementally refined throughout planning and execution as each ISR effort provides information and permits more specific focus in subsequent efforts. The size and complexity of the urban environment require that the ISR effort center strictly on decisive points or centers of gravity (COGs). Therefore, the overall ISR effort will have two major focuses. The first is to uncover and confirm information on the decisive points and COG. The second is to determine the approaches (physical, psychological, or both) leading to the decisive points. The first focus will likely drive ISR in support of major operations. The second focus will likely provide the impetus for tactical ISR efforts. For example, special operations reconnaissance might focus on a major command center that controls the entire urban area and that is one of the higher-echelon command's CCIR. Tactical reconnaissance might focus on the nature of the defense along a particular avenue of approach to the objective.

Integration

5-54. Another important aspect of urban ISR is integration. All reconnaissance capabilities provide both distinctive information as well as information that confirms and adds to that coming from other sources. Essential to urban ISR is the link between all of these sources, either directly or through an integrating headquarters.

5-55. ISR operations must be vertically and horizontally linked. Vertical links ensure that ISR operations among the various levels of command are complementary and that the information flow between these levels is rapid. Horizontal links ensure that forces operating in close proximity (particularly adjacent units), where areas of interest overlap, can rapidly share results of their individual ISR efforts. Together, this helps ensure that all Army forces share a common operational picture and permits the greatest flexibility and survivability of ISR resources.

5-56. ISR operations also are integrated into the planning system, especially the targeting process. As part of targeting, positioned reconnaissance and surveillance elements may become the trigger and terminal control for applying precision fires when appropriate and after considering the risks of compromise of the position or platform. ISR operations and plans must also be synchronized with the air tasking order, as many surveillance platforms are coordinated on the same timeline.

Flexibility

5-57. The urban ISR effort must be more flexible than in other operations. This flexibility permits the ISR effort to meet unforeseen circumstances and to deal with the challenges of the urban environment. As indicated previously, the urban environment is particularly difficult to penetrate. The practical effects of this characteristic are that—

- The initial ISR effort may not be as successful as in other operations.
- More intelligence requirements may be discovered later while executing ISR operations than otherwise.
- The threat may be more successful in active counterreconnaissance because of the concealment advantages of the urban environment (hiding in structures as well as among the urban population).

Therefore, tactical and operational commanders should consider requesting greater than usual ISR support from higher headquarters. Higher headquarters should be proactive in augmenting units conducting urban operations with additional ISR assets. Additionally, ISR assets remaining under the control of the higher headquarters must respond more quickly to the CCIR of supported commanders. Sequencing reconnaissance missions over time provides flexibility by creating uncommitted reconnaissance assets.

5-58. Time sequencing of ISR assets is essential to flexibility. It makes ISR assets more survivable and allows the intelligence cycle to mature the CCIR. It also creates a ready ISR capability to augment committed forces in critical areas if required or diverts them around centers of threat resistance. If not required, the original ISR tasks can be executed as envisioned in planning. Cueing allows a high-value ISR asset to be able to respond to multiple targets based on an ongoing assessment of the overall reconnaissance effort and the changing CCIR. Redundancy permits the effort to overcome line of sight restrictions, the destruction of an ISR asset, and the ability to combine ISR resources to create combat power if required. Maximizing the ISR effort requires applying all available ISR assets to support the urban operation. Additionally, assets—such as air defense artillery and field artillery radars and engineer reconnaissance teams—are integrated into the ISR effort. In urban operations, units will also commit infantry and armor elements (plus their organic reconnaissance elements) into the tactical reconnaissance effort. These units increase the dismount capability and the ability of reconnaissance elements to fight for information and fight through security zones.

INFORMATION OPERATIONS

5-59. Information operations are an integral part of all Army operations and a critical component in creating and maintaining information superiority. The information environment is the sum of individuals, organizations, or systems that collect, process, and disseminate information; it also includes the information itself. In UO, the information environment is extremely dense due to the proliferation of INFOSYS and widespread access to those systems. In urban operations, commanders must consider how the urban environment, particularly the human component, uniquely relates to executing IO. They must also anticipate the threat's information operation campaign and preclude successful operations on its part. Overall in regard to the urban society, commanders must convince the urban inhabitants of the inevitable success of UO, the legitimacy of Army actions, and the beneficial effects that Army success will eventually generate.

5-60. Because urban operations are likely to be joint, interagency, and multinational, commanders must guard against various forms of information fratricide as a multitude of actors plan and conduct IO in the urban area. For example, the lead governmental agency's key IO objective may be to establish the legitimacy of newly formed civilian authorities. In an attempt to develop close relationships with the civilian populace, Army commanders may continue to work closely with traditional, informal leaders to the exclusion of the new authority. These actions, while they are often conducted out of practical and immediate necessity, may run counter to the lead agency's goal. Overall, Army commanders must nest their IO campaign objectives and themes within those of the lead agency, aggressively coordinate with other governmental agencies and coalition partners, and synchronize activities down to the tactical level to prevent working at odds and avoid information fratricide. However, the process must not be so centralized and rigid that subordinate units lose the flexibility necessary to develop their own products and themes essential to addressing their own AO and the particular situation that confronts them.

5-61. IO are executed using core and supporting elements and related activities (see Figure 5-3 and FM 3-13). The elements of IO are employed in either an offensive or defensive role. Many aspects of IO are not affected differently in an urban environment from any other environment. The following sections outline some IO considerations unique to urban operations.

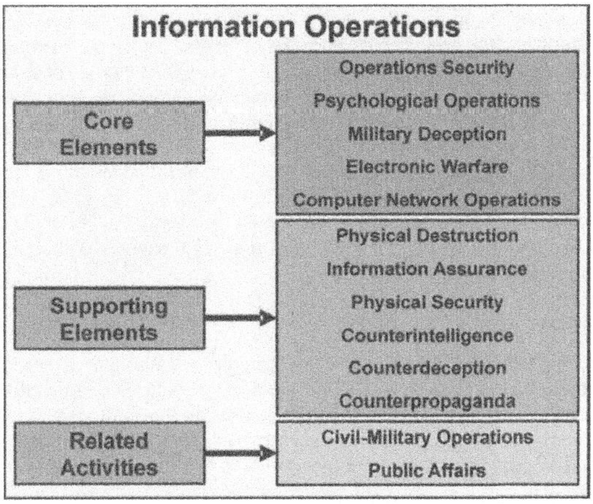

Figure 5-3. IO elements and related activities

Operations Security

5-62. In the urban environment, Army forces can leverage existing urban infrastructure, including the communications and information infrastructure, to enhance Army operations. The danger in integrating these systems is violating OPSEC. Commands ensure that Army forces use only approved systems and proper safeguards exist. Commands also supervise subordinate units for inadvertent breaches of OPSEC policies when using existing urban systems. Finally, established OPSEC procedures must be reasonable and practical. Overly complicated and stringent procedures increase the likelihood that unintentional OPSEC violations may occur as Army forces seek to rapidly accomplish the mission.

5-63. Of particular concern are computers (see also the discussion of Computer Network Operations below). During longer-term UO, commanders will constantly be upgrading and improving the quality of life for their soldiers through the establishment of well-resourced forward operating and sustainment bases. This will likely include the provision of unclassified access to the Internet and e-mail allowing Soldiers to communicate with friends and families around the world. E-mail addresses, facsimile numbers, cell phone numbers, photographs, and other sensitive (but unclassified) information can be valuable sources of threat information. Soldiers must be periodically trained and reminded of everyday procedures to combat potential threat exploitation of these unclassified sources (such as not opening e-mails from unknown sources as they may harbor Trojan programs and viruses).

5-64. To compound concerns, commanders will also establish protected systems to disseminate tactical lessons learned to individual soldiers. This creates the potential that Soldiers can inadvertently release classified information about ongoing operations to threat intelligence systems. In addition to OPSEC training and persistent awareness, commanders must establish procedures to shutdown unclassified Internet and e-mail access immediately, both before and after critical events, to mitigate potential OPSEC violations. (Commanders may also need to randomly shutdown Internet and e-mail access as part of their overall OPSEC measures to preclude such information blackouts serving as a potential warning of impending operations.) Technical surveillance countermeasures—the identification of technical collection activities conducted by threat intelligence entities—will be an important CI service to identify this and other technical OPSEC concerns.

5-65. The close proximity of Army operations to a civil population, particularly in stability and civil support operations, makes Army activities themselves an additional OPSEC concern. Hostile civilians or other threats integrated into the urban population may have more chances to observe Army activities closely. Such observations can provide insight into tactics, techniques, and procedures (TTP) and expose operational vulnerabilities. However, threats may coerce even friendly civilians to provide a threat with EEFI. Therefore, commanders in an urban environment must ensure that civilians cannot observe critical TTP. Any observable patterns and TTP vary and are supplemented with deception efforts. Physical security is increasingly important in urban areas to control civilians' access. Although many urban operations require close coordination with NGOs, commanders should screen information provided to these organizations to protect EEFI. Release of EEFI to NGOs is controlled and done with full recognition and understanding of potential consequences—the benefits must far outweigh the risks involved. (Even the best-intentioned NGO might inadvertently compromise security. For example, by revealing that a commander has closed a particular route the next day, the NGO may unwittingly reveal the objective of a pending offensive operation.)

Psychological Operations

5-66. PSYOP aim to influence the behavior and attitude of foreign audiences, both military and noncombatant, in the urban environment. PSYOP are a force multiplier and contribute in many ways to mission success (see FM 3-05.30). Effective PSYOP focuses on transmitting selected messages to specific individuals and groups in order to influence their actions. Their ability to influence the attitudes and disposition of the urban population cannot be overstated. While the complexity of the societal component of the urban environment can make PSYOP challenging, the urban society also offers many options and resources. Potentially, PSYOP (with other political and economic actions) may help limit or preclude the use of military force in urban areas. In some circumstances, UO may be relevant to the major operation only in terms of their psychological effect.

5-67. The positive influence created by PSYOP is often useful in developing an effective HUMINT capability particularly in an urban area where many civilians actively or passively support the threat. Persuading and influencing a few to support friendly forces may pay great dividends. These few supporters may allow Army forces to penetrate the urban area and obtain essential information. Such information can apply to threat capabilities, threat intentions, and even the urban environment itself.

5-68. PSYOP, combined with other elements of offensive IO, aid in isolation of a threat—a critical shaping action for any urban operation. For example, commanders may use PSYOP to inform civilians about new food distribution points located away from urban combat operations. This action supports the UO fundamental of separating combatants from noncombatants and helps to further isolate the threat (both physically and psychologically) from the civilian populace. Aside from projecting a positive image of friendly forces over threat forces, PSYOP can also isolate the threat by identifying and exploiting ethnic, cultural, religious, and economic differences between the elements of the civilian populace and threat forces as well as the differences among supportive and unsupportive civilian factions. The complexity of the urban environment enables quick changes in opinion or attitude. Commanders must continually evaluate the results of PSYOP for mission relevance.

Military Deception

5-69. Urban operations present numerous challenges to tactical commanders; however, higher-level commanders may help to mitigate some challenges through effective military deception Commanders can use military deception efforts designed to mislead threat decision makers as to friendly force disposition, capabilities, vulnerabilities, and intentions. Military deception actions may allow commanders to achieve tactical surprise or improve relative combat power at a selected location. For example, allowing the threat to observe certain activities on a selected avenue of approach may cause the threat to shift his forces (and effort) to the area perceived to be threatened. (This movement may also aid in determining the overall disposition of threat forces and intentions.) Repositioned forces or effort to activities or locations that are not decisive to the achievement of friendly objectives, combined with other IO designed to overwhelm threat information and intelligence systems, may create the force and tempo differential necessary to achieve success. Commanders should tailor urban deception plans to the specific urban area, paying close attention to the societal characteristics of the target population.

Electronic Warfare

5-70. Electronic warfare (EW) includes all actions that use electromagnetic or directed energy weapons to control the electromagnetic spectrum or to attack a threat. Conducting EW in urban areas seeks to achieve much the same results as in other environments. A major consideration in urban areas is collateral effects on portions of the urban infrastructure that rely on the electromagnetic spectrum for service. Thus, precision is a major factor in planning for EW operations. For example, EW attacking a threat's television broadcasts avoids affecting the television broadcasts of neutral or friendly television. Likewise, EW attacking military communications in a large urban area avoids adversely affecting the area's police and other emergency service communications. Urban offensive and defensive operations will have the least restrictions on EW operations while urban stability may have significant constraints on using EW capabilities.

Computer Network Operations

5-71. Computer network operations (CNO) include computer network attack (CNA), computer network defense (CND), and computer network exploitation (CNE). CNO are not applicable to units at corps and below. Echelons above corps (EAC) units will conduct CNA and CNE. If tactical units require either of these network supports, they will request it of EAC units.

Computer Network Defense

5-72. In urban operations, CND will require extreme measures to protect and defend the computers and networks from disruption, denial, degradation, or destruction. The nature of the urban environment and configuration of computer networks provides the threat with many opportunities to interdict local area networks (LANs) unless monitored by military forces. LANs controlled by military forces are normally more secure than the civilian infrastructure. Commanders should prepare for opportunities by the threat to insert misinformation.

Computer Network Attack

5-73. Considerations regarding the execution of CNA in urban operations are similar to those of EW: CNAs that do not discriminate can disrupt vital civilian systems. However, possible adverse effects on the civilian infrastructure can be much larger—potentially on a global scale. In the short term, CNAs may serve to enhance immediate combat operations but have a debilitating effect on the efficiency of follow-on urban stability operations. Because of these far-reaching effects, tactical units do not execute CNA. CNA is requested of EAC units. EAC units will receive all requests from lower echelons, carefully consider second- and third-order effects of CNA, and work to ensure its precise application.

Computer Network Exploitation

5-74. CNE consists of enabling operations and intelligence collection to gather data from target or adversary automated INFOSYS or networks. Tactical units do not have the capability for CNE. CNE contributes to intelligence collection at EAC. In UO, CNE will be centrally controlled.

Physical Destruction

5-75. Physical destruction includes those actions—including direct and indirect fires from air, land, sea, space, and Special Forces—taken with, to augment, or supplement IO actions. Like many other IO elements, major concerns with employing physical destruction in UO are precision and follow-on effects. Thus, commanders using physical destruction to support IO must adhere to the same constraints as all other urban fires.

Information Assurance

5-76. Information assurance in UO takes on an added dimension. As with other operations, availability of information means timely, reliable access to data and services by authorized users. In UO, the timeliness of information may be restricted because structures block the transmission waves. The need for retransmission facilities may overwhelm the signal community. The reliability can be questioned because of the blockage between units and communications nodes. Unauthorized users may intercept the communications and input misinformation or disinformation. Commanders must protect the integrity of all information from unauthorized changes, including destruction. INFOSYS with integrity operate correctly, consistently, and accurately. The authentication of information may be accomplished by sophisticated electronic means. However, it is more likely that communications-electronics operating instructions authentication tables will authenticate the information. Commanders should consider the confidential nature of all information in UO and establish procedures to protect the information from unauthorized disclosure. Of additional concern, the density of the infrastructure in urban areas may inhibit receipt by the intended individual or unit. Therefore, commanders must develop alternative means to determine if the message was received.

Physical Security

5-77. Physical security consists of those actions and measures to safeguard personnel; to prevent unauthorized access to equipment, installations, material, and documents; and to safeguard them against espionage, sabotage, damage, and theft. The joint, interagency, and multinational nature of urban operations increases the need for physical security measures, especially with organizations that do not emphasize physical and operational security within their own organizations. The potential for locating sustainment bases in or near urban areas and the use of indigenous personnel on these facilities to accomplish a multitude of tasks mandates that commanders carefully consider the effectiveness of their physical security measures. In some stability operations where Army forces must recruit and train security or police forces, and then conduct joint operations and patrols with them, commanders must examine how best to establish close working relations while maintaining essential physical (and operations) security.

Counterintelligence

5-78. The urban environment, particularly in stability operations, is ideal for espionage, other intelligence activities, sabotage, or assassination. Threats can approach, conduct reconnaissance, and escape under the concealment offered by the urban environment. The dense environment and electromagnetic clutter serves to hide technological collection means. To counter threat intelligence collection activities, the Army employs CI forces. They provide analysis of the threat's human, imagery, signals, and measurement and signature intelligence capabilities to develop countermeasures against them (see FM 34-60).

5-79. In conjunction with military police, engineers, medical personnel, and other as required, CI help create urban threat and vulnerability assessments necessary in developing comprehensive force protection measures and evaluating their effectiveness. CI personnel can also supplement the screening of linguists and other critical host-nation personnel employed by Army forces. As Soldiers operate in close proximity to the urban society, threats will attempt to use three, time-tested tools to influence them: money, sex, and drugs. Periodic CI refresher training remains crucial to decrease the effectiveness of these threat tactics.

Counterdeception

5-80. In UO, threat forces can easily accomplish deception operations. The force that controls the area above and below ground will have freedom of movement. Deception aimed at friendly commanders will cause them to deploy combat power at the wrong place and the wrong time. Counterdeception by friendly commanders will identify and exploit threat attempts to mislead friendly forces. Counterdeception is difficult. Cultures of certain rhetoric and actions are more predisposed to deception than others. Knowing a threat's previous deception methods is important. Dismissing tactical indicators because they conflict with preconceptions may allow a hostile deception operation that plays on the preconception to succeed.

Counterpropaganda

Four hostile newspapers are more to be feared than a thousand bayonets.

Napoleon Bonaparte

5-81. Because propaganda is aimed at both combatants and noncombatants, UO are especially concerned with its use. Propaganda can rapidly and dramatically affect the attitudes of the urban population and will probably occur (or increase significantly) after urban operations have begun. Thus, it can create situations in the human dimension of the environment quite different from those discovered in the pre-operations assessment. Counterpropaganda is, therefore, essential to urban operations. To negate, deflect, or destroy the threat's propaganda capability, counterpropaganda requires—

- Monitoring and evaluating the effectiveness the threat's propaganda efforts.
- Determining innovative methods using all Army force capabilities, especially PSYOP and PA units.

5-82. In particular, humanitarian assistance, when coupled with an effective information operations campaign, can be a powerful counterpropaganda tool to sway public support. For example, media coverage of Army forces directly providing or coordinating NGO humanitarian efforts to feed, shelter, and provide medical care to the urban population can help counter threat efforts to discredit Army forces as an evil occupying force. This type of coverage can be used to turn opinion against the threat and could also prove an effective tool in gaining favorable international support.

Public Affairs

5-83. A closely related activity to IO is public affairs (PA). PA influences urban operations by transmitting information through the media to internal (urban Army forces as well as the urban civilian populace) and external audiences. At higher levels of command, PA can help maintain popular national support for the urban operation by clarifying the links between strategic goals and operational objectives. At both the operational and tactical levels, it links Army units, the urban inhabitants, the U.S. and international public, and the media. The media will often focus on the negative actions that are ongoing in an AO; hence, the urban commander's challenge will be to publicize his unit's good news stories. Commanders should not wait for media inquiries and interview requests. Instead, they should aggressively plan public affairs to get the Army's story out to the media and, ultimately, the public.

5-84. PA can help determine potential media issues that may influence planned UO. It can also aid commanders in understanding the impact of UO on the environment (particularly its citizens) and other agencies and organizations operating in the urban area. PA also helps to counter rumors, uncertainty, fear, loneliness, confusion, and other factors that cause stress (to both soldiers and civilians) and undermine effective UO. If the populace does not understand the mission or the Army forces' capabilities, false expectations may be created that Army forces may not be able to meet. PA can help prepare the American public for the possibility of high casualty rates. Overall, PA supports urban commanders in their goals to achieve information superiority and preserve public support.

5-85. The density of information sources and reporters in UO ensures that all Army activities will be subject to media and public scrutiny. Many reporters will congregate in cities for their own comfort and take advantage of established communications networks. Urban areas are densely populated and, together with Army forces and NGOs operating there, will present the greatest number of human-interest stories.

The local urban or host-nation media, however, will often have their own agendas developed over a longer period of time. This local media may also have a greater influence over the urban population than the international media. The indigenous media may not follow international norms. Commanders are responsible for understanding the media (particularly the local media), its role, and its potential for negative or positive influence.

5-86. Commanders cannot allow themselves to be intimidated by the media. Media sources are going to get and report their story one way or another. Therefore, commanders should support open and independent reporting and grant access to their units as early and as far forward as the situation permits. Successful units often go beyond simply accommodating the media to actually integrating them into their units. In the future, embedding of media may become the standard for reporting Army operations. Other considerations for improving urban media relations include:

- Encouraging the media to report on all ongoing aspects of the unit's urban operations—not exclusively on combat operations. This may create a unit responsibility to protect journalists that participate in covering suggested activities.
- Hosting media luncheons or other events for key journalists. This provides commanders the ability to get the whole story out to the press from their unit's perspective. To gain media interest and unit credibility, this will often require key leadership participation.
- Identifying Soldiers that speak the local and international journalists' native language. Even if journalists speak English as a second language, they may overlook or misunderstand important points due to language subtleties. Additionally, people tend to be friendlier to those that speak their own language.
- Identifying selected Soldiers at lower tactical levels to act as PA representatives for their unit as an additional duty. However, commanders must provide them with appropriate training and resources to accomplish specific PA tasks.
- Writing articles for local newspapers that emphasize the unit's good news. Depending on the situation, subordinate units may be tasked to write stories about the good accomplishments in their AO. Commanders, with appropriate OPSEC considerations, may hire locals to assist in this effort. However, commanders must take extra precautions to ensure that these efforts are not perceived as covert propaganda or as undermining free and independent news reporting.

5-87. PA does not distort, direct, or manipulate information. Its effectiveness stems directly from establishing and maintaining credibility with the urban population and media. PA protects the overall integrity of the Army as an institution by ensuring that the Army is recognized as the most reliable source for official information among all other competing sources. Commanders must synchronize PA with the core and supporting elements of IO (particularly PSYOP and counterpropaganda) to ensure that all Army sources send only one message. Urban commanders must plan for the media and integrate PA into their decision-making and (through IO) targeting processes. Similar to lethal targeting, commanders should consider target audiences and coverage areas for the various new agencies and mediums. They also pay attention to print media circulation statistics and television and radio broadcast times. For example, an urban populace may tend to read morning newspapers vice late editions and watch or listen to nightly news instead of morning or daytime programming. Additionally, low literacy rates among the urban population may indicate that television and radio are better mediums to use to influence the urban population than print media such as newspapers, magazines, leaflets, and posters.

5-88. The PA principles listed in Figure 5-4 and addressed in FM 46-1 summarize PA. They serve as useful guides toward planning and executing PA operations regardless of the environment. However, the principles of "practice security at the source," and "truth is paramount" particularly apply to the urban environment. The compartmented nature of most UO impede commanders' and their PA officers' ability to be at all places where the media will likely be. Friendly inaccuracies, intentional or not, will help fuel the threat information campaign. Therefore, all soldiers are trained, provided with clear and understandable PA guidance, and prepared to accurately communicate to the civilian media. The keys at all levels are understanding, prepared acceptance, and truthfulness tempered with an essential concern for OPSEC.

- Truth is Paramount
- If News is Out, It is Out
- Public Affairs Must be Deployed Early
- Not All News is Good News
- Practice Security at the Source
- Media are Not the Enemy
- Telling Our Story is Good for the Army
- Soldiers and Families Come First

Figure 5-4. Public affairs principles

Civil-Military Operations

5-89. Civil-military operations (CMO) are critical and will be an essential, implied task of any urban operation. They are included here as a closely related activity of IO. Commanders use CMO to establish, maintain, influence, or exploit relations to achieve operational objectives. These relations are among military forces, civilian authorities and organizations (both governmental and nongovernmental), and the civilian population. CMO range from support to combat operations to assisting in establishing political, economic, and social stability. In some instances, commanders may need to be prepared to assume temporary responsibility for the functions and capacities of the urban government. Like PA, effective CMO is based on establishing and maintaining credibility and trust with the urban populace and civilian organizations operating in the urban environment. Because of the potential importance of CMO, commanders may consider placing an assistant commander, executive officer, or some other senior leader to monitor, guide, liaise with other units and organizations, and, overall, ensure that the appropriate command emphasis is placed on these activities.

Civil Affairs

5-90. Civil affairs (CA) units and teams will be critical during UO. While any military force can conduct CMO, CA units are specifically organized, trained, and equipped to conduct activities in support of CMO. They have experience in planning and conducting CMO, a regional focus (which includes enhanced cultural awareness and language training), civilian technical expertise that parallel common urban government and administration functions, and experience in negotiation and mediation. Such experience ensures relevant support to commanders conducting urban operations. (As it is likely that there will not be enough CA forces to cover all requirements, commanders should survey their Soldiers to determine the whether any possess civilian expertise that may be beneficial to CMO.) CA units organize their skills into six functional specialty areas (see Figure 5-5). Commanders should use these skills, often unfamiliar to most military personnel, to—

- Develop their situational understanding of the urban environment (particularly the infrastructure and society).
- Plan CMO to support UO.
- Achieve many of the fundamentals of UO.

Note: To ensure optimal integration, commanders must establish clear command and support relationships between CA and maneuver forces.

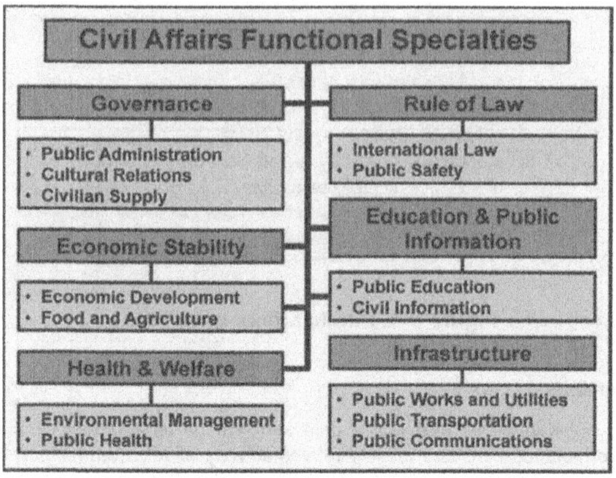

Figure 5-5. Civil affairs functional specialties

5-91. In addition to providing essential information for understanding the urban environment, CA personnel and activities help shape the battlefield, engage a civil problem, consolidate gains, and transition to a legitimate civil authority. Specifically, CA teams and CMO help urban commanders—

- Minimize civilian interference with UO and the impact of urban operations on the populace and infrastructure. CA personnel can help establish and run a civil-military operations center (CMOC), discussed later in this chapter, to coordinate UO with civilian agencies (both governmental and nongovernmental), other services, and multinational partners.
- Provide advice and assistance to restore or rehabilitate portions of the infrastructure.
- Plan, supervise, and execute necessary populace and resources control measures (in close coordination with military police units) until no longer required or the urban operation is completed.
- When requested or when military necessity or legitimate directives require, establish all or portions of the urban civil administration.
- Determine available supplies and services in the urban area and if necessary assist in negotiating their acquisition. CA also help commanders assess the capability, dependability, and willingness of urban sources to provide and sustain identified needs as well as to calculate the impact of using them on other aspects of the urban environment.
- In conjunction with the SJA, fulfill the Army's responsibilities toward the urban population under international, host-nation, and U.S. law.
- Plan and conduct the transition of control for the urban area or operation to another military or civilian governmental or nongovernmental organization or agency.

5-92. As part of CMO, stability or civil support operations will be critical as commanders begin to restore essential services to meet the emergency needs of a potentially devastated urban population. Besides repairing or restoring vital services and other critical infrastructure, these and other projects may also stimulate the urban area's economy and provide work for the unemployed. (Employed workers are less likely to be co-opted by terrorists, insurgents, and criminal gangs.) As such, they can be a critical stabilizing tool for commanders as they attempt to transition control and responsibility to other agencies or to legitimate civilian control. These projects may include—

- Providing clean water.
- Restoring waste management functions (less-extensive sewer repairs and trash collection and disposal activities).
- Improving medical care.
- Rehabilitating schools.
- Assisting in the establishment of local governments.
- Improving security and emergency services (police, fire, and rescue).
- Providing electrical generators for hospitals, clinics, and factories.
- Restoring irrigation to surrounding farmland.
- Repairing important cultural infrastructure.

5-93. As part of the initial planning process, CA units conduct an *area assessment*, which can provide commanders with essential information about the environment (see FM 41-10). Commanders should integrate this initial assessment into the overall urban-focused IPB process. To help analyze civil considerations in any environment, commanders and staffs can consider many characteristics such as areas, structures, capabilities, organizations, people, and events (ASCOPE). These characteristics easily align with an urban area's three main characteristics of terrain, society, and infrastructure; and, like them, they are overlapped and interdependent (see FM 6-0). Of these, the category of events is a critical addition to urban characteristics previously categorized and discussed. Significant spikes (and lulls) to threat and civilian activity can be expected to occur in conjunction with important religious, historic, and political events (for example elections). Therefore, commanders should determine, track, and pay particular attention to these events when planning operations and managing transitions (see also the Key Event Chart discussion in Appendix B, paragraph B-43).

5-94. Similar to PA operations, CMO are related to information operations. The nature of CMO and the need for CA personnel to develop and maintain close relationships with the urban population put CA personnel in a favorable position to collect information. CA personnel work daily with civilians, their equipment, and their records that may be prime sources of information. If used correctly, CA personnel can complement the intelligence collection process necessary to understand the dynamic societal component of the urban environment and detect significant changes. However, CA personnel are not, and cannot appear as, intelligence agents; otherwise, it will undermine their ability to interact with the civilian community. Examples of information available to CA units include government documents, libraries, and archives; files of newspapers and periodicals; industrial and commercial records; and technical equipment, blueprints, and plans.

5-95. Overall, CA expertise is critical to a commander's understanding of the complexities of the infrastructure and societal components of the urban area. These components (together with the terrain or physical component of the urban area) interconnect. CA forces help identify and understand the relationships and interactions between these urban components. From this understanding, commanders can anticipate how specific military actions affect the urban environment and the subsequent reactions. CA personnel help commanders predict and consider the second- and third-order effects and reactions as well as the long-term consequences. Understanding these long-term consequences helps in achieving stability and ensuring a smoother and speedier transition of the urban area back to civilian responsibility and control. Oppositely, an unconsidered short-term solution may, in the end, turn out to be the solution requiring the greatest amount of time. Poor solutions may exacerbate the situation. Others may contribute little to conflict resolution and waste precious resources (particularly time) until a more reasoned approach is formulated—one that should have been the first solution.

Civil-Military Reconstruction Efforts

5-96. Reconstruction efforts and activities, an integral part of CMO, should complement NGO efforts and are conducted in concert with larger urban area and nationwide projects often headed by the U.S. Army Corps of Engineers (USACE) and the U.S. Agency for International Development (USAID). For example, extensive, large-scale sewer projects or major repairs to the urban area's electrical infrastructure would be left to the USACE and USAID. When planning and conducting urban reconstruction, commanders should—

- **Keep expectations realistic.** With so much to do, it is easy for Army forces to become overcommitted. Commanders must focus on the achievable, build off successes, and create constructive forward process.
- **Seek opportunities.** While commanders think long-term, they should look for catalytic investment of resources. Minimal resources provided at the right time can generate significant activities resulting in increased momentum toward the desired end state.
- **Align resources with local needs.** In analyzing the merit of a project, commanders must check to determine if it is supported and desired by the urban populace. Local participation in the process reinforces ownership and acceptance and eventual transition of responsibility.
- **Balance speed and quality.** Commanders must balance the achievement of rapid response with quality construction and repairs.
- **Ensure efforts do not undermine locals.** Army reconstruction efforts should not undermine the growth and legitimacy of local institutions and authorities.
- **Integrate in the targeting process.** Commanders should routinely include approval of civil-military reconstruction products as part of the information operation's contribution to the overall targeting process.

INTEGRATION OF CONVENTIONAL AND SPECIAL OPERATIONS FORCES

5-97. One important Army and joint resource that commanders of a major operation can use to influence urban operations is SOF. Several types of these forces exist (including the CA forces discussed above), each with unique and complementary capabilities. They can be extremely valuable in UO for their ability to execute discrete missions with a higher degree of precision than conventional forces, to provide information, and to enhance cultural understanding. However, the challenges of using SOF include command and control; integration; coordination with conventional forces that will normally command, control, and conduct the bulk of UO tasks; and the imprudent inclination to use SOF forces for conventional purposes. The density and complexity of UO make close coordination and synchronization of conventional forces and SOF essential to mission success. The nature of the environment dictates that both forces will work in close proximity to each other; the separation in space and time between SOF and conventional forces will often be much less in urban areas than in other environments. Overall, the nature of the environment demands a synergistic combination of capabilities to achieve effects on the threat and mission success.

5-98. Successfully integrating SOF occurs with proper integration into, or coordination with, the command structure of the force conducting the UO. SOF within a theater (less CA and tactical PSYOP) ordinarily falls under joint command and control. Therefore, the commander of the major operation responsible for an urban area, if he is not a JFC, will have to coordinate through the JFC to integrate SOF capabilities into the UO. Examples of critical coordination elements include:

- Boundaries.
- No-fire areas. (If assigned, SOF liaison officers (LNOs) must be routinely included in targeting meetings.)
- Coordination points.
- The exchange of key intelligence and information.
- Requirements to support personnel recovery contingencies.

> **Conventional and Special Forces Integration**
>
> On 20 April 2003 during OPERATION IRAQI FREEDOM (OIF), an Army Special Forces (SF) operational detachment alpha (ODA) was patrolling through villages and secondary roads east-southeast of An Najaf. A local Iraqi approached the team and volunteered the location of a senior Ba'ath Party official in the nearby town of Ghamas. Unfortunately, he did not provide an exact location but rather an approximate location and the name of the family dwelling. The ODA coordinated with their company commander and nearby elements of the 101st Airborne Division to conduct a raid on the residence at dawn the next day. Joint planning, preparation, and rehearsals among the units occurred throughout the night.
>
> At 0430, the raid convoy departed its base en route to Ghamas. The small task force consisted of an assault team, a security team, a command and control element, and three blocking forces composed of Soldiers from a scout platoon and antiarmor company of the 101st. At 0515, the convoy hit the release point outside Ghamas. The 101st vehicles moved to their blocking positions at three bridges surrounding the town. The remainder of the task force moved into the town. An interpreter quickly found a local guide who knew where the house was located. He took the SF team to a walled two-story dwelling with a courtyard. The security team isolated the objective and provided overwatch while the assault team forcibly seized the house and apprehended three adult males. Among them was Abd Hamden, the target of the raid and a senior Ba'ath Party official from Baghdad.
>
> After a tactical interrogation, one of the men provided the location of a Fedayeen major nearby. The SF and conventional force raided this house minutes later but only found the major's relatives. As the Soldiers left the town with their captives, they were cheered by the locals throughout the town. Although only one small example, the extensive coordination between SOF and conventional forces during urban operations was a tremendous success during OIF.

SYNCHRONIZATION OF ACTIVITIES WITH OTHER AGENCIES

5-99. The population density of the urban environment, its economic and political importance, and its life-supporting infrastructure attracts many types of organizations. These organizations include—

- Other U.S. governmental agencies.
- International governmental organizations.
- Allied and neutral national governments.
- Allied and coalition forces.
- Local governmental agencies and politicians.
- NGOs.

5-100. Even in a major operation or campaign, many organizations operate in the area as long as possible before combat or as soon as possible after combat. Therefore, coordination with these organizations sharing the urban AO will be essential to achieve synchronization; however, effective synchronization is challenging, time consuming, and manpower intensive. The staffs of larger headquarters (divisions or higher) normally have the breadth of resources and experience to best conduct the necessary synchronization. They can effectively use or manage the organizations interested in the urban area and mitigate their potential adverse effects on UO. By taking on as much of the synchronization requirements as possible, the operational headquarters permits its tactical subordinates to remain focused on accomplishing their tactical missions. The higher headquarters should assume as much of the burden of synchronization as possible. However, the density of the urban environment will often require that smaller tactical units coordinate and synchronize their activities with other agencies and the local civilian leadership (formal and informal) simply because of their physical presence in the units' AOs. In urban stability and civil support operations, mission accomplishment will require effective civil-military coordination and synchronization activities and measures at all levels as either a specified or implied task.

Civil-Military Operations Centers

5-101. To coordinate activities among the varied agencies and organizations operating in an urban area and the local population, urban commanders can establish a CMOC. The CMOC synchronizes Army activities and resources with the efforts and resources of all others involved (see FM 41-10). This can be particularly important in stability and civil support operations where combat operations are not the dominant characteristic of the major operation. CMOCs can be established at all levels of command. Hence, more than one CMOC may exist in an AO, particularly large urban areas. CMOCs may be organized in various ways and include representatives from as many agencies as required to facilitate the flow of information among all concerned parties. Commanders must still ensure that force protection and OPSEC requirements are not compromised. Effective CMOCs can serve as clearinghouses for the receipt and validation of all civilian requests for support, can aid in prioritizing efforts and eliminating redundancy, can decrease the potential for inappropriate displays of wealth by one or more of the participating organizations, and, most importantly, can reduce wasting the urban commander's scarce resources.

Liaison Officers

5-102. LNOs—sufficiently experienced and adequately trained in liaison duties and functions—are necessary to deal with the other agencies that have interests in the urban area. Army LNOs work with the lead agency or other organizations (including local civilian agencies such as police) that the commander has identified as critical to mission success. Together they work to rapidly establish unity of effort and maintain coordination, often before a CMOC is established. The additional coordination afforded by the physical presence of LNOs within these organizations may be required even after the CMOC is fully functional. When commanders lack enough LNOs to meet requirements, they should prioritize and often assign a single LNO to several organizations. That LNO will then share his time and presence to those organizations based on the situation and his commander's guidance.

Commander's Personal Involvement

5-103. Overall, establishing a close relationship with other agencies and the urban civilian population will often be a major, positive factor in successful mission accomplishment, particularly in urban stability operations. Despite internally established command and control relationships, commanders that develop a direct and personal relationship with the leaders and staff of other agencies can often avoid conflict, win support, foster trust, and help eliminate the "us versus them" mentality that can frustrate cooperation among Army forces and civilian organizations.

Chapter 6
Foundations for Urban Operations

Utilities such as electricity and water are as much weapons of war as rifles, artillery pieces or fighter aircraft. . . . In the case of Manila, where there was a noncombatant, civilian population of one million in place, it was the attacker's aim to capture the utilities which the defender planned to destroy.

The Battle for Manila

Commanders conducting major urban operations (UO) use their ability to visualize how doctrine and military capabilities are applied within the context of the urban environment. An operational framework is the basic foundation for this visualization. In turn, this visualization forms the basis of operational design and decision making. To accurately visualize, describe, and direct the conduct of UO, commanders and their staffs must understand the basic fundamentals applicable to most UO.

URBAN OPERATIONAL FRAMEWORK

6-1. Army leaders who have an urban area in their area of operations (AO) or are assigned missions in an urban area follow an urban operational framework. They identify the portion of the urban area essential to mission success, shape the area, precisely mass the effects of combat power to rapidly dominate the area, protect and strengthen initial gains without losing momentum, and then transition control of the area to another agency. This framework divides into five essential components: understand, shape, engage, consolidate, and transition. These five components provide a means for conceptualizing the application of Army combat power and capabilities in the urban environment.

6-2. The urban operational framework assists commanders in visualizing urban operations. This framework is simply an aid to the commander. It is not sequential, nor is it a planner's tool for phasing an operation. Commanders should combine the urban operational framework with—

- The principles of war.
- The tenets of Army operations.
- The components of operational design.
- Considerations for stability operations and civil support operations.
- Sustainment characteristics.
- Running estimates.
- Commander's critical information requirements (CCIR).
- Each commander's experience.

The framework contributes to the visualizing, describing, and directing aspects of leadership that make commanders the catalysts of the operations process (see Figure 6-1). In the same manner, the urban operational framework contributes to the overall operations process (see FM 3-0).

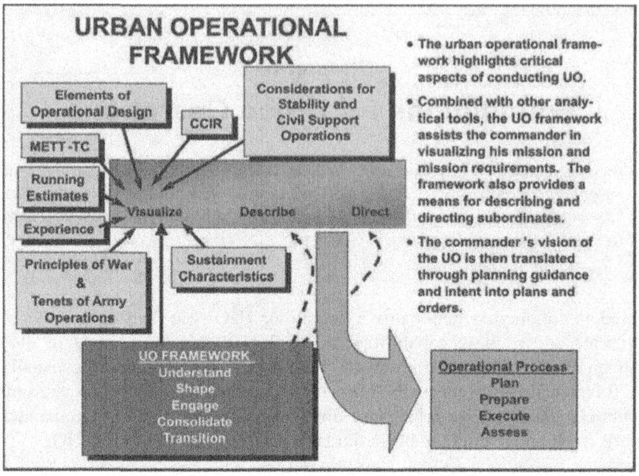

Figure 6-1. The urban operational framework and battle command

UNDERSTAND

6-3. Understanding requires continuous assessment—throughout planning, preparation, and execution—of the current situation and progress of an operation, and the evaluation of it against measures of effectiveness to make decisions and adjustments. Commanders use visualization, staff officers use running estimates, and all use the intelligence preparation of the battlefield (IPB) process to assess and understand the urban environment. Commanders and staffs begin to understand by observing and then collecting information about the situation. They observe and learn about the urban environment (terrain, society, and infrastructure), and other factors of METT-TC—mission, enemy, weather, troops and support available, and time available. They use intelligence, surveillance, and reconnaissance means; information systems (INFOSYS); and reports from other headquarters, services, organizations, and agencies. Then they orient themselves to the situation and achieve situational understanding based on a common operational picture (COP) and continuously updated CCIR. Largely, the ability to rapidly and accurately achieve a holistic understanding of the urban environment contributes to the commanders' abilities to seize, retain, and exploit the initiative during UO.

Disproportionately Critical

6-4. The Army operations process requires continuous assessment and understanding; it precedes and guides every activity. In UO, however, in-depth understanding is disproportionately critical for several reasons. First, each urban environment is unique. Other environments can be studied and their characteristics quantified in a general manner with accuracy. This is fundamentally not true of different urban areas. The characteristics and experience in one urban area often have limited value and application to an urban area elsewhere. This characteristic sets UO apart from operations in other environments.

Extremely Dynamic

6-5. The urban environment is also extremely dynamic. Either deliberate destruction or collateral damage can quickly alter physical aspects of the urban environment. The human aspect is even more dynamic and potentially volatile. A friendly civil population, for example, can become hostile almost instantaneously.

These dynamics (combined with initial difficulty of understanding and describing this unique environment) make it difficult for commanders and staffs to initially develop and maintain a COP and establish situational understanding. Furthermore, public reaction to media coverage of the urban operation and political changes influence national security strategy and objectives. Such changes can affect the basic nature of an operation, especially after it has commenced. Anticipating these potential effects and developing appropriate branches and sequels based on an accurate understanding often determines how quickly commanders can achieve the desired end state.

Risk Assessment

6-6. As in any environment, UO pose both tactical and accident risks. However, the level of uncertainty, ambiguity, and friction can often be higher than that of many other environments. Such challenges increase the probability and severity of a potential loss due to the presence of the enemy, a hostile civilian group, or some other hazardous condition within the urban environment (see Determining the Necessity of Urban Operations in Chapter 5). Therefore, commanders must—

- Identify, assess, and understand hazards that may be encountered in executing their missions.
- Develop and implement clear and practical control measures to eliminate unnecessary risk.
- Continuously supervise and assess to ensure measures are properly understood, executed, and remain appropriate as the situation changes.

6-7. Risk decisions are commanders' business. However, staffs, subordinate leaders, and even individual Soldiers must also understand the risk management process and continuously look for hazards at their level or within their area of expertise. Any risks identified (with recommended risk reduction measures) must be quickly elevated to the appropriate level within the chain of command (see FM 3-100.12 and FM 5-19).

Complex and Resource Intensive

6-8. The urban environment is the most complex of all the environments in which the Army conducts operations. It is often comprised of a diverse civil population and complex, ill-defined physical components. A sophisticated structure of functional, social, cultural, economic, political, and informational institutions unites it. Thus, the analysis to understand the environment is also complex and time and resource intensive. The nuances of the urban environment can take years to uncover. Hence, constant analysis of the environment requires greater command attention and resources. Accurately understanding the environment is a prerequisite to shaping it, and both understanding and shaping activities are crucial to effectively engage the elements of the urban environment critical to success.

SHAPE

6-9. Shaping operations, part of all Army operations, are essential to successful UO. They set the conditions for decisive operations at the tactical level in the urban area. Rapid action, minimum friendly casualties, and acceptable collateral damage distinguish this success when the AO is properly shaped. Failure to adequately shape the urban AO creates unacceptable risk. The commander of a major urban operation has several resources with which to begin shaping the AO. Important capabilities include—

- Fires.
- Information operations.
- Special operations capabilities.
- The maneuver of major subordinate units.

6-10. Critical urban shaping operations may include actions taken to achieve or prevent isolation, understand the environment, maintain freedom of action, protect the force, develop cooperative relationships with the urban population, and train Army forces for sustained UO.

Isolation

6-11. Isolation of an urban environment is often the most critical component of shaping operations. Commanders who's AO includes operationally significant urban areas often conduct many shaping operations to isolate, or prevent isolation of, those areas from other parts of the AO. Likewise, commanders operating in the urban area focus on isolating decisive points and objectives in the urban area or averting isolation of points that are critical to maintaining their own freedom of action. Isolation is usually the key shaping action that affects UO. It applies across full spectrum operations. Most successful UO have effectively isolated the urban area. Failure to do so often contributed to difficult or failed UO. In fact, the relationship between successful isolation and successful UO is so great that the threat often opposes external isolation actions more strongly than operations executed in the urban area (or critical areas within). In some situations, the success of isolation efforts has been decisive. This occurs when the isolation of the urban area compels a defending enemy to withdraw or to surrender before beginning or completing decisive operations. In UO that are opposed, Army forces attempt to isolate the threat three ways: physically, electronically, and, as a resultant combination of these first two, psychologically (see Figure 6-2).

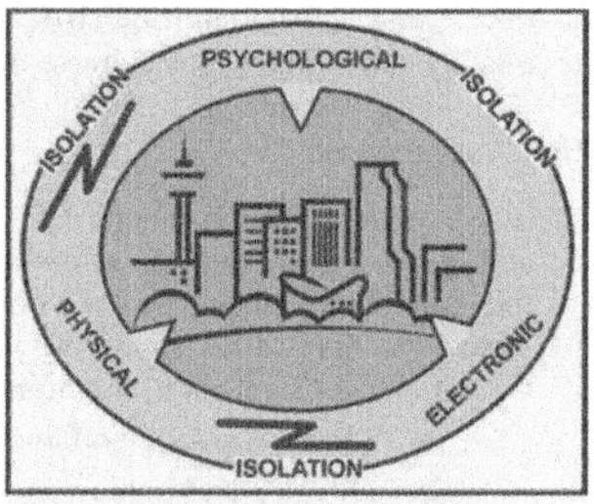

Figure 6-2. Urban isolation

Physical Isolation

6-12. In offensive UO, physical isolation keeps the threat from receiving information, supplies, and reinforcement while preventing him from withdrawing or breaking out. Conversely, a defending Army force attempts to avoid its own physical isolation. Simultaneously, this force conducts operations to isolate the threat outside, as they enter, or at selected locations in the urban area. Physical isolation can occur at all levels. In many situations, the commander of a major combat operation may attempt to isolate the entire urban area and all enemy forces defending or attacking it. At the tactical level, forces isolate and attack individual decisive points often using a cordon technique. In stability operations, physical isolation may be more subtly focused on isolating less obvious decisive points, such as a hostile civilian group's individual leaders. In many operations, isolation may be temporary and synchronized to facilitate a decisive operation elsewhere. To effectively isolate an urban area, air, space, and sea forces are often necessary additions to the capabilities of ground forces.

Electronic Isolation

6-13. Electronic isolation is achieved through offensive information operations (IO). Electronic warfare (particularly two of its components: electronic warfare support and electronic attack) and computer network attack are critical to electronic isolation (see FM 3-13 and the Information Operations discussion in Chapter 5). At the operational level, offensive IO aims to quickly and effectively control the information flow into and out of an urban area. This isolation separates the threat's command and control (C2) system in the urban area from its operational and strategic leadership outside the urban area. Offensive IO also focuses on preventing the threat from communicating with civilians through television, radio, telephone, and computer systems. At the tactical level, IO aim to isolate the threat's combat capability from its C2 and leadership within the urban area, thus preventing unity of effort within the urban area. Defensive IO can prevent isolation of friendly forces defending in an urban area.

Psychological Isolation

6-14. Psychological isolation is a function of physical actions, electronic warfare, and other forms of IO, especially military deception and psychological operations. Psychological isolation denies the threat political and military allies. It separates the enemy or hostile civilian group from the friendly population, nongovernmental organizations (NGOs) operating in the urban area, and from political leaders who may consider supporting Army forces. Psychological isolation destroys the morale of individual enemy soldiers or hostile civilians. It creates a feeling of isolation and hopelessness in the mind of the threat. It undermines the threat leadership's ability to think, plan, decide, and act. As such, it undermines the confidence of the threat in their leadership. On the other hand, IO, as well as the disciplined conduct of Army personnel, can help to forge legitimacy for Army operations. In stability operations, psychologically isolating the threat can result in the friendly urban population and NGOs positively supporting Army operations.

Other Shaping Actions

6-15. Other shaping actions can include the proper sequencing and deployment of forces, reconnaissance operations, and force protection. These actions contribute equally to the success of any urban operation. Commanders must understand how the urban environment affects their ability to accomplish these shaping actions. However, civil-military operations (CMO), another closely related activity of IO, are important to shaping the urban operational environment for decisive operations. The specific civil-military task can vary greatly and may include affecting a cooperative relationship with the civil political system, protecting portions of the civil population or infrastructure, or establishing refugee camps or safe areas for noncombatants. This is most true in stability and civil support operations. Successful CMO also can contribute to the psychological isolation of the threat. (See Civil-Military Operations in Chapter 5 for a more detailed discussion.)

Training and Education

6-16. Finally, Army commanders must consider that critical shaping actions often occur prior to the urban operation in the form of professional education, and home-station and in-theater training. Commanders can enhance training through joint, interagency, multinational, and combined arms exercises and effective rehearsals.

ENGAGE

6-17. In UO, Army forces engage by appropriately applying their full range of capabilities against decisive points leading to centers of gravity. Successful engagement takes advantage of the Army force's superior training, leadership, and, within the constraints of the environment, equipment and technology. Successful engagement also requires the establishment of necessary levels of control and influence over all or portions of the urban environment until responsibilities can be transferred to other legitimate military or civilian control. Engagement may range from the overwhelming, yet precise, application of combat power to defeat a threat, to large-scale humanitarian assistance operations, to unobtrusive advice and assistance to the urban security forces. It may require brief engagement in a relatively small portion of an urban area or with small segments of the population to long-term engagement throughout multiple, large urban areas and with vast sectors of the population. The commander's ability to engage in a specific urban operation depends, as always, on the situation and the assigned mission.

Offense: Attack Decisive Points

6-18. In offensive UO, forces successfully engage by striking at the enemy's center of gravity using multiple offensive actions from unexpected directions and throughout all dimensions. Army forces aim to engage identifiable decisive points. Successful efforts against decisive points lead to effects on the center of gravity. The center of gravity will differ in each offensive situation. It may be an individual enemy leader, the enemy's combat power, the enemy's communications capability, or a physical structure of cultural, political, or economic significance.

Defense: Deny Vital Functions and Critical Infrastructure

6-19. In defensive UO, engage translates into denying the enemy control of the vital functions and critical infrastructure of the urban area. Forces achieve this by leveraging the defensive advantages of the urban terrain, defending essential areas in depth, using economy of force in nonessential areas, controlling the enemy direction of attack with natural and man-made obstacles, and retaining the initiative through counterattacks.

Stability and Civil Support: Apply Innovation and Imagination

6-20. The ability to engage in urban stability operations hinges on the type of stability operation commanders execute. In a noncombatant evacuation operation, forces limit engagement to finite geographic areas and times. In contrast, a peace operation may require engagement of a large urban area for an extended time. In this operation, engage is defined as using the array of Army capabilities to create specific conditions among the belligerents. Thus, the techniques used for engagement in stability operations vary according to the situation and as situations mature during long-term operations.

6-21. In urban civil support operations, engagement is accomplished by innovative and subtle application of Army capabilities. Since Army forces usually support other agencies that lead the operation, effective engagement results from carefully and discretely applying Army capabilities to the tasks assigned by the lead agency. In a humanitarian relief situation, Army forces may be tasked to transport supplies in the urban area. Successful engagement in this activity then becomes the goal of Army forces and may be achieved by providing, managing, and protecting transportation assets and clearing and maintaining sufficient lines of communications (LOCs).

CONSOLIDATE

6-22. Army forces consolidate to protect and strengthen initial gains and ensure retention of the initiative. Consolidation includes actions taken to eliminate or neutralize isolated or bypassed threat forces (including the processing of enemy prisoners and civilian detainees) to increase security and protect LOCs. It includes the rapid repositioning and, as necessary, reorganization of forces and intelligence, reconnaissance, and surveillance (ISR) assets. Consolidation may also include activities in support of the civilian population such as relocating displaced civilians, reestablishing law and order, humanitarian assistance and relief operations, and restoration of key urban infrastructure. A significant consolidation effort may be the reduction or elimination numerous explosive hazards resulting from previous urban combat operations. During consolidation, commanders often coordinate with NGOs to establish unity of effort in the execution of stability or civil support operations. Consolidation begins with the initiation of the urban operation and follows each critical activity.

TRANSITION

6-23. When planning UO, commanders ensure that they plan, prepare for, and manage transitions. Transitions are movements from one phase of an operation to another or a change in responsibility for all or portions of the urban environment from one unit, organization, or agency to another. They involve significant changes in the type of operation, concept of the operation, mission, situation, task organization, forces, resource allocation and support arrangements, or C2. Transitions occur in all operations, but in UO they occur with greater frequency and intensity, are more complex, and often involve agencies other than U.S. military organizations. For example, a successful attack may transition to a defend mission that includes not only defense tasks but also stability tasks. All operations will likely include a transition of responsibility for some aspect of the urban environment to (or back to) a legitimate civilian authority. Unless planned and executed effectively, transitions can reduce the tempo of the urban operation, slow its momentum, and cede the initiative to the threat.

Mental and Physical Preparation

6-24. Transitions occur as conditions warrant. They can be carefully planned and controlled, or they can be quick and dramatic, such as the swift transformation of a stability operation into offense or defense—and back again. Units prepare mentally and physically to address rapid transitions. Accordingly, plans include branches and sequels that address anticipated or possible transition points. When the dominant type of operation changes from an offense to stability or civil support, the types of units originally conducting the UO may no longer be appropriate. A large mobile reserve may permit increased flexibility to react to unplanned transition requirements. Operations or units in one part of an urban area may transition before operations or units in a different part of the same urban area. This will require commanders to exercise careful command and control and execute various types of operations and associated tasks simultaneously.

Transition to Legitimate Civilian Authorities or Agencies

6-25. Combat operations are not an end to themselves. They are inevitably a means of transition, an activity for moving from an unsatisfactory state of affairs to an improved end (as determined by national objectives). In UO, a distinct aspect of transition is the requirement to quickly and efficiently transition the major portions of Army responsibilities to civil agencies. Some tasks to which units will transition are not traditional combat tasks but rather stability tasks more closely associated with CMO. In stability or civil support operations this is often a near-term critical mission objective. In these operations, commanders aim to alleviate the circumstances requiring Army forces and ensure that other civilian agencies assume the functions provided by Army forces. Following combat operations, civilian agencies are assisted and encouraged to quickly resume specific support activities—such as providing sanitary services, food services, law enforcement, and health services—because of their high demand on Army resources.

6-26. Transitioning responsibility for the urban environment (or selected portions) to legitimate civilian agencies or authorities requires commanders to make an accurate assessment and develop an in-depth understanding of the civilian organization's ability to accept and handle the responsibility. Commanders consider not only resources available but whether the organization has the appropriate leadership and technical expertise as well as a vision in consonance with U.S. objectives. Commanders should avoid forcing a transition only to alleviate the burden on Army forces. On the other hand, they may need to encourage acceptance of responsibility while continuing to provide "behind-the-scenes" advice, assistance, and resources. Transitioning responsibility prematurely may alleviate the commander's immediate resource concerns but may, in the longer term, negatively impact the achievement of strategic objectives.

Clearly Visualize and Describe the End State

6-27. Army UO conclude when Army forces depart and have no further mission requirements in the urban area. At the outset, commanders visualize and describe the intended end state of a unit's execution of UO. Commanders then clarify and update this visualization as the political or strategic situation is refined or changes. Importantly, commanders must create a command climate and procedures that allow their staffs and subordinate commanders the ability to ask questions that will enable all to share in an accurate perception of the commander's visualization. This enables subordinate units to identify likely transitions and ensures that current operational planning takes into account second- and third-order effects. As long as an active Army AO contains an urban area, some type of urban operation will exist. After urban combat successfully ends, combat forces may move on. Support forces conducting sustaining operations may then occupy the area and continue to conduct a different form of UO.

Applying the Urban Operational Framework: Panama – 1989

The U.S. conducted OPERATION JUST CAUSE in December 1989 to remove the illegal ruler of Panama, Manuel Noriega, and to restore that country to a democracy. It also conducted the operation to ensure the safety of a substantial number of U.S. personnel as well as the security of U.S. interests in Panama. The major focus of JUST CAUSE was in Panama City, the country's capital. Most operations occurred in this large urban area, one of the numerous smaller urban areas, or the urban-like military bases. These bases increased the AO and were directly linked to operations in the capital city. This successful operation illustrates how commanders can apply the urban operational framework to visualize, describe, and direct the critical aspects of urban operations.

Understand

The synchronization achieved during the operation may have obscured the challenges faced in the initial assessment process in Panama. However, it was not as simple as it may have seemed. Using the framework of the urban environment, U.S. forces required details of the physical characteristics of the environment, the infrastructure, and the human dimension including the capabilities of the Panamanian military.

Because Army forces had a long history in Panama, commanders clearly understood the physical challenges and layout of critical urban areas (see Figure 6-3), particularly Panama City. They also understood how the infrastructure in each urban area functioned and which parts was a key element to success. Examples of key portions of the infrastructure included the Madden Dam, which controlled the water flow through the Panama Canal, and the Cerro Azul television tower, which was the main Panamanian broadcast tower.

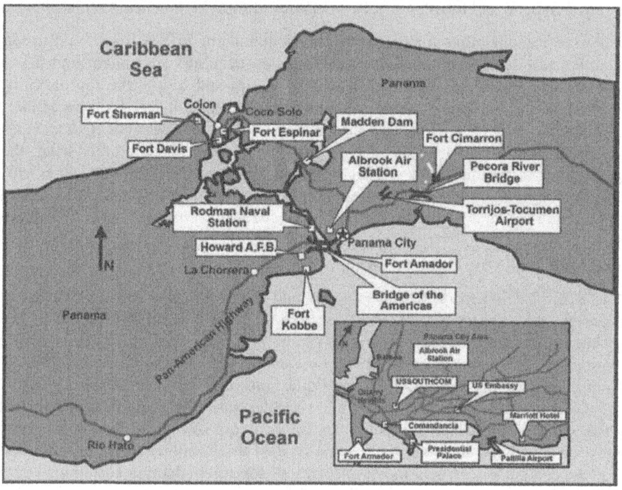

Figure 6-3. Panama

Collecting information and developing intelligence on the human elements of the urban environment was critical to operational success and a challenge. Because the Panamanian Defense Force (PDF) had traditionally been an ally of the United States, Army forces did not have a systemic database that adequately depicted their order of battle and their true capabilities. Additionally, much of the situation in Panama was colored in political terms making it more difficult for traditional military sources to evaluate the status of PDF forces. For example, Army planners needed to know if PDF military units (when faced with a formidable U.S. force) would fight at all for Noriega and if they did fight, how hard and long would they fight. The answers depended largely on their political loyalty to Noriega and on the individual loyalty of the unit officers to the Panamanian president. Thus, Army commanders needed to understand the military characteristics of PDF units and their political affiliations and tendencies.

Because transition from combat to noncombat tasks would be critical to achieving all objectives, particularly the restoration of democracy, Army forces also needed an accurate assessment of the political opposition to Noriega—including that opposition's capabilities and vulnerabilities. Again, Army forces were required to make assessments outside those needed solely for combat operations. Ultimately, assessing the political opposition's vulnerabilities led to assigning Army units to protect them throughout the operation so that they could serve as a foundation for a new democratic government.

Finally, the commander's assessment included an evaluation (often subjective) of the attitudes and disposition of the Panamanian people. Human intelligence (HUMINT) was the primary source of information on the population. Army forces had good access to the population because of their close proximity and historical ties to Panama. Many Soldiers were married to Panamanians, and the Army had total access to local media and to prominent individuals.

National imagery and special operations forces (SOF) also contributed to the ability of Army forces to assess the urban environments of JUST CAUSE. All units executing operations had detailed satellite photos of objective areas. Additionally, key objectives were placed under SOF surveillance well in advance. This surveillance revealed unit readiness, vulnerabilities, detailed disposition, and other patterns critical to mission success. The combination of the two capabilities allowed units to plan and achieve the synchronization necessary for such a complex urban operation.

Shape

During OPERATION JUST CAUSE, commanders conducted numerous shaping operations to establish the conditions for the decisive operations. Many operations were designed to control information, such as an assault on the Azul television tower identified during the assessment of the infrastructure. Planners designed many shaping operations to isolate various garrisons and PDF units. An example of tactical isolation was the plan for the Pacora River Bridge to prevent reinforcements from reaching the garrison at Torrijos-Tocumen Airport.

Operational isolation was achieved through the Ranger Regiment's and 82nd Airborne Division's assault on targets at Rio Hato in the west and Fort Cimarron in the east. These actions in conjunction with the securing of Maddam Dam had the primary objective of isolating Panama City. They were also the largest of the major actions occurring during OPERATION JUST CAUSE. The airborne assault was also the largest airborne operation conducted by U.S. forces since World War II. This large-scale shaping operation demonstrates that shaping operations are critical to mission success and can be more resource intensive than the actual operations that achieve domination.

Engage

U.S. Army forces achieved successful engagement in OPERATION JUST CAUSE by establishing unchallenged military control over Panama City and eliminating Noriega's capability to challenge that control. Toward this end, the operation attacked two decisive points. The first was the assault on the PDF headquarters located in Panama City: the Comandancia. The second was the operation undertaken to locate and seize Noriega himself.

Three battalions of task force (TF) Bayonet (5-87th Infantry, 1-508 Infantry [Airborne], and 4-6th Infantry [Mechanized]) executed the attack on the Comandancia and Fort Amador. They were also tasked to protect the American Embassy in downtown Panama City. To execute these missions, they moved from various staging areas located throughout the city to their assigned objectives using air assault, mounted, and dismounted approaches. The ground movement through the city proved to be the most difficult and hazardous part of the mission due to the vulnerability of the troops in their armored personnel carriers and trucks. The dismounted movement was slower than the mounted movement but allowed the Soldiers greater cover and concealment.

The strongest opposition to TF Bayonet occurred at the Comandancia. Elements of three PDF companies and two public order companies held out for three hours. The troops moving to Comandancia were subject to a large volume of sniper fire, and in the assault, unidentified indirect fire caused significant casualties among the mechanized forces. TF Bayonet forces, supported by airborne armored reconnaissance vehicles and Hellfire missiles from Apache helicopters, captured the

Comandancia. Commanders noted in particular the precision of the supporting fires from attack helicopters. The assault by fire from supporting AC 130 gunships severely damaged much of the reinforced Comandancia building.

Simultaneously, SOF attacked several targets where Noriega might be located. These initial attacks were unsuccessful. However, many subsequent actions neutralized Noriega's influence and eventually resulted in his apprehension on 3 January 1990. These actions included the well-organized and relentless manhunt conducted by SOF units, the isolation of Panama City itself, population control efforts, sophisticated IO, and cooperation with other U.S. agencies.

Consolidate and Transition

OPERATION JUST CAUSE demonstrated the vital need for a thought-out plan that adequately addresses consolidation and the transition from combat to noncombat before commanders initiate operations. Normally in complex UO, commanders cannot leave the details of transition until after the operation has begun without unacceptable risk to overall mission accomplishment. The follow-on stability operation, OPERATION BLIND LOGIC (later renamed OPERATION PROMOTE LIBERTY) began 24 hours after the initial assault and thus both operations were occurring simultaneously. This simultaneity of different types of operations is typical in major operations conducted in a large urban area. The stability operation involved more time than the combat operation and continued well after the close of OPERATION JUST CAUSE and after most of the major combat units had redeployed. It involved significant resources without the same level of risk to U.S. forces as the combat operations.

Civil affairs (CA) were a dominant part of the transition from combat to stability operations. The 96th Civil Affairs battalion was central to this operation. CA forces established a civil police force, emergency food distribution, property protection, production and distribution of newspapers, cleanup of the city, and building support for a new civil government. Most tasks were coordinated through Army CA forces and executed by other Army forces under the supervision of CA.

IO were also a major aspect of affecting a stable transition and successful postcombat operations. These operations built support for the U.S. operation among the population. They emphasized that the U.S. conflict was with Noriega and not the Panamanian people and that the U.S. forces would depart as soon as a new Panamanian government could take over.

Other U.S. agencies played critical roles in stability operations in Panama. The U.S. Drug Enforcement Agency and Justice Department were important to the negotiations that led to Noriega's capture. The U.S. State Department helped to negotiate for Noriega and develop military policies and plans during stability operations. The American Embassy advised commanders regarding the large diplomatic community that existed in Panama City.

FUNDAMENTALS OF URBAN OPERATIONS

6-28. UO often differ from one operation to the next. However, some fundamentals apply to UO regardless of the mission, geographical location, or level of command. Some of these fundamentals are not exclusive to urban environments. Yet, they are particularly relevant to an environment dominated by man-made structures and a dense noncombatant population (see figure 6-4). Appendix A provides an historical example of how many of these fundamentals applied to an actual conflict situation. Vitally, these fundamentals help to ensure that every action taken by a commander operating in an urban environment contributes to the desired end-state of the major operation.

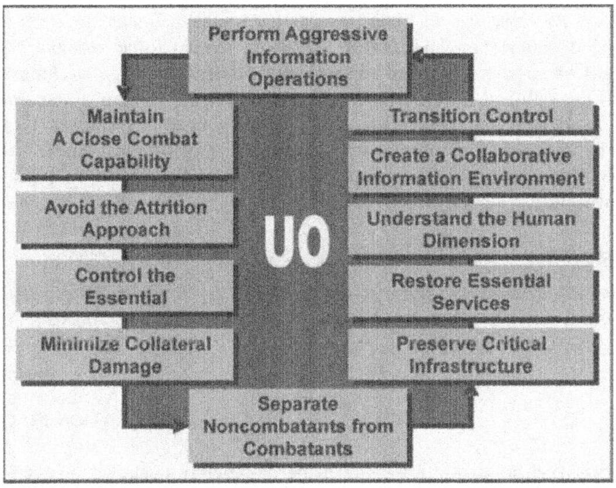

Figure 6-4. Fundamentals of urban operations

PERFORM AGGRESSIVE INFORMATION OPERATIONS

6-29. Information operations aimed at influencing non-Army sources of information are critical in UO across the spectrum of operations. Because of the density of noncombatants and information sources, the media, the public, allies, coalition partners, neutral nations, and strategic leadership will likely scrutinize how Army forces participate in UO. The proliferation of cell phones, video cameras, Internet capability, and media outlets ensure close observation of the activities of Army forces. With information sources rapidly expanding, public information of Army operations will often be available faster than the internal military INFOSYS can process it. Army forces should aggressively integrate IO into every facet and at all levels of the operation to manage perceptions and mitigate unintended consequences. Under media scrutiny, the actions of one Soldier may have significant strategic implications. IO aim to make the information accurate; placed in the proper context of the Army's mission; and available to all interested parties: the public, the media, and other agencies. Ensuring that the urban population understands how they and their urban area fit within the commander's vision, intent, and end state will be a critical task.

MAINTAIN A CLOSE COMBAT CAPABILITY

6-30. Close combat is required in all offensive and defensive UO. This core capability must also be present and visible in urban stability operations and may be required in urban civil support operations. Close combat in any urban operation is resource intensive, requires properly trained and equipped forces, and has the potential for high casualties. However, the ability to decisively close with and destroy enemy forces as a combined arms team remains essential. Potential threats respect strength and, instead, seek to attack ill-prepared or complacent Army forces—not necessarily forces that are better protected or armored, but those less likely or capable of inflicting casualties against them. In stability and civil support operations, a lack of respect and fear of Army forces can hinder recovery as much as the ill-advised use of forces. Hence, all Soldiers, regardless of branch or military occupational specialty, must be properly equipped and trained to fight in an urban environment. This ability allows Army forces to deter aggression, compel compliance, morally and physically dominate a threat and destroy his means to resist, and terminate urban conflicts on the Army commander's terms.

Avoid the Attrition Approach

6-31. Previous Army doctrine was inclined towards a systematic linear approach to urban combat. This approach emphasized standoff weapons and firepower. Army force structure does not support this approach towards UO. It can result in significant collateral damage, a lengthy operation, and an inconsistency with the political situation and strategic objectives. Enemy forces that defend urban areas want Army forces to adopt this approach because of the likely costs in resources. Commanders should only consider this approach to urban combat as an exception and justified by unique circumstances. Instead, commanders should seek to achieve precise, intended effects against multiple decisive points that overwhelm a threat's ability to react effectively.

Control the Essential

I'm talking about attacking those things from which the regime draws power but being very careful about it so that we don't get large bodies of young Americans caught up in house-to-house Berlin, World War II-type scenario.

Lt. Gen. William C. Wallace
Commander, V Corps
OPERATION IRAQI FREEDOM

6-32. Many modern urban areas are too large to be completely occupied or even effectively controlled without an enormous force. Therefore, Army forces focus their efforts on controlling only the essentials to mission accomplishment. At a minimum, this requires control of key terrain. Key terrain is terrain whose possession or control provides a marked advantage to one side or another. In the urban environment, commanders determine key terrain based on its functional, political, economic, or social significance. A power station or a place of worship may be key terrain.

6-33. All principles of war can apply to UO. The principle of mass and the principle of economy of force (in addition to the principle of unity of command discussed in the previous chapter) are particularly important in guiding UO and providing mission focus. Army forces mass combat power only to control those requirements essential for mission success. This permits conservation of combat power. It also implies economy of force and associated risk in those areas where Army forces choose not to exercise control.

Minimize Collateral Damage

6-34. Forces should integrate precision fires, IO, and nonlethal tactical systems consistent with mission accomplishment while decreasing the potential for collateral damage. Commanders develop unique rules of engagement (ROE) for each urban operation and provide necessary firepower constraints. IO and nonlethal systems may compensate for some restrictions, especially in stability or civil support operations. Commanders continually assess the short- and long-term effects of operations and firepower on the population, infrastructure, subsequent missions, and national and strategic objectives. They must also consider what, if any, provisions should be made to amend or address potential collateral damage. Overall, commanders must balance restraint and precision with speed and overwhelming combat power. By avoiding unnecessary harm to all elements of the urban environment, commanders retain the moral high ground and help sustain legitimacy for their operations. Minimization of collateral damage allows civilians to continue to provide for their own needs or the rapid return of the urban area to civilian self-sufficiency.

Separate Noncombatants from Combatants

6-35. Promptly separating noncombatants from combatants (psychologically and physically) may make the operation more efficient and diminish some of the threat's potential asymmetrical advantages. This separation also may reduce restrictions on the use of firepower, enhance force protection, and strip the threat from its popular support base. This important task becomes more difficult when the threat is an unconventional force that can mix with civilians.

6-36. In more recent operations, threats have sought to integrate their military capabilities as closely as possible into the civilian population and infrastructure. In these conditions, commanders must increase their efforts to discriminate between the two. Soldiers managing violence in this setting require the highest level of individual and organizational discipline and judgment. Soldiers will require the mental agility to separate their aggression toward threats from the noncombatant civilian population. The training, effort, and command emphasis in this area is as important as fully successful results. Such efforts strongly impact national and international perceptions of the operation.

PRESERVE CRITICAL INFRASTRUCTURE

6-37. Commanders should analyze the urban area to identify critical infrastructure. They should attempt to preserve and protect the critical elements for postcombat sustainment operations, stability or civil support operations, or the overall health and well-being of the indigenous population. Urban areas remain in the AO after combat operations have ceased. Postcombat UO are unavoidable. Different from simply avoiding collateral damage, Army forces may have to initiate actions to prevent an enemy or a hostile civilian group from removing or destroying critical infrastructure and assets. This may include cultural infrastructure such as religious and historical places. In some cases, preserving the infrastructure and the urban society's sources of economic and cultural wealth may be the assigned objective of the urban operation.

RESTORE ESSENTIAL SERVICES

6-38. Army forces plan to restore essential services that may fail to function before or during an operation. Essential services include power, food, water, sewage, medical care, and security and law enforcement. When planning for and conducting Army UO, units can use nonlethal and less destructive munitions and capabilities to keep potentially vital infrastructure intact. Initially, Army forces may be the only force able to restore or provide essential services and commanders must plan accordingly. Failure to do so can result in serious health problems for the civilians, which can affect the health of Army forces and negatively impact overall mission success. Army forces seek to transfer responsibility for providing essential services to other agencies, NGOs, or the local government as quickly and effectively as possible. Despite the potential causes for the failure or destruction of essential services, commanders also seek to ensure that civilians continually perceive Army restoration activities as assistance rather than an Army requirement. Otherwise, the civilian populace may be slow to accept or resume responsibility for their urban area.

UNDERSTAND THE HUMAN DIMENSION

> *Do not try to do too much with your own hands. Better the Arabs do it tolerably than that you do it perfectly. It is their war, and you are to help them, not to win it for them. Actually, also, under the very odd conditions of Arabia, your practical work will not be as good as, perhaps, you think it is. It may take them longer and it may not be as good as you think, but if it is theirs, it will be better.*
>
> T. E. Lawrence
> "Twenty-Seven Articles"

6-39. Commanders carefully consider and manage the perceptions, allegiance, and morale of the civilians. Their assessment of the environment needs to accurately identify the attitudes of the people toward Army forces. Operational guidance to subordinates—including ROE, protection, sustainment operations, and fraternization—is based on this assessment. Commanders expect and consider the demographic variance in the attitudes of an urban population. They cannot inadvertently apply Western cultural norms to a non-Western urban population. Commanders can only make reliable assessments based on a thorough understanding and appreciation of the local society and their culture. Developing this ability also requires the additional ability to effectively share information among all echelons of command.

6-40. Sound policies, proper discipline, adequate consideration for local culture, and rapid engagement of local urban leaders will positively affect the attitudes of the population toward Army forces. Additionally, well-conceived and executed IO will enhance the position of Army forces relative to the urban population. Even during high-intensity urban combat, heightened awareness of and sensitivity toward the civilians can lead to a better postcombat situation than if civil considerations were unobserved or diminished in importance. An improved postcombat situation enhances transition. As the environment of conflict becomes more complex, the human dimension (and associated moral aspects) takes on greater importance and may have the greatest potential for affecting the successful outcome of UO. Therefore, the human aspect and development of cultural acuity creates a discrete planning factor.

Create a Collaborative Information Environment

6-41. The complexity of the urban environment, particularly the human dimension, requires rapid information sharing—providing and receiving—from the national level to the tactical level, among Army headquarters at each echelon, with sister services and coalition partners, and with participating governmental and (at appropriate times) nongovernmental agencies. The analysis of urban information into the relevant intelligence necessary to refine and deepen a commander's understanding of the urban environment and its infrastructure of systems also demands collaboration among the various information sources and consumers. Therefore, commanders seek to establish streamlined procedures, develop commonality among databases, and make best use of existing information systems to disseminate and receive the necessary intelligence and relevant information for subordinates and partner organizations and agencies to exercise effective leadership and decision making and establish synergistic unity of effort in this multifaceted environment.

Transition Control

6-42. Because UO are resource intensive, commanders should plan to end them as quickly as possible, yet consistently with successful mission accomplishment. (Depending on the factors of METT-TC, successful transition may take a few days or many years to achieve.) The end state of all UO transfers control of the urban area to another agency or returns it to legitimate civilian control and responsibility. Rapid transition releases Army resources for use elsewhere and improves the civilian morale and disposition toward Army forces. This requires the successful completion of the Army force mission and a thorough transition plan. The transition plan may include returning control of the urban area to another agency a portion at a time as conditions permit. A successful transition plan should consider early alignment of military capabilities with existing urban governmental and administrative organizations, agencies, structures, and districts. Transition planning must be conducted before the onset of operations and continually adjusted as the situation develops.

Chapter 7
Urban Offensive Operations

... Capture Suez City "provided it does not become a Stalingrad situation."

Order to the Adan Armored Division
prior to its 1973 attack on Suez City
On the Banks of the Suez

Offensive urban operations (UO) are one of the most challenging operations that military forces can undertake. Campaigns and wars have sometimes hinged on their success or failure. Costly in resources, even when successful, they are not lightly entered into. Once engaged, they are executed rapidly and decisively. For reasons already discussed, threat forces defending in UO may gain advantages from the environment while Army force capabilities may diminish. Despite the challenges, Army forces conduct successful urban offensive operations by combining the Army's existing offensive doctrine with a thorough understanding of the environment.

PURPOSE OF URBAN OFFENSIVE OPERATIONS

7-1. Like all offensive operations, urban offensive operations are designed to impose the will of commanders on the threat. The urban offense often aims to destroy, defeat, or neutralize a threat force. However, the purpose may be to achieve some effect relating to the population or infrastructure of the urban area. Army forces may conduct offensive operations to secure a port or a communications center, to eliminate a threat to a friendly government or the urban population, or to deny the threat use of urban infrastructure. No matter the purpose, commanders must use a combined arms approach for successful urban offensive operations.

OFFENSIVE CHARACTERISTICS APPLIED TO URBAN OPERATIONS

7-2. All offensive operations, including those in urban areas, contain the characteristics of surprise, concentration, tempo, and audacity (see FM 3-0). Commanders consider and incorporate these characteristics in their offensive UO plans.

SURPRISE

7-3. Army forces can achieve offensive surprise at two levels: operational and tactical. In urban offensive operations, operational surprise can be decisive. The goal is to attack the urban area before the threat expects it, from a direction he doesn't expect, or in a manner he doesn't expect. In UO, this requires an attack against urban areas that the threat believes will provide sanctuary from the technological advantages of Army forces. Usually, urban areas that meet this criterion are not easily accessible. Army forces launch an attack against these urban complexes in different ways: through a vertical assault using airborne or air assault forces, through an amphibious assault, or through a penetration followed by a rapid and deep advance. All three attacks aim to achieve surprise and to deny the threat time to prepare and establish a defense. Surprise in a major urban operation prevents a threat from falling back to occupy prepared positions in and around an urban area.

7-4. At lower tactical levels, forces achieve surprise by attacking asymmetrically. An asymmetric method attacks the threat so he cannot respond effectively. This may be achieved by using special operations forces

(SOF) against a threat prepared for a conventional attack, by attacking decisively with heavy forces when the threat expects an effort by light forces or SOF, or by leveraging Army forces' extensive information operations (IO) capability. Offensive IO—primarily using IO elements of deception, electronic warfare, and operations security (OPSEC)—can help achieve surprise at all levels (see Chapter 4). Attacking at night surprises the threat and maximizes the Army forces' training, command and control (C2), and technological advantages. Attacking from unexpected or multiple directions achieves surprise by leveraging Army information systems (INFOSYS) and superior synchronization of combat power and capabilities.

CONCENTRATION

7-5. In UO, the attacking force creates a major advantage by concentrating the effects of combat power at the point and time of its choosing. The area and its compartmented effects naturally disperse and dissipate combat capability. Often commanders must position considerable amounts of stay-behind forces to protect hard-fought gains. The environment also hinders repositioning forces rapidly. Such effects can work equally against defending and attacking forces. However, in a well-prepared defense, the defender often has the advantage of interior lines. The defender can reinforce or reposition forces more quickly using covered and concealed routes (such as, sewers, tunnels, or prepared holes made in walls). Successful UO need synchronized air and ground maneuver with overwhelming effects from fires at decisive points on the urban battlefield. To achieve proper synchronization and precise effects, commanders should consider the unique time and distance relationships set by the environment.

TEMPO

7-6. Tempo is the rate of military action; it differs between UO and operations in more open terrain. Often, the primary purpose of the threat's urban defense is to take advantage of that differential to disrupt the rapidity and overall tempo of the Army's major operation. Additionally, the complexity and the potential for increased risk in an urban environment may invoke a cautious and methodical response on the part of commanders and their staffs thereby exacerbating tempo differences. While careful and meticulous preparation and planning are required in UO, commanders conducting major operations that include urban areas must strive to maintain an active tempo in offensive operations through synchronization of combat power and anticipation of threat reactions. A high tempo allows Army forces to achieve surprise and quickly gain positions of advantage. Controlling operational tempo and not allowing the different tempo of UO to adversely affect other operations is a challenge that commanders of major operations must overcome.

Operational Context of Urban Operations: Brittany Ports – 1944

The plan for the invasion of Normandy, France, in June 1944 was meticulously developed. The plan not only addressed the invasion itself, but also contained detailed planning for the campaign to follow. A major concern of the detailed campaign planning was logistics. To address this critical concern, and specifically the problem of ports to supply the allied armies once ashore, the pre-invasion planning called for the major ports of the French province of Brittany—Brest, Lorient, and Saint Nazaire—to be objectives of General Patton's Third Army, once it was activated.

Early August 1944, almost two months after the successful Normandy invasion, the operational situation significantly differed from that envisioned by the D-Day planners. General Montgomery's Twenty-first Army Group was still fighting in the Bocage of Normandy. In contrast, General Bradley's Twelfth Army Group had just achieved a major breakthrough at Saint Lo, secured the Cotentin Peninsula, and reached the city of Avranches. Here was a decision point. Bradley and Eisenhower had to decide whether to adhere to the original plan and turn west with Patton's forces to secure the peninsula or to take advantage of the breakout at Saint Lo and turn east to exploit the disruption of the German defenses.

Ultimately they reached a compromise. General Middleton's VIII Corps was tasked to secure the peninsula, and the bulk of Patton's Army, three Army corps, was turned northeast to exploit the operational collapse of the main German defenses. (See figure 7-1.)

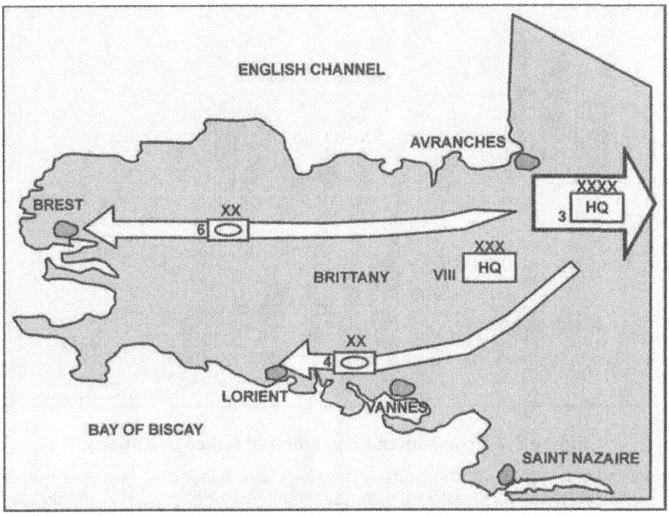

Figure 7-1. Initial attack in Brittany

Middleton's corps sprinted into the peninsula with the 4th and 6th Armored Divisions leading the way. However, poor communications, disagreements between commands, and contradictory orders caused the corps to hesitate before pushing the two divisions to continue to exploit toward the ports. The result: the 6th Armored Division missed an opportunity to seize Brest against light resistance by one day. The 4th Armored Division, after capturing the smaller port of Vannes, was also frustrated on the approaches to Lorient. The American reaction to the inability to rapidly seize the ports demonstrated an understanding of changing circumstances. The 6th Armored Division turned the attack at Brest to the 8th Infantry Division and then relieved the 4th Armored Division at Lorient. The 4th Armored was moved to rejoin the rest of Third Army exploiting to the east and north. Ultimately Brest fell to VIII Corps on 19 September after a 43-day siege by three infantry divisions. The victory yielded 36,000 German prisoners of war (POWs). However, the German defense and demolitions of the port left the port without an impact on the logistic situation of the allies. Brest cost the U.S. Army almost 10,000 casualties and the commitment of significant supplies. The experience convinced commanders to surround and bypass the other major Brittany ports. Lorient and Saint Nazaire remained under German control, deep in allied territory, until the war ended ten months later (see figure 7-2).

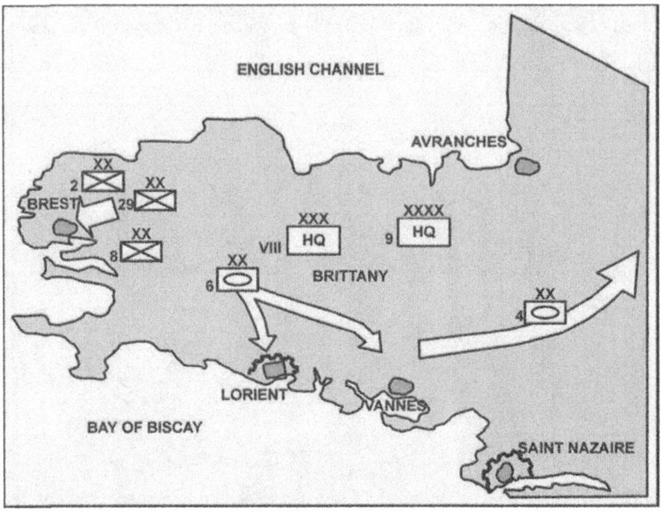

Figure 7-2. Subsequent disposition of forces in Brittany

The operational lessons of the Brittany campaign are numerous. First, commanders are responsible to continually assess assumptions and decisions made during planning based on the changing circumstances of the battlefield. This includes the planning decision to conduct urban offensive operations. When the allies arrived at the Brittany Peninsula, the focus of the operational maneuver was no longer securing logistics facilities but exploiting the breakthrough at Saint Lo and the disintegrating the German defense. The bulk of Third Army then was turned to the north and east rather than west into the peninsula.

The Brest experience also demonstrates that the costs of urban offensive operations are continually assessed against the operational value of the objective. This lesson was applied to the cities of Lorient and Saint Nazaire. The cities were never seized from the Germans because their logistic value failed to warrant the required resources to take them. German retention of the ports had no major adverse effect on the overall campaign.

Another lesson is that commanders cannot allow urban operations to disrupt the tempo of other offensive operations. One German goal of defending the ports was to disrupt the rapid tempo of the U.S. exploitation. They failed to achieve this goal because General Bradley continued the exploitation with the bulk of Third Army and executed the original plan with only a single corps.

Finally, commanders cannot allow emotion to color their decision to conduct or continue UO. The failure of 6th Armored Division to seize Brest rapidly caused some commanders to believe that Brest had to be captured because the prestige of the Army was committed to the battle. Costs of the continuing combat operations to seize Brest were significant. These resources might have been better committed elsewhere in the theater.

7-7. Tactical tempo is also important in urban combat. Because of the complex terrain, defending forces can rapidly occupy and defend from a position of strength. Once Army forces initiate tactical offensive operations, they cannot allow the threat to set the tempo of the operation. Instead, attacking forces seek to maintain a high tempo of operations. However, the tactical tempo of urban operations differs from operations in other terrain. Not necessarily slow, it requires a careful balance of preparation, speed, and security. In terms of unit fatigue, resource consumption, and contact with the threat, the tempo of most urban offensive operations may be rated as very high. On the other hand, in distances traveled and time consumed to achieve objectives, the tempo of many urban offensive operations might be rated as slow. The urban battlefield's density concentrates activity and consumes resources in a relatively small area. The lack of terrain seized or secured is not to be construed to mean a low tempo in the battle. Although the tempo may seem excruciatingly slow at higher levels of command and exceeding fast at lower, tactical levels, in reality, the natural tempo of urban operations is not faster or slower than other types of operations, merely different. Creating and operating at a tempo faster than an opponent can maintain, however, can favor forces which are better led, trained, prepared, and resourced.

7-8. A high tactical tempo in urban offensive operations challenges logisticians to provide for the increased consumption of munitions and degrades soldiers' physical capabilities. Commanders must anticipate these challenges and develop the means and abilities to overcome them. In the past, these challenges forced commanders to conduct urban offensives cyclically. They used night and other periods of limited visibility to resupply, rest, and refit forces. The environment influenced the tempo of their operations. This type of "battle rhythm" resulted in the forces spending each new day attacking a rested threat that was in a well-prepared position.

7-9. Army forces must maintain the tempo. Offensive operations continue even during darkness. Moreover, Army forces increase the tempo of operations at night to leverage the limited visibility capabilities, increased situational understanding, training, and INFOSYS that give an advantage to Army forces in all environments. To overcome the physical impact of the environment on soldiers, commanders can retain a large reserve to rotate, continuing offensive operations at night. The force that fights in daylight becomes the reserve, rests, and conducts sustaining operations while another force fights at night. Army forces can then maintain the tempo of operations and leverage technological advantages in urban offensive combat.

7-10. Tempo in UO does not necessarily mean speed. Offensive operations balance speed, security, and adequate firepower. Commanders must plan for the complex tactical environment and the requirements to secure flanks and airspace as the operation progresses. Mission orders allow subordinate units to make the most of tactical advantages and fleeting opportunities.

AUDACITY

7-11. Audacity is a simple plan of action, boldly executed. Superb execution and calculated risk exemplify it. In an urban attack, a thorough understanding of the physical terrain can mitigate risk. The terrain's complexity can be studied to reveal advantages to the attacker. Audacity can also be embodied in an operation by inventively integrating and coordinating the direct action tasks of SOF throughout the operation. Combining SOF actions with conventional attacks can asymmetrically unhinge a threat's defensive plan. Well trained Soldiers confident in their ability to execute urban offensive operations foster audacity.

URBAN OFFENSIVE OPERATIONS AND BATTLEFIELD ORGANIZATION

7-12. Urban offensive operations, like all operations, are arranged using the overall battlefield organization of decisive, shaping, and sustaining operations. Each operation is essential to the success of an urban offensive, and usually two or more of these operations occur simultaneously. Decisive operations are attacks that conclusively determine the outcome of UO. These attacks strike at a series of decisive points and directly lead to neutralizing the threat's center of gravity. Shaping operations in urban offensive operations create the conditions for decisive operations. In UO, much of the shaping effort focuses on isolation, which is critical in both major operations and tactical battles and engagements. Sustaining operations in urban offensive operations ensure freedom of action. They occur throughout the area of operations (AO) and for the duration of the operation.

Decisive Operations

7-13. A tactical commander fights decisive urban combat, whereas commanders conducting a larger major operation influence urban combat by setting the conditions for tactical success. Higher commanders may directly influence urban offensive operations by operational maneuver, by coordinating joint fires, by closely coordinating conventional forces, or with SOF.

7-14. Tactical urban offensive operations quickly devolve into small-unit tactics of squads, platoons, and companies seizing their objectives. The compartmented effect of the terrain and the obstacles to command and control of small units, especially once they enter close combat inside buildings or underground, often restricts the higher commander's ability to influence operations. Commanders can influence the actions of subordinates by clearly identifying the decisive points leading to the center of gravity; using mission orders (as discussed in Chapter 5); developing effective task organizations; synchronizing their decisive, shaping, and sustaining operations; and managing transitions.

7-15. Like all operations, successful decisive operations in UO depend on identifying the decisive points so the forces can destroy or neutralize the threat's center of gravity. Seizing a key structure or system that makes the threat's defense untenable; interdicting a key resupply route that effectively isolates the threat force from his primary source of support; or isolating the threat so that his force can no longer influence friendly activity may be more effective than his outright destruction.

7-16. Commanders must select the right subordinate force for the mission and balance it with appropriate attachments. Higher commanders do not direct how to organize the small tactical combined arms teams, but they must ensure that subordinates have the proper balance of forces from which to form these teams. Successful urban offensive operations require small tactical combined arms teams. Urban offensive operations require abundant infantry as the base of this force. However, successful urban combat requires a combined arms approach adjusted for the conditions of the environment. In urban offensive operations, it will normally not be a question of whether include armor and mechanized elements but, rather, how best to accomplish this task organization. Precision-capable artillery systems generally support urban operations better than unguided rocket artillery.

7-17. Divisions entering urban combat may require additional resources. These resources include military intelligence support in the form of linguists, human intelligence (HUMINT) specialists, and unmanned aircraft systems (UAS). Engineering assets will be at a premium; the task organization of a task force executing the decisive operation may require a one-to-one ratio of engineer units to combat units. Corps and higher engineering support may be necessary to meet these requirements and to repair vital and specialized infrastructure. A tailored and dedicated support battalion or group may need to assist in providing anticipated support to a displaced and stressed civil population. Finally, divisional CA units may require augmentation to deal with NGOs and civilian government issues.

7-18. Successfully conducting decisive operations in the urban environment requires properly synchronizing the application of all available combat power. Army forces have a major advantage in the command and control of operations. Commanders can use this advantage to attack numerous decisive points simultaneously or in rapid succession. They also use it to attack each individual decisive point from as many directions and with as many different complementary capabilities as possible. Commanders must completely understand urban environmental effects on warfighting functions to envision and execute the bold and imaginative operations required. Significantly, these operations require that C2 systems account for the mitigating effects of the environment as execution occurs.

7-19. Properly synchronized actions considerably enhance the relative value of the combat power applied at the decisive points. They present to the threat more requirements than he has resources with which to respond. Synchronized IO and multiple maneuver actions paralyze the threat's decision-making capacity with information overload combined with attacks on his C2 systems. Additionally, well-synchronized actions limit the time the threat has to make decisions and forces him into bad decisions. In the urban environment, these effects are enhanced because C2 systems are already strained, poor decisions are harder to retrieve, and units that do not react effectively are more easily isolated and destroyed.

SHAPING OPERATIONS

7-20. Shaping operations that support the urban attack separate into those focused on isolating the threat and all others. Army forces isolate the threat to ensure successful urban offensive operations. Depending on the threat reaction to isolation efforts and the nature of the threat center of gravity, this task may become decisive. Other shaping operations include those common to all offensive operations and others unique to urban operations. Unique urban shaping operations may include securing a foothold in a well-fortified defensive sector, securing key infrastructure, or protecting noncombatants. Because of the nature of UO, shaping operations may consume a much larger proportion of the force than during other operations and may take place both inside and outside the urban area (see Applying the Urban Operational Framework: Panama in Chapter 5). By successfully isolating a threat force, the force needed to conduct the decisive operation may be relatively small.

SUSTAINING OPERATIONS

7-21. Commanders conducting urban offensive operations must ensure security of the sustaining operation and bases; in many situations, protecting lines of communications (LOCs) and sustaining operations may be the greatest vulnerability of the attacking force. Those supporting an urban offensive are tailored to the urban environment and are well forward. Ideally, the supporting forces closely follow the combat forces and move within or just outside the urban area as soon as they secure an area. Operating in the urban area during offensive operations allows the sustaining operation to take advantage of the defensive attributes of the environment for security purposes.

7-22. Counterattacks against sustaining operations may take the form of special operations activities aimed at LOCs leading to or within the urban area. Choke points—such as bridges, tunnels, and mountain passes—are vulnerable to these attacks and may require combat forces to protect them. Threat forces attack the LOC to blunt the Army's combat power advantage in the urban area.

7-23. Attacks against the LOC into the urban area may also attempt to isolate the attacking Army forces from its sustainment base. Isolated forces in an urban area are greatly disadvantaged. Commanders must plan and aggressively execute strong measures to protect their LOC, even if it requires reduced combat power to execute their offensive operation.

7-24. Sustaining operations anticipate the volume and unique logistics requirements of urban operations. Specialized individual equipment—such as grappling hooks, ladders, and pads—is identified and provided to troops in quantity before they are needed. Forces stockpile and distribute their attacking units' special munitions requirements including small arms, explosives, and grenades of all types, precision artillery munitions, and mortar ammunition. Forces also supply transport to move the resources rapidly forward, both to and through the urban environment. Sustaining operations cannot rely on "operational pauses" to execute their tasks. Commanders must plan to continuously supply resources and capabilities to the most forward combatants as offensive operations advance.

7-25. Sustaining operations also anticipate the growth of sustainment requirements as Army forces secure and take responsibility for large portions of the urban area. The success of Army urban offensive operations will often uncover the civil population in former threat occupied areas. It may attract the civil population from sections of the urban area where the Army is not operating to areas occupied by Army forces. Rural populations may migrate to the urban area as the result of successful Army offensive operations.

7-26. Army forces may be required to take initial responsibility to provide for the urban population. This consideration is integrated into logistics planning and organization from the start of the planning process. To be successful and efficient in such a situation, sustainment planning includes Army civil affairs (CA) specialists and local government representatives. It also integrates and consults with the international community and nongovernmental organizations (NGOs) that might augment or supplement Army sustainment capabilities.

FORMS AND TYPES OF URBAN OFFENSE

7-27. Traditional forms of offensive maneuver include envelopment, turning movement, infiltration, penetration, and frontal attack. Traditional types of offensive operations are movement to contact, attack, exploitation, and pursuit. These traditional forms and types listed apply to urban combat. Some have greater application to an urban environment than others do. Moreover, success will belong to commanders who imaginatively combine and sequence these forms and types throughout the depth, breadth, and height of the urban battlefield. This is true at the lowest tactical level and in major operations.

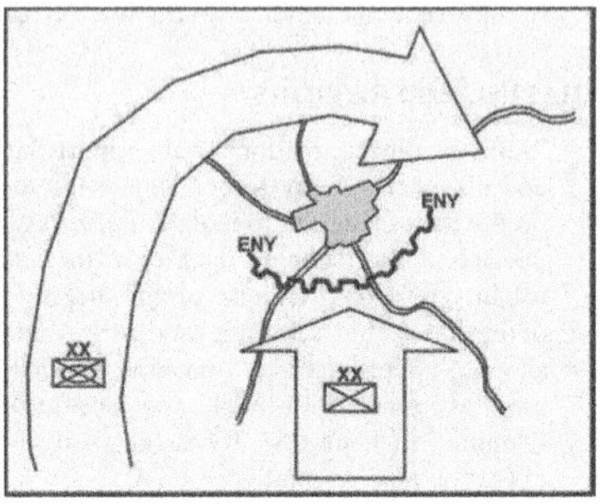

Figure 7-3. Envelopment isolates an urban area

FORMS OF OFFENSIVE MANEUVER

Envelopment

7-28. The envelopment is the ideal maneuver for isolating threat elements in the urban area or isolating the area itself. A deep envelopment effectively isolates the defending forces and sets the conditions for attacking the urban area from the flank or rear. Yet, enveloping an objective or threat force in the urban area is often harder since achieving speed of maneuver in the environment is so difficult (see figure 7-3). Vertical envelopment, however, works effectively if Army fires can effectively suppress or neutralize the threat air defense.

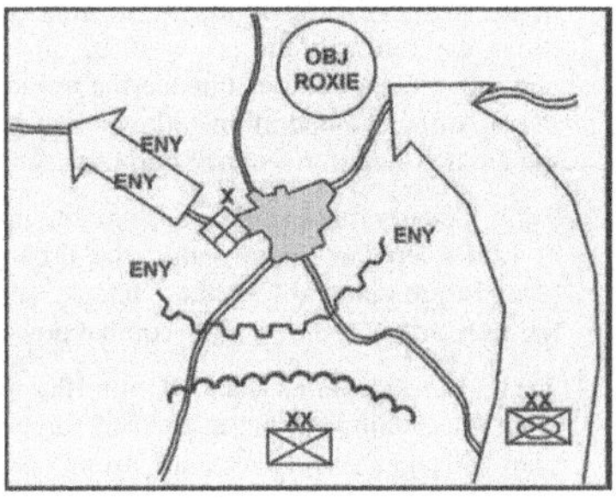

Figure 7-4. Turning movement

Turning Movement

7-29. Turning movements can also be extremely effective in major operations (see figure 7-4). By controlling key LOCs into the urban area, Army forces can force the threat to abandon the urban area entirely. These movements may also force the threat to fight in the open to regain control of LOCs.

Infiltration

7-30. Infiltration secures key objectives in the urban area while avoiding unnecessary combat with threat defensive forces on conditions favorable to them (see figure 7-5). This technique seeks to avoid the threat's defense using stealthy, clandestine movement through all dimensions of the urban area to occupy positions of

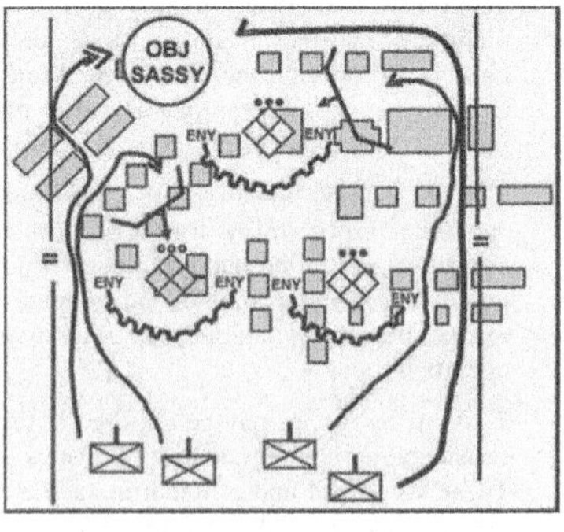

F

igure 7-5. Infiltration

advantage in the threat's rear (or elsewhere).It depends on the careful selection of objectives that threaten the integrity of the threat's defense and a superior common operational picture (COP). Well-planned and resourced deception operations may potentially play a critical role in masking the movement of infiltrating forces. The difficulty of infiltration attacks increases with the size and number of units involved. It is also more difficult when Army forces face a hostile civilian population. Under such circumstances, infiltration by conventional forces may be impossible. Armored forces are generally inappropriate for infiltration operations. However, they may infiltrate large urban areas if the threat is not established in strength and had insufficient time to prepare defenses.

Penetration

7-31. Penetration is often the most useful form of attack against a prepared and comprehensive urban defense (see figure 7-6). It focuses on successfully attacking a decisive point or on segmenting or fragmenting the defense—thereby weakening it and allowing for piecemeal destruction. (The decisive point may be a relatively weak or undefended area that allows Army forces to establish a foothold for attacks on the remainder of the urban area.) Ideally in urban combat, multiple penetrations in all dimensions are focused at the same decisive point or on several decisive points simultaneously. In urban combat, the flanks of a penetration attack are secure, and resources are positioned to exploit the penetration once achieved. Although always a combined arms team, rapid penetrations are enhanced by the potential speed, firepower, and shock action of armored and mechanized forces.

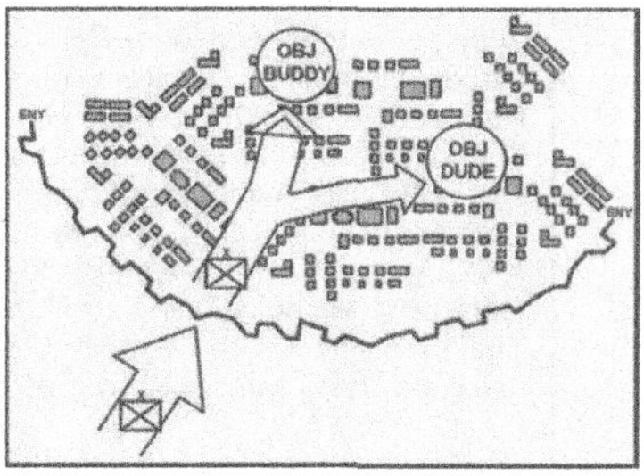

Figure 7-6. Penetration

7-32. Importantly, commanders must consider required actions and resources that must be applied following a successful (or unsuccessful) penetration. A penetration may result in the rapid collapse and defeat of the threat defense and complete capitulation. On the other hand, success may cause threat forces to withdraw but leave significant stay-behind forces (or disperse into the urban population as an insurgent-type force) which may necessitate methodical room-to-room clearance operations by significant dismounted forces. Additionally, securing portions (or all) of an urban area requires occupation by Army forces to prevent re-infiltration of threat forces thereby further increasing manpower requirements. Based on the factors of the situation, commanders may conclude that over time methodical clearance operations conducted from the outset, while frontloading time and resource requirements, will be less costly than a penetration followed by systematic clearing operations.

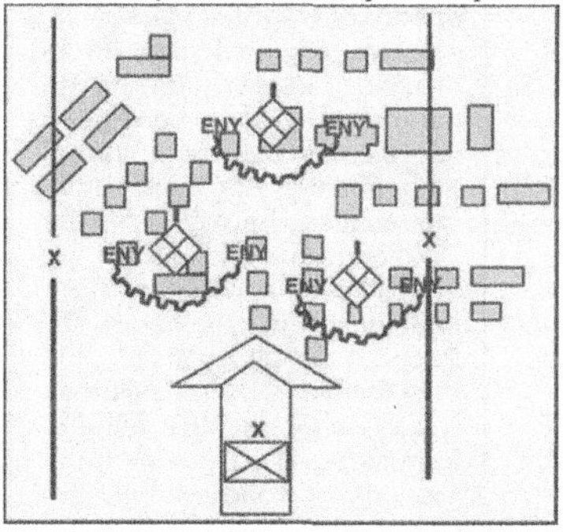

Figure 7-7. Frontal attack

Frontal Attack

7-33. For the commander of a major operation, the frontal attack is generally the least favorable form of maneuver against an urban area unless the threat is at an obvious disadvantage in organization; numbers, training, weapons capabilities, and overall combat power (see figure 7-7). Frontal attacks require many resources to execute properly, risk dispersing combat power into nonessential portions of the area, and risk exposing more of the force than necessary to threat fires. In urban offensive combat, forces most effectively use the frontal attack at the lowest tactical level once they set conditions to ensure that they have achieved overwhelming combat power. Then the force of the frontal attack overwhelms the threat with speed and coordinated and synchronized combat power at the point of attack. The assigned frontage for units conducting an attack on an urban area depends upon the size and type of the buildings and the anticipated threat disposition. Generally, a company attacks on a one- to two-block front and a battalion on a two- to four-block front (based on city blocks averaging 175 meters in width).

Forms of Attack in the Urban Offense: Metz – 1944

In November 1944, the U.S. Third Army launched its final effort to take the French city of Metz from the defending Germans. This was the Army's third attempt. The first attempt had been a surprise, mounted attack. This was followed by a series of piecemeal infantry assaults on the surrounding fortresses. Finally, a deliberate effort was made to take the city in a coordinated effort by XX Corps.

The initial failures stemmed from a shortage of resources on the U.S. side, to include fuel and units—especially infantry. This added to the ad-hoc nature of the first two efforts. The third effort, though more deliberate, was still constrained by resources. XX Corps could only muster three nearly full-strength infantry divisions to attack the German's defending with four under-strength divisions.

The third attempt to take Metz demonstrates how a corps operates with multiple divisions using various forms of attack to achieve its objective in urban offensive operations.

The opening phase of the Metz battle had attacks by the 90th and 5th Infantry Divisions to envelop the city from the north and south (see figure 7-8). This isolated the city and ensured the garrison could not escape nor be reinforced. The garrison was under orders from Hitler not to conduct a breakout and to resist to the last man; thus, German forces strongly counterattacked both wings of the envelopment to prevent isolation.

The second phase of the operation was the penetration of the city defenses from multiple directions. The 5th Infantry Division penetrated into the city with the 10th and 11th Infantry Regiments from the south. The 95th Infantry Division penetrated into the city from the north with the 377th Infantry Regiment and TF Bacon. Simultaneously, the 95th Infantry Division infiltrated battalions through the string of fortress positions guarding the western approaches into the city, isolated, and bypassed these positions with its other two regiments.

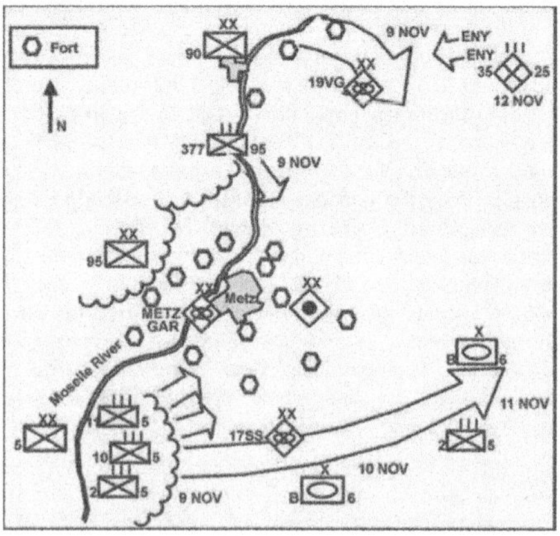

Figure 7-8. Metz envelopment

The final reduction of the defense was a series of battalion frontal attacks, which took place against the last remaining strongholds within the city. Even in these final engagements, however, the infantry battalions isolated, bypassed, and then attacked from the flanks and rear whenever possible (see figure 7-9).

The city was declared secured on 19 November. However, at that point more than a half-dozen of the fortresses had yet to be reduced. The 95th Infantry Division, after a four-day rest, quickly moved forward to rejoin the still rapidly advancing corps forward elements. Elements of the 5th Infantry Division remained in siege posture around the remaining strong points, the last of which surrendered on 19 December 1944 when it ran out of food. U.S. forces made no efforts to attempt to assault these bypassed fortresses although extensive psychological operations (PSYOP) were used.

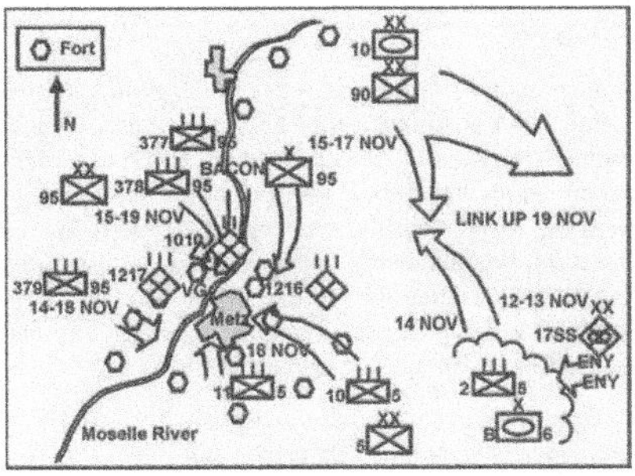

Figure 7-9. Metz final assault

TYPES OF OFFENSIVE OPERATIONS

Movement to Contact

7-34. In an urban area where the threat situation is vague, Army forces will often conduct a movement to contact to establish or regain threat contact and develop the situation. A movement to contact in an urban area occurs as both sides try to establish their influence or control over a contested urban area. The situation determines whether the movement to contact or its specific technique, the search and attack, is appropriate. A conventional force-oriented movement to contact will likely take place when friendly and threat conventional forces attempt to establish control simultaneously. Initially, neither side is defensive. The friendly force aims to quickly locate and fix the threat while establishing control of the urban area and its key infrastructure. The search and attack technique works well when a smaller threat has established a noncontiguous defense in an urban area. This operation is characterized by the friendly point defense of key infrastructure, robust reconnaissance, and rapidly concentrated combat power to fix and defeat or destroy threat resistance once located.

7-35. A meeting engagement often results from the movement to contact. It occurs when a moving force that is partially deployed for battle collides with and engages a threat at an unexpected time and place. In a meeting engagement in an urban area, the unit that reacts most quickly and decisively will likely win. Rapid and accurate decision making depends heavily on understanding the nature of the urban area and its impact on operations. Thus, in a meeting engagement, commanders should quickly assess the impact and role of all components of the urban environment (terrain, infrastructure, and society) on the operation. To this end, responsive reconnaissance is important. This permits accurate decision making regarding where to attack, where to defend, and how to allocate resources. Situational understanding enhanced by robust digital INFOSYS that provide an accurate COP also facilitates the rapid reaction of Army units and a synchronized response. This reaction and response allow Army forces to seize the initiative and dominate the threat.

Attack

7-36. The attack is the most common and likely offensive operation that Army forces will conduct in an urban environment. Commanders conducting major operations and commanders of large tactical units usually execute deliberate attacks. In the urban environment, units larger than battalion-size rarely conduct hasty attacks. Hasty attacks are common below company level as units use their initiative to take advantage of tactical opportunities. However, larger units will conduct hasty attacks when threat defenses are disrupted or unprepared, to take advantage of an unexpected situation, and to prevent the threat from establishing or re-establishing a coherent defense.

Exploitation

7-37. Exploitation follows a successful attack to disrupt the threat in depth. Commanders of major operations should consider focusing exploitation attacks on urban areas. A threat defeated in an attack will attempt to rally units, reinforce with reserves, and reorganize his defense. With its information and communications capability, transportation network, and defensive attributes, the urban area is the natural focal point to reestablish a disrupted defense. By establishing urban centers as the objectives of the exploitation, commanders deny the threat the sanctuary needed to reorganize and reestablish his defense. The exploitation focuses on the urban area as well as on the remnants of the threat. A successful exploitation to seize an urban area works efficiently because the attack preempts the defense and denies the threat the full advantages of urban terrain.

7-38. Commanders conducting exploitation must acknowledge the vulnerability of their forces to counterattack and ambush in urban areas. An urban area provides ideal cover and concealment to hide threat reserves, reinforcements, or reorganized forces. Constrictions of routes into and through the urban area make exploitation forces a potentially dense target and limit maneuver options. Robust and well-coordinated reconnaissance, tactical dispersal, and use of advance guard security forces protect against this threat (see Defensive Combat Power: Suez City vignette in Chapter 8).

Pursuit

7-39. The pursuit is designed to destroy threat forces attempting to escape. It focuses on the threat and not on urban areas. When conducting a pursuit, Army forces move through undefended urban areas and, if possible, bypass those in which threat forces successfully take refuge. The threat will likely attempt to use urban areas to disrupt the pursuit and permit the threat main body to escape. Commanders can prevent escape by denying the threat the time to establish forces in urban areas that cannot be bypassed. The agility of Army aviation forces for attack, reconnaissance, and transportation is essential to execute a successful pursuit around and through urban areas.

URBAN OFFENSIVE CONSIDERATIONS

7-40. The urban operational framework (understand, shape, engage, consolidate, and transition) provides a structure for developing considerations unique to urban offensive operations. The considerations vary depending on the situation and scale of the operation. Some considerations applicable to major operations that include an urban area will also be considerations at the tactical level focused in the urban area. However, no set rules exist. All urban operations are unique. Issues addressed at the operational level in one situation may be addressed in a new situation only at the tactical level. Under the right circumstances, a consideration may become an operational issue, a tactical issue, or a combination of the two. The following identifies some planning and execution issues that commanders conducting major operations should address.

UNDERSTAND

7-41. The first requirement, and a continuing requirement throughout the conduct of urban operations, is the assessment and understanding of the situation. Commanders should base this understanding on detailed information regarding the particular urban area. Since the threat will likely dominate or control most of the urban area during the planning phase of offensive operations, achieving an accurate understanding of the urban environment will be difficult. A comprehensive intelligence, surveillance, and reconnaissance (ISR) effort in support of a rigorous intelligence preparation of the battlefield (IPB) process overcomes this obstacle.

Integrated Intelligence, Surveillance, and Reconnaissance

7-42. The commander of a major operation that includes an urban area, unlike his subordinate commanders, can target reconnaissance deep into his AO and area of interest. This begins the application of ISR resources against the urban area that may lead to decisive ground operations. This ISR effort and the understanding it supports continue as long as the urban area remains in the AO. Commanders of major operations initially direct ISR assets on those information requirements that support determining whether or not to conduct urban offensive operations. Once decided, ISR resources shift to support the planning and execution of the operation in the urban area.

7-43. The first resources that a senior commander can use are national and strategic sensors. He requests them through the appropriate joint force commander. The commander aggressively pursues full use of these systems to begin building an initial database for analyzing the significant aspects of the terrain; key infrastructure considerations; the status and disposition of the population; and the size, type, disposition, and intentions of threat forces in the area.

7-44. Simultaneously, multiple intelligence sources contribute to the database. The sources collect, process, store, display, and disseminate the relevant information on large urban areas through open and classified resources. These information sources include—

- Historical research.
- Travel brochures that include cultural information and recent maps.
- Classified debriefings of diplomats, businesses, DOD personnel, and allies.
- Military maps and special geospatial products of the urban area.
- Previous intelligence assessments of the country, government, and population.
- Reachback to appropriate economic, political, cultural, and infrastructure subject matter experts not in the commanders AO.

The gathering and analysis of HUMINT plays a critical part of this process and assists commanders in understanding ethnic, cultural, religious, economic, political, and other societal and infrastructural facets of the environment.

7-45. As the intelligence and the national reconnaissance and surveillance efforts progress, commanders will insert, if available and feasible, Special Forces reconnaissance assets into the urban environment. These elements will seek to confirm or deny the information received from imagery intelligence (IMINT), signals intelligence (SIGINT), and HUMINT sources. Among many factors, using SOF depends on their availability, the particular urban area, the area's ethnic composition, and the relationship between the urban population and the threat. Other joint operational reconnaissance and surveillance assets that higher-echelon commanders may have available might include the Joint Surveillance Target Attack Radar System, Guard Rail targeting aircraft, UAS, and space-based systems.

7-46. The commander's staff will use all sources of information—IMINT and SIGINT sensors, HUMINT, historical research, and reconnaissance—to refine his ability to understand the urban environment. Digitally linking subordinate commanders with information sources helps to develop a COP essential to their situational understanding of the urban environment. The IPB process guides this assessment. As operations progress, additional reconnaissance and surveillance assets may become available. These may include UAS, long-range reconnaissance and surveillance units, counterfire radar, and air and ground cavalry. As these assets are employed, they are linked into the net of sources sharing information and further refine a common situational understanding of the environment.

Focused Assessment Efforts

7-47. In urban offensive operations, the tactical commander's assessment focuses on defeating the threat in the urban area within the constraints of the environment. Toward this end, identifying and assessing decisive points to attack is a commander's priority assessment task. Some unique aspects of the urban environment also require the focus of the commander's assessment efforts. These include the character of the urban defense, collateral damage considerations, and the effects of the environment on friendly and threat courses of action.

Character of the Urban Defense

7-48. To be both efficient and effective, Army urban offensive operations focus on what is decisive. Decisive points for an urban attack depend primarily on the mission within the urban area. They can vary widely in composition and size. Since commanders only focus on the essential, they may determine the decisive point to be a single building or a limited sector of an urban area. It could be an entire system within the urban infrastructure such as communications and information, or a limited subsystem of the transportation and distribution infrastructure such as a single airfield. Sometimes what is decisive in the urban area is the threat military capability, but even this large an objective, when carefully analyzed, may not require destruction of all threat forces or control of the entire urban area. Decisive points relate directly to the threat's center of gravity and to mission success. Some decisive points related to the urban threat's center of gravity may be physically located outside the urban area.

7-49. To gain specifics on threat dispositions within the urban area requires reconnaissance capability to see into the depths of the area and the intelligence capability to determine the threat's likely defensive course of action. With this information, commanders can determine decisive points and apply Army combat power discretely against them. Effective urban offensive operations require detailed situational understanding of an area of interest that extends well beyond the perimeter of the urban area.

7-50. Commanders see throughout the depth of the urban area using several actions (see figure 7-10). First, they evaluate sensor data and imagery. This guides targeting of special reconnaissance. Simultaneously, HUMINT is conducted using any persons who might know the urban area and threat. This includes civilians (allies, aides, neutrals, obstacles, and hostiles) and POWs. Finally, tactical conventional reconnaissance assets including reconnaissance forces, aviation, artillery radar, signals intelligence, and UAS are directed at the urban area. All these sources and data are linked through digital INFOSYS to provide commanders and their subordinates with improved situational understanding and a COP.

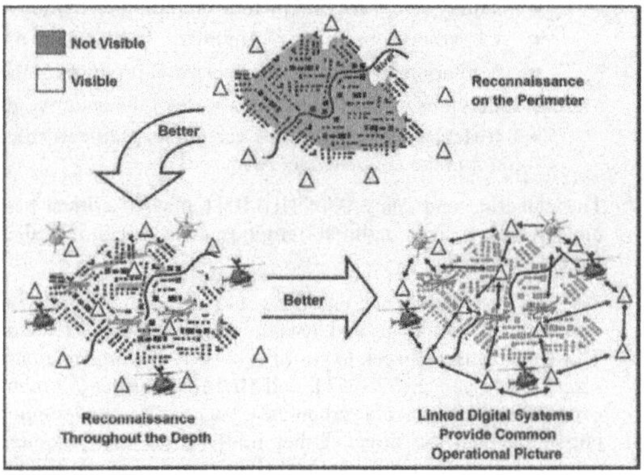

Figure 7-10. Required urban reconnaissance capabilities

Collateral Damage Considerations

7-51. Commanders must also assess the collateral damage risks that their operations may include. This assessment helps to initially determine the viability of a course of action. However, commanders should reassess their courses of action at frequent intervals in urban offensive operations based on known information to determine if the original evaluations remain valid. This reassessment minimizes potential collateral damage from a change in mission or a change in a course of action. Many aspects of the environment can change during mission execution.

The Environment's Effects on Courses of Action

7-52. The urban environment's unique aspects can significantly impact the course of action chosen by Army forces and the threat. Commanders must assess these effects in planning, but they must also verify and monitor these effects as forces execute offensive missions. In particular, commanders will want to confirm the civilian population's locations, beliefs, and actions and to monitor any changes. They will need to validate terrain considerations and monitor the effects of any changes due to rubble and other damage to structures. In urban terrain, dead space, cover, and concealment can only be identified physically and may change considerably as operations are executed.

SHAPE

7-53. Commanders of major operations have a primary contribution to urban operations: the planning and conduct of effective shaping operations that set the conditions for subordinate tactical success. In urban operations, isolation will be a critical condition. Effective isolation will require persistent, continuous surveillance and reconnaissance, innovative use of fires and maneuver (including effective force allocation decisions), and well-established sensor-to-shooter links. These efforts—combined and synchronized with SOF direct actions, IO that minimize noncombatant influences, and necessary shaping attacks (particularly the seizure of a foothold)—establish the conditions necessary for the subsequent offensive domination of the area.

Isolation is Essential

7-54. One key to success in the history of urban operations has been the effective isolation of the threat force (see figure 7-11). This applies today and equally well to major urban offensive operations as it does to smaller-unit attacks. This isolation not only denies access to the urban area from outside but also contains threat forces within. In a modern metropolis or megalopolis, this can appear a daunting task. Operational isolation requires dominating all physical and electronic contact between the threat in the urban area and supporting threat forces outside the urban area. This does not necessarily require physically encircling the urban area, but it does require that Army forces be able to exert control over the area's entire perimeter, as well as decisive points within. For a sprawling urban area, successful isolation may require the commitment of a large amount of resources.

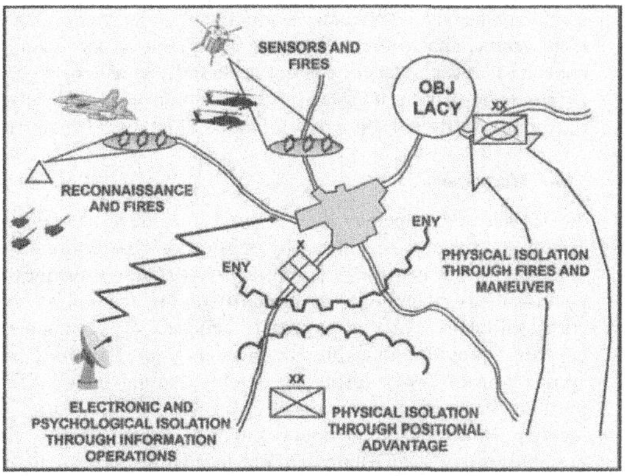

Figure 7-11. Shaping through isolation

7-55. Successful isolation of the urban area depends as much on the nature of the threat as it does on any other factor. A conventional threat in a large urban area may be much easier to isolate than an insurgent threat in a much smaller urban area. The forces needed in the former situation may be less than those needed in the latter. The more the characteristics of the threat are conventional in nature generally the easier it will be to isolate him using standard combat methods and equipment. Isolating a more unconventional force requires many of the same techniques as used against conventional forces. However, it also requires a much greater ability to simultaneously conduct offensive IO, to integrate CA units and civil-military operations (CMO), and to work with allies, nongovernmental organizations, and local authorities. Fundamentally, isolating a less conventional threat puts increased emphasis on separating combatants from noncombatants.

Offensive Isolation Objectives

7-56. Isolation seeks to achieve two primary objectives with respect to defeating a threat's urban defense:

- Weaken the overall coherence of his defense.
- Limit or manipulate his maneuver options.

Isolating the threat in the urban area from external support, as well as isolating him from sources of support within the urban area, weakens his overall defense. The defense is weakened through a combination of attrition (the threat cannot replace his losses) and the diversion of his combat power from the defense to operations to counter the isolation effort. Isolation can also prevent the threat from shifting his forces to reinforce decisive points in the urban area or to conduct counterattacks.

7-57. Commanders may choose not to isolate the urban area completely—or at least make it appear so to threat forces. Instead, they may afford the threat an apparent means of escape, create the conditions for its use through effective fire and maneuver against the defenders, and then destroy the threat through various ambush methods. While friendly forces may be able to move undetected to appropriate ambush sites, it is more likely that this technique will necessitate rapidly mobile air and ground forces moving along carefully chosen routes through the urban area. Commanders must consider maintaining the ability to complete the isolation of the urban area to prevent reinforcement and escape of urban threat forces—particularly if the ambush attack does not achieve desired effects.

Persistent Surveillance

7-58. Persistent surveillance of the urban area is essential to all types of actions used to isolate an urban area and as complete as resources will allow. Surveillance of the urban area relies on either reconnaissance forces or sensors continuously observing or monitoring urban avenues of approach. This network of ISR assets updates the commander's situational understanding and provides the means to quickly identify and, if necessary, attack threat elements as they move. However, particularly with sensors, commanders must consider that each detection is not necessarily an enemy to be attacked. Noncombatant activity clutters the environment making it easier for threats to disguise themselves and increases the burden (and the number of resources required) on Army forces to distinguish friend from foe.

Fires and Maneuver

7-59. Fires and maneuver may be used to achieve isolation, either singly or in combination. (As always, effective obstacles, monitored by sensors or observation, are integral to any isolation technique.) First, attacking forces can pre-position themselves along avenues of approach to deny entry and exit through positional advantage. Relying primarily on this method of isolation, particularly around a large urban area with multiple avenues of approach, can be overly resource intensive. Instead, the pairing of fires and maneuver provides attacking commanders more flexibility and allows them to isolate several avenues of approach with fewer resources. Highly mobile attack helicopters, operating outside threat-controlled portions of the urban area, are ideal for this purpose. Inside threat-controlled areas, it is more difficult to identify, eliminate, or effectively suppress the air defense threat. The threats may have numerous man-portable air defense weapons and enhanced effects of small arms used for air defense. Therefore, the risk to using this equipment may outweigh the potential benefits. However, mobile ground units—such as an air assault (subject to the same air defense threat considerations as attack aviation), armored, or mechanized forces—can also rapidly move to attack and destroy a threat moving in or out of an urban area. Potential disadvantages of the combined, fires and maneuver, option are that—

- Critical assets, on standby and dedicated to isolation efforts, may be unavailable for other missions.
- The attacking force may not locate the threat in time to complete its mission (an inherent risk to any attack).

7-60. Another alternative relies on indirect or joint fires alone to destroy the threat force. Its disadvantage is that fires alone often cannot completely destroy or stop a determined force from moving into or out of an urban area. Although targets and avenues of approach will require continual surveillance, it is usually a less resource-intensive option than those that include maneuver. It also does not normally require fires assets to remain on standby to accomplish the mission. However, fires must be able to reliably and quickly respond. For Army field artillery units and naval gunfire, the units must also be in range, which requires careful positioning. A skilled threat can avoid interdiction fires by using the geometry of the area to identify gaps due to obstructing terrain or the firing unit's range limitations. It can also use concealment and weather to avoid observation. However, effective sensor-to-shooter links throughout the urban battlefield will reduce the threat's ability to hide (see figure 7-12). A resolute threat may risk significant losses to fires to prevent isolation or may attempt to use noncombatants as a shield. Ultimately, commanders should use innovative combinations of all techniques discussed. Some units will physically block key avenues of approach. Surveillance will monitor less important routes and avenues. Artillery fires, joint fires, and maneuver units will then respond to the results of surveillance depending on the circumstances.

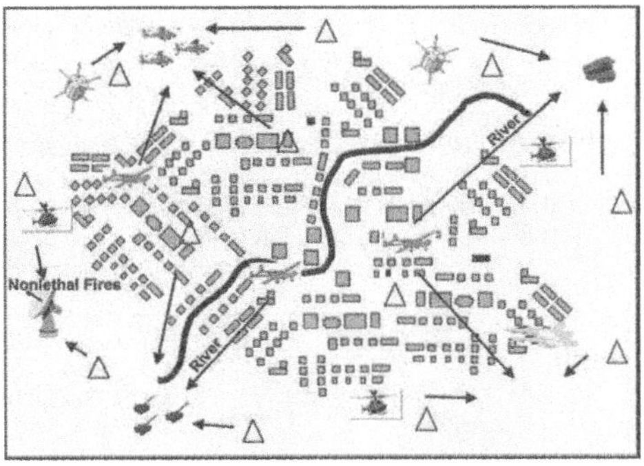

Figure 7-12. Critical sensor-to-shooter links

Threat Reactions

7-61. The reaction of the threat to the effects of isolation will depend on his mission, morale, force structure, and overall campaign plan. The threat may recognize isolation actions early and withdraw from the urban area before isolation is completed instead of risking destruction. On the other hand, the threat, based on a different or flawed assessment (perhaps a perception shaped by the Army force commander), may choose to—

- Continue to defend (or hide) and conduct local ambushes and counterattacks.
- Attack to break into the urban area or infiltrate forces and supplies in.
- Attack to break out of the urban area or infiltrate forces out.
- Execute any combination of the above (see figure 7-13).

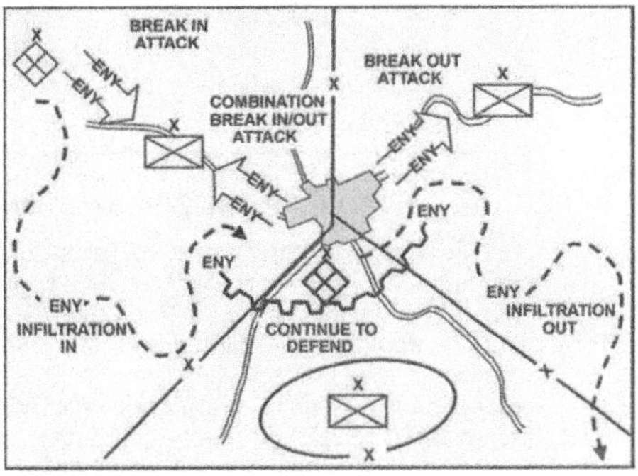

Figure 7-13. Reactions to isolation

7-62. Attacking commanders must consider how the threat leadership's subsequent actions will affect the continuance of overall offensive operations. They deliberate many considerations, to include—

- The allocation of more forces to the shaping operations to isolate the urban area.
- The allocation of more combat power to achieve rapid penetration and seizure of objectives to take advantage of developing threat dispositions in the urban area.

Isolating an Urban Area: Hue, Vietnam – 1968

On 31 January 1968, the 4th and 6th North Vietnamese Army (NVA) regiments and attached NVA and Vietcong (VC) sapper battalions attempted to seize control of Hue from the South Vietnamese Army's (ARVN) 1st Division as part of the North Vietnamese Tet Offensive. The attack, which was launched with complete surprise, successfully established temporary control over most of the city and facilitated the occupation of strong defensive positions within the city's ancient fortress known as the Citadel. However, it did not lead to the capture of the headquarters of the ARVN 1st Division located in the Citadel nor the U.S. military assistance command—Vietnam compound located in the southeast part of the city.

Both ARVN and U.S. forces swiftly reacted to the NVA attack. Within 24 hours, ARVN infantry and cavalry units counterattacked to recapture the Citadel. South of the Perfume River, the 1st Marine Division began attacking to clear the southern portion of the city (see figure 7-14).

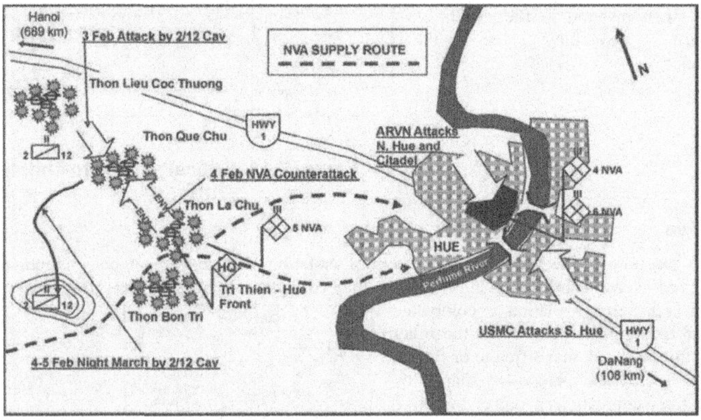

Figure 7-14. Initial attack to isolate Hue

On 2 February, the first Army unit was committed to Hue. The 2-12th Cavalry (an infantry battalion) was ordered to attack southeast along Highway 1 and clear enemy interdicting that route. On 3 February, 2-12th Cavalry began their attack moving along the south side of the highway. The plan was to clear a series of small hamlets in succession en route to the city itself. The first hamlet, Thon Lieu Coc Thuong, was cleared easily. Moving toward the second hamlet, Thon Que Chu, the battalion encountered a strong enemy force that was well entrenched. Air and artillery supported the battalion as it attacked and captured the hamlet against strong resistance. The battalion dug in for the night and prepared to resume the attack against the third hamlet, Thon La Chu, on 4 February.

Unknown to the 2-12th Cavalry, Thon La Chu was the headquarters of the Tri Thien-Hue Front, and it was defended by the NVA's 5th Infantry Regiment. Also important, the hamlet sat astride the NVA's primary supply route to the regiments fighting in Hue. The NVA was determined that 2-12th Cavalry's attack would not succeed; the NVA was determined to destroy the threat to its LOCs.

As the 2-12th Cavalry prepared to resume the attack on 4 February, the NVA fiercely counterattacked with all three battalions of the 5th NVA Regiment. It soon became apparent that 2-12th Cavalry could not continue to attack. As the day continued and the NVA pressure steadily increased, the battalion began to lose its position in Thon Que Chu. To avoid destruction, the battalion broke contact.

As darkness fell on 4 February, 2-12th Cavalry finally broke contact with the NVA. However, instead of retreating north back to its start line, the battalion moved west and then south to good defensive high ground deeper in the NVA AO. Eleven hours later the battalion was set in its new defensive position.

At dawn on 5 February, 2-12th Cavalry was established 5 kilometers west of Hue. The battalion soon observed enemy forces and supplies moving toward Hue. From its high ground position, the battalion directed artillery and air strikes against the NVA forces. By its bold move to bypass the 5th NVA Regiment, the battalion now held excellent position to direct fires on the primary NVA supply line into Hue. These fires were the first step toward isolating the NVA in Hue.

The fires controlled by the 2-12th Cavalry shut down the NVA LOCs into Hue during the daytime. However, under the cover of darkness supplies and reinforcements were still entering the city (see figure 7-15). It became apparent that the isolation of the NVA in Hue would require the capture of Thong La Chu. The problem facing American forces was concentrating combat power against the NVA. All U.S. units at this time were actively engaged against the numerous NVA attacks that constituted the NVA's 31 January Tet Offensive.

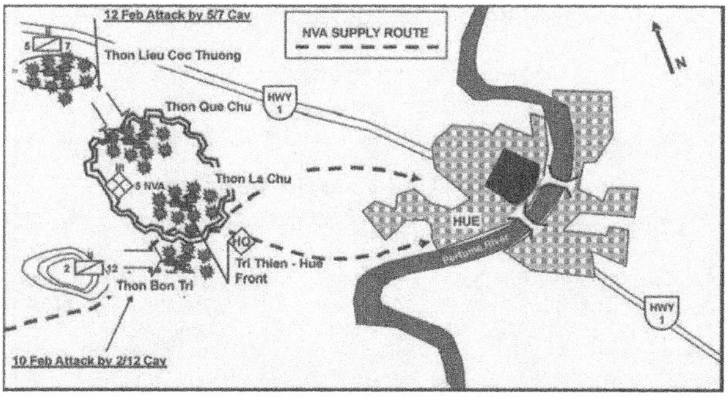

Figure 7-15. Subsequent attack to isolate Hue

The first additional American unit was not available until 12 February when the 5 7th Cavalry attacked Thong Que Chu much like the 2-12th Cavalry had attacked previously. The 5-7th Cavalry had even less success against the alert 5th NVA. The 5-7th Cavalry was forced to occupy defensive positions in Thon Lieu Coc Thuong and await the build up of combat power before it could continue to attack. In the interval, 2-12th Cavalry had moved off the high ground and captured the hamlet of Thon Bon Tri, south of the 5th NVA Regiment.

On 21 February, the 1st Cavalry Division had moved enough resources to the area to launch an effective attack to isolate Hue (see figure 7-16). In addition to the 5-7th and 2-12th Cavalry, the 1-7th Cavalry arrived in the AO and the 2-501st Airborne Infantry of the 101st Airborne Division was attached. On 21 February, after a combined artillery, air, and naval gunfire bombardment, the four battalions attacked the Thon La Chu stronghold. Elements of the 5th NVA Regiment were either destroyed in place or fled northeast. The next day resistance in Hue was noticeably lighter. U.S. Marine and ARVN units began the last phase of fighting to recapture the Imperial Palace. On 26 February, the North Vietnamese flag was removed from the Citadel and the ARVN I Corps declared the city secured.

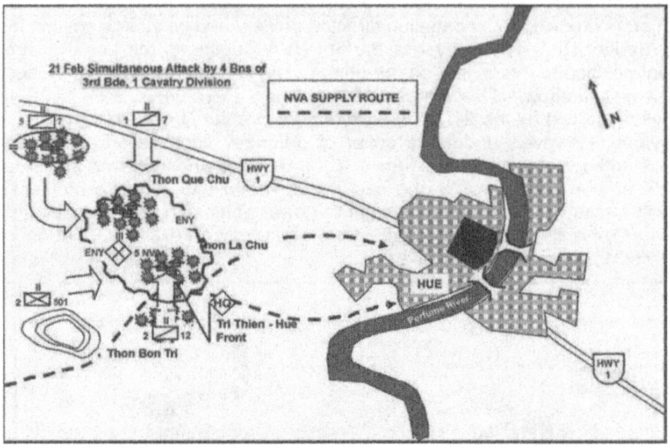

Figure 7-16. Final attack to isolate Hue

> The actions of the 1st Cavalry Division forces northwest of Hue demonstrated the importance and the difficulty of isolating an enemy fighting in an urban area. Isolating Hue was difficult not only because of the dispersion and surprise with which the Tet Offensive caught U.S. forces, but also because of the tenacity of the NVA. At least one-third of the combat power of the NVA in the Hue AO was focused on maintaining access to the city.
>
> Although only one aspect of this urban battle, Hue's isolation had an immediate and important, if not decisive, impact on operations. It not only resulted in restriction and then elimination of supplies and reinforcements, but it also immediately impacted the conduct of the defending NVA forces. Isolation caused an immediate drop in NVA morale and changed the nature of the defense. Once the enemy was isolated from external support and retreat, the objective of the NVA in the city changed from defending to avoiding destruction and attempting to infiltrate out of the city.

Civilian Reactions

7-63. Commanders must also consider the potential effects on and reactions and perceptions of the population living in the urban areas that they choose to isolate and bypass—either as a direct effect or as a response of the threat force being isolated. Isolation to reduce the threat's ability to sustain itself will likely have similar (and worse) effects on the civilian population remaining in the isolated area. (If food and water are in short supply, threat forces may take from noncombatants to satisfy their needs, leaving civilians to starve.) Isolation may also create a collapse of civilian authority within an urban area as it becomes apparent that the military arm of their government is suffering defeat. Due to their isolation, elements of the population may completely usurp the governmental and administrative functions of the former regime and establish their own local control, or the population may lapse into lawlessness. Returning later, Army commanders may find that these self-governing residents are proud of their accomplishments and, in some instances, less willing to allow Army forces to assume control since they may be perceived as having done nothing to earn that privilege. Alternatively, as witnessed in some urban areas during OPERATION IRAQI FREEDOM in 2003, a power vacuum may lead to intra-urban conflicts among rival factions coupled with general public disorder, looting, and destruction of the infrastructure (see also the discussion of Competing Power Structures in Chapter 3).

Direct Action by Special Operations Forces

7-64. Although SOF in urban offensive operations will likely conduct essential reconnaissance, they also have a direct action capability to shape the offensive operation (see figure 7-17). Special Forces and Rangers can use direct action capabilities to attack targets to help isolate the urban area or to directly support decisive actions subsequently or simultaneously executed by conventional forces. Successful attacks against urban infrastructure, such as transportation or communications centers, further the area's physical and electronic isolation. Direct action against command centers, logistics bases, and air defense assets can contribute to the success of conventional attacks by destroying or disrupting key threat capabilities. Direct action can also secure key targets such as airports, power stations, and television stations necessary for subsequent operations. Direct action by Special Forces and Rangers in these operations can help achieve precision and reduce potential damage to the target or noncombatant casualties.

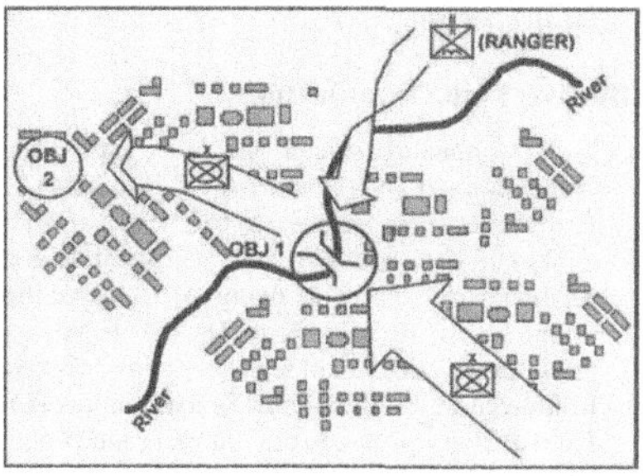

Figure 7-17. Coordination of SOF and conventional capabilities

Information Operations

7-65. Regardless of how Army forces physically isolate the urban area, they combine physical isolation with offensive IO to electronically isolate the threat and undermine his morale. Electronic isolation will cut off communications between forces in the urban area from their higher command to deny both from knowing the other's status. IO combined with isolation may persuade the threat's higher command or leadership that its forces located in the urban area are defeated. Thus, the command or leadership's intentions to break through to the besieged threat forces may be affected. PSYOP can undermine the morale of the threat in the area and reinforce electronic isolation and perceptions of abandonment. IO can be used to reduce any loyalty the civil population may have to the threat. IO can also ensure that civilians have the information that minimizes their exposure to combat and, as a result, overall noncombatant casualties. In addition, IO aim to deceive the threat regarding the time and place of Army force operations and intentions.

Detailed Leader Reconnaissance

7-66. Army commanders must clearly see the urban environment to understand the challenges facing their brigades, battalions, companies, platoons, and squads. Urban terrain can be deceptive until viewed from the soldier's perspective. Commanders are responsible to intimately know the conditions to allocate resources effectively to subordinate units. Often, particularly at battalion level and above, commanders will not be able to command and control dispersed forces from positions forward, but be forced by the terrain to rely on semifixed command posts. Detailed leader reconnaissance of the AO by commanders, their staff, and their subordinates before the mission can compensate for this challenge. This reconnaissance will give commanders a personal feel for the challenges of the terrain and will facilitate more accurate planning and better decision making during operations.

Mission Orders

7-67. Before contact, commanders can mitigate some terrain challenges to effective C2 using mission orders. Subordinates have mission orders to take advantage of opportunities before C2 systems can adversely impact the environment. To see the battle and provide effective and timely direction, tactical leaders will follow closely behind units as they assault buildings, floors, and rooms. Thus, only the most mobile INFOSYS can accompany tactical leaders into combat, and they will suffer the degrading effects of the environment. Mission orders permit rapid and decisive execution without commanders intervening at battalion level and above. Higher-level commanders can facilitate mission orders through their subordinates by articulating their desired end state, clearly stating their intent, and building flexibility into the overall plan.

Effective Task Organization

7-68. Commanders can shape urban offensive operations through effective and innovative task organization. Combined arms, often starting with an infantry base, are essential to success and may be the Army's asymmetric means of defeating an urban threat. Urban attacks will quickly break down into noncontiguous firefights between small units. To achieve the tactical agility for mission success in this nonlinear environment, many Army capabilities are task organized down to the company, platoon, and squad levels. Infantry provides the decisive capability to enter buildings and other structures to ensure threat destruction. Tanks, gun systems, and fighting vehicles provide additional speed and mobility, direct firepower, and protection. Combat engineers provide specialized breaching and reconnaissance capability. Field artillery provides the indirect (and if necessary, direct) firepower. Such mobility and firepower create the conditions necessary for the dismounted infantry to close with and destroy a covered threat in an urban defense. When a threat skillfully uses the urban area to limit ground maneuver, vertical envelopment or aerial attack using precision-guided munitions from Army aviation may circumvent his defenses and achieve necessary effects. Generally, ground systems used within the urban area will not be able to operate independently from dismounted infantry. The infantry will be required to protect armor and mechanized systems from close antiarmor weapons, particularly when those weapons are in well-prepared positions located throughout the urban area but especially on rooftops and in basements.

7-69. In urban offensive operations, direct fire support can be critical. Armor vehicle munitions types do not always achieve decisive effects against some urban structures. In some cases, field artillery high explosive munitions work better than armor for direct fire support of infantry. Large caliber (105 or 155mm) high explosives directly fired at a structure often produce a more severe shock effect than tank and fighting vehicle cannon and machine guns produce. Artillery is also able to achieve higher elevation than armor and engage threats located at greater heights.

7-70. However, commanders must view artillery as not just a weapon but a weapon system. As such, artillery should normally be placed under tactical control (TACON) of maneuver commanders, such as a platoon of three guns TACON to a company or a battery to a battalion, not just one gun to a company or other maneuver unit. Self-propelled artillery has some of the mobility characteristics of armor; however, it provides minimal ballistic protection from fragmentation for the crew. Although these systems seem formidable, they provide less crew protection than a Bradley fighting vehicle, for example, and contain large amounts of onboard ammunition and propellant. They are susceptible to catastrophic destruction by heavy automatic weapons, light cannon, and antitank fire. Therefore, infantry units must carefully secure and protect these systems (even more so than armored vehicles) when employed in urban offensive operations, particularly when forward in the direct fire role.

Creative Task Organization: Using Artillery in the Direct Fire Role

Task organizing artillery to permit its use in a direct fire role demonstrates the innovative task organization required for urban operations. The following provides three historical examples of task organizing and using field artillery for a direct fire role.

In 1944, U.S. Army units of the 1st Infantry Division were assigned to attack and seize the German city of Aachen. The city's internal defense included bunkers designed to serve as air raid shelters. These positions, buildings of stone, seemed impervious to direct fire tank weapons, demolitions, and small arms. To reduce the positions, the 1st Infantry Division relied on the artillery's direct fire.

Field artillery used this way had physical and psychological effects on the defenders. The 26th Regimental Combat Team's history of the battle describes the German reaction to the artillery pieces:

The chief shock to the defenders, Colonel Wilck (Aachen defense commander) said, came from the self-propelled 155s and tanks. The colonel spoke with considerable consternation of the 155mm self-propelled rifles. A shell from one of the guns, he said, pierced three houses completely before exploding and wrecking a fourth.

The 26th Infantry Regiment also described how the artillery, one piece attached to each assaulting infantry battalion, helped the infantry to penetrate buildings.

With solid blocks of buildings comprising most of the city, there wasn't any easy way to get at the Germans in the buildings. The eight-inch gun solved the problem. Beginning on the eastern outskirts the gun would plow a round into the side of the built up block of buildings at about ground level. One shell would usually open an entrance into the first tier of floors, i.e. the first building. Then several more shells were fired through the first hole. Thus a tunnel would be rapidly made all the way to the next cross street. Soldiers could then rush the newly formed entrance, clear the upper floors with hand grenades and rifles and then move on to the next building to repeat the process. When a block or square, was thus completely cleared of Germans—soldiers, skulkers, or even snipers—the next square was treated in the same way, working forward square by square, right and left, thereby avoiding nearly all exposure in the streets.

In 1982, Israeli forces invaded southern Lebanon to destroy base camps of the Palestine Liberation Organization (PLO). This operation involved significant fighting in urban areas including major operations in Beirut. Artillery, firing in a direct fire role, played a major part of the tactical solution. Artillery was particularly effective in the 33-day siege of Beirut. During this siege, Israeli forces used artillery in its traditional role as well as in the direct fire role.

The Israeli army was committed to a policy of disproportionate response during the Beirut siege. When fired on with small arms, crew-served weapons, tanks, or indirect artillery, the Israeli forces responded with intense, high-caliber direct and indirect fire from tanks and artillery positioned around the city. Many firing positions were on heights to the south and southwest that dominated much of the city. These positions had almost unrestricted fields of view. Israeli artillery fired from these positions directly into high-rise buildings concealing PLO gunners and snipers. The artillery, using direct fire, destroyed entire floors, collapsed floors on top of each other, and completely removed some upper floors. Such a response, as in Aachen in 1944, had as much a psychological impact as it did a physical impact on the PLO defenders.

In the early hours of 20 December 1989, the United States launched OPERATION JUST CAUSE. One of this operation's objectives was removing the Panamanian dictator, Manuel Noriega. U.S. forces carefully planned using all fires before the operation to minimize casualties and collateral damage. Part of this detailed fire planning called for applying artillery in a direct fire role.

The Panamanian Defense Force's (PDF) 5th Rifle Company based at Fort Amador was one of the key objectives of U.S. forces at the start of hostilities. This unit was high priority because it was the closest PDF unit to Noriega's headquarters, the Comandancia. U.S. forces expected the 500-man company to react first to

> OPERATION JUST CAUSE by reinforcing the defense of the Comandancia. It also posed a threat to U.S. military dependents housed at Fort Amador.
>
> To quickly neutralize this force, the United States assembled a three-company force composed of A, B, and headquarters elements of 1-508th Infantry (Airborne), supported by 105mm towed howitzers of 320th Field Artillery and M113 armored personnel carriers. The howitzers and the personnel carriers were covertly pre-positioned at the fort. At approximately 0100, helicopters transported the two airborne rifle companies into position. The howitzers then suppressed any personnel in the PDF-controlled buildings on Fort Amador while demonstrating the firepower of the U.S. task force. They used direct fire into the PDF barracks. The impact of the 105mm high explosives and .50-caliber fire from the M113s convinced the PDF infantry to give up after token resistance. Following the direct fire, U.S. infantry assaulted and cleared the dozen PDF buildings, finding that most occupants had fled or surrendered. For more details of OPERATION JUST CAUSE, see Applying the Urban Operational Framework: Panama in Chapter 5.
>
> ---
>
> The three examples cited indicate the importance of the situation-dependent and innovative task organization of artillery and its use in the direct fire role. Using artillery helps overcome some challenges of offensive operations in the urban environment, and it has an important psychological effect on a defending threat. Such task organization takes a traditional tool of a higher-level tactical commander and uses it to directly influence the company-level battle. This philosophy of task organization can be applied to other types of forces—not just artillery. PSYOP teams, interpreters, CA specialists, armor, engineers, and reconnaissance teams may require task organization different from traditional organization. The compartmented urban environment drives the requirements for these assets lower in the tactical scheme than in open operations. Consequently, commanders must understand and account for more of these assets for UO than for operations in less restrictive environments.

7-71. Army aviation may also be inventively task organized. It can support urban operations with lift, attack, and reconnaissance capabilities. Tactical commanders down to company may use all these capabilities to positively influence ground close combat. Army attack and reconnaissance aircraft can provide flank security for attacking ground forces. Attack aircraft may also provide direct fire support to individual platoons or squads. Lift may move entire battalions as part of brigade operations, or it may move single squads to a position of advantage (such as a roof) as part of a small unit assault. Army aviation can assist with C2 by providing airborne retransmission capability, airborne command posts, and the confirmed status and position of friendly forces. However, Army aviation is a limited and high-value asset; commanders should review its use in innovative task organizations. It is particularly vulnerable to urban air defense threats unless used over terrain secured by ground forces. From these positions, aircraft can use enhanced sensors to conduct reconnaissance and use precision weapons with standoff capability.

Shaping Attacks

7-72. In a large urban area, the defending threat cannot be strong everywhere. Shaping operations can also take the form of attacks against vulnerable positions to force the threat to maneuver and redeploy his forces in the urban area. This prevents him from merely defending from prepared positions. Forcing the threat to move negates many of the defensive advantages of urban terrain, confirms his dispositions, exposes vulnerable flanks, and permits target acquisition and engagement with precision standoff fires.

7-73. A critical shaping operation in urban offensive operations is usually an initial attack to seize a foothold. Once Army forces establish this foothold, they accrue some of the defensive advantages of urban terrain. From this protected location, Army forces continue offensive operations and have a position of advantage against neighboring threat defensive positions.

ENGAGE

7-74. Commanders may employ several methods to decisively engage elements of the urban area during offensive operations. These include—

- Rapid maneuver.
- Appropriate use of SOF.
- Precise application of fires and effects.
- Proper balance of speed and security.

7-75. None is unique to UO. Their effective execution, however, allows Army commanders to dominate in this challenging environment by effectively using resources with the least amount of collateral damage. Overall, decisive engagement results from urban offensive operations when forces achieve the objective of the assigned mission and establish preeminent control over the necessary terrain, population, and infrastructure. Largely, the Army commander's ability to engage is based on superior situational understanding and the correct application of unit strengths to the challenges found in the urban environment.

Rapid and Bold Maneuver

7-76. Commanders of major operations may have or create the opportunity to seize an urban area with rapid and bold maneuver. Such maneuver requires striking while the area remains relatively undefended—essentially preempting an effective defense. This opportunity occurs when the urban area is well to the rear of defending threat forces or before the onset of hostilities. Under such conditions, an attack requires striking deep behind threat forces or striking quickly with little time for the threat to make deliberate preparations. Attacks under such conditions may entail significant risk; however, the potential benefit of audacious offensive operations may be well worth possible losses. Three potential ways to accomplish such attacks (and their combinations) are:

- Airborne or air assault.
- Amphibious assault.
- Rapid penetration followed by an exceptionally aggressive exploitation, for example, a heavy force using shock, armor protection, and mobility.

7-77. Commanders should analyze all potential urban operations to seek an opportunity or advantage to apply rapid and bold operational maneuver to the task. Using operational maneuver to avoid urban combat against an established threat defense potentially marks a significant operational achievement and can have decisive strategic consequences. Just influencing the threat's morale can positively affect all future operations. However, commanders must evaluate the challenges of such a course of action. These challenges may include—

- Sustaining the operation.
- Avoiding isolation and piecemeal destruction.
- Successfully conducting shaping attacks.
- Achieving the necessary tactical, operational, and strategic surprise.

7-78. Commanders can also build on the shaping effects of isolating the urban area internally and externally by attacking urban decisive points from multiple directions. They can attack multiple decisive points either simultaneously or in a systematic, synchronized manner. This complicates the threat's situational understanding of the urban environment, further impedes his decision making, and allows Army commanders to dictate the tempo.

Bold Operational Maneuver to Seize an Urban Area: Inchon and Seoul, Korea – 1950

In August 1950, UN forces in Korea were desperately attempting to stave off defeat by establishing a viable defense in southern Korea. This defense, known as the Pusan Perimeter, was the focus of the world's attention. The commander of U.S. Forces Far East, General Douglas MacArthur, was not focused on the U.S. Eighth Army in Pusan. Instead, he focused on how to dramatically and decisively transition to the offense—the recapture of the South Korean capital of Seoul.

Seoul was more than just the South Korean political capital, although that was an important consideration. Seoul was the key to the Korean transportation system, the hub of the national communications system, and the center of the country's economy and culture. It was also strategically placed astride maneuver corridors to the north, south, and northeast. It had been Korea's capital since 1394 when King Taejo Yi Sung-Gye established it as the center of the Chosun dynasty. It had been captured by the surprise communist invasion of the south in June 1950, and the key to success in Korea was its recapture. The question was how.

The obvious military solution to the Korean situation in August 1950 was to build military strength within the Pusan Perimeter and then, when sufficiently strong, to breakout and counterattack north and recapture Seoul. The disadvantage of this course of action was that the counterattack would be through hundreds of miles of mountainous terrain, through several other major cities, and across numerous rivers. The North Korean People's Army (NKPA) would be able to fall back from one mountain defense and one river line to another and would have weeks or months to prepare the defenses of Seoul.

MacArthur's solution was to delay a conventional counterattack, boldly maneuver by sea deep into the flanks of the NKPA, land on the Korean west coast at Inchon, and rapidly seize Seoul before the enemy could react (see figure 7-18). This course of action had numerous disadvantages: achieving surprise; assembling a landing force trained in amphibious operations; few good landing sites; supplying a large force once landed; and needing to simultaneously continue to defend and prepare a counterattack with the Eighth Army from Pusan. MacArthur understood the challenges of the operation but believed that the potential value compensated for the risk.

The potential value of a rapid assault on Seoul through a deep amphibious operation was immense. The operation would trap the bulk of the NKPA in South Korea and facilitate its destruction. It would also capture Seoul before the NKPA could prepare a defense. The quick recapture would immeasurably impact the morale of UN forces and South Korea. MacArthur felt the payoff of success justified the risks and challenges of the operation.

On 15 September, elements of the 1st Marine Division landed in Inchon taking the NKPA completely by surprise. On D+1 they were joined by elements of the 7th Infantry Division, both divisions under the control of U.S. X Corps. Naval gunfire and Marine close air support flying from Navy aircraft carriers supported the landings. Seven days later, the 5th Marine Regiment had battled 25 miles to the outskirts of Seoul and began attacking the city from the north. Three days later, on 25 September, the 1st Marine Division and elements of the 7th Infantry Division had defeated the 10,000 defenders of the NKPA's 18th Rifle Division.

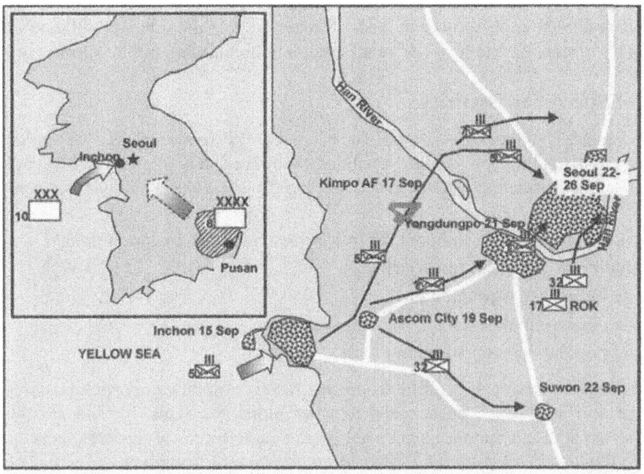

Figure 7-18. Inchon-Seoul campaign, September 1950

> Most of MacArthur's predictions for the Inchon-Seoul campaign were accurate. Superb staff work, excellent logistics operations, and unmatched training permitted the UN forces to assemble a trained landing force, land it over one of the most difficult shores in the history of military amphibious operations, and keep it supported. More importantly, the operation achieved total strategic and operational surprise. Seoul was recaptured quickly and, although resistance was fierce, the NKPA could not react fast enough to influence the outcome. In conjunction with the Eighth Army counterattack from Pusan, 75 percent of the NKPA was destroyed. The Inchon-Seoul campaign was an important turning point in the war, and had not the Chinese People's Army intervened two months later, it would have been the decisive campaign of the war.

Appropriate Use of Special Operations Forces

7-79. Sometimes Army forces can decisively engage (instead of simply shape) the urban area using the direct action capability of SOF. When the threat fails to develop a comprehensive defense and does not possess large, capable conventional forces, then Army forces can achieve operational surprise. Commanders, by synchronizing conventional and SOF effects, may actively control offensive operations to dominate the area. Importantly, however, conventional ground forces must be available to quickly assume the mission because SOF units acting as the primary striking force have limited logistic capability to sustain long-term operations. OPERATION JUST CAUSE offers several examples of this type of synchronization (see Applying the Urban Operational Framework: Panama in Chapter 6).

Precise Application of Fires and Effects

7-80. Precisely applied fires and the massed effects of combat power characterize successful urban attacks. The fires can be direct fire from combined heavy or light ground teams; direct or indirect fires from supporting Army aviation standing off from the target and any possible air defense threat; precision indirect fires from conventional tube artillery; or direct and indirect fires from supporting joint assets including Air Force and naval assets. All efforts strive to reduce collateral damage around the point of attack, consistent with mission success. Forces use fires to deny the threat the ability to maneuver in the urban area and to destroy the threat when he attempts to maneuver. When the threat exposes himself by moving, the environment no longer protects him, and fires can effectively engage him. Overall, precise fires and effects demonstrate the power of Army forces and help the urban population understand that only legitimate military targets are the focus of attacks, potentially building public support of Army UO.

Proper Balance of Speed and Security

7-81. Attacking units balance speed and security. Forces secure flanks as units advance, control dominating terrain (buildings), evacuate civilians, and keep the integrity and synchronization of the combined arms team. Obstacles are anticipated and rapidly breached. Commanders can choose avenues of approach to—

- Provide cover and concealment for following aviation and support units.
- Permit travel by all classes of vehicles.
- Easily defend from counterattack.
- Avoid nonessential centers of threat resistance.
- Avoid population concentrations.

7-82. Army aviation is a critical resource to protect flanks. Another important resource is engineers who seal off surface and subsurface entries and avenues along the flanks of the attack. Finally, as in all offensive operations, ground and air cavalry are ideal mobile forces to perform security in an economy of force role along flanks allowing decisive forces more freedom of maneuver.

CONSOLIDATE

7-83. Commanders, at all levels, must consolidate to strengthen their position during urban offensive operations without loss of momentum. They take the steps necessary to make permanent any temporary battlefield successes while maintaining relentless pressure on threat forces. Consolidation provides security and protection, facilitates reorganization, and allows forces to prepare for counterattack.

Repositioning of Forces

7-84. Following seizure of the objective, commanders normally consolidate by adjusting and repositioning forces. While UO are likely to be noncontiguous, commanders reposition ISR assets (including observation posts and reconnaissance patrols) to maintain contact with the threat, to establish contact with nearby friendly units and ensure that there are no exploitable gaps or seams, and to help maintain freedom of action. Physical occupation of the terrain (as well as continued reconnaissance) will provide commanders with a fuller understanding of the urban environment. With this enhanced understanding, commanders may need to adjust boundaries and other control measures to better adapt to the effects of urban terrain features such as canals, subway tunnels, raised roadways, and tall buildings. As necessary, commanders may also need to reposition communications assets and C2 facilities to enable subsequent operations.

7-85. Immediately after the conduct of successful urban operations, units must also be alert to the potential for rapid and violent counterattacks. Defenders may launch a quick counterattack to regain terrain before the offensive forces have consolidated and fully assumed the defensive advantages of the urban terrain. Delaying a counterattack in UO, even for a few minutes, permits the environment's advantages to shift to the successful attacker. Thus, attacking units anticipate this reaction and reposition forces such that they are prepared to defeat it.

7-86. As necessary, commanders should reposition armor and artillery (and other fire support assets) to account for the changing situation and battlefield geometry. Commanders should also consider consolidating and repositioning armored and artillery forces in positions—either inside or outside the urban area—to add significant combat power to a hasty defense, to defeat threat counterattacks, or to allow for quick resumption of the attack. If integrated into a hasty defense inside the urban area, these forces will require continued infantry protection. As a mobile counterattack force positioned inside the urban area, armored forces will also require careful selection of attack positions and counterattack routes. Damage to buildings and infrastructure may limit maneuver and the use of direct and indirect weapon systems.

Repositioning forces outside the urban area may contribute to strengthening or reestablishing the isolation of the urban area and will make better use of range and standoff capability, allow them to take full advantage of their speed and mobility, and decrease their need for additional protective support.

Eliminating Pockets of Resistance

7-87. Focusing on controlling the essential and attacking decisive points means that attacking units will have bypassed some elements of the threat's defense. Within the bounds of the initial plan and current situation, commanders should consider whether it is necessary to devote the time and resources to immediately clear and eliminate remaining threat forces and pockets of resistance with all or parts of the attacking force or leave the task to follow-on forces. In their deliberations, commanders should determine if remaining threat forces are capable of consolidating and mounting effective counterattacks before follow-on units can engage the remaining threat and take advantage of their dispersion and disunity. As another part of their deliberations, commanders should also consider if remaining threat forces will likely (and immediately) endanger—

- The urban inhabitants.
- Lines of communications.
- Critical resources within the urban area necessary to support the inhabitants or subsequent operations.

Reorganization

7-88. Reorganization includes all measures taken by the commander to maintain combat effectiveness or return to a specified level of combat capability. As necessary, these actions include—

- Redistributing or cross-leveling supplies and equipment until resupply can be accomplished.
- Replacing key personnel and combining units or crews to form mission-capable ones.
- Integrating replacement soldiers and systems into the unit and matching operational systems with available crews.
- Recovering, treating, and evacuating casualties, enemy prisoners of war, and damaged equipment.
- Conducting training and disseminating critical lessons learned.
- Other actions to reestablish unit cohesion.

TRANSITION

7-89. Effective transitioning allows commanders to continue full spectrum operations in the urban area and elsewhere in the AO without unnecessary delays. Commanders can accomplish this by thorough planning—including appropriate branches and sequels (revised as the situation changes)—that gives adequate consideration to post-offensive organizational, training, psychological, and civilian requirements. If properly prepared, commanders can anticipate rather than react to potential mission changes.

Early and Concurrent Transition Planning

7-90. Commanders can ensure smooth transitions of urban offensive operations by planning for post-offensive operations early. Based on the mission envisioned, they determine which subordinates and what type of force structure to use. Post-offensive missions, like all urban operations, encompass full spectrum operations. At the successful conclusion of offensive operations, Army forces transition to some type of stability or civil support operation. Commanders may leave the subordinate unit in place to execute the new mission, may reorganize the subordinate unit for the mission, or relieve the unit that just completed offensive operations with a new unit.

Changes to Task Organization

7-91. Commanders should consider the organization of forces following offensive operations. Hostile civilians may require significant combat forces or military police forces to maintain stability. On the other hand, friendly civilians may require a minimum of military police or combat forces, but significant logistic support. Commanders must carefully consider the urban situation before deciding how to use combat forces that recently participated in a high-intensity offensive operation.

Training and Psychological Considerations

7-92. Many Army combat tasks may not support follow-on stability or civil support operations without considerable modification (if at all). Often, noncombat skills—not normally part of a unit's mission essential task list such as negotiating or mediating skills—will be required. However, the greatest modification required applies to each soldier's mental outlook. Forces that transition directly from combat to stability operations may not be psychologically prepared for a rapid and drastic change of mission. Commanders cannot expect troops who have just completed high-intensity offensive operations to rapidly adjust and exercise the sensitivity and judgment required in most stability operations. This especially applies if the population is hostile to Army forces. If possible, combat forces assisting in stability operations, particularly in hostile civilian situations, should not have had recent experience in high-intensity urban operations and they should have trained for the mission.

Return to Civil Agencies

7-93. Commanders of major operations also have the critical role of transitioning aspects of the urban offensive operation to civil, allied, NGOs, and other agencies as appropriate. Transition planning is detailed and aims to quickly return as much civil control of the area as is feasible after the attack. Beyond local civil control, outside civil agencies and NGOs assume tasks as completely and as rapidly as possible. These organizations are consulted and integrated into the planning process as early as possible. Commanders should begin planning for transition simultaneously with planning for offensive operations. They consider the feasibility of relinquishing control of urban areas to civil government, law enforcement, or NGOs even before completing offensive operations. During the conduct of urban operations, these transition operations are closely synchronized with the execution of the attack.

Transition to a New Mission

7-94. In urban offensive operations, like other offensive missions, the change in mission after a successful urban attack may be to a hasty defense or a continuation of offensive operations outside the area. However, in urban offensive operations the mission will just as likely rapidly change to a civil support or stability mission. This is particularly true if the unit has had special training and is task organized for urban operations. Transition to stability or civil support operations is often accompanied by a transition in roles from supported to supporting.

7-95. Even more challenging than transition at the end of the mission is transition during the conduct of the mission. Soldiers may have a difficultly transitioning from stability or civil support to offense and defense, and back again multiple times during an urban offensive operation. Soldiers may apply the tactics, techniques, and procedures of urban offensive operations directly to the stability or civil support missions with potentially disastrous results. Commanders need to segregate missions in time and space, and, if sufficient forces exist, by unit. To this end, commanders should consider permanently designating specific units to conduct civil-military and humanitarian support tasks and avoid rapid mission changes that rotate units (particularly at company level and below) between violent and nonviolent tasks. However, commanders may not have that luxury and may need to rely heavily on preparatory training (including the inculcation of Army values) and strong unit leadership to mitigate potential difficulties.

Chapter 8
Urban Defensive Operations

Generally, a modern city magnifies the power of the defender and robs the attacker of his advantages in firepower and mobility. A city can ingest an invading army, paralyze it for weeks on end, and grind it down to a state of ineffectiveness.

"Military Operations on Urbanized Terrain:
The 2d Battalion, 26th Infantry, at Aachen, October 1944"

The skillful defense of an urban area can decisively affect a campaign. The urban area offers many advantages to defending forces. An adroit defender can use the advantages of the urban environment to negate combat power disparities, blunt the tempo of an attack, attrit threat forces, and sap the morale of attacking troops. The defender gains an opportunity to concentrate resources, reconstitute attrited units, and transition to the offense. A successful defense of an urban area can also deny the threat vital resources. Defense in the urban environment is an essential Army capability and can significantly affect the outcome of entire campaigns and the achievement of national objectives.

PURPOSE OF URBAN DEFENSIVE OPERATIONS

8-1. Army forces defend urban areas for various reasons: defeating a threat attack, buying time, economizing forces, protecting an ally's political institutions and economic infrastructure, protecting an urban population, shaping conditions for decisive offensive operations, and shaping conditions for executing stability or civil support operations. During force projection operations, urban areas may be used as initial lodgment areas that Army commanders may need to defend at the outset until they build sufficient combat power. Usually two or more of these purposes apply to the urban defense. Urban defensive operations provide commanders great opportunities to turn the environment's characteristics to the advantage of Army forces. Urban areas are ideal for defensive operations and greatly enhance the combat power of defending units.

DEFENSIVE CHARACTERISTICS APPLIED TO URBAN OPERATIONS

8-2. There are five general characteristics of the successful defense: preparation, security, disruption, massing effects, and flexibility. All apply to the successful urban defense and to the higher commander supporting a subordinate defending in the urban area.

PREPARATION

8-3. The urban area suits the defense since the area's physical characteristics naturally enhance the combat power of defending units. These characteristics include protection, obstacles, and concealment. Urban terrain provides superb defensive positions with minimum preparation. With deliberate preparation, urban defensive positions can rapidly become strong points.

8-4. One primary characteristic of urban terrain that enhances the defense is protection. With little or no advance preparation, buildings, subsurface structures, and walls protect Soldiers from direct and indirect fire, interdict indirect fire, limit observation, and limit engagement ranges (requiring Soldiers to be skilled at combat in close quarters and quick-fire techniques). Nearly all buildings provide some ballistic protection from direct and indirect fire. Mason and stone buildings with basements and cellars can protect Soldiers from most fires except the largest caliber or tonnage bomb. Minimal additional preparation turns them into formidable, defensive strong points.

8-5. Buildings in urban areas, because of their height and close proximity, also can protect Soldiers by masking them from indirect fire. The height of a building may interdict the flight path of an artillery round, rocket, missile, or bomb at a point short of the intended target. Masking protects static defending forces and protects forces moving along routes bordered with tall buildings that form urban "canyons". These protected routes can be used for sustainment, counterattacks, and maneuver.

8-6. Structurally significant buildings in an urban area can create major obstacles to maneuver. These obstacles immediately canalize maneuver into existing streets and routes without any preparation by the defense. These obstacles then become kill zones for well-positioned and sited defensive forces. Minimal obstacle construction as point obstacles blocking streets and routes can further restrict the maneuver options of the attacking force. Rubble from structures collapsing into streets after fires (intentional or unintentional) can also block routes.

8-7. Buildings also conceal the location, disposition, and intent of the defense. They limit visual observation to the external perimeter of the urban area. They degrade radar and electronic position identifiers and decrease the utility of overhead imagery. The physical aspect of the urban environment greatly enhances the defense by degrading the opposition's intelligence, surveillance, and reconnaissance (ISR) capabilities. Buildings can conceal static defensive positions and the maneuver of defensive forces in the urban area. Although the environment constrains defensive mobility in much the same manner as offensive mobility, the defender has the time and opportunity to conduct careful reconnaissance and select and prepare routes. This gives the defender the ability to move reserves, maneuver counterattack forces, and plan sustainment without observation. Careful preparation provides the defender a mobility advantage over attacking forces.

SECURITY

8-8. The urban area can be an advantage or a disadvantage to the security of defending forces. This largely depends on the nature of the human dimension of the environment. If the population is evacuated or allied with Army forces, then the environment may assist in the security of defending Army forces. However, if the population is present and hostile, then the environment may make security difficult.

8-9. The physical aspects of the urban environment, uninfluenced by the human dimension, may assist in the security of defending Army forces. The combat power of small security forces manning observation posts is greatly enhanced. Forces can more easily restrict and monitor avenues of approach for threat reconnaissance. Defending forces positioned mostly in structures are difficult to locate.

8-10. The physical aspects of the environment may also present some security challenges, primarily with observation. The compartmented terrain limits the field of observation from any one point. The defense may require more security forces to adequately observe the mounted and dismounted avenues to prevent infiltration. Threat forces that successfully infiltrate will be more difficult to locate. These forces will gain numerous hide positions for small reconnaissance units in complex terrain and the effect the terrain has in masking electronic signatures.

8-11. Friendly civilians in the urban area can help identify threat forces attempting to conduct reconnaissance. Civilian activity will also help to mask defense preparations. However, a hostile element of the population may pass intelligence information to the threat. They may assist threat reconnaissance to infiltrate the urban area or provide guides, manpower, or resource support for threat forces. Commanders should take measures to ensure strict control of hostile populations. If resources permit, commanders may consider removing potentially hostile civilians from the area.

DISRUPTION

8-12. The urban environment's attributes assist defending Army forces to disrupt the attacker. It does this through compartmentalization, inhibiting command and control (C2), and facilitating counterattacks.

8-13. The physical aspects of the urban area force the attacking threat into compartmented urban canyons that make mutual support between attacking threat columns difficult. Shifting resources from one portion of the threat attack to another is also difficult. Physically, the urban area disrupts tactical communications making synchronization of combat power difficult.

8-14. The urban terrain hinders the mobility capabilities of the defense. However, careful planning, preparation, and rehearsals can facilitate more rapid movement of larger forces. Defending forces can assemble counterattacks undetected, move them along covered and concealed routes, and achieve surprise at the point of the counterattack. Attacking forces, using the compartmented terrain, often leave forward elements in position to be isolated or expose long and vulnerable flanks to friendly counterattack and interdiction.

MASSING EFFECTS

8-15. The urban environment allows defenders to better protect their centers of gravity and decisive points. The restrictive terrain reduces the attacker's maneuver options. Defenders can position forces in protected and mutually supportive positions oriented on deadly engagement areas. Relatively few well-positioned defenders can generate significant combat power. Without the positional advantage and the corresponding protective effects of the terrain, attacking forces often mass numbers to achieve the necessary combat power.

8-16. Knowing the complex terrain permits defending forces to plan engagement areas that maximize the effects of their combat power. Defending forces can remove fences, walls, rooftops, and even entire buildings to facilitate fields of fire and unmask indirect fire flight paths. Forces carefully choose firing positions for indirect fire systems so that flight paths travel between buildings into engagement areas. By leveraging this knowledge of the terrain, numerically inferior defenders can synchronize devastating fires on offensive forces that are forced by terrain and reinforcing obstacles to mass in confined spaces where fires can have the greatest effect.

FLEXIBILITY

8-17. Defensive flexibility results from detailed planning and mental agility. Commanders develop defensive flexibility by ensuring that plans adequately address branches and sequels that include alternate and subsequent positions and emphasize counterattack options. The urban area facilitates defensive flexibility because the urban terrain can be quickly adapted for defensive operations with little or no preparation. The effect is similar to having multiple, prepared positions on nearly every possible approach. The urban area can also permit rapid, covered movement on interior lines. This permits swift movement to and occupation of strong defensive positions with little or no preparation. The defense also has more flexibility since defenders often know and better understand the urban terrain's effects on operations. Normally, defenders will not get lost as easily, will know complex lines of sight and masking effects, and will best understand the ballistic characteristics of individual structures.

8-18. Mental agility allows commanders to see that the best urban defense may actually be to defend outside of the area. Such a defense mitigates the danger to the urban population and potentially reduces collateral damage. It takes advantage of Army long-range engagement capabilities and denies the threat the opportunity to "hug" Army forces or noncombatants as protection from fires. This defense may be appropriate when Army forces have enough resources to defend more open terrain, when time permits deploying extensive obstacles and constructing protected positions, and when natural terrain such as river obstacles aids the defense.

URBAN DEFENSIVE OPERATIONS AND BATTLEFIELD ORGANIZATION

8-19. Urban defensive operations are organized within the overall battlefield organization of sustaining, shaping, and decisive operations. The success of urban defense depends on each operation, but commanders must synchronize these simultaneous operations as one action. Sustaining operations in defensive urban operations (UO) ensure freedom of action. Critically, urban sustaining operations ensure security of the lines of communications and establish effective movement control. Shaping operations in defensive UO create the conditions for decisive operations. Shaping operations vary greatly depending on the type of defense. For example, in a mobile defense the shaping operation may be the fixing force. In contrast, in an area defense the fixed defense may be the decisive operation. In the urban defense, decisive operations focus on accomplishing the commander's mission. The decisive operation may not defeat the threat's main effort, and it may not prevent threat occupation of large portions of the urban area if those tasks are not essential to mission accomplishment. For example, if the defense's objective is to protect a critical communications node, then, depending on the commander's overall intent, threat actions to secure an airfield elsewhere may not be important.

TYPES OF URBAN DEFENSE

8-20. Commanders can view urban area defensive operations two ways: as conducting a major defensive operation with an urban area in their area of operations (AO) and as defending entirely in an urban area.

AREA DEFENSE

8-21. At the operational level, an area defense may include both urban areas and open maneuver areas. The most common defense in an urban area and the most suitable for the characteristics of this distinct environment is the area defense. As a defensive operation, the area defense concentrates on denying threat forces access to designated terrain for a specific time rather than destroying the threat outright. Although an area defense in an urban area does not directly seek to destroy or defeat attacking threat forces, as an objective it does aim to force culmination of the threat's attack. The urban area defense often works effectively to exhaust threat resources and shape conditions for a transition to offensive operations. The urban area may also be used as a strong point to force threat movement in a different direction or to fix threat forces as part of a large, mobile defense taking place in the AO outside the urban area.

MOBILE DEFENSE

8-22. A mobile defense can operate in an urban area but only under specific conditions. It focuses on destroying or defeating the threat through a decisive attack by a striking force. It requires the defender to have greater mobility than the attacker. To shape a mobility advantage, the urban defender effectively uses the terrain and correctly task organizes his forces' mobility. The principles of applying the mobile defense in the urban area remain the same: a small fixing force stops the threat and limits any ability to maneuver while a striking force quickly maneuvers and counterattacks to destroy him.

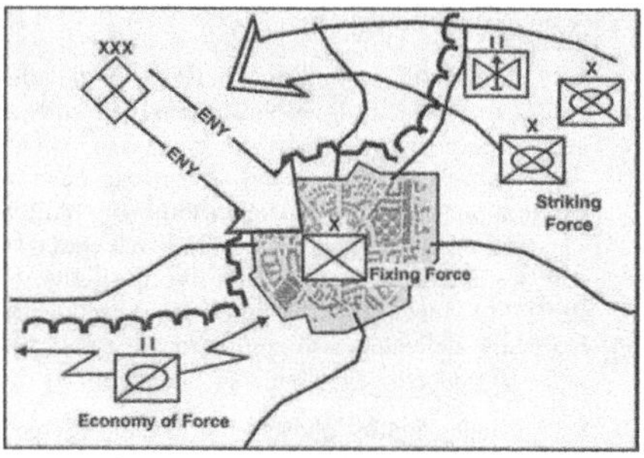

Figure 8-1. An urban area incorporated into a larger mobile defense

8-23. One key to executing a mobile defense in the urban area is to entice a threat force into the depths of the urban area where it begins to lose mobility options. A well-placed fixing force augmented with man-made obstacles and taking advantage of the naturally constrictive terrain can stop a much larger force. If the attacking force is largely mounted and armored, its mobility in the urban area may be reduced to less than that of dismounted infantry. In addition, if the attacking force's movement into the urban area is mounted and rapid, the commander's situational understanding also diminishes. Then the striking force, consisting of dismounted infantry forces, can execute the counterattack with surprise from multiple directions and dimensions (subsurface, surface, supersurface, and airspace). Man-portable antiarmor weapons—firing from flanks and top down and supported by precision indirect fires from both organic and joint systems—can rapidly destroy the threat.

8-24. From the perspective of commanders of the major operation, the urban environment can help defending forces achieve a mobility advantage over an attacker in a broader sense. Defending commanders can attempt to shape the battlefield so that the attacker commits significant resources into an urban area, where his maneuver capabilities are reduced (see figure 8-1). A disproportionately small defending force, which relies on the defensive combat power advantages of the urban environment, reduces and fixes the attacker's maneuver capabilities. Other defending forces mass outside the urban area then strike the threat with a combined mobility and firepower advantage.

Urban Defense in a Major Operation: Stalingrad – 1942 to 1943

The German and Soviet battle for Stalingrad in late 1942 and into early 1943 illustrates how a tactical urban area defense integrates into a larger mobile defense. Stalingrad was a battle fought on a huge scale: it involved army groups on both sides and thousands of square kilometers. Though the city was relatively small, it remained the focus of both German and Soviet offensive and defensive operations during the six-month battle.

In the summer of 1942, the Germans launched a strategic offensive in southern Russia. Its goal was the valuable oil fields of the Caucasus. German forces turning south into the Caucasus exposed a vulnerable flank to Soviet forces positioned between the Don and Volga Rivers. For the German Caucasus operation to succeed, it had to destroy Soviet forces between the Don and Volga, establish a good defensive line, and capture Stalingrad. This city would anchor the German defense and interdict the critical flow of supplies from the Caspian Sea via the Volga River into central Russia. Stalingrad, by virtue of its name, also had important political and cultural value to the Germans and Soviets.

The opening phases of the German offensive were successful; German forces—the 6th Army and 4th Panzer Army—entered the outskirts of Stalingrad in late August 1942 (see figure 8-2).

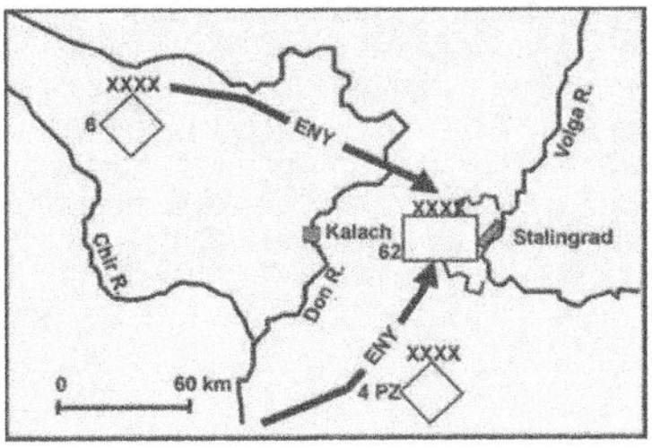

Figure 8-2. German attacks to seize Stalingrad

After a month of intense fighting, the Germans possessed nearly 90 percent of the city. At this point, the 6th Army commanded all German forces in and around Stalingrad. The Soviet 62nd Army's defense was reduced to a front only a few hundred meters deep and a couple of kilometers long on the banks of the Volga. The Soviet defenses hinged on fortress-like concrete industrial buildings and the fanatical bravery and tenacity of soldiers and civilians fighting in the city's remains (see figure 8-3 on page 8-6).

Beginning in mid-September, the Soviet command began looking at how to convert the defense of Stalingrad into an operational opportunity. During October and November, the 62nd Army held on to its toehold in Stalingrad. While maintaining the defense of the 62nd Army, the Soviets secretly began to build up strength on both flanks of the German 6th Army. The Germans increased their vulnerability by securing the German 6th Army's flanks with less capable Romanian, Hungarian, and Italian armies. Also, the 6th Army moved powerful German divisions into the city and rotated with German divisions that were exhausted by urban combat.

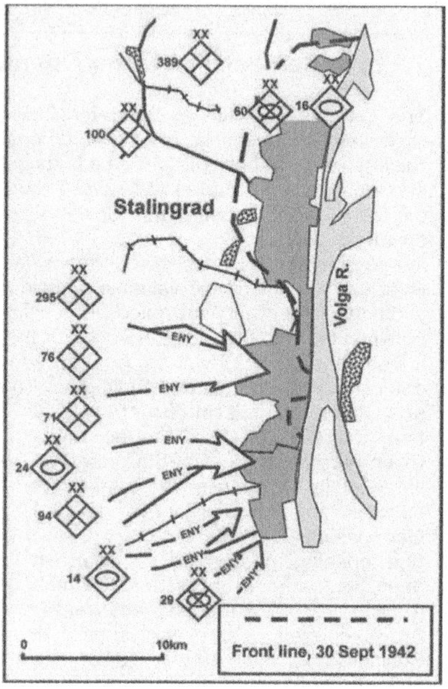

Figure 8-3. German attacks to seize Stalingrad, September 1942

On 19 November, the Soviets launched Operation Uranus that attacked two Romanian armies with seven Soviet armies. Simultaneously, the 8th Russian Army attacked to aid the 62nd Army in further fixing the German 6th Army. Within five days, the Soviet armies of the Don Front, Southwest Front, and Stalingrad Front met near the city of Kalach and sealed the fate of the German 6th Army's 300,000 troops in Stalingrad (see figure 8-4).

On the third day of the Soviet offensive, when encirclement seemed inevitable but not yet complete, the 6th Army commander asked permission to withdraw from the trap. The German high command denied permission believing that the Army could be supplied by air and then a renewed offensive could break through to the city. On 12 December, the German LVII Panzer Corps launched an offensive north to break through to Stalingrad. This offensive made progress until another Soviet offensive on 16 December forced its cancellation. This ended any hope of recovering Stalingrad and the 6th Army. On 31 January 1943, the 6th Army surrendered after sustaining losses of almost two-thirds of its strength. The Soviets took over 100,000 prisoners.

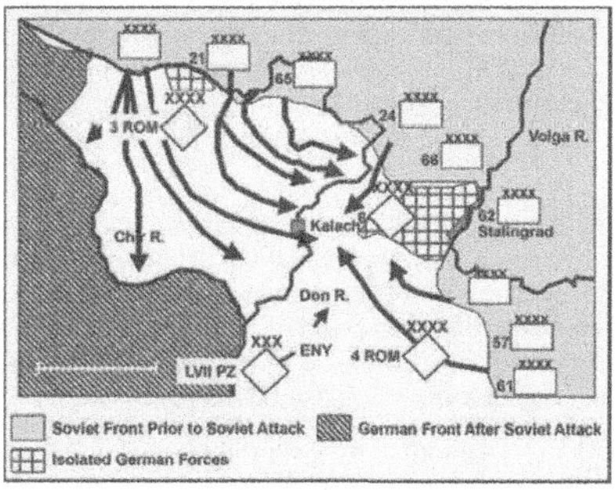

Figure 8-4. Soviet attacks trap German 6th Army

Many lessons emerge from the successful defense of Stalingrad. Tactically, the defense showed how using the terrain of a modern industrial city wisely could increase the combat power of an inferior, defending force and reduce the maneuver options of a mobile, modern attacking force. Another element in the Soviet's tactical success was the Germans' inability to isolate the defenders. The Germans never threatened the Soviet supply bases east of the Volga and, despite German air superiority, the Soviets continuously supplied and reinforced the 62nd Army across the Volga River. Also, Soviet artillery west of the river was able to fire in support of Soviet forces and was never threatened with ground attack.

At the operational level, the Soviets demonstrated a keen understanding of using an urban area within the context of a mobile defense. The 62nd Army's stubborn area defense of Stalingrad drew the bulk of the German combat power into the urban area where they were fixed by a smaller and quantitatively inferior defending force. This allowed the Soviets to build combat power outside the urban area. The Soviets set the conditions for a mobile defense by positioning powerful Soviet armor forces in open terrain outside the urban area against quantitatively inferior German allied forces. In OPERATION URANUS, the mobile defense's strike force destroyed the enemy outside the urban area and trapped the greater part of the best enemy formations inside the urban area. The trapped units were then subjected to dwindling resources and extensive psychological operations, further isolated into pockets, and defeated in detail.

RETROGRADE

8-25. A retrograde involves organized movement away from the threat. Retrograde operations include withdrawals, delays, and retirements. These defensive operations often occur in an urban environment. The urban environment enhances the defending force's ability to conduct retrograde operations successfully (see figure 8-5).

8-26. The cover and concealment afforded by the urban environment facilitates withdrawals where friendly forces attempt to break contact with the threat and move away. The environment also restricts threat reconnaissance, which is less able to detect friendly forces moving out of position, and presents excellent opportunities for deception actions. Finally, a small security force's ability to remain concealed until contact in the urban environment significantly slows threat attempts to regain contact once Army forces have broken contact and begun to move.

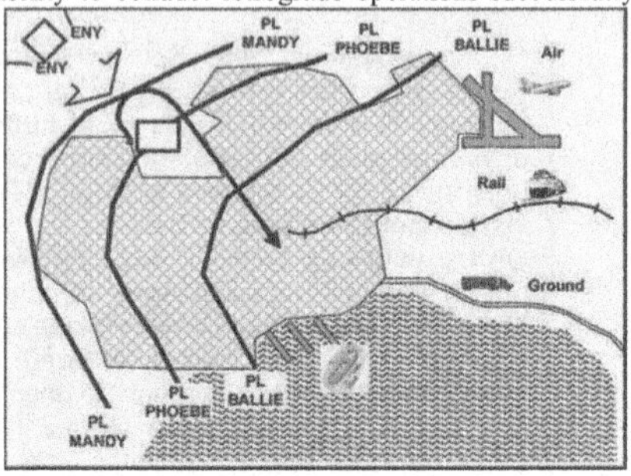

Figure 8-5. Retrograde through an urban area

8-27. The urban environment's natural cover and concealment, as well as the compartmented effects, facilitates delays. Delays can effectively draw the threat into the urban area for subsequent counterattack or as an integral part of a withdrawal under threat pressure. Delaying units can quickly displace from one covered and concealed position to another; the repositioning options are vast. Compartmented effects force the attacking threat to move on well-defined and easily interdicted routes and limit the threat's ability to flank or bypass delaying positions.

8-28. The urban area's transportation and distribution network facilitates retiring forces that are not in contact. Properly used, the urban transportation system can quickly move large forces and associated resources, using port facilities, airfields, railheads, and well-developed road networks.

URBAN DEFENSIVE CONSIDERATIONS

What is the position about London? I have a very clear view that we should fight every inch of it, and that it would devour quite a large invading army.

Winston Churchill
War in the Streets

8-29. The urban operational framework—understand, shape, engage, consolidate, and transition—provides structure to developing considerations for defensive operations. The considerations can vary depending on the level of war at which the operation is conducted, the type of defense, and the situation. Most issues discussed may, in the right circumstances, apply to both commanders conducting major UO and commanders at lower tactical levels of command.

UNDERSTAND

8-30. The commander defending in the urban area must assess many factors. His mission statement and guidance from higher commanders help him focus his assessment. If the mission is to deny a threat access to port facilities in an urban area, the commander's assessment will be focused much differently than if the mission is to deny the threat control over the entire urban area. The mission, enemy, terrain and weather, troops and support available, time available, civil considerations (METT TC) structure guides the commander's assessment. Of these, the impacts of the threat and environment—to include the terrain, weather, and civil considerations—are significant to the commander's understanding of urban defensive operations.

The Threat

8-31. In the urban defense, a key element is the commander's understanding of the threat. One of his primary concerns is to determine the attacker's general scheme, methodology, or concept. Overall, the attacker may take one of two approaches. The most obvious would be a direct approach aimed at seizing the objectives in the area by a frontal attack. A more sophisticated approach would be indirect and begin by isolating Army forces defending the urban area. Innumerable combinations of these two extremes exist, but the threat's intentions toward the urban area will favor one approach over another. The defending Army commander (whose AO includes but is not limited to the urban area) conducts defensive planning, particularly his allocation of forces, based on this initial assessment of threat intentions. This assessment determines whether the commander's primary concern is preventing isolation by defeating threat efforts outside the area or defeating a threat attacking the urban area directly. For the higher commander, this assessment determines how he allocates forces in and outside the urban area. For the commander in the urban area, this assessment clarifies threats to sustainment operations and helps shape how he arrays his forces.

The Environment's Defensive Characteristics

8-32. A second key assessment is the defensive qualities of the urban environment. This understanding, as in any defensive scenario, is based on mission requirements and on a systemic analysis of the terrain in terms of observation and fields of fire, avenues of approach, key terrain, obstacles, and cover and concealment (OAKOC). It is also based on potential chemical, biological, radiological, nuclear and fire hazards that may be present in the urban area. This understanding accounts for the unique characteristics of urban terrain, population, and infrastructure as discussed in Chapter 2.

8-33. Generally, units occupy less terrain in urban areas than more open areas. For example, an infantry company, which might occupy 1,500 to 2,000 meters in open terrain, is usually restricted to a frontage of 300 to 800 meters in urban areas. The density of building in the urban area, building sizes and heights, construction materials, rubble, and street patterns will dictate the actual frontage of units; however, for initial planning purposes, figure 8-6 provides approximate frontages and depths for units defending in an urban area.

UNIT	Frontage (Blocks*)	Depth (Blocks*)
Battalion	4 – 8	3 – 6
Company	3 – 4	2 – 3
Platoon	1 – 2	1
*Average block is 175 meters		

Figure 8-6. Approximate defensive frontages and depths

SHAPE

8-34. Commanders of a major operation shape the urban battle according to the type of defense they are attempting to conduct. If conducting an area defense or retrograde, they use shaping actions like those for any defensive action. Important shaping actions that apply to all defensive UO include—

- Preventing or defeating isolation.
- Separating attacking forces from supporting resources.
- Creating a mobility advantage.
- Applying economy of force measures.
- Effectively managing the urban population.
- Planning counterattacks.

Preventing or Defeating Isolation

8-35. Failure to prevent isolation of the urban area can rapidly lead to the failure of the entire urban defense. Its importance cannot be overstated. In planning the defense, commanders must anticipate that the threat will attempt to isolate the urban area. Defensive planning addresses in detail defeating threat attacks aimed at isolation of the urban area. Commanders may defeat this effort by allocating sufficient defending forces outside the urban area to prevent its isolation. Defensive information operations (IO) based on deception can also be used to mislead the threat regarding the defensive array in and outside the urban area. Such information can convince the threat that a direct attack against the urban area is the most favorable approach.

8-36. If the threat has successfully isolated the urban area, commanders of a major operation have several courses of actions. Three options are an exfiltration, a breakout attack by forces defending the urban area, or an attack by forces outside the urban area to relieve the siege. A fourth option combines the last two: counterattacks from both inside and outside the urban area to rupture the isolation (see breakout operations in FM 3-90). Time is critical to the success of either operation. Commanders should plan for both contingencies to ensure rapid execution if necessary. Delay permits threat forces surrounding the urban area to prepare defenses, reorganize their attacking force, retain the initiative, and continue offensive operations. The passage of time also reduces the resources of defending forces and their ability to breakout. Therefore, commanders and staff of a major operation must vigilantly avoid isolation when Army forces are defending urban areas in their AO.

Separating Attacking Forces from Supporting Resources

8-37. Commanders of the major operation primarily use fires and IO for separating in space and time threat forces attacking the urban area from echelons and resources in support. The purpose of this shaping action is the same as for any conventional area defense. It aims to allow the defending forces to defeat the threat piecemeal as they arrive in the urban area without support and already disrupted by deep fires and IO against information systems. This separation and disruption of the threat also sets the conditions for a mobile defense if commanders choose to execute that type of defense. These operations also prevent the threat commander from synchronizing and massing his combat power at the decisive point in the close battle.

8-38. If the urban area is part of a major mobile defense operation, the urban defense becomes the fixing force. Commanders can shape the defense to encourage the threat to attack into the urban area. They can lure the threat using a combination of techniques depending on the situation. They may make the urban area appear only lightly defended while other alternative courses of action appear strongly defended by friendly forces. Placing the bulk of the defending forces in concealed positions well within the urban area and positioning security forces on the periphery of the urban area portray a weak defense. In other situations, the opposite is true. If the urban area is an important objective to the threat, friendly forces can make the urban area appear heavily defended, thus ensuring that he commits sufficient combat power to the urban area to overwhelm the defense. Both cases have the same objective: to cause a major commitment of threat forces in the urban area. Once this commitment is made, the mobile defense striking force attacks and defeats the threat outside the urban area. This isolates the threat in the urban area and facilitates its destruction.

8-39. In the urban tactical battle, many shaping actions mirror those in all defensive operations. The size and complexity of the urban area prevent defending forces from being strong everywhere; shaping operations designed to engage the threat on terms advantageous to the defense have particular importance. Shaping actions include reconnaissance and security operations, passages of lines, and movement of reserve forces prior to their commitment. In addition, shaping operations critical to urban defense include mobility and countermobility operations, offensive IO, economy of force operations, and population management operations.

Creating a Mobility Advantage

8-40. In urban terrain, countermobility operations can greatly influence bringing the threat into the engagement areas of defending forces. Countermobility operations—based on understanding the urban transportation system, design, and construction characteristics—can be unusually effective (see Chapter 2). Demolitions can have important implications for creating impassable obstacles in urban canyons as well as for clearing fields of fire where necessary. Careful engineer planning can make the already constrictive terrain virtually impassable to mounted forces where appropriate, thus denying the threat combined arms capabilities. Countermobility operations in urban terrain drastically increase the defense's ability to shape the attacker's approach and to increase the combat power ratios in favor of the defense. As with all aspects of UO, countermobility considers collateral damage and the second- and third-order effects of obstacle construction.

8-41. Well-conceived mobility operations in urban terrain can provide defending forces mobility superiority over attacking forces. This is achieved by carefully selecting routes between primary, alternate, and subsequent positions, and for moving reserves and counterattack forces. These routes are reconnoitered, cleared, and marked before the operation. They maximize the cover and concealment characteristics of the terrain. Using demolitions, lanes, and innovative obstacles denies the defense of these same routes.

Applying Economy of Force Measures

8-42. Economy of force is extremely important to effective tactical urban defense. A megalopolis is too large and too easily accessible for defending forces to be strong everywhere. Economy of force enables the defending force to mass effects at decisive points. Forces used in an economy of force role execute security missions and take advantage of obstacles, mobility, and firepower to portray greater combat power than they actually possess. They prevent the threat from determining the actual disposition and strength of the friendly defense. If, contrary to expectations, they are strongly attacked, their mobility—stemming from a mounted maneuver capability, planning, and an intimate knowledge of the terrain—allows them to delay until reserves can meet the threat. Security forces in an economy of force role take position in parts of the urban area where the threat is less likely to attack.

Defensive Combat Power: Suez – 1973

At the end of October, the Israeli Army was in the midst of effective counterattack against the Egyptian Army. The Israelis had success attacking west across the Suez Canal. Their armored divisions were attempting to achieve several objectives, to include destroying Egyptian air defense sites and completing the encirclement of the Egyptian 3rd Army, which was trapped on the canal's east side.

To completely encircle the Egyptian 3rd Army, the Israelis had to seize all possible crossing sites to it from the canal's west bank and the Red Sea. Also, as international negotiations towards a cease-fire progressed, the Israeli government wanted to capture as much Egyptian territory as possible to improve their negotiating position after hostilities.

Consequently, the Israeli Adan Armored Division was tasked to seize the Egyptian Red Sea port of Suez on the morning of 24 October. A cease-fire was to begin at 0700, and the Israeli intent was to be decisively engaged in the city by that time and then consolidate their position as part of the cease-fire compliance.

The Adan Division plan to seize Suez was a two-part operation. Each of the division's armored brigades would have a role. The 460th Brigade would attack west of the city and complete the city's encirclement. Simultaneously, the 217th Brigade would attack in columns of battalions through the city to seize three key intersections in the city. This was in accordance with standard Israeli armored doctrine for fighting in an urban area. The 217th Brigade would seize its objectives through speed, firepower, and shock action. Once the objectives were seized, infantry and armored teams would continue attacking from the secured objectives to mop up and destroy pockets of resistance. The Israeli commanders expected to demoralize the defending Egyptians—two infantry battalions and one antitank company—by this rapid attack. The armored division commander was specifically advised by his commander to avoid a "Stalingrad" situation.

The attack got off to an ominous beginning as mist greatly inhibited a scheduled aerial bombardment in support of the attack. The 217th Brigade began its attack without infantry support and was quickly stopped by antitank missiles and antitank fire. Infantry was quickly integrated into the brigade and the attack resumed.

At the first objective, the Israelis encountered their first problems. A withering barrage of small arms, antitank missiles, and antitank fire hit the lead tank battalion, including direct fire from SU-23 anti-aircraft guns. Virtually all the officers and tank commanders in the tank battalion were killed or wounded, and several tanks were destroyed. Disabled vehicles blocked portions of the road, and vehicles that turned on to secondary roads were ambushed and destroyed. The battalion, however, successfully fought its way through the first brigade objective and on to the final brigade objective.

Hastily attached paratroop infantry in company strength were next in column following the tanks. They were traveling in buses and trucks. As the lead tank battalion took fire, the paratroopers dismounted, and attempted to secure adjacent buildings. The tank battalion's action of fighting through the objective caused the paratroopers to mount up and also attempt to move through the objective. Because of their soft skinned vehicles the paratroopers were unable to remain mounted and again dismounted, assaulted, and secured several buildings that they could defend. Once inside the buildings, the paratroopers found they were cut off, pinned down, and unable to evacuate their considerable casualties, which included the battalion commander. The paratroopers were on the initial brigade objective but were unable to maneuver and were taking casualties.

A second paratroop company also dismounted and quickly became stalled in house-to-house fighting. The brigade reconnaissance company in M113 personnel carriers brought up the rear of the brigade column and lost several vehicles and was also unable to advance.

By 1100 the Israeli attack culminated. Elements of the 217th Brigade were on all three of the brigade's objectives in the city. However, the armored battalion, which had achieved the deepest penetration, was without infantry support and under severe antitank fire. Both paratroop companies were isolated and pinned down. In addition, an attempt to link up with the paratroopers had failed. At the same time, the civilian population of the city began to react. They erected impromptu barriers, ambushed isolated Israeli troops, and carried supplies and information to Egyptian forces.

The Israeli division commander ordered the brigade to break contact and fight its way out of the city. The armored battalion was able to fight its way out in daylight. The paratroop companies were forced to wait until darkness and then infiltrated out of the city carrying their wounded with them. Israeli casualties totaled 88 killed and hundreds wounded in addition to 28 combat vehicles destroyed. Egyptian casualties were unknown but not believed to be significant.

The fight for Suez effectively demonstrates numerous urban defensive techniques. It also vividly demonstrates the significant effect on defensive combat power of the urban environment.

The Egyptian defense demonstrates how the compartmented urban terrain restricts the mobility and the massing of firepower of armored forces. Trapped in column on the road, the Israelis were unable to mass fire on particular targets nor effectively synchronize and coordinate their fires. The short-range engagement, also a characteristic of urban combat, reduced the Israeli armor protection and eliminated the Israeli armor's ability to keep out of small arms range. Thus, hand-held antiarmor weapons were more effective in an urban area. Additionally, Egyptian small arms and sniper fire critically affected Israeli C2 by successfully targeting leaders.

The Egyptian defenders effectively isolated the mounted Israelis by defending and planning engagement areas in depth. The Egyptians synchronized so that they engaged the entire Israeli force simultaneously. This forced the Israelis to fight in multiple directions. It also separated the Israeli infantry from the armor and prevented the formation of combined arms teams necessary for effective urban offensive operations.

Suez also demonstrated how civilians come to the advantage of the defense. After the battle was joined, the population—by threatening isolated pockets of Israelis and building barricades—helped prevent the Israelis from reorganizing while in contact and hindered the Israelis breaking contact. The population was also a valuable source of intelligence for the Egyptians and precluded covert Israeli movement in daytime.

Suez shows the ability of a well-placed defense in depth to fix a superior force in an urban area. Despite the Israeli commander's caution to avoid a "Stalingrad," the Israeli division, brigade, and battalion commanders were quickly trapped and unable to easily break contact. Even a successful defense on the perimeter of the city would not have been nearly as effective, as the Israelis would have easily broken contact once the strength of the defense was recognized.

Another key to the success of the Egyptian defense was the Israelis' inadequate reconnaissance. While the Israelis knew the approximate size of the defending forces, they had no idea of the Egyptian dispositions. In this case, time prevented adequate reconnaissance. Key to a successful defense is adequate security to obscure defense dispositions, which permits surprise and shock effect.

The Suez defense was a decisive defeat of elite Israeli forces by regular infantry units inferior in training, morale, and numbers. Total disaster was averted only because of the professionalism of the Israeli armored forces and paratroopers that permitted them to continue to fight and eventually exfiltrate the urban trap. The Israeli forces thus escaped total destruction. Suez strongly demonstrates how the enhancing effects of the urban environment on defensive combat power are significant enough to permit inferior regular forces to defeat elite formations. Since

> the 1973 Suez battle, U.S. forces in Mogadishu, Somalia, and Russian forces in Grozny, Chechnya have faced similar urban defensive ambushes.

Effectively Managing the Urban Population

8-43. Another way to shape the urban defensive battle is population management. In most cases, defending force commanders are in the urban area before combat. This time gives them the chance to manage civilians. Consequently, they can better manage and protect the population (a legal requirement) and gain more freedom of action for their forces.

8-44. Managing the civilians during the defense is a function of the size, disposition, and needs of the population and the resources available to the commander. Requesting higher support or coordinating with nongovernmental organizations and the local civil leadership for support may make up shortages of resources. Resources devoted to population management are carefully weighed against availability, military mission requirements, and possible collateral damage affecting tactical, operational, or strategic success. It may prove impractical to evacuate an urban area's population; still, commanders should attempt to create and move most civilians to protected areas. Moving the population allows defending forces to more liberally apply fires, emplace obstacles, and relieve combat units and support units of requirements to continue life support for civilians while executing combat operations. Overall, effective civil-military operations can turn a friendly (or a neutral) population into an effective force multiplier providing support to every warfighting function.

Planning Counterattacks

8-45. Counterattacks are also an important tool in shaping the battlefield for defensive success. Counterattacks as a shaping tool have two applications: retaining the initiative and separating forces. However, opportunity for effective counterattacks will be brief and, therefore, timing will be critical. If conducted too soon, the counterattack may expend resources required later; if conducted too late, it may not be effective. Commanders should understand the effect of the urban environment on time-distance relationships; otherwise, the timing of the attack may be upset and the operation desynchronized. Additionally, commanders should develop plans beyond the counterattack to exploit potential success.

ENGAGE

8-46. Engaging the urban area in a defensive operation requires decisively defeating the threat's attacks. Defensive forces use the terrain to their advantage, employ precision supporting fires, and use direct fire from protected positions aligned against carefully selected avenues of approaches and kill zones. The combat power of the defense augmented by shaping actions and the characteristics of urban terrain force culmination of the threat attack. Like urban offensive operations, effective engagement in urban defensive operations typically results from successful actions at the tactical level of war. These actions include—

- Performing aggressive ISR.
- Creating depth.
- Executing an effective obstacle plan.
- Conducting coordinated counterattacks.

Performing Aggressive ISR

8-47. ISR efforts of the defender are focused initially on identifying relevant information about the location and nature of the threat's main effort. Once identified, the defender's ISR focus shifts to assessing the rate at which the threat attack moves to its culminating point. Indicators of culmination may be physical fatigue of Soldiers, a breakdown in C2 capability, difficulty providing logistics support, or the increasing time required to reorganize small units to attack. When that culmination is achieved, friendly forces counterattack before the threat has a chance to transition to a hasty defense.

Creating Depth

8-48. Depth in the defense is the key to forcing the threat to culminate. The urban defense cannot allow itself to be penetrated nor permit forward elements to be destroyed. The defense is designed with the greatest depth possible. Defending forces weaken the threat to the fullest extent possible by attack from each position but not permit themselves to be destroyed by fires or close assault. Instead, as threat combat power builds up against individual positions, the use of mission orders permits subordinate leaders to disengage on their own initiative and move on preplanned routes to subsequent positions. Positions are designed to be mutually supporting—withdrawing from one position to a subsequent one while supporting positions cover by fires. The attacker is constantly forced to deploy and reorganize without being able to achieve decisive effects against the defender.

Executing an Effective Obstacle Plan

8-49. Obstacles in the urban defense are designed to break up the threat's combined arms capability. Separating dismounted forces from mounted forces disrupts the cohesion of the attacker and reduces his combat power. It also exposes his individual elements to the effects of asymmetric counterattack. The leading threat dismounted force can be effectively counterattacked by a friendly combined arms element while the threat armored force remains vulnerable to antiarmor attack by dismounted forces.

Conducting Coordinated Counterattacks

8-50. The counterattack is one of the key actions of the urban defense. However, commanders do not counterattack unless there is a reasonable chance of success. As the attacker moves into the depth of the urban area, his forces become fatigued, attrited, and increasingly disorganized. He likely also creates an increasingly long and exposed flank. At all levels, forces defending in urban terrain look for opportunities to counterattack. As the offensive force reaches the culmination point where it can no longer continue to attack with the available forces, the defensive commander executes a planned and coordinated counterattack. The counterattack aims to regain the initiative and to make the threat fight in multiple directions. Infiltration using superior knowledge of the terrain (including supersurface and subsurface capabilities) and the skillful use of stay-behind forces permits attacking the threat throughout the depth of his formations. Small-scale counterattacks focus on C2 and sustainment capabilities. These counterattacks can set the conditions for a deliberate attack leading to the ultimate destruction of the attacking threat force.

CONSOLIDATE

8-51. Consolidation is as important to urban defensive operations as to offensive operations. Many of the same consolidation considerations for the urban offense apply equally well to the defense. Commanders should reinforce or reposition maneuver forces and fire support assets on the urban battlefield based on weaknesses uncovered during rehearsals and opportunities discovered during actual execution. While maximizing the many advantages of the urban defense, the commander of the urban defense aggressively seeks ways of attriting and weakening enemy forces before they enter the urban area and close combat is initiated. Commanders should combine the static and mobile elements of the defense to strengthen their positions in relation to the enemy while seeking every opportunity to transition to urban offensive operations. As in urban offensive operations, commanders should conduct any necessary reorganization actions that they were unable to accomplish during execution.

TRANSITION

8-52. Transitions in urban defensive operations occur at all levels. As with offensive operations, commanders of major operations should address which units are assigned to continue to operate in the area after defensive operations have ceased. In defensive UO, this task is not as challenging as an occupation mission during urban offensive operations. The psychology of troops defending an urban area differs from those attacking into it. Defending forces become accustomed to the environment, having experience in the environment before combat. In terms of training, it is easier for follow-on missions to be assigned to a unit that has successfully defended the urban area. This course of action takes advantage of the defending unit's experience in the area and its relationships with other agencies—agencies that were operating alongside the units before and possibly during the defense. In defensive operations, regardless of the civilians' attitudes, policies regarding that population are established before the successful defense, and the command likely has experience executing operations with civil authorities and other agencies. Thus, these relationships are neither new nor as significant an issue as in offensive operations. Therefore, commanders must be prepared to execute various stability or civil support operations or use a successful defense to springboard into more decisive offensive operations elsewhere in the commanders' AO.

Transition Emphasis to Stability or Civil Support Operations

8-53. At the end of a successful urban defense, operational commanders should generally expect civil authority, control, and jurisdiction to increase. Additionally, the civil population may be anxious to return. Defensive combat will require virtually complete military control of the urban area; however, after the successful defense, a rapid transition will occur from military control to civil or joint military and civil control afterward. Although the full spectrum of Army operations will often be conducted simultaneously in UO, commanders will transition emphasis to stability or civil support operations. Important transition tasks will include demilitarizing munitions, clearing obstacles, and searching for isolated threat pockets of resistance. Conclusion of the defensive operations will also require transition to joint civil-military tasks, such as evaluating structures for safety, restoring essential services, and possibly creating joint law enforcement. Commanders of major operations, using a civil-military operations center and G9, should anticipate these requirements and begin early preparations to ensure a smooth, successful transition.

Transition to Offensive Operations

8-54. Units that have successfully defended the urban area may then transition to offensive operations. A rapid transition to offensive operations will require identification, preparation, and training of units designated to assume missions as the defending units leave the urban area. This preparation emphasizes continuity of policies and relationships already established. A relief in place occurs. The new occupying units provide not only a continuity of policy, but also a continuity of attitude toward the urban area, its population, and its institutions.

Chapter 9

Urban Stability and Civil Support Operations

The fundamental shared aims between stability and civil support operations and any operation conducted in an urban environment are the increased significance and influence of the civil population and nonmilitary organizations. Often, no military victory is to be achieved. The center of gravity for these operations normally cannot be attacked through military means alone; the Army (and the military in general) is often but one tool supporting a larger, civil-focused effort. When conducted, military defeat of an armed adversary is only a phase in a larger campaign, the overall objectives of which will be to bring about a more satisfactory political, social, diplomatic, economic, and military situation than existed prior to the conflict. Without a tightly coordinated civil-military effort, overall success will be difficult or impossible to achieve. Commanders who can understand and cope with the complexities of stability or civil support operations gain insights that directly apply to executing any urban operation. Urban stability or civil support operations may complement urban offensive and defensive operations, or may dominate the overall operation. Army forces may need to conduct offensive and defensive operations to defend themselves or destroy urban threats seeking to prevent the decisive stability or civil support mission. During hostilities, urban stability operations may keep armed conflict from spreading, encourage coalition and alliance partners, and secure the civilian population's support in unstable urban areas (and the surrounding rural areas under their influence). Following hostilities, urban stability operations may provide a secure environment for civil authorities to rebuild. Urban civil support operations can range from transporting, feeding, and sheltering the population made homeless as a result of man-made or natural disasters to conducting civil search and rescue and providing urgent medical care.

PURPOSE OF URBAN STABILITY AND CIVIL SUPPORT OPERATIONS

9-1. Army forces conduct stability and civil support operations to deter war, resolve conflict, promote peace, strengthen democratic processes, retain U.S. influence or access abroad, assist U.S. civil authorities, and support moral and legal imperatives. Stability operations promote and sustain regional and global stability. Civil support operations meet the urgent needs of designated U.S. civilians until civil authorities can accomplish these tasks without military assistance. Nearly every urban operation will involve some type or form of stability operation or civil support operation combined and sequenced with offensive and defensive operations.

CHARACTERISTICS OF URBAN STABILITY AND CIVIL SUPPORT OPERATIONS

9-2. Worldwide urbanization, migration trends from rural to urban areas, and more centralized populations in urban areas increase the chance that Army forces will conduct stability or civil support operations in or near urban areas. Simply put, many people live in urban areas, and their welfare will be the primary reason for conducting these operations. Urban areas that serve as economic and government centers (the ideal location for U.S. and allied embassies) are often the focal point for many threat activities. Therefore, Army forces may need to conduct stability operations in these foreign cities to counter those threats. Additionally, urban areas may contain the resources and infrastructure to support both types of operations, regardless of whether the overall focus is in urban or rural areas. Repairing or restoring the infrastructure may be a critical task in accomplishing either type of operation. Supported governmental and nongovernmental agencies are not as logistically self-sufficient as the Army. As such, these agencies will often need to center their operations in and around urban areas to use the area's infrastructure to support themselves and their objectives. These agencies may require military protection to accomplish their missions. Figure 9-1 lists some defining characteristics of these wide-ranging operations.

- Long or Short Duration
- Joint and Interagency
- Unilateral or Multinational
- Increased Civil-Military and Legal Considerations
- Greater Potential for Ambiguity
- Increased Constraints Necessitating More Restrictive ROE
- Amplified Need for Cultural and Political Sensitivity

Figure 9-1. Characteristics of stability and reconstruction and support operations

9-3. Stability and civil support operations are diverse, varied in duration, and unilateral or multinational. Stability operations are conducted overseas as part of a campaign or major operation, while civil support operations are conducted within the United States as part of homeland security. Like all urban operations (UO), they are usually joint. Unlike urban offensive and defensive operations, they are more often interagency operations and require more restrictive rules of engagement (ROE). The multiplicity of actors involved usually increases the scope and scale of required coordination and communication. In urban stability or civil support operations, adverse conditions arising from natural or man-made disasters or other endemic conditions—such as human suffering, disease, violations of human rights, or privation—will significantly modify the urban environment. Unresolved political issues and tenuous agreements, difficulties discriminating combatants from noncombatants or between parties of a dispute, and the absence of basic law and order all serve to complicate an already complex and uncertain environment. Civil-military and legal considerations take on added significance in all urban operations (see Civil-Military Operations and Legal Support in Chapters 5 and 10 respectively), but even more so in urban stability and civil support operations. Finally, recognizing, defining, and achieving the desired end state is often more difficult than in offensive and defensive operations.

9-4. Overall, commanders of major operations involving urban stability and civil support operations should not expect clear guidance. They must learn, adapt, and live with ambiguity. They cannot expect to operate in a political vacuum (even commanders at the tactical level); rather, they should expect to work alongside both governmental and nongovernmental leaders and organizations. Commanders conducting UO should not expect an easily identifiable enemy located across a clearly demarcated line. In fact in many peace operations, they and their Soldiers must resist the need to have an enemy—difficult at best when one side or another (or both) may be sniping at them. Commanders should also expect changing and additional missions and tasks, without being allowed to use every means at their disposal to carry out those missions. Many tasks required may be ones for which their units have never, or rarely, trained. Finally, commanders should expect to show restraint with a keen sensitivity to political considerations and to alien cultures, either or both of which they might find confusing or even repugnant.

URBAN STABILITY AND CIVIL SUPPORT OPERATIONS AND BATTLEFIELD ORGANIZATION

9-5. Each urban stability operation or civil support operation will be distinct from one another. These operations differ even more when applied to a specific urban area. Due to the complexity of the environment, commanders must carefully arrange their forces and operations according to purpose, time, and space to accomplish the mission. In most UO the terrain, the dense population (military and civilian), and the participating organizations will further complicate this arrangement.

9-6. The support and assistance that Army forces will provide during these operations is only temporary (although it may be of long duration) and this facet requires particular emphasis. Commanders should plan and execute both of these types of operations with that essential consideration always in mind. Eventually, the government and administration—either foreign or domestic—must secure and support their population by themselves. Therefore, commanders must envision and set the conditions that allow for the transition of control and responsibility to shift to legitimate civilian authorities. While commanders may only provide assistance and support based on specific and well planned civilian requests, more often commanders will need to determine requirements in collaboration with competent civilian authorities and agencies or, in some cases, with little or no initial civilian assistance at all. Ultimately, transition planning should occur as an integral part of the overall operational planning and include collaboration with appropriate civilian agencies and organization as early as possible to ensure a seamless transition to civilian control without major setbacks and loss of forward momentum.

STABILITY OPERATIONS

9-7. Stability operations establish, sustain, and exploit security and control over foreign areas, populations, and resources. Urban areas will be decisive to accomplishing many types of stability operations because urban areas are the centers of population, culture, economy, and government. Much of the support provided by Army forces will aim to assist local, regional, or national governments to restore essential services and infrastructure and reestablish civil order and authority. The location of civilian authorities in urban areas will, by necessity, be a dominating factor in accomplishing the mission. As importantly, many stability operations—enforcing peace in Bosnia for example—will require interacting with, influencing, controlling, or protecting all or parts of the civilian population. Assessing, understanding, and gaining the support of civilians in key economic, cultural, or political urban areas may influence surrounding regions (smaller urban areas and the rural countryside) and may be decisive to achieving overall stability objectives.

9-8. Stability operations may involve both coercive and cooperative actions. They are conducted in situations in which legitimate civil authority is unable to provide the necessary security and control for the urban population as a result of—

- Deliberate operations to influence a regime change.
- Offensive or defensive operations or natural disasters resulting in ineffective civil authorities, conditions beyond the capabilities of the foreign urban government, or both.

CIVIL SUPPORT OPERATIONS

9-9. Civil support operations are an integral part of homeland security which aim to protect U.S. citizens and infrastructure from conventional and unconventional threats and mitigate the effects of man-made and natural disasters. (The other part to homeland security is homeland defense.) Army forces conduct civil support operations by providing Army resources, expertise, and capabilities in support of a lead agency—which may be to Department of Defense—to address the consequences of man-made or natural accidents beyond the capabilities of local or regional U.S. civilian authorities. Such a situation could result from a chemical, biological, radiological, nuclear, and high-yield explosive (CBRNE) incident, hurricane, flood, or civil disturbance in a domestic urban area.

CBRNE INCIDENTS

9-10. CBRNE incidents can be disastrous and are of particular concern for both stability and civil support UO. In urban areas, the potential for catastrophic loss of life and property is enormous. The Army categorizes CBRNE incidents separately from other natural and man-made disasters because it has specific expertise with these weapons. CBRNE incidents usually result from a military or terrorist threat (adding a law enforcement dimension to the disaster). Other urban CBRNE considerations include the following:

- Subways and other subsurface areas offer ideal areas for limited chemical or biological attacks. Nuclear attack (and high-yield explosives) can produce tragic results due to the effects of collapsing structures, flying debris, and fires. Dispersion patterns are affected by the urban terrain and are more difficult to predict and monitor. Large-scale incidents may produce hundreds of thousands of casualties, but even a limited attack may require evacuating and screening large numbers of civilians. Requirements for medical support, basic life support, and, if necessary, decontamination may quickly overwhelm the Army force's capabilities even with augmentation.
- Similar to natural disasters, panic and disorder may accompany the CBRNE event. Fleeing civilians may clog elements of the transportation and distribution infrastructure. Physical destruction may also affect other components of the infrastructure of critical and immediate concern, such as energy and administration and human services (water, sanitation, medical, fire fighting, and law enforcement). Because all elements of the infrastructure may be affected, the overall recovery time may be lengthened and the effects broadened to include much of the surrounding area. The effects of a single urban CBRNE event potentially could be felt nationally or globally.

DECISIVE OPERATIONS

9-11. In urban stability operations, decisive operations may take many years and include multiple actions before achieving the desired end state. This particularly applies to the strategic and operational levels. Oppositely, decisive operations involved in an urban civil support operation for mitigating or reducing disease, hunger, privation, and the effects of disasters normally achieve faster results. However, any operation that attacks the underlying cause and seeks to prevent or relieve such conditions will usually take longer. In urban areas, establishing law and order to protect critical infrastructure and the inhabitants from lawlessness and violence is often critical and also the decisive operation.

SHAPING OPERATIONS

9-12. Shaping operations establish and maintain the conditions for executing decisive operations. In urban stability or civil support operations, shaping operations always include information operations (IO) that influence perceptions and maintain legitimacy. Often, various participants, and their potentially divergent missions and methods, are involved. Army commanders must coordinate their planning and efforts (early and continuously) to ensure that their decisive, shaping, or sustaining operations are not working against other agencies' efforts and operations—agencies that may have the lead role in the operation. Thus, a critical shaping operation may be to establish the coordination to help develop a common purpose and direction among agencies, particularly those that may experience continuous personnel turnover during the conduct of a lengthy operation. In some instances and with some organizations and agencies, particularly nongovernmental organizations (NGOs), genuine unity of effort may not be achievable; however, recognizing the differences in aims and goals will allow Army commanders to conduct operations with less friction. Commanders should actively request and include NGOs and appropriate governmental agencies in mission readiness exercises or any other training for stability or civil support operations.

SUSTAINING OPERATIONS

9-13. Sustaining operations enable decisive and shaping operations and include logistics, rear area and base security, movement control, terrain management, and infrastructure development. Sustainment bases, especially those located in urban areas, become an attractive target for hostile civilians; therefore, commanders must actively and aggressively protect these bases as well as lines of communications (see Chapter 10).

CONSIDERATIONS FOR URBAN STABILITY AND CIVIL SUPPORT OPERATIONS

In wars of intervention the essentials are to secure a general who is both a statesman and a soldier; to have clear stipulations with the allies as to the part to be taken by each in the principal operations; finally, to agree upon an objective point which shall be in harmony with the common interests.

Lieutenant General Antoine-Henri,
Baron de Jomini

9-14. The urban operational framework (understand, shape, engage, consolidate, and transition) provides a structure for developing considerations unique to urban stability and civil support operations. Many considerations presented in urban offensive and defensive operations apply to urban stability and civil support operations, particularly those that address how to understand the urban and overall operational environment. Because the situations in which stability and civil support operations normally occur share strong similarities with any urban environment, many of these considerations are closely linked to the urban fundamentals presented in Chapter 5. Taken together, commanders will often find them useful in conducting UO throughout full spectrum operations. Appendix C has a historical example of how to apply the urban operational framework to a stability operation.

UNDERSTAND

9-15. In urban stability or civil support operations, commanders must carefully assess and understand the political dimension of the operational environment, as well as their role and the media's part in managing information. These operations are inherently tied to the exercise of diplomatic power. All operations in urban areas are often the focus of the media and thus gain considerable public and political attention. Therefore, military objectives in urban stability or civil support operations are more directly linked with political objectives. The relationship between the levels of war—strategic, operational, and tactical—is often closer than in urban offensive and defensive operations. Military objectives must be carefully nested within political objectives. Commanders must ensure that the ways and means to accomplish their objectives, to include security and force protection measures, will hold up to media scrutiny and are appropriate for the situation and environment. All levels of command must understand the link between political and military objectives, to include a basic understanding at the Soldier level. One uncoordinated, undisciplined, or inappropriate action, even at the lowest level, could negate months or years of previous, disciplined effort. Commanders must balance security and force protection measures with mission accomplishment. Ineffective measures can put Soldiers at too great a risk and jeopardize the mission. Conversely, overly stringent measures may make it difficult for forces to interact with the population closely—essential in many of these operations. Finally, commanders will need a thorough assessment of the governmental and nongovernmental organizations and agencies that will be operating in or near urban areas that fall within their area of operations.

Political and Military Objectives

9-16. Commanders must translate political objectives into military objectives that are clear and achievable (clear tasks and purposes) and can lead to the desired end state. (In most stability operations, defeat of a threat is not the ultimate desired end state.) Political objectives may be vague making it difficult for commanders to conduct their mission analysis. This applies to tactical- and even operational-level commanders, unskilled at higher level, strategic political-military assessments. Each type of stability operation or civil support operation is distinct, often unfamiliar to the executing unit, and unique to the specific situation. These factors often make it difficult to confidently determine the specific tasks that will lead to mission success. Therefore, commanders must also establish measures of effectiveness that aid in understanding and evaluating progress and help gauge mission accomplishment.

> ***Note***: Commanders should consult the US Agency for International Development's (USAID) Field Operations Guide for Disaster Assessment and Response when conducting their assessments and developing measures of effectiveness for many urban relief operations.

9-17. These measures should be quantifiable (in some circumstances, a qualitative assessment may be most appropriate) and link cause with effect. They help determine the changes required and are essential to the assessment cycle required for urban stability or civil support operations. In a humanitarian relief operation to aid the starving, commanders could determine that the decisive effort is delivering safe food to the urban area. To judge success or effectiveness, they could determine that the appropriate measure is the number of food trucks dispatched daily to each distribution site; the more trucks, the more effective the efforts. However, this measure must correlate with the overarching measure of effectiveness: decline in the mortality rate. If no significant decrease in deaths due to starvation occurs, they may need to reassess and modify the tasks or measures of effectiveness. A better measure may be to track the amount of food consumed by those in need instead of simply counting the number of trucks dispatched. Areas around which measures of effectiveness can be formed for many stability or civil support operations (including the example above) and which will help return most societies to some degree of normalcy and self-sufficiency include:

- Restoring law and order.
- Decreasing morbidity and mortality rates.
- Securing safe food and water.
- Restoring critical infrastructure.
- Resettling the population.
- Reestablishing economic activity.

Commanders can often develop measures of effectiveness to address these areas in terms of establishing or restoring security and providing logistics. (Figure 9-2 provides example measures of effectiveness from the strategic to tactical levels for a possible stability operation.)

9-18. Political objectives are fluid and modified in response to new domestic and international events or circumstances. Thus, assessment is continuous, and commanders must adjust their own objectives and subsequent missions accordingly. In urban stability or civil support operations, commanders often develop military objectives that support or align with the objectives of another agency that has overall responsibility for the urban operation. In this supporting role, commanders may receive numerous requests for Soldier and materiel assistance from the supported agency and other supporting agencies operating in the urban area (to include elements of the urban population). With such unclear lines of authority and areas of responsibility, they must ensure that the tasks, missions, or requested Army resources fall clearly in the intended scope and purpose of the Army's participation in the operation. They must not develop or execute missions based on inadequate or false assumptions, misinterpreted intent, or well-meaning but erroneously interpreted laws or regulations by any organization, to include even the lead agency. When missions appear outside their scope, commanders must quickly relay their assessment to their higher headquarters for immediate resolution. The commander's goal is not to limit or slow military participation but to contribute as intended and in consonance with political objectives and the law.

Security and Force Protection Measures

9-19. Commanders must plan for and continually assess the security of their forces operating in an urban area as well as constantly review protection measures. Establishing a robust intelligence—particularly human intelligence (HUMINT)—network that can determine the intentions and capabilities of the threat and the urban populace is the basis for establishing force protection for Army forces operating in the urban environment. However, many such operations, particularly stability operations, require extra time to forge a lasting change. Over time, and particularly in peacetime when objectives center on helping others and avoiding violence, even the complex urban environment may seem benign. Without continued, aggressive command emphasis, Soldiers may become lulled into complacency. Also, during periods of transition or transfer of authority from one unit or organization to another, departing Soldiers' focus often shifts to redeployment activities and away from force protection concerns. It is usually then that Army forces are most vulnerable to terrorist and insurgent tactics, such as bombings, kidnappings, ambushes, raids, and other forms of urban violence.

Measures of Effectiveness: The Need to Measure Progress

Strategic Level Criteria	Operational Level Criteria	Tactical Level Criteria
Accountable to the American people for defining and measuring progress towards defeating terrorism and meeting national security goals.	Accountable to the strategic level for measuring operational success and providing linkage to strategic goals.	Responsible to the operational level for measuring tactical success and providing linkage to operational goals.
Examples: • Prevention of the insurgency from receiving aid or resources from other international groups. • A functioning national government. • Amount of international support and aid to reconstruction. • Number of nations contributing manpower to coalition forces.	Examples: • Host-nation security forces trained and equipped. • Denial of the merging of insurgent forces with terrorist groups. • Amount and distribution of— • Electricity. • Liquid propane gas. • Gasoline. • Functioning provincial governments.	Examples: • Number of insurgent forces in the AO neutralized. • Reduced attacks on coalition forces in the AO. • Reduced civilian-on-civilian violence in the AO. • Host-nation security force recruitment goals met. • Host-nation security force training goals met. • Number of reliable human intelligence walk-ins. • Amount of unexploded explosive ordnance cleared. • Functioning neighborhood and district advisory councils.

Figure 9-2. Example measures of effectiveness

9-20. Although force protection will not ensure successful urban stability or civil support operations, improper assessment and inadequate force protection measures can cause the operation to fail. In either operation, keeping a neutral attitude toward all elements of the urban population, while maintaining the appropriate defensive posture, enhances security. One threat principle discussed in Chapter 3 was that threats would seek to cause politically unacceptable casualties. An improper threat assessment and a lapse in security at the tactical level could result in casualties. That result could affect strategy by influencing domestic popular support and subsequently national leadership decisions and policy.

9-21. Emphasizing security and force protection measures does not mean isolating Soldiers from contact with the urban population. On the contrary, commanders should balance survivability with mobility according to the factors of METT-TC—mission, enemy, terrain and weather, troops and support available, time available, civil considerations. Survivability measures—such as sandbagging, hardening, or fortifying buildings and installations, particularly where large numbers of Soldiers are fed and billeted—will be necessary and require considerable command attention. On the other hand, mobility operations are essential in preserving freedom of action and denying a threat the opportunity to observe, plan, and attack urban forces. Continual Army presence in the urban area may provide the urban population a sense of security and allow Soldiers to develop a detailed knowledge of the "patterns of life" in their assigned area of operation (AO). Armed with this knowledge, they can detect the absence of the normal or the presence of the abnormal that might indicate a potential threat. Overall, mission degradation and increased risk to the force can result if force protection measures prevent Army forces from conducting prudent missions and establishing an active and capable presence.

Assessment of Security and Force Protection: Belfast, Northern Ireland

Since 1969, Belfast has significantly affected the British military campaign for stabilizing the area. British operations in Belfast illustrate the difficulty of balancing the security and protection of forces with maintaining the stabilizing presence necessary to uphold law and order, minimize violence, and control the urban population.

British successes in protecting Belfast's infrastructure and government facilities from terrorist attacks compelled various terrorist cells, especially the Irish Republican Army and the Provisional Irish Republican Army, to attack more military targets.

At the time, British soldiers and bases presented relatively unprotected targets to these factions, and attacks against them solidified their legitimacy as an "army." In response, British commanders implemented extreme security and force protection measures—from ballistic protection vests and helmets to fortress-like operational bases and large unit patrols. These protection measures successfully decreased the violence against British soldiers in Belfast. However, they also decreased the soldiers' interaction with the population and their ability to stabilize the city. The large patrols protected the soldiers, but also inhibited effective saturation of neighborhoods. These patrols, coupled with fortress-like bases and bulky protective clothing, created an "us-versus-them" mentality among civilians and soldiers.

As force protection increased and stabilizing effects decreased, the terrorists were provided more targets of opportunity among the civilians and infrastructure. British commanders reassessed the situation, identified this "see-saw" effect, and adapted to strike a better balance between force protection and effective presence patrols. For example, British forces switched to four-man patrols to enable greater mobility and wore berets instead of helmets to appear less aggressive.

While the above historical example illustrates a critical aspect of stability and civil support operations, it subtly reveals another important lesson for Army commanders. Many of our future coalition partners and assisting governmental and nongovernmental organizations will bring key insights to UO (particularly dealing with

> urban societies) based on their historical and operational experience. Army commanders and their staffs should seek out and remain open to others' unique understandings that are based on hard-earned and well-analyzed lessons learned.

Participating Organizations and Agencies

9-22. Across the spectrum of urban operations, but more so in these operations, numerous NGOs may be involved in relieving adverse humanitarian conditions. Dense populations and infrastructure make an urban area a likely headquarters location for them. (In 1994 during OPERATION UPHOLD DEMOCRACY, for example, over 400 civilian agencies and relief organizations were operating in Haiti.) Therefore, commanders should assess all significant NGOs and governmental agencies operating (or likely to operate) in or near the urban area to include their—

- Functions, purposes, or agendas.
- Known headquarters and operating locations.
- Leadership or senior points of contact (including telephone numbers).
- Communications capabilities.
- Potential as a source for critical information.
- Financial abilities and constraints.
- Logistic resources: transportation, energy and fuel, food and water, clothing and shelter, and emergency medical and health care services.
- Law enforcement, fire fighting, and search and rescue capabilities.
- Refugee services.
- Engineering and construction capabilities.
- Other unique capabilities or expertise.
- Previous military, multinational, and interagency coordination experience and training.
- Rapport with the urban population.
- Relationship with the media.
- Biases or prejudices (especially towards participating U.S. or coalition forces, other civilian organizations, or elements of the urban society).

Commanders can then seek to determine the resources and capabilities that these organizations may bring and the possible problem areas to include resources or assistance they will likely need or request from Army forces. These organizations will be critical to meeting the population's immediate needs and minimizing the effects of collateral damage or disaster. However, commanders should consider whether a close relationship with any of these organizations will compromise the organization's appearance of neutrality (particularly threat perceptions) and adversely affect their ability to assist the population.

SHAPE

9-23. Commanders conduct many activities to shape the conditions for successful decisive operations. In urban stability and civil support operations, two activities rise to the forefront of importance: aggressive IO and security operations.

Vigorous Information Operations

9-24. IO, particularly psychological operations (PSYOP) and the related activities of civil affairs (CA) and public affairs, are essential to shape the urban environment for the successful conduct of stability or civil support operations. Vigorous IO can influence the perceptions, decisions, and will of the threat, the urban population, and other groups in support of the commander's mission. IO objectives are translated to IO tasks that are then executed to create the commander's desired effects in shaping the battlefield. These operations can isolate an urban threat from his sources of support; neutralize hostile urban populations or gain the support of neutral populations; and mitigate the effects of threat IO, misinformation, rumors, confusion, and apprehension. Developing effective measures of effectiveness is essential to a good IO campaign strategy. One of the most valuable methods for obtaining data for use in this process is face-to-face encounters with targeted audiences by unit patrols and HUMINT, PSYOP, and CA teams. A valuable technique may be to conduct periodic, unbiased surveys or opinion polls of the civilian population to determine changes in their perceptions and attitudes.

Security Operations

Protecting Civilians and Critical Infrastructure

9-25. Security for NGOs and civilians may also be an important shaping operation, particularly for stability and civil support operations. Commanders may need to provide security to civil agencies and NGOs located near or operating in the urban area so that these agencies can focus their relief efforts directly to the emergency. Commanders may also need to protect the urban population and critical infrastructure to maintain law and order if the urban area's security or police forces are nonexistent or incapacitated, or the urban area's security situation has undergone drastic change (such as the result of a natural disaster) and security or police forces require additional augmentation. (See also the discussion of the legal aspects of Civilians Accompanying the Force in Chapter 10.)

Preserving Resources

9-26. Just as forces may be at risk during urban stability or civil support operations, so may their resources. In urban areas of great need, supplies and equipment are extremely valuable. Criminal elements, insurgent forces, and people in need may try to steal weapons, ammunition, food, construction material, medical supplies, and fuel. Protecting these resources may become a critical shaping operation. Otherwise, Army forces and supporting agencies may lack the resources to accomplish their primary objectives or overall mission.

Prioritize Resources and Efforts

9-27. During UO, commanders will always face limited resources with which to shape the battlefield, conduct their decisive operations, and accomplish their objectives. They must continually prioritize, allocate, and apply those resources to achieve the desired end state. To this end, they may develop an order of merit list for proposed projects and constantly update it over time. To some degree, the local urban population will usually need to be protected and sustained by Army forces. Hence, commanders must tailor their objectives and shape their operations to achieve the greatest good for the largest number. Commanders first apply the urban fundamental of preserving critical infrastructure to reduce the disruption to the residents' health and welfare. Second, they apply the urban fundamental of restoring essential services, which includes prioritizing their efforts to provide vital services for the greatest number of inhabitants possible. In operations that include efforts to alleviate human suffering, the criticism for any participating organization is likely to be there is not enough being done or Army forces are not being responsive enough. Therefore, commanders must develop clear measures of effectiveness not only to determine necessary improvements to operational plans but also to demonstrate their Soldiers' hard work and sacrifice and U.S. commitment to the operation.

Improve the Urban Economy

9-28. When conducting reconstruction and infrastructure repair, commanders should consider using such activities to simultaneously improve the urban economy. Hiring local civilians and organizations to do reconstruction work helps satisfies urban job requirements, may inspire critical elements of the urban society to assume responsibility for the success or failure of urban restoration efforts, and may potentially reduce threat influence by diminishing their civilian sources of aid. Hiring indigenous personnel for short-term projects does not replace the need for long-term economic planning and the development of stable jobs. The overall reconstruction effort must be guided by the commander's vision of the end state. However, the commander's guidance and intent should be broad and expansive enough to allow responsive decision making by subordinate commanders based on their analysis of the urban population's needs in their assigned AO.

ENGAGE

> *If there is any lesson to be derived from the work of the regular troops in San Francisco, it is that nothing can take the place of training and discipline, and that self-control and patience are as important as courage.*
>
> Brigadier General Frederick Funston
> commenting on the Army's assistance following
> the 1906 San Francisco earthquake and fire

9-29. The focus of the Army is warfighting. Therefore, when Army commanders conduct many urban stability and civil support operations, they must adjust their concept of success. Commanders will most often find themselves in a supporting role and less often responsible for conducting decisive operations. They must accept this supporting function and capitalize on the professional values they have instilled in each Soldier, particularly the sense of duty to do what needs to be done despite difficulty, danger, and personal hardship. Commanders must also put accomplishing the overall mission ahead of individual desires to take the lead—desire often fulfilled by being the supported rather than supporting commander. In many stability and civil support operations, success may be described as settlement and compromise rather than victory. Yet, the Army's professionalism and values—combined with inherent adaptability, aggressive coordination, perseverance, and reasonable restraint—will allow Army forces to engage purposefully and dominate during complex urban stability or civil support operations.

Adaptability

9-30. Adaptability is particularly critical to urban stability or civil support operations because these operations relentlessly present complex challenges to commanders for which no prescribed solutions exist. Commanders often lack the experience and training that provide the basis for creating the unique solutions required for these operations. Since the primary purpose for the Army is to fight and win the nation's wars, the challenge then is to adapt urban warfighting skills to the unique stability or civil support situation.

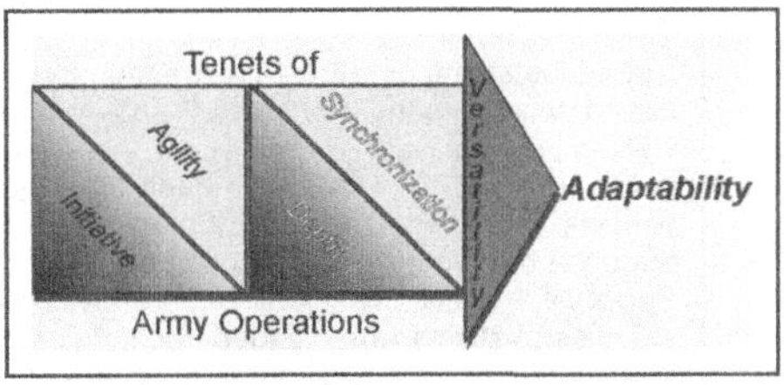

Figure 9-3. Adaptability

9-31. Doctrine (joint and Army) provides an inherent cohesion among the leaders of the Army and other services. Still, Army commanders conducting urban stability or civil support operations will often work with and support other agencies that have dissimilar purposes, methods, and professional languages. Army commanders must then capitalize on three of the five doctrinal tenets of Army operations: initiative, agility, and versatility (see figure 9-3 and FM 3-0). Commanders must bend as each situation and the urban environment demand without losing their orientation. They must thoroughly embrace the mission command philosophy of command and control addressed in FM 6-0 and Chapter 5 to encourage and allow subordinates to exercise creative and critical thinking required for planning and executing these UO. Commanders must also be alert to and recognize "good ideas," and effective tactics, techniques, and procedures, regardless of their source—other services, coalition partners, governmental and nongovernmental organizations, and even the threat—and adapt them for their own purposes in UO.

9-32. Adaptability also springs from detailed planning that carefully considers and realistically accounts for the extent of stability and civil support operations. Although no plan can account for every contingency and completely eliminate the unexpected, good plans—which include detailed civil considerations—provide platforms from which to adjust course more readily. Adequate planning allows commanders not only to react more effectively, but also to be forward-thinking and take actions that favorably steer the course of events.

Aggressive Coordination and Synchronization

9-33. In urban stability and civil support operations, the increased number of participants (both military and nonmilitary) and divergent missions and methods create a significant coordination and synchronization challenge. Significant potential for duplicated effort and working at cross-purposes exists. The success of UO often depends on establishing a successful working relationship with all groups operating in the urban area. The absence of unity of command among civil and military organizations does not prevent commanders from influencing other participants not under his direct command through persuasion, good leadership, and innovative ideas.

9-34. Commanders may consider establishing, as necessary, separate organizations for combat operations and for stability operations or civil support operations to increase coordination and enhance local, NGO, and international support. Further, aligning the unit or subordinate units with NGOs may contribute to establishing popular legitimacy for the operation and place greater pressure on threat forces. In some instances, commanders may consider organizing part of their staff around government, administrative, and infrastructure functions that mirror the urban area in which their forces are operating. Development of a mirror urban area organization may give greater legitimacy to the urban government or administration and ease transition of responsibility once the end state is achieved. Commanders must be mindful, however, that local groups seen as allying themselves with Army or coalition authorities will likely experience pressure to demonstrate their independence as established dates for redeployment or other critical events approach. In some instances that demonstration of independence may be violent.

Civil Support and Coordination with Civilian Authorities: Los Angeles – 1992

During the spring of 1992, Soldiers from the 40th Infantry Division, California National Guard were among the forces deployed to Los Angeles County to assist the California Highway Patrol, Los Angeles County Sheriffs, and civilian law enforcement. They worked to quell the riots that were sparked by the "not guilty" verdicts concerning four police officers who, following a lengthy high-speed chase through Los Angeles, were accused of brutally beating Rodney King.

Successful accomplishment of this civil support operation was attributed to the exercise of strong Army leadership and judgment at lower tactical levels, particularly among the unit's noncommissioned officers. An essential component of combat power, it was especially critical in executing noncontiguous and decentralized operations in the compartmented terrain of Los Angeles. As important, however, was the clear understanding that Army forces were to support civilian law enforcement—and not the other way around. The 40th Infantry Division aligned its area of operations with local law enforcement boundaries and relied heavily on police recommendations for the level at which Soldiers be armed (the need for magazines to be locked in weapons or rounds chambered).

One incident emphasized the need for coordination of command and control measures with civilian agencies even at the lowest tactical levels. To civilian law enforcement and Army forces, the command "Cover me" was interpreted the same: be prepared to shoot if necessary. However, when a police officer responding to a complaint of domestic abuse issued that command to an accompanying squad of Marines, they responded by immediately providing a supporting base of fire that narrowly missed some children at home. However, the Marines responded as they had been trained. This command meant something entirely different to them than for Army Soldiers and civilian law enforcement. Again, coordination at all levels is critical to the success of the operation (see also the vignette in Appendix B).

9-35. In the constraints imposed by METT-TC and operations security (OPSEC), commanders should seek to coordinate all tactical stability operations with other agencies and forces that share the urban environment. Importantly, they should seek to coordinate appropriate information and intelligence sharing with participating organizations. Commanders must strive to overcome difficulties, such as mutual suspicion, different values and motivations, and varying methods of organization and execution. Frequently, they must initiate cooperative efforts with participating civilian agencies and determine where their objectives and plans complement or conflict with those agencies. Commanders can then match Army force capabilities to the needs of the supported agencies. In situations leading to many urban civil support operations, confusion may initially make it difficult to ascertain specific priority requirements. Reconnaissance and liaison elements—heavily weighted with CA, engineers, and health support personnel—may need to be deployed immediately to determine what type of support Army forces provide. Overall, aggressive coordination will foster trust and make unity of effort possible in urban stability or civil support operations where unity of command is difficult or impossible to achieve.

Perseverance

9-36. The society is a major factor responsible for increasing the overall duration of urban operations. This particularly applies to urban stability operations where success often depends on changing people's fundamental beliefs and subsequent actions. Modifying behavior requires influence, sometimes with coercion or control, and perseverance. The urban population often must be convinced or persuaded to accept change. Necessary change may take as long as or longer than the evolution of the conflict. Decades of problems and their consequences cannot be immediately corrected. Frequently, the affected segments of the urban society must see that change is lasting and basic problems are being effectively addressed.

9-37. In most stability operations, success will not occur unless the host nation, not Army forces, ultimately prevails. The host urban administration must address the underlying problem or revise its policies toward the disaffected portions of the urban population. Otherwise, apparent successes will be short lived. The UO fundamental of understanding the human dimension is of paramount importance in applying this consideration. After all Army forces, particularly commanders and staff of major operations, understand the society's history and culture, they can begin to accurately identify the problem, understand root causes, quickly engage and assist key civilian leadership, and, overall, plan and execute successful Army UO.

Reasonable Restraint

9-38. Unlike offensive and defensive operations where commanders seek to apply overwhelming combat power at decisive points, restraint is more essential to success in urban stability and civil support operations. It involves employing combat power selectively, discriminately, and precisely (yet still at decisive points) in accordance with assigned missions and prescribed legal and policy limitations. Similar to the UO fundamentals of minimizing collateral damage and preserving critical infrastructure, restraint entails restrictions on using force. Commanders of major operations should issue or supplement ROE to guide the tactical application of combat power. Excessively or arbitrarily using force is never justified or tolerated by Army forces. Even unintentionally injuring or killing inhabitants and inadvertently destroying their property and infrastructure lessens legitimacy and the urban population's sympathy and support. Collateral damage may even cause some inhabitants to become hostile. In urban stability or civil support operations, even force against a violent opponent is minimized. Undue force often leads to commanders applying ever-increasing force to achieve the same results.

9-39. Although restraint is essential, Army forces, primarily during urban stability operations, must always be capable of decisive (albeit in some circumstances, limited) combat operations. This is in accordance with the UO fundamental of maintaining a close combat capability. This capability must be present and visible, yet displayed in a nonthreatening manner. A commander's intent normally includes demonstrating strength and resolve without provoking an unintended response. Army forces must be capable of moving quickly through the urban area and available on short notice. When necessary, Army forces must be prepared to apply combat power rapidly, forcefully, and decisively to prevent, end, or deter urban confrontations. Keeping this deterrent viable requires readiness, constant training, and rehearsals. It also requires active reconnaissance, superb OPSEC, a combined arms team, and timely and accurate intelligence.

Restraint: An Najaf, Iraq – 2003

On 31 March 2003, during OPERATION IRAQI FREEDOM, The 101st Airborne (Air Assault) Division transitioned from isolating An Najaf—among the holiest places for Shia Muslims in Iraq and the home of one of their leading holy men, the Grand Ayatollah Ali Hussein Sistani—to an attack to clear Iraqi fighters from the town. The decision to attack was made in response to increasing attacks against 101st forces situated near An Najaf and due to alarming reports that the Fedayeen were killing Iraqi family members to force adult males to fight coalition forces. Since An Najaf had an airfield, control of the town would also allow the 101st to obtain needed hardstands for their aircraft.

Ultimately, the division used two brigades to attack and clear the town. The 2nd Brigade Combat Team (BCT) attacked from the north and the 1st BCT—responsible for the initial penetration—attacked from the southwest. Using tanks, infantry fighting vehicles, and light infantry, the commander of the 1st BCT formed effective combined arms teams supported by artillery and air that successfully fought their way into the city.

On 3 April 2003, Soldiers of TF 2-327 Infantry turned a street corner to face a group of civilian men blocking their way and shouting in Arabic, "God is great." The crowd quickly grew into hundreds as they mistakenly thought the Soldiers were trying to capture Sistani and seize the Imam Ali Mosque. Someone in the crowd threw a rock at the Soldiers, which started a hailstorm of rocks; even the battalion commander was hit on the head, chest, and the corner of his sunglasses.

Based on information from a Free-Iraqi Fighter accompanying the battalion and his own observations during the attack, the commander of TF 2-327 believed that most of the people in An Najaf neither wanted to fight nor obstruct his unit's efforts. Consequently, he ordered his Soldiers to position themselves on one knee, smile, and point their weapons toward the ground. At this gesture, many of the Iraqis backed off and sat down, enabling the commander to identify the true troublemakers. He identified eight. In case these agitators were to produce weapons and start to shoot, the commander wanted to make sure that the remainder of the crowd would know from where the shooting would originate. Next, he ordered his Soldiers to withdraw to allow the tension to subside. With his own rifle pointed toward the ground, the battalion commander bowed to the crowd and led his Soldiers away. When tempers had calmed, the Grand Ayatollah Sistani issued a decree (fatwa) calling on the people of Najaf to welcome the American Soldiers.

The attack on An Najaf by the 101st provides an excellent example of well-led Soldiers capable of understanding cultural differences and possessing the discipline and mental flexibility to transition from the aggressive mindset required for high-intensity urban combat to the restraint essential for the initiation of stability operations. In his weekly radio address, President George W. Bush commented: "This gesture of respect helped defuse a dangerous situation and made our peaceful intentions clear."

CONSOLIDATE

9-40. As urban operations will often be full spectrum, many of the consolidation activities necessary to secure gains in urban offensive and defensive operations will be applicable to urban stability and civil support operations. However, the greatest obstacles to attaining strategic objectives will come after major urban combat operations. Therefore, emphasis will appropriately shift from actions to ensure the defeat of threat forces to those measures that address the needs of the urban population, manage their perceptions, and allow responsibility to shift from Army forces to legitimate indigenous civilian control (or the intermediate step to other military forces, governmental agencies, and organizations).

Continued Civilian and Infrastructure Protection

9-41. Following urban offensive or defensive operations, there will likely be a need to secure and protect the civilian population and much of the civilian infrastructure from the civilians themselves. After having minimized collateral damage and preserved critical infrastructure, commanders must implement measures to preclude looting and destruction of critical and essential infrastructure by the urban population and civilian-on-civilian violence. This may be as relatively simple as allowing the urban police force to return to work or may be as difficult as hiring, vetting, and training an indigenous police force. In the latter case commanders may need to determine—

- Number and operability of police stations.
- Responsibility for recruiting, hiring, training, and equipping the urban security or police force. The accountable unit, organization, or agency will need to consider a vetting process, suitable salaries and wages, and appropriate training standards.
- The appropriate responses toward those civilians who threaten, oppose, or harm the new police force.

9-42. Alternatively, civilian security firms (from inside or outside the urban area or country) can provide supplemental protection until indigenous police forces are fully functional. Further, the commander of the major urban operation may plan to manage expected instability primarily with Army forces. Often, this will require larger numbers of infantry, military police, and dismounted forces. Other populace and resource control measures such as curfews may also assist in civilian (including NGOs) and infrastructure protection. Previous shaping operations aimed at improving the local economy can also assist in this regard.

Resolute Legitimacy

9-43. Closely linked to the restraint described previously is legitimacy or the proper exercise of authority for reasonable purposes. Achieving or maintaining legitimacy during urban stability or civil support operations is essential in gaining and maintaining the support of the urban population. Commanders can ensure legitimacy by building consent among the population, projecting a credible force, and appropriately using that force. Perceptions play a key role in legitimacy, and skillful IO can shape perceptions. Commanders must send messages that are consistent with the actions of their forces. Generally, the urban population will accept violence for proper purposes if that force is used impartially. Perceptions that force is excessive or that certain groups are being favored over others can erode legitimacy and generate resentment, resistance, and, in some situations, violent acts of revenge.

9-44. In stability operations against an elusive insurgent threat, commanders may need to explain to urban residents why damage was necessary, apologize, or make near-instant restitution for some unsuccessful Army operations that may have been planned based on inaccurate or incomplete intelligence. (Soldiers may even make it a point to thank homeowners for allowing the search of their homes during cordon and search operations.) During urban operations, a single Soldier's misbehavior can significantly degrade a commander's ability to project an image of impartiality and legitimacy. Fortunately, disciplined Soldiers can contribute immeasurably to gaining and maintaining legitimacy, mitigating ill will, or otherwise winning the urban population's trust and confidence. In stability operations, the greater fight will often be to win the battle of perceptions and ideas instead of one to seize terrain and triumph over an enemy. Inconsistencies in message and behavior will provide threats with raw material for their propaganda and precipitate doubt in the minds of the urban populace who might otherwise support Army objectives.

TRANSITION

9-45. Commanders of major operations are the focal point for synchronizing tactical stability operations or civil support operations with strategic diplomatic and political issues. They are also the critical links between national intelligence resources and the tactical commander. Because strategic, diplomatic, and political changes can quickly transition the type of urban operation, they must keep subordinate tactical commanders abreast of changes in intelligence, policy, and higher decisions. The potential to rapidly transition to urban combat operations emphasizes the need to maintain the capability to conduct close, urban combat. Failure to recognize changes and transition points may lead to UO that do not support the attainment of the overall objective and needlessly use resources, particularly Soldiers' lives. Therefore, Army forces on the ground in an urban stability operation must be more aware of the strategic environment than the threat and the civilian population, each of whom will have their own means of monitoring the national and international situation.

Legitimate and Capable Civilian Control

I met immediately with the [special operations forces liaison officer] to discuss how we would establish a new government that would be able to make instant progress. We decided to select a city council with the following departments: commerce, public works, social services, health, emergency services, education, public relations, and agriculture. A city council director would lead the council, and each department would also have a director in charge of its respective services. A critical step in selecting the city council was ensuring that there was a balance on the council between Altun Kupri's three major ethnic groups. An ethnic imbalance on the council or perceived favoritism would have decreased the legitimacy of the Coalition forces' efforts in the city and served as fuel for the fire of competition for power between the Turkomen and Kurds.

Captain Jeffrey B. Van Sickle
"Stability Operations in Northern Iraq"

9-46. Commanders must maintain or enhance the credibility and legitimacy of the government and police of the urban area and, in the case of stability operations, of the host nation's military forces operating there. In accordance with the urban fundamental of transitioning control, urban commanders must conclude UO quickly and successfully, often to use assets elsewhere in their area of operations. This entails returning the control of the urban area back to civilian responsibility as soon as feasible. The host nation's military and the urban area's leadership and police are integrated into all aspects of the urban stability or civil support operations to maintain their legitimacy. They must be allowed (or influenced) to take the lead in developing and implementing solutions to their own problems. This requires commanders to transition from "leading from the front" to "leading from behind" in an advisory and assistance role. Effective transition to civilian control and responsibility requires commanders at all echelons who understand the basic operation of civil governments and the administration and management of key urban infrastructure.

9-47. If the host nation's leadership, military, and police are not up to the task, commanders can take steps to increase its capabilities through training, advice, and assistance by CA units or by other nongovernmental or governmental organizations and agencies. Sometimes, new leadership and a restructured police force may be required, particularly when corrupt and no longer trusted by the population. This candid assessment of the urban leadership's ability to govern, protect, and support itself is made early in the planning process. Only then can commanders ensure that resources and a well thought-out and coordinated plan (particularly with civilian organizations and agencies) are available for a speedy transition. IO will be paramount in these instances to ensure that the urban population sees the training and rebuilding process itself as legitimate. Throughout urban stability or civil support operations, commanders shape the conditions to successfully hand over all activities to urban civilian authorities.

Longer-Term Commitment

Maintain the Focus

9-48. Many stability operations often require perseverance and a longer-term U.S. commitment requiring operational endurance and the establishment of a sustainable battle rhythm. Lengthy operations will also require a transitional rotation of Army units into the area of operations to continue the mission. Considerations for these transitions are similar to a relief in place and battle handover (see FM 3-90) combined with considerations for deployment and redeployment. FM 41-10 contains comprehensive guidance on transition planning and coordination activities applicable to UO. In addition to any threat considerations, planning for urban transitions between units will often include emphasis and understanding of—

- Formal and informal civilian leadership and relationships.
- Government institutions and administrative functions.
- On-going reconstruction projects.
- The urban economy.
- Participating governmental and nongovernmental organizations and established relationships and cooperation activities, particularly information sharing.
- Significant key events affecting (or likely to affect) operations.
- Significant cultural lessons learned as a result of the outgoing unit's operations.

9-49. The commander of the major operation must ensure that the incoming unit also understands the political and strategic objectives behind the tasks that they will be required to accomplish. Otherwise, the new unit may begin to plan operations that are similar to those conducted by the previous unit without achieving the desired end state or accomplishing the mission.

Transition Trust

9-50. If units are successful in establishing and cultivating relationships with the urban populace and their leadership, and have built an effective level of trust, their anticipated departure from the AO as part of a unit rotation schedule may have detrimental effects on working relationships. As a result, local leaders may develop deep anxieties about the replacement force and their ability to "be as good" as the current command. Therefore, incumbent forces must take steps to identify successful and ongoing efforts and activities in which the new unit can continue to achieve success. In some instances, the current unit may delay a project or activity whose scheduled completion is close to the date of relief or the transfer of authority so as to allow the new unit immediate success. In other instances, projects or activities can be backward planned to ensure that the completion date is scheduled soon after the new unit arrives. The goal for the outgoing unit is to ensure that the new unit is readily accepted and that both units execute a seamless relief in place.

Modify Objectives to Match the Current Environment

9-51. While it is important that incoming commanders create a seamless transition with their outgoing counterparts, they must make full use of the opportunity to review the current political, strategic, and local urban environment to determine potential modifications to logical lines of operations and unit objectives. Otherwise, units may fall into the trap of executing a series of six-month to one-year rotations that do not significantly contribute to solidifying the conditions required to ultimately transition the urban environment back to legitimate civilian control. In operations that involve longer-term commitments, the potential exists for the same unit to conduct two (or more) rotations back into the same AO. Commanders must resist the desire to simply continue with previous lines of operations that, at the time, were deemed successful. The urban environment will change in response to Army actions and this change will necessitate subsequent operational adaptations.

Provide an End State Not an End Date

9-52. Though it may not be feasible, due to political constraints, commanders of major urban operations should fight the public desire to establish a fixed transition timetable. Instead, they should plan urban stability operations based on an attainable end state and not on a specified withdrawal or end date. Progress toward strategic (and operational) goals is susceptible to many changes and delay, particularly in coalitions which will likely be a part of most future urban operations. As such, an end date will only serve to unduly constrain commanders in the attainment of their urban operational objectives. Having set and announced a firm timetable, American and coalition publics will expect withdrawal to commence as scheduled and will not easily be convinced of the need to stay longer even if the situation changes drastically. In general, the public can understand and accept a logical and clearly articulated end state as long as commanders can demonstrate adequate progress based on well developed measures of effectiveness. As important, a timetable allows threats to adjust their plans to the anticipated withdrawal date. Instead of yielding to relentless diplomatic, informational, military, and economic pressure backed by firm political resolve, threats will likely exercise strategic and operational patience.

This page intentionally left blank.

Chapter 10

Urban Sustainment

Sustainment activities exist to enable the Army to initiate and sustain full spectrum operations. Inseparable from decisive and shaping operations, logistics is the central component of sustaining operations and provides the means for commanders to generate and maintain combat power and extend the operational reach of the force. In offensive and defensive operations, sustainment activities are not by themselves likely to be decisive or shaping, yet significantly contribute to those operations. In stability and civil support operations, when the critical objectives may be restoring the infrastructure and the welfare of civilians, sustainment activities will often be decisive. Their success will allow Army forces to dominate this complex environment. However, like all urban operations (UO), sustainment activities affect and are affected by the environment. The urban terrain, infrastructure, and existing resources, coupled with supportive civilians, may facilitate sustainment activities. In contrast, a poorly designed or damaged infrastructure and a hostile population may severely hamper sustainment activities. In the latter case, critical Army resources required elsewhere in the area of operations (AO) may be diverted to repair facilities and control and support the inhabitants of the urban areas.

URBAN SUSTAINMENT CHARACTERISTICS

10-1. Fundamental sustainment characteristics (see figure 10-1) guide prudent sustainment planning regardless of the environment. They provide commanders an excellent framework to analyze and help develop urban sustainment requirements, assess the impact of the environment on the provision of logistics, and gauge the effectiveness of urban sustainment activities.

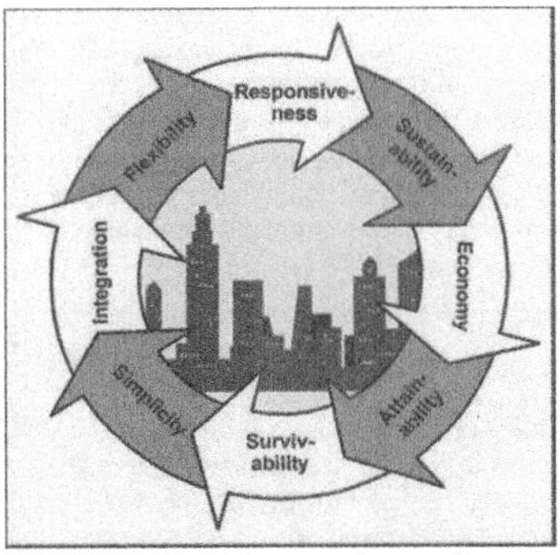

Figure 10-1. Sustainment characteristics

Responsiveness and Sustainability

10-2. UO require responsiveness and sustainability to establish and maintain the tempo necessary for success. Responsiveness—providing the right support in the right place at the right time—is the essential sustainment characteristic. It requires that sustainment commanders and planners accurately forecast urban operational requirements. Continuous urban operations will drain personnel, equipment, and supplies at rates vastly different than in other environments. Therefore, sustainability—the ability to maintain continuous support throughout all phases of the operation—will be a significant concern. Anticipation is critical to both responsiveness and sustainability. It requires that sustainment commanders and planners comprehend the potential effects that components of the urban environment (terrain, infrastructure, and society) may have on operations and sustainment, either benefiting or impeding UO. Effective urban operational and sustainment planning cannot be accomplished separately. Operational and sustainment planners, as well as sustainment operators, must be closely linked to aid in synchronizing and attaining responsiveness and sustainability. The key is often the right balance between efficiency and effectiveness.

Economy and Attainability

10-3. A thoughtful assessment and understanding of the urban environment can also help determine how specific urban areas can contribute to or frustrate the achievement of economy and attainability. Economy is providing the most efficient support at the least cost to accomplish the mission. Attainability means generating the minimum essential supplies and services necessary to begin operations. If available, obtaining support in the AO costs less than purchasing the supplies outside the area and then transporting them there. Critical resources may be available in urban areas to support the operation. However, relying on sources outside the established military sustainment system may create conflict with other sustainment characteristics. A strike by longshoremen, for example, may shut down port operations (at least temporarily) lowering responsiveness and sustainability.

Survivability

10-4. Survivability is being able to protect support functions from destruction or degradation. Commanders often choose to locate sustainment functions in an urban area because the buildings may better protect and conceal equipment, supplies, and people. Urban industrial areas and airports are frequently chosen as support areas because they offer this protection as well as sizeable warehouses, large parking areas, and materials handling equipment (MHE). Such areas facilitate the storage and movement of equipment and supplies. They may also provide readily available water, electricity, and other potentially useful urban resources and infrastructure. However, these areas may also contain toxic industrial materials (TIM) (see Chapter 2). These materials and chemicals in close proximity to support areas may unjustifiably increase the risk to survivability. Sustainment activities in any environment will always be targeted by threat forces. Furthermore, sustainment activities located in any type of confined urban area can offer lucrative targets for terrorists, insurgents, or even angry crowds and mobs. Therefore, no sustainment activity should be considered safe from attack during UO. (During OPERATION JUST CAUSE, Panamanian paramilitary forces and deserters even attacked marked ambulances). Although host-nation support may include assets to assist in defending sustainment units, bases, and lines of communications (LOCs), sustainment commanders must carefully consider if adequate protection measures can ensure survivability. The sustainment commander's greatest challenge to force protection may be complacency born of routine.

Base Security: Tan Son Nhut, Vietnam – 1968

Colonel Nam Truyen was the commander of the 9th Vietcong Division who planned and conducted the attack on the U.S. airbase at Tan Son Nhut during the 1968 Tet Offensive. He had previously entered the airbase during the 1967 Christmas cease-fire using forged identity papers to conduct his own personal and highly thorough reconnaissance.

> UO increase the likelihood that the indigenous population will be utilized as a workforce to supplement Army sustainment activities. Hence, strict operations security will be paramount. Commanders will need to carefully screen potential workers to determine their relative trustworthiness. Human intelligence (HUMINT) and counterintelligence (CI) assets will likely be required to assist in this effort. However, higher-level commanders should seek to take on as much of this burden as possible so that lower-level, tactical commanders can use their limited intelligence assets to accomplish other critical intelligence collection and analysis activities.
> All commanders with base protection responsibilities should implement random, changing force protection measures that focus on civilians—even those civilians who have, over time, earned a perceived measure of trust. (Civilians can be co-opted and coerced by threat forces.) As such, all civilian activity that seems out of the ordinary should be promptly investigated regardless of whether a specific force protection measure is in effect. Although leaders and Soldiers seek to establish mutual trust and respect between themselves and the local populace, commanders should normally err on the side of protecting their Soldiers.

SIMPLICITY

10-5. Simplicity is required in both planning and executing sustainment activities in this complex environment. Developing standard procedures among the Army, other services, and especially civilian governmental and nongovernmental agencies, liaison and open channels of communication, simple plans and orders, and extensive rehearsals contribute immeasurably to attaining this necessary characteristic.

INTEGRATION

10-6. The need for sustainment integration increases in UO due to its joint and multinational nature and greater numbers of other governmental and nongovernmental organizations (NGOs) and agencies operating in or near urban areas. More NGOs will likely exist because urban areas often contain most of a region's population. Most NGOs focus on people. Army forces and other military and nonmilitary groups must cooperate and coordinate their actions. Much of their coordination will revolve around sustainment activities. Cooperation and coordination will take advantage of each group's sustainment capabilities, help to avoid duplicated effort (contributing to economy), and create sustainment synergy. It will also help to curtail competition for the same urban resources and assist in developing a unified list of priorities. Such coordination will help ensure that other operations by one force or agency will not disrupt or destroy portions of the urban infrastructure critical to another's sustainment operations and the overall mission.

10-7. Success in UO requires combined arms and innovative task organizations. As such, UO will often result in the integration of heavy and light units necessitating a simultaneous integrated approach to sustainment. Gaining units must integrate forces into their maintenance systems and carefully consider potentially increased resource requirements. (For example, armor units consume greater amounts of Class III—bulk and petroleum—than a similar-sized light unit.) When integrating into a light unit, armored and mechanized units should bring as much of their critical parts and supplies as possible and immediately integrate into the gaining unit's logistical reporting system to ensure adequate resupply.

FLEXIBILITY

10-8. Lastly, commanders must develop flexibility. Although they and their staffs must thoroughly understand the urban environment essential to planning sustainment operations, they cannot possibly anticipate every eventuality. Urban commanders must possess the ability to exploit fleeting opportunities. Knowledge of the environment, particularly its infrastructure, can aid in developing innovative solutions to sustainment acquisition and distribution problems. Flexibility enables sustainment personnel to remain adaptive and responsive to the force commander's needs. Key to maintaining flexibility is the constant assessment of the sustainment situation and the readiness to modify or change procedures to adapt to the current conditions confronting Army forces operating within the urban environment.

10-9. Maneuver and sustainment commanders should consider and prioritize these characteristics as they visualize UO. Each characteristic does not affect every operation and urban area in the same way. The sustainment characteristics seldom exert equal influence, and their importance varies according to mission, enemy, terrain and weather, troops and support available, time available, civil considerations (METT-TC). Like the principles of war, commanders must not ignore the potential impact of sustainment characteristics and how their influence changes as the operation evolves (see FM 4-0).

LOGISTICS PREPARATION OF THE THEATER

[Sustainment] planning accounts for increased consumption, increased threats to lines of communications, and anticipated support to noncombatants...Urban operations place a premium on closely coordinated, combined arms teams and carefully protected [sustainment activities]. Urban operations are [logistic]-intensive, demanding large quantities of material and support for military forces and noncombatants displaced by operations.

FM 3-0

10-10. A thorough logistics preparation of the theater (LPT) is critical for an adaptable UO support plan. Sustainment planners conduct the LPT to understand the situation from a sustainment perspective and determine how best to support the maneuver commander's plan. However, decisions that impact the urban environment and the political situation require a combined operational and sustainment perspective. Sustainment planners must understand the urban environment, the fundamentals of UO, the proposed course of action, and the urban environment's effects on sustainment (as well as the other warfighting functions). Such knowledge allows the planners to develop a detailed estimate of support requirements. A thorough LPT helps commanders determine the most effective method of providing adequate, responsive support to meet support estimates while minimizing the sustainment footprint. Overall, it helps tie together UO requirements with acquisition and distribution. As with all operations, but particularly in a dynamic urban environment, this assessment process is continuous since requirements will change as the urban operation unfolds and matures.

SUPPORT TO INTELLIGENCE

10-11. The LPT resembles and runs parallel to the intelligence preparation of the battlefield (IPB). Products generated under IPB may be useful in the sustainment analysis. Conversely, the LPT may contribute to the IPB by identifying critical resources and infrastructure and assessing their potential to influence (positively or negatively) the operational plan. This information may warrant a course of action that includes offensive or defensive operations to seize, secure, or destroy those critical resources. In UO initially planned for other than sustainment reasons, the information may require altering the plan or imposing additional constraints to protect the identified resources. While these resources may or may not be critical to current operations, they are usually important to set or shape the conditions necessary for Army forces to consolidate and transition to subsequent missions or redeploy. This close relationship between IPB and LPT underscores the need to quickly and continuously involve sustainment personnel for their sustainment expertise and perspective in planning UO.

10-12. As sustainment units are among the most dispersed and omnipresent in any AO, sustainment Soldiers are potentially valuable intelligence collectors. Soldiers working in and around urban sustainment bases and executing frequent convoys through urban areas quickly become very familiar with the urban terrain and the routine associated with that terrain. As with all Soldiers, sustainment personnel must be incorporated into a comprehensive human intelligence collection system.

URBAN SUSTAINMENT INFORMATION

10-13. Figure 10-2 illustrates that a thorough analysis of the key components of urban areas in the commander's AO provides the data for an accurate LPT and subsequent UO sustainment plan (see Chapter 2 and Appendix B). Analyzing the urban terrain and infrastructure helps to determine—

- Geographic influences on consumption factors and on the provision of support (weather, climate, and topography).
- The availability of supplies, such as safe food, potable water, petroleum, electrical energy, barrier material, and compatible repair parts.
- The location of facilities, such as warehouses, cold-storage sites, water treatment facilities, manufacturing plants, hospitals, hotels for billeting, and waste treatment facilities.
- Transportation information, such as seaport and harbor facilities, airfields, rail and road networks, traffic flow patterns, choke points, and control problems.
- Locations and accessibility of maintenance facilities and equipment, and machine works for the possible fabrication of parts.
- The available general skills among the urban population, such as linguists, drivers, mechanics, MHE operators, longshoremen, and other vital trade skills.

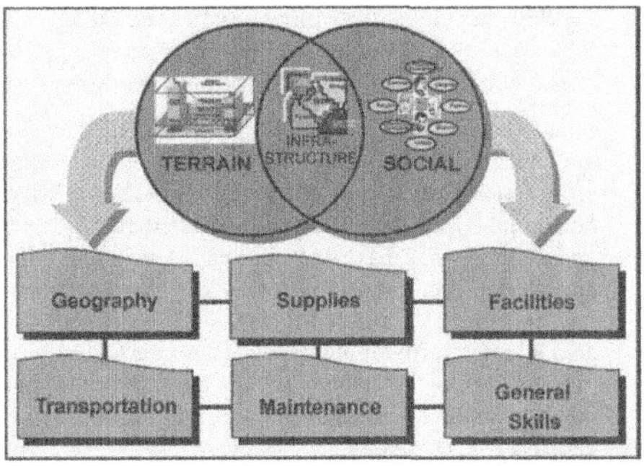

Figure 10-2. Urban environment and essential elements of sustainment information

POTENTIAL RESTRICTIONS

10-14. Commanders must be aware of restrictions that apply to the use of some non-U.S. resources. Security and requirements for U.S. national control dictate that only U.S. assets may perform certain services and functions. Therefore, some foreign urban-area capabilities, even if abundantly available, may not be used. These might include—

- Command and control of medical supply, service, maintenance, replacements, and communications.
- Triage of casualties for evacuation.
- Treatment of chemical, biological, radiological, nuclear, and nuclear (CBRN) casualties, as well as the decontamination of U.S. equipment, personnel, and remains.
- Identification and burial of deceased U.S. personnel.
- Veterinary subsistence inspection.
- Law and order operations over U.S. forces and U.S. military prisoner confinement operations, as well as accountability and security of enemy prisoners of war in U.S. custody.

URBAN SOCIETAL CONSIDERATIONS

10-15. As in all aspects of UO, the urban society is a critical element of the LPT analysis. Sustainment planners must assess whether they can acquire and use urban resources and infrastructure without overly disrupting the urban society and their environment. If the resources are only sufficient for the urban area's inhabitants (and dependent populations in outlying areas), and the facilities cannot increase production to accommodate the needs of Army forces, then commanders may not be able to rely on those resources to support their operations. In fact, the opposite may be true. The effects of UO on the inhabitants, particularly during offensive and defensive operations, may place increased burdens on the Army's resources. For example, commanders may require civilians to evacuate all or parts of an urban area to safeguard them from the effects of planned combat operations. This precautionary measure may require Army forces to support civilian evacuees with temporary housing, food, water, clothing, and medical and sanitary facilities. It will also require carefully planned evacuation routes to ensure that the movement does not interfere with sustainment of the fight.

10-16. Sustainment planners must take into account the potential effects that the provision of supplies to the urban population may have on the operational environment (provided by Army forces or not). For example, an embargo on petroleum products was lifted after Army forces entered Haiti in the 1990s and a less repressive regime assumed governmental control. Few civilian vehicles had moved on the streets of the Port-au-Prince, the capital city, when fuel was scarce and the movement of mounted forces through the urban area was virtually unimpeded. This changed almost overnight once fuel supplies became available, initially making it very difficult for forces to move where required (unless transported by helicopter).

10-17. Sustainment planners must also consider the urban society's ability to restore their own facilities and provide for themselves (if necessary with assistance from Army forces). Throughout this analysis, civil affairs (CA) units can advise and assist in identifying and assessing urban supply systems, services, personnel, resources, and facilities. Critically, commanders must understand that purchasing local goods and services may have the unintended negative consequence of financially sustaining the most disruptive and violent factions in the area. Army forces must seek to purchase urban products and services that will not contribute to prolonging the conflict or crisis. In many stability or civil support operations, they must also attempt to distribute the contracts for goods and services purchased locally as fairly as possible among urban factions and ethnic groups to maintain impartiality and legitimacy. As part of their coordination efforts, commanders should attempt to achieve the cooperation of relief agencies and other NGOs in this endeavor.

10-18. Finally, sustainment planners must also identify potential threats and increased protection requirements that the urban society (criminals, gangs, and riotous mobs) may present, particularly when support units and activities are positioned in urban areas. The disposition or allegiance of the urban population is also important to consider. The infrastructure of an urban area may exhibit great potential to support the sustainment efforts of Army forces, but if the population is hostile or unreliable, use of the facilities and resources may be unfeasible.

Urban Support Areas and Bases

10-19. A major influence on the operation plan and its subsequent execution is often the proper identification and preparation of support areas and bases. The LPT helps commanders determine the need, advantages, and disadvantages of using urban areas in the AO as bases from which to provide support and conduct distribution operations. Ideally, these areas should support reception, staging, onward movement, and integration operations. They should also allow easy sea and air access, offer adequate protection and storage space, increase the throughput of supplies and equipment, and be accessible to multiple LOCs. Consequently, commanders often establish support areas and bases near seaports and airports that are part of a larger urban area. However, threats recognize the Army's need for ports and airfields and may devote substantial resources and combat power to defend them. Army commanders also face the challenge of integrating airfield and port defensive operations with air and naval component commanders to deny threats the ability to conduct stand-off attacks against aircraft and ships. Only an integrated, joint approach can ensure persistent air and sea support– essential not only to sustainment but to firepower as well—is not degraded by the threat. Therefore, planners may determine during the LPT that the risks of seizing or establishing urban lodgment areas may be too high (see Chapter 4). Instead, they may recommend building an airfield, conducting logistics over-the-shore operations, or constructing sustainment bases in more isolated—but more easily secured—locations.

Overall Assessment

10-20. As shown above, the LPT process and analysis help to determine if urban areas in the AO—

- Are suitable as areas and bases for support.
- Can contribute sufficient quantities of and are a dependable source for resources for the overall operation.
- May additionally drain the supported commander's resources.

10-21. The results of this process often serve as a basis for reviewing requirements for civilian contract support and host-nation support and for developing sustainment input into time-phased force and deployment data. This chapter focuses on the effects urban areas may have on accomplishing sustainment functions and related activities, particularly when sustainment units and activities are in urban areas.

ACHIEVING FORCE AGILITY

10-22. To maintain the responsiveness necessary to support full spectrum operations against a wide array of threats and operational environments, sustainment forces must remain agile, possessing the mental and physical agility to transition within or between types of operation with minimal augmentation, no break in contact, and no significant additional training. Responsiveness, flexibility, and economy are key characteristics that enable sustainment forces to support a dynamic combat force; they represent the capability of sustainment forces to provide the resources to initiate, sustain, and extend UO. Agile sustainment forces are characterized by modular force designs, the ability to tailor sustainment organizations for the supporting mission, and the ability to conduct split-based operations. These sustainment aspects are critical to achieving the agility necessary for Army forces to succeed in UO.

MODULAR DESIGN

10-23. Sustainment units are structured as modular organizations, or must possess the ability to modularize in order to conduct decentralized sustainment operations required in UO. At the company level, modularity enables each major company sub-element to assume a cross-section of the company's total capabilities, allowing commanders to employ individual modules to provide a support function, while the rest of the unit remains operational. Modularity enhances responsiveness while enabling support forces to project capabilities as far forward as possible into urban areas, yet with the minimum forward footprint necessary to sustain operations.

FORCE TAILORING

10-24. Sustainment force tailoring refers to the process of determining and deploying the right mix of capabilities to support the force or mission. Sustainment commanders may task organize specific capabilities into functional elements in order to maximize effectiveness and efficiency, and to minimize the sustainment footprint. During UO, the sustainment commander can tailor the support element required to accomplish a specific mission or task, thereby mitigating the risk associated with deploying a larger, more robust capability package forward into the urban area.

SPLIT-BASED OPERATIONS

10-25. Split-based operations refer to performing certain sustainment administrative and management functions outside the area of operations. During UO, Soldiers and civilians can perform personnel, materiel, and distribution management functions without deploying forward into the urban area if the information systems are adequate. This is essential to minimizing risk to support forces and the sustainment footprint in the urban area, and still ensures that all support requirements are met.

SUSTAINMENT FUNCTIONS

- Supply and Field Services
- Maintenance Support
- Transportation Support
- Force Health Protection
- Explosive Ordnance Disposal Support
- Human Resources Support
- Financial Management Support
- Religious Support
- Legal Support to Operations
- General Engineering Support

Figure 10-3. Sustainment functions

SUSTAINMENT FUNCTIONS

10-26. Sustainment consists of eleven functions that must be carefully planned, managed, and synchronized to provide responsive and efficient support to maneuver commanders (see figure 10-3).

Similar to the components of the urban environment, particularly its infrastructure, they overlap and are interdependent. The success of one function depends on the success of several others. Like urban infrastructure, they have two components: a physical component (supplies, equipment, and facilities) and a human component (the Soldiers and civilians who execute these functions). Like city mayors, commanders must plan, manage, and synchronize these functions to provide responsive and efficient sustainment to UO.

10-27. Commanders and planners should consider two essential aspects when addressing these sustainment functions. One aspect looks outward and one looks inward. The first aspect is how these functions can best support full-spectrum UO—the outward analysis. The second aspect is how the urban area affects the conduct of sustainment functions, particularly when those functions are located or performed in an urban area—the inward analysis.

SUPPLY AND FIELD SERVICES

10-28. The supply function involves acquiring, managing, receiving, storing, protecting, maintaining, salvaging, and distributing all classes of supply (except Class VIII) required to equip and sustain Army (and joint) forces. For UO, commanders may need to decide early in the planning cycle whether to stockpile resources forward or to rely on distribution-based sustainment management to satisfy requirements. Some specialized items identified below may not be available through the normal military supply system and may take sustainment personnel much longer to obtain or fabricate. Operation planners must quickly identify the special equipment and increased supply requirements for UO to give sustainment personnel the time necessary to acquire them.

10-29. Field services are those essential services required to enhance a Soldier's quality of life. They consist of clothing exchange, laundry and shower support, textile repair, mortuary affairs, preparation for aerial delivery, food services, billeting, and sanitation. The urban commander determines the need and priority of each service after careful METT-TC analysis. Some facilities such as shower, laundry, and cold storage may be available in the urban area through theater support contracting or provided by external support contracts such as the logistics civilian augmentation program (LOGCAP). Additionally, requirements to care for the urban population will increase requirements for field services immensely. In some circumstances, most notably urban stability or civil support operations, adequately protected field service units or activities will be critical and may be the only support provided.

Increased Urban Supply Requirements

Water

10-30. Water production and distribution will become more difficult for sustainment forces during UO. (Many parts of an urban area can become virtual deserts if public water supplies are cut off or contaminated due to deliberate military action or natural disasters.) Consequently, sustainment managers will be called upon to find more efficient and effective means of moving and distributing potable water, especially when bulk storage of water within the urban area may not be possible or feasible. In sustainment bases that do not have access to urban or other water sources, the transportation and storage of non-potable water may be as great a concern as potable water. Unless adequate storage is available for both potable and non-potable water, units may be forced to rely solely on potable water for all bulk requirements to mitigate health risk. Due to the compartmented nature of many urban operations, planners may need to ensure that additional containers for water (and fuel) are available to support dispersed stock at small-unit level, and that man-portable systems are available in sufficient quantities to ensure timely replenishment.

Class II

10-31. Clothing and individual equipment exposed to the urban environment (concrete, glass, and steel) experience wear and damage at accelerated rates. Units will also require items such as rope, grappling hooks, crowbars, ladders, chain saws, elbow- and kneepads, special vehicle and personnel body armor, fire-fighting equipment, packboards, carpentry tools, and other specialized items to conduct UO. Planners must recognize this and plan accordingly to ensure operations are not interrupted due to equipment shortfalls.

Class III (Bulk)

10-32. With the potential exception of aviation fuel, bulk Class III requirements for UO generally decrease at the maneuver unit level. However, increased fuel requirements for engineer and power-generation equipment attached to or operating with forward units may offset these decreases. Generally, fuel sources within the urban area (refinement facilities, gas stations, garages, and airfields) should not be utilized unless tested and approved for use by a fuel laboratory.

Class IV

10-33. Requirements for construction and barrier material (and the power tools to cut it) during UO will likely far exceed the capability of the sustainment force to provide. Commanders must prioritize distribution, defensive projects, and construction and repair efforts in order to maximize the efficient use of available Class IV. Every effort should be made to obtain supplemental Class IV material from the local economy, as long as that effort does not adversely affect operations or the restoration and improvement of critical urban infrastructure. Of note, specialized, prefabricated road barriers will also be in great demand to block roads, create checkpoints, reinforce defensive positions, and protect headquarters and sustainment activities.

Class V

10-34. Urban combat operations will often increase requirements for Class V. Ammunition consumption rates for urban environments have been as much as five to ten times greater than operations in other environments. Urban offensive and defensive operations will often increase overall requirements for certain types of ammunition (crew-served and small arms ammunition, mortars, antitank rounds, grenades, mines, and demolitions) while others may typically decrease (tank, artillery, attack helicopter). Since forces operating on urban terrain may not be capable of storing ammunition, an established system of distribution is vital to ensuring a constant flow of ammunition necessary to sustain continuous operations and retain the initiative.

Class IX

10-35. Requirements for repair parts generally decrease in UO, with the notable exception of those systems that experience increased ammunition consumption. The heavy use of crew-served and small arms, mortar, and antitank systems in the urban area causes wear and failure at significantly increased rates. Additionally, wheeled-vehicle tires and armored-vehicle track pads and road arms, road wheels, center guides, wedge bolts, and track blocks may wear rapidly on urban streets, rubble, and debris. Based on the tempo of operations, it may be more feasible to replace whole sets of tracks rather than have crews replace individual track pads. Hence, commanders may consider resourcing all wheeled vehicles with spare tires and creating containerized and rapidly transportable sets of track and track parts for speedy delivery to urban sustainment bases. On the whole, planners should consider the use of repair parts packages specifically configured for UO that can be transported by tailored, modular sustainment elements that directly support operations within the urban environment. During long-term stability operations, outgoing units may leave large portions of their prescribed load list in theater to ensure that relieving units have adequate supplies to maintain the tempo of operations.

Storage and Distribution

10-36. UO present challenges to storage and distribution that may not be experienced outside of the urban environment. Forces operating within these areas may not be static and, therefore, cannot store equipment and resources beyond what they can transport organically. Even forces conducting operations from secure locations within the urban area may not be able to store materiel, necessitating a continuous, coordinated distribution system to ensure forces are adequately resourced. During UO, sustainment forces may need to utilize unit or throughput distribution techniques to provide the level of support required to sustain the force.

10-37. Urban commanders must also consider the benefits of stockpiling resources forward against relying on the Army's distribution-based sustainment system. Stockpiling brings supplies close to urban forces and helps ensure supplies are immediately available to support potentially increased consumption rates. However, this method may burden the support structure that moves, handles, and protects large quantities of resources often on a repetitive basis. The Army's normal distribution-based sustainment system reduces this burden significantly; however, available transportation assets (sea, air, and ground) impact delivery response times. These transportation assets often combine military assets overlaid on the host nation's (and urban area's) transportation and distribution infrastructure. Although the civilian infrastructure may initially support the Army's distribution system, later effects of UO, such as destruction of equipment and facilities or loss of civilian workers, may degrade the system. Moreover, Army forces may have to share these assets with other military, multinational force, or civilian organizations participating in the urban operation, as well as with the indigenous civilian population at large. This shared system may put at risk the timely delivery of critical supplies to Army forces.

10-38. Sustainment planners must understand the urban environment and its effects on the proposed method of distribution. They must also understand how urban operations (to include sustainment UO) may affect the urban environment. Storing bulk fuel or ammunition in or near an urban area, for example, may increase the risk of fire and explosive hazard to civilians and Army forces. With this awareness, planners present the commander with an estimate that considers both risks and benefits. Depending on the particular area and other METT-TC factors, they may recommend one method of distribution then transition later to another or a combination of methods. Preconfigured resupply loads delivered as close as possible to where they are needed is vital to responsiveness. Only an agile distribution-based sustainment system will allow Army forces to be responsive and operationally effective across full spectrum operations in an urban environment.

Food Preparation

10-39. The daily feeding standard for Soldiers operating in urban areas remains the same: three quality meals per Soldier. UO creates an increased energy demand on Soldiers, requiring a caloric intake of about 5,000 calories each day. Since producing and delivering prepared meals to forward elements may be difficult or even impossible due to the urban environment, the innovative use of meal supplements can be critical to meeting the dietary requirements of the force. Energy drinks and bars, rich in electrolytes and carbohydrates, are an essential supplement to prepared meals and should be provided in quantities that support or exceed the requirements of the forces operating in the urban area.

10-40. Food (and water) may be available in the urban area; however, local sources must be tested, carefully monitored, and medically approved before consumption. Garbage disposal may be an important consideration in the urban area. Improper trash disposal may leave a signature trail (particularly during urban defensive operations) that may degrade sanitary conditions leading to increased disease and nonbattle injuries (DNBIs). Commanders must understand that food operations, if not properly positioned and secured, can become a focal point for the urban population. Strict policy regarding distribution and control of any Class I supplies (including waste products) will be enforced. Black marketers will be attracted to Army food service activities as well (even during offensive and defensive operations).

Water Purification

Vigilant Monitoring

10-41. Urban areas will often have a ready source of water to support the urban inhabitants and its infrastructure. However, this water may not be suitable for U.S. and allied forces (though the urban population may have developed immunity toward its microorganisms). The higher concentration of TIM in urban areas compounds this problem, as ground water is highly susceptible to chemical contamination, even supplies located miles away from the source of contamination. U.S. Soldiers should be trained and cautioned against using water from an urban area (to include ice and bottled water) until preventive medicine and veterinary personnel can determine its quality. When water quality is unknown, commanders ensure use of tactical water purification equipment that will upgrade it to the Army's water quality standards. Even if initial testing indicates that urban water sources are safe for Army forces, personnel must continuously monitor the water quality. However, Army water purification, storage, and transportation requirements for UO can be greatly reduced if the existing urban water supply can be integrated into sustainment operations. An early assessment of the feasibility of this course of action is critical to sustainment planning.

Potentially Greater Requirements

10-42. Offensive and defensive UO are often intense and can produce more casualties, including civilians. Consequently, medical facilities, already consumers of large volumes of water, may require even more water. Water purification, particularly in the urban areas of developing nations and during urban disaster relief operations, will be a critical and constant concern for Army forces.

Possible Key Terrain

10-43. Sites that can control the water of the urban area may be key terrain, providing not only a resource for Army forces, but also a means to control the threat, the civilians, or both. These sites may be the sources of the water—the river, lake, reservoir, or storage tanks—or the means that process and transport the water—pipelines, pumping stations, or treatment facilities. Many sites may be outside the urban area, as many large urban areas draw water from distant sources. The seizure of a pumping station or pipeline may make it possible for commanders to control water supplies without expending resources required to enter the urban area. To preserve critical infrastructure, commanders may increase security to protect these locations from contamination or destruction. Engineers also may need to restore, maintain, or operate existing water facilities damaged by the threat or disaster and to drill new wells and construct new water facilities.

Mortuary Affairs

10-44. Mortuary affairs provide the necessary care and disposition of deceased personnel. It supports the Army across the spectrum of operations. It may directly and suddenly impact (positively or negatively) the morale of the Soldiers and the American public and may influence relations with the civilian population in the AO. It can also affect the health of Soldiers and the urban populace. Overall, commanders must carefully plan evacuation routes and temporary collection and internment sites, adhering to local customs and traditions to lessen potentially negative consequences.

10-45. Units are responsible for recovering the remains of their own fatalities and evacuating them to the closest mortuary affairs collection point, usually located at the nearest support area. Urban governmental authorities, family, and representatives of appropriate nongovernmental organization are responsible (and should always be relied on to the greatest extent possible) for recovering and handling indigenous civilian remains. However, because of the sheer density of noncombatants in UO, commanders may also find themselves responsible for civilian remains. Deaths of civilians under Army control, such as urban evacuees and refugees at Army-operated sites, often obligate the Army to care for their remains including medical certification and records of death. High-intensity urban combat may result in civilian deaths, and health concerns will require Army forces to deal with civilian remains expeditiously. Commanders should consult local religious leaders, the Staff Judge Advocate (SJA), CA personnel, and chaplains to verify that they are abiding by law and customs. Overall, commanders must ensure that forces treat all deceased, including civilians, with dignity and respect (another important aspect of adhering to the urban fundamental of understanding the human dimension).

Aerial Delivery

10-46. Aerial delivery is the movement by fixed- or rotary-wing aircraft and delivery by the use of parachute or sling load of Soldiers, supplies, and equipment. As a vital link in the distribution system, it contributes to flexibility and provides the capability of supplying the force even when urban ground LOCs are disrupted. Forces use aerial delivery to deliver supplies and equipment when not possible or timely by other means. During urban stability or civil support operations, it is used extensively to move supplies to meet the urgent needs of a population in crisis. In all UO, delivery aircraft are highly vulnerable to small arms, rockets, and air defense systems. A threat may further decrease an already limited number of urban drop zones (aircraft may be able to avoid air defense systems, but ground forces may not be able to secure the drop zone and retrieve the cargo). Equipment and supplies transported by helicopter sling-load lessen the latter disadvantage. There are usually more available sites to deposit sling loads, such as rooftops (engineers may need to determine if they have sufficient structural integrity), parking lots, and athletic fields. However, load instability during flight may restrict a helicopter's airspeed and maneuver capabilities making it more vulnerable to small arms and man-portable air defense systems. For these reasons, aerial delivery of supplies in UO may be much less efficient than in many other environments.

MAINTENANCE SUPPORT

10-47. Maintenance entails keeping materiel in operating condition, returning it to service, or updating and upgrading its capability. It includes recovering and evacuating disabled equipment; replacing forward; performing preventive maintenance checks and services (PMCS); increasing battle damage assessment and repair (BDAR); and determining potential maintenance resources in the urban area.

Rapid Recovery Essential

10-48. Disabled vehicles produce obstacles to movement during UO, blocking narrow thoroughfares and creating vulnerable targets for threat forces. This makes rapid recovery operations essential to avoid clogging limited LOCs and mounted avenues with vehicle evacuation operations. Hence, maintenance personnel should be proficient in hastily securing unit maintenance collection sites near the damaged equipment, ensuring route security to and from the disabled vehicle, and methods of egress under fire. The increased potential for the task organization of armored units into smaller attachments during UO (often platoon-sized or smaller) will strain limited recovery assets and increase the need for sections to be well trained and resourced (to include adequate tow-bars and chains) for self-recovery operations. Units must maintain centralized and responsive control over these potentially critical recovery assets and position them as far forward as the situation allows.

Replace Forward

10-49. One of the guiding maintenance principles is to replace forward and fix rear. Maintenance activities, with a forward focus on system replacement, task and use the distribution and evacuation channels to push components and end items to the sustainment level for repair. However, the urban environment may make distribution and evacuation difficult or even impossible. Repairing equipment as far forward as possible will often be vital to maintaining the tempo necessary to succeed in UO. When recovery is required, equipment is moved only as far back as the point where repairs can be made. When selecting a maintenance site, commanders consider:

- Security and force protection measures.
- Sufficient area around equipment for lift or recovery vehicles.
- Availability of existing maintenance facilities or garages.

10-50. The unforgiving urban terrain will invariably increase damage to man-portable weapons and equipment, particularly electronic equipment sensitive to jarring. More frequent jarring can also translate into more frequent calibration requirements. Although by definition man-portable weapons and equipment are easier to transport, evacuating these systems may prove as difficult as evacuating vehicles and larger, heavier equipment. Therefore, field support maintenance support teams (MSTs) will frequently need to repair equipment at (or as near as possible to) the point where it was damaged. Equipment operators are responsible to properly diagnose the fault or damage. Such action ensures that the correct repair parts and maintenance personnel are sent forward to complete necessary repairs. In UO, particularly offensive and defensive operations, units may need to replace rather than repair equipment. Subsequently, sustainment personnel may need to plan for increased replacement of what might normally be repairable equipment, as well as plan for increased repair parts for man-portable items.

Add-On Protection Increases Wear

10-51. To increase protection against small arms, mines, rocket-propelled grenades, and lightweight antiarmor weapons, units may need to attach additional armor and improvised wire-mesh screens and cages to both wheeled and tracked vehicles operating in an urban environment. (Units can also sandbag vehicles to achieve a degree of increased protection.) These modifications, however, increase wear on brakes, springs, suspension, and tires, all of which are already vulnerable to the increased amount of debris present in the urban environment. Over time, this wear will begin to increase failure rates in major assemblies, such as engines, transmissions, and transfers. To offset the increase in wear and, subsequently repair parts requirements, commanders conducting UO should consider emphasizing these aspects during daily operator PMCS and increasing the frequency of scheduled services.

Increased Battle Damage Assessment and Repair

10-52. In UO, operators, crews, MSTs, and recovery teams execute BDAR far more than in other environments. BDAR quickly restores minimum essential combat capabilities for a specific mission (normally of short duration) or allows the equipment to self-recover by expediently fixing or bypassing components. Commanders may need to authorize supervised battlefield cannibalization and controlled exchange when units lack critical parts or cannot bring them forward. Sustainment commanders should consider the use of BDAR kits, specifically designed for the systems they support. The use of these kits, though providing only temporary repair, will extend the availability of combat systems until full repair capability is accessible.

Potential Urban Maintenance Resources

10-53. Although urban areas can complicate maintenance, they may contribute to this sustainment function. Analyzing the urban area in the commander's AO may reveal potential sources of parts, tools, equipment, and facilities, and the expertise necessary to fix equipment, repair components, and fabricate critical parts. Sites of potential value may include automotive repair shops, foundries, car dealerships, parts stores, junkyards, scrap-metal dealers, and machine shops. Urban areas may also serve as key sources for parts and facilities (and theater support contractor personnel) to repair automation and network communication equipment.

Host Nation Repair Operations: OPERATION IRAQI FREEDOM

During operations in Mosul, Iraq in the summer months of 2003, elements of the 2nd Brigade Combat Team (BCT), 101st Airborne Division (Air Assault) suffered from a decline in combat power that outpaced the ability of the distribution-based sustainment system to provide repair parts. As readiness across the BCT threatened to degrade operational capability, sustainment personnel from the brigade's forward support battalion (FSB)—drawing on lessons learned from the 1990-1991 Gulf War—scoured the local populace for repair parts, facilities, and expertise in an attempt to offset the losses suffered due to supply shortfalls.

Initially, local retailers provided limited repair parts and facilities to the FSB, but the effect was significant and readiness rates began to stabilize. As time passed, however, the continued wear on equipment operating in the urban environment, coupled with the harsh summer weather, began to take a toll on reparable components and major assemblies. With limited availability in theater, and order-ship times as long as 60 days, requisitions would not be filled before the effect on combat fleet readiness reached a critical point.

Recognizing a trend that threatened the success of the BCT's urban operations in Mosul, the FSB established a contract maintenance facility in the brigade support area, used its contingency contracting officer to hire a group of local mechanics certified through General Motors, and opened the first host nation repair facility in Iraq. The group repaired major assemblies and reparable components and systems, maintaining a failure rate consistent with major CONUS rebuild facilities. The effect was immediate: during the first four months of the program, a full two-thirds of the combat platforms repaired and returned to service operated with components or assemblies rebuilt in the host nation facility.

Over time, the program was expanded to support other divisional units in northern Iraq. This effort provided maneuver commanders with responsive maintenance support and demonstrated the effect of sustainment flexibility and sustainability in urban operations. Furthermore, this action met the division commander's intent of infusing much-needed funds into the local economy while strengthening American-Iraqi relations in the region.

TRANSPORTATION SUPPORT

10-54. Transportation supports the concept of the urban operation by moving and transferring units, Soldiers, equipment, and supplies. Transportation incorporates military, commercial, host-nation, and urban area capabilities to establish a flexible system that expands to meet the needs of the force. Transportation includes movement control, terminal operations, and mode operations.

10-55. Urban areas are often critical to transportation operations. These areas may serve as a lodgment or support area for entry of Army forces and sustainment supplies. The existing transportation and distribution infrastructure may be essential to reception, staging, and onward movement. Theater support contracts, as well as host-nation support agreements, may greatly increase the ability of Army forces to use the urban area's facilities, which may include docks, airfields, warehouses, and cargo handling equipment. Urban support may also include skilled urban workers, such as longshoremen, truck drivers, and MHE operators.

US Army Photo

Urban Terminals

10-56. In addition to serving as major seaports and aerial ports of debarkation, urban areas may provide additional terminals in the AO. Forces may use these terminals for further staging, loading, discharging, and transferring the handling between various inland transportation modes and carriers (motor, air, rail, and water). These urban terminals—with synchronized movement control—permit commanders to rapidly shift transportation modes and carriers. This increases flexibility and ensures the continued forward movement of equipment and supplies to support ongoing operations. Movement control, particularly in urban areas, relies heavily on support from military police in their maneuver and mobility support role. Without this support, urban LOCs may become congested, hinder movement and maneuver, and degrade force effectiveness (see FM 3-19.4). Urban commanders may need to establish multiple checkpoints, roadblocks, and traffic control points; restrict selected roads to military traffic; and reroute movement to unaffected road networks when civil support and refugee control operations compete for available routes. Again, military police operations are critical in this regard and will require continuous, close coordination with urban civilian police (if available).

Obstacles to Ground Transportation

10-57. Although urban areas can contribute to transportation operations, rubble and other damage can become obstacles to ground movement. Even in an undamaged urban environment, road and bridge weight restrictions, barriers and medians, steep embankments, large drainage systems, and other urban structures may limit transportation operations. Urban route maintenance, to include reinforcing bridges and constructing bypasses, may become a priority task for engineer units. Bypassed pockets of resistance and ambushes pose a constant threat along urban supply routes and routes in close proximity to urban areas. Urban convoy and resupply operations will often require increased security in the form of continuous route security operations, regular (daily, if necessary) mine and improvised explosive device (IED) clearance operations, numerous observation posts, attack helicopter support, a larger, more mobile tactical combat force, and adequate communications capabilities among convoy vehicles and with supporting forces and higher headquarters. While the level of security and force protection may vary, convoy operations are never considered an "administrative" operation (see FM 4-01.45). Security needs may increase manpower requirements for sustaining operations and potentially reduce resources from decisive operations. Moving critical supplies may require heavily armed convoys and armored vehicles. Drivers must be well trained in convoy operations, rehearsed, and alert. They must recognize and avoid potential mines and IEDs (stationary and vehicle-borne) and be able to react rapidly to ambushes. Additionally, commanders may need to institute deception measures such as randomly closing and opening supply routes to confuse threat pattern analysis. As before, aerial resupply can alleviate problems due to ground obstacles, but the air defense threat and proximity of threat forces may preclude their routine use.

Population Effects

10-58. The ability of Army forces to use vital urban transportation facilities depends largely on the civilians and the threat. The civilian population can affect the transportation system if they do not support the goals of Army operations. Urban transportation systems—such as ports, railroads, and rivers—require many specialists to operate. Without these specialists, the system's utility is degraded and may not function at all. In urban stability or civil support operations, Army forces will share the system with civilians and other agencies. Civilian authorities may refuse to allow Army forces to use any portions of an urban area's transportation system. Negotiating for access to that system under these circumstances then becomes a command priority.

Threat Effects

10-59. The threat can significantly affect urban transportation systems. Many are composed of smaller subsystems. Each subsystem is vulnerable to attack, which in turn often shuts down the whole system. A large canal system, for example, may have entrance and exit facilities, the canal itself, a means to pull the vessel along such as a locomotive engine, and the civilians that run each of these subsystems. Both an attacker and defender understand the components of the particular transportation system. If important to current or subsequent operations, defending forces then develop plans and allocate forces to protect these subsystems. Attacking forces, on the other hand, often avoid collateral damage to the system, while simultaneously preventing enemy destruction of the facilities.

FORCE HEALTH PROTECTION

10-60. Force health protection (FHP) consists of all services performed, provided, or arranged to promote, improve, conserve, or restore the mental or physical well-being of personnel in the Army and, as directed, for other services, agencies, and organizations. FHP conserves the force by—

- Preventing DNBIs and controlling combat stress.
- Clearing the urban area of casualties.
- Providing forward medical treatment and hospitalization and en route care during medical evacuation.
- Providing required veterinary, dental, and laboratory services.
- Ensuring that adequate Class VIII supplies, medical equipment, and blood are available.

10-61. FHP operations minimize the effects of wounds, injuries, disease, urban environmental hazards, and psychological stresses on unit effectiveness, readiness, and morale. Effective UO require acclimated Soldiers trained in specific urban tactics, techniques, and procedures. FHP helps maintain the health of urban forces, thereby conserving that trained manpower. Historically, urban combat operations have generated three- to six-times greater casualty rates than operations in any other type environment. (Greater enemy wounded and a need to provide care for them will also exacerbate this situation.) FHP operations that maintain the health and readiness of Army forces reduce the strain on the replacement and evacuation systems. Such care allows Soldiers to concentrate on the task at hand instead of the increased risks associated with UO.

10-62. As part of the overall LPT, commanders and medical planners must analyze and continuously assess the urban area. They must identify medical threats, required medical resources, and the quality and availability of medical facilities and resources (to include civilian medical personnel). This assessment prevents duplicated services and permits more effectively and efficiently organized medical resources.

10-63. FHP analysis also identifies available hospitals, clinics, medical treatment facilities, and medical supplies and equipment (including production facilities) in the urban area. It may also identify NGOs capable of providing medical services and supplies. (However, stringent federal regulations, standards of medical care, and a need for unavailable advanced technologies may limit their use by Army forces. But, this analysis will still be crucial in determining whether urban medical facilities can support civilian requirements.) FHP personnel keep abreast of the operational situation and its impact on FHP and the urban population. Peace operations, for example, may rapidly transition to high-intensity offensive and defensive operations requiring medical support able to handle potential mass casualty scenarios which may include civilians.

Care of Civilians

10-64. During combat operations, the military normally does not provide injured civilians with medical care, instead relying on civilian medical personnel to provide this service. However, based on METT-TC and requirements under Geneva Conventions, commanders may need to recover, evacuate, and treat numerous civilians (particularly in urban civil support or stability operations) until local civilian medical personnel and facilities can be reconstituted and supplied. In urban civil support operations involving weapons of mass destruction, the primary focus of Army support may be FHP.

10-65. If commanders provide civilian medical support, they adhere to the UO fundamental of transitioning control and transferring responsibilities for medical care of civilians to another agency or into civilian hands as soon as is practical. In UO, commanders consider and address the medical treatment of civilians (enemy and friendly) early in the planning process. Any initial assessment or survey teams therefore contain medical representatives. Of significant concern, commanders may need to ensure that adequate supplies and equipment are on hand to treat the urban population's children as combat health service support is normally focused on the treatment of adult Soldiers. Additionally, follow-on military care for civilians after treatment at a combat support hospital is often nonexistent or extremely limited to a case-by-case basis necessitating the quick restoration of civilian hospital capabilities. Finally, military health practitioners must also assess and consider cultural factors related to civilian medical treatment. (For example, male doctors may not be able to examine or treat indigenous women.) To alleviate confusion and preclude avoidable negative repercussions and unrealistic expectations, commanders may need to develop specific medical policies, directives, and standing operating procedures. These "medical rules of engagement" (ROE) ensure that subordinates know how much medical care they may provide to—

- The urban population.
- Other host-nation and third-country civilians.
- Coalition and host-nation forces.
- Contractor personnel.

10-66. Commanders should also consider that medical aid to the local urban population and other civilians can be a powerful influential tool. Properly planned and executed, its potential benefits for Army forces include enhanced security and intelligence collection created by increasing the willingness of civilians to interact with Soldiers. These effects cannot be left to happenstance. They should be a conscious part of the information operations campaign nested within the overall UO plan. Consideration must also be given, however, to mitigating the potential negative consequences involved in the inevitable reduction or curtailing of medical assistance by Army FHP personnel. As with any aspect of UO, providing care without considering the longer-term consequences could result in far greater disadvantages than benefits.

Disease and Nonbattle Injury

10-67. DNBI is a major medical threat during all operations and presents additional risk during UO. Urban pollution hazards and potential exposure to TIM may increase the risk to soldier health. Some urban areas, particularly those in developing countries, are already large sources of communicable diseases, such as tuberculosis, cholera, typhus, hepatitis, malaria, dengue, and acquired immune deficiency syndrome (AIDS). Other factors that increase the number of disease vectors and the potential for the transmission of infectious diseases might include—

- Physical damage or deterioration of urban infrastructure—such as electricity, water, and sewage services.
- Industries that use or produce hazardous materials.
- Increased habitats for rodents and other pests due to rubble, debris, and uncollected trash.
- Inability for the threat or local authorities to recover and dispose of human remains and animal carcasses.
- Greater threat of animal diseases transmittable to humans due to larger numbers of animals wounded or injured by the effects of UO.
- Greater susceptibility of urban civilians to infectious diseases—increasing the potential of person-to-person disease transmission—due to declining public health capabilities and nutrition levels.

10-68. The density of the environment extends these urban medical risks to Army forces. Commanders should establish a medical epidemiological surveillance system early to assess the health of the force and promptly identify unusual or local occurrences that may signal preventive medicine problems or the influence of biological or chemical agents. These potential hazards, particularly the release of TIM, will influence the type of medical supplies needed by medical personnel and will also necessitate critical planning and preparation for potential mass casualties (civilian as well as military).

10-69. Adhering to the UO fundamental of preserving critical infrastructure will likely prevent surges in DNBI. Preventive medicine personnel must identify the diseases and recommend control and preventive measures. In urban areas, they may also conduct civilian health screening, health education, and immunization programs. Medical screening of military personnel, particularly multinational forces, may be required to prevent introducing new diseases (especially drug-resistant strains) into an urban area. A new disease may tax the medical system and introduce a new medical problem into an area already in crisis. Lastly, field sanitation training (to include a general emphasis on hand washing; training in the use of barrier protection such as latex gloves when rendering care to any person; and animal, rodent, and pest control), equipment, and supplies are part of overall preventive medicine measures and considerations.

Combat Stress

10-70. Stress occurs in every operation and type of environment; while limited stress levels are beneficial, too much is harmful. Controlled combat stress can invoke positive, adaptive reactions such as loyalty, selflessness, and heroism. On the other hand, uncontrolled combat stress can result in negative, harmful behavior and combat stress casualties. Such behaviors and casualties—battle fatigue, misconduct stress behaviors, and post-traumatic stress disorder—can interfere with the unit's mission. Physical and mental factors leading to combat stress result from the environment and the deliberate enemy actions aimed at killing, wounding, or demoralizing Soldiers. However, many stressors are also generated from the Soldier's own leaders and mission demands.

Combat Stress: Chechnya – 1994 to 1996

Russia's 1994-1996 conflict with Chechnya, a republic in the southwestern part of the Russian Federation, produced an increased number of psychological trauma and combat stress casualties. One medical survey found 72 percent of the soldiers screened had some sort of psychological disorder symptoms. Of that, 46 percent exhibited asthenic depression (a weak, apathetic, or retarded motor state). The other 26 percent exhibited psychotic reactions such as high states of anxiety, excitement, or aggressiveness and a deterioration of moral values or interpersonal relations. The statistics showed more troops experienced combat stress disorders than during their 1980s war in Afghanistan. One primary difference was that in Chechnya, Russian forces conducted combat mostly in cities rather than in mountains, valleys, and other rural areas.

Combat always invokes fear in soldiers. However, poor training and planning, uncertainty in their cause, and urban populations that resented their presence exacerbated the psychological climate for the Russian forces in Chechnya. Acts of subversion and terrorism by Chechen guerrillas kept the Russians in a constant, high state of readiness and anxiety; the soldiers viewed every civilian—young or old, male or female—as a potential enemy. This psychological pressure was not simply a by-product, but an objective of information operations and a prime reason for taking the fight into the close confines of Chechnya's urban areas. Torture and mutilation of prisoners; immediate execution of captured pilots; imitative electromagnetic deception (Chechens mimicked Russian radio transmissions and directed Russian close air support against their own forces); and propaganda to convince civilians that Russia's actions had a religious bias against Muslims and Islam were conducted to exert intense, unremitting psychological pressure on Russian forces—with great success.

The characteristics of urban areas combined with Chechen insurgent activities and information operations, civilians that did not welcome foreigners, and an unpopular and poorly supported conflict with an open-ended mission reinforced the need for a national will to strengthen and support the urban fight. These characteristics also reinforce the need for clear objectives, proper leadership and training, and available medical assets able to properly diagnose and treat combat stress casualties.

10-71. During UO, compartmented urban combat leads to physical isolation while difficulties in transmitting radio signals lead to communications isolation; together, they create an overwhelming sense of loneliness. Snipers, mines, and IEDs combined with the closeness and high intensity of urban combat contribute to an unremitting fear of attack from any quarter that further increases stress casualties. Additionally, seeing and perhaps accidentally inflicting casualties on civilians (especially women and children) increase battle fatigue. If civilians are hostile or a threat uses the population as cover and concealment, then the potential for misconduct stress behaviors often increases. Urban areas may provide temptations for looting, alcohol and substance abuse, black marketeering, and harmful social interactions; these temptations may increase misconduct stress behaviors. Leaders can prevent or rapidly identify, successfully manage, and treat stress-related casualties (see FM 22-51 and FM 6-22.5) and prevent misconduct stress behaviors as well as potential violations of the law of war. To reduce the number of stress casualties, commanders must—

- Provide tough, realistic training in urban combat (and combat stress management).
- Plan for rotation of forces and adequate rest and relaxation opportunities.
- Establish effective ROE.
- Create unit cohesion.
- Ensure adequate mental health support is available.
- Provide strong, capable leadership. Leaders, forward on the battlefield, will determine whether the necessary discipline perseveres.

Evacuation

10-72. Transportation restrictions may preclude the medical evacuation (MEDEVAC) of urban casualties. MEDEVAC by air ambulance remains the preferred method; however, it may not be feasible. The threat from small arms fire, rocket-propelled grenades, and other shoulder-fired weapons may prohibit the employment of helicopters. Aeromedical evacuation can also be hampered by the lack of landing zones due to narrow streets, heavily damaged buildings, and barricaded or booby-trapped roofs. If the employment of air ambulances is not feasible, a ground evacuation operation will be necessary. In UO, ground ambulances should often be—

- Hardened to improve the safety of the crew and patients as well as to increase the survivability of the evacuation platform.
- Field-sited as close to the operation as is tactically and physically possible.
- Maneuverable in tight spaces and narrow streets.

10-73. Evacuation by ground ambulance may have their own set of obstacles to surmount such as rubble; debris; barricades; mines, booby-traps, and IEDs; damaged or destroyed vehicles, and flooded or snow-filled streets. If enough ground ambulances are unavailable due to increased casualties or the tactical situation does not allow, units will need to execute casualty evacuation (CASEVAC) operations.

10-74. In UO, CASEVAC may entail evacuation by litter and the expedient use of a vehicle of opportunity or other military vehicle not dedicated for medical evacuation. Based on the threat situation, CASEVAC may also be limited to armored vehicles, conducted only during limited visibility, or both. For evacuation by litter, units may require more litter bearers to move the injured to a point where they can be further evacuated by ground or air ambulance. Lengthy evacuation routes require more litter bearers, as multiple litter relay teams will be necessary to conserve energy and expedite evacuation. However, unless augmented or relieved of this responsibility by another unit, these litter bearers will come from the casualties' own unit thereby diminishing the unit's strength levels necessary to accomplish its primary mission. Instead, tactical commanders may need to wait until a lull in or a cessation of the battle occurs. Depending on projected casualty rates, higher-level commanders may augment units with additional personnel to perform evacuation or may assign maneuver units this mission. Evacuation personnel will require specific training in urban evacuation techniques (moving casualties from subsurface and supersurface levels to and along the surface level).

10-75. Overall, urban commanders at all echelons must develop detailed casualty evacuation plans that include the establishment of treatment facilities as far forward as the situation allows. Engineers are critical to clear routes for medical resupply and evacuation. Based on the threat situation, additional armored vehicles may need to accompany ambulances in a multi-vehicle convoy not only to and from the patient pickup point but also through to a higher level of care. Army forces may also need to plan for and acquire unique capabilities, equipment, and skills now typically seen in civilian urban search-and-rescue teams to clear debris and search for urban casualties. For example, specially trained dogs may play a vital role in locating victims among urban rubble.

Treatment

10-76. First aid training will have increased significance in UO. The compartmented nature of UO, transportation restrictions, communications difficulties, and the low density of trauma specialists (combat medics) may limit the urban casualty's initial treatment administered by nonmedical personnel or to self-treatment measures. To mitigate this risk, units identify and train combat lifesavers (CLS) to perform in the absence of trauma specialists. Since the likelihood of Army forces performing UO continues to increase, commanders strive to meet or exceed Army standards for the number of fully trained CLS required for their specific unit. CLS (and trauma specialist) training must be realistic, hands-on, and conducted at appropriate intervals to ensure CLS can perform lifesaving tasks to include starting intravenous lines. This increase in self, buddy, and CLS care, as well as longer delays in evacuation, may also increase requirements for additional first aid, CLS, and trauma specialist-carried supplies. (Ultimately, commanders may have to consider weighting the main effort with greater overall FHP support for units conducting UO than for units conducting operations in other environments.)

10-77. The increased potential for delayed evacuation during UO mandates that Army trauma specialists be skilled in prolonged casualty care. During the 3-4 October 1993 battle in Somalia, for example, seven trauma specialists managed 39 casualties for more than 14 hours before they could be evacuated. Evacuation delays significantly increase the potential for infection; therefore, trauma specialists should also be trained to administer antibiotics on the battlefield.

10-78. FHP personnel must be capable of recognizing and treating injuries due to IEDs and incendiary or thermobaric weapons (fuel-air explosives)—favored, urban-oriented threat weapons (see Chapter 3). Thermobaric weapons explode; create a cloud of volatile gases, liquids, or powders; and then ignite, creating an immense fireball consuming oxygen and creating enormous overpressure. When employed in an urban structure, the blast wave or overpressure of any explosive is greatly amplified. Injuries resulting from these weapons are massive burns, broken or crushed bones, concussions and traumatic brain injuries, missile injuries, and internal injuries. To counter this threat, medical personnel must be aware of the capabilities of these blast weapons, wounding patterns, and treatment methods.

10-79. The increased use of body armor during UO will help prevent penetrating chest and abdomen wounds. As such, FHP personnel should expect more groin, pelvis, head, and extremity injuries. Furthermore, when fighting soldiers that are known to use body armor, a threat (particularly snipers) can be expected to target the head and face more often than other anatomic areas resulting in more head injuries. Medical personnel must also anticipate and be prepared to treat an increase in hearing loss due to the firing of weapons (particularly recoilless) in enclosed spaces with little ventilation. Medical personnel (and engineers) should also be trained on how to properly remove rubble that has fallen on a Soldier or civilian so as not to increase the extent of his injuries. Lastly, FHP personnel will need to research and develop unique treatments for injuries due to an ever-increasing arsenal of nonlethal weapons.

Explosive Ordnance Disposal Support

10-80. Explosive ordnance disposal (EOD) support provides the capability to neutralize domestic or foreign conventional explosive hazards which include unexploded explosive ordnance (UXO), booby traps, IEDs, captured enemy ammunition (CEA), and bulk explosives. Such material and devices threaten military operations and military and civilian facilities, equipment, and personnel. Explosive hazards create a much greater risk during UO than operations in any other environment. Confined spaces, hard surfaces, and more personnel (both Soldiers and civilians) in the vicinity may magnify the detonating effects of explosives. Dense terrain makes all types of explosive hazards more difficult to locate. Moreover, terrorists select urban settings for their booby traps and IEDs to potentially kill and destroy more, thus gaining greater visibility for their message.

10-81. 10-81. EOD units detect, mark, identify, render safe, and dispose of explosive hazards. Importantly, EOD specialists will work closely with unit intelligence personnel to conduct explosive forensics—technical intelligence collection and exploitation—to help identify the makers of the devices as well as their ever-evolving methods of construction, placement, concealment, and detonation. Chemical reconnaissance with their array of chemical sensors will also work closely with EOD units to reduce the uncertainty of the chemical threat. This integrated forensic approach is a proactive method to take the advantage away from a terrorist or insurgent threat that seeks to employ these devices as a primary tactic.

10-82. Urban operations will increase EOD specialists' role as advisors and instructors on explosive hazards, protection measures, and disposal techniques. They will advise and train other Army forces, other services, multinational partners, and civilian authorities. EOD specialists will also work closely with public affairs and psychological operations personnel to increase public awareness of the horrific effects of these munitions and devices. This education process is designed not only to teach the urban population to identify and avoid explosive hazards, especially UXO, booby traps, and IEDs, but to gain civilian assistance in reporting their locations. (In this regard, rewards for information leading to the discovery of unexploded ordnance may be beneficial. However, the largest reward should normally be reserved for information leading to the arrest or capture of bomb makers.)

10-83. In an urban environment where there are large amounts of UXO, IEDs, and CEA, the sheer amount of explosives may overwhelm the capabilities of available EOD personnel. Because of their training in demolitions and explosives, commanders will often use combat engineers to augment the destruction and clearance effort. Engineers are normally given responsibility for UXO clearance and disposal while EOD teams use their technical skills for IEDs. Commanders may also consider embedding EOD expertise within engineer units to provide the knowledge necessary to determine the correct amount and type of explosive necessary to detonate UXO and to understand potential risks and effects of detonation of various munitions. For example, EOD specialists would better understand the second-order effects and risk reduction measures necessary to detonate a large missile containing numerous submunitions.

10-84. Unfortunately, there is no single device or technique that can eliminate every explosive hazard encountered. Clearance operations are dangerous and time-consuming. Based on METT-TC, commanders may consider the following options when dealing with these explosive hazards:

- Report, mark, and bypass the hazard.
- Employ tactical breaching procedures.
- Self-extract to alternate routes or positions.
- Restrict routes or positions from further use.
- Clear specific areas or positions, or conduct an extensive route clearance.
- Lastly, accept risk of casualties and continue with the assigned mission.

10-85. Effective tactics, techniques, and procedures (TTP) for addressing explosive hazards must be captured and shared with friendly forces and coalition partners. Commanders recognize, however, that adaptive threats will continually modify their own techniques to counter the unit's response. See FM 3-34.119, FM 20-32, FM 3-100.38, FM 4-30.51, and FM 4-30.16.

Human Resources Support

10-86. Human resources support (HRS) encompasses the following functions: manning the force, personnel support, and personnel services. These activities include personnel readiness; replacement operations; personnel accounting; personnel information management; casualty operations; essential personnel services; postal operations; morale, welfare, and recreation; and band support provided to Soldiers, their families, Department of Army civilians, and contractors.

10-87. Successful UO require HRS functions. Clearly, HRS focuses on caring for the needs of people: Soldiers and civilians who support them. Since a critical component of the urban environment is the population, these activities, when required, may also support them. For example, HRS elements, in conjunction with mortuary affairs, military police, and CA, may provide administrative support to assist in the accountability of displaced persons and civilian internees and the establishment of other populace and resources control measures. This support may include the provision of identification and registration cards, travel permits, and other documents, and administrative support to casualty operations. Human resources managers may also assist with community activities and coordination with the Red Cross. If necessary, a postal platoon may assist urban officials in training and reestablishing civilian postal operations. Overall, this type of assistance and support helps to care for the immediate needs of the urban population and, ultimately, speed transition back to legitimate civilian responsibility and control.

10-88. In offensive and defensive UO, HRS personnel will need to account for more casualties and more frequent reconstitution. In all UO, success relies on training individual replacements. Urban combat requires Soldiers skilled in specific urban combat TTP. They understand the societal aspects of the urban population and have training in crowd control to avoid escalating potentially volatile situations. Training these replacements prior to their actual deployment or while in the replacement system frees urban commanders from having to do so. It also helps reduce Soldier (and supporting civilian) isolation, anxiety, and fear associated with urban areas and operations.

Financial Management Support

10-89. Money and its management are often critical to the success of UO. As such, financial management support makes resources available when and where they are needed and assist the urban commander in maintaining fiscal responsibilities. Financial management operations are necessary to conduct the contracting necessary to restore and rebuild infrastructure and provide real-time information, accounting, and services. Financial management operations ensure that urban operational policies and procedures adhere to laws and regulations, develop command resource requirements, and leverage appropriate fund sources to meet them.

10-90. UO will likely include other U.S. military services, governmental agencies, and contractors; multinational forces; NGOs; and various elements of the host nation. So many actors complicate financial accounting, resource management, and cost capturing. Yet, these activities are vital to accomplishing UO and maintaining legal requirements. Army forces will receive support from and provide support to these participants. Financial managers provide crucial expertise to the urban commander on the obligation authority. They also furnish the documentation necessary to obtain and pay for local goods and services using contract or commercial vendor services. However, commanders (with financial management assistance) must assess the economic impact of UO on the economy. An appropriate analysis includes how well the urban economics and commerce infrastructure can support the deployed force and how an influx of U.S. currency may affect the overall economy either positively or negatively.

Labor, Reward, and Solatia Programs

10-91. Financial management elements can also provide joint pay support and non-U.S. pay support for host-nation employees and day laborers supporting UO, as well as for civilian internees. These units can support reward programs and solatia (financial compensation). Commanders can use reward programs to purchase weapons, radios, information, and other items from the urban inhabitants. However, they must use caution when paying for critical human intelligence. First, they must ensure that multiple collection agencies in the command avoid paying the same source for the same information and interpreting these repeated inputs as validation that information is accurate and reliable. Second, they must establish a price or pay scale so subordinate units (particularly in multinational UO) equally compensate their sources for information. Commanders may also need to make solatia to alleviate grief, suffering, and anxiety resulting from injuries and property or personal loss resulting from some aspect of UO. At other times, commanders make solatia to meet cultural expectations of the urban population. These payments are nominal in amount and made without an admission of liability by the Army. Significantly, they may help maintain the legitimacy of the operation and preserve the support of the urban populace.

Financial Support to Civil-Military Projects

> *...[I]n an endeavor like that in Iraq, money is ammunition. In fact, depending on the situation, money can be more important than real ammunition—and that has often been the case in Iraq since early April 2003 when Saddam's regime collapsed and the focus rapidly shifted to reconstruction, economic revival, and restoration of basic services. Once money is available, the challenge is to spend it effectively and quickly to rapidly achieve measurable results. This leads to a related observation that the money needs to*

> *be provided as soon as possible to the organizations that have the capability and capacity to spend it in such a manner.*
>
> Lieutenant General David H. Petraeus
> Commander, 101st Airborne Division (Air Assault) and, later,
> Commander, Multi-National Security Transition Command-Iraq
> during OPERATION IRAQI FREEDOM

10-92. Other financial resources separate from operational funds may be appropriated to support civil-military operations (CMO) projects during stability operations. (During OPERATION IRAQI FREEDOM, these funds were referred to as commander's emergency response program [CERP] funds.) These funds are typically disbursed as bulk currency to tactical units, where the funds are allocated against lower-cost, high-impact projects that support the immediate needs of civil authorities and their citizens. As such, commanders must be prepared to receive, account, secure, and properly disburse these resources in accordance with established procedures. Commanders have an inherent responsibility for ensuring the proper use and disbursement of these funds and must execute appropriate judgment, fiscal responsibility, and supervisory oversight of any resources appropriated in support of CMO. While commanders establish appropriate guidelines to accomplish this fiscal responsibility, they must provide their subordinate commanders—who have the need, capacity, and capability to put these funds to use—reasonable flexibility in how they spend at least a portion of the money, so that they can address emerging needs within their AO.

Contracting Expertise

10-93. Due to the nature of UO, tactical-level units will be likely required to plan and initiate contracts not only for reconstruction projects but to supplement their own sustainment needs. Therefore, higher-level commands must be prepared to assist tactical-level units with contracting expertise. This expertise is required to overcome difficulties in delineating the scope of work, defining construction standards, and establishing viable delivery dates. For small projects and purchases, a standard contractual format may be sufficient. However, for large projects, commanders must provide their subordinates with knowledgeable contracting personnel, who can write contracts with clearly-understood specifications, create purchase orders that specify quality and delivery times, and help monitor work for quality control and assurance. Payment guidelines and standards among all services and members of a coalition are essential to eliminate competition and potential overpayment. Translators will be also essential to accurately translate requirements and standards in addition to trusted local agents to assist in selecting quality engineering and construction firms and workers. Overall, subordinate commanders need adequate contracting support and expertise to meet civilian needs as well as their expectations for timeliness and quality. Without it, commanders may risk losing critical opportunities to positively influence civilian perceptions. (See also Legal Support to Operations below.)

RELIGIOUS SUPPORT

10-94. Religious support entails providing and performing religious support operations for commanders to protect the free exercise of religion for Soldiers, family members, and authorized civilians. This includes personal delivery of rites, sacraments, ordinances, pastoral and spiritual care, and religious education. Such support also consists of:

- Advising the commander on matters of religion, morals, and morale as affected by religion.
- The impact of local religions on threat courses of action, the urban society, and, subsequently, the mission.
- The ethical impact of command decisions.

Moral and Ethical Climate

10-95. Chaplains can help commanders sense the moral and ethical climate in Army units and understand potential moral dilemmas associated with planned UO. The urban environment affects Soldiers' mental health and can increase combat stress casualties, especially misconduct stress behaviors. Chaplains are an important part of combat stress support. The chaplain's presence and faith can sustain Soldiers throughout periods of great trauma. Religious support contributes to the total well being of Soldiers and aids in their return to combat readiness. Chaplains help bolster Soldiers' moral and ethical behaviors through spiritual fitness training so that Soldiers may better cope with ambiguous moral and ethical situations. Chaplains can help identify ethical concerns before they become critical command problems. To this end, their observations help develop and modify ROE, which—due to ambiguity or overly restraining rules—may be part of the problem.

Influence of Local Religions

10-96. Chaplains advise commanders on matters of religion as it affects the Soldiers within their units. They also explain the influences of local religions on threat courses of action and the urban populace and, subsequently, their potential effects on Army forces and UO. Religion is a crucial aspect of assessing the societal component of the urban environment. Understanding the major tenets and concepts of the religions and the impact of faith on civilians' lives may help commanders understand what motivates the threat and the urban populace. (Historically, when religion was a critical factor in armed conflict, the level of lethality has been significantly higher.) To understand threat strategy and tactics, commanders understand their ideology, which, in many cases, involves strong religious beliefs. This understanding also helps commanders to appreciate the inhabitants' attitudes toward other races, religions, and cultures and to identify unacceptable kinds of social interaction (particularly between Soldiers and civilians). Failure to recognize and respect religious beliefs can rapidly erode the perceived legitimacy of the mission and increase the length of the operation. In addition to its influence on threat courses of action, a thorough analysis of the urban environment also includes the degree of influence religion and religious leaders have on the area's government and economy.

Legal Support to Operations

10-97. Legal support provides operational law support in all legal disciplines (including military justice, international law, civil law—composed of contract, fiscal, and environmental law, claims, and legal assistance). This support assists in command and control, sustainment, and HRS of UO. Legal considerations are important in any operation; they take on added significance during UO. They form the foundation for establishing ROE and are critical in the targeting process. (For example, with the increased likelihood that the threat will use the media in an attempt to categorize Army attacks as unlawful, legal support will be critical in determining the protected status of targets.) Adequate legal instruction and training may help subordinate commanders and their Soldiers better understand some aspects ROE and, with this understanding, help them make the rapid decisions often required in a complex urban environment.

10-98. Legal considerations also affect how units acquire goods and services from urban areas and provide support to other agencies and organizations operating in an area. The environment's complex nature requires commanders and their staffs to review and closely consider applicable legal constraints when developing and executing friendly courses of action. Most urban areas have a highly developed legal system nested within county, provincial, state, and national systems. The SJA support to commanders addresses this urban system in conjunction with other legal systems and their potential to affect, positively or negatively, the success of UO.

10-99. International, host-nation, and U.S. law and other regulatory guidelines may vary in their applicability by time and place; actions permissible in one situation may be prohibited in another. These exceptions and complexities increase requirements for SJA, often working with CA personnel, to identify and resolve technical legal issues. Therefore, the SJA must actively advise and participate in all aspects of UO from predeployment training and initial planning through transition and redeployment. FM 27-100 contains detailed legal guidelines affecting UO.

International and Host-Nation Law

10-100. International law consists primarily of agreements, treaties, and customary law to include the law known as the law of war (see FM 27-10). The law of war consists of four general principles applicable when conducting any operation but requiring particular attention during UO. Figure 10-4 lists and describes the four principles: military necessity, discrimination (or distinction), unnecessary suffering (or humanity), and proportionality.

10-101. International law may affect urban operational issues, such as the right of entry, base operations, use of urban infrastructure, and overflight and landing rights. Status-of-forces agreements (SOFAs) exist or can be negotiated to resolve legal issues, such as the status of Soldiers (and, as necessary, contractors) operating in foreign areas to include criminal and civil jurisdictions, taxation concerns, and claims for damages and injuries. Unless a SOFA or other convention exists, Soldiers operating in foreign urban areas have the same legal status as tourists; they are subject to the laws and judicial process of the host nation. (During armed conflict, however, Soldiers always have the rights afforded to them by the Geneva Conventions). Commanders are responsible for understanding the international and host-nation agreements and laws that influence foreign UO. If local law hinders the operation, commanders may be able to inform the local U.S. diplomatic mission and request that it negotiate a solution.

10-102. Commanders may encounter civilian resistance groups whose actions may range from providing enemies with sustainment support to actively fighting Army and coalition forces. Members of such resistance groups must be dealt with in accordance with applicable provisions of the law of war. Commanders should seek legal guidance concerning detention and disposition of persons participating in various acts harmful to friendly forces and detrimental to the mission.

U.S. Law

10-103. Commanders conducting UO must also comply with U.S. law whether it is in the form of a statute, executive order, regulation, or other directive from a federal branch or agency. U.S. law influences UO by governing the acquisition of supplies and services for Army forces, regulating the assistance that can be rendered to foreign nations, and controlling intelligence activities. The Posse Comitatus Act, for example, makes it a crime for Army forces to enforce civil law. Similarly, portions of the Foreign Assistance Act prohibit Soldiers from performing law enforcement activities in foreign urban areas. However, circumstances—expressly authorized by the Constitution, acts of Congress, and other exceptions to these statutes—exist that allow the Army to support civilian law enforcement. Although not nearly all-inclusive, the above demonstrates how U.S. law further complicates urban operations, particularly stability operations, and increases the need for proactive SJA advice and counsel in all facets of UO.

Military Necessity: The principle that justifies the use of measures not forbidden by international or domestic law necessary to rapidly achieve military objectives.

Discrimination: The principle of distinguishing between combatants (who may be attacked) and noncombatants (who may not be attacked).

Unnecessary Suffering: The principle that prohibits the use of weapons, projectiles, or other materials in a manner calculated to cause superfluous injury and unnecessary suffering.

Proportionality: The principle that injury to persons and damage to property incidental to military action, in the circumstances ruling at the time, must not be excessive in relation to the concrete and direct military advantage anticipated.

Figure 10-4. General principles of the law of war

Legal Aspects of Nonlethal Force

10-104. Nonlethal capabilities can augment the means of deadly force and extend urban firepower options. They enhance the commander's ability to apply force in proportion to the threat and to allow discrimination in its use. The range of nonlethal capabilities includes offensive information operations, smoke and obscurants, irritants (such as chemical riot control agents), nonpenetrating projectiles, high-pressure water devices, and directed-energy weapons. These continually expanding capabilities give commanders more options to confront situations that do not warrant deadly force but require Soldiers to employ overwhelming decisive power. However, nonlethal capabilities are subject to the same legal constraints as lethal force (in fact, some nonlethal capabilities can cause serious injury and death, particularly if not employed properly) and undergo the same legal review. Like lethal force, nonlethal capabilities show military necessity, distinguish between combatants and noncombatants, distinguish between military objectives and protected property, are used proportionally, and do not result in unnecessary suffering.

10-105. Of special note, commanders cannot employ chemical herbicides or riot control agents as a method of warfare without prior Presidential approval. However, the Secretary of Defense and regional combatant commanders have authority to use herbicides and riot control agents in limited circumstances during armed conflict and in peacetime operations. Considering the complexity of this issue, commanders of major operations must streamline the approval process so that subordinate commanders do not discount use of these agents simply because of the high-level approval authority. Subordinate commanders must, however, be able to anticipate and clearly justify their potential use in UO. Also, any approval does not mean that these agents must be employed.

Civilians Accompanying the Force

10-106. LOGCAP personnel and other civilian support contractors provide a variety of sustainment functions for the Army. These functions may range from providing unskilled labor, transportation support, and health care to technical support of sophisticated equipment and weapons systems. Commanders must ensure that civilians providing support in their AO are not placed in positions of jeopardy, but if they are, these civilians must understand the risks that they assume when they engage in activities that might be misconstrued as direct or active participation in hostilities.

10-107. This last caution indicates an important legal consideration for commanders. They must try to ensure that threat forces can distinguish civilian support contractors from combatants and thereby remove any justification for them to be targeted or attacked. (As Army forces seek to work closely with the members of governmental and nongovernmental agencies and organizations in urban stability operations, commanders may need to take steps necessary to mitigate the risk of attack caused by their involvement with Army forces.) In the event that threat forces capture civilian support contractors, this same concern helps to ensure that, as lawful civilians accompanying the force, they will be afforded prisoner of war status. As the role of contractor support to Army operations continues to expand, commanders must stay abreast of changes in the law (to include evolving interpretative distinctions) as it affects the use of contractors on the urban battlefield. Other important considerations for commanders with civilian contractors supporting their force are—

- **Minimize Exposure to Harm**. Civilian contractors should be assigned duties at locations that minimize their exposure to harm. Commanders will need to periodically review these locations since they are likely to change as the urban operation evolves and transitions.
- **Prohibit the Performance of Combat Functions**. Civilian support personnel should not be working in situations that involve combat operations where they might be conceived as combatants. As an example, they should not perform force protection for Army forces such as fortification construction and checkpoint security.
- **Prohibit the Use of Arms**. Civilians should not be armed, unless approved by the combatant or service component commander on a limited, by-exception basis for personal defense.
- **Provide Identification Cards**. Civilians should be provided, and carry on their person, a Geneva Convention identification card identifying them as civilians authorized to accompany military forces and entitling them to be treated, if captured, as prisoners of war.
- **Restrict the Wear of Military Uniforms**. Civilians should not normally wear distinctive U.S. military uniforms unless the combatant or service component commander authorizes them to do so. Regardless of their clothing, however, civilians will wear a symbol that establishes their civilian status.

General Engineering Support

10-108. General engineering support will be essential during UO. This support helps assess, construct, maintain, and restore essential LOCs and urban facilities to sustain Army forces, the urban population, or both. Using civilian resources and investing Army general engineering resources requires careful consideration by commanders and staff planners. Since all elements of the urban infrastructure interconnect, general engineering support touches each category to some degree. Figure 10-5 illustrates how urban-specific, general engineering tasks align primarily with the transportation and distribution, energy, and administration and human services components of the urban infrastructure. These engineering tasks are significant and readily apply to UO. The last two, providing fire fighting support and waste management, have not been previously addressed and require more specific consideration.

Construct, Maintain, or Restore

- Roads and Highways
- Over-the-Shore Facilities
- Ports
- Railroad Facilities
- Airports and Heliports
- Fixed Bridges
- Electric Power Facilities
- Petroleum Pipelines and Storage Facilities
- Water Facilities

and Provide

- Mobile Electric Power
- Construction Support
- Fire Fighting Support
- Waste Management

Figure 10-5. General engineering support tasks

The Use or Investment of Resources

10-109. During urban offensive and defensive operations, Army engineer units accomplish tasks to sustain or improve movement and mobility, protection, and sustainment of U.S. and allied forces. These units should maximize the existing urban facilities, host-nation support, civilian contractors, and joint engineer assets. However, commanders should consider how using urban facilities to support military forces may negatively affect the population. On the other hand, construction and repair may benefit both Army units and the urban inhabitants. Restoring the urban transportation network not only improves military LOCs, but may also allow needed commerce to resume. Repairing urban airfields or ports increases throughput capabilities for military supplies, facilitates medical evacuation operations to the support base, accelerates needed relief efforts, and allows international commerce to proceed. Commanders may first invest resources and conduct general engineering tasks to restore facilities for civilian use. Such actions stem future drains on operational resources or facilitate later transition of control back to civilian authorities. For example, repairing police stations, detention facilities, and marksmanship ranges may help urban governments reestablish law and order after completing urban offensive or defensive operations. During stability or civil support operations, the focus of general engineering will often be to support and assist the urban population rather than Army forces.

Fire Fighting Support

10-110. Fire protection and prevention, as well as fire fighting, takes on added significance during UO, particularly offensive and defensive operations. Most ordnance produces heat and flame. This, coupled with an abundance of combustible material (buildings, furniture, gasoline, oil, and propane), poses a serious risk to Soldiers, civilians, and the urban operation. Large shantytowns can exacerbate this problem. In highly combustible areas, commanders may even need to limit or preclude the use of small-arms tracer ammunition. Fire threats to urban areas can be categorized as—

- **Isolated Fires**. These are restricted to a single structure or a specific area within a structure.
- **Area Fires**. These fires consume two or more structures and may extend to encompass an entire block. Generally, streets will serve as firebreaks and help contain the fire within a single block.
- **Firestorms**. These are the most violent and dangerous fires capable of rapidly consuming large areas by creating windstorms and intense heat. Firestorms are often inextinguishable until they have consumed all available combustible materials.

- **Explosive Hazards.** These are present in areas containing fuels and chemicals as well as military explosive hazards.

10-111. When analyzing the administration and human services component of the infrastructure, commanders must determine the adequacy of existing civilian fire fighting support. A deteriorated or nonexistent infrastructure that cannot support the urban area will likely fail to handle the increased risk due to military operations. Commanders may need to provide fire fighting teams to support their own forces and civilians.

10-112. A military force task organized with multiple fire fighting teams (even with maximum use of available civilian fire-fighting assets) will only be able to fight some fires in the AO. Water distribution systems damaged during operations, chemical and other TIM, and hostile activities will further complicate and limit fire fighting capabilities. Commanders must develop priorities for equipment, facility, and infrastructure protection. All Soldiers need training in fire prevention and initial or immediate response fire fighting. Such training includes planning covered and concealed movement, withdrawal, and evacuation routes. Soldiers should be trained to identify and remove ignition and fuel sources and provided additional fire fighting material such as extinguishers, sand, and blankets (see FM 5-415 and the U.S. Department of Transportation's current version of the Emergency Response Guidebook).

Waste Management

10-113. Management of all forms of waste, particularly human, putrescible (such as food), and medical, may become a critical planning consideration for Army forces. This particularly applies if the urban waste management infrastructure was previously inadequate or damaged during UO or natural disasters, the Army force is operating in the urban area for an extended period of time, and a significant number of the urban population remains. Failure to adequately consider this aspect, possibly coupled with the decaying remains of humans and animals and an inadequate or tainted water supply (such as may be caused by flooding), may create unacceptable sanitary and hygiene conditions and subsequently increase DNBIs as well as civilian casualties.

Appendix A

Siege of Beirut: An Illustration of the Fundamentals of Urban Operations

The [Israeli Defense Forces] IDF had neither the strategy nor the experience nor the configuration of forces to fight and sustain a house-to-house campaign in Beirut.

Richard A. Gabriel

Operation Peace for Galilee: The Israeli-PLO War in Lebanon

OVERALL STRATEGIC SITUATION

A-1. In 1982, Israel launched OPERATION PEACE FOR GALILEE designed to destroy the Palestine Liberation Organization (PLO) presence in southern Lebanon. On 1 June, the IDF launched a massive assault across the border into southern Lebanon. The Israeli attack focused on the PLO, but the operations quickly involved major ground and air combat between Israel and Syrian forces.

A-2. In the first few weeks, Israeli forces quickly pushed back both the Syrians and the PLO. However, except for some PLO forces isolated in bypassed urban areas, such as Tyre and Sidon, most of the PLO fell back into Beirut (see Figure A-1). By 30 June, Israeli forces had reached the outskirts of southern Beirut, occupied East Beirut, isolated the city from Syria and the rest of Lebanon, and blockaded the sea approaches to the city. Even so, with most of the PLO intact inside and with significant military and political capability, the Israelis had yet to achieve the objective of OPERATION PEACE FOR GALILEE. The Israeli command had to make a decision. It had three choices: permit the PLO to operate in Beirut; execute a potentially costly assault of the PLO in the city; or lay siege to the city and use the siege to successfully achieve the objective. The Israelis opted for the latter.

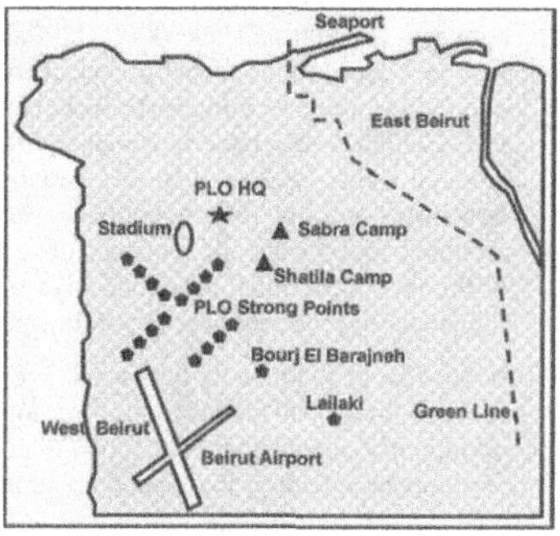

Figure A-1. The city of Beirut

Israeli Military Position

A-3. The Israelis had an excellent position around Beirut. They occupied high ground to the south and west, virtually dominating the entire city. Israeli naval forces controlled the seaward approaches to Beirut. The Israelis' position was also strong defensively, capable of defeating any attempt to break out of or into the city from northern Lebanon or Syria. The Israeli air force had total and complete air superiority. The Israelis controlled the water, fuel, and food sources of West Beirut. Although the PLO forces had stockpiles of food and supplies, the Israelis regulated the food, water, and generating power for the civil population.

A-4. Despite the superior positioning of Israeli forces, the IDF faced significant challenges to include the combat power of the PLO, Syria, and other threats in Beirut. Israeli doctrine and training did not emphasize urban operations. Additionally, Israel was constrained by its desire to limit collateral damage and friendly and noncombatant casualties. Organizationally, the Israeli army was not optimized to fight in urban terrain.

Armor and self-propelled artillery formations dominated the Israeli forces, and most Israeli infantry was mechanized. The Israeli forces had only a few elite formations of traditional dismounted infantry.

PLO MILITARY POSITION

A-5. Despite being surrounded and cut off from support, the PLO position in Beirut offered numerous advantages in addition to the characteristic advantages of urban defense. The PLO had long anticipated an Israeli invasion of southern Lebanon; it had had months to prepare bunkers, obstacles, and the defensive plan of Beirut and other urban areas. Approximately 14,000 Arab combatants in West Beirut readied to withstand the Israeli siege. This was done with the advice of Soviet, Syrian, and east European advisors. The preparation included stockpiling essential supplies in quantities sufficient to withstand a six-month siege. Also, the PLO fighters integrated into the civil populations of the urban areas. Often their families lived with them. The civil population itself was friendly and provided both information and concealment for PLO forces. PLO fighters were experienced in urban combat and knew the urban terrain intimately. PLO forces had been involved in urban fighting against Syrian conventional forces and Christian militias in Beirut several years prior to the Israeli invasion. Finally, the organization of the PLO—centered on small teams of fighters armed with machine-guns and antitank weapons, and trained in insurgent, hit-and-run tactics—was ideally suited to take maximum advantage of the urban environment.

ROLE OF CIVILIANS

A-6. Various ethnic and religious groups make up the civil population of southern Lebanon. However, West Beirut's population was heavily Palestinian and Lebanese. The civil population of West Beirut was between 350,000 and 500,000. The Palestinian population supported the PLO. The Lebanese population may be described as friendly neutral to the Israelis. Although unhappy under Palestinian dominance, this population was unwilling to actively support Israel. The civilian population was a logistic constraint on the PLO, which would have become significant had the siege lasted longer. The civilians in West Beirut were an even larger constraint on the Israelis. The presence of civilians significantly limited the ability of the Israelis to employ firepower. However, the Palestine refugee camps located in West Beirut were both civilian centers and military bases. The Israeli constraints on artillery and other systems against these parts of the city were much less restrictive than in other parts of West Beirut where the population was mostly Lebanese and where fewer key military targets existed.

A-7. The PLO knew of the Israeli aversion to causing civil casualties and purposely located key military centers, troop concentrations, and logistics and weapons systems in and amongst the population—particularly the refugee Palestinian population in the southern part of West Beirut. Tactically, they used the civilians to hide their forces and infiltrate Israeli positions.

A-8. The friendly Palestinian population provided intelligence to the PLO while the friendly Lebanese population provided intelligence for the IDF. Throughout the siege, the IDF maintained a policy of free passage out of Beirut for all civilians. This policy was strictly enforced and permitted no weapons to leave the city. Some estimates are that as many as 100,000 refugees took advantage of this policy.

INFORMATION OPERATIONS

A-9. The siege of Beirut involved using information operations (IO) to influence the media. PLO information operations were aimed at controlling the media and hence the international perception of the operation. This was done by carefully cultivating a select group of pro-PLO media years before hostilities even began. Once hostilities started, only these media sources were permitted to report from the besieged portions of the city, and they were only shown activities that portrayed the IDF negatively. The IDF did not vigorously counter the PLO plan. In fact, the IDF contributed to it by limiting media access to their activities. The PLO information operations had a successful impact. The international community was constantly pressuring the Israeli government to end hostilities. This put pressure on the IDF to conduct operations rapidly and to limit firepower and casualties.

CONDUCT OF THE URBAN OPERATIONS

A-10. The siege of Beirut began 1 July (see Figure A-2). By 4 July, Israeli forces occupied East Beirut, the Green Line separating East and West Beirut, and dominating positions south of the airport. IDF naval forces also controlled the sea west and north of Beirut. On 3 and 4 July, IDF artillery and naval fire began a regular campaign of firing on military targets throughout West Beirut. On 4 July, the IDF cut power and water to the city.

A-11. From 5 to 13 July, the Israeli fires continued to pound PLO targets in West Beirut. The PLO gave one significant response, firing on an Israeli position south of the city and causing several casualties. On 7 July, reacting to international pressure, the IDF returned power and water to West Beirut's civil population. On 11 July, the IDF launched its first attack, probing the southern portion of the airport with an armored task force (see Figure A-3). The PLO repulsed this attack and destroyed several IDF armored vehicles.

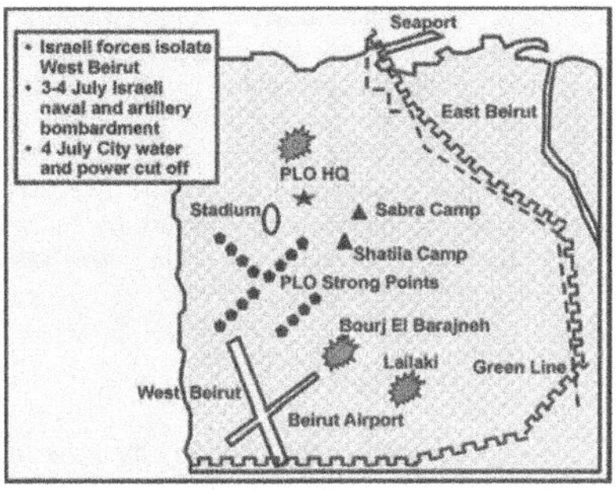

Figure A-2. Initial conduct of the urban operation

A-12. On 13 July, both sides entered into a cease-fire that lasted until 21 July. They began negotiations, mediated by international community, to end the siege. The PLO used this period to continue to fortify Beirut. The Israelis used the time to train their infantry and other arms in urban small unit tactics in Damour, a town the Israeli paratroopers had captured.

A-13. The cease-fire ended on 21 July as PLO forces launched three attacks on IDF rear areas. The Israelis responded with renewed and even more vigorous artillery, naval, and air bombardment of PLO positions in the city. The IDF attacks went on without respite until 30 July. On 28 July, the IDF renewed its ground attack in the south around the airport (see Figure A-4). This time IDF forces methodically advanced and captured a few hundred meters of ground establishing a toehold.

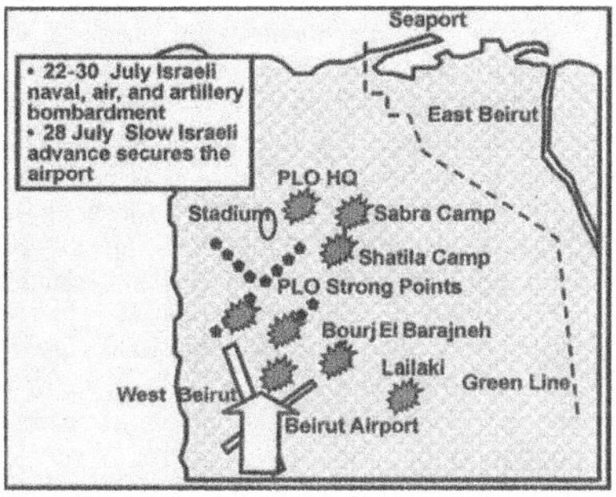

Figure A-3. Israeli probe of PLO defenses

A-14. The Israeli bombardment stopped on 31 July. However, on 1 August the IDF launched its first major ground attack, successfully seizing Beirut airport in the south (see Figure A-5). Israeli armored forces began massing on 2 August along the green line, simultaneously continuing the attack from the south to the outskirts of the Palestinian positions at Ouzai. On 3 August, the Israeli forces continued to reinforce both their southern attack forces and forces along the green line to prepare for continuing offensive operations. On 4 August, the IDF attacked at four different places. This was the much-anticipated major Israeli offensive.

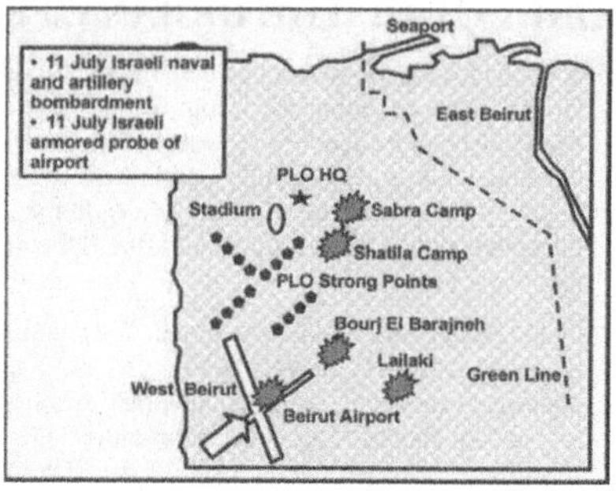

Figure A-4. Initial Israeli attack

A-15. The Israeli attack successfully disrupted the coherence of the PLO defense. The southern attack was the most successful: it pushed PLO forces back to their camps of Sabra and Shatila and threatened to overrun PLO headquarters. Along the green line the IDF attacked across three crossing points. All three attacks made modest gains against stiff resistance. For this day's offensive, the Israelis suffered 19 killed and 84 wounded, the highest single day total of the siege, bringing the total to 318 killed. Following the major attacks on 4 August, Israeli forces paused and, for four days, consolidated their gains and prepared to renew the offensive. Skirmishes and sniping continued, but without significant offensive action. On 9 August, the IDF renewed air and artillery attacks for four days. This activity culminated on 12 August with a massive aerial attack that killed over a hundred and wounded over 400—mostly civilians. A cease-fire started the next day and lasted until the PLO evacuated Beirut on 22 August.

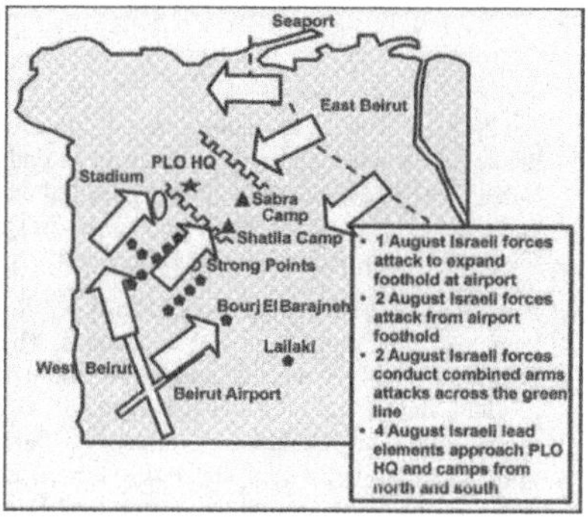

Figure A-5. Final Israeli attack

LESSONS

A-16. The Israeli siege of West Beirut was both a military and a political victory. However, the issue was in doubt until the last week of the siege. Military victory was never in question; the issue in doubt was whether the Israeli government could sustain military operations politically in the face of international and domestic opposition. On the other side, the PLO faced whether they could last militarily until a favorable political end could be negotiated. The answer was that the PLO's military situation became untenable before the Israeli political situation did.

A-17. This favorable military and political outcome stemmed from the careful balance of applying military force with political negotiation. The Israelis also balanced the type of tactics they employed against the domestic aversion to major friendly casualties and international concern with collateral damage.

Perform Aggressive Information Operations

A-18. The PLO devoted considerable resources and much planning on how to use IO to their best advantage. They chose to focus on media information sources as a means of influencing international and domestic opinion.

A-19. The PLO's carefully orchestrated misinformation and control of the media manipulated international sentiment. The major goal of this effort was to grossly exaggerate the claims of civilian casualties, damage, and number of refugees—and this was successfully accomplished. Actual casualties among the civilians were likely half of what the press reported during the battle. The failure of the IDF to present a believable and accurate account of operations to balance PLO efforts put tremendous pressure on the Israeli government to break off the siege. It was the PLO's primary hope for political victory.

A-20. In contrast to the overall weak performance in IO, the IDF excelled in psychological operations. IDF psychological operations attacked the morale of the PLO fighter and the Palestinian population. They were designed to wear down the will of the PLO to fight while convincing the PLO that the IDF would go to any extreme to win. Thus defeat was inevitable. The IDF used passive measures, such as leaflet drops and loudspeaker broadcasts. They used naval bombardment to emphasize the totality of the isolation of Beirut. To maintain high levels of stress, to deny sleep, and to emphasize their combat power, the IDF used constant naval, air, and artillery bombardment. They even employed sonic booms from low-flying aircraft to emphasize the IDF's dominance. These efforts helped to convince the PLO that the only alternative to negotiation on Israeli terms was complete destruction.

Maintain a Close Combat Capability

A-21. The ground combat during the siege of Beirut demonstrated that the lessons of tactical ground combat learned earlier in the twentieth century were still valid. Small combined arms teams built around infantry, but including armor and engineers, were the key to successful tactical combat. Artillery firing in direct fire support of infantry worked effectively as did the Vulcan air defense system. The Israeli tactical plan was sound. The Israelis attacked from multiple directions, segmented West Beirut into pieces, and then destroyed each individually. The plan's success strongly influenced the PLO willingness to negotiate. Tactical patience based on steady though slow progress toward decisive points limited both friendly and noncombatant casualties. In this case, the decisive points were PLO camps, strong points, and the PLO headquarters.

A-22. The Israeli willingness to execute close combat demonstrated throughout the siege, but especially in the attacks of 4 August, was decisive. Decisive ground combat was used sparingly, was successful and aimed at decisive points, and was timed carefully to impact on achieving the political objectives in negotiations. The PLO had hoped that their elaborate defensive preparations would have made Israeli assaults so costly as to convince the Israelis not to attack. That the Israelis could successfully attack the urban area convinced the PLO leadership that destruction of their forces was inevitable. For this reason they negotiated a cease-fire and a withdrawal on Israeli terms.

Avoid the Attrition Approach

A-23. The Israelis carefully focused their attacks on objects that were decisive and would have the greatest impact on the PLO: the known PLO headquarters and refugee centers. Other areas of West Beirut were essentially ignored. For example, the significant Syrian forces in West Beirut were not the focus of Israeli attention even though they had significant combat power. Selectively ignoring portions of the urban area allowed the Israelis to focus their combat power on the PLO and limit both friendly casualties and collateral damage.

Control the Essential and Preserve Critical Infrastructure

A-24. The Israeli siege assured Israeli control of the essential infrastructure of Beirut. The initial Israeli actions secured East Beirut and the city's water, power, and food supplies. The Israelis also dominated Beirut's international airport, closed all the sea access, and controlled all routes into and out of the city. They controlled and preserved all that was critical to operating the city and this put them in a commanding position when negotiating with the PLO.

Minimize Collateral Damage

A-25. The Israeli army took extraordinary steps to limit collateral damage, preserve critical infrastructure, and put in place stringent rules of engagement (ROE). They avoided randomly using grenades in house clearing, limited the use of massed artillery fires, and maximized the use of precision weapons. With this effort, the Israelis extensively used Maverick missiles because of their precise laser guidance and small warheads.

A-26. The strict ROE, however, conflicted with operational guidance that mandated that Israeli commanders minimize their own casualties and adhere to a rapid timetable. The nature of the environment made fighting slow. The concern for civilian casualties and damage to infrastructure declined as IDF casualties rose. They began to bring more field artillery to bear on Palestinian strong points and increasingly employed close air support. This tension underscores the delicate balance that Army commanders will face between minimizing collateral damage and protecting infrastructure while accomplishing the military objective with the least expenditure of resources—particularly soldiers. ROE is but one tool among many that a commander may employ to adhere to this UO fundamental.

Understand the Human Dimension

A-27. The Israelis had a noteworthy (although imperfect and at times flawed) ability to understand the human dimension during their operations against the PLO in Beirut. This was the result of two circumstances. First, the PLO was a threat with which the Israeli forces were familiar after literally decades of conflict. Second, through a close alliance and cooperation with Lebanese militia, the Israelis understood a great deal regarding the attitudes and disposition of the civil population both within and outside Beirut.

Separate Noncombatants from Combatants

A-28. Separating combatants from noncombatants was a difficult but important aspect of the Beirut operation. The Israelis made every effort to positively identify the military nature of all targets. They also operated a free passage system that permitted the passage of all civilians out of the city through Israeli lines. The need to impose cease-fires and open lanes for civilians to escape the fighting slowed IDF operations considerably. Additionally, Israeli assumptions that civilians in urban combat zones would abandon areas where fighting was taking place were incorrect. In many cases, civilians would try to stay in their homes, leaving only after the battle had begun. In contrast, the PLO tied their military operations closely to the civilian community to make targeting difficult. They also abstained from donning uniforms to make individual targeting difficult.

A-29. Earlier in OPERATION PEACE FOR GALILEE when the IDF attacked PLO forces located in Tyre, Israeli psychological operations convinced 30,000 Lebanese noncombatants to abandon their homes and move to beach locations outside the city. However, the IDF was subsequently unable to provide food, water, clothing, shelter, and sanitation for these displaced civilians. IDF commanders compounded the situation by interfering with the efforts by outside relief agencies to aid the displaced population (for fear that the PLO would somehow benefit). Predictably, many civilians tried to return to the city complicating IDF maneuver and targeting—that which the separation was designed to avoid. IDF commanders learned that, while separation is important, they must also adequately plan and prepare for the subsequent control, health, and welfare of the noncombatants they displace.

Restore Essential Services

A-30. Since essential services were under Israeli command, and had been since the beginning of the siege, the Israelis had the ability to easily restore these resources to West Beirut as soon as they adopted the cease-fire.

Transition Control

A-31. In the rear areas of the Israeli siege positions, the Israeli army immediately handed over civic and police responsibility to civil authorities. This policy of rapid transition to civil control within Israeli lines elevated the requirement for the Israeli army to act as an army of occupation. The Israeli army believed the efficient administration of local government and police and the resulting good will of the population more than compensated for the slightly increased force protection issues and the increased risk of PLO infiltration.

A-32. Upon the cease-fire agreement, Israeli forces withdrew to predetermined positions. International forces under United Nations (UN) control supervised the evacuation of the PLO and Syrian forces from Beirut. These actions were executed according to a meticulous plan developed by the Israeli negotiators and agreed to by the PLO. Israeli forces did not take over and occupy Beirut as a result of the 1982 siege (an occupation did occur later but as a result of changing situations).

SUMMARY

A-33. The Israeli siege of West Beirut demonstrates many of the most demanding challenges of urban combat. Apart from Israel's poor understanding of the strategic influence of the media on operations (a lesson that they learned and subsequently applied to future operations), the IDF's successful siege of Beirut emerged from their clear understanding of national strategic objectives and close coordination of diplomatic efforts with urban military operations. A key part of that synchronization of capabilities was the understanding that the efforts of the IDF would be enhanced if they left an escape option open to the PLO. This option was the PLO's supervised evacuation that occurred after the siege. Although the PLO was not physically destroyed, the evacuation without arms and to different host countries effectively shattered the PLO's military capability. Had Israel insisted on the physical destruction of the PLO in Beirut, the overall operation might have failed because that goal may not have been politically obtainable in view of the costs in casualties, collateral damage, and international opinion.

This page intentionally left blank.

Appendix B

An Urban Focus to the Intelligence Preparation of the Battlefield

Maneuvers that are possible and dispositions that are essential are indelibly written on the ground. Badly off, indeed, is the leader who is unable to read this writing. His lot must inevitably be one of blunder, defeat, and disaster.

Infantry in Battle

The complexity of the urban environment and increased number of variables (and their infinite combinations) increases the difficulty of providing timely, relevant, and effective intelligence support to urban operations (UO). Conducted effectively, however, the intelligence preparation of the battlefield (IPB) allows commanders to develop the situational understanding necessary to visualize, describe, and direct subordinates in successfully accomplishing the mission.

IPB is the systematic process of analyzing the threat and environment in a specific geographic area—the area of operations (AO) and its associated area of interest (see figure B-1). It provides the basis for intelligence support to current and future UO, drives the military decision-making process, and supports targeting and battle damage assessment. The procedure (as well as each of its four steps) is performed continuously throughout the planning, preparation, and execution of an urban operation.

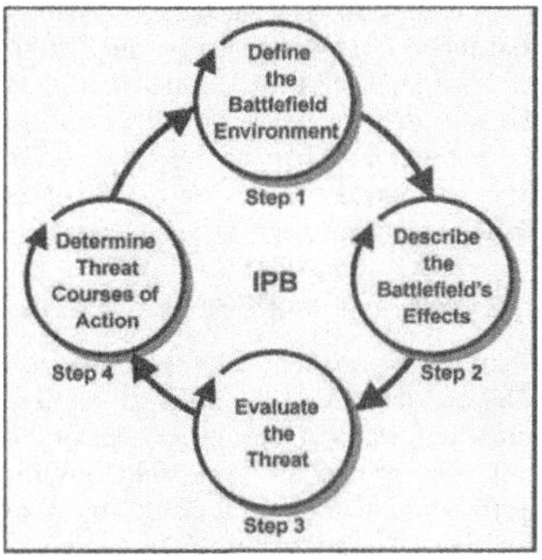

Figure B-1. The steps of IPB

UNAFFECTED PROCESS

B-1. The IPB process is useful at all echelons and remains constant regardless of the operation or environment. However, an urban focus to IPB stresses some aspects not normally emphasized for IPBs conducted for operations elsewhere. The complex urban mosaic is composed of the societal, cultural, or civil dimension of the urban environment; the overlapping and interdependent nature of the urban infrastructure; and the multidimensional terrain. This mosaic challenges the conduct of an urban-focused IPB. There is potential for full spectrum operations to be executed near-simultaneously as part of a single major operation occurring in one urban area with multiple transitions. Such multiplicity stresses the importance of a thorough, non-stop IPB cycle aggressively led by the commander and executed by the entire staff. Overall, the art of applying IPB to UO is in properly employing the steps to the specific operational environment. In UO, this translates to analyzing the significant characteristics of the environment, the role that its populace has in threat evaluation, and understanding how these affect the planning and execution of UO. FM 2-01.3 details how to conduct IPB; FM 34-3 has the processes and procedures for producing all-source intelligence. This appendix supplements the information found there; it does not replace it.

INCREASED COMPLEXITY

B-2. Uncovering intricate relationships takes time, careful analysis, and constant refinement to determine actual effects on friendly and threat courses of action (COAs). These relationships exist among—

- Urban population groups.
- The technical aspects of the infrastructure.
- The historical, cultural, political, or economic significance of the urban area in relation to surrounding urban and rural areas or the nation as a whole.
- The physical effects of the natural and man-made terrain.

B-3. A primary goal of any IPB is to accurately predict the threat's likely COA (step four—which may include political, social, religious, informational, economic, and military actions). Commanders can then develop their own COAs that maximize and apply combat power at decisive points. Understanding the decisive points in the urban operation allows commanders to select objectives that are clearly defined, decisive, and attainable.

Reducing Uncertainty and Its Effects

B-4. Commanders and their staffs may be unfamiliar with the intricacies of the urban environment and more adept at thinking and planning in other environments. Therefore, without detailed situational understanding, commanders may assign missions that their subordinate forces may not be able to achieve. As importantly, commanders and their staffs may miss critical opportunities because they appear overwhelming or impossible (and concede the initiative to the threat). They also may fail to anticipate potential threat COAs afforded by the distinctive urban environment. Commanders may fail to recognize that the least likely threat COA may be the one adopted precisely because it is least likely and, therefore, may be intended to maximize surprise. Misunderstanding the urban environment's effect on potential friendly and threat COAs may rapidly lead to mission failure and the unnecessary loss of Soldiers' lives and other resources. A thorough IPB of the urban environment can greatly reduce uncertainty and contribute to mission success.

Training, Experience, and Functional Area Expertise

B-5. Not all information about the urban environment is relevant to the situation and mission—hence the difficulty and the reason for IPB and intelligence analysis. Although it may appear daunting, institutional education, unit training, and experience at conducting intelligence support to UO will improve the ability to rapidly sort through all the potential information to separate the relevant from merely informative. (This applies to any new or difficult task.) The involvement and functional expertise of the entire staff will allow commanders to quickly identify the important elements of the environment affecting their operations. Fortunately, IPB as part of the entire intelligence process (plan, prepare, collect, process, and produce) is comprehensive enough to manage the seemingly overwhelming amounts of information coming from many sources. Accomplished properly, it allows commanders to recognize opportunities often without complete information.

B-6. As in any operational environment, tension exists between the desire to be methodical and the need to create the tempo necessary to seize, retain, and exploit the initiative necessary for decisive UO. Quickly defining the significant characteristics of the urban environment requiring in-depth evaluation (not only what we need to know but what is possible to know) allows rapid identification of intelligence gaps (what we know versus what we don't know). Such identification leads to information requirements and will drive the intelligence, surveillance, and reconnaissance (ISR) plan (how will we get the information we need). Commanders must carefully consider how to develop focused priority information requirements (PIR) to enable collectors to more easily weed relevant information from the plethora of information. Commanders can then make better decisions and implement them faster than urban threats can react.

AMPLIFIED IMPORTANCE OF CIVIL (SOCIETAL) CONSIDERATIONS

B-7. The Army focuses on warfighting. The experiences in urban operations gained at lower echelons often center on the tactics of urban offensive and defensive operations where the influences of terrain and enemy frequently dominate. At higher echelons, the terrain and enemy are still essential considerations, but the societal component of the urban environment is considered more closely and throughout the operational process. Moreover, the human or civil considerations gain importance in civil support or stability operations regardless of the echelon or level of command. In addition to the echelon and the type of operation, a similar relationship exists between the key elements of the urban environment and other situational factors. These factors can include where the operation lies within the range of operations or the level of war and the conventional or unconventional nature of the opposing threat. Figure B-2 graphically represents the varying significance of these elements to urban IPB. Overall, population effects are significant in how they impact the threat, Army forces, and overall accomplishment of strategic and operational goals.

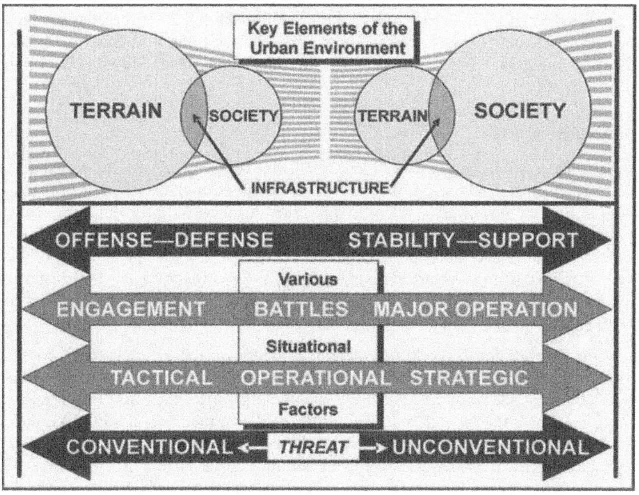

Figure B-2. Relativity of key urban environment elements

B-8. Describing the battlefield's effects—step two in IPB—ascribes meaning to the characteristics analyzed. It helps commanders understand how the environment enhances or degrades friendly and threat forces and capabilities. It also helps commanders understand how the environment supports the population. It also explains how changes in the "normal" urban environment (intentional or unintentional and because of threat or friendly activities) may affect the population. Included in this assessment are matters of perception. At each step of the IPB process, commanders must try to determine the urban society's perceptions of ongoing activities to ensure Army operations are viewed as intended. Throughout this process, commanders, staffs, and analysts cannot allow their biases—cultural, organizational, personal, or cognitive—to markedly influence or alter their assessment (see FM 34-3). This particularly applies when they analyze the societal aspect of the urban environment. With so many potential groups and varied interests in such a limited area, misperception is always a risk.

SIGNIFICANT CHARACTERISTICS

B-9. Urban intelligence analysis must include consideration of the urban environment's distinguishing attributes—man-made terrain, society, and infrastructure—as well as the underlying natural terrain (to include weather) and the threat. Because the urban environment is so complex, it is often useful to break it into categories. Then commanders can understand the intricacies of the environment that may affect their operations and assimilate this information into clear mental images. Commanders can then synthesize these images of the environment with the current status of friendly and threat forces and develop a desired end state. Then they can determine the most decisive sequence of activities that will move their forces from the current state to the end state. Identifying and understanding the environment's characteristics (from a friendly, threat, and noncombatant perspective) allows commanders to establish and maintain situational understanding. Then they can develop appropriate COAs and rules of engagement that will lead to decisive mission accomplishment.

B-10. Figures B-3, B-4, and B-5 are not intended to be all-encompassing lists of urban characteristics. Instead, they provide a starting point or outline useful for conducting an urban-focused IPB and analysis that can be modified to fit the specific operational environment and meet the commander's requirements. Commanders and staffs can compare the categories presented with those in the civil affairs area study and assessment format found in FM 41-10 and the IPB considerations for stability and civil support operations found in ST 2-91.1.

INTERCONNECTED SYSTEMS

B-11. Since the urban environment comprises an interconnected "system of systems," considerations among the key elements of the environment will overlap during an urban intelligence analysis. For example, boundaries, regions, or areas relate to a physical location on the ground. Hence, they have urban terrain implications. These boundaries, regions, or areas often stem from some historical, religious, political, administrative, or social aspect that could also be considered a characteristic of the urban society. Overlaps can also occur in a specific category, such as infrastructure. For instance, dams are a consideration for their potential effects on transportation and distribution (mobility), administration and human services (water supply), and energy (hydroelectric).

B-12. This overlap recognition is a critical concern for commanders and their staffs. In "taking apart" the urban environment and analyzing the pieces, commanders and staffs cannot lose perspective of how each piece interacts with any other and as part of the whole. Otherwise, their vision will be shortsighted, and they will fail to recognize the second- and third-order effects of their proposed COAs; the actual end state differing dramatically from the one envisioned by the commander. The increased density of combatants and noncombatants, infrastructure, and complex terrain means that a given action will likely have unintended consequences—positive or negative. Those consequences will be more widely felt and their impact will spread in less time than in other environments. These unintended results may have important strategic and operational consequences. The multiple ways these dynamic urban elements and characteristics combine make it necessary to approach each urban environment as a unique challenge for intelligence analysis.

URBAN TERRAIN AND WEATHER

Terrain

B-13. Earlier admonitions that civil considerations are more closely considered in UO do not necessarily mean that consideration for urban terrain is de-emphasized. In every urban operation, terrain and its effects on both threat and friendly forces must be assessed and understood. Then commanders can quickly choose and exploit the terrain (and weather conditions) that best supports their missions. Effective terrain analysis thoroughly assesses urban structures as well as the ground on which they stand (see figure B-3 and FM 5-33). An analysis of urban terrain first considers broader urban characteristics and effects and then progresses to a more detailed examination.

Natural Terrain

B-14. The natural terrain features beyond the urban area and beneath urban structures significantly influence unit operations. They dictate where buildings can be constructed, the slope and pattern of streets, and even the broad urban patterns that develop over longer periods of time, thereby influencing a unit's scheme of maneuver. The military aspects of terrain—observation and fields of fire, avenues of approach, key terrain, obstacles, and cover and concealment (OAKOC)—remain critical to the analysis of natural terrain in, under, and around urban areas. Fortunately, commanders and their staffs are normally accustomed to this type of analysis.

Forms and Functions

- Cores
- Industrial areas
  - Toxic industrial material production and storage facilities
  - Standard signs and markings for toxic chemicals
- Outlying high-rise areas
- Residential areas and shantytowns
- Commercial ribbon areas
- Forts and military bases

Broad Urban Patterns

- Types
  - Satellite
  - Network
  - Linear
  - Segment
- Dominant or central hub (if any)
- Area covered (square miles)

Street Patterns

- Basic Types
  - Radial
  - Grid
  - Irregular (planned and unplanned)
- Variations
  - Rayed
  - Radial-Ring
  - Contour-forming
  - Combined
- Widths

Construction and Placement

- Construction
  - Mass or framed
  - Light or heavy clad
  - Material (dirt, wood, stone, brick, cinder block, concrete, steel, and glass)
  - Density and thickness (roofs, floors, and interior and exterior walls)
  - Load bearing walls and columns
  - Height (floors)
  - Doors, windows, fire escapes, and other openings
  - Interior floor plan (including crawl spaces, elevators, and stairs)
- Placement
  - Random
  - Close-orderly block
  - Dispersed
- Ownership

Military Aspects of Terrain: OAKOC

- Observation and fields of fire
  - Smoke (fire), dust (explosions), and flying debris
  - Rubble
  - Engagement ranges (including minimum safe distances and backblast factors) and obliquity/angle (ricochets) considerations
  - Elevation and depression considerations
  - Lasers and reflective concerns
- Avenues of approach (mobility corridors)
  - Airspace
  - Surface
  - Supersurface
  - Subsurface
- Key terrain
  - Landmarks
  - Buildings of significant cultural, social, political, historical, or economic significance
- Obstacles
  - Rubble and vehicles
  - Steep embankments
  - Medians
  - Inadequate bridges and overpasses (destroyed, weight-restricted, or narrow)
  - Tunnels and underpasses (destroyed or narrow)
  - Mines and roadside improvised explosive devices
  - Masking of fires
  - Burning buildings or other fire hazards
  - Rivers and lakes
- Cover and concealment
  - Building protection
  - Weapon penetration (single shot and multiple rounds) considerations
  - Rubble and vehicles

Figure B-3. Significant urban terrain characteristics

Analysis of an Urban Area's Underlying Terrain: Mitrovica, Kosovo – 1999

An urban area's underlying terrain provides many clues into its history, economy, society, and current situation. Mitrovica, Kosovo is an illustrative example. The Ibar River creates a natural line of communications through the middle of the city as well as an obstacle that bisects the urban area. This bisection naturally divides the two resident ethnic groups: Albanians and Serbs. The separation became significant at both the strategic and tactical levels during 1999 deployments to Kosovo. Army forces had to ensure that the Orthodox Church located south of the Ibar was accessible to Serbs residing in the north. North Atlantic Treaty Organization (NATO)

> peacekeepers built a footbridge across the river that allowed reliable, safe passage. The natural feature separating the two groups assisted NATO troops in maintaining stability in the region.

Man-Made Terrain

B-15. Building composition, frontages, placement, forms and functions, size, floor plans, and window and door locations affect maneuver, force positioning, and weapons deployment considerations. Angles, displacement, surface reflection, and antenna locations influence command and control. Structures also influence ISR operations. The increased density and volume created by man-made structures increases how much information commanders and their staffs collect and assess as well as the number of forces required. Building materials and construction will also influence force structures to include weapons and equipment required. Heavily constructed buildings combined with hot and cold extremes may affect target identification for thermal sights. Thick walls, for example, may make combat vehicle identification difficult by distorting hotspots. Additionally, the increased use of heaters and warming fires may clutter thermal sights with numerous hotspots. The ability to maneuver through the urban dimensions—airspace, supersurface, surface, and subsurface—and shoot through walls, ceilings, and floors also creates increased psychological stress. The physical characteristics of man-made terrain can also be analyzed using OAKOC.

Weather

B-16. Weather and its effects are often considered when examining the military aspects of terrain. Military aspects of weather include temperature (heat and cold), light conditions, precipitation (cloud cover, rain, snow, fog, and smog), and wind. Their military effects during UO are similar to any operational environment (see FM 34-81 and FM 34-81-1). Extremes of heat and cold affect weapon systems and the Soldiers that operate them. The extra luminescence provided by the ambient light of an urban area, unless controlled, may affect night vision capabilities and the ability of the Army to "own the night." Precipitation affects mobility and visibility. Smog inversion layers are common over cities. An inversion layer may trap smoke and chemicals in the air and increase ground and air temperatures to the detriment of Soldiers' health. (If the conditions are severe enough, it might require the use of protective masks.) Changes in temperature as a result of air inversions can also affect thermal sights during crossover periods of warm to cold and vice versa. Winds, which may increase as they are funneled through urban canyons, may—

- Increase other weather effects (for example, wind chill).
- Decrease visibility (blowing debris, sand, rain, and snow).
- Spread radiation, biological, and chemical hazards.
- Adversely affect low-altitude air mobility and airborne ISR platforms.

B-17. Commanders also analyze weather for its potential effect on civilians and civilian infrastructure as well as Soldiers and military equipment. Rain might create sewage overflow problems in urban areas with collapsed sewage infrastructure and in refugee camps, increasing disease and even creating panic. Rain and flooding may also make some subsurface areas impassable or extremely hazardous to civilian and military forces alike. Chemical agents and TIM may be washed into underground systems creating toxic hotspots. These effects become more pronounced as chemical agents or TIM are absorbed by brick and unsealed concrete sewers. Other weather effects on UO can include—

- Heavy snowfall in an urban area that may paralyze area transportation and distribution infrastructure, hindering the urban administration's ability to provide vital human services (police, fire fighting, medical, and rescue). Heavy rains and flooding may have similar effects especially on poorly designed and constructed roads or roads that have been damaged by tracked vehicles.
- Extreme hot and cold weather climates that increase the dependence (and military significance) of many elements of the infrastructure. For example, the energy infrastructure may be critical; without it, civilians may not be adequately cooled or heated or they may not be able to cook their food.

- In urban areas located in tropical regions, it can rain at the same time each day during the wet season. Threat forces may attack during these periods knowing aircraft will have difficulty responding. Bad weather also reduces the effectiveness of surveillance, direct and indirect fire, and logistic support.
- Inclement weather may preclude demonstrations or rallies by threats. Good weather may mean a maximum turnout of civilians for events such as festivals, sporting events, and other social, cultural, or religious activities.
- Severe weather may affect psychological, civil-military and humanitarian assistance operations. Heavy rains and severe dust storms may disrupt leaflet drops, construction projects, food and water distribution, and medical and veterinary assistance programs.

URBAN SOCIETY

B-18. This manual shows that societal considerations take on added importance in UO. Critical to operational success is knowing and understanding which groups live in an urban area, what relationships exist among them, and how each population group will respond to friendly and threat activities. Often determining any of this is very difficult. Cultural acuity is also essential in helping commanders and their staffs to view the urban area as the residents view it. The demographics presented depict what conditions exist, while the other categories help to explain the root causes or why conditions exist (see figure B-4 on page B-8). Other categories besides basic demographics that are important to gain this understanding include health, history, leadership, ethnicity and culture, religion, and government and politics.

From Cultural Awareness to Understanding and Competence

B-19. This societal understanding must go beyond a superficial awareness of gestures and taboos to developing a thorough understanding of the organizing principles that make the urban society hold together and function as a coherent entity—or not. In UO, particularly longer-term stability operations, commanders use this knowledge and insight to design and plan operations that work effectively within that society. For example in OPERATION IRAQI FREEDOM, Army units, as they shifted to long-term stability operations, were required to recruit, organize, equip, and train indigenous security forces in order to be able to transition security responsibility from U.S. forces to the burgeoning Iraqi government. Army commanders were successful—to a point. The first units that they built and trained attained a remarkable level of unity and tactical ability in a relatively short period of time. The newly formed units effectively fought alongside Army forces. However when prodded into conducting operations with minimal Army supervision, many Iraqi units lost unity and coherence. Army commanders initially failed to account for a basic Iraqi societal consideration: their society is patriarchal—each member or organization in Iraqi society always seeks a patron or sponsor to support them and provide for their needs. When Iraqi units trained and operated with Army forces, this relationship was satisfied. Operating on their own, however, Iraqi commanders and units lost this patronage without an equivalent Iraqi replacement. In hindsight, the establishment of an effective and more robust ministry of defense, either as a precursor or simultaneously with the establishment of tactical units, may have provided the Iraqi security forces the patronage that their culture required.

Resource the Societal Analysis

B-20. Aside from friendly and threat forces, the society is the only thinking component of the urban environment able to rapidly impact the urban operation. (Even people going about their daily routines can unwittingly hamper the mission.) Urban residents create conditions for restrictive rules of engagement, increase stress on Soldiers and logistic capabilities, and confuse threat identification (see Threat Considerations in this appendix). Demographic, health, safety, ethnic, and cultural concerns will be essential considerations in most UO. Other situational factors—the mission, enemy, and time available—dictate the balance between the level of detail and analysis to support the overall urban operation with the level of detail that commanders and their staffs can achieve. However, an analysis that fails to devote enough time and resources to understanding the urban society can find large elements of the population turned against the Army force. Detailed analysis of the urban society will help save time and ISR resources. While not necessarily a resource saver in the short term, a thorough understanding of the urban society and its culture ultimately contributes to faster achievement of strategic goals.

Population Demographics

- General population size
  - Village
  - Town
  - City
  - Metropolis
  - Megalopolis
- Group size based on race age, sex, political affiliation, economics, religion, tribe, clan, gang, criminal activities, or other significant grouping
  - Significant US or allied populations
  - Distribution, densities, and physical boundaries and overlaps
  - Majority, minority, and dominant groups
- Increasing or decreasing migration trends
  - Dislocated civilians
- Nongovernmental organizations (NGOs)
  - Local
  - National
  - International
- Languages (distribution, dialects, relationship to social structure)
- Educational levels and literacy rates
- Crime rates
- Birth and death rates
- Labor statistics and considerations
  - Skilled and unskilled
  - Imported and exported
  - Unemployment
  - Standard wages and per capita income
  - Workday and workweek norms

Health

- Diseases
- Nutritional deficiencies
- Local standards of care
- Pollution and environmental hazards (air, water, food, and soil)
- Health workers (types, numbers, and degree of skill)

History

- General and for a specific group
  - Internal or external
  - Recent conflicts
- Relationship with US, allies, and other participating multinational forces
- Applicable international treaties
- Status-of-forces agreements
- Antagonists/protagonists
- Heroes
- Events, facts, and dates considered important or celebrated
- Urban area's historical importance

Leadership and Prominent Personalities

- Identification, location, and prioritization of influential leaders (exploitation, evacuation, protection, etc.)
- Affiliation (ethnic, religion, military, government, industry, criminal, or entertainment)
- Education attained
- Organization and distribution of power
- Associations among different leaders and groups

Ethnicity and Culture

- Values, moral codes, taboos, and insults (verbal and non-verbal)
- Attitudes towards age, sex, and race (including same-sex interaction)
- Role of the clan, tribe, or family
- Biases between ethnic groups
- Privacy and individuality
- Recreation, entertainment, and humor
- Fatalism or self-determination
- Exchanges of gifts
- Displays of emotion
- Lines of authority
- Dating and marriage
- Greetings, leave-takings, and gestures
- Visiting practices
- Alcohol and drug use
- Important holidays, festivals, sporting, or entertainment events
- Eating and dietary practices
- Significance of animals and pets
- Urban-rural similarities and differences
- Driving habits
- Clothing

Religion

- Sects, divisions, and overlaps
- Religious biases and problems
- Relationship and influence on government, politics, economics, and education
- Impact on ethnic and cultural beliefs
- Key events or celebrations (daily, weekly, monthly, or annually)
- Funeral and burial practices

Government and Politics

- Present and past forms
- Organization and powers (executive, legislative, judicial, and administrative divisions)
- Scheduled elections and historical turnouts
- Degree of control over the population
  - Identification required
  - Border-crossing procedures
- Relationship with US or multinational governments, national governments, and criminal elements
- Political factions and boundaries
- Political traditions
- Grievances
- Censorship
- Nepotism and other clan, tribal, or social ties
- Civil defense and disaster preparedness (organization, plans, training, equipment, and resources)
  - Evacuation routes
- Legal system
  - System of laws
  - Applicable treaties
  - Courts and tribunals
  - Procedures
  - Records (birth and deeds)
- Property control
- Monetary system (formal and informal)
- Domestic and foreign trade
  - Taxation and tariffs
  - Customs requirements
  - Rationing and price controls
  - Economic performance and contribution to gross national product
- Economic aid
- Perception of relative deprivation
- Trade unions
- Competition with the black market and organized crime

Figure B-4. Significant urban societal characteristics

Transportation and Distribution

- Water
    - Shipyards and other port and harbor facilities
    - Inland waterways, canals, and locks
    - Offshore pipeline berths
    - Cargo storage and handling
    - Types and number of ships, boats, and ferries
    - Dams
- Streets and roads
    - Bridges and fords
    - Over- and underpasses
    - Raised embankments, tunnels, culverts, and other subterranean features (widths and clearances)
    - Parking areas (surface, subsurface, and supersurface)
    - Weight restrictions
    - Traffic light operations
    - Traffic patterns
    - Widths
    - Surface materials
- Rail
    - Lines
    - Terminals
    - Switchyards and junctions
    - Subways, bridges, elevated rail lines, and underpasses (clearances)
    - Track gauges
    - Types and number of rolling stock
    - Electrification
- Air
    - Airfields and runways (including capabilities)
    - Heliports and helipads (including rooftop)
    - Types and number of aircraft
    - Cargo storage and handling
- Trucking companies and delivery services
- Available material-handling equipment
- Rush hour and market time considerations
- Seasonal (weather) effects
- Rubble effects
- Impact of dislocated civilians and migration patterns
- Likely population congregation points
- Identifiable primary and alternate lines of communication

Economics and Commerce

- Industries
    - Types and locations
    - Important companies (including US or allied)
    - Military production facilities
- Sources of raw materials
- Use of toxic industrial materials and biological agents
    - Agriculture (insecticides, herbicides, and fertilizers)
    - Manufacturing
    - Cleaning
    - Research
- Food types, quantities, and sources
    - Requirements and availability
    - Storage and processing
    - Cleanliness standards
- Stores, shops, restaurants, hotels, and strip malls
- Recreation facilities
    - Outdoor and amusement parks
    - Stadiums and other sports facilities
- Machine shops
- Brick and lumber yards
- Banking and investment institutions

Administration and Human Services

- Police and fire protection
    - Headquarters, station, and key facilities locations
    - Organization and strengths
    - Equipment
    - Functions, authority, and jurisdictional boundaries
    - Contract guard services
- Welfare and public assistance
    - Monetary assistance
    - Orphanages
    - Elderly care facilities
- Water supply systems
    - Water sources and storage (lakes, reservoirs, cisterns, pools, and public baths)
    - Water treatment and quality
    - Pumping stations and other distribution methods (trucks, bottles)
    - Hydrant locations
- Snow removal capabilities
- Street light operations
- Health facilities
    - Hospitals
    - Emergency medical services
    - Mental institutions
    - Medical supplies and equipment
    - Research and pharmaceutical buildings
    - Blood banks
- Governmental buildings
    - Embassies
    - Capitol building
    - Legislative, judicial, and ministry buildings
    - Hall of records
- NGOs providing assistance
- Effects of military control measures on providing vital human services
- Waste and sanitation
    - Types (solid, sewage, and toxic)
    - Collection, processing, and disposal
    - Dumps or landfills
    - Drainage systems

Cultural

- Religious buildings and places of worship
- Shrines, monuments and other historical structures
- Schools
- Museums
- Theaters
- Libraries

Energy

- Types
    - Electric
    - Oil
    - Coal
    - Natural gas
    - Nuclear
    - Solar
    - Hydroelectric
    - Geothermal
- Facilities
    - Production and processing
    - Storage
- Distribution
    - Pipelines (above and below ground)
    - Power lines (overhead and underground)
    - Water, rail, and road
    - Potential hazards

Communication and Information

- Print media
    - Newspapers, periodicals, and pamphlets
    - Billboards and posters
    - Postal facilities
- Telephone facilities
    - Wire or wireless
    - Facsimile machines
- Telegraph facilities
- Radio facilities
- Police, fire, and rescue systems
- Security systems
- Television facilities
- Computers and the Internet
- Antennas, towers, relay stations, and lines (surface and subsurface)
- Integration of space-based capabilities
- Public forums and speech
- Low-technology media (car horns, drums, graffiti, and burning fires)
- Key media organizations and reporters
    - Local
    - International
    - US

Figure B-5. Significant urban infrastructure characteristics

Urban Infrastructure

B-21. Functional and analytical interdependence is readily apparent when examining urban infrastructures (see figure B-5 on page B-9). They are composed of physical structures or facilities and people. Hence, much of the analysis conducted for terrain and society can apply when assessing the urban infrastructure. For example, commanders, staffs, and analysts could not effectively assess the urban economic and commercial infrastructure without simultaneously considering labor. All aspects of the society relate and can be used to further analyze the urban work force since they are a sub-element of the urban society. Similarly, the OAKOC aspects used to evaluate terrain may also apply to the urban infrastructure, especially considerations of key terrain. Depending on the mission, elements of the urban infrastructure may become key terrain. While infrastructure analysis is not an engineer-specific task, Army engineers may provide the general and geospatial engineering expertise necessary to better understand the urban terrain and infrastructure.

THREAT CONSIDERATIONS

... [T]he adversaries of freedom ... send arms, agitators, aid, technicians and propaganda to every troubled area.... [S]ubversives and saboteurs and insurrectionists . ... [possess] the power to conscript talent and manpower for any purpose. and long experience in the techniques of violence and subversion....It is a contest of will and purpose as well as force and violence—a battle for minds and souls as well as lives and territory.

John F. Kennedy
Message to Congress, 5 May 1961

B-22. Chapter 3 outlines the instability and uncertainty of the strategic environment. Commanders, staffs, and analysts identify and analyze the threat's composition, strength, disposition, leadership, training, morale, weapons and capabilities, vulnerabilities, internal logistics and external support, doctrine (if any), strategy or modus operandi, and tactics. The threat can take a variety of forms:

- Conventional military forces.
- Paramilitary forces.
- Guerrillas and insurgents.
- Terrorists.
- Militia or special police organizations.

B-23. A general study of guerrilla and insurgent tactics, techniques, and procedures may prove beneficial to many types of operations regardless of the actual composition or type of threat forces. Insurgent strategies and tactics may work especially well in this complex environment and will likely be a part of any threat COA. Particularly, commanders should understand how a threat might restrict itself by the laws of land warfare and similar conventions, or exploit the use of these conventions to its own gain. Commanders can refer to FM 31-20-3 and FMI 3-07.22 for more information. For many of the above threats, no doctrinal templates may exist. Intelligence staffs will need to identify and track common threat patterns and tactics, techniques, and procedures as they develop and change during the course of the operation. Understanding culture will be critical to determining threat motives that drive decisions. Comman415081ders, staffs, and analysts must evaluate, update (or create), and manage threat databases early (and continuously) during the operational process.

Urban Environmental Threats

B-24. While threats vary, they share a common characteristic: the capability and intent to conduct violence against Army forces to negatively influence mission accomplishment. These threats are often the most recognizable for forces trained for warfighting—these are often the enemy. In order for Army units to be able to conduct full spectrum urban operations, commanders broaden their concept of the threat when analyzing the urban environment's terrain, societal, and infrastructure characteristics. This analysis includes many environmental dangers (potentially affecting both sides of a conflict as well as noncombatants) such as—

- Natural disasters (earthquakes, tornados, hurricanes, volcanic eruptions, fires, floods, tidal waves or tsunamis, and heavy snows).
- Food shortages.
- Water shortages (both quantity and quality).
- Rampant disease.
- Pollution and toxic industrial materials.

B-25. A critical difference between the latter forms of threat and the former is the lack of intent to do harm. The latter may stand alone as threats, or these conditions may be created, initiated, or used by the enemy or a hostile civilian group as a weapon or tool. Threat analysis includes identifying and describing how each relevant characteristic of the area of operations can hinder mission accomplishment. This analysis, particularly during stability or civil support operations, will require extensive coordination and cooperation with urban civil authorities, law enforcement, and numerous governmental and nongovernmental organizations.

CIVILIANS

B-26. In a major combat operation or campaign where offensive and defensive operations are conducted against a conventional enemy, threat identification is more readily accomplished. However, adaptation may be necessary to even further broaden the threat concept to include specific elements of the urban society and, in some instances, nongovernmental organizations (NGOs) and other civilian agencies working in the urban area. In many stability operations, this modification can account for opposing armed forces that are not an enemy but are a threat to the mission. As discussed in Chapter 9, Army forces in many stability or civil support operations must avoid classifying or thinking of these threats as the enemy. Further, commanders ensure that Soldiers do not shift the animosity that they feel toward urban threats onto the civilian population. In OPERATION IRAQI FREEDOM, for example, some Soldiers began to refer to Iraqi civilians as "hajjis" in reference to the annual pilgrimage—the Hajj—that forms one of the tenets or pillars of Islamic faith. Those who have completed the pilgrimage are entitled to add the phrase al-Hajj or hajji (pilgrim) to their name. However, the Soldiers were not using the term in its proper context but rather to denigrate. In response, commanders prohibited the inappropriate use of this term as one of many measures to stress and develop cultural understanding and respect for the people that Army forces were there to assist.

Need for More Accurate Categories

B-27. Urban commanders recognize that threat evaluation is not solely a straightforward assessment of the capabilities of a known, armed enemy. Instead, they develop a system of categories that mitigates potential situational uncertainty and better indicates the level of threat (or utility) that civilian groups pose for Army forces conducting UO. Categories can range from a simple construct—threat, neutral, or friendly—to a more refined one (see figure B-6 on page B-12). In most UO, a more intricate approach is necessary to detect and monitor shifts in key relevant relationships. While this analysis is important in any urban operations, it is especially critical when Army forces are opposed by urban insurgent and guerrilla forces.

B-28. In any system, commanders consider that the classification of a group is relative to the perspective from which it is viewed. This is an especially important consideration in multinational and interagency UO. During operations in Somalia, for example, U.S. Army forces may have viewed a particular clan as a hostile element. The United Nations' Italian contingent, with their colonial background in the region, may have considered the same clan as neutral or even an ally.

Shifting and Overlapping Groups

B-29. Commanders recognize that no system of categorization will precisely classify any given group; no system can reflect the overall nature and complexity of the urban society. A single group may fit in a particular category. It may also have components in two or more categories simultaneously. Often, it can shift among categories during an operation. A given group may have individuals in it who have interests identical to or different from that group and these individual interests may change over time.

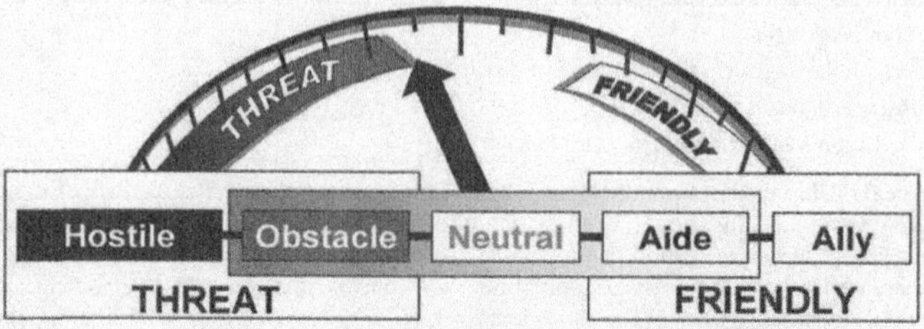

Hostile:	**An element (*individual, group, or organization*) that has the intent to negatively influence mission accomplishment by supporting enemy forces (or another hostile element) or exploiting friendly vulnerabilities but is not engaging in combat operations—otherwise, they lose any civilian status and become an enemy force.**
Obstacle:	**An element whose presence or activities serve to unintentionally impede, disrupt, or constrain friendly operations by imposing additional losses in personnel, time, equipment, and resources. Its interests may or may not be compatible with friendly objectives.**
Neutral:	**An element whose presence, activities, and overall interests do not conflict with friendly, enemy, or hostile elements.**
Aide:	**An element whose *present* intentions are—either wittingly or unwittingly—compatible with friendly force objectives.**
Ally:	**An element whose *fundamental* interests and intent align with friendly force objectives.**

Figure B-6. A refined approach to civilian allegiance

B-30. A peace enforcement operation illustrates the varying nature of groups. An identified criminal group might be classified as an obstacle to the commander's mission because its illegal activities impede unit progress. Its compelling interest, however, is to make money rather than resist the actions of friendly forces. In the same operation, one of the armed belligerents may be intent on disrupting the peace process and would be, therefore, classified as a hostile. (Again, they are not an enemy unless the belligerents are engaged in planned combat operations against the peacekeeping force. Commanders should consider, however, that under international law, if any civilian takes a direct or active role in hostilities, that civilian can be subject to attack.) The belligerent force may finance the criminal organization to assist in further obstructing the peace mission. In this case, the criminal organization moves from being an obstacle to being a hostile.

B-31. This classification effort, therefore, is not a one-time undertaking; commanders must constantly review and update it. Groups or individuals can be influenced into assisting either the friendly or opposing force. People will also act opportunistically, shifting support and alliances as perceived advantages arise. Even seemingly passive and law abiding members of the urban society may conduct themselves in unexpected ways given the right conditions (mob violence, for example).

> **Shifting Civilian Interests and Intent**
>
> Among other applicable lessons (see also the vignette in Chapter 9), the Los Angeles riots of 1992 illustrate how urban population groups can shift their relative positions due to changing conditions in an urban AO. Several gangs exist in the Los Angeles area. Usually, these gangs are hostile to one another. During the riots, however, several rival gangs formed a "united front" against what was seen as a larger obstacle to their own interests: law enforcement. As a result, the hostile gangs became one another's aides during that time.

Similarities, Differences, Capabilities, and Vulnerabilities

B-32. Similarities and differences in interests and interdependencies between groups are often focal points for analysis (and the allocation of ISR assets). They may indicate how commanders may influence, coerce, or align civilian interests and intentions with mission objectives. Simultaneously, commanders should consider an analysis of the civilian element's (individual, group, or organization) capability to influence the accomplishment of friendly objectives. They should also consider civilians' vulnerabilities and dependencies. If a civilian group's fundamental interests align with friendly objectives and this group has the intent to assist friendly forces, it is clearly an ally. However, with limited or no capability, a specific group will not help move the commander any closer to achieving his desired end state and accomplishing the mission. Then the commander would normally limit the resources expended on this group to those necessary for maintaining their commitment to common goals and objectives.

B-33. In contrast, commanders may provide a group with resources to enhance or create the capability to assist in mission accomplishment. They may do this if they felt that the assistance gained (or reduction to threat support) exceeds the potential diminishment of their own force's capabilities from losing those same resources. Commanders would also consider the group's dependencies (such as food, infrastructure, and protection) and overall vulnerability to outside influence. If vulnerable to friendly influence or control (understanding urban societal considerations and matters of perception are critical in this regard), then forces are likely susceptible to enemy or hostile manipulation. Even if commanders can generate extra resources (and not significantly affect their own force's capabilities), they should still conduct this same cost-benefit analysis to determine which civilian group (if any) should receive resources.

Greatest Potential Nearer the Center

B-34. The most critical population sectors often are those nearest the middle of the spectrum shown in figure B-6, particularly if their capabilities (or potential) significantly degrade or enhance mission accomplishment. These are the obstacle, neutral, or aide categories. If their beliefs and interests can be adequately understood, commanders may have great chance to influence the population segment and significantly contribute to mission accomplishment. In the fight for information superiority, persuading a neutral population to become an aide may provide critical information required about the urban environment and threat forces operating within.

Political and Strategic Concerns

B-35. The aide category may be of significant political or strategic concern. An aide group may be invaluable for accomplishing intermediate objectives but become a vulnerability to accomplishing a larger stability operation. (Even an urban offensive or defensive operation is likely to transition to a stability operation.) For example, commanders may provide resources to a criminal organization to assist in defeating insurgent forces during urban combat operations. Once these forces have been subdued, the interests (monetary gain and defeat of the threat) linking friendly forces with this criminal element disappear. What may remain is a criminal organization with more power than a reconstituted or newly established law enforcement agency and a truly destabilizing force. This also illustrates the second- and third-order thinking that will be required of commanders and their staffs during UO.

URBAN INTELLIGENCE TOOLS AND PRODUCTS

B-36. Developing tools and products to assist with situational understanding is not the task of intelligence sections alone. Commanders and all elements of their staffs must develop and adapt products and tools suited to their particular requirements. Listed below are some of the analytic tools and products that may help meet those requirements in a complex environment (see figure B-7 on page B-14 and FMI 2-91.4). Standard tools and products include: modified combined obstacle overlays and doctrinal, situation, event, and decision support templates or matrices. In addition to these standards aids, staffs and analysts may develop or produce other innovative tools to assist commanders in their situational understanding of the complex urban environment. These tools may contain overlapping information as different types of information are compared to determine patterns among them. Staffs and analysts may also initiate requests for products (or information) from their higher headquarters or other agencies with the technical means or control over assets when the capability lies outside the Army force's means. The tools that may be developed or requested include—

- Imagery.
- Three-dimensional representations.
- Infrastructure blueprints.
- Hydrographic surveys.
- Psychological profiles.
- Matrices, diagrams, or charts.
- Various urban overlays.

B-37. There are many software applications available to the Army that can be used to conduct intelligence analysis and create many of the relevant products described above. These applications currently range from such programs as the Analyst Notebook and Crimelink which have link analysis, association matrix, and pattern analysis software tools to the Urban Tactical Planner developed by the Topographic Engineering Center as an operational planning tool. The focus of the following information, however, is the various types of tools that can be developed and used to understand the urban environment rather than the constantly evolving and changing software or hardware that may be used to create them.

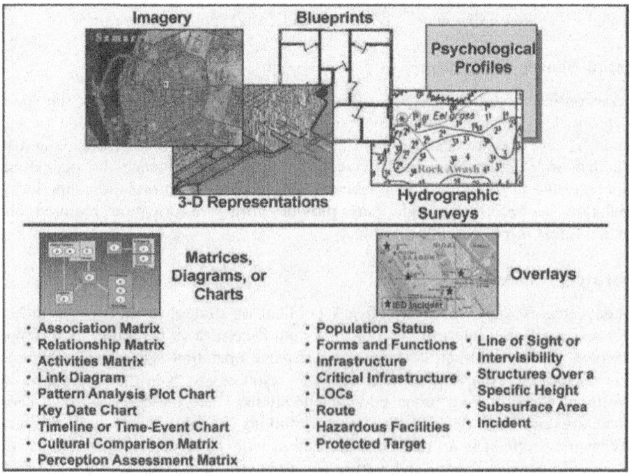

Figure B-7. Urban intelligence tools and products

IMAGERY

B-38. Recent satellite imagery or aerial photography will be required for most types of UO. Such images clarify vague and inaccurate maps and other graphic representations. Satellite assets provide responsive data input into the geographic information systems (GIS). (The National Geospatial Intelligence Agency [NGA] is one important source of imagery and map data.) GIS will often form the basis for creating the three-dimensional representations and the various overlays described below. Frequently updated (or continuous real-time) satellite or aerial imagery may be required for detailed pattern analysis and maintaining accurate situational understanding. For example, imagery taken during an area's rainy season may appear significantly altered during the summer months. Finally, digital, hand-held still and video cameras, particularly at the tactical level, will also be critical in developing and attaining situational understanding in a complex urban environment.

THREE-DIMENSIONAL REPRESENTATIONS

B-39. Often, physical or computer-generated (virtual) three-dimensional representations may be required to achieve situational understanding. These models or computer representations include specific sections of the urban area or specific buildings or structures. Such detail is particularly important for special operating forces and tactical-level units. These units require detail to achieve precision, increase the speed of the operation, and lessen friendly casualties and collateral damage.

INFRASTRUCTURE BLUEPRINTS

B-40. Urban police, fire, health, public utilities, city engineer, realty, and tourist agencies, and other urban organizations often maintain current blueprints and detailed maps. Such documents may prove useful to update or supplement military maps or to clarify the intricacies of a specific infrastructure. They may prove critical in operations that require detailed information to achieve the speed and precision required for success. Without such detail, analysts determine interior configurations based on a building's outward appearance. That task generally becomes more difficult as the building size increases.

HYDROGRAPHIC SURVEYS

B-41. Many urban areas are located along the world's littoral regions and major rivers. Therefore, commanders may need current hydrographic surveys to support amphibious, river crossing, riverine, and sustainment operations.

PSYCHOLOGICAL PROFILES

B-42. Psychological profiles analyze how key groups, leaders, or decision makers think or act—their attitudes, opinions, and views. They include an analysis of doctrine and strategy, culture, and historical patterns of behavior. The degree to which the attitudes, beliefs, and backgrounds of the military either reflect or conflict with the urban populace's (or civilian leadership's) core values is extremely important in this analysis. Psychological profiles help to assess the relative probability of a threat (or noncombatant group) adopting various COAs as well as evaluating a threat's vulnerability to deception. These profiles are derived from open-source intelligence as well as signals and human intelligence.

MATRICES, DIAGRAMS, AND CHARTS

B-43. Matrices, diagrams, and charts help to identify and understand key relationships among friendly and threat forces and other significant elements of the urban environment. While similar, each looks at information in a different way to uncover hidden patterns and connections useful in understanding the complex urban environment, particularly the threat and the urban society. These tools and products might include—

- **Association Matrix**. The association matrix helps identify the nature and relationship between individuals. Association matrices also help to identify those personalities and associations needing a more in-depth analysis necessary to establish the degree of relationship, contacts, or knowledge between individuals. Threat organizational structure and functions are uncovered as connections between personalities are made.

- **Relationship Matrix**. Relationship matrices are an extension of association matrices described above. They are used to examine the relationship between groups and organizations—threat forces, friendly forces, NGOs, the media, and the various elements of the urban population. Urban terrain and infrastructure may also be included as analytical factors to help expose relationships. These matrices graphically depict how human elements of the urban area interact with the physical elements to promote or degrade mission success. A significant relationship matrix may be a comparison of cultural perspectives—ideology, politics, religion, acceptable standards of living, and mores—between urban population groups and Army (and multinational) forces to help understand and accurately predict a civilian element's actions. The relationships discovered may suggest ways that commanders can shape the environment. For example, conflicting ideals or issues between two disparate threat forces may be emphasized to cause threats to focus their resources against each other rather than Army forces. Conversely, stressing common ideals identified between opposing noncombatant organizations may cause these groups to unite and work harmoniously toward a goal reflecting their commonalities.
- **Activities Matrix**. An activities matrix connects individuals to organizations, professions, events, activities, or addresses. Information from this matrix (combined with information from association and relationship matrices) also helps to link personalities, uncover the structure of an organization (threat or otherwise), and recognize differences, similarities, and dependencies for possible exploitation.
- **Link Diagram**. This tool graphically depicts many of the relationships and associations described above—people, groups, events, locations, or other factors deemed significant in any given situation. It helps commanders and analysts visualize how people and other factors are interrelated to determine key links.
- **Pattern Analysis Plot Chart**. This chart depicts the times and dates of a selected activity (such as ambushes, sniper and mortar attacks, bombings, and demonstrations) to search for patterns of activity for predictive purposes as well as to discern intent. This analysis can be conducted using a time-event wheel as well as by plotting events on maps using multiple historical overlays (analog or digital). (Compare with Incidents Overlay covered later in this chapter.) During pattern analysis, commanders and analysts consider not only what is occurring but also what is not occurring (or ceases to occur).
- **Key Event Chart**. In many urban operations, particularly stability operations, key holidays, historic events, and other significant cultural or political events can be extremely important for commanders to understand. More than just dates, these charts depict what can be expected to happen on each particular event. In Bosnia, for example, weddings are often held on Fridays and celebratory fire is a common occurrence. Understanding this cultural phenomenon could reduce collateral damage and accompanying civilian deaths.
- **Timeline or Time-Event Chart**. Timelines are a list of significant historical dates with relevant information and analysis that provide the commander with a record of past activities necessary to understand current operational conditions. A timeline may highlight a specific feature of the present situation such as population movements or political shifts or outline the general chronological record of the urban area, perhaps highlighting the activities of a certain population sector. Understanding past events also helps commanders predict future reactions to proposed COAs. Similarly to link diagrams, time-event charts display large amounts of information in a small space. These charts may help commanders discover larger-scale patterns of activity and relationships.
- **Cultural Comparison Matrix**. Commanders must avoid ethnocentricity—assuming that only one cultural perspective exists (often their own) and using that as the single lens with which to view the situation. To avoid this common obstacle to an accurate understanding of the urban society, commanders use this matrix to compare local ideology, attitudes, beliefs, and acceptable standards of living with associated U.S. and Army norms and values. This matrix not only highlights differences but also similarities.

- **Perception Assessment Matrix**. All of the above tools contribute to developing the perception assessment matrix; they allow commanders to see the urban environment from the perspective of its inhabitants. The perception assessment matrix is another tool that uses this refined perspective and the framework outlined in the matrix to predict the urban population's perception of a proposed COA. Although perceptions are not actions, they—more than reality—drive decision-making and resultant civilian activity. (As an example, legitimacy is a critical factor in developing many COAs that affect the civilian population, particularly during stability operations. This legitimacy must be viewed from the perspective of the inhabitants and not based on the Army commander's own culturally-shaped perceptions. Moreover, the legitimacy of an operation may need to be considered from other perspectives as well. These might include those of the American public, coalition partners and their publics, and those of neighboring states in the region.) This same matrix can later be used as a foundation to track and gauge the effectiveness of the chosen COA with favorable perception as one criterion of success. Commanders must be wary that perceptions are not fixed. They may change based on factors that the commander may not be able to control or influence. Therefore, commanders must continually monitor perceptions for deviation.

VARIOUS URBAN OVERLAYS

B-44. Staffs can produce various map overlays. These overlays depict physical locations of some aspect critical to the planning and conduct of the urban operation. Given adequate lead time, NGA can produce many overlays as an integrated map product (including satellite imagery). These overlays can include the—

- **Population Status Overlay**. This tool depicts the physical location of various groups identified by any significant social category such as religion, tribe, or language. During offensive and defensive operations, it may simply be where significant numbers of people are "huddled" or located throughout the battlefield. Population dispersal can vary significantly through the day, particularly at night, and must be considered as part of the overall analysis leading to the development of this tool.
- **Forms and Functions Overlay**. Based on the urban model, this overlay depicts the urban core or central business district, industrial areas, outlying high-rise areas, commercial ribbon areas, and residential areas, to include shantytowns.
- **Infrastructure Overlay**. This overlay is actually a series of overlays. It depicts identifiable subsystems in each form of urban infrastructure: communications and information, transportation and distribution, energy, economics and commerce, administration and human services, and cultural. Each subsystem can be broken down into more detail. Infrastructure data may be used to develop three other overlays—

  ▪ **Critical Infrastructure Overlay**. This tool displays specific elements of the urban infrastructure that, if harmed, will adversely affect the living conditions of the urban society to the detriment of the mission. These elements may include power generation plants, water purification plants and pumping stations, and sewage treatment plants. This information could be coded as part of the overall infrastructure overlay.

  ▪ **Lines of Communications (LOCs) Overlay**. The LOCs overlay highlights transportation systems and nodes, such as railways, road, trails, navigable waterways, airfields, and open areas for drop zones and landing zones. It also includes subsurface areas and routes such as sewage, drainage, and tunnels and considers movement between supersurface areas. The LOCs overlay and the route overlay (below) consider traffic conditions, times, and locations, to include potential points where significant portions of the urban population may congregate.

  ▪ **Route Overlay**. This overlay emphasizes mobility information to assist commanders and planners in determining what forces and equipment can move along the urban area's mobility corridors. Pertinent data includes street names, patterns, and widths; bridge, underpass, and overpass locations; load capacities; potential sniper and ambush locations (which may be its own overlay); and key navigational landmarks such as major roads and highways, rivers and canals, cemeteries, bridges, stadiums, and churches. The structures over a specific height overlay and subsurface overlay may assist in its development. As with the LOCs overlay, commanders, planners, and analysts think in all dimensions.

- **Line of Sight or Intervisibility Overlay**. This product creates a profile view (optical or electronic) of the terrain from the observer's location to other locations or targets. It can show trajectory or flight-line masking as well as obstructed or unobstructed signal pathways.
- **Structures over a Specific Height Overlay**. This level of detail may also be critical to communications, fires, and Army airspace command and control (air mobility corridors especially low-level flight profiles). Incorporated as part of this overlay, it may include floors or elevations above limitations for particular weapon systems at various distances from the structure.
- **Subsurface Area Overlay**. As an alternate to the building or structure height overlay, this product provides the locations of basements, underground parking garages, sewers, tunnels, subways, naturally-occurring subterranean formations, catacombs, and other subsurface areas. Similar to elevation "dead spaces," this overlay may show areas that exceed depression capabilities of weapon systems and potential threat ambush locations—again, affecting maneuver options.
- **Urban Logistic Resources Overlay**. This product identifies the locations of urban logistic resources that may contribute to mission accomplishment. It may contain specific warehouse sites, hospitals and medical supply locations, viable food stores, building material locations, fuel storage areas, car or truck lots, maintenance garages, and appliance warehouses. (NGO locations, taken from an NGO relationship matrix, may be an essential, overlapping element of this overlay.)
- **Hazardous Facilities Overlay**. This overlay identifies urban structures with known or suspected chemical, biological, or radiological features, such as nuclear power plants, fertilizer plants, oil refineries, pharmaceutical plants, and covert locations for producing weapons of mass destruction. These locations are critical to maneuver and fire planning.
- **Protected Target Overlay**. This overlay depicts terrain that should not be destroyed or attacked based on restrictions due to international, host-nation, or U.S. law and subsequent rules of engagement. These may include schools, hospitals, historical or other culturally significant monuments, and religious sites. This overlay may incorporate no-fire areas, such as special operations forces locations, critical infrastructure, logistic sources, and hazardous sites that must be protected as part of the commander's concept of the operation.
- **Incident Overlay**. Similar to the pattern analysis plot chart, this product depicts the location of different threat actions and types of tactics employed to uncover recurring routines, schemes, methods, tactics, or techniques and overall threat interests, objectives, or the desired end state.

B-45. The above IPB tools and products constitute a small sampling of what staffs and analysts can produce. They are limited only by their imaginations and mission needs (not all tools presented above may be relevant or necessary to every operation). Many products can be combined into a single product or each can generate further products of increasing level of detail. This is similar to transparent overlays positioned one atop another on a map. As discussed earlier, technology—software and hardware—will allow for more urban data to be combined, compared, analyzed, displayed, and shared—ultimately allowing scaleable products that can expand and contract to meet the intelligence requirements of any echelon of command down to the individual Soldier. The challenge will remain, however, to provide timely, accurate, complete, and relevant information in an understandable, usable, and share-capable form—share-capable among different levels of command as well as among multinational partners and various governmental and (as required) nongovernmental organizations—without overloading the commander.

Appendix C

Operations in Somalia: Applying the Urban Operational Framework to Stability Operations

It's impossible for an American mother to believe that a Somali mother would raise children to avenge the clan.

Major General Thomas M. Montgomery

GENERAL SITUATION

C-1. Following decades of political unrest and the fall of Somali dictator Siad Barre, a civil war broke out as 14 clans vied for power. The resulting nation composed of hostile social factions was held together by weak political alliances—none strong enough to unite and lead the country to national reconciliation. An ongoing drought led to famine and compounded the ethnic tensions and political instability. This volatile situation rapidly led to a phased U.S. involvement (see figure C-1). Army forces combined, sequenced, and proportionally emphasized the different types of operations to accomplish changing political objectives. Throughout all operations in Somalia, urban areas were critical to achieving mission success.

SOMALI OPERATIONS

INITIAL UN RESPONSE

C-2. The United Nations (UN) initially responded to requests for assistance from international relief organizations by sending supplies and other forms of humanitarian aid to Somalia. However, widespread looting, fighting between gangs, and other lawlessness prevented supplies from reaching the hungry and sick. Only 20 percent of the food entering the country reached the people who needed it. An estimated 25 percent of Somalia's 6 million people died of starvation or disease. In April 1992, the UN issued Security Council Resolution (UNSCR) 751 which authorized 50 unarmed observers, but the action had little effect. Under continuing pressure for additional measures to ensure the delivery of supplies and relief, the UN authorized 500 armed peacekeepers (furnished by Pakistan and transported by U.S. sea- and airlift) to protect humanitarian workers. The battalion's limited mission, designated UN Operations in Somalia I (UNOSOM I), encompassed safeguarding the unloading of ships and providing convoy security.

PROVIDE RELIEF (UNOSOM I)

C-3. In July 1992, the UN requested an increased airlift of supplies and the United States quickly responded. U.S. Central Command (USCENTCOM) activated joint task force (JTF) OPERATION PROVIDE RELIEF. Based on careful mission analysis, USCENTCOM limited the JTF's actions to—

- Deploying a humanitarian assistance survey team to assessing relief requirements.
- Providing an emergency airlift of supplies.
- Using Air Force cargo aircraft for daily relief sorties into Somalia.

C-4. USCENTCOM restricted the sorties to flying during daylight hours and to locations that would provide a permissive and safe environment. In mid-September 1992, the United States prudently expanded its role by stationing the amphibious ready group, USS Tarawa, offshore to provide support to the Pakistani security battalion and to provide security for U.S. airlift operations. The 11th Marine expeditionary unit (MEU) was on board the USS Tarawa to rapidly respond to any change in mission (see Appendix D for a description and the capabilities of a MEU).

Operation	Dates	UN Security Council Resolution	Relative Proportionality Between Types of Operations
PROVIDE RELIEF (UNOSOM I)	Aug 92 – Dec 92	UNSCR# 751 24 Apr 92	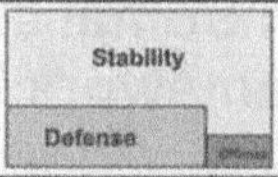 Stability Defense Offense
RESTORE HOPE (UNITAF)	Dec 92 – May 93	UNSCR# 794 3 Dec 92	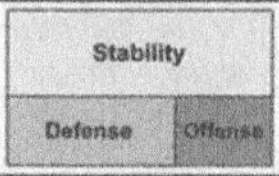 Stability Defense Offense
CONTINUED HOPE (UNOSOM II)	May 93 – Mar 94	UNSCR# 814 26 Mar 93	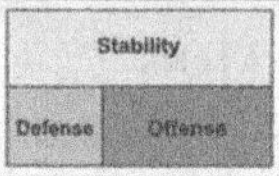 Stability Defense Offense

Figure C-1. Relative proportionality between types of operations

RESTORE HOPE (UNITAF)

C-5. By November 1992, the magnitude of the task, UN organizational deficiencies, and a continued lack of security precluded delivery of sufficient supplies to the needy. Notably, a ship laden with relief supplies was fired on in the harbor at Mogadishu, forcing its withdrawal before the supplies could be brought ashore, and a Pakistani peacekeeper was shot when his car was hijacked. Subsequently, the United States offered to provide forces and lead a UN-sponsored operation to reopen the flow of food to where it was needed most. In December 1992, the UN issued UNSCR 794, which authorized member states "to use all necessary means to establish a secure environment for humanitarian relief operations in Somalia" and demanded all "factions in Somalia immediately cease hostilities." To allay concerns of colonialism by a number of African countries, the UN Secretary-General was given oversight of the operation. The resolution also required soldiers to be withdrawn once order was restored; however, it provided no exit strategy. As clearly as possible, the USCENTCOM mission statement (below) for OPERATION RESTORE HOPE reflected the UN mandate:

> *When directed by the [President or the Secretary of Defense], [Commander, USCENTCOM] will conduct joint/combined military operations in Somalia to secure the major air and sea ports, key installations and food distribution points, to provide open and free passage of relief supplies, provide security for convoys and relief organization operations, and assist UN/[nongovernmental organizations] in providing humanitarian relief under UN auspices. Upon establishing a secure environment for uninterrupted relief operations, [Commander, USCENTCOM] terminates and transfers relief operations to UN peacekeeping forces.*

C-6. Mogadishu was the largest port in the country and the focal point of previous humanitarian relief activities of nongovernmental organizations (NGOs). It was also the headquarters of the coalition of 20 nations and over 30 active humanitarian relief organizations. As such, Mogadishu became the entry point for the operational buildup of the multinational force known as Unified Task Force (UNITAF) and the key logistic hub for all operations in Somalia. UNITAF immediately gained control over the flow of relief supplies into and through Mogadishu and stabilized the conflict among the clans. In less than a month, UNITAF forces expanded control over additional ports and interior airfields. They secured additional distribution sites in other key urban areas in the famine belt to include Baidoa, Baledogle, Gialalassi, Bardera, Belet Uen, Oddur, Marka, and the southern town of Kismayo (see figure C-2). With minimal force, the U.S.-led UNITAF established a secure environment that allowed relief to reach those in need, successfully fulfilling its limited—yet focused—mandate.

CONTINUED HOPE (UNOSOM II)

C-7. In March 1993, the UN issued UNSCR 814 establishing a permanent peacekeeping force, UNOSOM II. However, the orderly transition from UNITAF to UNOSOM II was repeatedly delayed until May 1993. (The UN Secretary-General urged the delay so that U.S. forces could effectively disarm bandits and rival clan factions in Somalia.) This resolution was significant in two critical aspects:

- It explicitly endorsed nation building with the specific objectives of rehabilitating the political institutions and economy of Somalia.
- It mandated the first ever UN-directed peace enforcement operation under the Chapter VII enforcement provisions of the Charter, including the requirement for UNOSOM II to disarm the Somali clans. The creation of a peaceful, secure environment included the northern region that had declared independence and had hereto been mostly ignored.

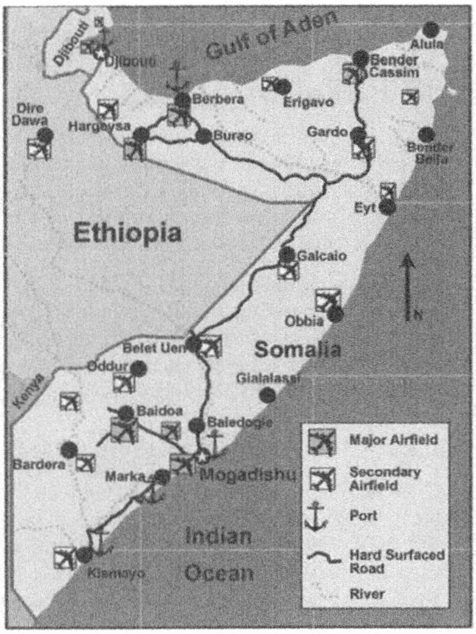

Figure C-2. Map of Somalia

C-8. These far-reaching objectives exceeded the limited mandate of UNITAF as well as those of any previous UN operation. Somali clan leaders rejected the shift from a peacekeeping operation to a peace enforcement operation. They perceived the UN as having lost its neutral position among rival factions. A more powerful clan leader, General Mohammed Farah Aideed (leader of the Habr Gidr clan), aggressively turned against the UN operation and began a radio and propaganda campaign. This campaign characterized UN soldiers as an occupation force trying to recolonize Somalia.

C-9. The mounting crisis erupted in June 1993. Aideed supporters killed 24 Pakistani soldiers and wounded 57 in an ambush while the soldiers were conducting a short-warning inspection of one of Aideed's weapons arsenals. UNSCR 837, passed the next day, called for immediately apprehending those responsible and quickly led to a manhunt for Aideed. The United States deployed 400 rangers and other special operations forces (SOF) personnel to aid in capturing Aideed, neutralizing his followers, and assisting the quick reaction force (QRF), composed of 10th Mountain Division units, in maintaining the peace around Mogadishu.

PHASED WITHDRAWAL

C-10. On 3 October 1993, elements of Task Force (TF) Ranger (a force of nearly 100 Rangers and SOF operators) executed a raid to capture some of Aideed's closest supporters. Although tactically successful, 2 helicopters were shot down, 75 soldiers were wounded, and 18 soldiers were killed accomplishing the mission. The U.S. deaths, as well as vivid scenes of mutilation to some of the soldiers, increased calls to Congress for withdrawing U.S. forces from Somalia. The President then ordered reinforcements to protect U.S. Forces, Somalia (USFORSOM) as they began a phased withdrawal with a 31 March deadline. The last contingent sailed from Mogadishu on 25 March, ending OPERATION CONTINUED HOPE and the overall U.S. mission in Somalia.

C-11. Although U.S. forces did not carry out the more ambitious UN goals of nation building, they executed their missions successfully, relieving untold suffering through humanitarian assistance with military skill and professionalism. Operations in Somalia occurred under unique circumstances, yet commanders may glean lessons applicable to future urban stability operations, as well as civil support operations. In any operations, commanders balance changing mission requirements and conditions.

UNDERSTAND

C-12. Although accomplished to varying degrees, U.S. forces failed to adequately assess and understand the urban environment, especially the society. Somali culture stresses the unity of the clan; alliances are made with other clans only when necessary to elicit some gain. Weapons, overt aggressiveness, and an unusual willingness to accept casualties are intrinsic parts of the Somali culture. Women and children are considered part of the clan's order of battle.

C-13. Early in the planning for OPERATION RESTORE HOPE, U.S. forces did recognize the limited transportation and distribution infrastructure in Mogadishu. The most notable was the limited or poor airport and harbor facilities and its impact on the ability of military forces and organizations to provide relief. Therefore, a naval construction battalion made major improvements in roads, warehouses, and other facilities that allowed more personnel, supplies, and equipment to join the relief effort faster.

Understanding the Clan (The Human Dimension)

C-14. During OPERATION RESTORE HOPE, the UNITAF worked with the various clan leaders as the only recognized leadership remaining in the country. The UNITAF was under the leadership of LTG Robert B. Johnston and U.S. Ambassador to Somalia, Robert Oakley. In addition, UNITAF forces also tried to reestablish elements of the Somali National Police—one of the last respected institutions in the country that was not clan-based. This reinstated police force manned checkpoints throughout Mogadishu and provided crowd control at feeding centers. Largely because of this engagement strategy, the UNITAF succeeded in its missions of stabilizing the security situation and facilitating humanitarian relief. Before its termination, the UNITAF also worked with the 14 major Somali factions to agree to a plan for a transitional or transnational government.

C-15. The UN Special Representative of the Secretary-General, retired U.S. Navy Admiral Jonathon Howe, worked with the UNOSOM II commander, Turkish General Cevik Bir. During OPERATION CONTINUED HOPE, Howe and General Bir adopted a philosophy and operational strategy dissimilar to their UNITAF predecessors. Instead of engaging the clan leaders, Howe attempted to marginalize and isolate them. Howe initially attempted to ignore Aideed and other clan leaders in an attempt to decrease the warlord's power. Disregarding the long-established Somali cultural order, the UN felt that, in the interest of creating a representative, democratic Somali government, they would be better served by excluding the clan leadership. This decision ultimately set the stage for strategic failure.

Threat Strategy and Tactics

C-16. During OPERATION RESTORE HOPE, U.S. forces also failed to properly analyze and understand their identified threat's intent and the impact that the urban environment would have on his strategy, operations, and tactics. The UN began to view eliminating Aideed's influence as a decisive point when creating an environment conducive to long-term conflict resolution. Aideed's objective, however, remained to consolidate control of the Somali nation under his leadership—his own brand of conflict resolution. He viewed the UN's operational center of gravity as the well-trained and technologically advanced American military forces, which he could not attack directly. He identified a potential American vulnerability—the inability to accept casualties for an operation not vital to national interests—since most Americans still viewed Somalia as a humanitarian effort. If he could convince the American public that the price for keeping troops in Somalia would be costly, or that their forces were hurting as many Somalis as they were helping, he believed they would withdraw their forces. If U.S. forces left, the powerless UN would leave soon after, allowing Aideed to consolidate Somalia under his leadership.

Vulnerability and Risk Assessment

C-17. U.S. forces failed to assess, understand, and anticipate that Aideed would adopt this asymmetric approach and attack the American public's desire to remain involved in Somalia. By drawing U.S. forces into an urban fight on his home turf in Mogadishu, he could employ guerrilla insurgency tactics and use the urban area's noncombatants and its confining nature. Such tactics made it difficult for the U.S. forces to employ their technological superiority. If U.S. forces were unwilling to risk harming civilians, his forces could inflict heavy casualties on them, thereby degrading U.S. public support for operations in Somalia. If, on the other hand, the U.S. forces were willing to risk increased civilian casualties to protect themselves, those casualties would likely have the same effect.

C-18. However, an assessment and understanding of the Somali culture and society should have recognized the potential for Aideed's forces to use women and children as cover and concealment. Accordingly, the plan should have avoided entering the densely populated Bakara market district with such restrictive rules of engagement. As legitimacy is critical to stability operations, TF Ranger should have been prepared and authorized to employ nonlethal weapons, to include riot control gas, as an alternative to killing civilians or dying themselves.

C-19. U.S. forces also failed to assess and understand the critical vulnerability of their helicopters in an urban environment and the potential impact on their operations. TF Ranger underestimated the threat's ability to shoot down its helicopters even though they knew Somalis had attempted to use massed rocket-propelled grenade (RPG) fires during earlier raids. (Aideed brought in fundamentalist Islamic soldiers from Sudan, experienced in downing Russian helicopters in Afghanistan, to train his men in RPG firing techniques). In fact, the Somalis had succeeded in shooting down a UH-60 flying at rooftop level at night just one week prior to the battle. Instead, TF Ranger kept their most vulnerable helicopters, the MH-60 Blackhawks, loitering for forty minutes over the target area in an orbit that was well within Somali RPG range. The more maneuverable AH-6s and MH-6s could have provided the necessary fire support. Planning should have included a ready ground reaction force, properly task organized, for a downed helicopter and personnel recovery contingency.

C-20. Information operations considerations apply throughout the entire urban operational framework; however, operations security (OPSEC) is critical to both assessment and shaping. OPSEC requires continuous assessment throughout the urban operation particularly as it transitions among the spectrum of operations. As offensive operations grew during OPERATION CONTINUED HOPE, U.S. forces did little to protect essential elements of friendly information. Combined with the vulnerability of U.S. helicopters, Aideed's followers used U.S. forces' inattention to OPSEC measures to their advantage. The U.S. base in Mogadishu was open to public view and Somali contractors often moved about freely. Somalis had a clear view both day and night of the soldiers' billets. Whenever TF Ranger would prepare for a mission, the word rapidly spread through the city. On 3 October 1993, Aideed's followers immediately knew that aircraft had taken off and, based on their pattern analysis of TF Ranger's previous raids, RPG teams rushed to the rooftops along the flight paths of the task force's Blackhawks.

SHAPE

C-21. One of the most critical urban shaping operations is isolation. During OPERATION CONTINUED HOPE, U.S. forces largely discounted other essential elements of friendly information and did not establish significant public affairs and psychological operations (PSYOP) initiatives. In fact, Army forces lacked a public affairs organization altogether. Consequently, Aideed was not isolated from the support of the Somali people. This failure to shape the perceptions of the civilian populace coupled with the increased use of lethal force (discussed below) allowed Aideed to retain or create a sense of legitimacy and popular support. Ironically, the focus on Aideed helped create a popular figure among the warlords through which many noncompliant, hostile, and even neutral Somalis could rally behind.

C-22. During OPERATION RESTORE HOPE, Aideed conducted propaganda efforts through "Radio Aideed"—his own radio station. UNITAF countered these efforts with radio broadcasts. This technique proved so effective that Aideed called MG Anthony C. Zinni, UNITAF's director of operations, over to his house on several occasions to complain about UNITAF radio broadcasts. General Zinni responded, "If he didn't like what we said on the radio station, he ought to think about his radio station and we could mutually agree to lower the rhetoric." This approach worked.

ENGAGE

C-23. The complexity of urban operations requires unity of command to identify and effectively strike the center of gravity with overwhelming combat power or capabilities. Complex command and control relationships will only add to the complexity and inhibit a commander's ability to dominate and apply available combat power to accomplish assigned objectives. Stability operations as seen in Somalia required commanders to dominate within a supporting role and, throughout, required careful, measured restraint.

UNITY OF COMMAND (EFFORT)

C-24. During OPERATION RESTORE HOPE, UNITAF successfully met unity of command challenges through three innovations. First, they created a civil-military operations center (CMOC) to facilitate unity of effort between NGOs and military forces. Second, UNITAF divided the country into nine humanitarian relief sectors centered on critical urban areas that facilitated both relief distribution and military areas of responsibility. Third, to establish a reasonable span of control, nations that provided less than platoon-sized contingents were placed under the control of the Army, Marine Corps, and Air Force components.

C-25. On the other hand, during OPERATION CONTINUED HOPE, UNOSOM II command and control relationships made unity of command (effort) nearly impossible. The logistic components of USFORSOM were under UN operational control, while the QRF remained under USCENTCOM's combatant command—as was TF Ranger. However, the USCENTCOM commander was not in theater. He was not actively involved in planning TF Ranger's missions or in coordinating and integrating them with his other subordinate commands. It was left to TF Ranger to coordinate with the QRF as needed. Even in TF Ranger, there were dual chains of command between SOF operators and the Rangers. This underscores the need for close coordination and careful integration of SOF and conventional forces (see Chapter 4). It also emphasizes overall unity of command (or effort when command is not possible) among all forces operating in a single urban environment.

C-26. Following TF Ranger's 3 October mission, the command structure during OPERATION CONTINUED HOPE was further complicated with the new JTF-Somalia. This force was designed to protect U.S. forces during the withdrawal from Somalia. JTF-Somalia came under the operational control of USCENTCOM, but fell under the tactical control of USFORSOM. Neither the JTF nor USFORSOM controlled the naval forces that remained under USCENTCOM's operational control. However unity of effort (force protection and a rapid, orderly withdrawal) galvanized the command and fostered close coordination and cooperation among the semiautonomous units.

MEASURED RESTRAINT

C-27. During OPERATIONS PROVIDE RELIEF and RESTORE HOPE, U.S. forces dominated within their supporting roles. Their perseverance, adaptability, impartiality, and restraint allowed them to provide a stable, secure environment. Hence, relief organizations could provide the food and medical care necessary to reduce disease, malnourishment, and the overall mortality rate. However, during OPERATION CONTINUED HOPE, U.S. operations became increasingly aggressive under the UN mandate. However, peace enforcement also requires restraint and impartiality to successfully dominate and achieve political objectives. The increased use of force resulted in increased civilian casualties, which in turn reduced the Somalis' perception of U.S. legitimacy. As a result, most moderate Somalis began to side with Aideed and his supporters. Many Somalis, without effective information operations to shape their perceptions, felt that it was fine for foreign military forces to intervene in their country to feed the starving and even help establish a peaceful government, but not to purposefully target specific Somali leaders as criminals.

CONSOLIDATE AND TRANSITION

C-28. Across the spectrum of conflict, Army forces must be able to execute full spectrum operations not only sequentially but, as in the case of operations in Somalia, simultaneously. OPERATION PROVIDE RELIEF began primarily as a stability operation in the form of foreign humanitarian assistance. Later this operation progressed to include peacekeeping (another stability operation), defensive operations to protect

UN forces and relief supplies, and minimum offensive operations. As operations transitioned to OPERATION RESTORE HOPE, it became apparent that while foreign humanitarian assistance was still the principal operation, other operations were necessary. More forceful peacekeeping, show of force, and arms control (all forms of stability operations), and offensive and defensive operations grew increasingly necessary to establish a secure environment for uninterrupted relief operations. In the final phase of U.S. involvement during OPERATION CONTINUED HOPE, major changes to political objectives caused a major transition in the form of the stability operation to peace enforcement with an even greater increase in the use of force, offensively and defensively, to create a peaceful environment and conduct nation building. Finally, the TF Ranger raid demonstrated the need to maintain a robust, combined arms force capable of rapidly transitioning from stability operations (primarily peace operations) to combat operations.

SUMMARY

C-29. OPERATIONS PROVIDE RELIEF and RESTORE HOPE were unquestionably successes. Conversely, during OPERATION CONTINUED HOPE, the 3-4 October battle of Mogadishu (also known as the "Battle of the Black Sea") was a tactical success leading to an operational failure. TF Ranger succeeded in capturing 24 suspected Aideed supporters to include two of his key lieutenants. Arguably, given the appropriate response at the strategic level, it had the potential to be an operational success. After accompanying Ambassador Oakley to a meeting with Aideed soon after the battle, MG Zinni described Aideed as visibly shaken by the encounter. MG Zinni believed Aideed and his subordinate leadership were tired of the fighting and prepared to negotiate. Unfortunately, the U.S. strategic leadership failed to conduct the shaping actions necessary to inform and convince the American public (and its elected members of Congress) of the necessity of employing American forces to capture Aideed. The President was left with little recourse after the battle of Mogadishu but to avoid further military confrontation.

C-30. Despite this strategic failing, the operational commanders might have avoided the casualties, and any subsequent public and Congressional backlash, had they better communicated among themselves and worked with unity of effort. Recognizing the separate U.S. and UN chains of command, the UN Special Representative, along with the USCENTCOM, USFORSOM, and TF Ranger commanders, should have established the command and control architecture needed. This architecture would have integrated planning and execution for each urban operation conducted. These commanders failed to "operationalize" their plan. They did not properly link U.S. strategic objectives and concerns to the tactical plan. The TF Ranger mission was a direct operational attempt to obtain a strategic objective in a single tactical action. Yet, they failed to understand the lack of strategic groundwork, the threat's intent and capabilities, and the overall impact of the urban environment, to include the terrain and society, on the operation. Such an understanding may not have led to such a high-risk course of action but instead to one that de-emphasized military operations and emphasized a political solution that adequately considered the clans' influence.

This page intentionally left blank.

Appendix D

Joint and Multinational Urban Operations

[Joint force commanders] synchronize the actions of air, land, sea, space, and special operations forces to achieve strategic and operational objectives through integrated, joint campaigns and major operations. The goal is to increase the total effectiveness of the joint force, not necessarily to involve all forces or to involve all forces equally.

JP 3-0

As pointed out earlier, Army forces, brigade size and larger, will likely be required to conduct operations in, around, and over large urban areas in support of a joint force commander (JFC). The complexity of many urban environments, particularly those accessible from the sea, requires unique leveraging and integration of all the capabilities of U.S. military forces to successfully conduct the operation. This appendix discusses many of these capabilities; JP 3-06 details joint urban operations.

PURPOSE

D-1. In some situations, a major urban operation is required in an inland area where only Army forces are operating. Army commanders determine if the unique requirements of the urban environment require forming a joint task force (JTF) or, if not, request support by joint capabilities from the higher joint headquarters. Sometimes the nature of the operation is straightforward enough or the urban operation is on a small enough scale that conventional intraservice support relationships are sufficient to meet the mission requirements.

D-2. Most major urban operations (UO), however, require the close cooperation and application of joint service capabilities. A JTF may be designated to closely synchronize the efforts of all services and functions in an urban area designated as a joint operations area (JOA). If a large urban area falls in the context of an even larger ground force area of operations, a JTF dedicated to the urban operation may not be appropriate. These situations still require joint capabilities. In such cases, the responsible JFC designates support relations between major land units and joint functional commands. The major land units can consist of Army forces, Marine Corps forces, or joint forces land component command. The joint functional commands can consist of the joint special operations task force, joint psychological operations task force, or joint civil-military operations task force.

D-3. This appendix describes the roles of other services and joint combatant commands in UO. It provides an understanding that enables Army commanders to recommend when to form a JTF or to request support from the JFC. It also provides information so commanders can better coordinate their efforts with those of the JFC and the commanders of other services or components conducting UO. Lastly, this appendix describes some considerations when conducting UO with multinational forces.

SISTER SERVICE URBAN CAPABILITIES

D-4. Army forces conducting UO rely on other services and functional joint commands for specialized support in the urban environment. These capabilities are requested from and provided through the commanding JFC. Army forces request the assets and capabilities described in this annex through their higher headquarters to the joint command. The JFC determines if the assets will be made available, the appropriate command relationship, and the duration of the support. Army forces prepare to coordinate planning and execution with other services and to exchange liaison officers. These capabilities can greatly increase the Army's ability to understand, shape, engage, consolidate, and transition within the context of UO.

AIR FORCE

D-5. Air Force support is an important aspect of the Army force concept for urban operations. Air Force elements have a role to play in UO across the spectrum of operations.

D-6. Air Force intelligence, surveillance, and reconnaissance (ISR) systems contribute significantly to understanding the urban environment. These ISR systems include the E-8 Joint Surveillance Target Attack Radar System (JSTARS) (see figure D-1), U2 Dragon Lady, RC-135 Rivet Joint, or RQ-4 Global Hawk and RQ/MQ-1 Predator unmanned aerial systems. Air Force ISR systems can provide vital data to help assess threat intentions, threat dispositions, and an understanding of the civilian population. These systems also can downlink raw information in real-time to Army intelligence processing and display systems, such as the common ground station or division tactical exploitation system.

Figure D-1. USAF E-8 JSTARS platform

D-7. Air interdiction (AI) can be a vital component of shaping the urban operational environment. Often, AI of the avenues of approach into the urban area isolates the threat by diverting, disrupting, delaying, or destroying portions of the threat force before they can be used effectively against Army forces. AI is especially effective in major combat operations where restrictions on airpower are limited and the threat is more likely to be a conventionally equipped enemy. In 1991 during OPERATION DESERT STORM, AI helped prevent the Iraqi 5th Mechanized Division from reaching Khafji.

D-8. Precision-guided munitions (PGMs) delivered from combat aircraft can positively and directly affect the conduct of Army close combat actions in the urban area. Special munitions designed to penetrate reinforced concrete can provide unique support to land forces executing UO. Although laser-guided bombs are extremely accurate with an effective laser lock-on, environmental factors such as smoke, haze, and weather can adversely affect successful laser designation. If the launching aircraft can achieve a successful laser designation and lock-on, these weapons have devastating effects, penetrating deep into reinforced concrete before exploding with great force. If launched without a lock-on, or if the laser spot is lost, these weapons are unpredictable and can travel long distances before they impact. However, continual advancements in tactics and weapons systems, such as joint direct attack munition (JDAM) weapons, help decrease the adverse effects of these environmental factors. (For example, in 2003 during OPERATION IRAQI FREEDOM, the Air Force supported the Army's attack on An Najaf with JDAM-equipped F 16s and B 1s during a blinding sandstorm. With near zero visibility, JDAM weapons provided significant fire support, allowing Army forces to complete their mission and continue the attack north.) When PGMs are unavailable, general-purpose bombs from 500 to 2,000 pounds can also prove effective. However, there may be risk factors associated with their use such as collateral damage due to decreased accuracy, threat exposure to the delivering aircraft due to lower delivery altitudes, or reduced weapons effectiveness due to shallower penetration angles. (See FM 3-06.1 for a more detailed description of various aircraft weapons and capabilities.)

D-9. In addition to shaping and engaging the UO through firepower, commanders can use Air Force capabilities to improve and augment the urban transportation and distribution infrastructure. Air Force units can repair or improve airfields, revitalize civil aviation maintenance facilities, manage air-delivered cargo, and control civil and military air traffic. These latter capabilities particularly enhance urban stability and civil support operations. These capabilities may even be decisive. Air-delivered cargo and air traffic management, for example, were the decisive factors in U.S. forces' successful resistance of the Soviet blockade of Berlin in 1948. In OPERATION RESTORE HOPE, from December 1992 to May 1993, Air Force operations in Mogadishu were critical to airlifting and staging supplies and forces. The Air Force determined the ultimate success of the humanitarian assistance operation (see Appendix C).

D-10. In unique situations, such as the Berlin Blockade and OPERATION RESTORE HOPE, exercising Air Force urban capabilities may be the decisive action of the operation. Most often, however, Air Force capabilities will play a shaping role, sustaining role, or both in joint urban operations because of the inherent requirement to occupy terrain and interface with the population.

MARINE CORPS

D-11. The Marine Corps can assault across water obstacles into a defended urban environment. This capability is an invaluable tactical and operational tool. The mere threat of this capability can divert many enemy forces from other avenues of approach and obscure the true nature of an attack. The impact of the threat of amphibious assault was vividly demonstrated during OPERATION DESERT STORM where embarked Marine Forces diverted several Iraqi divisions to defensive positions along the coast and near Kuwait City.

D-12. The presence of Marine amphibious equipment, apart from Marine infantry, provides Army forces unique capabilities. In UO, the amphibious operation is often not an assault from the sea, but rather an assault river crossing. In 1950, the 7th U.S. Infantry Division used amphibious tractor support from the 1st Marine Division to conduct an assault river crossing of the Han River into downtown Seoul. See FM 3-31.1.

D-13. The worldwide deployment of Marine air-ground task forces (MAGTFs) enables a short notice response into any urban areas accessible from the sea. Typically, a deployed MAGTF is a Marine expeditionary unit (special operations capable) (MEU[SOC]). The MEU(SOC) can perform forcible entry operations, seize lodgments, and may execute these tasks anticipating reinforcement by Army or joint forces. They are also well positioned and equipped to rapidly reinforce Army forces already deployed in theater. The special-operations-capable training that these units accomplish before deploying includes urban warfare training and contributes to their value in UO. The MEU(SOC) is relatively small (its core unit is a Marine infantry battalion), is forward deployed, and has a wide spectrum of organic capabilities. It is an important asset in crisis stability and civil support operations.

D-14. The Marine expeditionary brigade is the MAGTF between a MEU and a Marine expeditionary force (the largest MAGTF). All MAGTFs include an aviation combat element, which is particularly skilled at providing accurate and timely close air support. The responsiveness and accuracy of Marine close air support aptly suits it to UO where the lethality of combat and the close range of engagements demand accurate and responsive fires.

D-15. Marine forces that conduct UO work well in littoral urban areas because of their unique relationship with naval forces and thus their capability to closely integrate land and sea operations. A supporting arms liaison team may be attached to Army forces at battalion level to provide ship-to-shore communications and coordination for naval gunfire support.

NAVY

D-16. Many major urban areas are accessible from the sea. Therefore, Army commanders should understand how sea power can influence and support UO. The Navy brings several major capabilities to UO. These include naval gunfire support, naval air support, and port and coastal security. In addition, the Navy can offer secure command post sites and meeting areas; provide limited emergency shelter and critical care for Soldiers and, as necessary, civilians; and store and transport critical supplies ashore providing flexibility in meeting support requirements. The ability to provide support from "over the horizon" not only enhances force protection but also the legitimacy of urban and host nation authorities during transition.

D-17. Naval surface fire support (which includes naval gunfire support) particularly applies to forcible entry operations in littoral urban areas. It provides an initial indirect fire support capability until Army forces land ashore. However, naval gunfire support lacks a precision munitions capability. In defensive operations, naval gunfire support can also add major fires, especially during retrograde operations through an urban area. Naval gunfire support of the 3rd Infantry Division was the primary fire support for the last two days of the X Corps retrograde operation out of the port of Hungnam, North Korea, in December 1950.

D-18. Destroyers and cruisers, which mount one and two 127mm MK45 lightweight gun systems respectively (see figure D-2), are capable of providing naval gunfire support. This gun system can provide a rate of fire of 20 rounds per minute per gun to a range of approximately 23 kilometers. Destroyers, cruisers, and attack submarines are capable of precision strikes with Tomahawk cruise missiles. Depending on the location of the urban area, naval air can provide responsive aviation support to UO alleviating the need for fixed-wing aviation bases.

Figure D-2. USN MK45 lightweight gun system

D-19. Due to its flat trajectory, terrain masking affects naval gunfire more than field artillery. Naval gunfire also results in large range probable errors (the dispersion pattern of the naval gun is roughly elliptical with the long axis in the direction of fire). Hence, coverage of targets such as roads and airfields is most effective when the gun-target line (GTL) coincides with the long axis of the target. Very close supporting fire can be delivered when the GTL is parallel to the front line of troops. Oppositely, a GTL perpendicular to the front trace can endanger friendly forces. Within the limits of hydrographic conditions, the ship can maneuver to achieve a better GTL, but ship movement also makes it difficult to adjust fire. Overall, naval and air threats, bad weather, and large range probable errors make naval gunfire difficult and can cause cancellation of supporting fires.

D-20. Water terminals located in urban areas are usually the debarkation points for the bulk of Army forces as well as a joint force theater logistics requirement. Army forces are responsible for water terminal operations while naval capabilities protect these strategic and operationally vital facilities.

D-21. Naval coastal warfare (NCW) is the responsibility of the JFC and is often exercised through the Navy component commander. He may assign a naval coastal warfare commander for an appropriate geographic area. NCW includes coastal sea control, port security, and harbor defense. While coastal sea control is conducted in the environment of the open seas, port security and harbor security include the urban environment. Port security is the safeguarding of vessels, harbors, ports, waterfront facilities, and cargo from internal threats. It includes destruction, loss, or injury from sabotage or other subversive acts; accidents; thefts; or other causes of similar nature. The Navy's role in protecting essential urban infrastructure is often key. This task is also important when executing stability operations if the threats against urban infrastructure will likely be unconventional.

D-22. Harbor defense protects harbor approaches, harbors, anchorages, and ports from external threats. Harbor defense focuses on the conventional defense of port infrastructure. It is a task appropriate in a major operation or campaign and often includes port security as a subtask. The JFC executes NCW (focused on harbor defense and port security) using combined Navy capabilities including surface warfare, aviation, and naval special operations. See also discussions in this appendix on transportation command and special operations command.

COAST GUARD

D-23. The Coast Guard, like the Navy, can significantly influence the conduct of UO when the urban area is accessible from the sea. The U.S. Coast Guard (USCG) is the federal authority for port security and harbor defense of domestic facilities. When directed by the President, the USCG can augment the Navy in operations overseas. Historically, the entire USCG was under Navy control during both World Wars I and II. USCG elements deployed overseas and operated under Navy control during the Vietnam War and during OPERATIONS DESERT SHIELD and DESERT STORM.

D-24. The USCG uses surface warfare systems and aerial reconnaissance systems to conduct its missions. Its air systems are unarmed, whereas its surface systems are armed for self-defense and law enforcement operations. The USCG is experienced and adept at supporting other agencies, local governments, and law enforcement. Its capabilities can best support Army UO in civil support operations or the civil support aspects of an operation. The USCG works effectively against an unconventional threat and threats with on-water capability.

URBAN FUNCTIONAL COMBATANT COMMAND CAPABILITIES

D-25. Three commands provide urban functional combatant command capabilities. Transportation Command manages the global transportation system. Strategic Command integrates plans for and employs space forces ad assets. Special Operations Command is responsible for providing special operations forces.

TRANSPORTATION COMMAND

D-26. U.S. Transportation Command (USTRANSCOM) provides strategic air, land, and sea transportation for the Department of Defense to deploy, employ, sustain, and redeploy U.S. military forces worldwide. USTRANSCOM provides global transportation management, using an integrated transportation system across the spectrum of operations through its transportation component commands (TCCs). The TCC consists of Air Mobility Command (AMC), Military Sealift Command (MSC), and Surface Deployment and Distribution Command (SDDC).

D-27. During urban operations, USTRANSCOM, through its TCC, can provide common-user terminal services in support of strategic transportation movements to a theater of operations. AMC provides common-user airlift, air refueling, and aeromedical evacuation services. It is the worldwide aerial port manager and, where designated, the operator of common-user aerial ports of embarkation and aerial ports of debarkation. MSC provides common-user sealift services between seaports of embarkation (SPOEs) and seaports of debarkation (SPODs). SDDC provides common-user ocean terminal service and, where designated, serves as the single port manager at SPOEs and SPODs. The urban transport system, in the form of railheads, ports, and airfields, is integral to many urban operations across the spectrum of Army operations and often the objective of Army UO. Army forces planning, preparing, and executing urban operations engage the supported geographic commander of a combatant command or his component commands in all aspects of conducting UO when the objective is transport related.

D-28. In offensive operations, commanders use transportation expertise to identify the urban transportation infrastructure (both the terrain and social [human] aspects) that is secured or that can affect current and future operations. This analysis also includes second- and third-order support systems. USTRANSCOM is consulted regarding the degree of acceptable damage that the system can sustain and still meet mission requirements. During execution, USTRANSCOM units may integrate into the operation so they can begin operating the transportation systems as early as possible. During the Inchon landing of September 1950, Army forces had begun rail operations on D+1 and port operations under way by D+3 of the forcible entry.

D-29. In defensive and stability or civil support operations, USTRANSCOM units safeguard and prevent disruption of the transport system by the conduct of defensive operations or stability tasks.

USTRANSCOM advises Army commanders of the impact of defensive tactics, techniques, and procedures (TTP) on USTRANSCOM operations as well as security requirements for USTRANSCOM facilities. The JFC provides guidance to deconflict any issues that may arise from these potentially divergent missions and tasks.

STRATEGIC COMMAND

D-30. The United States Strategic Command (USSTRATCOM) is the combatant command that integrates and synchronizes space operations, information operations, integrated missile defense, global command and control (C2); ISR, global strike, and strategic deterrence to provide. A unified source for greater understanding of specific threats around the world and the means to rapidly respond to those threats including weapons of mass destruction. USSTRATCOM operates assigned forces through its service component commands: U.S. Army Space Command/Army Forces Strategic Command; Air Force Space Command, and Naval Net Warfare Command.

D-31. USSTRATCOM provides critical space systems that offer global coverage and potential for real-time and near real-time support to urban operations. Space systems are unconstrained by political boundaries. Commanders can use space systems during peacetime or times of crisis to monitor an urban area before inserting friendly forces. Space systems enhance operations and assist commanders in overcoming some of the physical challenges of the urban environment. These enhancement operations include ISR; positioning and navigation; environmental monitoring; and communications. Space-based imagery and sensors are important ISR capabilities that contribute to situational understanding throughout the depth, breadth, and height of the urban area. ISR systems can provide route and target information for mission planning, locate presurveyed missile launch sites, detect camouflage, assess threat operations and movements, and warn of hostile acts and reconnaissance. Positioning and navigation systems also assist situational understanding through links to digital information systems (INFOSYS) while assisting tactical navigation in ambiguous terrain found in some urban areas. These enhancements can include precise location and position information for urban fires, ingress and egress routes, and rendezvous coordination. Environmental monitoring systems can provide weather and ionospheric information needed to assess weapon selection, air routes, ground and water trafficability, and communications. Communications systems provide secure, survivable links between elements of Army and joint forces to disseminate plans, orders, and warnings. These systems may form a critical link in the INFOSYS that transmit data to assessment centers and intelligence to key decision makers. However, some of the same environmental influences and degradation as ground-based systems may affect space-based systems. Commanders should refer to FM 3-14 for more information on space support to Army operations.

SPECIAL OPERATIONS COMMAND

D-32. U.S. Special Operations Command (USSOCOM) exercises combatant command of all active and reserve special operations forces stationed in the United States. USSOCOM also provides trained and combat-ready special operations forces (SOF) to the geographical combatant commanders and, when directed by the President or the Secretary of Defense, command-designated special operations.

D-33. SOF provide commanders with capabilities critical to success in the urban environment. The density of this environment in both space and time requires the careful integration of SOF and conventional forces. Army forces conducting UO have a clear and unambiguous C2 relationship with the SOF in the urban area to ensure coordination, massing of effects, and unity of effort.

D-34. Each service has unique special operations (SO) capabilities. For example, SOF can identify and seize or destroy key terrain or infrastructure in denied areas; secure or capture key personnel; identify supportive civilians (aides and allies) who can facilitate future infiltrations and the cache of critical supplies; counter urban insurgencies; and conduct unconventional warfare in enemy-held urban areas. SOF can also emplace covert sensors and surveillance devices, provide clandestine intelligence collection, and provide target acquisition information or emplace beacons (for weapons guidance or navigation) in the highly restrictive terrain of the urban environment. In multinational UO, they can provide coalition support teams with trained, culturally aware, language proficient, military liaison personnel with organic communications connectivity. Army leaders understand the SO capabilities available to ensure that they request the right support for Army UO and to ensure unity of effort within the urban JOA.

Army SOF

D-35. The Army provides five types of SOF units to USSOCOM. Two of these, psychological operations (PSYOP) and civil affairs, are discussed in Chapter 4.

Ranger Forces

D-36. The U.S. Army 75th Ranger Regiment can rapidly deploy light infantry forces from company through regimental size. Rangers specialize in direct action (DA) SO missions and focus on airfield seizure and raids. Typically, Ranger units turn over their objectives to conventional units upon mission completion. The Ranger DA capability especially applies to UO because of the many critical infrastructures that often prompt Army forces to engage in UO. Ranger capabilities are ideal for seizing critical facilities to preempt their defense or destruction. The size and combat power of Ranger units permit execution of offensive and defensive operations against enemy conventional units for periods of limited duration; austere combat service support capabilities limit the Ranger regiment's ability to sustain combat action without extensive augmentation.

Special Forces

D-37. U.S. Army Special Forces Command (Airborne) trains and prepares Army Special Forces (SF) to deploy and execute operational requirements for geographical combatant commanders. SF units are small and capable of extended operations in remote and hostile locations. SF units execute seven basic missions: foreign internal defense (FID), unconventional warfare (UW), counterproliferation of weapons of mass destruction, special reconnaissance (SR), DA, counterterrorism, and information operations. Important collateral activities consist of coalition support, combat search and rescue, counterdrug activities, humanitarian demining activities, humanitarian assistance, security assistance, and special activities.

D-38. SR, DA, and coalition support are particularly important in UO. SF units may physically penetrate an urban area to conduct SR to determine threat strengths, dispositions, and intentions. In some situations, SF units, due to their language and cultural training, can gather accurate information regarding the disposition and attitudes of the population. SF special reconnaissance can also determine or verify the functional status of urban infrastructure as well as conduct target acquisition, area assessment, and poststrike reconnaissance.

D-39. SF teams can execute terminal guidance and with qualified joint tactical air controllers can control operations for Army and joint precision fires. Although these teams possess limited organic combat power for their short-duration, DA missions, they can bring significant effects to bear against high-payoff targets. Yet, SF units have no capability to conduct a sustained defense of such targets.

D-40. SF units can advise, train, and assist urban indigenous movements already in existence to conduct unconventional warfare and possibly accompany these groups into combat. The upsurge in urban insurgency and terrorism has caused worldwide concern since it is not confined to developing countries. Present day dissident groups are well aware of this situation. These groups realize that to be successful, they must center the insurgent activities on the major cities or political center of their countries.

Special Operations Aviation Forces

D-41. Army special operations aviation provided by the 160th Special Operations Aviation Regiment operates primarily to support SO missions. They execute insertion, extraction, and resupply missions to support SO. Similar to conventional aviation, this capability may be more vulnerable to concealed air defenses when operating over hostile or unsecured urban terrain than in many other environments.

Navy SOF

Sea-Air-Land Teams

D-42. Navy sea-air-land teams (SEALs) specialize in water approaches to targets. They operate in small, squad-size teams and have many of the same capabilities as Army Special Forces (see above). Navy SEALs do not typically have the cultural and language training of Army Special Forces. Their capability to insert from sea gives them a unique ability to penetrate into urban areas that are accessible from the sea.

Special Boat Units

D-43. Special boat units (SBUs) employ, operate, and maintain various surface combatant craft (see figure D-3) to conduct and support naval and joint special operations, riverine warfare, and coastal patrol and interdiction. The SBU can infiltrate and exfiltrate forces; provide small-caliber gunfire support; and conduct coastal patrol, surveillance, harassment, and interdiction of maritime lines of communications. These units are ideal in UO that include ports and rivers. They can assist in port security, conduct river patrols, and participate in harbor defense. They are well suited for preventing sea infiltration by unconventional threats.

Figure D-3. USN MK V special

Operations CraftAir Force SOF

D-44. The Air Force has three primary SOF elements: special operations air units, battlefield Airmen, and combat aviation advisory units.

Special Operations Wings and Groups

D-45. Air Force special operations aviation elements operate both fixed-wing and rotary-wing aircraft (which include variants of both C-130s and MH-53s). Air Force SOF conduct core missions which include specialized air mobility, precision aerospace fires, aerospace surface interface, personnel recovery operations, specialized refueling, combat aviation advisory, information operations, and PSYOP. As a result, Air Force SOF conduct infiltrations, exfiltrations, resupplies, close air support of SOF ground elements, electronic warfare, and aerial C2 support.

D-46. The AC-130 U and AC-130 H model aircraft (see figure D-4) are designed specifically to provide close air support to ground SOF. However, they will be an invaluable asset to any ground force operating in an urban environment. Hence, Army commanders seek opportunities to train with AC-130 gunships prior to the conduct of UO. AC-130s are armed with one 40mm autocannon and one 105mm howitzer. The AC-130 U is also armed with one 25mm autocannon. This aircraft has night capability and is extremely accurate. Its fires are responsive and can be decisive in close urban combat. Its cannon and howitzer are accurate enough to concentrate fire onto a single spot to create a rooftop breach that allows fire to be directed deeper into the building. The aircraft can provide excellent covert illumination with its infrared spotlight. The AC-130 is very vulnerable to air defense systems, which friendly forces must suppress or destroy to effectively use this system. (During OPERATION DESERT STORM in 1991, an AC-130 was shot down over Kuwait.)

Battlefield Airmen

D-47. Battlefield Airmen consist of special tactics teams (STTs) and special operations weather teams (SOWTs). Combat controllers and pararescue personnel constitute STTs. STTs are specially tailored to meet mission criteria and may vary from a small three-man team to a larger twenty-five-man element. STTs support the UO Army commanders by—

- Performing air-land-sea personnel recovery operations.
- Providing terminal attack control or guidance.
- Establishing and operating navigational aids and beacons.
- Providing liaison to ground commanders.
- Providing visual flight rules and limited instrument flight rules air traffic control.
- Providing positive control of the terminal objective area aviation environment during SOF operations.

Note: SOWTs are normally attached to Army SOF to provide weather observation and limited tactical forecasting.

Combat Aviation Advisory (CAA) Operations

D-48. CAA operations are part of an interagency process aimed toward freeing and protecting the host nation's population from subversion, lawlessness, and insurgency. Specifically, CAA operations are tailored to assess, train, advise, and assist foreign aviation forces in air operations employment and sustainability. CAA operations support geographical combatant commanders throughout the spectrum of conflict primarily by facilitating the availability, reliability, safety and interoperability, and integration of friendly and allied aviation forces supporting joint, interagency, and multinational forces and organizations. CAA operations provide assistance in the interrelated areas of FID, coalition support, UW, and humanitarian and disaster relief. CAA operations also include a liaison role in coalition support.

MULTINATIONAL CONSIDERATIONS

D-49. Army UO in foreign urban areas will often be joint and likely have a multinational component. When properly executed, integrating multinational forces into UO greatly enhances the operation's military (as well as political) effectiveness. Properly integrating multinational forces into UO requires a thorough understanding of both the urban environment and the nature of individual national forces. This understanding includes the political, cultural, and historical characteristics of the other national forces. Such understanding also includes the national force's doctrine and military capabilities, strengths, and weaknesses. In UO, these considerations are critical because these factors will alter the urban population's attitude toward multinational forces and the behavior of such forces as they interact with the urban population. Combining this understanding with effective C2 and an equal understanding of the urban environment results in effective multinational UO (see FM 100-8).

CULTURAL COMPATIBILITY

D-50. When evaluating and assigning UO tasks, Army commanders also consider the degree of interaction with the civilian population. A national force from a Muslim-majority country may work better with a civilian population that is also Muslim than a force having a different religion. In such a situation, the national force with the same religion as the urban population may be assigned tasks that require close relations with civilians. Army forces, in this situation, may be assigned tasks that are more remote from the population. National forces that have a national history of animosity to the civil population (or that sympathize with antagonists within the civilian population) are not used in tasks requiring diplomacy and close cooperation with the civilians or government.

Military Capabilities and Restrictions

D-51. Army forces are responsible to understand the military capabilities of national forces with which they work. Some national forces, as part of their normal capabilities, are adept at police functions that enable them to operate with little training in a law enforcement role. Other national forces specialize in small-unit, light infantry patrolling. These forces may be ideal in a stability operation. In contrast, a national force composed of conscripts and trained primarily in conventional warfare techniques may best work as a reinforcing force or may require extensive training before mission execution in an urban environment or a stability situation. Army commanders also consider the type of weapon systems with which participating multinational forces are equipped; they may be more or less effective in an urban environment. For example, some countries may still possess the Vulcan antiaircraft gun system (or similar weapon) that can be very useful in urban offensive and defensive operations.

D-52. Commanders in multinational UO must also understand nation-specific rules of engagement (ROE). Some nations will severely limit their participating military's use of lethal and nonlethal fires. On the other hand, other nations may not be as constrained as Army units. Ideally, commanders attempt to create ROE that are the same across the coalition but recognize that this may not be achievable. Hence, in an environment where minimization of collateral damage will be paramount, Army commanders will need to carefully consider national limitations on the employment of forces when planning and executing UO. Relatedly, commanders take into account each nation's ability to achieve precision effects with their available forces and systems.

Intelligence Sharing

D-53. Due to classification issues, sharing intelligence with coalition partners may be challenging; the United States may have close intelligence ties with some countries within the coalition and few or none with others. In many cases intelligence personnel from other countries have unique skills (and cultural perspectives) that can significantly contribute to the friendly intelligence effort. Reports from some coalition members may fill intelligence gaps for Army forces and the coalition as a whole. On the other hand, few countries have the sophisticated intelligence collection assets available to Army forces and information that Army units may provide could be critical to a coalition partner's force protection and overall mission success.

Information Management

D-54. National capabilities to collect, process, store, display, and disseminate information and establish a common operational picture vary considerably between coalition partners. As the capabilities of Army information systems increase, those of our future partners may not keep pace. Even when capabilities are similar, equipment may not be compatible. The complex nature of the urban environment and coalition language differences (and linguist shortages) will only serve to exacerbate incompatibilities and limitations. Therefore commanders must understand capabilities and limitations to ensure that orders, control measures, and information can be appropriately disseminated, displayed, and understood. Significantly, they must understand the coalition's ability to accurately locate and track friendly forces as well as identify friend from foe. Commanders must take the appropriate steps to ensure that essential intelligence and relevant information is provided to coalition partners in a form (and language) that they can accept and use. This will be critical to establishing the common operational picture necessary to facilitate rapid decision making and effective command and control across the coalition.

Logistic Requirements

D-55. Logistically, agreement among the multinational forces should include support consolidation whenever possible. Creating a multinational logistics office or coordination center works best when accomplished early. This office can coordinate local contracts as well as already agreed upon host-nation support. This coordination among participating nations will reduce the competition for local assets that could otherwise have detrimental effects on one or more participating nations. Additionally, many nations providing forces for a coalition effort may not be able to support them logistically and will rely almost entirely on the United States or another coalition army for logistic support. Finally, pooled resources such as transportation assets and maintenance personnel could potentially reduce deployment requirements for one or more coalition partners. As the coalition matures, the role of this office or center may be expanded to include command activities.

This page intentionally left blank.

Source Notes

These are sources quoted or paraphrased in this publication. They are listed by page number. Quotations are identified by the first few word of the quote. Where a quote is embedded within a paragraph, the paragraph number is listed. Boldface indicates the title of historical vignettes.

Chapter 1–The Urban Outlook

1-1 "Today's security environment demands more...": FM 1, *The Army*, 14 June 2005: 1-19.

1-4 **Rome: A Microcosm of Urban Warfare**: R. Ernest Dupuy and Trevor N. Dupuy, *The Encyclopedia of Military History: From 3500 B.C. to the Present* (San Francisco: Harper & Row, 1977).

1-6 **The Three Block War**: as quoted in Russell W. Glenn, Steven Hartman, and Scott Gerwehr, *Urban Combat Service Support Operations: The Shoulders of Atlas* (Santa Monica, CA: RAND, 2003): 26.

1-8 **Winning the Peace in Iraq: The Requirement for Full Spectrum Operations**: Peter W. Chiarelli and Patrick R. Michaelis, "Winning the Peace: The Requirement for Full-Spectrum Operations," *Military Review* (July/August 2005): 4.

Chapter 2–Understanding the Urban Environment

2-1 "From a planning perspective, commanders view cities....": FM 3-0, *Operations*, 14 June 2001: 6-19.

2-16 "Me and Somalia Against the World": cited in Andrew S. Natsios, "Food Through Force: Humanitarian Intervention and U.S. Policy," *The Washington Quarterly* (winter 1994): 136.

2-18 **Understanding the Effects of Unit and Soldier Actions in Iraq**: *Initial Impressions Report: Operation Iraqi Freedom—Stability Operations-Support Operations,* Center for Army Lessons Learned (CALL) Newsletter No. 04-13 (Fort Leavenworth, KS: Center for Army Lessons Learned, U.S. Training and Doctrine Command [TRADOC], 2004): ii.

2-21 "Whoever coined the phrase... ": General Sir Rupert Smith, "Wars in Our Time–A Survey of Recent and Continuing Conflicts," *World Defence Systems*, volume 3:2 (London: Royal United Services Institute for Defence Studies, 2001): 121.

2-25 **Understanding the Urban Environment: Paris – 1944**: Russell W. Glenn, "Urban Combat is Complex," *Proceedings* (February 2002): 65. Russell W. Glenn, Steven Hartman, and Scott Gerwehr, *Urban Combat Service Support Operations: The Shoulders of Atlas* (Santa Monica, CA: RAND, 2003): 10-11.

Chapter 3–Understanding the Urban Threat

3-1 "...[T]he United States could be forced to intervene...": *Quadrennial Defense Review Report*, 30 September 2001: 6.

3-3 "...Iraq made no direct effort...": Gregory Fontenot, E. J. Degen, and David Tohn, *On Point: US Army in Operation IRAQI FREEDOM* (Fort Leavenworth, KS: Combat Studies Institute Press, 2004): 388.

3-4 **Tempo**: 26th Infantry Regimental Association, *Aachen. Military Operations in Urban Terrain* (Lititz, PA: 26th Infantry Regimental Association, 1999); Irving Werstein, *The Battle of Aachen* (New York: Thomas Y. Crowell Company, 1962); Charles Whiting, *Bloody Aachen* (New York: Stein and Day, 1976).

3-6 "Chechen fighters sometimes disguised...": Olga Oliker, *Russia's Chechen Wars 1994–2000: Lessons from Urban Combat* (Santa Monica, CA: RAND, 2001): 21.

3-7 **Identifying Threats from Noncombatants**: Olga Oliker, *Russia's Chechen Wars 1994–2000: Lessons from Urban Combat* (Santa Monica, CA: RAND, 2001).

3-9 **Information and the Media**: Eric Hammel, *Fire in the Streets: The Battle for Hue, Tet 1968* (New York: Dell Publishing, 1991).

3-12 **Cultural and Religious Instability**: Sid Heal, "Crowds, Mobs and Nonlethal Weapons," *Military Review* (March/April 2000): 45–50.

3-13 **Food and Water Shortages**: Leif R. Rosenberger, "The Strategic Importance of the World Food Supply," *Parameters* (spring 1997): 84–105.

3-15 **Insurgencies and the Urban Society**: Roger Trinquier, *Modern Warfare: A French View of Counterinsurgency* (New York: Praeger Publisher, 1964): 16-17.

3-16 **Crime and Criminal Organizations**: Eugene Linden, "The Exploding Cities of the Developing World," *Foreign Affairs* (January/February 1996): 52–65; Andrew S. Natsios, "Commander's Guidance: A Challenge of Complex Humanitarian Emergencies," *Parameters* (summer 1996): 50–66.

Chapter 4–Understanding the Urban Environment's Effects on Warfighting Functions and Tactics

4-1 "War is, above all things, an art...": Francis V. Greene, "The Important Improvements in the Art of War During the Past Twenty Years and Their Probable Effect on Future Military Operations," *Journal of the Military Service Institution of the United States* 4, no. 13 (1883): 41.

4-13 "Fighting in a city is much...": Vasili I. Chuikov, *The Battle for Stalingrad* (New York: Holt, Rinehart, and Winston, 1964): as cited by JP 3-06, *Doctrine for Urban Operations*, 16 September 2002: 1-8.

4-14 "I heard small-arms fire and...": Mark A. B. Hollis, "Platoon Under Fire," *Infantry* (January/April 1998): 29–30.

4-17 **Example of Simple Communications Innovation: Israel's Six-Day War – 1967**: Trevor N. Dupuy and Paul Martell, *Flawed Victory: The Arab-Israeli Conflict and the 1982 War in Lebanon* (Fairfax, VA: Hero Books, 1986); Richard A. Gabriel, *Operation Peace for Galilee: The Israeli-PLO War in Lebanon* (New York: Hill and Wang, 1984).

Chapter 5–Contemplating Urban Operations

5-1 "We based all our further calculations...": S. M. Shtemenko, *The Soviet General Staff at War, 1941-1945: Book One* (Moscow: Progress Publishers, 1981): 317–318.

5-19 "Four hostile newspapers...": Napoleon Bonaparte, Justin Wintle (ed.), *The Dictionary of War Quotations* (New York: The Free Press, 1989): 73.

5-24 **Conventional and Special Forces Integration**: Gregory Fontenot, E. J. Degen, and David Tohn, *On Point: US Army in Operation IRAQI FREEDOM* (Fort Leavenworth, KS: Combat Studies Institute Press, 2004): 403.

Chapter 6–Foundations for Urban Operations

6-1 "Utilities such as electricity…": Richard Connaughton, John Pimlott, and Duncan Anderson, *The Battle for Manila: The Most Devastating Untold Story of World War II* (Novata, CA: Presidio Press, 1995): 103.

6-7 **Applying the Urban Operational Framework: Panama – 1989**: Ronald H. Cole, *Operation Just Cause* (Washington DC: Joint History Office, 1995); Thomas Donnelly, Margaret Roth, and Caleb Baker, *Operation Just Cause: The Storming of Panama* (New York: Lexington Books, 1991); John T. Fishel, *The Fog of Peace: Planning and Executing the Restoration of Panama* (Carlisle, PA: Strategic Studies Institute, 1992); Malcolm McConnell, *Just Cause: The Real Story of America's High-Tech Invasion of Panama* (New York: St. Martin's Press, 1991); John Embry Parkerson, Jr., "United States Compliance with Humanitarian Law Respecting Civilians During Operation Just Cause," *Military Law Review* (1991): 31–140; Jennifer Morrison Taw, *Operation Just Cause: Lessons for Operations Other Than War* (Santa Monica, CA: RAND, 1996).

6-12 "I'm talking about attacking...": William S. Wallace quoted in Michael R. Gordon, "Baghdad Targets Picked if Hussein Holes Up There," *New York Times* (7 March 2003): A11.

6-13 "Do not try to..." T. E. Lawrence, "Twenty-Seven Articles," *The Arab Bulletin* (20 August 1917): as cited in David H. Petraeus, "Learning Counterinsurgency: Observations from Soldiering in Iraq," *Military Review* (January/February 2006): 4–5.

Chapter 7–Urban Offensive Operations

7-1 "…Capture Suez City...": Abraham Adan, *On the Banks of the Suez* (Novato, CA: Presidio Press, 1980): 409.

7-2 **Operational Context of Urban Operations: Brittany Ports – 1944**: Kent H. Butts, "The Strategic Importance of Water," *Parameters* (spring 1997): 65–83; Paul Carell, *Stalingrad: The Defeat of the German 6th Army* (Atglen, PA: Schieffer Publishing, 1993); R. Ernest Dupuy and Trevor N. Dupuy, *The Encyclopedia of Military History: From 3500 B.C. to the Present* (San Francisco: Harper & Row, 1977); Ralph Peters, "Our Soldiers, Their Cities," *Parameters* (spring 1996): 43–50; John J. Peterson, *Into the Cauldron* (Clinton, MD: Clavier House, 1973).

7-10 **Forms of Attack in the Urban Offense: Metz – 1944**: John Colby, *War From the Ground Up: The 90th Division in WWII* (Austin, TX: Nortex Press, 1991); Anthony Kemp, *The Unknown Battle: Metz, 1944* (New York: Stein and Day, 1981).

7-17 **Isolating an Urban Area: Hue, Vietnam – 1968**: James R. Arnold, *Tet Offensive 1968: Turning Point in Vietnam* (London: Osprey, 1990); Eric Hammel, *Fire in the Streets: The Battle for Hue, Tet 1968* (New York: Dell Publishing, 1991).

7-21 "Often what seems to be...": Russell W. Glenn, *An Attack on Duffer's Downtown* (Santa Monica, CA: RAND, 2001): 28.

7-22 **Creative Task Organization: Using Artillery in the Direct Fire Role**: 26th Infantry Regimental Association, *Aachen. Military Operations in Urban Terrain* (Lititz, PA: 26th Infantry Regimental Association, 1999); Ronald H. Cole, *Operation Just Cause* (Washington DC: Joint History Office, 1995); Thomas Donnelly, Margaret Roth, and Caleb Baker, *Operation Just Cause: The Storming of Panama* (New York: Lexington Books, 1991); David Eshel, *Mid-East Wars: The Lebanon War 1982* (Hod Hasharon, Israel: Eshel-Dramit, 1983); Richard A. Gabriel, *Operation Peace for Galilee: The Israeli-PLO War in Lebanon* (New York: Hill and Wang, 1984); Malcolm McConnell, *Just Cause: The Real Story of America's High-Tech Invasion of Panama* (New York: St. Martin's Press, 1991); Irving Werstein, *The Battle of Aachen* (New York: Thomas Y. Crowell Company, 1962); Charles Whiting, *Bloody Aachen* (New York: Stein and Day, 1976); Bruce Allen Watson, *Sieges: A Comparative Study* (Westport, CT: Praeger, 1993).

7-26 **Bold Operational Maneuver to Seize an Urban Area: Inchon and Seoul, Korea – 1950**: Robert Debs Heinl, Jr., *Victory at High Tide: The Inchon-Seoul Campaign* (Annapolis, MD: Nautical & Aviation Publishing Company of America, 1979); Douglas MacArthur, *Reminiscences* (New York: McGraw-Hill, 1964); Shelby L. Stanton, *Ten Corps in Korea, 1950* (Novato, CA: Presidio Press, 1989).

Chapter 8–Urban Defensive Operations

8-1 "Generally, a modern city magnifies...": Christopher R. Gabel, "Military Operations on Urbanized Terrain: The 2d Battalion, 26th Infantry, at Aachen, October 1944," *Urban Combat Operations: tactics, Techniques and Procedure.* Center for Army Lessons Learned (CALL) Newsletter No. 99-16 (Fort Leavenworth, KS: Center for Army Lessons Learned, U.S. Training and Doctrine Command [TRADOC], November 1999): 1-9.

8-5 **Urban Defense in a Major Operation: Stalingrad –1942 to 1943**: Anthony Beevor, *Stalingrad* (New York: Penguin Books, 1998); Vasili I. Chuikov, *The Battle for Stalingrad* (New York: Holt, Rinehart, and Winston, 1964); William Craig, *Enemy at the Gates: The Battle for Stalingrad* (New York: Reader's Digest Press, 1973).

8-8 "What is the position...": Michael Dewar, *War in the Streets: The Story of Urban Combat from Calais to Khafji* (New York: Sterling Publishers, 1992): 8.

8-10 **Defensive Combat Power: Suez – 1973**: Abraham Adan, *On the Banks of the Suez* (Novato, CA: Presidio Press, 1980).

Chapter 9–Urban Stability and Civil Support Operations

9-1 "The Rangers were bound...": Mark Bowden, *Black Hawk Down: A Story of Modern War* (New York: Atlantic Monthly Press, 1999): 18.

9-5 "In wars of intervention...": Antoine-Henri Jomini, "The Art of War," in *Roots of Strategy: Book 2* (Harrisburg, PA: Stackpole Books, 1987): 441.

9-8 **Assessment of Security and Force Protection: Belfast, Northern Ireland**: Michael Dewar, *British Army in Northern Ireland* (New York: Sterling Publishers, 1985); Desmond Hamill, *Pig in the Middle: The Army in Northern Ireland, 1969-1984* (London: Methuen, 1985).

9-11 "If there is any lesson...": Frederick Funston, "How the Army Worked to Save San Francisco," *Cosmopolitan Magazine* (July 1906). Available at http: //www.sfmuseum.org/1906/cosmo.html.

9-12 **Civil Support and Coordination with Civilian Authorities: Los Angeles – 1992**: James D. Delk, *Fires & Furies: The LA Riots* (Palm Springs, CA: Etc. Publications, 1995); William W. Mendel, *Combat in Cities: The LA Riots and Operation Rio* (Fort Leavenworth, KS: Foreign Military Studies Office, 1996).

9-14 **Restraint: An Najaf, Iraq – 2003**: Gregory Fontenot, E. J. Degen, and David Tohn, *On Point: US Army in Operation IRAQI FREEDOM* (Fort Leavenworth, KS: Combat Studies Institute Press, 2004).

9-16 "I met immediately with the...": Jeffrey B. Van Sickle, "Stability Operations in Northern Iraq: Task Force Altun Kupri," *Infantry* (January/February 2005): 26.

Chapter 10–Urban Sustainment

10-1 "Even supply is different…" Ralph Peters, "Our Soldiers, Their Cities," *Parameters* (spring 1996): 48.

10-2 **Base Security: Tan Son Nhut, Vietnam – 1968**: James R. Arnold, *Tet Offensive 1968: Turning Point in Vietnam* (London: Osprey, 1990); Keith William Nolan, *The Battle for Saigon: Tet 1968* (New York: Pocket Books, 1996).

10-4 "[Sustainment] planning accounts for...": FM 3-0, *Operations*, 14 June 2001: 6-19.

10-8 "Greater friendly force density would appear...": Russell W. Glenn, Steven Hartman, and Scott Gerwehr, *Urban Combat Service Support Operations: The Shoulders of Atlas* (Santa Monica, CA: RAND, 2003): 4.

10-13 **Host Nation Repair Operations: OPERATION IRAQI FREEDOM**: Steven Leonard, personal experience written specifically for FM 3-06.

10-18 **Combat Stress: Chechnya – 1994 to 1996**: Timothy L. Thomas and Charles P. O'Hara, "Combat Stress in Chechnya: 'The Equal Opportunity Disorder'," *Army Medical Department Journal* (January/March 2000): 46–53.

10-22 "…[I]n an endeavor like that in Iraq…": David H. Petraeus, "Learning Counterinsurgency: Observations from Soldiering in Iraq," *Military Review* (January/February 2006): 4–5.

Appendix A–Siege of Beirut: An Illustration of the Fundamentals of Urban Operations

A-1 **Siege of Beirut: An Illustration of the Fundamentals of Urban Operations**: Trevor N. Dupuy and Paul Martell, *Flawed Victory: The Arab-Israeli Conflict and the 1982 War in Lebanon* (Fairfax, VA: Hero Books, 1986); David Eshel, *Mid-East Wars: The Lebanon War 1982* (Hod Hasharon, Israel: Eshel-Dramit, 1983); Richard A. Gabriel, *Operation Peace for Galilee: The Israeli-PLO War in Lebanon* (New York: Hill and Wang, 1984).

A-1 "The [Israeli Defense Forces] IDF had...": Richard A. Gabriel, *Operation Peace for Galilee: The Israeli-PLO War in Lebanon* (New York: Hill and Wang, 1984): 128.

Appendix B–An Urban Focus to the Intelligence Preparation of the Battlefield

B-1 "Maneuvers that are possible...": *Infantry in Battle, 2d ed.* (Richmond, VA: Garret & Massie, 1939; Reprint, Fort Leavenworth, KS: U.S. Army Command and General Staff College, 1981): 69.

B-5 **Analysis of an Urban Area's Underlying Terrain: Mitrovica, Kosovo – 1999**: Jamison Jo Medby and Russell W. Glenn, *Street Smart: Intelligence Preparation of the Battlefield for Urban Operations* (Santa Monica, CA: RAND, 2002).

B-10 "…[T]he adversaries of freedom...": John F. Kennedy, Message to Congress, 25 May 1961. Available at <http://www.jfklibrary.org/Historical+Resources/Archives>.

B-12 **Shifting Civilian Interests and Intent**: James D. Delk, *Fires & Furies: The LA Riots* (Palm Springs, CA: Etc. Publications, 1995); Jamison Jo Medby and Russell W. Glenn, *Street Smart: Intelligence Preparation of the Battlefield for Urban Operations* (Santa Monica, CA: RAND, 2002).

Appendix C–Operations in Somalia: Applying the Urban Operational Framework to Stability Operations

C-1 **Operations in Somalia: Applying the Urban Operational Framework to Stability and Reconstruction Operations**: Kenneth Allard, *Somalia Operations: Lessons Learned* (Washington, DC: National Defense University Press, 1995).

C-1 "It's impossible for an American mother...": Thomas M. Montgomery, interview, "Ambush in Mogasdishu," *Frontline*, WGBH Boston, 29 September 1998, <http://www.pbs.org/wgbh/pages/frontline/programs/1998.html>.

C-2 C-5 …" to use all necessary…": *United Nations Security Council Resolution 794* (1992) (New York: United Nations, 1992), <http://daccessdds.un.org/doc/UNDOC/GEN/N92/772/11/PDF>.

C-2 C-5 "When directed by the…": CENTCOM mission statement as cited in Kenneth Allard, *Somalia Operations: Lessons Learned* (Washington, DC: National Defense University Press, 1995): 13-14.

Appendix D–Joint and Multinational Urban Operations

D-1 "[Joint force commanders] synchronize the actions…" JP 3-0, *Doctrine for Joint Operations*, 10 September 2001: II-4.

This page intentionally left blank.

Glossary

SECTION I – ACRONYMS AND ABBREVIATIONS

AI	air interdiction
AIDS	acquired immune deficiency syndrome
AMC	Air Mobility Command
AO	area of operations
ARVN	Army of the Republic of Viet Nam
ASCOPE	areas, structures, capabilities, organizations, people, and events
BCT	brigade combat team
BDAR	battle damage assessment and repair
BENELUX	Belgium, Netherlands, and Luxembourg
C2	command and control
CA	civil affairs
CAA	combat aviation advisory
CASEVAC	casualty evacuation
CBRN	chemical, biological, radiological, nuclear
CBRNE	chemical, biological, radiological, nuclear, high-yield explosive
CCIR	commander's critical information requirement
CI	counterintelligence
CEA	captured enemy ammunition
CERP	commander's emergency response program
CLS	combat lifesaver
CMO	civil-military operations
CMOC	civil-military operations center
CNA	computer network attack
CND	computer network defense
CNE	computer network exploitation
CNO	computer network operations
COA	course of action
COG	center of gravity
CONUS	continental United States
COP	common operational picture
DA	direct action
DNBI	disease and nonbattle injury
DOTMLPF	doctrine, organization, training, materiel, leadership and education, personnel, and facilities
EAC	echelons above corps
EEFI	essential elements of friendly information

EOD	explosive ordnance disposal
EW	electronic warfare
FHP	force health protection
FID	foreign internal defense
FM	field manual; frequency modulation
FMI	field manual, interim
FSB	forward support battalion
G-9	assistant chief of staff, civil affairs
GIS	geographic information system
GTL	gun-target line
HRS	human resources support
HUMINT	human intelligence
IDF	Israeli Defense Forces
IED	improvised explosive device
IMINT	imagery intelligence
INFOSYS	information system
IO	information operations
IPB	intelligence preparation of the battlefield
ISR	intelligence, surveillance, and reconnaissance
JDAM	Joint Direct Attack Munition
JFC	joint force commander
JOA	joint operations area
JP	joint publication
JTF	joint task force
JUO	joint urban operations
LAN	local area network
LNO	liaison officer
LOC	line of communications
LOGCAP	logistics civilian augmentation program
LOS	line of sight
LPT	logistics preparation of the theater
MAGTF	Marine air-ground task force
MANPADS	man-portable air defense system
MEDEVAC	medical evacuation
METT-TC	mission, enemy, terrain and weather, troops and support available, time available, civil considerations
MEU	Marine expeditionary unit
MEU(SOC)	Marine expeditionary unit (special operations capable)
MHE	materials handling equipment
MLRS	Multiple Launch Rocket System
MP	military police

MSC	Military Sealift Command
MST	maintenance support team
NATO	North Atlantic Treaty Organization
NCW	naval coastal warfare
NGA	National Geospatial-Intelligence Agency
NGO	nongovernmental organization
NKPA	North Korean People's Army
NVA	North Vietnamese Army
OAKOC	observation and fields of fire, avenues of approach, key terrain, obstacles, and cover and concealment
ODA	operational detachment alpha
OIF	Operation Iraqi Freedom
OPSEC	operations security
PA	public affairs
PDF	Panamanian Defense Force
PGM	precision-guided munitions
PLO	Palestinian Liberation Organization
PMCS	preventive maintenance checks and services
POW	prisoner of war
PSYOP	psychological operations
QRF	quick reaction force
ROE	rules of engagement
RPG	rocket-propelled grenade
SBU	special boat unit
SDDC	Surface Deployment and Distribution Command
SEAL	sea-air-land (team)
SF	special forces
SIGINT	signals intelligence
SJA	Staff Judge Advocate
SO	special operations
SOF	special operations forces
SOFA	status-of-forces agreement
SOP	standing operating procedure
SOWT	special operations weather team
SPOD	seaport of debarkation
SPE	seaport of embarkation
SR	special reconnaissance
STT	special tactics team
SWET	sewer, water, electricity, trash
SWEAT	sewer, water, electricity, academics trash
SWEAT-MS	sewer, water, electricity, academics trash-medical, security

TACON	tactical control
TC	training circular
TCC	transportation component command
TF	task force
TIM	toxic industrial material
TTP	tactics, techniques, and procedures
UAS	unmanned aircraft system
UN	United Nations
UNITAF	Unified Task Force
UNOSOM	United Nations Operations in Somalia
UNSCR	UN Security Council resolution
UO	urban operations
USACE	United States Army Corps of Engineers
USAID	United States Agency for International Development
USCENTCOM	United States Central Command
USFORSOM	United States Forces, Somalia
USCG	United States Coast Guard
USTRANSCOM	United States Transportation Command
UXO	unexploded ordnance
UW	unconventional warfare
VC	Viet Cong
WMD	weapons of mass destruction

SECTION II – Terms and Definitions

agility – The ability to move and adjust quickly and easily. (FM 3-0)

air interdiction – Air operations conducted to destroy, neutralize, or delay the enemy's military potential before it can be brought to bear effectively against friendly forces at such distance from friendly forces that detailed integration of each air mission with the fire and movement of friendly forces is not required. (JP 1-02)

area defense – A type of defensive operation that concentrates on denying enemy forces access to designated terrain for a specific time rather than destroying the enemy outright. (FM 3-0)

area of interest – That area of concern to the commander, including the area of influence, areas adjacent thereto, and extending into enemy territory to the objectives of current or planned operations. This area also includes areas occupied by enemy forces who could jeopardize the accomplishment of the mission. (JP 1-02)

area of operations – An operational area defined by the joint force commander for land and naval forces. Areas of operations do not typically encompass the entire operational area of the joint force commander, but should be large enough for component commanders to accomplish their missions and protect their forces. (JP 1-02)

ARFOR – The senior Army headquarters and all Army forces assigned or attached to a combatant command, subordinate joint force command, joint functional command, or multinational command. (FM 3-0)

assessment – (Army) The continuous monitoring of the current situation and progress of an operation.(FMI 5-0.1)

asymmetry – Dissimilarities in organization, equipment, doctrine, and values between other armed forces (formally organized or not) and US forces. Engagements are symmetric if forces, technologies, and weapons are similar; they are asymmetric if forces, technologies, and weapons are different, or if a resort to terrorism and rejection of more conventional rules of engagement are the norm. (FM 3-0)

attack – An offensive operation that destroys or defeats enemy forces, seizes and secures terrain, or both. (FM 3-0)

attainability - One of the eight characteristics of combat service support: generating the minimum essential supplies and services necessary to begin operations. (FM 4-0)

attrition - The reduction of the effectiveness of a force caused by loss of personnel and materiel. (JP 1-02)

avenue of approach – An air or ground route of an attacking force of a given size leading to its objective or to key terrain in its path. (JP 1-02)

axis of advance – (Army) The general area through which the bulk of a unit's combat power must move. (FM 3-90)

battle – A set of related engagements that lasts longer and involves larger forces than an engagement. (FM 3-0)

battlefield organization – The allocation of forces in the area of operations by purpose. It consists of three all-encompassing categories of operations: decisive, shaping, and sustaining. (FM 3-0)

battle rhythm – The sequencing of command and control activities within a headquarters and throughout the force to facilitate effective command and control. (FMI 5-0.1)

branch – (Army) A contingency plan or course of action (an option built into the basic plan or course of action) for changing the mission, disposition, orientation, or direction of movement of the force to aid success of the current operation, based on anticipated events, opportunities, or disruptions caused by enemy actions. Army forces prepare branches to exploit success and opportunities, or to counter disruptions caused by enemy actions. (FM 3-0)

breakout – An operation conducted by an encircled force to regain freedom of movement or contact with friendly units. It differs from other attacks only in that a simultaneous defense in other areas of the perimeter must be maintained. (FM 3-90)

buffer zone – A defined area controlled by a peace operations force from which disputing or belligerent forces have been excluded. A buffer zone is formed to create an area of separation between disputing or belligerent forces and reduce the risk of renewed conflict. Also called **area of separation** in some United Nations operations. (JP 1-02)

bypass – A tactical mission task in which the commander directs his unit to maneuver around an obstacle, position, or enemy force to maintain the momentum of the operation while deliberately avoiding combat with an enemy force. (FM 3-90)

campaign – A series of related military operations aimed at accomplishing a strategic or operational objective within a given time and space. (JP 1-02)

canalize – (Army) A tactical mission task in which the commander restricts enemy movement to a narrow zone by exploiting terrain coupled with the use of obstacles, fires, or friendly maneuver. (FM 3-90)

capability - The ability to execute a specified course of action (a capability may or may not be accompanied by an intention) (JP 1-02)

casualty evacuation – (Army) A term used by nonmedical units to refer to the movement of casualties aboard nonmedical vehicles or aircraft. (FM 8-10-6)

centers of gravity – Those characteristics, capabilities, or sources of power from which a military force derives its freedom of action, physical strength, or will to fight. (JP 1-02)

civil affairs – Designated Active and Reserve component forces and units organized, trained, and equipped specifically to conduct civil affairs activities and to support civil-military operations. (JP 1-02)

civil considerations – The influence of manmade infrastructure, civilian institutions, and attitudes and activities of the civilian leaders, populations, and organizations within an area of operations on the conduct of military operations. (FM 6-0)

civil disturbance – Group acts of violence and disorder prejudicial to public law and order. (JP 1-02)

civil-military operations – The activities of a commander that establish, maintain, influence, or exploit relations between military forces, governmental and nongovernmental civilian organizations and authorities, and the civilian populace in a friendly, neutral, or hostile operational area in order to facilitate military operations, to consolidate and achieve US objectives. Civil-military operations may include performance by military forces of activities and functions normally the responsibility of the local, regional, or national government. These activities may occur prior to, during, or subsequent to other military actions. They may also occur, if directed, in the absence of other military operations. Civil-military operations may be performed by designated civil affairs, by other military forces, or by a combination of civil affairs and other forces. (JP 1-02)

civil-military operations center – An ad hoc organization, normally established by the geographic combatant commander or subordinate joint force commander, to assist in the coordination of activities of engaged military forces, and other United States Government agencies, nongovernmental organizations, and regional and international organizations. There is no established structure, and its size and composition are situation dependent. (JP 1-02)

civil support operations – Domestic operations that address the consequences of man-made or natural accidents and incidents beyond the capabilities of civilian authorities. (FM 1)

clear – (Army) 1. A tactical mission task that requires the commander to remove all enemy forces and eliminate organized resistance within an assigned area. (FM 3-90) 2. The total elimination or neutralization of an obstacle that is usually performed by follow-on engineers and is not done under fire. (FM 3-34.2)

close air support – Air action by fixed- and rotary-wing aircraft against hostile targets that are in close proximity to friendly forces and that require detailed integration of each air mission with the fire and movement of those forces. (JP 1-02)

close combat – Combat carried out with direct fire weapons, supported by indirect fire, air-delivered fires, and nonlethal engagement means. Close combat defeats or destroys enemy forces or seizes and retains ground. (FM 3-0)

coalition – An ad hoc arrangement between two or more nations for common action. (JP 1-02)

collateral damage – Unintentional or incidental injury or damage to persons or objects that would not be lawful military targets in the circumstances ruling at the time. Such damage is not unlawful so long as it is not excessive in light of the overall military advantage anticipated from the attack. (JP 1-02)

combatant command – A unified or specified command with a broad continuing mission under a single commander established and so designated by the President, through the Secretary of Defense and with the advice and assistance of the Chairman of the Joint Chiefs of Staff. Combatant commands typically have geographic or functional responsibilities. (JP 1-02)

combat power – The total means of destructive and/or disruptive force which a military unit/formation can apply against the opponent at a given time. (JP 1-02)

combined arms – The synchronized or simultaneous application of several arms—such as infantry, armor, field artillery, engineers, air defense, and aviation—to achieve an effect on the enemy that is greater than if each arm were used against the enemy separately or in sequence. (FM 3-0)

combined arms team – (Army) Two or more arms mutually supporting one another, usually consisting of infantry, armor, cavalry, aviation, field artillery, air defense artillery, and engineers. (FM 3-90).

command and control system – (Army) The arrangement of personnel, information management, procedures, and equipment and facilities essential for the commander to conduct operations. (FM 6-0)

commander's critical information requirements – (Army) – Elements of information required by commanders that directly affect decisionmaking and dictate the successful execution of military operations. (FM 3-0)

commander's intent – (Army) A clear, concise statement of what the force must do and the conditions the force must meet to succeed with respect to the enemy, terrain, and civil considerations that represent the operation's desired end state. (FMI 5-0.1)

commander's visualization – The mental process of developing situational understanding, determining a desired end state, and envisioning how the force will achieve that end state. (FMI 5-0.1)

common operational picture – (Army) An operational picture tailored to the user's requirements, based on common data and information shared by more than one command. (FM 3-0)

computer network attack – Actions taken through the use of computer networks to disrupt, deny, degrade, or destroy information resident in computers or computer networks, or the computers and networks themselves. (JP 1-02)

computer network defense – Actions taken through computer networks to protect, monitor, analyze, detect and respond to unauthorized activity within Department of Defense information systems and computer networks. (JP 1-02)

concept of operations – (Army) How commanders see the actions of subordinate units fitting together to accomplish the mission. As a minimum, the description includes the scheme of maneuver and concept of fires. The concept of operations expands the commander's selected course of action and expresses how each element of the force will cooperate to accomplish the mission. (FM 3-0)

constraint – A restriction placed on the command by a higher command. A constraint dictates an action or inaction, thus restricting the freedom of action a subordinate commander has for planning. (FM 5-0)

control – (Army) 1. The regulation of forces and warfighting functions to accomplish the mission in accordance with the commander's intent. (FMI 5-0.1) 2. A tactical mission task that requires the commander to maintain physical influence over a specified area to prevent its use by an enemy. (FM 3-90)

controlled exchange – The removal of serviceable parts, components, or assemblies from unserviceable, economically reparable equipment and their immediate reuse in restoring a like item of equipment to a combat operable or serviceable condition. (FM 4-30.3)

control measure – A means of regulating forces or warfighting functions. (FMI 5-0.1)

counterattack – (Army) A form of attack by part or all of a defending force against an enemy attacking force with the general objective of denying the enemy his goal in attacking. See FM 3-0.

counterdeception – Efforts to negate, neutralize, diminish the effects of, or gain advantage from a foreign deception operation. Counterdeception does not include the intelligence function of identifying foreign deception operations. (JP 1-02)

counterdrug – Those active measures taken to detect, monitor, and counter the production, trafficking, and use of illegal drugs. (JP 1-02)

counterinsurgency – Those military, paramilitary, political, economic, psychological, and civic actions taken by a government to defeat insurgency. (JP 1-02)

counterintelligence – Information gathered and activities conducted to protect against espionage, other intelligence activities, sabotage, or assassinations conducted by or on behalf of foreign governments or elements thereof, foreign organizations, or foreign persons, or international terrorist activities. (JP 1-02)

countermobility operations – The construction of obstacles and emplacement of minefields to delay, disrupt, and destroy the enemy by reinforcement of the terrain. The primary purpose of countermobility operations is to slow or divert the enemy, to increase time for target acquisition, and to increase weapon effectiveness. (JP 1-02)

counterpropaganda – Programs of products and actions designed to nullify propaganda or mitigate its effects. (FM 3-05.30)

counterterrorism – Operations that include the offensive measures taken to prevent, deter, preempt, and respond to terrorism. (JP 1-02)

course of action – 1. Any sequence of activities that an individual or a unit may follow. 2. A possible plan open to an individual or a commander that would accomplish or is related to the accomplishment of a mission. 3. The scheme adopted to accomplish a job or mission. 4. A line of conduct in an engagement. 5. A product of the Joint Operation Planning and Execution System concept development phase. (JP 1-02)

cover – (Army) 1. Protection from the effects of fires. (FM 6-0) 2. A form of security operation whose primary task is to protect the main body by fighting to gain time while also observing and reporting information and preventing enemy ground observation of and direct fire against the main body. Unlike a screening or guard force, the covering force is a self-contained force capable of operating independently of the main body. (FM 3-90)

dead space – 1. An area within the maximum range of a weapon, radar, or observer, which cannot be covered by fire or observation from a particular position because of intervening obstacles, the nature of the ground, the characteristics of the trajectory, or the limitations of the pointing capabilities of the weapon. 2. An area or zone which is within range of a radio transmitter, but in which a signal is not received. 3. The volume of space above and around a gun or guided missile system into which it cannot fire because of mechanical or electronic limitations. (JP 1-02)

debarkation – The unloading of troops, equipment, or supplies from a ship or aircraft. (JP 1-02)

decision point – (Army) An event, area, or point in the battlespace where and when the friendly commander will make a critical decision. (FM 5-0)

decision support matrix – An aid used by the commander and staff to make battlefield decisions. This matrix is a staff product of the wargaming process that lists the decision point, location of the decision point, the criteria to be evaluated at the point of decision, the action or operations to occur at the decision point, and the unit or element that is to act and has responsibility to observe and report the information affecting the criteria for the decision. (FM 5-0)

decision support template – (Army) A staff product initially used in the wargaming process that graphically represents the decision points and projected situations, and indicates when, where, and under what conditions a decision is most likely to be required to initiate a specific activity or event. (FM 5-0)

decisive engagement – In land and naval warfare, an engagement in which a unit is considered fully committed and cannot maneuver or extricate itself. In the absence of outside assistance, the action must be fought to a conclusion and either won or lost with the forces at hand. (JP 1-02)

decisive operation – The operation that directly accomplishes the task assigned by the higher headquarters. Decisive operations conclusively determine the outcome of major operations, battles, and engagements. (FM 3-0)

decisive point – A geographic place, specific key event, critical system or function that allows commanders to gain a marked advantage over an enemy and greatly influence the outcome of an attack. (JP 1-02)

defeat – A tactical mission task that occurs when an enemy force has temporarily or permanently lost the physical means or the will to fight. The defeated force's commander is unwilling or unable to pursue his adopted course of action, thereby yielding to the friendly commander's will, and can no longer interfere to a significant degree with the actions of friendly forces. Defeat can result from the use of force or the threat of its use. (FM 3-90)

defeat in detail – Concentrating overwhelming combat power against separate parts of a force in sequence rather than defeating the entire force at once. (FM 3-90)

defense in depth – The siting of mutually supporting defense positions designed to absorb and progressively weaken attack, prevent initial observations of the whole position by the enemy, and to allow the commander to maneuver the reserve. (JP 1-02)

defensive information operations – The integration and coordination of policies and procedures, operations, personnel, and technology to protect and defend information and information systems. Defensive information operations ensure timely, accurate, and relevant information access while denying adversaries the opportunity to exploit friendly information and information systems for their own purposes. (FM 3-0)

defensive operations – Operations that defeat an enemy attack, buy time, economize forces, or develop conditions favorable for offensive operations. Defensive operations alone normally cannot achieve a decision. Their purpose is to create conditions for a counteroffensive that allows Army forces to regain the initiative. (FM 3-0)

delay – A form of retrograde in which a force under pressure trades space for time by slowing the enemy's momentum and inflicting maximum damage on the enemy without, in principle, becoming decisively engaged. (FM 3-90)

depth – The extension of operations in time, space, and resources. (FM 3-0)

destroy – 1. A tactical mission task that physically renders an enemy force combat-ineffective until it is reconstituted. 2. To damage a combat system so badly that it cannot perform any function or be restored to a usable condition without being entirely rebuilt. (FM 3-90)

direct action – Short-duration strikes and other small-scale offensive actions conducted as a special operation in hostile, denied, or other politically sensitive environments and which employ specialized military capabilities to seize, destroy, capture, exploit, recover, or damage designated targets. Direct action differs from conventional offensive actions in the level of physical and political risk, operational techniques, and the degree of discriminate and precise use of force to achieve specific objectives. (JP 1-02)

direct approach – The application of combat power directly against the enemy center of gravity or the enemy's principal strength. (FM 3-90)

direct fire – Fire delivered on a target using the target itself as a point of aim for either the weapon or the director. (JP 1-02)

disinformation – Information disseminated primarily by intelligence organizations or other covert agencies designed to distort information or deceive or influence United States decisionmakers, United States forces, coalition allies, key actors, or individuals via indirect or unconventional means. (FM 3-13)

displaced person – (Army) A civilian who is involuntarily outside the national boundaries of his or her country or as an internally displaced person is a civilian involuntarily outside his area or region within his country. (FM 2-0)

display – (Army) An information management activity: to represent relevant information in a usable, easily understood audio or visual form tailored to the needs of the user that conveys the common operational picture for decisionmaking and exercising command and control functions (FM 6-0)

disposition – 1. Distribution of the elements of a command within an area, usually the exact location of each unit headquarters and the deployment of forces subordinate to it. (JP 1-02)

disrupt – 1. A tactical mission task in which a commander integrates direct and indirect fires, terrain, and obstacles to upset an enemy's formation or tempo, interrupt his timetable, or cause his forces to commit prematurely or attack in piecemeal fashion. (FM 3-90) 2. An engineer obstacle effect that focuses fire planning and obstacle effort to cause the enemy to break up his formation and tempo, interrupt his timetable, commit breaching assets prematurely, and attack in a piecemeal effort. (FM 90-7)

distribution system – That complex of facilities, installations, methods, and procedures designed to receive, store, maintain, distribute, and control the flow of military materiel between the point of receipt into the military system and the point of issue to using activities and units. (JP 1-02)

doctrinal template – A model based on known or postulated adversary doctrine. Doctrinal templates illustrate the disposition and activity of adversary forces and assets conducting a particular operation unconstrained by the effects of the battlespace. They represent the application of adversary doctrine under ideal conditions. Ideally, doctrinal templates depict the threat's normal organization for combat, frontages, depths, boundaries and other control measures, assets available from other commands, objective depths, engagement areas, battle positions, and so forth. Doctrinal templates are usually scaled to allow ready use with geospatial products. (JP 1-02)

doctrine – Fundamental principles by which the military forces or elements thereof guide their actions in support of national objectives. It is authoritative but requires judgment in application. (JP 1-02)

economy – One of the eight characteristics of combat service support: providing the most efficient support at the least cost to accomplish the mission. (FM 4-0)

economy of force – One of the nine principles of war: allocate minimum essential combat power to secondary efforts. (FM 3-0)

effect – (Army) A result, outcome, or consequence of an action. (FMI 5-0.1)

electromagnetic deception – The deliberate radiation, re-radiation, alteration, suppression, absorption, denial, enhancement, or reflection of electromagnetic energy in a manner intended to convey misleading information to an enemy or enemy electromagnetic-dependent weapons, thereby degrading or neutralizing the enemy's combat capability. Among the types of electromagnetic deception are: a. **manipulative electromagnetic deception**—Actions to eliminate revealing, or convey misleading, electromagnetic telltale indicators that may be used by hostile forces; b. **simulative electromagnetic deception**—Actions to simulate friendly, notional, or actual capabilities to mislead hostile forces; c. **imitative electromagnetic deception**—The introduction of electromagnetic energy into enemy systems that imitates enemy emissions. (JP 1-02)

electromagnetic spectrum – The range of frequencies of electromagnetic radiation from zero to infinity. It is divided into 26 alphabetically designated bands. (JP 1-02)

electronic attack—That division of electronic warfare involving the use of electromagnetic energy, directed energy, or antiradiation weapons to attack personnel, facilities, or equipment with the intent of degrading, neutralizing, or destroying enemy combat capability and is considered a form of fires. Electronic attack includes: 1. actions taken to prevent or reduce an enemy's effective use of the electromagnetic spectrum, such as jamming and electromagnetic deception, and 2. employment of weapons that use either electromagnetic or directed energy as their primary destructive mechanism (lasers, radio frequency weapons, particle beams), or antiradiation weapons. (JP 1-02)

electronic warfare – Any military action involving the use of electromagnetic and directed energy to control the electromagnetic spectrum or to attack the enemy. The three major subdivisions within electronic warfare are: electronic attack, electronic protection, and electronic warfare support. (JP 1-02)

electronic warfare support—That division of electronic warfare involving actions tasked by, or under direct control of, an operational commander to search for, intercept, identify, and locate or localize sources of intentional and unintentional radiated electromagnetic energy for the purpose of immediate threat recognition, targeting, planning, and conduct of future operations. Thus, electronic warfare support provides information required for immediate decisions involving electronic warfare operations and other tactical actions such as threat avoidance, targeting, and homing. Electronic warfare support data can be used to produce signals intelligence, provide targeting for electronic or destructive attack, and produce measurement and signature intelligence. (JP 1-02)

embarkation – The process of putting personnel and/or vehicles and their associated stores and equipment into ships and/or aircraft. (JP 1-02)

encirclement – An operation where one force loses its freedom of maneuver because an opposing force is able to isolate it by controlling all ground lines of communications. (FM 3-0)

end state – (Army) At the operational and tactical levels, the conditions that, when achieved, accomplish the mission. At the operational level, these conditions attain the aims set for the campaign or major operation. (FM 3-0)

engagement – A tactical conflict, usually between opposing lower echelon maneuver forces. (JP 1-02)

envelopment – (Army) – A form of maneuver in which an attacking force seeks to avoid the principal enemy defenses by seizing objectives to the enemy rear to destroy the enemy in his current positions. At the tactical level, envelopments focus on seizing terrain, destroying specific enemy forces, and interdicting enemy withdrawal routes. (FM 3-0).

essential elements of friendly information – (Army) The critical aspects of a friendly operation that, if known by the enemy, would subsequently compromise, lead to failure, or limit success of the operation, and, therefore, must be protected from enemy detection. (FM 3-13)

essential task – A specified or implied task that must be executed to accomplish the mission. Essential tasks are always included in the unit's mission statement. (FM 5-0)

exfiltration – The removal of personnel or units from areas under enemy control by stealth, deception, surprise, or clandestine means. (JP 1-02)

exploitation – 1. Taking full advantage of success in military operations, following up initial gains, and making permanent the temporary effects already achieved. 2. An offensive operation that usually follows a successful attack and is designed to disorganize the enemy in depth. (JP 1-02)

explosive ordnance disposal – The detection, identification, on-site evaluation, rendering safe, recovery, and final disposal of unexploded explosive ordnance. It may also include explosive ordnance which has become hazardous by damage or deterioration. (JP 1-02)

firepower – 1. The amount of fire which may be delivered by a position, unit, or weapon system. 2. Ability to deliver fire. (JP 1-02)

fires – The effects of lethal or nonlethal weapons. (JP 1-02)

fire support coordinating measure – A measure employed by land or amphibious commanders to facilitate the rapid engagement of targets and simultaneously provide safeguards for friendly forces. (JP 1-02)

fix – (Army) 1. A tactical mission task where a commander prevents the enemy from moving any part of his force from a specific location for a specific period of time. 2. An engineer obstacle effect that focuses fire planning and obstacle effort to slow an attacker's movement within a specified area, normally an engagement area. (FM 3-90)

flexibility – One of the eight characteristics of combat service support: being able to adapt combat service support structures and procedures to changing situations, missions, and concepts of operations. (FM 4-0)

force projection – The ability to project the military element of national power from the continental United States (CONUS) or another theater in response to requirements for military operations. Force projection operations extend from mobilization and deployment of forces to redeployment to CONUS or home theater. (JP 1-02)

force protection – Actions taken to prevent or mitigate hostile actions against Department of Defense personnel (to include family members), resources, facilities, and critical information. These actions conserve the force's fighting potential so it can be applied at a decisive time and place and incorporates the coordinated and synchronized offensive and defensive measures to enable the effective employment of the joint force while degrading opportunities for the enemy. Force protection does not include actions to defeat the enemy or protect against accidents, weather, or disease. (JP 1-02)

force tailoring – The process of determining the right mix and sequence of units for a mission. (FM 3-0)

foreign humanitarian assistance – Programs conducted to relieve or reduce the results of natural or manmade disasters or other endemic conditions such as human pain, disease, hunger, or privation that might present a serious threat to life or that can result in great damage to or loss of property. Foreign humanitarian assistance (FHA) provided by US forces is limited in scope and duration. The foreign assistance provided is designed to supplement or complement the efforts of the host nation civil authorities or agencies that may have primary responsibility for providing FHA. FHA operations are those conducted outside the United States, its territories, and possessions. (JP 1-02)

foreign internal defense – Participation by civilian and military agencies of a government in any of the action programs taken by another government to free and protect its society from subversion, lawlessness, and insurgency. (JP 1-02)

fratricide – The unintentional killing or wounding of friendly personnel by friendly firepower. (FM 3-0)

frontal attack – (Army) A form of maneuver in which the attacking force seeks to destroy a weaker enemy force or fix a larger enemy force in place over a broad front. (FM 3-0)

full spectrum operations – The range of operations Army forces conduct in war and military operations other than war. (FM 3-0)

graphic control measure – A symbol used on maps and displays to regulate forces and warfighting functions. (FMI 5-0.1)

gun-target line – An imaginary straight line from gun to target. (JP 1-02)

hazardous material – Any substance which has a human health hazard associated with it. Special storage, use, handling, and shipment safety procedures and protocols must be followed to help protect against accidental exposure. Hazardous materials are specifically identified under federal law. (FM 3-100.4)

high angle fire – (Army) Fire delivered to clear an obstacle (such as a hill) that low-angle fire cannot, or fire delivered to attack targets on the reverse side of an obstacle (such as a hill) that cannot be attacked with low-angle or direct fire. (FM 6-30).

host nation – A nation that receives the forces and/or supplies of allied nations, coalition partners, and/or NATO organizations to be located on, to operate in, or to transit through its territory. (JP 1-02)

host-nation support – Civil and/or military assistance rendered by a nation to foreign forces within its territory during peacetime, crises or emergencies, or war based on agreements mutually concluded between nations. (JP 1-02)

human intelligence – A category of intelligence derived from information collected and provided by human sources. (JP 1-02)

imagery intelligence – Intelligence derived from the exploitation of collection by visual photography, infrared sensors, lasers, electro-optics, and radar sensors such as synthetic aperture radar wherein images of objects are reproduced optically or electronically on film, electronic display devices, or other media. (JP 1-02)

implied task – A task that must be performed to accomplish a specified task or mission, but is not stated in the higher headquarters order. (FM 5-0)

indirect fire – Fire delivered on a target that is not itself used as a point of aim for the weapons or the director. (JP 1-02)

infiltration – (Army) 1. A form of maneuver in which an attacking force conducts undetected movement through or into an area occupied by enemy forces to occupy a position of advantage in the enemy rear while exposing only small elements to enemy defensive fires. (FM 3-90)

information – (Army) 1. In the general sense, the meaning humans assign to data. 2. In the context of the cognitive hierarchy, data that have been processed to provide further meaning. (FM 6-0)

information fratricide – The results of employing information operations elements in a way that causes effects in the information environment that impede the conduct of friendly operations or adversely affect friendly forces. (FM 3-13)

information management – The provision of relevant information to the right person at the right time in a usable form to facilitate situational understanding and decisionmaking. It uses procedures and information systems to collect, process, store, display, and disseminate information. (FM 3-0)

information operations – (Army) The employment of the core capabilities of electronic warfare, computer network operations, psychological operations, military deception, and operations security, in concert with specified supporting and related capabilities, to affect and defend information and information systems and to influence decisionmaking. (FM 3-13)

information requirements – (Army) All information elements the commander and staff require to successfully conduct operations, that is, all elements necessary to address the factors of METT-TC. (FM 6-0)

information superiority – The operational advantage derived from the ability to collect, process, and disseminate an uninterrupted flow of information while exploiting or denying an adversary's ability to do the same. (JP 1-02)

information system – (Army) The equipment and facilities that collect, process, store, display, and disseminate information. This includes computers—hardware and software—and communications, as well as policies and procedures for their use. (FM 3-0)

insurgency – An organized movement aimed at the overthrow of a constituted government through the use of subversion and armed conflict. (JP 1-02)

integration – One of the eight characteristics of combat service support: the total inclusion of Army combat service support into the operations process (plan, prepare, execute, assess), as well as into other logistics comnponents of the unified force. (FM 4-0).

intelligence – 1. The product resulting from the collection, processing, integration, analysis, evaluation, and interpretation of available information concerning foreign countries or areas. 2. Information and knowledge about an adversary obtained through observation, investigation, analysis, or understanding. (JP 1-02)

intelligence preparation of the battlefield – The systematic, continuous process of analyzing the threat and environment in a specific geographic area. Intelligence preparation of the battlefield (IPB) is designed to support the staff estimate and military decisionmaking process. Most intelligence requirements are generated as a result of the IPB process and its interrelation with the decisionmaking process. (FM 34-130)

intelligence, surveillance, and reconnaissance – An activity that synchronizes and integrates the planning and operation of sensors, assets, and processing, exploitation, and dissemination systems in direct support of current and future operations. This is an integrated intelligence and operations function. (JP 1-02)

intelligence, surveillance, and reconnaissance plan – An integrated plan for collection of information from all available sources and analysis of that information to produce intelligence to meet requirements. Specifically, a logical plan for transforming priority intelligence requirements into orders or requests to reconnaissance and surveillance assets to collect pertinent information within a required time limit. (FM 34-3)

interagency – Activities or operations conducted by or through coordination with two or more agencies or an agency and one or more Services of the same nation. (FM 3-07)

interdict – A tactical mission task where the commander prevents, disrupts, or delays the enemy's use of an area or route. (FM 3-90)

interior lines – A force operates on interior lines when its operations diverge from a central point. (FM 3-0)

isolate – A tactical mission task that requires a unit to seal off—both physically and psychologically—an enemy from his sources of support, deny an enemy freedom of movement, and prevent an enemy unit from having contact with other enemy forces. (FM 3-90)

joint force – A general term applied to a force composed of significant elements, assigned or attached, of two or more Military Departments, operating under a single joint force commander. (JP 1-02)

joint force commander – A general term applied to a combatant commander, subunified commander, or joint task force commander authorized to exercise combatant command (command authority) or operational control over a joint force. (JP 1-02)

joint operations – A general term to describe military actions conducted by joint forces, or by Service forces in relationships (e.g., support, coordinating authority), which, of themselves, do not create joint forces. (JP 1-02)

joint task force – A joint force that is constituted and so designated by the Secretary of Defense, a combatant commander, a subunified commander, or an existing joint task force commander. (JP 1-02)

key terrain – Any locality, or area, the seizure or retention of which affords a marked advantage to either combatant. (JP 1-02)

landing zone – Any specified zone used for the landing of aircraft. JP 1-02)

law of war – That part of international law that regulates the conduct of armed hostilities. (JP 1-02)

liaison – That contact or intercommunication maintained between elements of military forces or other agencies to ensure mutual understanding and unity of purpose and action. (JP 1-02)

line of communications – A route, either land, water, and/or air, that connects an operating military force with a base of operations and along which supplies and military forces move. (JP 1-02)

lines of operations – (Army) A line that defines the orientation of the force in time and space, or purpose, in relation to an enemy or objective. (FMI 5-0.1).

logistics – The science of planning and carrying out the movement and maintenance of forces. In its most comprehensive sense, those aspects of military operations which deal with: a. design and development, acquisition, storage, movement, distribution, maintenance, evacuation, and disposition of materiel; b. movement, evacuation, and hospitalization of personnel; c. acquisition or construction, maintenance, operation, and disposition of facilities; and d. acquisition and furnishing of services. (JP 1-02)

logistics-over-the-shore operations – The loading and unloading of ships without the benefit of deep draft-capable, fixed port facilities; or as a means of moving forces close to tactical assembly areas dependent on threat force capabilities. (JP 1-02)

logistics preparation of the theater – Actions taken by combat service support personnel to optimize means—force structure, resources, and strategic lift—of supporting the joint force commander's plan. (FM 4-0)

major operation – A series of tactical actions (battles, engagements, strikes) conducted by various combat forces of a single or several Services, coordinated in time and place, to accomplish operational and, sometimes, strategic objectives in an operational area. These actions are conducted simultaneously or sequentially in accordance with a common plan and are controlled by a single commander. (JP 1-02)

maneuver – One of the nine principles of war: place the enemy in a disadvantageous position through the flexible application of combat power. (FM 3-0)

Marine air-ground task force – The Marine Corps principal organization for all missions across the range of military operations, composed of forces task-organized under a single commander capable of responding rapidly to a contingency anywhere in the world. The types of forces in the Marine air-ground task force (MAGTF) are functionally grouped into four core elements: a command element, an aviation combat element, a ground combat element, and a combat service support element. The four core elements are categories of forces, not formal commands. The basic structure of the MAGTF never varies, though the number, size, and type of Marine Corps units comprising each of its four elements will always be mission dependent. The flexibility of the organizational structure allows for one or more subordinate MAGTFs to be assigned. In a joint or multinational environment, other Service or multinational forces may be assigned or attached. (FM 1-02)

Marine expeditionary force – The largest Marine air-ground task force (MAGTF) and the Marine Corps principal warfighting organization, particularly for larger crises or contingencies. It is task-organized around a permanent command element and normally consists of one or more Marine divisions, Marine aircraft wings, and Marine force service support groups. The Marine expeditionary force is capable of missions across the range of military operations, including amphibious assault and sustained operations ashore in any environment. It can operate from a sea base, a land base, or both. In a joint or multinational environment, it may contain other Service or multinational forces assigned or attached to the MAGTF. (FM 1-02)

Marine expeditionary unit – A Marine air-ground task force (MAGTF) that is constructed around an infantry battalion reinforced, a helicopter squadron reinforced, and a task-organized combat service support element. It normally fulfills Marine Corps' forward sea-based deployment requirements. The Marine expeditionary unit provides an immediate reaction capability for crisis response and is capable of limited combat operations. In a joint or multinational environment, it may contain other Service or multinational forces assigned or attached to the MAGTF. (FM 1-02)

mass – One of the nine principles of war: the effects of combat power at the decisive place and time. (FM 3-0)

mass casualty – Any large number of casualties produced in a relatively short period of time, usually as the result of a single incident such as a military aircraft accident, hurricane, flood, earthquake, or armed attack that exceeds logistical support capabilities. (JP 1-02)

measure of effectiveness – (Army) A criterion used to assess changes in system behavior, capability, or operational environment that is tied to measuring the attainment of an end state, achievement of an objective, or creation of an effect. (FMI 5-0.1)

medical evacuation – The timely and efficient movement of the wounded, injured, or ill while providing en route medical care to and between medical treatment facilities. (FM 4-02)

meeting engagement – A combat action that occurs when a moving force engages an enemy at an unexpected time and place. (FM 3-0)

METT-TC – A memory aid used in two contexts: (1) In the context of information management, the major subject categories into which relevant information is grouped for military operations: mission, enemy, terrain and weather, troops and support available, time available, civil considerations. (2) In the context of tactics, the major factors considered during mission analysis. (FM 6-0)

military deception – Actions executed to deliberately mislead adversary military decisionmakers as to friendly military capabilities, intentions, and operations, thereby causing the adversary to take specific actions (or inactions) that will contribute to the accomplishment of the friendly mission. (JP 1-02)

military decisionmaking process – A process that integrates the activities of the commander, staff and subordinate commanders in developing and operation plan or order. It establishes procedures for analyzing a mission; developing, analyzing, and comparing courses of action; selecting the best course of action; and producing an operation plan or order (FMI 5-0.1).

misinformation – Incorrect information from any source that is released for unknown reasons or to solicit a response or interest from a nonpolitical or nonmilitary target. (FM 3-13)

mission – 1. The task, together with the purpose, that clearly indicates the action to be taken and the reason therefor. 2. In common usage, especially when applied to lower military units, a duty assigned to an individual or unit; a task. (JP 1-02)

mission command – The conduct of military operations through decentralized execution based upon mission orders for effective mission accomplishment. Successful mission command results from subordinate leaders at all echelons exercising disciplined initiative within the commander's intent to accomplish missions. It requires an environment of trust and mutual understanding. (FM 6-0)

mission orders – A technique for completing combat orders that allows subordinates maximum freedom of planning and action in accomplishing missions and leaves the "how" of mission accomplishment to the subordinate. (FM 6-0)

mobile defense – (Army) A type of defensive operation that concentrates on the destruction or defeat of the enemy through a decisive attack by a striking force. (FM 3-0)

mobility corridor – Areas where a force will be canalized due to terrain restrictions. They allow military forces to capitalize on the principles of mass and speed and are therefore relatively free of obstacles. (JP 1-02)

mobility operations – Obstacle reduction by maneuver and engineer units to reduce or negate the effects of existing or reinforcing obstacles. The objective is to maintain freedom of movement for maneuver units, weapon systems, and critical supplies. (FM 3-34)

modified combined obstacle overlay – A joint intelligence preparation of the battlespace product used to portray the effects of each battlespace dimension on military operations. It normally depicts militarily significant aspects of the battlespace environment, such as obstacles restricting military movement, key geography, and military objectives. (JP 1-02)

monitoring – (Army) Continuous observation of the current situation to identify opportunities for the force, threats to the force, gaps in information, and progress according to the plan or order. (FMI 5-0.1)

movement to contact – A form of the offensive designed to develop the situation and to establish or regain contact. (JP 1-02)

naval coastal warfare – Coastal sea control, harbor defense, and port security, executed both in coastal areas outside the United States in support of national policy and in the United States as part of this Nation's defense. (JP 1-02)

naval gunfire support – Fire provided by Navy surface gun systems in support of a unit or units tasked with achieving the commander's objectives. A subset of naval surface fire support. (JP 1-02)

naval surface fire support – Fire provided by Navy surface gun and missile systems in support of a unit or units. (JP 1-02)

neutral – (Army) 1. An individual, group of individuals, organization, or nation that is not hostile to or in any way supportive of only one belligerent force in a hostile environment. (FM 3-07)

no-fire area – A land area, designated by the appropriate commander, into which fires or their effects are prohibited. (JP 1-02)

noncombatant – 1. An individual in an area of combat operations who is not armed and is not participating in any activity in support of any of the factions or forces involved in combat. 2. An individual, such as chaplain or medical personnel, whose duties do not involve combat. (FM 3-07)

noncombatant evacuation operations – Operations directed by the Department of State, the Department of Defense, or other appropriate authority whereby noncombatants are evacuated from foreign countries when their lives are endangered by war, civil unrest, or natural disaster to safe havens or to the United States. (JP 1-02)

nongovernmental organization – A private, self-governing, not-for-profit organization dedicated to alleviating human suffering; and/or promoting education, health care, economic development, environmental protection, human rights, and conflict resolution; and/or encouraging the establishment of democratic institutions and civil society. (JP 1-02)

nonlethal fires – Any fires that do not directly seek the physical destruction of the intended target and are designed to impair, disrupt, or delay the performance of enemy operational forces, functions, and facilities. Psychological operations, electronic warfare (jamming), and other command and control countermeasures are all nonlethal fire options. (FM 6-20)

nonlethal weapons – Weapons that are explicitly designed and primarily employed so as to incapacitate personnel or materiel, while minimizing fatalities, permanent injury to personnel and undesired damage to property and the environment. a. Unlike conventional lethal weapons that destroy their targets through blast, penetration, and fragmentation, nonlethal weapons employ means other than gross physical destruction to prevent the target from functioning. b. Nonlethal weapons are intended to have one, or both, of the following characteristics: (1) They have relatively reversible effects on personnel or materiel. (2) They affect objects differently within their area of influence. (JP 1-02)

objective – (Army) 1. One of the nine principles of war: direct every military operation toward a clearly defined, decisive and attainable objective. (FM 3-0) 2. A location on the ground used to orient operations, phase operations, facilitate changes of direction, and provide for unity of effort. (FM 3-90)

offensive – One of the nine principles of war: seize, retain, and exploit the initiative. (FM 3-0)

offensive information operations – The integrated use of assigned and supporting capabilities and activities, mutually supported by intelligence, to affect enemy decisionmakers or to influence others to achieve or promote specific objectives. (FM 3-0)

offensive operations – Operations which aim at destroying or defeating an enemy. Their purpose is to impose US will on the enemy and achieve decisive victory. (FM 3-0)

operation – 1. A military action or the carrying out of a strategic, operational, tactical, service, training, or administrative military mission. 2. The process of carrying on combat, including movement, supply, attack, defense, and maneuvers needed to gain the objectives of any battle or campaign. (JP 1-02)

operational control – Command authority that may be exercised by commanders at any echelon at or below the level of combatant command. Operational control is inherent in combatant command (command authority) and may be delegated within the command. When forces are transferred between combatant commands, the command relationship the gaining commander will exercise (and the losing commander will relinquish) over these forces must be specified by the Secretary of Defense. Operational control is the authority to perform those functions of command over subordinate forces involving organizing and employing commands and forces, assigning tasks, designating objectives, and giving authoritative direction necessary to accomplish the mission. Operational control includes authoritative direction over all aspects of military operations and joint training necessary to accomplish the missions assigned to the command. Operational control should be exercised through the commanders of subordinate organizations. Normally this authority is exercised through subordinate joint force commanders and Service and/or functional component commanders. Operational control normally provides full authority to organize commands and forces and to employ those forces as the commander in operational control considers necessary to accomplish assigned missions; it does not, in and of itself, include authoritative direction for logistics or matters of administration, discipline, internal organization, or unit training. (JP 1-02)

operational framework – The arrangement of friendly forces and resources in time, space, and purpose with respect to each other and the enemy or situation. It consists of the area of operations, battlespace, and battlefield organization. (FM 3-0)

operations process – The major command and control activities performed during operations: planning, preparation, execution with continuous assessment. These activities occur continuously throughout an operation, overlapping and recurring as required. (FMI 5-0.1)

operations security – A process of identifying critical information and subsequently analyzing friendly actions attendant to military operations and other activities to: a. identify those actions that can be observed by adversary intelligence systems; b. determine indicators hostile intelligence systems might obtain that could be interpreted or pieced together to derive critical information in time to be useful to adversaries; and c. select and execute measures that eliminate or reduce to an acceptable level the vulnerabilities of friendly actions to adversary exploitation. (JP 1-02)

paramilitary forces – Forces or groups distinct from the regular armed forces of any country, but resembling them in organization, equipment, training, or mission. (JP 1-02)

patrol – A detachment of ground, sea, or air forces sent out for the purpose of gathering information or carrying out a destructive, harassing, mopping-up, or security mission. (JP 1-02)

peace enforcement – Application of military force, or the threat of its use, normally pursuant to international authorization, to compel compliance with resolutions or sanctions designed to maintain or restore peace and order. (JP 1-02)

peacekeeping – Military operations undertaken with the consent of all major parties to a dispute, designed to monitor and facilitate implementation of an agreement (ceasefire, truce, or other such agreement) and support diplomatic efforts to reach a long-term political settlement. (JP 1-02)

peace operations – A broad term that encompasses peacekeeping operations and peace enforcement operations conducted in support of diplomatic efforts to establish and maintain peace. (JP 1-02)

penetration – (Army) A form of maneuver in which an attacking force seeks to rupture enemy defenses on a narrow front to disrupt the defensive system. (FM 3-90).

perceptions – Mental images the commander wants the deception target to believe are real. (FM 3-13)

phase line – A line utilized for control and coordination of military operations, usually an easily identified feature in the operational area. (JP 1-02)

physical destruction – The application of combat power to destroy or degrade adversary forces, sources of information, command and control systems, and installations. It includes direct and indirect forces from ground, sea, and air forces. Also included are direct actions by special operations forces. (FM 3-13)

physical security – 1. That part of security concerned with physical measures designed to safeguard personnel; to prevent unauthorized access to equipment, installations, material, and documents; and to safeguard them against espionage, sabotage, damage, and theft. 2. In communications security, the component that results from all physical measures necessary to safeguard classified equipment, material, and documents from access thereto or observation thereof by unauthorized persons. (JP 1-02)

pickup zone – A geographic area used to pick up troops or equipment by helicopter. (FM 90-4)

plan – A design for a future or anticipated operation. (FM 5-0)

planning – The process by which commanders (and staffs, if available) translate the commander's visualization into a specific course of action for preparation and execution, focusing on the expected results. (FMI 5-0.1)

port of debarkation – (DOD) The geographic point at which cargo or personnel are discharged. This may be a seaport or aerial port of debarkation; for unit requirements, it may or may not coincide with the destination. (JP 1-02)

port of embarkation – (DOD) The geographic point in a routing scheme from which cargo and personnel depart. This may be a seaport or aerial port from which personnel and equipment flow to a port of debarkation; for unit and nonunit requirements, it may or may not coincide with the origin. (JP 1-02)

principles of war – Principles that provide general guidance for conducting war and military operations other than war at the strategic, operational, and tactical levels. The nine principles of war are: objective, offensive, mass, economy of force, maneuver, unity of command, security, surprise, and simplicity (FM 3-0)

propaganda – Any form of communication in support of national objectives designed to influence the opinions, emotions, attitudes, or behavior of any group in order to benefit the sponsor, either directly or indirectly. (JP 1-02)

psychological operations – Planned operations to convey selected information and indicators to foreign audiences to influence their emotions, motives, objective reasoning, and ultimately the behavior of foreign governments, organizations, groups, and individuals. The purpose of psychological operations is to induce or reinforce foreign attitudes and behavior favorable to the originator's objectives. (JP 1-02)

public affairs – Those public information, command information, and community relations activities directed toward both the external and internal publics with interest in the Department of Defense. (JP 1-02)

pursuit – An offensive operation designed to catch or cut off a hostile force attempting to escape, with the aim of destroying it. (JP 1-02)

push – In logistics, the delivery of a predetermined amount of supplies to a user on a scheduled basis without the user requesting them. (FM 4-0)

railhead – A point on a railway where loads are transferred between trains and other means of transport. (JP 1-02)

reachback – The process of obtaining products, services, and applications, or forces, or equipment, or material from organizations that are not forward deployed. (JP 1-02)

rear area – (Army) For any particular command, the area extending forward from its rear boundary to the rear of the area assigned to the next lower level of command. This area is provided primarily for the performance of support functions, and is where the majority of the echelon's sustaining functions occur. (FM 3-0)

refugee – A person who, by reason of real or imagined danger, has left his home country or country of nationality and is unwilling or unable to return. (JP 1-02)

relevant information – All information of importance to commanders and staffs in the exercise of command and control. (FM 3-0)

relief in place – (Army) An tactical enabling operation in which, by direction of higher authority, all or part of a unit is replaced in an area by the incoming unit. (FM 3-90)

reorganization – Action taken to shift internal resources within a degraded unit to increase its level of combat effectiveness. (FM 100-9)

reserve – Portion of a body of troops which is kept to the rear or withheld from action at the beginning of an engagement, in order to be available for a decisive movement. (JP 1-02)

responsiveness – One of the eight principles of combat service support: providing the right support at the right place at the right time. (FM 4-0)

retrograde – A type of defensive operation that involves organized movement away from the enemy. (FM 3-0)

riot control agent – (Army) A substance that produces temporary irritating or disabling physical effects that disappear within minutes of removal from exposure. There is no significant risk of permanent injury, and medical treatment is rarely required. (FM 3-11.11)

risk – (DOD) Probability and severity of loss linked to hazards. (JP 1-02)

risk management – The process of identifying, assessing, and controlling risk arising from operational factors, and making decisions that balance risk cost with mission benefits. (JP 1-02)

rules of engagement – Directives issued by competent military authority that delineate the circumstances and limitations under which United States forces will initiate and/or continue combat engagement with other forces encountered. (JP 1-02)

running estimate – A staff section's continuous assessment of current and future operations to determine if the current operation is proceeding according to the commander's intent and if future operations are supportable. (FMI 5-0.1)

search and attack – A technique of conducting a movement to contact that shares many of the characteristics of an area security mission. (FM 3-0)

secure – (Army) 1. A tactical mission task that involves preventing a unit, facility, or geographical location from being damaged or destroyed as a result of enemy action. (FM 3-90)

security – One of the nine principles of war: never permit the enemy to acquire an unexpected advantage (FM 3-0).

seize – (Army) A tactical mission task that involves taking possession of a designated area using overwhelming force. (FM 3-90)

sequel – (Army) An operation that follows the current operation. It is a future operation that anticipates possible outcomes – success, failure, or stalemate – of the current operation. (FM 3-0).

shaping operation – An operation at any echelon that creates and preserves conditions for the success of decisive operations. (FM 3-0)

show of force – An operation designed to demonstrate US resolve that involves increased visibility of US deployed forces in an attempt to defuse a specific situation that, if allowed to continue, may be detrimental to US interests or national objectives. (JP 1-02)

signals intelligence – 1. A category of intelligence comprising either individually or in combination all communications intelligence, electronic intelligence, and foreign instrumentation signals intelligence, however transmitted. 2. Intelligence derived from communications, electronics, and foreign instrumentation signals. (JP 1-02)

simplicity – One of the nine principles of war: prepare clear, uncomplicated plans, and clear, concise orders to ensure thorough understanding. (FM 3-0) [Note: "simplicity" is also one of the eight characteristics of combat service support.]

situational understanding – (Army) The product of applying analysis and judgment to the common operational picture to determine the relationship among the factors of METT-TC. (FM 3-0)

space operations – The employment of space system capabilities that provide the means to enhance command and control, facilitate the maneuver of forces, reduce the commander's uncertainty, and improve fire support, air defense, intelligence collection, and combat service support operations which will support strategic, operational, and tactical missions across the operational continuum in the near, mid, and far term. (FM 3-14)

special operations – Operations conducted in hostile, denied, or politically sensitive environments to achieve military, political, economic, and/or informational objectives employing military capabilities for which there is no broad conventional force requirement. These operations often require covert, clandestine, or low visibility capabilities. Special operations are applicable across the rnage of military operations. They can be conducted independently or in conjunction with operations of conventional forces or other government agencies and may include operations through, with, or by indigenous or surrogate forces. Special operations differ from conventional operations in degree of physical and political risk, operational techniques, mode of employment, independence from friendly support, and dependence on detailed operational intelligence and indigenous assets. (JP 1-02)

special operations forces – Those Active and Reserve Component forces of the Military Services designated by the Secretary of Defense and specifically organized, trained, and equipped to conduct and support special operations. (JP 1-02)

special reconnaissance – Reconnaissance and surveillance actions conducted by special operations forces in hostile, denied, or politically sensitive environments to collect or verify information of strategic or operational significance, employing military capabilities not normally found in conventional forces. These forces provide an additive capability for commanders and supplement other conventional reconnaissance and surveillance actions. (JP 1-02)

stability operations – Operations that promote and protect US national interests by influencing the threat, political, and information dimensions of the operational environment through a combination of peacetime developmental, cooperative activities and coercive actions in response to crisis. (FM 3-0)

status-of-forces agreement – An agreement that defines the legal position of a visiting military force deployed in the territory of a friendly state. Agreements delineating the status of visiting military forces may be bilateral or multilateral. Provisions pertaining to the status of visiting forces may be set forth in a separate agreement, or they may form a part of a more comprehensive agreement. These provisions describe how the authorities of a visiting force may control members of that force and the amenability of the force or its members to the local law or to the authority of local officials. To the extent that agreements delineate matters affecting the relations between a military force and civilian authorities and population, they may be considered as civil affairs agreements. (JP 1-02)

strategy – The art and science of developing and employing instruments of national power in a synchronized and integrated fashion to achieve theater, national and/or multinational objectives. (JP 1-02)

striking force – A committed force organized to conduct the decisive attack in a mobile defense. It normally comprises the maximum combat power available to the commander at the time of the attack. (FM 3-0)

strong point – (Army) A heavily fortified battle position tied to a natural or reinforcing obstacle to create an anchor for the defense or to deny the enemy decisive or key terrain. (FM 3-90)

surprise – One of the nine principles of war: strike the enemy at a time or place or in a manner for which he is unprepared. (FM 3-0)

survivability – (Army) 1. Concept which includes all aspects of protecting personnel, weapons, and supplies while simultaneously deceiving the enemy. Survivability tactics include building a good defense; employing frequent movement; using concealment, deception, and camouflage; and constructing fighting and protective positions for both individuals and equipment. Encompasses planning and locating position sites, designing adequate overhead cover, analyzing terrain conditions and construction materials, selecting excavation methods, and countering the effects of direct and indirect fire weapons. (FM 5-103) 2. One of the eight characteristics of combat service support: being able to shield support functions from destruction or degradation. (FM 4-0)

sustainability - One of the eight characteristics of combat service support: the ability to maintain continuous support, rear area and base security, movement control, terrain management, and infrastructure development. (FM 4-0)

sustaining operation – An operation at any echelon that enables shaping and decisive operations by providing combat service support, rear area and base security, movement control, terrain management, and infrastructure development. (FM 3-0)

synchronization – 1. The arrangement of military actions in time, space, and purpose to produce maximum relative combat power at a decisive place and time. See FM 3-0. 2. In the intelligence context, application of intelligence sources and methods in concert with the operation plan. (JP 1-02)

tactical combat force – A combat unit, with appropriate combat support and combat service support assets, that is assigned the mission of defeating Level III threats. (JP 1-02)

tactical control – Command authority over assigned or attached forces or commands, or military capability or forces made available for tasking, that is limited to the detailed direction and control of movements or maneuvers within the operational area necessary to accomplish missions or tasks assigned. Tactical control is inherent in operational control. Tactical control may be delegated to, and exercised at any level at or below the level of combatant command. When forces are transferred between combatant commands, the command relationship the gaining commander will exercise (and the losing commander will relinquish) over these forces must be specified by the Secretary of Defense. Tactical control provides sufficient authority for controlling and directing the application of force or tactical use of combat support assets within the assigned mission or task. (JP 1-02)

tactical level of war –The level of war at which battles and engagements are planned and executed to accomplish military objectives assigned to tactical units or task forces. Activities at this level focus on the ordered arrangement and maneuver of combat elements in relation to each other and to the enemy to achieve combat objectives. (JP 1-02)

target acquisition – The detection, identification, and location of a target in sufficient detail to permit the effective employment of weapons. (JP 1-02)

targeting – The process of selecting and prioritizing targets and matching the appropriate response to them, taking account of operational requirements and capabilities. (JP 1-02)

target reference point – An easily recognizable point on the ground (either natural or man-made) used to initiate, distribute, and control fires. Target reference points (TRPs) can also designate the center of an area where the commander plans to distribute or converge the fires of all his weapons rapidly. They are used by task force and below, and can further delineate sectors of fire within an engagement area. TRPs are designated using the standard target symbol and numbers issued by the fire support officer. Once designated, TRPs also constitute indirect fire targets. (FM 3-90)

task organization – A temporary grouping of forces designed to accomplish a particular mission. (FM 3-0)

task-organizing – (Army) The process of allocating available assets to subordinate commanders and establishing their command and support relationships. (FM 3-0)

tempo – The rate of military action. (FM 3-0)

terminal control – 1. The authority to direct the maneuver of aircraft which are delivering ordnance, pasengers, or cargo to a specific location or target. Terminal control is a type of air control. 2. Any electronic, mechanical, or visual control given to aircraft to facilitate target acquisition and resolution. (JP 1-02)

terrorism – The calculated use of unlawful violence or threat of unlawful violence to inculcate fear; intended to coerce or to intimidate governments or societies in the pursuit of goals that are generally political, religious, or ideological. (JP 1-02)

terrorist – An individual who uses violence, terror, and intimidation to achieve a result. (JP 1-02)

throughput distribution – The bypassing of one or more intermediate supply echelons in the supply system to avoid multiple handling. (FM 4-0)

time-phased force and deployment data – The Joint Operation Planning and Execution System database portion of an operation plan; it contains time-phased force data, non-unit-related cargo and personnel data, and movement data for the operation plan, including the following: a. In-place units; b. Units to be deployed to support the operation plan with a priority indicating the desired sequence for their arrival at the port of debarkation; c. Routing of forces to be deployed; d. Movement data associated with deploying forces; e. Estimates of non-unit related cargo and personnel movements to be conducted concurrently with the deployment of forces; and f. Estimate of transportation requirements that must be fulfilled by common-user lift resources as well as those requirements that can be fulfilled by assigned or attached transportation resources. (JP 1-02)

toxic industrial material – A generic term for toxic radioactive compounds in solid, liquid, aerosolized or gaseous form. These may be used, or stored for use, for industrial, commercial, medical, military, or domestic purposes. Toxic industrial materials may be chemical, biological, or radioactive and described as toxic industrial chemical, toxic industrial biological, or toxic industrial radiological. (FM 3-11.4)

track – 1. To display or record the successive positions of a moving object. 2. To keep a gun properly aimed, or to point continuously a target-locating instrument at a moving target. (JP 1-02)

troop leading procedures – A dynamic process used by small unit leaders to analyze a mission, develop a plan, and prepare for an operation. (FM 5-0)

turning movement – (Army) A form of maneuver in which the attacking force seeks to avoid the enemy's principal defensive positions by seizing objectives to the enemy rear and causing the enemy to move out of his current positions or divert major forces to meet the threat. (FM 3-0)

unconventional warfare – A broad spectrum of military and paramilitary operations, normally of long duration, predominantly conducted through, with, or by indigenous or surrogate forces who are organized, trained, equipped, supported, and directed in varying degrees by an external source. It includes, but is not limited to, guerrilla warfare, subversion, sabotage, intelligence activities, and unconventional assisted recovery. (JP 1-02)

unexploded explosive ordnance – Explosive ordnance which has been primed, fused, armed, or otherwise prepared for action, and which has been fired, dropped, launched, projected, or placed in such a manner as to constitute a hazard to operations, installations, personnel, or material, and remains unexploded either by malfunction or design or for any other cause. (JP 1-02)

unity of command – One of the nine principles of war: for every objective, ensure unity of effort under one responsible commander.

unity of effort – Coordination and cooperation among all military forces and other organizations toward a commonly recognized objective, even if the forces and nonmilitary organizations are not necessarily part of the same command structure. (FM 6-0)

unmanned aircraft system – A powered aerial vehicle that does not carry a human operator, uses aerodynamic forces to provide lift, can fly automomously or is remotely operated, can be expendable or recoverable, and can carry a lethal or nonlethal payload. Ballistic or semi-ballistic vehicles, cruise missiles, and artillery projectiles are not considered unmanned aircraft systems.

urban area – A topographical complex where manmade construction or high population density is the dominant feature. See also **urban environment**; **urban operations**. (FM 3-06)

urban environment – The physical urban area, as well as the complex and dynamic interaction among its key components—the terrain (natural and manmade), the population, and the supporting infrastructure—as an overlapping and interdependent system of systems. (FM 3-06)

urban operations – Offense, defense, stability, and support operations conducted in a topographical complex and adjacent natural terrain where manmade construction and high population density are the dominant features. (FM 3-0)

versatility – The ability of Army forces to meet the global, diverse mission requirements of full spectrum operations. (FM 3-0)

visualize – To create and think in mental images. (FM 6-0)

vulnerability – 1. The susceptibility of a nation or military force to any action by any means through which its war potential or combat effectiveness may be reduced or its will to fight diminished. 2. The characteristics of a system that cause it to suffer a definite degradation (incapability to perform the designated mission) as a result of having been subjected to a certain level of effects in an unnatural (manmade) hostile environment. 3. In information operations, a weakness in information system security design, procedures, implementation, or internal controls that could be exploited to gain unauthorized access to information or an information system. (JP 1-02)

warfighting function – A group of tasks and systems (people, organizations, information, and processes), united by a common purpose, that commanders use to accomplish missions and training objectives. The six warfighting functions are: a. **command and control** – the related tasks and systems that support the commander in exercising authority and direction; b. **fire support** – the related tasks and systems that provide collective and coordinated use of Army indirect fires, joint fires, and offensive information operations; c. **intelligence** – the related tasks and systems that facilitate understanding of the enemy, terrain, weather, and civil considerations; d. **movement and maneuver** – the related tasks and systems that move forces to achieve a position of advantage in relation to the enemy; e. **protection** – the related tasks and systems that preserve the force so the commander can apply maximum combat power; f. **sustainment** – the related tasks and systems that provide support and services to ensure freedom of action, extend operational reach, and prolon endurance. (FMI 5-0.1)

weapons of mass destruction – Weapons that are capable of a high order of destruction and/or of being used in such a manner as to destroy large numbers of people. Weapons of mass destruction can be high explosives or nuclear, biological, chemical, and radiological weapons, but exclude the means of transporting or propelling the weapon where such means is a separable and divisible part of the weapon. (JP 1-02)

This page intentionally left blank.

References

Military publications are listed by title. When a field manual has been published under a new number for the first time, the old number is provided in parenthesis after the new number.

REQUIRED PUBLICATIONS

These documents must be available to intended users of this publication.

JOINT PUBLICATIONS

Joint publications are available at <http://www.dtic.mil/doctrine/jel/>.

JP 1-02. *Department of Defense Dictionary of Military and Associated Terms*, 12 April 2001.

JP 3-06. *Doctrine for Joint Urban Operations.* 16 September 2002.

ARMY PUBLICATIONS

Army doctrinal publications are available at <http://www.apd.army.mil/>.

FM 1-02 (101-5-1). *Operational Terms and Graphics.* 21 September 2004.

FM 34-130. *Intelligence Preparation of the Battlefield.* 8 July 1994. FM 34-130 will be republished as FM 2-01.3

FM 3-0 (100-5). *Operations.* 14 June 2001.

FM 3-06.11 (90-10-1). *Combined Arms Operations in Urban Terrain.* 28 February 2002.

FM 3-07 (100-20). *Stability Operations and Support Operations.* 20 February 2003. When updated, FM 3-07 will be retitled *Stability Operations.*

FM 3-90. *Tactics.* 04 July 2001.

FM 5-19 (100-14). *Composite Risk Management.* 21 August 2006.

FM 7-0. *Training the Force.* 22 October 2002.

FM 7-1. *Battle Focused Training.* 15 September 2003.

FM 5-0 (101-5). *Army Planning and Orders Production.* 20 January 2005.

RELATED PUBLICATIONS

These sources contain relevant supplemental information.

JOINT PUBLICATIONS

Joint publications are available at <http://www.dtic.mil/doctrine/jel/>.

JP 3-0. *Doctrine for Joint Operations.* 10 September 2001.

JP 3-05. *Doctrine for Joint Special Operations.* 17 December 2003.

JP 3-08. *Interagency Coordination During Joint Operations* (2 Volumes). 9 October 1996.

JP 3-16. *Joint Doctrine for Multinational Operations.* 05 April 2000.

JP 4-0. *Doctrine for Logistic Support of Joint Operations.* 06 April 2000.

JP 5-00.2. *Joint Task Force Planning Guidance and Procedures.* 13 January 1999.

Joint Military Operations Historical Collection. 15 July 1997.

ARMY PUBLICATIONS

Army doctrinal publications are available at <http://www.apd.army.mil/>.

Army Regulations

FM 1. *The Army.* 14 June 2005.

FM 2-0 (34-1). *Intelligence.* 17 May 2004.

FM 2-22.3 (34-52). *Human Intelligence Collector Operations*. 6 September 2006.

FM 3-05 (100-25). *Army Special Operations Forces*. 20 September 2006.

FM 3-05.20 (31-20). *Special Forces Operations*. 1 April 2004.

FM 3-05.30 (33-1). *Psychological Operations*. 15 April 2005.

FM 3-05.105. *Multiservice Tactics, Techniques, and Procedures for Special Operations Forces in Nuclear, Biological, and Chemical Environments*. 28 September 2001.

FM 3-05.212 (TC 31-25). *Special Forces Waterborne Operations*. 31 August 2004.

FM 3-05.301 (33-1-1). *Psychological Operations Tactics, Techniques, and Procedures*. 31 December 2003.

FM 3-05.401 (41-10-1). *Civil Affairs Tactics, Techniques, and Procedures*. 23 September 2003.

FM 3-06 (90-10). *Urban Operations*. 01 June 2003.

FM 3-06.1. *Aviation Urban Operations: Multi-service Tactics, Techniques, and Procedures for Aviation Urban Operations*. 09 July 2005.

FM 3-07.2. *Antiterrorism*. (Future Publication).

FM 3-11.3 ( ). *Multiservice Tactics, Techniques, and Procedures for Chemical, Biological, Radiological, and Nuclear Contamination Avoidance*. 2 February 2006.

FM 3-11.9 (3-9). *Potential Military Chemical/Biological Agents and Compounds*. 10 January 2005.

FM 3-11.14 (3-14). *Multiservice Tactics, Techniques, and Procedures for Nuclear, Biological, and Chemical Vulnerability Assessment*. 28 November 2004.

FM 3-11.19 (3-19). *Multiservice Tactics, Techniques, and Procedures for Nuclear, Biological, and Chemical Reconnaissance*. 30 July 2004.

FM 3-11.21 (3-21). *Multiservice Tactics, Techniques, and Procedures for Nuclear, Biological, and Chemical Aspects of Consequence Management*. 12 December 2001.

FM 3-11.22. *Weapons of Mass Destruction Civil Support Team Tactics, Techniques, and Procedures*. 06 June 2003.

FM 3-13 (100-6). *Information Operations: Doctrine, Tactics, Techniques, and Procedures*. 28 November 2003.

FM 3-14 (100-18). *Space Support to Army Operations*. 18 May 2005.

FM 3-19.4 (19-4). *Military Police Leaders' Handbook*. 04 March 2002.

FM 3-19.15 (19-15). *Civil Disturbance Operations*. 18 April 2005.

FM 3-19.17. *Military Working Dogs*. 22 September 2005.

FM 3-19.40 (19-40). *Military Police Internment/Resettlement Operations*. 01 August 2001.

FM 3-19.50. *Police Intelligence Operations*. 21 July 2006.

FM 3-22.40 (90-40). *NLW: Tactical Employment of Nonlethal Weapons*. 15 January 2003.

FM 3-28. *Civil Support Operations*. (Future Publication).

FM 3-31. *Joint Force Land Component Commander Handbook (JFLCC)*. 13 December 2001.

FM 3-31.1 (90-31). *AMCI: Army and Marine Corps Integration in Joint Operations*. 21 November 2001.

FM 3-34 (5-100, 5-114). *Engineer Operations*. 02 January 2004.

FM 3-34.2. *Combined Arms Breaching Operations*. 31 August 2000. FM 3-34.2 will be republished as FM 3-90.11

FM 3-34.230 (5-105). *Topographic Operations*. 03 August 2000.

FM 3-61.1. *Public Affairs Tactics, Techniques and Procedures*. 01 October 2000.

FM 3-90.12 (90-13). *Combined Arms Gap Crossing Operations*. TBP.

FM 3-97.6 (90-6). *Mountain Operations*. 28 November 2000.

FM 3-100.12. *Risk Management: Multiservice Tactics, Techniques, and Procedures for Risk Management*. 15 February 2001.

FM 3-100.38. *UXO: Multi-service Tactics, Techniques and Procedures for Unexploded Ordnance Operations*. 18 August 2005.

FM 4-0 (100-10). *Combat Service Support*. 29 August 2003.

FM 4-01.45. *Tactical Convoy Operations: Multiservice Tactics, Techniques, and Procedures for Tactical Convoy Operations*. 24 March 2005.

FM 4-02.51 (8-51). *Combat and Operational Stress Control*. 6 June 2006.

FM 4-30.16. *EOD: Multiservice Procedures for Explosive Ordnance Disposal in a Joint Environment*. 27 October 2005.

FM 4-30.51. *Unexploded Ordnance (UXO) Procedures*. 13 July 2006.

FM 5-01.12 (90-41). *JTF LNO Integration: Multiservice Tactics, Techniques, and Procedures for Joint Task Force (JTF) Liaison Officer Integration*. 27 January 2003.

FM 5-33. *Terrain Analysis*. 8 September 1992. FM 5-33 will be republished as FM 3 34.330.

FM 5-103. *Survivability*. 10 June 1985. FM 5-103 will be republished as FM 5-34.112.

FM 5-104. *General Engineering*. 12 November 1986. FM 5-104 will be republished as FM 3-34.250.

FM 5-170. *Engineer Reconnaissance*. 05 May 1988. FM 5-170 will be republished as FM 3 34.212.

FM 5-415. *Fire-Fighting Operations*. 09 February 1999. FM 5-415 will be republished as FM 3-34.281.

FM 6-0. *Mission Command: Command and Control of Army Forces*. 11 August 2003.

FM 6-22.5 (22-9). *Combat Stress*. 23 June 2000.

FM 7-15. *The Army Universal Task List (Incl. changes 2)*. 6 July 2006.

FM 20-32. *Mine/Countermine Operations*. 1 April 2005.

FM 22-51. *Leaders' Manual for Combat Stress Control*. 29 September 1994. FM 22-51 will be republished as FM 4-02.22.

FM 22-100. *Army Leadership: Be, Know, Do*. 31 August 1999. FM 22-100 will be republished as FM 6-22.

FM 23-10. *Sniper Training*. 17 August 1994. FM 23-10 will be republished as FM 3-22.10.

FM 27-10. *The Law of Land Warfare*. 18 July 1956. FM 27-10 will be republished as FM 1-04.10.

FM 27-100. *Legal Support to Operations*. 01 March 2000. FM 27-100 will be republished as FM 1-04.

FM 31-20-3. *Foreign Internal Defense Tactics, Techniques, and Procedures for Special Forces*. 20 September 1994. FM 31-20-3 will be republished as FM 3-05.202.

FM 31-50. *Combat in Fortified and Built-Up Areas*. March 1964.

FM 34-3. *Intelligence Analysis*. 15 March 1990. FM 34-3 will be republished as FM 2-33.4.

FM 34-60. *Counterintelligence*. 03 October 1995. FM 34-60 will be republished as FM 2-01.2.

FM 34-81. *Weather Support for Army Tactical Operations*. 31 August 1989. FM 34-81 will be republished as FM 2-33.2.

FM 34-81-1. *Battlefield Weather Effects*. 23 December 1992. FM 34-81-1 will be republished as FM 2-33.201.

FM 41-10. *Civil Affairs Operations*. 14 February 2000. FM 41-10 will be republished as FM 3-05.40.

FM 44-8. *Combined Arms for Air Defense*. 01 June 1999. FM 44-8 will be republished as FM 3-01.8.

FM 46-1. *Public Affairs Operations*. 30 May 1997. FM 46-1 will be republished as FM 3-61.

FM 90-8. *Counterguerrilla Operations*. 29 August 1986. FM 90-8 will be republished as FM 3-07.11.

FM 90-10. *Military Operations on Urbanized Terrain (MOUT) (How to Fight)*. 15 August 1979.

FM 100-5. *Operations*. 1944.

FM 100-8. *The Army in Multinational Operations*. 24 November 1997. FM 100-8 will be republished as FM 3-16.

FM 100-11. *Force Integration*. 15 January 1998. FM 100-11 will be republished as FM 3-100.11.

FMI 2-91.4. *Intelligence Support to Operations in the Urban Environment*. 30 June 2005. FMI 2-91.4 will be republished as FM 2-91.4.

FMI 3-07.22. *Counterinsurgency Operations*. 01 October 2004. FMI 3-07.22 will be republished as FM 3-07.22.

FMI 3-34.119. *Improvised Explosive Device Defeat*. 21 September 2005.

FMI 4-30.50. *Modular Explosive Ordnance Disposal Operations*. 31 July 2006.

FMI 5-0.1. *The Operations Process*. 31 March 2006.

TC 3-15. *Nuclear Accident and Incident Response and Assistance (NAIRA)*. 27 December 1988.

TC 5-230. *Army Geospatial Guide for Commanders and Planners*. 28 November 2003.

TC 7-98-1. *Stability and Support Operations Training Support Package*. 05 June 1997.

TC 9-21-01. *Soldiers Improvised Explosive Device (IED) Awareness Guide: Iraq and Afghanistan Theaters of Operation*. 28 May 2004.

TC 90-1. *Training for Urban Operations*. 01 April 2002.

Department of Defense Publications

Defense Intelligence Agency Manual 58-11. *Defense Human Resources Intelligence Collection Management*. 20 August 1968.

Quadrennial Defense Review Report. 30 September 2001.

U.S. Army Training and Doctrine Command (TRADOC), Deputy Chief of Staff for Intelligence (DCSINT). *TRADOC DCSINT Handbook No. 1: A Military Guide to Terrorism in the Twenty-First Century* (Version2.0). Fort Leavenworth, KS: U.S. Government Printing Office, 2004.

Nonmilitary Publications

26th Infantry Regimental Association. *Aachen: Military Operations in Urban Terrain*. Lititz, PA: 26th Infantry Regimental Association, 1999.

Adan, Abraham. *On the Banks of the Suez*. Novato, CA: Presidio Press, 1980.

Akers, Frank H., Jr. and George B. Singleton. "Task Force Ranger: A Case Study Examining the Application of Advanced Technologies in Modern Urban Warfare." Oak Ridge, TN: National Security Program Office, November 2000.

Allard, Kenneth. *Somalia Operations: Lessons Learned*. Washington, DC: National Defense University Press, 1995.

Arnold, James R. *Tet Offensive 1968: Turning Point in Vietnam*. London: Osprey, 1990.

Bass, Bernard M. "Leading in the Army after Next." *Military Review* (March/April 1998): 46–58.

Beevor, Anthony. *Stalingrad*. New York: Penguin Books, 1998.

Bowden, Mark. *Black Hawk Down: A Story of Modern War*. New York: Atlantic Monthly Press, 1999.

Butts, Kent H. "The Strategic Importance of Water." *Parameters* (spring 1997): 65–83.

Carell, Paul. *Stalingrad: The Defeat of the German 6th Army*. Atglen, PA: Schieffer Publishing, 1993.

Celeski, Joseph D. "Joint Urban Operations: Special Forces in Urban Campaign Planning." *Special Warfare* (fall 2001): 2–13.

Chiarelli, Peter W. and Patrick R. Michaelis. "Winning the Peace: The Requirement for Full-Spectrum Operations." *Military Review* (July/August 2005): 4–17.

Chiarelli, Peter W., Patrick R. Michaelis, and Geoffrey A. Norman. "Armor in Urban Terrain: The Critical Enabler." *Armor* (March/April 2005): 7–12.

Chuikov, Vasili I. *The Battle for Stalingrad.* New York: Holt, Rinehart, and Winston, 1964.

Colby, John. *War From the Ground Up: The 90th Division in WWII.* Austin, TX: Nortex Press, 1991.

Cole, Ronald H. *Operation Just Cause.* Washington, DC: Joint History Office, 1995.

Connaughton, Richard, John Pimlott, and Duncan Anderson. *The Battle for Manila: The Most Devastating Untold Story of World War II.* Novata, CA: Presidio, 1995.

Craig, William. *Enemy at the Gates: The Battle for Stalingrad.* New York: Reader's Digest Press, 1973.

Delk, James D. *Fires & Furies: The LA Riots.* Palm Springs, CA: Etc. Publications, 1995.

Dewar, Michael. *British Army in Northern Ireland.* New York: Sterling Publishers, 1985.

______. *War in the Streets: The Story of Urban Combat from Calais to Khafji.* New York: Sterling Publishers, 1992.

Donnelly, Thomas, Margaret Roth, and Caleb Baker. *Operation Just Cause: The Storming of Panama.* New York: Lexington Books, 1991.

Dupuy, R. Ernest and Trevor N. Dupuy. *The Encyclopedia of Military History: From 3500 B.C. to the Present.* San Francisco: Harper & Row, 1977.

Dupuy, Trevor N. and Paul Martell. *Flawed Victory: The Arab-Israeli Conflict and the 1982 War in Lebanon.* Fairfax, VA: Hero Books, 1986.

Edwards, Sean J. A. *Freeing Mercury's Wings: Improving Tactical Communications in Cities.* Santa Monica, CA: RAND, 2001.

Eshel, David. *Mid-East Wars: The Lebanon War 1982.* Hod Hasharon, Israel: Eshel-Dramit, 1983.

Fishel, John T. *The Fog of Peace: Planning and Executing the Restoration of Panama.* Carlisle, PA: Strategic Studies Institute, 1992.

Fontenot, Gregory, E. J. Degen, and David Tohn. *On Point: US Army in Operation IRAQI FREEDOM.* Fort Leavenworth, KS: Combat Studies Institute Press, 2004.

Funston, Frederick. "How the Army Worked to Save San Francisco." *Cosmopolitan Magazine* (July 1906): <http://www.sfmuseum.org/1906/cosmo.html>.

Gabel, Christopher R. "Military Operations on Urbanized Terrain: The 2d Battalion, 26th Infantry, at Aachen, October 1944." *Urban Combat Operations: tactics, Techniques and Procedure.* Center for Army Lessons Learned (CALL) Newsletter No. 99-16. Fort Leavenworth, KS: Center for Army Lessons Learned, U.S. Training and Doctrine Command [TRADOC], November 1999: 1-9–1-13.

Gabriel, Richard A. *Operation Peace for Galilee: The Israeli-PLO War in Lebanon.* New York: Hill and Wang, 1984.

Glenn, Russell W. *An Attack on Duffer's Downtown.* Santa Monica, CA: RAND, 2001.

______. *Heavy Matter: Urban Operations' Density of Challenges.* Santa Monica, CA: RAND, 2000.

______. "Urban Combat is Complex." *Proceedings* (February 2002): 62–65.

______. "...We Band of Brothers": *The Call for Joint Urban Operations Doctrine.* Santa Monica, CA: RAND, 1999.

Glenn, Russell W., Steven Hartman, and Scott Gerwehr. *Urban Combat Service Support Operations: The Shoulders of Atlas.* Santa Monica, CA: RAND, 2003.

Gordon, Michael R. "Baghdad Targets Picked if Hussein Holes Up There." *New York Times* (7 March 2003): 1A2.

Grau, Lester W. "Something Old, Something New: Guerrillas, Terrorists, and Intelligence Analysis." *Military Review* (July/August 2004): 42–49.

Grau, Lester W. and Timothy Smith. *A 'Crushing' Victory: Fuel-Air Explosives and Grozny 2000.* Fort Leavenworth, KS: Foreign Military Studies Office, 2000.

Hamill, Desmond. *Pig in the Middle: The Army in Northern Ireland, 1969–1984.* London: Methuen, 1985.

Hammel, Eric. *Fire in the Streets: The Battle for Hue, Tet 1968*. New York: Dell Publishing, 1991.

Heal, Sid. "Crowds, Mobs and Nonlethal Weapons." *Military Review* (March/April 2000): 45–50.

Heinl, Robert Debs, Jr. *Victory at High Tide: The Inchon-Seoul Campaign*. Annapolis, MD: Nautical & Aviation Publishing Company of America, 1979.

Herzog, Cham. *The War of Atonement, October, 1973*. Boston: Little, Brown and Company, 1975.

Higham, Robin. *Bayonets in the Streets: The Use of Troops in Civil Disturbances*. Lawrence, KS: University Press of Kansas, 1969.

Hoar, Joseph P. "A CINC's Perspective." *JFQ Forum* (autumn 1993): 56–63.

Hollis, Mark A. B. "Platoon Under Fire." *Infantry* (January/April 1998): 27–34.

Infantry in Battle. 2d ed. Garret & Massie: Richmond, VA, 1939. Reprint, Fort Leavenworth, KS: U.S. Army Command and General Staff College, 1981.

Initial Impressions Report: Operation Iraqi Freedom—Stability Operations-Support Operations Center for Army Lessons Learned (CALL) Newsletter No. 04-13 (Fort Leavenworth, KS: Center for Army Lessons Learned, U.S. Army Training and Doctrine Command [TRADOC], 2004).

Initial Impressions Report: Operation Iraqi Freedom—Stability Operations-Support Operations. Center for Army Lessons Learned (CALL) Newsletter No. 03 (Fort Leavenworth, KS: Center for Army Lessons Learned, U.S. Army Training and Doctrine Command [TRADOC], 2003).

Jomini, Antoine-Henri. "The Art of War" in *Roots of Strategy: Book 2*. Harrisburg, PA: Stackpole Books, 1987.

Kemp, Anthony. *The Unknown Battle: Metz, 1944*. New York: Stein and Day, 1981.

Kennedy, John F. Message to Congress, 25 May 1961. Available at <http://www.jfklibrary.org/Historical+Resources/Archives>.

Linden, Eugene. "The Exploding Cities of the Developing World." *Foreign Affairs* (January/February 1996): 52–65.

Mabry, Robert L. et al. "United States Army Rangers in Somalia: An Analysis of Combat Casualties on an Urban Battlefield." *Journal of Special Operations Medicine* (fall 2001): 24–40.

MacArthur, Douglas. *Reminiscences*. New York: McGraw-Hill, 1964.

Manwaring, Max G. *Street Gangs: The New Urban Insurgency*. Carlisle, PA: Strategic Studies Institute, 2005.

Maxwell, Mark D. "The Law of War and Civilians on the Battlefield: Are We Undermining Civilian Protections?" *Military Review* (September/October 2004): 17–25.

McConnell, Malcolm. *Just Cause: The Real Story of America's High-Tech Invasion of Panama*. New York: St. Martin's Press, 1991.

Medby, Jamison Jo and Russell W. Glenn. *Street Smart: Intelligence Preparation of the Battlefield for Urban Operations*. Santa Monica, CA: RAND, 2002.

Mendel, William W. *Combat in Cities: The LA Riots and Operation Rio*. Fort Leavenworth, KS: Foreign Military Studies Office, 1996.

Monmonier, Mark. *How to Lie with Maps*. Chicago: University of Chicago Press, 1996.

Montgomery, Thomas M. Interview. "Ambush in Mogasdishu." *Frontline*, WGBH Boston. 29 September 1998. Available at <http://www.pbs.org/wgbh/pages/frontline/programs/1998.html>.

Moore, John C. "Sadr City: The Armor Pure Assault in Urban Terrain." *Armor* (November/December 2004): 31–37.

Natsios, Andrew S. "Commander's Guidance: A Challenge of Complex Humanitarian Emergencies." *Parameters* (summer 1996): 50–66.

______. "Food Through Force: Humanitarian Intervention and U.S. Policy." *The Washington Quarterly* (winter 1994): 136.

Nolan, Keith William. *The Battle for Saigon: Tet 1968*. New York: Pocket Books, 1996.

Oakley, Robert B. "An Envoy's Perspective." *JFQ Forum* (autumn 1993): 44–55.

Oliker, Olga. *Russia's Chechen Wars 1994–2000: Lessons from Urban Combat*. Santa Monica, CA: RAND, 2001.

Operation Restore Hope Lessons Learned Report: 3 December 1992 – 4 May 1993, Operations Other than War. Fort Leavenworth, KS: Center for Army Lessons Learned (CALL), U.S. Army Combined Arms Command (CAC), 1993. Reprint, Washington, DC: U.S. Government Printing Office, 1996.

Parkerson, John Embry, Jr. "United States Compliance with Humanitarian Law Respecting Civilians During Operation Just Cause." *Military Law Review* (1991): 31–140.

Peters, Ralph. "Our Soldiers, Their Cities." *Parameters* (spring 1996): 43–50.

Peterson, John J. *Into the Cauldron*. Clinton, MD: Clavier House, 1973.

Petraeus, David H. "Learning Counterinsurgency: Observations from Soldiering in Iraq." *Military Review* (January/February 2006): 2–12.

Ritchie, Elspeth C. and Robert Mott. "Caring for Civilians During Peace Keeping Missions: Priorities and Decisions." *Military Medicine* (Vol. 167, Supplement 3, 2002): 14–16.

Rosenberger, Leif R. "The Strategic Importance of the World Food Supply." *Parameters* (spring 1997): 84–105.

Schadlow, Nadia. "War and the Art of Governance." *Parameters* (autumn 2003): 85–94.

Shelton, H. Hugh and Timothy D. Vane. "Winning the Information War in Haiti." *Military Review* (November/December 1995): 3–9.

Shtemenko, S. M. *The Soviet General Staff at War, 1941-1945: Book One*. Moscow: Progress Publishers, 1981.

Smith, Rupert. "Wars in Our Time: A Survey of Recent and Continuing Conflicts." *World Defence Systems,* volume 3:2. London: Royal United Services Institute for Defence Studies, 2001.

Stanton, Shelby L. *Ten Corps in Korea, 1950*. Novato, CA: Presidio Press, 1989.

Taw, Jennifer Morrison. *Operation Just Cause: Lessons for Operations Other Than* War. Santa Monica, CA: RAND, 1996.

Thomas, Timothy L. "The Battle for Grozny: Deadly Classroom for Urban Combat." *Parameters* (summer 1999): 87–102.

Thomas, Timothy L. and Charles P. O'Hara. "Combat Stress in Chechnya: 'The Equal Opportunity Disorder'." *Army Medical Department Journal* (January/March 2000): 46–53.

Trinquier, Roger. *Modern Warfare: A French View of Counterinsurgency*. New York: Praeger Publisher, 1964.

United Nations Security Council Resolution 794 (1992). New York: United Nations, 1992. Available at <http://daccessdds.un.org/doc/UNDOC/GEN/N92/772/11/PDF>.

U.S. Agency for International Development, Bureau for Humanitarian Response, Office of Foreign Disaster Assistance. *Field Operations Guide for Disaster Assessment and Response, Version 3.0*. Washington, DC: U.S. Government Printing Office, 2003. Available at <http://www.usaid.gov/our_work/humanitarian_assistance/>.

U.S. Department of Transportation, Research and Special Programs Administration, Office of Hazardous Materials Initiative and Training. *2004 Emergency Response Guidebook*. Washington, DC: U.S. Government Printing Office, 2004. Available at <http://hazmat.dot.gov/pubs/erg/erg2004.pdf>.

U.S. Environmental Protection Agency, Office of Emergency and Remedial Response. *HAZMAT Team Planning Guidance*. Washington, DC: U.S. Government Printing Office, 1990.

Van Sickle, Jeffrey B. "Stability Operations in Northern Iraq: Task Force Altun Kupri." *Infantry* (January/February 2005): 25–29.

Watson, Bruce Allen. *Sieges: A Comparative Study*. Westport, CT: Praeger, 1993.

Werstein, Irving. *The Battle of Aachen.* New York: Thomas Y. Crowell Company, 1962.

Whiting, Charles. *Bloody Aachen.* New York: Stein and Day, 1976.

Wintle, Justin, ed. *The Dictionary of War Quotations.* New York: The Free Press, 1989.

Yates, Lawrence A. "Military Stability and Support Operations: Analogies, Patterns and Recurring Themes." *Military Review* (July/August 1997): 51–61.

RECOMMENDED READINGS

The following sources contain relevant supplemental information.

Alexander, John B. *Future War: Non-Lethal Weapons in Twenty-First-Century* Warfare. New York: St. Martin's Press, 1999.

Ashworth, G. J. *The City and War.* London: Routledge, 1991.

Barker, A. J. *Dunkirk: The Great Escape.* New York: David McKay, 1977.

Baumann, Robert F., Lawrence A. Yates, and Versalle F. Washington. "My Clan Against the World": *U.S. and Coalition Forces in Somalia, 1992-1994.* Fort Leavenworth, KS: Combat Studies Institute Press, 2004.

Birtle, Andrew J. *U.S. Army Counterinsurgency and Contingency Operations Doctrine: 1860-1941.* Washington, DC: Center of Military History, 2001.

Blaufarb, Douglas S. *The Counterinsurgency Era: U.S. Doctrine and Performance 1950 to Present.* New York: The Free Press, 1977.

Breuer, William B. *Hitler's Fortress Cherbourg: The Conquest of a Bastion.* New York: Stein and Day, 1984.

Brown, Stephen D. "PSYOP in Operation Uphold Democracy." *Military Review* (September/October 1996): 57–76.

Bulloch, Gavin. "Military Doctrine and Counterinsurgency: A British Perspective." *Parameters* (summer 1996): 4–16.

Urban Combat Operations: Tactics, Techniques and Procedures. Center for Army Lessons Learned (CALL) Newsletter No. 99-16 (Fort Leavenworth, KS: Center for Army Lessons Learned, U.S. Army Training and Doctrine Command [TRADOC], November 1999).

Cordesman, Anthony H. and Abraham R. Wagner. *The Lessons of Modern War—Volume II: The Iran-Iraq War.* Boulder, CO: Westview Press, 1990.

Corum, James S. and Wray R. Johnson. *Airpower in Small Wars: Fighting Insurgents and Terrorists.* Lawrence, KS: University Press of Kansas, 2003.

Desch, Michael C., ed. *Soldiers in Cities: Military Operations on Urban Terrain.* Carlisle Barracks, PA: Strategic Studies Institute, 2001.

Duffy, Christopher. *Fire & Stone: The Science of Fortress Warfare, 1660–1860.* Mechanicsburg, PA: Stackpole Books, 1996.

______. *The Fortress in the Age of Vauban and Frederick the Great, 1660–1789: Siege Warfare Volume I.* London: Routledge & Kegan Paul, 1985.

______. *Siege Warfare: The Fortress in the Early Modern World, 1494–1660.* London: Routledge & Kegan Paul, 1979.

Edwards, Sean J. A. *Mars Unmasked: The Changing Face of Urban Operations.* Santa Monica, CA: RAND, 2000.

Eikenberry, Karl W. "Take No Casualties." *Parameters* (summer 1996): 109–118.

Erickson, John. *The Road to Berlin: Continuing the History of Stalin's War with Germany.* Boulder, CO: Westview Press, 1983.

Gerwehr, Scott and Russell W. Glenn. *The Art of Darkness: Deception and Urban Operations.* Santa Monica, CA: RAND, 2000.

Glenn, Russell W. *Combat in Hell: A Consideration of Constrained Urban Warfare.* Santa Monica, CA: RAND, 1996.

______. *Marching Under Darkening Skies: The American Military and Impending Urban Operations Threat*. Santa Monica, CA: RAND, 1998.

Glenn, Russell W., ed. *Capital Preservation: Preparing for Urban Operations in the Twenty-First Century: Proceedings of the RAND Arroyo-TRADOC-MCWL-OSD Urban Operations Conference, March 22-23, 2000*. Santa Monica, CA: RAND, 2001.

Glenn, Russell W. et al. *Denying the Widow-Maker, Summary of Proceedings, RAND-DDBL Conference on Military Operations on Urbanized Terrain*. Santa Monica, CA: RAND, 1999.

______. *Getting the Musicians of Mars on the Same Sheet of Music: Army, Joint, Multinational, and Interagency C4ISR Interoperability*. Santa Monica, CA: RAND, 2000.

Glenn, Russell W. et al., eds. *The City's Many Faces: Proceedings of the RAND Arroyo-MCWL-J8 UWG Urban Operations Conference, April 13-14, 1999*. Santa Monica, CA: RAND, 2000.

Gouré, Leon. *The Siege of Leningrad*. Stanford, CA: Stanford University Press, 1962.

Grau, Lester W. and Jacob W. Kipp. "Urban Combat: Confronting the Specter." *Military Review* (July/August 1999): 9–17.

Groves, John R. "Operations in Urban Environments." *Military Review* (July/August 1998): 30–39.

Habeck, Mary R. Lecture. "Jihadist Strategies in the War on Terrorism." Washington, DC: The Heritage Foundation, 8 November 2004. Available at http://www.heritage.org/research/nationalsecurity/hl855.cfm.

Hahn, Robert F., II and Bonnie Jezior. "Urban Warfare and the Urban Warfighter of 2025." *Parameters* (summer 1999): 74–86.

Harman, Nicholas. *Dunkirk: The Patriotic Myth*. New York: Simon and Schuster, 1980.

Heckstall-Smith, Anthony. *Tobruk: The Story of a Siege*. New York: W. W. Norton, 1959.

Hibbert, Christopher. *The Battle of Arnhem*. London: Batsford, 1962.

House, John M. "The Enemy After Next." *Military Review* (March/April 1998): 22–28.

Jackson, W. G. F. *The Battle for Italy*. New York: Harper & Row, 1967.

Karslake, Basil. *1940 – The Last Act: The Story of the British Forces in France After Dunkirk*. Hamden, CT: Archon Books, 1979.

Kostof, Spiro. *The City Shaped: Urban Patterns and Meanings Through History*. London: Thames and Hudson, 1999.

______. *The City Assembled: The Elements of Urban Form Throughout History*. Boston: Little Brown and Company, 1999.

Kuehl, Daniel T. "Airpower vs. Electricity: Electric Power as a Target for Strategic Air Operations." *Journal of Strategic Studies* (March 1995): 237–266.

Larson, Eric. *Casualties and Consensus: The Historical Role of Casualties in Domestic Support of U.S. Military Operations*. Santa Monica, CA: RAND, 1996.

Lieven, Anatol. *Chechnya: Tombstone of Russian Power*. New Haven: Yale University Press, 1998.

______. "The World Turned Upside Down: Military Lessons of the Chechen War." *Armed Forces Journal International* (August 1998): 40–43.

McKee, Alexander. *Caen: Anvil of Victory*. New York: St. Martin's Press, 1964.

McLaurin, R. D. "Military Forces in Urban Antiterrorism." Aberdeen Proving Ground, MD: U.S. Army Human Engineering Laboratory, 1989.

McLaurin, R. D. and R. Miller. "Urban Counterinsurgency: Case Studies and Implications for U.S. Military Forces." Aberdeen Proving Ground, MD: U.S. Army Human Engineering Laboratory, 1989.

Mercier, Charles E. "Terrorists, WMD, and the U.S. Army Reserve." *Parameters* (autumn 1997): 98–118.

Milton, T. R. "Urban Operations: Future War." *Military Review* (February 1994): 37–46.

Moskin, J. Robert. *Among Lions: The Battle for Jerusalem, June 5–7, 1967.* New York: Ballantine Books, 1982.

Northam, Ray M. *Urban Geography.* New York: Wiley, 1975.

Peters, Ralph. "The Culture of Future Conflict." *Parameters* (winter 1995 96): 18–27.

______. "The Future of Armed Warfare." *Parameters* (autumn 1997): 50–59.

______. "The Human Terrain of Urban Operations." *Parameters* (spring 2000): 4–12.

Robertson, William G., ed. *Block by Block: The Challenges of Urban Operations.* Fort Leavenworth, KS: U.S. Army Command and Staff College Press, 2003.

Rosenau, William G. "'Every Room is a New Battle': The Lessons of Modern Urban Warfare." *Studies in Conflict and Terrorism* (October–December 1997): 371–394.

Scalard, Douglas. "People of Whom We Know Nothing: When Doctrine Isn't Enough." *Military Review* (July/August 1997): 4–11.

Scales, Robert H. "The Indirect Approach: How U.S. Military Forces Can Avoid the Pitfalls of Future Urban Warfare." *Armed Forces Journal International* (October 1998): 68–74.

Siegel, Adam B. *Eastern Exit: The Noncombatant Evacuation Operation (NEO) from Mogadishu, Somalia, in January 1991.* Alexandria, VA: Center for Naval Analyses, 1992.

Spiller, Roger J. *Sharp Corners: Urban Operations at Century's End.* Fort Leavenworth, KS: Combat Studies Institute, 2001.

Stech, Frank J. "Winning CNN Wars." *Parameters* (autumn 1994): 37–56.

Stevenson, Jonathan. *Losing Mogadishu: Testing U.S. Policy in Somalia.* Annapolis, MD: Naval Institute Press, 1995.

Taw, Jennifer Morrison and Bruce Hoffman. *The Urbanization of Insurgency: The Potential Challenge to U.S. Army Operations.* Santa Monica, CA: RAND, 1994.

Tully, Andrew. *Berlin: Story of a Battle.* Westport, CT: Greenwood Press, 1977.

United Nations Department of Economic and Social Affairs, Population Division. *World Urbanization Prospects: The 2003 Revision.* New York: United Nations, 2003.

Vick, Alan et al. *Aerospace Operations in Urban Environments: Exploring New Concepts.* Santa Monica, CA: RAND, 2000.

Wykes, Alan. *The Siege of Leningrad.* New York: Ballentine Books Inc., 1968.

Index

O

P-Q

R

S

T

U-V

W-Z

This page intentionally left blank.

FM 3-06
26 October 2006

By Order of the Secretary of the Army:

PETER J. SCHOOMAKER
General, United States Army
Chief of Staff

Official:

JOYCE E. MORROW
Administrative Assistant to the
Secretary of the Army
0627803

DISTRIBUTION:

Active Army, Army National Guard, and U. S. Army Reserve: To be distributed in accordance with initial distribution number 111233, requirements for FM 3-06.

PREFACE

Mountains exist in almost every country in the world and almost every war has included some type of mountain operations. This pattern will not change; therefore, soldiers will fight in mountainous terrain in future conflicts. Although mountain operations have not changed, several advancements in equipment and transportation have increased the soldiers' capabilities. The helicopter now allows access to terrain that was once unreachable or could be reached only by slow methodical climbing. Inclement weather, however, may place various restrictions on the capabilities of air assets available to a commander. The unit must then possess the necessary mountaineering skills to overcome adverse terrain to reach an objective.

This field manual details techniques soldiers and leaders must know to cope with mountainous terrain. These techniques are the foundation upon which the mountaineer must build. They must be applied to the various situations encountered to include river crossings, glaciers, snow-covered mountains, ice climbing, rock climbing, and urban vertical environments. The degree to which this training is applied must be varied to conform to known enemy doctrine, tactics, and actions. This FM also discusses basic and advanced techniques to include acclimatization, illness and injury, equipment, anchors, evacuation, movement on glaciers, and training.

This field manual is a training aid for use by *qualified* personnel in conjunction with FM 3-97.6, Mountain Operations, which is used for planning operations in mountainous terrain. Personnel using FM 3-97.61 should attend a recognized Department of Defense Mountain Warfare School for proper training. **Improper use of techniques and procedures by untrained personnel may result in serious injury or death.** Personnel should be certified as Level I, Basic Mountaineer; Level II, Assault Climber; or Level III, Mountain Leader before using FM 3-97.61 for training (see Appendix A).

The measurements in this manual are stated as they are used in training (either metric or standard). Appendix B contains a measurement conversion chart for your convenience.

The proponent of this publication is HQ TRADOC. Submit changes for improving this publication to doctrine@benning.army.mil or on DA Form 2028 (Recommended Changes to Publications and Blank Forms) and forward to the Commander, U.S. Army Infantry School, ATTN: ATSH-RBO, Fort Benning, GA 31905-5593.

Unless otherwise stated, whenever the masculine gender is used, both men and women are included.

FIELD MANUAL
No. 3-97.61

HEADQUARTERS
DEPARTMENT OF THE ARMY
WASHINGTON, DC, 26 August 2002

MILITARY MOUNTAINEERING

CONTENTS

DISTRIBUTION RESTRICTION: Approved for public release; distribution is unlimited.

---

*This publication supersedes TC 90-6-1, dated 26 April 1989.

Page

CHAPTER 1

MOUNTAIN TERRAIN, WEATHER, AND HAZARDS

Commanders must consider the effects terrain and weather will have on their operations, mainly on their troops and logistics efforts. Weather and terrain combine to challenge efforts in moving supplies to forward areas. Spring storms, which may deposit a foot of snow on dry roads, combined with unprepared vehicles create hazardous situations. Helicopters are a valuable asset for use in moving men and supplies, but commanders should not plan to use them as the only means of movement and resupply. Alternate methods must be planned due to the variability of weather. Units scheduled for deployment in mountainous terrain should become self-sufficient and train under various conditions. Commanders must be familiar with the restraints that the terrain can place on a unit.

Section I. MOUNTAIN TERRAIN

Operations in the mountains require soldiers to be physically fit and leaders to be experienced in operations in this terrain. Problems arise in moving men and transporting loads up and down steep and varied terrain in order to accomplish the mission. Chances for success in this environment are greater when a leader has experience operating under the same conditions as his men. Acclimatization, conditioning, and training are important factors in successful military mountaineering.

1-1. DEFINITION

Mountains are land forms that rise more than 500 meters above the surrounding plain and are characterized by steep slopes. Slopes commonly range from 4 to 45 degrees. Cliffs and precipices may be vertical or overhanging. Mountains may consist of an isolated peak, single ridges, glaciers, snowfields, compartments, or complex ranges extending for long distances and obstructing movement. Mountains usually favor the defense; however, attacks can succeed by using detailed planning, rehearsals, surprise, and well-led troops.

1-2. COMPOSITION

All mountains are made up of rocks and all rocks of minerals (compounds that cannot be broken down except by chemical action). Of the approximately 2,000 known minerals, seven rock-forming minerals make up most of the earth's crust: quartz and feldspar make up granite and sandstone; olivene and pyroxene give basalt its dark color; and amphibole and biotite (mica) are the black crystalline specks in granitic rocks. Except for calcite, found in limestone, they all contain silicon and are often referred to as silicates.

1-3. ROCK AND SLOPE TYPES

Different types of rock and different slopes present different hazards. The following paragraphs discuss the characteristics and hazards of the different rocks and slopes.

a. **Granite.** Granite produces fewer rockfalls, but jagged edges make pulling rope and raising equipment more difficult. Granite is abrasive and increases the danger of ropes or accessory cords being cut. Climbers must beware of large loose boulders. After a rain, granite dries quickly. Most climbing holds are found in cracks. Face climbing can be found, however, it cannot be protected.

b. **Chalk and Limestone.** Chalk and limestone are slippery when wet. Limestone is usually solid; however, conglomerate type stones may be loose. Limestone has pockets, face climbing, and cracks.

c. **Slate and Gneiss.** Slate and gneiss can be firm and or brittle in the same area (red coloring indicates brittle areas). Rockfall danger is high, and small rocks may break off when pulled or when pitons are emplaced.

d. **Sandstone.** Sandstone is usually soft causing handholds and footholds to break away under pressure. Chocks placed in sandstone may or may not hold. Sandstone should be allowed to dry for a couple of days after a rain before climbing on it—wet sandstone is extremely soft. Most climbs follow a crack. Face climbing is possible, but any outward pull will break off handholds and footholds, and it is usually difficult to protect.

e. **Grassy Slopes.** Penetrating roots and increased frost cracking cause a continuous loosening process. Grassy slopes are slippery after rain, new snow, and dew. After long, dry spells clumps of the slope tend to break away. Weight should be distributed evenly; for example, use flat hand push holds instead of finger pull holds.

f. **Firm Spring Snow (Firn Snow).** Stopping a slide on small, leftover snow patches in late spring can be difficult. Routes should be planned to avoid these dangers. Self-arrest should be practiced before encountering this situation. Beginning climbers should be secured with rope when climbing on this type surface. Climbers can glissade down firn snow if necessary. Firn snow is easier to ascend than walking up scree or talus.

g. **Talus.** Talus is rocks that are larger than a dinner plate, but smaller than boulders. They can be used as stepping-stones to ascend or descend a slope. However, if a talus rock slips away it can produce more injury than scree because of its size.

h. **Scree.** Scree is small rocks that are from pebble size to dinner plate size. Running down scree is an effective method of descending in a hurry. One can run at full stride without worry—the whole scree field is moving with you. Climbers must beware of larger rocks that may be solidly planted under the scree. Ascending scree is a tedious task. The scree does not provide a solid platform and will only slide under foot. If possible, avoid scree when ascending.

1-4. ROCK CLASSIFICATIONS

Rock is classified by origin and mineral composition.

a. **Igneous Rocks.** Deep within the earth's crust and mantle, internal heat, friction and radioactive decay creates magmas (melts of silicate minerals) that solidify into igneous rocks upon cooling. When the cooling occurs at depth, under pressure, and over time, the minerals in the magma crystallize slowly and develop well, making coarse-grained plutonic rock. The magma may move upward, propelled by its own lower density, either melting and combining with the overlying layers or forcing them aside. This results in an intrusive rock. If the melt erupts onto the surface it cools rapidly and the minerals form little or no crystal matrix, creating a volcanic or extrusive rock.

(1) ***Plutonic or Intrusive Rocks.*** Slow crystallization from deeply buried magmas generally means good climbing, since the minerals formed are relatively large and interwoven into a solid matrix. Weathering develops protrusions of resistant minerals, which makes for either a rough-surfaced rock with excellent friction, or, if the resistant

crystals are much larger than the surrounding matrix, a surface with numerous knobby holds. Pieces of foreign rock included in the plutonic body while it was rising and crystallizing, or clusters of segregated minerals, may weather differently than the main rock mass and form "chicken heads."

(a) Intrusions are named according to location and size. Large (100 square kilometers or larger) masses of plutonic rock are called "batholiths" and small ones "stocks." Most plutonic rock is in the granite family, differing only in the amounts of constituent minerals contained. A core of such batholiths is in every major mountain system in the world. In the Alps, Sierras, North Cascades, Rockies, Adirondacks, and most other ranges this core is at least partly exposed.

(b) Small plutonic intrusions are stocks, forced between sedimentary strata, and dikes, which cut across the strata. Many of these small intrusive bodies are quickly cooled and thus may look like extrusive rock.

(2) ***Volcanic or Extrusive Rocks.*** Explosive eruptions eject molten rock so quickly into the air that it hardens into loose aerated masses of fine crystals and uncrystallized glass (obsidian). When this ash consolidates while molten or after cooling, it is called "tuff," a weak rock that breaks down quickly and erodes easily. Quieter eruptions, where widespread lava flows from large fissures, produce basalt. Basaltic rocks are fine-grained and often sharp-edged.

(3) ***Jointing Rocks.*** In plutonic rocks, joints or cracks are caused by internal stresses such as contraction during cooling or expansion when overlying rock erodes or exfoliates. Some joints tend to follow a consistent pattern throughout an entire mountain and their existence can often be predicted. Therefore, when a ledge suddenly ends, the joint—and thus the ledge—may begin again around the corner. When molten rock extrudes onto the surface as a lava flow or intrudes into a cold surrounding mass as a dike or sill, the contraction from rapid cooling usually causes so much jointing that climbing can be extremely hazardous. Occasionally, this jointing is regular enough to create massed pillars with usable vertical cracks such as Devil's Tower in Wyoming.

b. **Sedimentary Rocks.** Sedimentary rocks are born high in the mountains, where erosion grinds down debris and moves it down to rivers for transportation to its final deposition in valleys, lakes, or oceans. As sediments accumulate, the bottom layers are solidified by pressure and by mineral cements precipitated from percolating groundwater. Gravel and boulders are transformed into conglomerates; sandy beaches into sandstone; beds of mud into mudstone or shale; and shell beds and coral reefs into limestone or dolomite.

(1) Though in general sedimentary rocks are much more friable than those cooled from molten magmas, pressure and cementing often produce solid rocks. In fact, by sealing up internal cracks cementing can result in flawless surfaces, especially in limestone.

(2) Most high mountain ranges have some sedimentary peaks. Ancient seafloor limestone can be found on the summits of the Himalayas and the Alps. The Canadian Rockies are almost exclusively limestone. With the exception of the Dolomites, in general sedimentary rocks do not offer high-angle climbing comparable to that of granite.

c. **Metamorphic Rocks.** These are igneous or sedimentary rocks that have been altered physically and or chemically by the tremendous heat and pressures within the earth. After sediments are solidified, high heat and pressure can cause their minerals to recrystallize. The bedding planes (strata) may also be distorted by folding and squeezing. Shale changes to slate or schist, sandstone and conglomerate into quartzite, and limestone to marble. These changes may be minimal, only slightly altering the sediments, or extensive enough to produce gneiss, which is almost indistinguishable from igneous rock.

(1) Metamorphic rocks may have not only joints and bedding, but cleavage or foliation, a series of thinly spaced cracks caused by the pressures of folding. Because of this cleavage, lower grades of metamorphic rocks may be completely unsuitable for climbing because the rock is too rotten for safe movement.

(2) Higher degrees of metamorphism or metamorphism of the right rocks provide a solid climbing surface. The Shawangunks of New York are an excellent example of high-grade conglomerate quartzite, which offers world class climbing. The center of the Green Mountain anticline contains heavily metamorphosed schist, which also provides solid climbing.

1-5. MOUNTAIN BUILDING

The two primary mechanisms for mountain-building are volcanic and tectonic activity. Volcanoes are constructed from lava and ash, which begin within the earth as magma. Tectonic activity causes plates to collide, heaving up fold mountains, and to pull apart and crack, forming fault-block mountain ranges.

a. **Plate Tectonics.** The massive slabs composing the outer layer are called tectonic plates. These plates are made up of portions of lighter, granitic continental crust, and heavier, basaltic oceanic crust attached to slabs of the rigid upper mantle. Floating slowly over the more malleable asthenosphere, their movement relative to each other creates earthquakes, volcanoes, ocean trenches, and mountain ridge systems.

b. **Mountain Structure.** The different horizontal and vertical stresses that create mountains usually produce complex patterns. Each type of stress produces a typical structure, and most mountains can be described in terms of these structures.

(1) ***Dome Mountains.*** A simple upward bulge of the crust forms dome mountains such as the Ozarks of Arkansas and Missouri, New York's Adirondacks, the Olympics of Washington, and the High Uintahs of Utah. They are usually the result of the upward movement of magma and the folding of the rock layers overhead. Erosion may strip away the overlying layers, exposing the central igneous core.

(2) ***Fault-Block Mountains.*** Faulting, or cracking of the crust into large chunks, often accompanies upwarp, which results in fault-block mountains. Many forms are created by the motion of these chunks along these faults.

(a) The ranges of the desert country of California, Nevada, and Utah provide the clearest display of faulting. The breakage extends to the surface and often during earthquakes—caused by slippage between the blocks—fresh scarps many feet high develop.

(b) Sometimes a block is faulted on both sides and rises or falls as a unit. More often, however, it is faulted on one side only. The Tetons of Wyoming and the Sierra Nevada display this—along the single zone of faults the range throws up impressive steep scarps, while on the other side the block bends but does not break, leaving a gentler slope from the base of the range to the crest. An example of a dropped block is California's Death Valley, which is below sea level and could not have been carved by erosion.

(3) ***Fold Mountains.*** Tectonic forces, in which continental plates collide or ride over each other, have given rise to the most common mountain form—fold mountains. Geologists call folds geosynclines. Upward folded strata are anticlines and downward folds are synclines. When erosion strips down the overburden of rock from folded mountain ranges, the oldest, central core is all that remains. The Alps and the Appalachians are examples of fold mountains. When the squeezing of a range is intense the rocks of the mountain mass first fold but then may break, and parts of the rocks are pushed sideways and override neighboring formations. This explains why older rocks are often found perched on top of younger ones. Isolated blocks of the over thrust mass may form when erosion strips away links connecting them with their place of origin. Almost every range of folded mountains in the world exhibits an over thrust of one sort or another.

(4) ***Volcanic Mountains.*** Along convergent plate boundaries volcanic activity increases. As it is forced underneath an overriding neighbor, continental crust melts and turns to magma within the mantle. Since it is less dense than the surrounding material it rises and erupts to form volcanoes.

(a) These volcanoes are found in belts, which correspond to continental margins around the world. The best known is the "Ring of Fire" encircling the Pacific Ocean from Katmai in Alaska through the Cascades (Mount Rainier and Mount Saint Helens) down through Mexico's Popocatepetl to the smokes of Tierra del Fuego. This belt then runs west down the Aleutian chain to Kamchatka, south to the volcanoes of Japan and the Philippines, and then east through New Guinea into the Pacific. Smaller volcanic belts are found along the Indonesian-SE Asian arc, the Caucasus region, and the Mediterranean.

(b) Volcanic activity also arises at boundaries where two plates are moving away from each other, creating deep rifts and long ridges where the crust has cracked apart and magma wells up to create new surface material. Examples of this are the Mid-Atlantic Ridge, which has created Iceland and the Azores, and the Rift Valley of East Africa with Kilimanjaro's cone.

(5) ***Complex Mountains.*** Most ranges are complex mountains with portions that have been subject to several processes. A block may have been simply pushed upward without tilting with other portions folded, domed, and faulted, often with a sprinkling of volcanoes. In addition, these processes occur both at the macro and the micro level. One massive fold can make an entire mountain peak; however, there are folds measured by a rope length, and tiny folds found within a handhold. A mountain front may be formed from a single fault, but smaller faults that form ledges and gullies may also be present.

1-6. ROUTE CLASSIFICATION

Military mountaineers must be able to assess a vertical obstacle, develop a course of action to overcome the obstacle, and have the skills to accomplish the plan. Assessment of a vertical obstacle requires experience in the classifications of routes and understanding the levels of difficulty they represent. Without a solid understanding of the difficulty of a chosen route, the mountain leader can place his life and the life of other soldiers in extreme danger. Ignorance is the most dangerous hazard in the mountain environment.

a. In North America the Yosemite Decimal System (YDS) is used to rate the difficulty of routes in mountainous terrain. The YDS classes are:

- Class 1—Hiking trail.
- Class 2—Off-trail scramble.
- Class 3—Climbing, use of ropes for beginners (moderate scrambling).
- Class 4—Belayed climbing. (This is moderate to difficult scrambling, which may have some exposure.)
- Class 5—Free climbing. (This class requires climbers to be roped up, belay and emplace intermediate protection.)

b. Class 5 is further subdivided into the following classifications:

(1) Class 5.0-5.4—Little difficulty. This is the simplest form of free climbing. Hands are necessary to support balance. This is sometimes referred to as advanced rock scrambling.

(2) Class 5.5—Moderate difficulty. Three points of contact are necessary.

(3) Class 5.6—Medium difficulty. The climber can experience vertical position or overhangs where good grips can require moderate levels of energy expenditure.

(4) Class 5.7—Great difficulty. Considerable climbing experience is necessary. Longer stretches of climbing requiring several points of intermediate protection. Higher levels of energy expenditure will be experienced.

(5) Class 5.8—Very great difficulty. Increasing amount of intermediate protection is the rule. High physical conditioning, climbing technique, and experience required.

(6) Class 5.9—Extremely great difficulty. Requires well above average ability and excellent condition. Exposed positions, often combined with small belay points. Passages of the difficult sections can often be accomplished under good conditions. Often combined with aid climbing (A0-A4).

(7) Class 5.10—Extraordinary difficulty. Climb only with improved equipment and intense training. Besides acrobatic climbing technique, mastery of refined security technique is indispensable. Often combined with aid climbing (A0-A4).

(8) Class 5.11-5.14—Greater increases of difficulty, requiring more climbing ability, experience, and energy expenditure. Only talented and dedicated climbers reach this level.

c. Additional classifications include the following.

(1) Classes are further divided into a, b, c, and d categories starting from 5.10 to 5.14 (for example, 5.10d).

(2) Classes are also further divided from 5.9 and below with +/- categories (for example, 5.8+).

(3) All class 5 climbs can also be designated with "R" or "X," which indicates a run-out on a climb. This means that placement of intermediate protection is not possible on portions of the route. (For example, in a classification of 5.8R, the "R" indicates periods of run-out where, if a fall was experienced, ground fall would occur.) Always check the local guidebook to find specific designation for your area.

(4) All class 5 climbs can also be designated with "stars." These refer to the popularity of the climb to the local area. Climbs are represented by a single "star" up to five "stars;" a five-star climb is a classic climb and is usually aesthetically pleasing.

d. Aid climb difficulty classification includes:

(1) A0—"French-free." This technique involves using a piece of gear to make progress; for example, clipping a sling into a bolt or piece of protection and then pulling up on it or stepping up in the sling. Usually only needed to get past one or two more difficult moves on advanced free climbs.

(2) A1—Easy aid. The placement of protection is straight forward and reliable. There is usually no high risk of any piece of protection pulling out. This technique requires etriers and is fast and simple.

(3) A2—Moderate aid. The placement of protection is generally straight forward, but placement can be awkward and strenuous. Usually A2 involves one or two moves that are difficult with good protection placement below and above the difficult moves. No serious fall danger.

(4) A3—Hard aid. This technique requires testing your protection. It involves several awkward and strenuous moves in a row. Generally solid placements which will hold a fall and are found within a full rope length. However, long fall potential does exist, with falls of 40 to 60 feet and intermediate protection on the awkward placements failing. These falls, however, are usually clean and with no serious bodily harm.

(5) A4—Serious aid. This technique requires lots of training and practice. More like walking on eggs so none of them break. Leads will usually take extended amounts of time which cause the lead climber to doubt and worry about each placement. Protection placed will usually only hold a climber's weight and falls can be as long as two-thirds the rope length.

(6) A5—Extreme aid. All protection is sketchy at best. Usually no protection placed on the entire route can be trusted to stop a fall.

(7) A6—Extremely severe aid. Continuous A5 climbing with A5 belay stations. If the leader falls, the whole rope team will probably experience ground fall.

(8) Aid climbing classes are also further divided into +/- categories, such as A3+ or A3-, which would simply refer to easy or hard.

e. Grade ratings (commitment grades) inform the climber of the approximate time a climber trained to the level of the climb will take to complete the route.

- I—Several hours.
- II—Half of a day.
- III—About three-fourths of a day.
- IV—Long hard day (usually not less than 5.7).
- V—1 1/2 to 2 1/2 days (usually not less than 5.8).
- VI—Greater than 2 days.

f. Climbing difficulties are rated by different systems. Table 1-1 shows a comparison of these systems.

- YDS (Yosemite Decimal System)—Used in the United States.
- UIAA (Union des International Alpine Association)—Used in Europe.
- British—The British use adjectives and numbers to designate the difficulty of climbs. This system can be confusing if the climber is not familiar with it.
- French—The French use numbers and letters to designate the difficulty of climbs.
- Brazilian—Brazil uses Roman Numerals and adjectives to designate difficulty.
- Australian—Australia uses only numbers to designate difficulty.

YDS	UIAA	BRITISH	FRENCH	BRAZIL	AUSTRALIA
Class 1	I	easy (E)			
Class 2	II	easy (E)			
Class 3	III	easy (E)	1a, b, c		
Class 4	III-	moderate (MOD)	1a, b, c		
5.0	III	moderate (MOD)	2a, b		4
5.1	III+	difficult (DIFF)	2a, b		5
5.2	IV-	hard difficult	2c, 3a		6
5.3	IV	very difficult	3b, c, 4a		7
5.4	IV+	hard very difficult	3b, c, 4a	II	8, 9
5.5	V-	mild severe	3b, c, 4a	IIsup	10, 11
5.6	V	severe, hard severe, 4a	4a, b, c	III	12, 13
5.7	V+	severe, hard severe, 4b	4a, b, c	IIIsup	14
5.8	VI-	hard severe, hard very severe, 4c	5a, b	IV	15
5.9	VI	5a	5b, c	IVsup	16, 17
5.10a	VII-	E1, 5b	5b, c	V	18
5.10b	VII	E1, 5b	5b, c	Vsup	19
5.10c	VII	E1, 5b	5b, c	VI	20
5.10d	VII+	E1/E2, 5b-5c	5b, c	VIsup	21
5.11a	VIII-	E3, 6a	6a, b, c	VII	22
5.11b	VIII	E3/E4, 6a	6a, b, c	VII	23
5.11c	VIII	E4, 6b	6a, b, c	VIIsup	24
5.12a	IX-	E5, E6/7, 6c	7a	VIII	26
5.12b	IX	E5, E6/7, 6c	7a	VIIIsup	27
5.12c	IX	E5, E6/7, 6c	7a		28
5.12d	IX+	E6/7, 7a	7a		29

Table 1-1. Rating systems.

g. Ice climbing ratings can have commitment ratings and technical ratings. The numerical ratings are often prefaced with WI (waterfall ice), AI (alpine ice), or M (mixed rock and ice).

(1) ***Commitment Ratings.*** Commitment ratings are expressed in Roman numerals.

- I—A short, easy climb near the road, with no avalanche hazard and a straightforward descent.
- II—A route of one or two pitches within a short distance of rescue assistance, with little objective hazard.
- III—A multipitch route at low elevation, or a one-pitch climb with an approach that takes about an hour. The route requires anywhere from a few hours to a long day to complete. The descent may require building rappel anchors, and the route might be prone to avalanche.
- IV—A multipitch route at higher elevations; may require several hours of approach on skis or foot. This route is subject to objective hazards, possibly with a hazardous descent.
- V—A long climb in a remote setting, requiring all day to complete the climb itself. Requires many rappels off anchors for the descent. This route has sustained exposure to avalanche or other objective hazards.

- VI—A long ice climb in an alpine setting, with sustained technical climbing. Only elite climbers will complete it in a day. A difficult and involved approach and descent, with objective hazards ever-present, all in a remote area.
- VII—Everything a grade VI has, and more of it. Possibly days to approach the climb, and objective hazards rendering survival as questionable. Difficult physically and mentally.

(2) ***Technical Ratings.*** Technical ratings are expressed as Arabic numerals.

- 1—A frozen lake or stream bed.
- 2—A pitch with short sections of ice up to 80 degrees; lots of opportunity for protection and good anchors.
- 3—Sustained ice up to 80 degrees; the ice is usually good, with places to rest, but it requires skill at placing protection and setting anchors.
- 4—A sustained pitch that is vertical or slightly less than vertical; may have special features such as chandeliers and run-outs between protection.
- 5—A long, strenuous pitch, possibly 50 meters of 85- to 90-degree ice with few if any rests between anchors. The pitch may be shorter, but on featureless ice. Good skills at placing protection are required.
- 6—A full 50-meter pitch of dead vertical ice, possibly of poor quality; requires efficiency of movement and ability to place protection while in awkward stances.
- 7—A full rope length of thin vertical or overhanging ice of dubious adhesion. An extremely tough pitch, physically and mentally, requiring agility and creativity.
- 8—Simply the hardest ice climbing ever done; extremely bold and gymnastic.

1-7. CROSS-COUNTRY MOVEMENT

Soldiers must know the terrain to determine the feasible routes for cross-country movement when no roads or trails are available.

a. A pre-operations intelligence effort should include topographic and photographic map coverage as well as detailed weather data for the area of operations. When planning mountain operations, additional information may be needed about size, location, and characteristics of landforms; drainage; types of rock and soil; and the density and distribution of vegetation. Control must be decentralized to lower levels because of varied terrain, erratic weather, and communication problems inherent to mountainous regions.

b. Movement is often restricted due to terrain and weather. The erratic weather requires that soldiers be prepared for wide variations in temperature, types, and amounts of precipitation.

(1) Movement above the timberline reduces the amount of protective cover available at lower elevations. The logistical problem is important; therefore, each man must be self-sufficient to cope with normal weather changes using materials from his rucksack.

(2) Movement during a storm is difficult due to poor visibility and bad footing on steep terrain. Although the temperature is often higher during a storm than during clear weather, the dampness of rain and snow and the penetration of wind cause soldiers to chill quickly. Although climbers should get off the high ground and seek shelter and

warmth, if possible, during severe mountain storms, capable commanders may use reduced visibility to achieve tactical surprise.

c. When the tactical situation requires continued movement during a storm, the following precautions should be observed:

- Maintain visual contact.
- Keep warm. Maintain energy and body heat by eating and drinking often; carry food that can be eaten quickly and while on the move.
- Keep dry. Wear wet-weather clothing when appropriate, but do not overdress, which can cause excessive perspiration and dampen clothing. As soon as the objective is reached and shelter secured, put on dry clothing.
- Do not rush. Hasty movement during storms leads to breaks in contact and accidents.
- If lost, stay warm, dry, and calm.
- Do not use ravines as routes of approach during a storm as they often fill with water and are prone to flash floods.
- Avoid high pinnacles and ridgelines during electrical storms.
- Avoid areas of potential avalanche or rock-fall danger.

1-8. COVER AND CONCEALMENT

When moving in the mountains, outcroppings, boulders, heavy vegetation, and intermediate terrain can provide cover and concealment. Digging fighting positions and temporary fortifications is difficult because soil is often thin or stony. The selection of dug-in positions requires detailed planning. Some rock types, such as volcanic tuff, are easily excavated. In other areas, boulders and other loose rocks can be used for building hasty fortifications. In alpine environments, snow and ice blocks may be cut and stacked to supplement dug-in positions. As in all operations, positions and routes must be camouflaged to blend in with the surrounding terrain to prevent aerial detection.

1-9. OBSERVATION

Observation in mountains varies because of weather and ground cover. The dominating height of mountainous terrain permits excellent long-range observation. However, rapidly changing weather with frequent periods of high winds, rain, snow, sleet, hail, and fog can limit visibility. The rugged nature of the terrain often produces dead space at midranges.

a. Low cloud cover at higher elevations may neutralize the effectiveness of OPs established on peaks or mountaintops. High wind speeds and sound often mask the noises of troop movement. Several OPs may need to be established laterally, in depth, and at varying altitudes to provide visual coverage of the battle area.

b. Conversely, the nature of the terrain can be used to provide concealment from observation. This concealment can be obtained in the dead space. Mountainous regions are subject to intense shadowing effects when the sun is low in relatively clear skies. The contrast from lighted to shaded areas causes visual acuity in the shaded regions to be considerably reduced. These shadowed areas can provide increased concealment when combined with other camouflage and should be considered in maneuver plans.

1-10. FIELDS OF FIRE

Fields of fire, like observation, are excellent at long ranges. However, dead space is a problem at short ranges. When forces cannot be positioned to cover dead space with direct fires, mines and obstacles or indirect fire must be used. Range determination is deceptive in mountainous terrain. Soldiers must routinely train in range estimation in mountainous regions to maintain their proficiency.

Section II. MOUNTAIN WEATHER

Most people subconsciously "forecast" the weather. If they look outside and see dark clouds they may decide to take rain gear. If an unexpected wind strikes, people glance to the sky for other bad signs. A conscious effort to follow weather changes will ultimately lead to a more accurate forecast. An analysis of mountain weather and how it is affected by mountain terrain shows that such weather is prone to patterns and is usually severe, but patterns are less obvious in mountainous terrain than in other areas. Conditions greatly change with altitude, latitude, and exposure to atmospheric winds and air masses. Mountain weather can be extremely erratic. It varies from stormy winds to calm, and from extreme cold to warmth within a short time or with a minor shift in locality. The severity and variance of the weather causes it to have a major impact on military operations.

1-11. CONSIDERATIONS FOR PLANNING

Mountain weather can be either a dangerous obstacle to operations or a valuable aid, depending on how well it is understood and to what extent advantage is taken of its peculiar characteristics.

a. Weather often determines the success or failure of a mission since it is highly changeable. Military operations plans must be flexible, especially in planning airmobile and airborne operations. The weather must be anticipated to allow enough time for planning so that the leaders of subordinate units can use their initiative in turning an important weather factor in their favor. The clouds that often cover the tops of mountains and the fogs that cover valleys are an excellent means of concealing movements that normally are made during darkness or in smoke. Limited visibility can be used as a combat multiplier.

b. The safety or danger of almost all high mountain regions, especially in winter, depends upon a change of a few degrees of temperature above or below the freezing point. Ease and speed of travel depend mainly on the weather. Terrain that can be crossed swiftly and safely one day may become impassable or highly dangerous the next due to snowfall, rainfall, or a rise in temperature. The reverse can happen just as quickly. The prevalence of avalanches depends on terrain, snow conditions, and weather factors.

c. Some mountains, such as those found in desert regions, are dry and barren with temperatures ranging from extreme heat in the summer to extreme cold in the winter. In tropical regions, lush jungles with heavy seasonal rains and little temperature variation often cover mountains. High rocky crags with glaciated peaks can be found in mountain ranges at most latitudes along the western portion of the Americas and Asia.

d. Severe weather may decrease morale and increase basic survival problems. These problems can be minimized when men have been trained to accept the weather by being self-sufficient. Mountain soldiers properly equipped and trained can use the weather to their advantage in combat operations.

1-12. MOUNTAIN AIR

High mountain air is dry and may be drier in the winter. Cold air has a reduced capacity to hold water vapor. Because of this increased dryness, equipment does not rust as quickly and organic material decomposes slowly. The dry air also requires soldiers to increase consumption of water. The reduced water vapor in the air causes an increase in evaporation of moisture from the skin and in loss of water through transpiration in the respiratory system. Due to the cold, most soldiers do not naturally consume the quantity of fluids they would at higher temperatures and must be encouraged to consciously increase their fluid intake.

a. Pressure is low in mountainous areas due to the altitude. The barometer usually drops 2.5 centimeters for every 300 meters gained in elevation (3 percent).

b. The air at higher altitudes is thinner as atmospheric pressure drops with the increasing altitude. The altitude has a natural filtering effect on the sun's rays. Rays are absorbed or reflected in part by the molecular content of the atmosphere. This effect is greater at lower altitudes. At higher altitudes, the thinner, drier air has a reduced molecular content and, consequently, a reduced filtering effect on the sun's rays. The intensity of both visible and ultraviolet rays is greater with increased altitude. These conditions increase the chance of sunburn, especially when combined with a snow cover that reflects the rays upward.

1-13. WEATHER CHARACTERISTICS

The earth is surrounded by an atmosphere that is divided into several layers. The world's weather systems are in the lower of these layers known as the "troposphere." This layer reaches as high as 40,000 feet. Weather is a result of an atmosphere, oceans, land masses, unequal heating and cooling from the sun, and the earth's rotation. The weather found in any one place depends on many things such as the air temperature, humidity (moisture content), air pressure (barometric pressure), how it is being moved, and if it is being lifted or not.

a. Air pressure is the "weight" of the atmosphere at any given place. The higher the pressure, the better the weather will be. With lower air pressure, the weather will more than likely be worse. In order to understand this, imagine that the air in the atmosphere acts like a liquid. Areas with a high level of this "liquid" exert more pressure on an area and are called high-pressure areas. Areas with a lower level are called low-pressure areas. The average air pressure at sea level is 29.92 inches of mercury (hg) or 1,013 millibars (mb). The higher in altitude, the lower the pressure.

(1) ***High Pressure.*** The characteristics of a high-pressure area are as follows:

- The airflow is clockwise and out.
- Otherwise known as an "anticyclone".
- Associated with clear skies.
- Generally the winds will be mild.
- Depicted as a blue "H" on weather maps.

(2) ***Low Pressure.*** The characteristics of a low-pressure area are as follows:

- The airflow is counterclockwise and in.
- Otherwise known as a "cyclone".
- Associated with bad weather.
- Depicted as a red "L" on weather maps.

b. Air from a high-pressure area is basically trying to flow out and equalize its pressure with the surrounding air. Low pressure, on the other hand, is building up vertically by pulling air in from outside itself, which causes atmospheric instability resulting in bad weather.

c. On a weather map, these differences in pressure are depicted as isobars. Isobars resemble contour lines and are measured in either millibars or inches of mercury. The areas of high pressure are called "ridges" and lows are called "troughs."

1-14. WIND

In high mountains, the ridges and passes are seldom calm; however, strong winds in protected valleys are rare. Normally, wind speed increases with altitude since the earth's frictional drag is strongest near the ground. This effect is intensified by mountainous terrain. Winds are accelerated when they converge through mountain passes and canyons. Because of these funneling effects, the wind may blast with great force on an exposed mountainside or summit. Usually, the local wind direction is controlled by topography.

a. The force exerted by wind quadruples each time the wind speed doubles; that is, wind blowing at 40 knots pushes four times harder than a wind blowing at 20 knots. With increasing wind strength, gusts become more important and may be 50 percent higher than the average wind speed. When wind strength increases to a hurricane force of 64 knots or more, soldiers should lay on the ground during gusts and continue moving during lulls. If a hurricane- force wind blows where there is sand or snow, dense clouds fill the air. The rocky debris or chunks of snow crust are hurled near the surface. During the winter season, or at high altitudes, commanders must be constantly aware of the wind-chill factor and associated cold-weather injuries (see Chapter 2).

b. Winds are formed due to the uneven heating of the air by the sun and rotation of the earth. Much of the world's weather depends on a system of winds that blow in a set direction.

c. Above hot surfaces, air expands and moves to colder areas where it cools and becomes denser, and sinks to the earth's surface. The results are a circulation of air from the poles along the surface of the earth to the equator, where it rises and moves to the poles again.

d. Heating and cooling together with the rotation of the earth causes surface winds. In the Northern Hemisphere, there are three prevailing winds:

(1) ***Polar Easterlies***. These are winds from the polar region moving from the east. This is air that has cooled and settled at the poles.

(2) ***Prevailing Westerlies***. These winds originate from approximately 30 degrees north latitude from the west. This is an area where prematurely cooled air, due to the earth's rotation, has settled to the surface.

(3) ***Northeast Tradewinds***. These are winds that originate from approximately 30° north from the northeast.

e. The jet stream is a long meandering current of high-speed winds often exceeding 250 miles per hour near the transition zone between the troposphere and the stratosphere known as the tropopause. These winds blow from a generally westerly direction dipping down and picking up air masses from the tropical regions and going north and bringing down air masses from the polar regions.

f. The patterns of wind mentioned above move air. This air comes in parcels called "air masses." These air masses can vary from the size of a small town to as large as a country. These air masses are named from where they originate:

- Maritime—over water.
- Continental—over land
- Polar—north of 60° north latitude.
- Tropical—south of 60° north latitude.

Combining these parcels of air provides the names and description of the four types of air masses:

- Continental Polar—cold, dry air mass.
- Maritime Polar—cold, wet air mass.
- Maritime Tropical—warm, wet air mass.
- Continental Tropical—warm, dry air mass.

g. Two types of winds are peculiar to mountain environments, but do not necessarily affect the weather.

(1) ***Anabatic Wind (Valley Winds).*** These winds blow up mountain valleys to replace warm rising air and are usually light winds.

(2) ***Katabatic Wind (Mountain Wind).*** These winds blow down mountain valley slopes caused by the cooling of air and are occasionally strong winds.

1-15. HUMIDITY

Humidity is the amount of moisture in the air. All air holds water vapor even if it cannot be seen. Air can hold only so much water vapor; however, the warmer the air, the more moisture it can hold. When air can hold all that it can the air is "saturated" or has 100 percent relative humidity.

a. If air is cooled beyond its saturation point, the air will release its moisture in one form or another (clouds, fog, dew, rain, snow, and so on). The temperature at which this happens is called the "condensation point". The condensation point varies depending on the amount of water vapor contained in the air and the temperature of the air. If the air contains a great deal of water, condensation can occur at a temperature of 68 degrees Fahrenheit, but if the air is dry and does not hold much moisture, condensation may not form until the temperature drops to 32 degrees Fahrenheit or even below freezing.

b. The adiabatic lapse rate is the rate at which air cools as it rises or warms as it descends. This rate varies depending on the moisture content of the air. Saturated (moist) air will warm and cool approximately 3.2 degrees Fahrenheit per 1,000 feet of elevation gained or lost. Dry air will warm and cool approximately 5.5 degrees Fahrenheit per 1,000 feet of elevation gained or lost.

1-16. CLOUD FORMATION

Clouds are indicators of weather conditions. By reading cloud shapes and patterns, observers can forecast weather with little need for additional equipment such as a barometer, wind meter, and thermometer. Any time air is lifted or cooled beyond its saturation point (100 percent relative humidity), clouds are formed. The four ways air gets lifted and cooled beyond its saturation point are as follows.

a. **Convective Lifting.** This effect happens due to the sun's heat radiating off the Earth's surface causing air currents (thermals) to rise straight up and lift air to a point of saturation.

b. **Frontal Lifting.** A front is formed when two air masses of different moisture content and temperature collide. Since air masses will not mix, warmer air is forced aloft over the colder air mass. From there it is cooled and then reaches its saturation point. Frontal lifting creates the majority of precipitation.

c. **Cyclonic Lifting.** An area of low pressure pulls air into its center from all over in a counterclockwise direction. Once this air reaches the center of the low pressure, it has nowhere to go but up. Air continues to lift until it reaches the saturation point.

d. **Orographic Lifting.** This happens when an air mass is pushed up and over a mass of higher ground such as a mountain. Air is cooled due to the adiabatic lapse rate until the air's saturation point is reached.

1-17. TYPES OF CLOUDS

Clouds are one of the signposts to what is happening with the weather. Clouds can be described in many ways. They can be classified by height or appearance, or even by the amount of area covered vertically or horizontally. Clouds are classified into five categories: low-, mid-, and high-level clouds; vertically-developed clouds; and less common clouds.

a. **Low-Level Clouds.** Low-level clouds (0 to 6,500 feet) are either cumulus or stratus (Figures 1-1 and 1-2, page 1-16). Low-level clouds are mostly composed of water droplets since their bases lie below 6,500 feet. When temperatures are cold enough, these clouds may also contain ice particles and snow.

(1) The two types of precipitating low-level clouds are nimbostratus and stratocumulus (Figures 1-3 and 1-4, page 1-17).

(a) Nimbostratus clouds are dark, low-level clouds accompanied by light to moderately falling precipitation. The sun or moon is not visible through nimbostratus clouds, which distinguishes them from mid-level altostratus clouds. Because of the fog and falling precipitation commonly found beneath and around nimbostratus clouds, the cloud base is typically extremely diffuse and difficult to accurately determine.

(b) Stratocumulus clouds generally appear as a low, lumpy layer of clouds that is sometimes accompanied by weak precipitation. Stratocumulus vary in color from dark gray to light gray and may appear as rounded masses with breaks of clear sky in between. Because the individual elements of stratocumulus are larger than those of altocumulus, deciphering between the two cloud types is easier. With your arm extended toward the sky, altocumulus elements are about the size of a thumbnail while stratocumulus are about the size of a fist.

Figure 1-1. Cumulus clouds.

Figure 1-2. Stratus clouds.

Figure 1-3. Nimbostratus clouds.

Figure 1-4. Stratocumulus clouds.

(2) Low-level clouds may be identified by their height above nearby surrounding relief of known elevation. Most precipitation originates from low-level clouds because rain or snow usually evaporate before reaching the ground from higher clouds. Low-level clouds usually indicate impending precipitation, especially if the cloud is more than 3,000 feet thick. (Clouds that appear dark at their bases are more than 3,000 feet thick.)

b. **Mid-Level Clouds.** Mid-level clouds (between 6,500 to 20,000 feet) have a prefix of alto. Middle clouds appear less distinct than low clouds because of their height. Alto clouds with sharp edges are warmer because they are composed mainly of water droplets. Cold clouds, composed mainly of ice crystals and usually colder than -30 degrees F, have distinct edges that grade gradually into the surrounding sky. Middle clouds usually indicate fair weather, especially if they are rising over time. Lowering middle clouds indicate potential storms, though usually hours away. There are two types of mid-level clouds, altocumulus and altostratus clouds (Figures 1-5 and 1-6).

(1) Altocumulus clouds can appear as parallel bands or rounded masses. Typically a portion of an altocumulus cloud is shaded, a characteristic which makes them distinguishable from high-level cirrocumulus. Altocumulus clouds usually form in advance of a cold front. The presence of altocumulus clouds on a warm humid summer morning is commonly followed by thunderstorms later in the day. Altocumulus clouds that are scattered rather than even, in a blue sky, are called "fair weather" cumulus and suggest arrival of high pressure and clear skies.

(2) Altostratus clouds are often confused with cirrostratus. The one distinguishing feature is that a halo is not observed around the sun or moon. With altostratus, the sun or moon is only vaguely visible and appears as if it were shining through frosted glass.

Figure 1-5. Altocumulus.

Figure 1-6. Altostratus.

c. **High-Level Clouds.** High-level clouds (more than 20,000 feet above ground level) are usually frozen clouds, indicating air temperatures at that elevation below -30 degrees Fahrenheit, with a fibrous structure and blurred outlines. The sky is often covered with a thin veil of cirrus that partly obscures the sun or, at night, produces a ring of light around the moon. The arrival of cirrus indicates moisture aloft and the approach of a traveling storm system. Precipitation is often 24 to 36 hours away. As the storm approaches, the cirrus thickens and lowers, becoming altostratus and eventually stratus. Temperatures are warm, humidity rises, and winds become southerly or south easterly. The two types of high-level clouds are cirrus and cirrostratus (Figure 1-7 and Figure 1-8, page 1-20).

(1) Cirrus clouds are the most common of the high-level clouds. Typically found at altitudes greater than 20,000 feet, cirrus are composed of ice crystals that form when super-cooled water droplets freeze. Cirrus clouds generally occur in fair weather and point in the direction of air movement at their elevation. Cirrus can be observed in a variety of shapes and sizes. They can be nearly straight, shaped like a comma, or seemingly all tangled together. Extensive cirrus clouds are associated with an approaching warm front.

(2) Cirrostratus clouds are sheet-like, high-level clouds composed of ice crystals. They are relatively transparent and can cover the entire sky and be up to several thousand feet thick. The sun or moon can be seen through cirrostratus. Sometimes the only indication of cirrostratus clouds is a halo around the sun or moon. Cirrostratus clouds tend to thicken as a warm front approaches, signifying an increased production of ice crystals. As a result, the halo gradually disappears and the sun or moon becomes less visible.

Figure 1-7. Cirrus.

Figure 1-8. Cirrostratus.

d. **Vertical-Development Clouds.** Clouds with vertical development can grow to heights in excess of 39,000 feet, releasing incredible amounts of energy. The two types of clouds with vertical development are fair weather cumulus and cumulonimbus.

(1) Fair weather cumulus clouds have the appearance of floating cotton balls and have a lifetime of 5 to 40 minutes. Known for their flat bases and distinct outlines, fair weather cumulus exhibit only slight vertical growth, with the cloud tops designating the limit of the rising air. Given suitable conditions, however, these clouds can later develop into towering cumulonimbus clouds associated with powerful thunderstorms. Fair weather cumulus clouds are fueled by buoyant bubbles of air known as thermals that rise up from the earth's surface. As the air rises, the water vapor cools and condenses forming water droplets. Young fair weather cumulus clouds have sharply defined edges and bases while the edges of older clouds appear more ragged, an artifact of erosion. Evaporation along the cloud edges cools the surrounding air, making it heavier and producing sinking motion outside the cloud. This downward motion inhibits further convection and growth of additional thermals from down below, which is why fair weather cumulus typically have expanses of clear sky between them. Without a continued supply of rising air, the cloud begins to erode and eventually disappears.

(2) Cumulonimbus clouds are much larger and more vertically developed than fair weather cumulus (Figure 1-9). They can exist as individual towers or form a line of towers called a squall line. Fueled by vigorous convective updrafts, the tops of cumulonimbus clouds can reach 39,000 feet or higher. Lower levels of cumulonimbus clouds consist mostly of water droplets while at higher elevations, where the temperatures are well below freezing, ice crystals dominate the composition. Under favorable conditions, harmless fair weather cumulus clouds can quickly develop into large cumulonimbus associated with powerful thunderstorms known as super-cells. Super-cells are large thunderstorms with deep rotating updrafts and can have a lifetime of several hours. Super-cells produce frequent lightning, large hail, damaging winds, and

tornadoes. These storms tend to develop during the afternoon and early evening when the effects of heating from the sun are the strongest.

Figure 1-9. Cumulonimbus.

e. **Other Cloud Types.** These clouds are a collection of miscellaneous types that do not fit into the previous four groups. They are orographic clouds, lenticulars, and contrails.

(1) Orographic clouds develop in response to the forced lifting of air by the earth's topography. Air passing over a mountain oscillates up and down as it moves downstream. Initially, stable air encounters a mountain, is lifted upward, and cools. If the air cools to its saturation temperature during this process, the water vapor condenses and becomes visible as a cloud. Upon reaching the mountain top, the air is heavier than the environment and will sink down the other side, warming as it descends. Once the air returns to its original height, it has the same buoyancy as the surrounding air. However, the air does not stop immediately because it still has momentum carrying it downward. With continued descent, the air becomes warmer then the surrounding air and accelerates back upwards towards its original height. Another name for this type of cloud is the lenticular cloud.

(2) Lenticular clouds are cloud caps that often form above pinnacles and peaks, and usually indicate higher winds aloft (Figure 1-10, page 1-22). Cloud caps with a lens shape, similar to a "flying saucer," indicate extremely high winds (over 40 knots). Lenticulars should always be watched for changes. If they grow and descend, bad weather can be expected.

Figure 1-10. Lenticular.

(3) Contrails are clouds that are made by water vapor being inserted into the upper atmosphere by the exhaust of jet engines (Figure 1-11). Contrails evaporate rapidly in fair weather. If it takes longer than two hours for contrails to evaporate, then there is impending bad weather (usually about 24 hours prior to a front).

Figure 1-11. Contrails.

f. **Cloud Interpretation.** Serious errors can occur in interpreting the extent of cloud cover, especially when cloud cover must be reported to another location. Cloud cover always appears greater on or near the horizon, especially if the sky is covered with cumulus clouds, since the observer is looking more at the sides of the clouds rather than between them. Cloud cover estimates should be restricted to sky areas more than 40 degrees above the horizon—that is, to the local sky. Assess the sky by dividing the 360 degrees of sky around you into eighths. Record the coverage in eighths and the types of clouds observed.

1-18. FRONTS

Fronts occur when two air masses of different moisture and temperature contents meet. One of the indicators that a front is approaching is the progression of the clouds. The four types of fronts are warm, cold, occluded, and stationary.

a. **Warm Front.** A warm front occurs when warm air moves into and over a slower or stationary cold air mass. Because warm air is less dense, it will rise up and over the cooler air. The cloud types seen when a warm front approaches are cirrus, cirrostratus, nimbostratus (producing rain), and fog. Occasionally, cumulonimbus clouds will be seen during the summer months.

b. **Cold Front.** A cold front occurs when a cold air mass overtakes a slower or stationary warm air mass. Cold air, being more dense than warm air, will force the warm air up. Clouds observed will be cirrus, cumulus, and then cumulonimbus producing a short period of showers.

c. **Occluded Front.** Cold fronts generally move faster than warm fronts. The cold fronts eventually overtake warm fronts and the warm air becomes progressively lifted from the surface. The zone of division between cold air ahead and cold air behind is called a "cold occlusion." If the air behind the front is warmer than the air ahead, it is a warm occlusion. Most land areas experience more occlusions than other types of fronts. The cloud progression observed will be cirrus, cirrostratus, altostratus, and nimbostratus. Precipitation can be from light to heavy.

d. **Stationary Front.** A stationary front is a zone with no significant air movement. When a warm or cold front stops moving, it becomes a stationary front. Once this boundary begins forward motion, it once again becomes a warm or cold front. When crossing from one side of a stationary front to another, there is typically a noticeable temperature change and shift in wind direction. The weather is usually clear to partly cloudy along the stationary front.

1-19. TEMPERATURE

Normally, a temperature drop of 3 to 5 degrees Fahrenheit for every 1,000 feet gain in altitude is encountered in motionless air. For air moving up a mountain with condensation occurring (clouds, fog, and precipitation), the temperature of the air drops 3.2 degrees Fahrenheit with every 1,000 feet of elevation gain. For air moving up a mountain with no clouds forming, the temperature of the air drops 5.5 degrees Fahrenheit for every 1,000 feet of elevation gain.

a. An expedient to this often occurs on cold, clear, calm mornings. During a troop movement or climb started in a valley, higher temperatures may often be encountered as altitude is gained. This reversal of the normal cooling with elevation is called temperature inversion. Temperature inversions are caused when mountain air is cooled by ice, snow, and heat loss through thermal radiation. This cooler, denser air settles into the valleys and low areas. The inversion continues until the sun warms the surface of the earth or a moderate wind causes a mixing of the warm and cold layers. Temperature inversions are common in the mountainous regions of the arctic, subarctic, and mid-latitudes.

b. At high altitudes, solar heating is responsible for the greatest temperature contrasts. More sunshine and solar heat are received above the clouds than below. The important effect of altitude is that the sun's rays pass through less of the atmosphere and more direct heat is received than at lower levels, where solar radiation is absorbed and reflected by dust and water vapor. Differences of 40 to 50 degrees Fahrenheit may occur between surface temperatures in the shade and surface temperatures in the sun. This is particularly true for dark metallic objects. The difference in temperature felt on the skin between the sun and shade is normally 7 degrees Fahrenheit. Special care must be taken to avoid sunburn and snow blindness. Besides permitting rapid heating, the clear air at high altitudes also favors rapid cooling at night. Consequently, the temperature rises fast after sunrise and drops quickly after sunset. Much of the chilled air drains downward, due to convection currents, so that the differences between day and night temperatures are greater in valleys than on slopes.

c. Local weather patterns force air currents up and over mountaintops. Air is cooled on the windward side of the mountain as it gains altitude, but more slowly (3.2 degrees Fahrenheit per 1,000 feet) if clouds are forming due to heat release when water vapor becomes liquid. On the leeward side of the mountain, this heat gained from the condensation on the windward side is added to the normal heating that occurs as the air descends and air pressure increases. Therefore, air and winds on the leeward slope are considerably warmer than on the windward slope, which is referred to as Chinook winds. The heating and cooling of the air affects planning considerations primarily with regard to the clothing and equipment needed for an operation.

1-20. WEATHER FORECASTING

The use of a portable aneroid barometer, thermometer, wind meter, and hygrometer help in making local weather forecasts. Reports from other localities and from any weather service, including USAF, USN, or the National Weather Bureau, are also helpful. Weather reports should be used in conjunction with the locally observed current weather situation to forecast future weather patterns.

a. Weather at various elevations may be quite different because cloud height, temperature, and barometric pressure will all be different. There may be overcast and rain in a lower area, with mountains rising above the low overcast into warmer clear weather.

b. To be effective, a forecast must reach the small-unit leaders who are expected to utilize weather conditions for assigned missions. Several different methods can be used to create a forecast. The method a forecaster chooses depends upon the forecaster's experience, the amount of data available, the level of difficulty that the forecast situation presents, and the degree of accuracy needed to make the forecast. The five ways to forecast weather are:

(1) ***Persistence Method.*** "Today equals tomorrow" is the simplest way of producing a forecast. This method assumes that the conditions at the time of the forecast will not

change; for example, if today was hot and dry, the persistence method predicts that tomorrow will be the same.

(2) ***Trends Method.*** "Nowcasting" involves determining the speed and direction of fronts, high- and low-pressure centers, and clouds and precipitation. For example, if a cold front moves 300 miles during a 24-hour period, we can predict that it will travel 300 miles in another 24-hours.

(3) ***Climatology Method.*** This method averages weather statistics accumulated over many years. This only works well when the pattern is similar to the following years.

(4) ***Analog Method.*** This method examines a day's forecast and recalls a day in the past when the weather looked similar (an analogy). This method is difficult to use because finding a perfect analogy is difficult.

(5) ***Numerical Weather Prediction.*** This method uses computers to analyze all weather conditions and is the most accurate of the five methods.

1-21. RECORDING DATA

An accurate observation is essential in noting trends in weather patterns. Ideally, under changing conditions, trends will be noted in some weather parameters. However, this may not always be the case. A minor shift in the winds may signal an approaching storm.

a. **Wind Direction.** Assess wind direction as a magnetic direction from which the wind is blowing.

b. **Wind Speed.** Assess wind speed in knots.

(1) If an anemometer is available, assess speed to the nearest knot.

(2) If no anemometer is available, estimate the speed in knots. Judge the wind speed by the way objects, such as trees, bushes, tents, and so forth, are blowing.

c. **Visibility in Meters.** Observe the farthest visible major terrain or man-made feature and determine the distance using any available map.

d. **Present Weather.** Include any precipitation or obscuring weather. The following are examples of present weather:

- Rain—continuous and steady liquid precipitation that will last at least one hour.
- Rain showers—short-term and potentially heavy downpours that rarely last more than one hour.
- Snow—continuous and steady frozen precipitation that will last at least one hour.
- Snow showers—short-term and potentially heavy frozen downpours that rarely last more than one hour.
- Fog, haze—obstructs visibility of ground objects.
- Thunderstorms—a potentially dangerous storm. Thunderstorms will produce lightning, heavy downpours, colder temperatures, tornadoes (not too frequently), hail, and strong gusty winds at the surface and aloft. Winds commonly exceed 35 knots.

e. **Total Cloud Cover.** Assess total cloud cover in eighths. Divide the sky into eight different sections measuring from horizon to horizon. Count the sections with cloud cover, which gives the total cloud cover in eighths. (For example, if half of the sections are covered with clouds, total cloud cover is 4/8.)

f. **Ceiling Height.** Estimate where the cloud base intersects elevated terrain. Note if bases are above all terrain. If clouds are not touching terrain, then estimate to the best of your ability.

g. **Temperature.** Assess temperature with or without a thermometer.

(1) With a thermometer, assess temperature in degrees Celsius (use Fahrenheit only if Celsius conversion is not available). To convert Fahrenheit to Celsius: C = F minus 32 times .55. To convert Celsius to Fahrenheit: F = 1.8 times C plus 32.

Example: 41 degrees F – 32 x .55 = 5 degrees C.
5 degrees C x 1.8 + 32 = 41 degrees F.

(2) Without a thermometer, estimate temperature as above or below freezing (0°C), as well as an estimated temperature.

h. **Pressure Trend.** With a barometer or altimeter, assess the pressure trend.

(1) A high pressure moving in will cause altimeters to indicate lower elevation.

(2) A low pressure moving in will cause altimeters to indicate higher elevation.

i. **Observed Weather.** Note changes or trends in observed weather conditions.

(1) Deteriorating trends include:

- Marked wind direction shifts. A high pressure system wind flows clockwise. A low pressure system wind flows counterclockwise. The closer the isometric lines are, the greater the differential of pressure (greater wind speeds).
- Marked wind speed increases.
- Changes in obstructions to visibility.
- Increasing cloud coverage.
- Increase in precipitation. A steady drizzle is usually a long-lasting rain.
- Lowering cloud ceilings.
- Marked cooler temperature changes, which could indicate that a cold front is passing through.
- Marked increase in humidity.
- Decreasing barometric pressure, which indicates a lower pressure system is moving through the area.

(2) Improving trends include:

- Steady wind direction, which indicates no change in weather systems in the area.
- Decreasing wind speeds.
- Clearing of obstructions to visibility.
- Decreasing or ending precipitation.
- Decreasing cloud coverage.
- Increasing height of cloud ceilings.
- Temperature changes slowly warmer.
- Humidity decreases.
- Increasing barometric pressure, which indicates that a higher pressure system is moving through the area.

j. **Update.** Continue to evaluate observed conditions and update the forecast.

Section III. MOUNTAIN HAZARDS

Hazards can be termed natural (caused by natural occurrence), man-made (caused by an individual, such as lack of preparation, carelessness, improper diet, equipment misuse), or as a combination (human trigger). There are two kinds of hazards while in the mountains—subjective and objective. Combinations of objective and subjective hazards are referred to as cumulative hazards.

1-22. SUBJECTIVE HAZARDS

Subjective hazards are created by humans; for example, choice of route, companions, overexertion, dehydration, climbing above one's ability, and poor judgment.

a. **Falling.** Falling can be caused by carelessness, over-fatigue, heavy equipment, bad weather, overestimating ability, a hold breaking away, or other reasons.

b. **Bivouac Site.** Bivouac sites must be protected from rockfall, wind, lightning, avalanche run-out zones, and flooding (especially in gullies). If the possibility of falling exists, rope in, the tent and all equipment may have to be tied down.

c. **Equipment.** Ropes are not total security; they can be cut on a sharp edge or break due to poor maintenance, age, or excessive use. You should always pack emergency and bivouac equipment even if the weather situation, tour, or a short climb is seemingly low of dangers.

1-23. OBJECTIVE HAZARDS

Objective hazards are caused by the mountain and weather and cannot be influenced by man; for example, storms, rockfalls, icefalls, lightning, and so on.

a. **Altitude.** At high altitudes (especially over 6,500 feet), endurance and concentration is reduced. Cut down on smoking and alcohol. Sleep well, acclimatize slowly, stay hydrated, and be aware of signs and symptoms of high-altitude illnesses. Storms can form quickly and lightning can be severe.

b. **Visibility.** Fog, rain, darkness, and or blowing snow can lead to disorientation. Take note of your exact position and plan your route to safety before visibility decreases. Cold combined with fog can cause a thin sheet of ice to form on rocks (verglas). Whiteout conditions can be extremely dangerous. If you must move under these conditions, it is best to rope up. Have the point man move to the end of the rope. The second man will use the first man as an aiming point with the compass. Use a route sketch and march table. If the tactical situation does not require it, plan route so as not to get caught by darkness.

c. **Gullies.** Rock, snow, and debris are channeled down gullies. If ice is in the gully, climbing at night may be better because the warming of the sun will loosen stones and cause rockfalls.

d. **Rockfall.** Blocks and scree at the base of a climb can indicate recurring rockfall. Light colored spots on the wall may indicate impact chips of falling rock. Spring melt or warming by the sun of the rock/ice/snow causes rockfall.

e. **Avalanches.** Avalanches are caused by the weight of the snow overloading the slope. (Refer to paragraph 1-25 for more detailed information on avalanches.)

f. **Hanging Glaciers and Seracs.** Avoid, if at all possible, hanging glaciers and seracs. They will fall without warning regardless of the time of day or time of year. One cubic meter of glacier ice weighs 910 kilograms (about 2,000 pounds). If you must cross these danger areas, do so quickly and keep an interval between each person.

g. **Crevasses.** Crevasses are formed when a glacier flows over a slope and makes a bend, or when a glacier separates from the rock walls that enclose it. A slope of only two to three degrees is enough to form a crevasse. As this slope increases from 25 to 30 degrees, hazardous icefalls can be formed. Likewise, as a glacier makes a bend, it is likely that crevasses will form at the outside of the bend. Therefore, the safest route on a glacier would be to the inside of bends, and away from steep slopes and icefalls. Extreme care must be taken when moving off of or onto the glacier because of the moat that is most likely to be present.

1-24. WEATHER HAZARDS

Weather conditions in the mountains may vary from one location to another as little as 10 kilometers apart. Approaching storms may be hard to spot if masked by local peaks. A clear, sunny day in July could turn into a snowstorm in less than an hour. Always pack some sort of emergency gear.

a. Winds are stronger and more variable in the mountains; as wind doubles in speed, the force quadruples.

b. Precipitation occurs more on the windward side than the leeward side of ranges. This causes more frequent and denser fog on the windward slope.

c. Above approximately 8,000 feet, snow can be expected any time of year in the temperate climates.

d. Air is dryer at higher altitudes, so equipment does not rust as quickly, but dehydration is of greater concern.

e. Lightning is frequent, violent, and normally attracted to high points and prominent features in mountain storms. Signs indicative of thunderstorms are tingling of the skin, hair standing on end, humming of metal objects, crackling, and a bluish light (St. Elmo's fire) on especially prominent metal objects (summit crosses and radio towers).

(1) Avoid peaks, ridges, rock walls, isolated trees, fixed wire installations, cracks that guide water, cracks filled with earth, shallow depressions, shallow overhangs, and rock needles. Seek shelter around dry, clean rock without cracks; in scree fields; or in deep indentations (depressions, caves). Keep at least half a body's length away from a cave wall and opening.

(2) Assume a one-point-of-contact body position. Squat on your haunches or sit on a rucksack or rope. Pull your knees to your chest and keep both feet together. If half way up the rock face, secure yourself with more than one point—lightning can burn through rope. If already rappelling, touch the wall with both feet together and hurry to the next anchor.

f. During and after rain, expect slippery rock and terrain in general and adjust movement accordingly. Expect flash floods in gullies or chimneys. A climber can be washed away or even drowned if caught in a gully during a rainstorm. Be especially alert for falling objects that the rain has loosened.

g. Dangers from impending high winds include frostbite (from increased wind-chill factor), windburn, being blown about (especially while rappelling), and debris being blown about. Wear protective clothing and plan the route to be finished before bad weather arrives.

h. For each 100-meter rise in altitude, the temperature drops approximately one degree Fahrenheit. This can cause hypothermia and frostbite even in summer, especially when combined with wind, rain, and snow. Always wear or pack appropriate clothing.

i. If it is snowing, gullies may contain avalanches or snow sloughs, which may bury the trail. Snowshoes or skis may be needed in autumn or even late spring. Unexpected snowstorms may occur in the summer with accumulations of 12 to 18 inches; however, the snow quickly melts.

j. Higher altitudes provide less filtering effects, which leads to greater ultraviolet (UV) radiation intensity. Cool winds at higher altitudes may mislead one into underestimating the sun's intensity, which can lead to sunburns and other heat injuries. Use sunscreen and wear hat and sunglasses, even if overcast. Drink plenty of fluids.

1-25. AVALANCHE HAZARDS

Avalanches occur when the weight of accumulated snow on a slope exceeds the cohesive forces that hold the snow in place. (Table 1-2, page 1-32, shows an avalanche hazard evaluation checklist.)

a. **Slope Stability.** Slope stability is the key factor in determining the avalanche danger.

(1) ***Slope Angle.*** Slopes as gentle as 15 degrees have avalanched. Most avalanches occur on slopes between 30 and 45 degrees. Slopes above 60 degrees often do not build up significant quantities of snow because they are too steep.

(2) ***Slope Profile.*** Dangerous slab avalanches are more likely to occur on convex slopes, but may occur on concave slopes.

(3) ***Slope Aspect.*** Snow on north facing slopes is more likely to slide in midwinter. South facing slopes are most dangerous in the spring and on sunny, warm days. Slopes on the windward side are generally more stable than leeward slopes.

(4) ***Ground Cover.*** Rough terrain is more stable than smooth terrain. On grassy slopes or scree, the snow pack has little to anchor to.

b. **Triggers.** Various factors trigger avalanches.

(1) ***Temperature.*** When the temperature is extremely low, settlement and adhesion occur slowly. Avalanches that occur during extreme cold weather usually occur during or immediately following a storm. At a temperature just below freezing, the snowpack stabilizes quickly. At temperatures above freezing, especially if temperatures rise quickly, the potential for avalanche is high. Storms with a rise in temperature can deposit dry snow early, which bonds poorly with the heavier snow deposited later. Most avalanches occur during the warmer midday.

(2) ***Precipitation.*** About 90 percent of avalanches occur during or within twenty-four hours after a snowstorm. The rate at which snow falls is important. High rates of snowfall (2.5 centimeters per hour or greater), especially when accompanied by wind, are usually responsible for major periods of avalanche activity. Rain falling on snow will increase its weight and weakens the snowpack.

(3) ***Wind.*** Sustained winds of 15 miles per hour and over transport snow and form wind slabs on the lee side of slopes.

(4) ***Weight.*** Most victims trigger the avalanches that kill them.

(5) ***Vibration.*** Passing helicopters, heavy equipment, explosions, and earth tremors have triggered avalanches.

c. **Snow Pits.** Snow pits can be used to determine slope stability.

(1) Dig the snow pit on the suspect slope or a slope with the same sun and wind conditions. Snow deposits may vary greatly within a few meters due to wind and sun variations. (On at least one occasion, a snow pit dug across the fall line triggered the suspect slope). Dig a 2-meter by 2-meter pit across the fall line, through all the snow, to the ground. Once the pit is complete, smooth the face with a shovel.

(2) Conduct a shovel shear test.

(a) A shovel shear test puts pressure on a representative sample of the snowpack. The core of this test is to isolate a column of the snowpack from three sides. The column should be of similar size to the blade of the shovel. Dig out the sides of the column without pressing against the column with the shovel (this affects the strength). To isolate the rear of the column, use a rope or string to saw from side to side to the base of the column.

(b) If the column remained standing while cutting the rear, place the shovel face down on the top of the column. Tap with varying degrees of strength on the shovel to see what force it takes to create movement on the bed of the column. The surface that eventually slides will be the layer to look at closer. This test provides a better understanding of the snowpack strength. For greater results you will need to do this test in many areas and formulate a scale for the varying methods of tapping the shovel.

(3) Conduct a Rutschblock test. To conduct the test, isolate a column slightly longer than the length of your snowshoes or skis (same method as for the shovel shear test). One person moves on their skis or snowshoes above the block without disturbing the block. Once above, the person carefully places one showshoe or ski onto the block with no body weight for the first stage of the test. The next stage is adding weight to the first leg. Next, place the other foot on the block. If the block is still holding up, squat once, then twice, and so on. The remaining stage is to jump up and land on the block.

d. **Types of Snow Avalanches.** There are two types of snow avalanches: loose snow (point) and slab.

(1) Loose snow avalanches start at one point on the snow cover and grow in the shape of an inverted "V." Although they happen most frequently during the winter snow season, they can occur at any time of the year in the mountains. They often fall as many small sluffs during or shortly after a storm. This process removes snow from steep upper slopes and either stabilizes lower slopes or loads them with additional snow.

(2) Wet loose snow avalanches occur in spring and summer in all mountain ranges. Large avalanches of this type, lubricated and weighed down by meltwater or rain can travel long distances and have tremendous destructive power. Coastal ranges that have high temperatures and frequent rain are the most common areas for this type of avalanche.

(3) Slab avalanches occur when cohesive snow begins to slide on a weak layer. The fracture line where the moving snow breaks away from the snowpack makes this type of avalanche easy to identify. Slab release is rapid. Although any avalanche can kill you, slab avalanches are generally considered more dangerous than loose snow avalanches.

(a) Most slab avalanches occur during or shortly after a storm when slopes are loaded with new snow at a critical rate. The old rule of never travel in avalanche terrain for a few days after a storm still holds true.

(b) As slabs become harder, their behavior becomes more unpredictable; they may allow several people to ski across before releasing. Many experts believe they are susceptible to rapid temperature changes. Packed snow expands and contracts with temperature changes. For normal density, settled snow, a drop in temperature of 10 degrees Celsius (18 degrees Fahrenheit) would cause a snow slope 300 meters wide to contract 2 centimeters. Early ski mountaineers in the Alps noticed that avalanches sometimes occurred when shadows struck a previously sun-warmed slope.

d. **Protective Measures.** Avoiding known or suspected avalanche areas is the easiest method of protection. Other measures include:

(1) ***Personal Safety.*** Remove your hands from ski pole wrist straps. Detach ski runaway cords. Prepare to discard equipment. Put your hood on. Close up your clothing to prepare for hypothermia. Deploy avalanche cord. Make avalanche probes and shovels accessible. Keep your pack on at all times—do not discard. Your pack can act as a flotation device, as well as protect your spine.

(2) ***Group Safety.*** Send one person across the suspect slope at a time with the rest of the group watching. All members of the group should move in the same track from safe zone to safe zone.

e. **Route Selection.** Selecting the correct route will help avoid avalanche prone areas, which is always the best choice. Always allow a wide margin of safety when making your decision.

(1) The safest routes are on ridge tops, slightly on the windward side; the next safest route is out in the valley, far from the bottom of slopes.

(2) Avoid cornices from above or below. Should you encounter a dangerous slope, either climb to the top of the slope or descend to the bottom—well out of the way of the run-out zone. If you must traverse, pick a line where you can traverse downhill as quickly as possible. When you must ascend a dangerous slope, climb to the side of the avalanche path, and not directly up the center.

(3) Take advantage of dense timber, ridges, or rocky outcrops as islands of safety. Use them for lunch and rest stops. Spend as little time as possible on open slopes.

(4) Since most avalanches occur within twenty-four hours of a storm and or at midday, avoid moving during these periods. Moving at night is tactically sound and may be safer.

f. **Stability Analysis.** Look for nature's billboards on slopes similar to the one you are on.

(1) ***Evidence of Avalanching.*** Look for recent avalanches and for signs of wind-loading and wind-slabs.

(2) ***Fracture Lines.*** Avoid any slopes showing cracks.

(3) ***Sounds.*** Beware of hollow sounds—a "whumping" noise. They may suggest a radical settling of the snowpack.

g. **Survival.** People trigger avalanches that bury people. If these people recognized the hazard and chose a different route, they would avoid the avalanche. The following steps should be followed if caught in an avalanche.

(1) Discard equipment. Equipment can injure or burden you; discarded equipment will indicate your position to rescuers.

(2) Swim or roll to stay on tope of the snow. FIGHT FOR YOUR LIFE. Work toward the edge of the avalanche. If you feel your feet touch the ground, give a hard push and try to "pop out" onto the surface.

(3) If your head goes under the snow, shut your mouth, hold your breath, and position your hands and arms to form an air pocket in front of your face. Many avalanche victims suffocate by having their mouths and noses plugged with snow.

(4) When you sense the slowing of the avalanche, you must try your hardest to reach the surface. Several victims have been found quickly because a hand or foot was sticking above the surface.

(5) When the snow comes to rest it sets up like cement and even if you are only partially buried, it may be impossible to dig yourself out. Don't shout unless you hear rescuers immediately above you; in snow, no one can hear you scream. Don't struggle to free yourself—you will only waste energy and oxygen.

(6) Try to relax. If you feel yourself about to pass out, do not fight it. The respiration of an unconscious person is more shallow, their pulse rate declines, and the body temperature is lowered, all of which reduce the amount of oxygen needed. (See Appendix C for information on search and rescue techniques.)

AVALANCHE HAZARD EVALUATION CHECKLIST

Critical Data PARAMETERS:	KEY INFORMATION	Hazard Rating G	Y	R
TERRAIN:	*Is the terrain capable of producing an avalanche?*			
	-Slope angle (steep enough to slide? prime time?)	☐	☐	☐
	-Slope aspect (leeward, shadowed, or extremely sunny?)	☐	☐	☐
	-Slope configuration (anchoring? shape?)	☐	☐	☐
	Overall Terrain Rating:	☐	☐	☐
SNOWPACK:	*Could the snow fall?*			
	-Slab Configuration (slab? depth and distribution?)	☐	☐	☐
	-Bonding Ability (weak layer? tender spots?)	☐	☐	☐
	-Sensitivity (how much force to fail? shear tests? clues?)	☐	☐	☐
	Overall Snowpack Rating:	☐	☐	☐
Weather:	*Is the weather contributing to instability?*			
	-Precipitation (type, amount, intensity? added weight?)	☐	☐	☐
	-Wind (snow transport? amount and rate of deposition?)	☐	☐	☐
	-Temperature (storm trends? effects on snowpack?)	☐	☐	☐
	Overall Weather Rating:	☐	☐	☐
Human:	*What are your alternatives and their possible consequences?*			
	-Attitude (toward life? risk? goals? assumptions?)	☐	☐	☐
	-Technical Skill Level (traveling? evaluating aval. hazard?)	☐	☐	☐
	-Strength/Equipment (strength? prepared for the worst?)	☐	☐	☐
	Overall Human Rating:	☐	☐	☐

Decision/Action:
Overall Hazard Rating/GO or NO Go? **GO** ☐ **or NOGO** ☐

***HAZARD LEVEL SYMBOLS:**
R = Red light (stop/dangerous)
G = Green light (go/OK)
Y = Yellow light (caution/potentially dangerous).

Table 1-2. Avalanche hazard evaluation checklist.

CHAPTER 2
MOUNTAIN LIVING

Units deploying to high elevations must receive advanced training to survive in the harsh mountain environment. Normal activities (navigation, communications, and movement) require specialized techniques. Training should be conducted as realistically as possible, preferably under severe conditions so the soldier gains confidence. Extended training exercises test support facilities and expose the soldier to the isolation common to mountain operations. Training should reflect the harsh mountain environment and should consider the following:

- *Temperature and altitude extremes.*
- *Hygiene and sanitation.*
- *Limited living space (difficulty of bivouac).*
- *Clothing requirements.*

Section I. SURVIVAL

The soldier trained to fight and survive in a mountain environment will have increased confidence in himself. Training should include: psychological preparation, locating water, shelter considerations, fire building, health hazards, and techniques for obtaining food (see FM 21-76).

2-1. WATER SUPPLY

Mountain water should never be assumed safe for consumption. Training in water discipline should be emphasized to ensure soldiers drink water only from approved sources. Fluids lost through respiration, perspiration, and urination must be replaced if the soldier is to operate efficiently.

a. Maintaining fluid balance is a major problem in mountain operations. The sense of thirst may be dulled by high elevations despite the greater threat of dehydration. Hyperventilation and the cool, dry atmosphere bring about a three- to four-fold increase in water loss by evaporation through the lungs. Hard work and overheating increase the perspiration rate. The soldier must make an effort to drink liquids even when he does not feel thirsty. One quart of water, or the equivalent, should be drunk every four hours; more should be drunk if the unit is conducting rigorous physical activity.

b. Three to six quarts of water each day should be consumed. About 75 percent of the human body is liquid. All chemical activities in the body occur in water solution, which assists in removing toxic wastes and in maintaining an even body temperature. A loss of two quarts of body fluid (2.5 percent of body weight) decreases physical efficiency by 25 percent, and a loss of 12 quarts (15 percent of body weight) is usually fatal. Salt lost by sweating should be replaced in meals to avoid a deficiency and subsequent cramping. Consuming the usual military rations (three meals a day) provides sufficient sodium replacement. Salt tablets are not necessary and may contribute to dehydration.

c. Even when water is plentiful, thirst should be satisfied in increments. Quickly drinking a large volume of water may actually slow the soldier. If he is hot and the water is cold, severe cramping may result. A basic rule is to drink small amounts often. Pure water should always be kept in reserve for first aid use. Emphasis must be placed on the three rules of water discipline:

- Drink only treated water.
- Conserve water for drinking. Potable water in the mountains may be in short supply.
- Do not contaminate or pollute water sources.

d. Snow, mountain streams, springs, rain, and lakes provide good sources of water supply. Purification must be accomplished, however, no matter how clear the snow or water appears. Fruits, juices, and powdered beverages may supplement and encourage water intake (do not add these until the water has been treated since the purification tablets may not work). Soldiers cannot adjust permanently to a decreased water intake. If the water supply is insufficient, physical activity must be reduced. Any temporary deficiency should be replaced to maintain maximum performance.

e. All water that is to be consumed must be potable. Drinking water must be taken only from approved sources or purified to avoid disease or the possible use of polluted water. Melting snow into water requires an increased amount of fuel and should be planned accordingly. Nonpotable water must not be mistaken for drinking water. Water that is unfit to drink, but otherwise not dangerous, may be used for other purposes such as bathing. Soldiers must be trained to avoid wasting water. External cooling (pouring water over the head and chest) is a waste of water and an inefficient means of cooling. Drinking water often is the best way to maintain a cool and functioning body.

f. Water is scarce above the timberline. After setting up a perimeter (patrol base, assembly area, defense), a watering party should be employed. After sundown, high mountain areas freeze, and snow and ice may be available for melting to provide water. In areas where water trickles off rocks, a shallow reservoir may be dug to collect water (after the sediment settles). Water should be treated with purification tablets (iodine tablets or calcium hypochlorite), or by boiling at least one to two minutes. Filtering with commercial water purification pumps can also be conducted. Solar stills may be erected if time and sunlight conditions permit (see FM 21-76). Water should be protected from freezing by storing it next to a soldier or by placing it in a sleeping bag at night. Water should be collected at midday when the sun thaw available.

2-2. NUTRITION

Success in mountain operations depends on proper nutrition. Because higher altitudes affect eating habits, precautions must be taken. If possible, at least one hot meal each day should be eaten, which may require personnel to heat their individual rations.

a. The following elements are characteristic of nutritional acclimatization in mountain operations:

- Weight loss during the first two to three days at high elevation.
- A loss of appetite with symptoms of mountain sickness.
- Loss of weight usually stops with acclimatization.
- At progressively higher elevations (greater than 14,000 feet), the tolerance of fatty/high-protein foods rapidly decreases. A high carbohydrate diet may lessen the symptoms of acute mountain sickness and is digested better than fat at high altitudes.

b. Increased fatigue may cause soldiers to become disinterested in eating properly. Decreased consumption may result in malnutrition because of the unpleasant taste of cold rations. Leaders should ensure that fuel tablets and squad stoves are available, or that natural flammable materials are used if possible. Although there is no physiological need for hot food, it does increase morale and a sense of well being. Loss of weight in the first few days occurs because of dehydration, metabolic changes, and loss of appetite. Carbohydrate-containing beverages, such as fruit juices and sports drinks, are an effective means of increasing carbohydrates, energy, and liquid intake when the normal appetite response is blunted at altitude.

c. Three major food components are required to maintain a well-functioning body: proteins, fats, and carbohydrates. These food components provide energy, amino acids, vitamins, fiber, and minerals. All three components must be provided in the correct proportions to maintain a healthy body.

(1) ***Protein***. Proteins consist of a large number of amino acid units that are linked together to form the protein. The amino acids, resulting from digestion of protein, are absorbed through the intestine into the blood, and are used to make or replace body proteins (muscle and body tissue). Sources of readily useable animal proteins include eggs, milk, cheese, poultry, fish, and meats. Other foods such as cereals, vegetables, and legumes also provide amino acids. These proteins are not as balanced in essential amino acid composition as meat, eggs, or milk proteins. The minimum daily protein requirement, regardless of physical activity, is 8 ounces for a 154-pound man. Since amino acids are either oxidized for energy or stored as fats, consuming excess protein is inefficient and may increase the water intake needed for urea nitrogen excretion. Protein requires water for digestion and may facilitate dehydration. Proteins provide the body about four kilocalories of energy per gram and require the most energy for the body to digest.

(2) ***Fats.*** Fats are the most concentrated form of food energy. Of the total daily caloric intake, 25 to 30 percent may be supplied as fats. Main sources of fats are meats, nuts, butter, eggs, milk, and cheese. Fats require more water and oxygen, and are harder to digest at higher altitudes. Fats are the body's natural stored source of energy. Fats provide the body around 9 kilocalories of energy per gram and require less energy for the body to digest than protein but more than carbohydrates.

(3) ***Carbohydrates.*** Carbohydrates are an important source of calories. In the form of glucose, carbohydrates are found in the most important energy-producing cycles in the body's cells. If carbohydrate intake exceeds energy needs, moderate amounts are stored in the muscles and liver. Larger amounts are converted into fats and stored in that form. Carbohydrates should compose up to 50 percent of the total daily caloric intake. Nutritionally, the most useful sources of carbohydrates are foods such as unrefined grains, vegetables, and fruit. Carbohydrates provide the body around four kilocalories of energy per gram and are the easiest to digest.

(4) ***Vitamins.*** Vitamins are classified into two groups on the basis of their ability to dissolve in fat or water. The fat-soluble vitamins include vitamins A, D, E, and K. The water-soluble vitamins include the B vitamins and vitamin C, which are found in cereals, vegetables, fruits, and meats. A well-balanced diet provides all of the required vitamins. Since most water-soluble vitamins are not stored, a proper diet is necessary to ensure adequate levels of these vitamins. If an improper and unbalanced diet is likely to occur

during a deployment, vitamin supplements should be considered, especially if this period is to exceed 10 days.

(5) ***Minerals.*** Mineral elements can be divided into two groups: those needed in the diet in amounts of 100 milligrams or more a day such as calcium, phosphorous, and magnesium; and trace elements needed in amounts of only a few milligrams a day such as iodine, iron, and zinc. Required minerals are contained in a balanced diet (meats, vegetables, fruits).

d. Eating a balanced diet provides the energy needed to conduct daily activities and to maintain the internal body processes. A balanced diet containing adequate amounts of vitamins and minerals ensures an efficient metabolism. Since climbing is a strenuous activity and demands high-energy use, a balanced diet is a necessity.

(1) The efficiency of the body to work above the basal metabolism varies from 20 to 40 percent, depending on the soldier. Over 50 percent of caloric intake is released as heat and is not available when the soldier works. (About 4,500 calories are expended for strenuous work and 3,500 calories for garrison activity.) Heat is a by-product of exertion. Exertion causes excessive bodily heat loss through perspiration and increased radiation. During inactivity in cold weather, the metabolism may not provide enough heat. The "internal thermostat" initiates and causes the muscles to shiver, thus releasing heat. Shivering also requires energy and burns up to 220 calories per hour (estimated for a 100-pound man).

(2) With an abrupt ascent to high altitudes, the soldier experiences physiological acclimatization. The circulatory system labors to provide the needed oxygen to the body. Large meals require the digestive system to work harder than usual to assimilate food. Large meals may be accompanied by indigestion, shortness of breath, cramps, and illness. Therefore, relatively light meals that are high in carbohydrates are best while acclimatizing at higher elevations. Personnel should eat moderately and rest before strenuous physical activity. Since fats and protein are harder to digest, less digestive disturbances may occur if meals are eaten before resting. A diet high in carbohydrates is not as dense in energy and may require eating more often. Carbohydrates, beginning in the morning and continuing through mid-afternoon, are important in maintaining energy levels.

(3) Extra food should be carried in case resupply operations fail. Food should be lightweight and easy to digest, and be eaten hot or cold. Meals-ready-to-eat (MREs) meet these criteria and provide all of the basic food groups. Commanders may consider supplementing MREs with breakfast bars, fruits, juices, candies, cereal bars, and chocolate. Bouillon cubes can replace water and salt as well as warming cold bodies and stimulating the appetite. Hot beverages of soup, juices, powdered milk, and cider should also be considered. Since coffee, tea, and hot chocolate are diuretics, the consumption of these beverages should not be relied upon for hydration.

(4) Warm meals should be provided when possible. When cooking, the heat source must be kept away from equipment and ammunition. At higher elevations, the cooking time may be doubled. To conserve fuel, stoves, fires, and fuel tablets should be protected from the wind. Extra fuel should be stored in tightly sealed, marked, metal containers. Use stoves and heat tabs for warming food and boiling water. Canteen cups and utensils should be cleaned after use. All food items and garbage are carried with the unit. If possible, garbage should be burned or deep buried. Caution must be taken to prevent animals from foraging through rucksacks, ahkios, and burial sites. As all missions are tactical, no trace of a unit should be detected.

(5) Certain drugs, medications, alcohol, and smoking have adverse effects on the circulation, perspiration, hydration, and judgment of soldiers. Therefore, they should be avoided when operating in extremely cold conditions or at high altitudes.

2-3. PERSONAL HYGIENE AND SANITATION

The principles of personal hygiene and sanitation that govern operations on low terrain also apply in the mountains. Commanders must conduct frequent inspections to ensure that personal habits of hygiene are not neglected. Standards must be maintained as a deterrent to disease, and as reinforcement to discipline and morale.

a. **Personal Hygiene.** This is especially important in the high mountains, mainly during periods of cold weather. In freezing weather, the soldier may neglect washing due to the cold temperatures and scarcity of water. This can result in skin infections and vermin infestation. If bathing is difficult for any extended period, the soldier should examine his skin and clean it often. Snow baths in lieu of a water bath are recommended. This helps reduce skin infections and aids the comfort of the soldier.

(1) Snow may be used instead of toilet paper. Soldiers should shave at rest periods in the shelter so that oils stripped in shaving will be replenished. A beard may mask the presence of frostbite or lice. Water-based creams and lotions should be avoided in cold environments since this will further dehydrate tissues and induce frostbite by freezing. The nonwater-based creams can be used for shaving in lieu of soap. Sunscreens and chap sticks should be used on lips, nose, and eyelids. Topical steroid ointments should be carried for rashes. The teeth must also be cleaned to avoid diseases of the teeth and gums. Underwear should be changed when possible, but this should not be considered a substitute for bathing. When operating in areas where resupply is not possible, each soldier should carry a complete change of clothing. If laundering of clothing is difficult, clothes should be shaken and air-dried. Sleeping bags must be regularly cleaned and aired.

(2) The principles of foot hygiene must be followed to protect the feet from cold injuries. The causes of such injuries are present throughout the year in high mountains. Boots should be laced tightly when climbing to provide needed support but not so tight as to constrict circulation. Socks should be worn with no wrinkles since this causes blisters on the feet. Feet should be washed daily, and kept as dry and clean as possible. If regular foot washing is impossible, socks should be changed often (at halts and rest periods or at least once a day) and feet massaged, dried, and sprinkled with foot powder. Talc or antifungal powder should be used when massaging; excess powder is brushed off to avoid clumping, which may cause blisters. Feet can be cleaned with snow, but must be quickly dried. Whenever changing socks, soldiers should closely examine their feet for wrinkles, cracks, blisters, and discoloration. Nails should be trimmed but not too short. Long nails wear out socks; short nails do not provide proper support for the ends of the toes. Medical attention should be sought for any possible problems.

(3) Feet should be sprayed two or three times a day with an aluminum chlorohydrate antiperspirant for a week and then once a day for the rest of the winter. If fissures or cracks occur in the feet, it is best to discontinue spraying until they are healed or to spray less often to control sweating. This process stops about 70 percent of the sweating in the feet.

(4) During periods of extreme cold, there is a tendency for the soldier to become constipated. This condition is brought about by the desire to avoid the inconvenience and

discomfort of defecating. Adequate water intake plus a low protein, high roughage diet can be helpful in preventing constipation.

b. **Sanitation.** In rocky or frozen ground, digging latrines is usually difficult. If latrines are constructed, they should be located downwind from the position and buried after use. In tactical situations, the soldier in a designated, downwind location away from water sources may dig "cat holes." Since waste freezes, it can be covered with snow and ice or pushed down a crevasse. In rocky areas above the timberline, waste may be covered with stones.

Section II. ACCLIMATIZATION AND CONDITIONING

Terrestrial altitude can be classified into five categories. Low altitude is sea level to 5,000 feet. Here, arterial blood is 96 percent saturated with oxygen in most people. Moderate altitude is from 5,000 to 8,000 feet. At these altitudes, arterial blood is greater than 92 percent saturated with oxygen, and effects of altitude are mild and temporary. High altitude extends from 8,000 to 14,000 feet, where arterial blood oxygen saturation ranges from 92 percent down to 80 percent. Altitude illness is common here. Very high altitude is the region from 14,000 to 18,000 feet, where altitude illness is the rule. Areas above 18,000 feet are considered extreme altitudes.

Soldiers deployed to high mountainous elevations require a period of acclimatization before undertaking extensive military operations. The expectation that freshly deployed, unacclimatized troops can go immediately into action is unrealistic, and could be disastrous if the opposing force is acclimatized. Even the physically fit soldier experiences physiological and psychological degradation when thrust into high elevations. Time must be allocated for acclimatization, conditioning, and training of soldiers. Training in mountains of low or medium elevation (5,000 to 8,000 feet) does not require special conditioning and acclimatization procedures. However, some soldiers will have some impairment of operating efficiency at these low altitudes. Above 8,000 feet (high elevation), most unacclimatized soldiers may display some altitude effects. Training should be conducted at progressively higher altitudes, starting at about 8,000 feet and ending at 14,000 feet. Attempts to acclimatize beyond 17,000 feet results in a degradation of the body greater than the benefits gained. The indigenous populations can out-perform even the most acclimatized and physically fit soldier who is brought to this altitude; therefore, employment of the local population may be advantageous.

2-4. SYMPTOMS AND ADJUSTMENTS

A person is said to be acclimatized to high elevations when he can effectively perform physically and mentally. The acclimatization process begins immediately upon arrival at the higher elevation. If the change in elevation is large and abrupt, some soldiers can suffer from acute mountain sickness (AMS), high-altitude pulmonary edema (HAPE), or high-altitude cerebral edema (HACE). Disappearance of the symptoms of acute mountain sickness (from four to seven days) does not indicate complete acclimatization. The process of adjustment continues for weeks or months. The altitude at which complete acclimatization is possible is not a set point but for most soldiers with proper ascent, nutrition and physical activity it is about 14,000 feet.

a. Immediately upon arrival at high elevations, only minimal physical work can be performed because of physiological changes. The incidence and severity of AMS symptoms vary with initial altitude, the rate of ascent, and the level of exertion and individual susceptibility. Ten to twenty percent of soldiers who ascend rapidly (in less than 24 hours) to altitudes up to 6,000 feet experience some mild symptoms. Rapid ascent to 10,000 feet causes mild symptoms in 75 percent of personnel. Rapid ascent to elevations of 12,000 to 14,000 feet will result in moderate symptoms in over 50 percent of the soldiers and 12 to 18 percent may have severe symptoms. Rapid ascent to 17,500 feet causes severe, incapacitating symptoms in almost all individuals. Vigorous activity during ascent or within the first 24 hours after ascent will increase both the incidence and severity of symptoms. Some of the behavioral effects that will be encountered in unacclimatized personnel include:

- Increased errors in performing simple mental tasks.
- Decreased ability for sustained concentration.
- Deterioration of memory.
- Decreased vigilance or lethargy.
- Increased irritability in some individuals.
- Impairment of night vision and some constriction in peripheral vision (up to 30 percent at 6,000 feet).
- Loss of appetite.
- Sleep disturbances.
- Irregular breathing.
- Slurred speech.
- Headache.

b. Judgment and self-evaluation are impaired the same as a person who is intoxicated. During the first few days at a high altitude, leaders have extreme difficulty in maintaining a coordinated, operational unit. The roughness of the terrain and the harshness and variability of the weather add to the problems of unacclimatized personnel. Although strong motivation may succeed in overcoming some of the physical handicaps imposed by the environment, the total impact still results in errors of judgment. When a soldier cannot walk a straight line and has a loss of balance, or he suffers from an incapacitating headache, he should be evacuated to a lower altitude (a descent of at least 1,000 feet for at least 24 hours).

2-5. PHYSICAL AND PSYCHOLOGICAL CONDITIONING

The commander must develop a conditioning/training program to bring his unit to a level where it can operate successfully in mountain conditions. Priorities of training must be established. As with all military operations, training is a major influence on the success of mountain operations.

a. U.S. forces do not routinely train in mountainous terrain. Therefore, extensive preparations are needed to ensure individual and unit effectiveness. Units must be physically and psychologically conditioned and adjusted before undertaking rigorous mountain operations. Units must be conditioned and trained as a team to cope with the terrain, environment, and enemy situation. Certain factors must be considered:

- What are the climatic and terrain conditions of the area of operations?
- How much time is available for conditioning and training?
- Will the unit conduct operations with other U.S. or Allied forces? Are there language barriers? What assistance will be required? Will training and conditioning be required for attached personnel?

- What additional personnel will accompany the unit? Will they be available for training and conditioning?
- What is the current level of physical fitness of the unit?
- What is the current level of individual expertise in mountaineering?
- What type of operations can be expected?
- What is the composition of the advance party? Will they be available to assist in training and acclimatization?
- What areas in the U.S. most closely resemble the area of operations?
- Are predeployment areas and ranges available?
- Does the unit have instructors qualified in mountain warfare?
- What type equipment will be required (to fit the season, mission, terrain)?
- Does the unit have enough of the required equipment? Do personnel know how to use the equipment? Will the equipment go with the advance party, with the unit, or follow after the unit's arrival?
- Does equipment require modification?
- Do weapons and equipment require special maintenance?

b. When the unit arrives in the area of operations, all personnel require a period of conditioning and acclimatization. The time schedule should allow for longer and more frequent periods of rest. The rigors of establishing an assembly area exhaust most unacclimatized personnel. Water, food, and rest must be considered as priorities, ensuring sufficient amounts while individual metabolisms and bodies become accustomed to functioning at higher elevations.

c. Since the acclimatization process cannot be shortened, and the absence of acclimatization hampers the successful execution of operations, deployment to higher elevations must consider the following:

(1) Above 8,000 feet, a unit should ascend at a rate of 1,000 to 2,000 feet per day. Units can leapfrog, taking an extended rest period.

(2) Units should not resort to the use of pharmaceutical pretreatment with carbonic anhydrase inhibitors such as acetazolamide (Diamox). These drugs have side effects that mimic the signs and symptoms of AMS. Inexperienced medics may have difficulty recognizing the differences between the side effects of the drug and a condition that could possibly be life threatening. Additionally, these drugs are diuretics, which results in higher hydration levels (at least 25 percent increase per man per day). These higher hydration levels create a larger logistical demand on the unit by requiring more water, time to acquire water, water purification supplies, and, if in a winter environment, fuels for melting snow and ice for water.

(3) Carbonic anhydrase inhibitors such as acetazolamide are effective in the treatment of mild and severe AMS. These drugs should accompany attached medical personnel because they can treat the soldier suffering the symptoms of AMS and, although rest may be required evacuation may not be needed.

(4) Do not move troops directly to high altitudes even if allowances can be made for inactivity for the first three to five days before mission commitment. Moving troops directly to high altitude can increase the probability of altitude sickness. Even if inactivity follows deployment, the incidence of altitude sickness is more likely than with a gradual ascent.

d. Training on high-altitude effects can prevent psychological preconceptions. Soldiers who have lived on flat terrain may have difficulty when learning to negotiate steep slopes or cliffs, developing a sense of insecurity and fear. They must be slowly introduced to the new terrain and encouraged to develop the confidence required to negotiate obstacles with assurance and ease. They must be taught the many climbing techniques and principles of mountain movement. They overcome their fear of heights by becoming familiar with the problem. The soldier cannot be forced to disregard this fear.

e. Regardless of previous training and the amount of flat cross-country movement practice, the untrained soldier finds mountain movement hard and tiring. A different group of muscles are used, which must be developed and hardened. A new technique of rhythmic movement must be learned. Such conditioning is attained through frequent marches and climbs, while carrying TOE and special equipment loads. This conditions the back and legs, which results in increased ability and endurance. At the same time, the men acquire confidence and ability to safely negotiate the terrain. The better the physical condition of the soldier, the better the chance of avoiding exhaustion. Proper physical conditioning ensures the soldier is an asset and not a liability. The body improves its capacity for exercise, the metabolism becomes more efficient, and blood and oxygen flow quickly and effectively.

f. A physical fitness training program that gradually increases in difficulty should include marches, climbing, and calisthenics. This increases the soldier's endurance. Through a sustained high level of muscular exertion, the soldier's capacity for exertion is increased. Physical conditioning should include long-distance running for aerobic conditioning; calisthenics and weight training to strengthen the heart, lungs, abdomen, legs, back, arms, and hands; a swimming program to increase lung efficiency; and road marches over mountainous terrain with all combat equipment. Upon deploying to high elevations, caution must be exercised by units that are in superior physical condition. The heart rate, metabolism, and lungs must become accustomed to the elevation and thinner air. A conditioning program must be set up on site and integrated in gradual stages where acclimatization, conditioning, and mountaineering skills are realized.

g. Conditioning should begin with basic climbing. It is equally important to instill the will to climb. Confidence goes hand in hand with physical conditioning and skill development. Repetitive practice, to the point of instinctive reaction, is key to learning and maintaining climbing proficiency and technical skills. There are no quick and easy methods to becoming acclimatized and conditioned. Training should gradually challenge the soldier over an extended period and reinforce learning skills.

Section III. MEDICAL CONSIDERATIONS

Improper acclimatization poses many problems for medical personnel. Facilities and supplies may be inadequate to treat all victims. After acclimatization, personnel can still become injured (sprains, strains, fractures, frostbite, hypothermia, and trench foot). Mountain sickness and other illnesses may also occur. Evacuation of the sick and wounded is compounded by the terrain and weather.

2-6. ILLNESS AND INJURY

Units operating in mountainous regions are exposed to varied types of injuries and illnesses not associated with other areas. Medical considerations are like those for other environments; however, there are some unique aspects of mountain operations to be considered if effective support is to be provided. Most injuries in the mountain environment are soft tissue injuries. These include sprains, strains, abrasions, contusions and fractures. As with any other injuries, the most life threatening are treated first with the emphasis on airway control, breathing management, and circulatory support. Skills in basic first aid are essential to the mountain leader and should be reinforced with regular sustainment training.

2-7. TREATMENT AND EVACUATION

In harsh mountain weather, the most important course of action is to provide injured soldiers with medical aid as soon as possible. Immediate first aid is given on site. Due to rough terrain, medical units can seldom reach unit aid stations by vehicle to evacuate casualties. Litter bearers are required to move casualties to the rear where they can be evacuated by ground or air to clearing stations. The victim is protected from the weather and shock during transportation. Rendezvous points are coordinated with medical units as far forward as possible. Training must be accomplished with all litter bearers on evacuation techniques and first aid. Lightly wounded personnel may need assistance to move over rough terrain.

2-8. SOLAR INJURIES

Solar injuries can happen in warm weather or in cold weather. These types of injuries can be just as incapacitating as most other injuries but usually are not fatal. The peak hours of ultraviolet (UV) radiation are between the hours of 1100 and 1500. Due to the long wavelengths of ultraviolet light, cloudy days can be more dangerous than sunny days. On sunny days the soldier takes more care due to the bright conditions. On cloudy days the soldier tends not to wear sunglasses or sunscreen.

a. **Sunburn.** Sunburn is the burning of exposed skin surfaces by ultraviolet radiation.

(1) Contributing factors include fair skin, improper use of para-amino benzoic acid (PABA)-based sunscreens, and exposure to intense ultraviolet rays for extended periods.

(2) Symptoms of sunburn are painful, burning, red or blistered skin with a slight swelling. The skin may be warm to the touch. In severe cases chills, fever, and headaches may occur.

(3) To treat sunburn, apply cool saline dressings to alleviate pain and swelling. Do not pop blisters. If blisters do break, wash thoroughly, bandage, and seek medical attention. A solution of vinegar (acetic) and water can be lightly applied with sterile gauze to alleviate burning. The tannic acid in used tea bags can also be applied to alleviate burning. Administer pain medication if needed.

(4) To prevent sunburn, skin should be covered with clothing or PABA-based sunscreens (at least sun protection factor [SPF] 15) should be applied liberally to exposed skin during the peak hours of UV exposure. The SPF means that you can stay exposed to the suns UV rays that many times longer than without it. (For example, an SPF of 15 means that skin can be exposed to UV rays 15 times longer than without sunscreen.) During sustained activity, the sunscreen should be regularly reapplied to maintain the SPF.

b. **Snowblindness.** Snowblindness is sunburn of the cornea of the eye caused by exposure to ultraviolet radiation.

(1) A contributing factor is the reflection of sunlight from all directions off the snow, ice, and water. Ultraviolet rays can cause vision problems even on cloudy days. They are less filtered at high altitudes than at low altitudes.

(2) Symptoms of snowblindness are painful, red, watery eyes; a gritty feeling; blurred vision; and a headache.

(3) To treat snowblindness, patch both eyes with cold compresses for 24 hours. Topical anesthetics such as Tetracaine Ophthalmic can be used to relieve pain. Avoid rubbing the eyes. If still painful, keep the victim's eyes patched and administer oral pain medication. Snowblindness will usually resolve in about 24 hours for mild to moderate cases. Victims are rarely in need of evacuation unless the case is unusually severe.

(4) To prevent snow blindness, use quality sunglasses even on cloudy days in snow-covered terrain. Proper sunglasses should provide 100 percent UVA and UVB protection and have hoods on the sides to prevent reflected light from entering the eye. (Currently, the U.S. Army does not have these types of "glacier" sunglasses in their inventory and they must be acquired from nonmilitary sources.) In an emergency, improvise slit glasses from materials such as cardboard or birch bark.

2-9. COLD-WEATHER INJURIES

Cold-weather injuries can occur during any season of the year. Death has resulted in temperatures as high as 10 degrees Celsius (50 degrees Fahrenheit). A loss of body heat combined with shock produces devastating results. However, most of these accidents can be prevented by proper planning to include: timely requisition and receipt of supplies and proper clothing; thorough training of personnel with respect to the hazards of cold weather; effective methods for the receipt, dissemination, and use of cold-weather data; periodic inspections of clothing, personnel, and equipment; and personnel receiving a balance of water, rest, and nutrition.

a. Soldiers must be prepared to survive, move, and fight in winter conditions. Intense cold affects the mind as well as the body. Simple tasks take longer to perform, and they take more effort than in a temperate climate. When weather conditions become extreme the problems of survival become more significant. Warmth and comfort become the top priorities. The effects of extreme cold and the probability of injury are magnified due to the lack of proper diet and sleep. The most important measure in the prevention of cold-weather injuries is the education of personnel and their leaders.

b. Cold injuries may be divided into two types: freezing and nonfreezing. The freezing type is known as frostbite. The nonfreezing type includes hypothermia, dehydration, and immersion foot. Cold injuries result from impaired circulation and the action of ice formation and cold upon the tissues of the body. Temperature alone is not a reliable guide as to whether a cold injury can occur. Low temperatures are needed for cold injuries to occur, but freezing temperatures are not. Wind speed can accelerate body heat loss under both wet and cold conditions. All commanders and subordinate leaders/instructors must be familiar with and carry GTA 5-8-12, which includes a wind chill equivalent temperature chart (Figure 2-1, page 2-12).

WIND CHILL FACTOR CHART												
COOLING POWER OF WIND EXPRESSED AS AN EQUIVALENT CHILL TEMPERATURE (UNDER CALM CONDITIONS)												
ESTIMATED WIND SPEED (IN MPH)	ACTUAL THERMOMETER READING (F)											
	50	40	30	20	10	0	-10	-20	-30	-40	-50	-60
	EQUIVALENT TEMPERATURES (F)											
Calm	50	40	30	20	10	0	-10	-20	-30	-40	-50	-60
5	48	37	27	16	6	-5	-15	-26	-36	-47	-57	-68
10	40	28	16	4	-9	-24	-33	-46	-58	-70	-83	-95
15	36	22	9	-5	-18	-32	-45	-58	-72	-85	-99	-112
20	32	18	4	-10	-25	-39	-53	-67	-82	-96	-110	-124
25	30	16	0	-15	-29	-44	-59	-74	-88	-104	-118	-133
30	28	13	-2	-18	-33	-48	-63	-79	-94	-109	-125	-140
35	27	11	-4	-21	-35	-51	-67	-82	-98	-113	-129	-145
40	26	10	-6	-21	-37	-53	-69	-85	-100	-116	-132	-148
Winds greater than 40 MPH have little additional effect.	LITTLE DANGER (For properly clothed person) Maximum danger of false sense of security.				INCREASING DANGER Danger from freezing of exposed flesh.			GREAT DANGER				
	Trench foot and immersion foot may occur at any point on this chart.											

Figure 2-1. Wind chill chart.

c. Many other factors in various combinations determine if cold injuries will occur.

(1) ***Previous Cold Injuries.*** If a soldier has had a cold injury before, he is at higher risk for subsequent cold injuries.

(2) ***Race.*** Blacks are more susceptible to cold-weather injuries than Caucasians.

(3) ***Geographic Origin.*** Personnel from warmer climates are more susceptible to cold injury than those from colder climates.

(4) ***Ambient Temperature.*** The temperature of the air (or water) surrounding the body is critical to heat regulation. For example, the body uses more heat to maintain the temperature of the skin when the temperature of the surrounding air is 37 degrees Fahrenheit than when it is 50 degrees Fahrenheit.

(5) ***Wind Chill Factor.*** The commander should know the wind chill factor. When the forecast gives a figure that falls within the increased danger zone or beyond, caution must be taken to minimize cold injury. The equivalent wind chill temperature is especially important when the ambient temperature is 0 degrees Celsius (32 degrees Fahrenheit) or less. Tissue can freeze if exposed for a prolonged period and if frequent warming is not practiced. The lower the wind chill, the faster tissue freezing can occur. Wind chill is the rate of cooling. Wind does not lower the ambient temperature. The ambient temperature alone determines freezing or nonfreezing injuries. Frostbite Wind chill may cause faster cooling due to increased convection, but not below the ambient temperature.

(6) ***Type of Mission.*** Combat action requiring prolonged immobility and long hours of exposure to low temperatures, or not having an opportunity to warm up increases the possibility of cold injuries.

(7) ***Terrain.*** Minimal cover and wet conditions increase the potential for cold injury.

(8) ***Clothing.*** Clothing for cold weather should be worn with the acronym **C.O.L.D.** in mind.

- **C**—Clothing should be clean since prolonged wear reduces its air-trapping abilities and clogs air spaces with dirt and body oils.
- **O**—Overheating. Avoid overheating. Appropriate measures should be taken when a change in weather or activity alters the amount of clothing needed to prevent overheating and, therefore, accumulation of perspiration.
- **L**—Loose and in layers (to trap air and to conserve body heat). The uniform should be worn completely and correctly to avoid injury to exposed body surfaces. The cold-weather uniform is complete when worn with gloves and inserts.
- **D**—Dry. Keep dry. Wet clothing loses insulation value.

(9) ***Moisture.*** Water conducts heat more rapidly than air (25 percent). When the skin or clothing becomes damp or wet, the risk of cold injury is greatly increased.

(10) ***Dehydration.*** The most overlooked factor causing cold injuries is dehydration. Individuals must retain their body fluids. In cold weather the human body needs special care, and the consumption of water is important to retain proper hydration.

(11) ***Age.*** Within the usual age range of combat personnel, age is not a significant factor.

(12) ***Fatigue.*** Mental weariness may cause apathy leading to neglect of duties vital to survival.

(13) ***Concomitant Injury.*** Injuries resulting in shock or blood loss reduce blood flow to extremities and may cause the injured individual to be susceptible to cold injury, which in turn can accelerate shock.

(14) ***Discipline, Training, and Experience.*** Well-trained and disciplined soldiers suffer less than others from the cold.

(15) ***Nutrition.*** Good nutrition is essential for providing the body with fuel to produce heat in cold weather. The number of calories consumed normally increases as the temperature becomes colder.

(16) ***Excess Activity.*** Excess activity (overheating) results in loss of large amounts of body heat by perspiration. This loss of body heat combined with the loss of insulation value provided by the clothing (due to perspiration dampening the clothing) can subject a soldier to cold injuries.

(17) ***Radical Changes in the Weather.*** Weather conditions in mountainous terrain are known to change considerably throughout the day. Weather can quickly change to extremely cold and wet conditions, especially in higher elevations.

d. Commanders should ensure that the following measures are taken.

(1) Soldiers' uniforms are kept as dry as possible and are protected from the elements.

(2) Soldiers are educated on proper use of clothing systems to avoid the effects of overheating and perspiration (layer dressing and ventilate).

(3) The buddy system is used to watch for early signs of cold-weather injuries.

(4) All soldiers waterproof their equipment.

(5) The rate of movement should be slow, deliberate, and careful. Soldiers should not move out at a force march pace and then be stationary after they have perspired heavily. Soldiers should not wear excessive cold-weather clothing while moving.

e. Medical procedures are needed when sickness and injuries occur. Leaders should—

- Assess the situation (tactical and environmental).
- Approach the victim safely (avoid rock or snow slide).
- Perform emergency first aid.
- Treat for shock (always assume that shock is present).
- Check for other injuries/cold injuries.
- Develop a course of action (decide on a means of evacuation).
- Execute the plan and monitor the victim's condition.

f. Body heat may be lost through radiation, conduction, convection, or evaporation.

(1) ***Radiation.*** The direct heat loss from the body to its surrounding atmosphere is called radiation heat loss. The head can radiate up to 80 percent of the total body heat output. On cold days, personnel must keep all extremities covered to retain heat. This accounts for the largest amount of heat lost from the body.

(2) ***Conduction.*** Conduction is the direct transfer of heat from one object in contact with another (being rained on or sitting in snow).

(3) ***Convection.*** Convection is the loss of heat due to moving air or water in contact with the skin. Wind chill is convection cooling. Clothing that ventilates, insulates, and protects must control the layer of warm air next to the skin.

(4) ***Evaporation.*** The evaporation of perspiration causes heat loss. Wet clothing can cause heat loss by conduction and evaporation. Dressing in layers allows soldiers to remove or add clothing as needed.

g. Some of the most common cold-weather injuries are described in the following paragraphs.

(1) ***Shock.*** Shock is the depressed state of vital organs due to the cardiovascular (heart) system not providing enough blood. Although shock is not a cold-weather injury, it is a symptom or a result of other injuries. Any illness or injury can produce shock, which increases the instance and severity of a cold-weather injury. Shock should be assumed in all injuries and treated accordingly. Even minor injuries can produce shock due to cold, pain, fear, and loss of blood.

(a) *Symptoms.* Initial symptoms of shock include apprehension, shortness of breath, sweating, cold skin, rapid and faint pulse, and excessive thirst. If the victim is not given adequate first aid immediately, his condition may digress into incoherence, slower heart beat, unconsciousness, and possibly death.

(b) *Treatment.* To treat shock, restore breathing and heart rate through artificial respiration or cardiopulmonary resuscitation. Treat the injury and control hemorrhaging. Make the victim as comfortable as possible and try to relieve the pain. Keep the victim warm but do not overheat him. Elevate the back and head, or feet. If the victim is conscious and has no abdominal injuries, administer water. The victim should receive proper medical attention as soon as possible.

(2) ***Dehydration.*** Dehydration is the loss of body fluids to the point that normal body functions are prevented or slowed. This is usually caused by overexertion and improper water intake. Dehydration precedes all cold-weather injuries and is a major symptom in acute mountain sickness. It contributes to poor performance in all physical activities—even more so than lack of food. Cold weather requirements for water are no different than in the desert. They may, in fact, exceed desert requirements because of the increased difficulty in moving with extra clothing and through the snow. At high altitudes, the air is dry. Combined

with a rapid rate of breathing, as much as two liters of liquid may be lost each day through respiration. A soldier needs about three to six quarts of water each day to prevent dehydration when living and performing physical labor in a cold or mountainous environment. Coffee and tea are diuretics and cause excessive urination and should be avoided. The adequacy of liquid intake can best be judged by the urine color and volume. Dark amber colored urine instead of light yellow or the absence of a need to urinate upon awakening from a night's sleep are indicators of dehydration. Thirst is not a good indicator of hydration.

(a) *Contributing Factors.* Factors that contribute to dehydration in cold weather are:

- The thirst mechanism does not function properly in cold weather.
- Water is often inconvenient to obtain and purify.
- The air in cold climates and at high altitudes lacks moisture.
- Cold causes frequent urination.

(b) *Symptoms.* Symptoms of dehydration include darkening urine, decreased amounts of urine being produced, dry mouth, tiredness, mental sluggishness, lack of appetite, headache, fainting, rapid heartbeat, dizziness, higher temperature, upset stomach, and unconsciousness. The symptoms of dehydration are similar to those of hypothermia. To distinguish between them, open the victim's clothes and feel the stomach. If the stomach is cold, the victim is probably hypothermic; if it is warm, he is probably dehydrated. However, this test is not conclusive since cold-weather dehydrating can also lead to total body cooling. The cold environment may act as a diuretic and impair the body's ability to conserve fluid (cold-induced diuresis and increased rate of urination).

(c) *Treatment.* Prevent dehydration by consuming three to six quarts of fluids each day (forced drinking in the absence of thirst is mandatory) and avoid caffeine and alcohol, which may chemically contribute to dehydration. Keep the victim warm and treat for shock. In advanced cases, administer fluids by mouth if the victim is conscious. Do not let him eat snow; eating snow uses body heat. Allow the victim to rest. If he fails to improve within one hour or is unconscious, evacuate him to a medical facility immediately.

(3) ***Hypothermia.*** Hypothermia is the lowering of the body core temperature at a rate faster than the body can produce heat. Hypothermia may be caused by exposure or by sudden wetting of the body such as falling into a lake or being sprayed with fuel or other liquid. Hypothermia can occur even on moderate days with temperatures of 40 to 50 degrees Fahrenheit with little precipitation if heat loss exceeds heat gain and the condition of the soldier is allowed to deteriorate. Hypothermia is classified as mild (core temperature above 90 degrees Fahrenheit or 32 degrees Celsius) or severe (core temperature below 90 degrees Fahrenheit or 32 degrees Celsius). An individual is considered to be "clinically hypothermic" when the core temperature is less than or equal to 95 degrees Fahrenheit.

(a) *Contributing Factors.* Factors that contribute to hypothermia are:

- Dehydration.
- Poor nutrition.
- Diarrhea.
- Decreased physical activity.
- Accidental immersion in water.
- Change in weather.
- High winds.
- Inadequate types or amounts of clothing.

(b) *Symptoms* The first symptom of hypothermia is when the body core (rectal) temperature falls to about 96 degrees Fahrenheit. Other symptoms include:

- Shivering, which may progress to an uncontrollable point making it hard for an individual to care for himself. Shivering begins after a drop in body temperature of one to two degrees. This is followed by clumsiness (stumbling or falling), slow reactions, mental confusion, and difficulty in speaking.
- Body temperature drop from 95 degrees Fahrenheit to 90 degrees Fahrenheit, which can cause sluggish thinking, irrational thought, apathy, and a false sense of warmth. The victim becomes cold and pale; cannot perform simple tasks; experiences amnesia and hallucinations; develops blueness of skin and decreased heart and respiratory rate with a weak pulse; pupils of the eyes dilate; speech becomes slurred; and visual disturbance occurs.
- Body temperature drop from 90 degrees Fahrenheit to 85 degrees Fahrenheit, which causes irrationality, incoherence, loss of contact with the environment, muscular rigidity, disorientation, and exhaustion. The soldier might stop shivering after his core temperature drops below 90 degrees Fahrenheit.
- Body temperature drop from 85 degrees Fahrenheit and below, which causes muscle rigidity, unconsciousness, comatose state, and faint vital signs. The pulse may be faint or impalpable, and breathing is too shallow to observe.

(c) *Prevention*. Prevent hypothermia by using the buddy system to watch each other for symptoms; consume adequate amounts of liquids daily; rest; and eat properly.

(d) *Avoidance*. Hypothermia can be avoided by dressing in layers, which permits easy additions or deletions to prevent overheating, becoming too cold, or getting wet or windblown. If the soldier is in a situation that precludes staying warm and dry, he should seek shelter. Sweets and physical activity help to produce body heat.

(e) *Treatment*. Treatment methods vary based on the severity of the hypothermia.

- Mild cases: If a soldier shows symptoms of hypothermia, prevent additional heat loss by getting the victim into a shelter; removing wet clothing and replacing it with dry, insulated clothing; insulating the victim from the ground; and sharing a sleeping bag (cover head) to transfer body heat. Make a diagnosis (rectal temperature). Rehydrate the victim with warm liquids, sweets, and food. If the tactical situation allows, build a fire. Above all else, keep the victim conscious until his vital signs are normal, and seek medical assistance. If possible, keep the victim physically active to produce body heat.
- Severe cases: If the victim is unconscious or appears dead without any obvious injury, prevent further heat loss. Rapid rewarming of an unconscious victim may create problems and should not be attempted. It is best to evacuate as soon as possible. At all times, the victim should be handled gently so as not to cause the cold blood from the extremities to rush to the heart. Do not allow the victim to perform ANY physical activity. Immediately transport the victim to the nearest medical facility. Field reheating is not effective and may be hazardous. Provide artificial respiration if breathing stops. If no pulse is detectable, be aware that in hypothermia there is often effective circulation for the victim's hypothermic state. In such a case, cardiac compression (such as CPR) may be fatal. The exception is acute hypothermia with near drowning.

- Breathing warm, moist air is the fastest way to warm the inside of the body. If breathing steam is not possible, place tubing under the rescuer's shirt so the victim will still breathe warm, moist air. This process can be done while on the move. In addition to breathing moist, warm air the victim must be gradually rewarmed using external heat sources. Padded hot water bottles or heated stones should be placed in the armpits.
- If conscious, the victim can be given warm, sweet drinks.
- The Hibler Pack is an improvised method of rewarming hypothermic victims in the field. This is used to heat the body core first so the vital organs are warmed and not the extremities. As the body warms up the warm blood will eventually warm all parts of the body. First lay out a blanket or sleeping bag and place a poncho or space blanket inside of it. The poncho or space blanket should go from the base of the skull to the base of the butt. This keeps the sleeping bag/blanket dry and acts like a vapor barrier. Lay the hypothermic patient inside the sleeping bag/blanket. Using a stove, warm water until it is hot to the touch (but not hot enough to burn the patient) and completely dampen any absorbable materials (such as T-shirt, towel, BDU top, and so on). Place the warm, wet items inside a plastic bag or directly in the armpits and chest of the patient. After the warm, wet item has been placed on the patient, wrap the patient tightly inside the poncho/space blanket and the blanket/sleeping bag. Continually check the temperature of the wet material and keep it warm.
- All bodily systems in hypothermia are brittle so treat the victim gently. As these attempts are being made, try to evacuate the victim. Severe complications may arise as the body temperature rises, which may result in cardiac arrest even though the victim seems to be doing well.

(4) ***Immersion or Trench Foot.*** This is damage to the circulatory and nervous systems of the feet that occurs from prolonged exposure to cold and wet at above freezing temperatures. This can happen wearing boots or not. A soldier may not feel uncomfortable until the injury has already begun.

(a) *Contributing Factors.* Factors that contribute to immersion or trench foot are:

- Stepping into water over the boot tops.
- Not changing socks often enough.
- Improper hygiene.
- Prolonged exposure (three to five days).

(b) *Symptoms.* Symptoms of immersion or trench foot include the sensation of tingling, numbness, and then pain. The toes are pale, and feel cold and stiff. The skin is wet and soggy with the color turning from red to bright red, progressing to pale and mottled, and then grayish blue. As symptoms progress and damage appears, the skin becomes red and then bluish or black. Swelling may occur. Because the early stages of trench foot are not painful, soldiers must be constantly aware to prevent it.

(c) *Treatment.* To prevent this condition, keep the feet dry and clean. Change socks often, drying the insides of boots, massaging the feet, and using foot powder. Drying the feet for 24 hours usually heals mild cases. Moderate cases usually heal within three to five days. The feet should be handled gently—NOT rubbed or massaged. They should be cleaned with soap and water, dried, elevated, and exposed to room temperature. The victim must stay off his feet and seek medical attention. Severe cases, when feet are not allowed to dry, are evacuated as a litter casualty.

(5) ***Blisters.*** When first noticed and before the formation of a blister, cover a hotspot with moleskin (over the area and beyond it). Use tincture benzoin to help the moleskin adhere to and toughen the skin. Once a blister has formed, cover it with a dressing large enough to fit over the blister, and then tape it. Never drain blisters unless they are surrounded by redness, or draining pus indicates infection. If this occurs, drain the blister from the side with a clean sterile needle. After cleaning with soap and water, gently press out the fluid leaving the skin intact. Make a doughnut of moleskin to go around the blister and apply to the skin. For toe blisters, wrap the entire toe with adhesive tape over the moleskin. (Toenails should be trimmed straight across the top, leaving a 90-degree angle on the sides. This provides an arch so that the corners do not irritate the skin.)

(6) ***Frostbite.*** Frostbite is the freezing or crystallization of living tissues due to heat being lost faster than it can be replaced by blood circulation, or from direct exposure to extreme cold or high winds. Exposure time can be minutes or instantaneous. The extremities are usually the first to be affected. Damp hands and feet may freeze quickly since moisture conducts heat away from the body and destroys the insulating value of clothing. Heat loss is compounded with intense cold and inactivity. With proper clothing and equipment, properly maintained and used, frostbite can be prevented. The extent of frostbite depends on temperature and duration of exposure. Frostbite is one of the major nonfatal cold-weather injuries encountered in military operations, but does not occur above an ambient temperature of 32 degrees Fahrenheit.

(a) *Categories of Frostbite.* Superficial (mild) frostbite involves only the skin (Figure 2-2). The layer immediately below usually appears white to grayish with the surface feeling hard, but the underlying tissue is soft. Deep (severe) frostbite extends beyond the first layer of skin and may include the bone (Figure 2-3). Discoloration continues from gray to black, and the texture becomes hard as the tissue freezes deeper. This condition requires immediate evacuation to a medical facility.

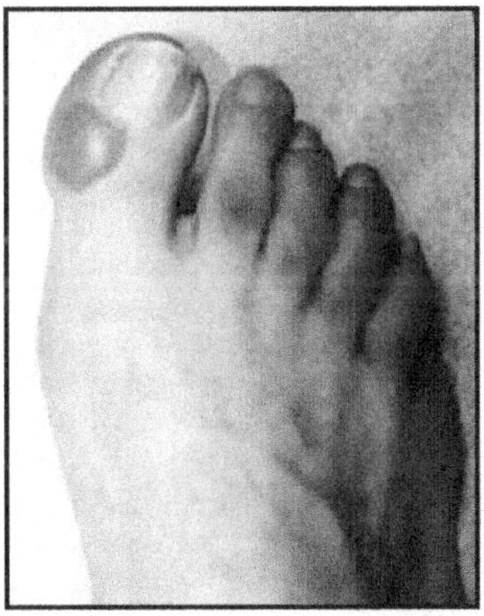

Figure 2-2. Superficial frostbite.

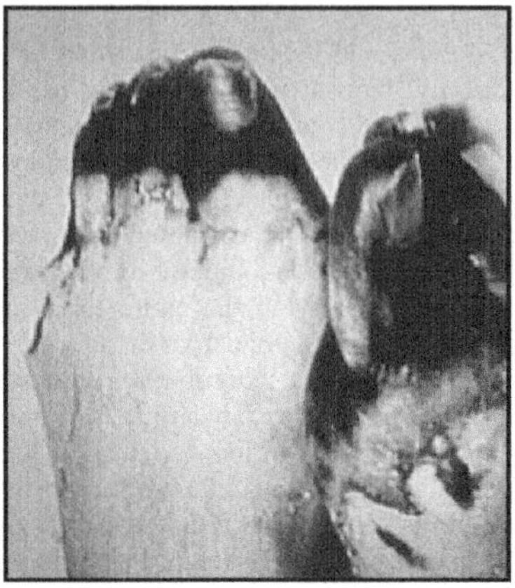

Figure 2-3. Deep frostbite.

(b) *Contributing Factors.* Factors that contribute to frostbite are:

- Dehydration.
- Below-freezing temperatures.
- Skin contact with super cooled metals or liquids.
- Use of caffeine, tobacco, or alcohol.
- Neglect.

(c) *Symptoms.* Symptoms of frostbite vary and may include a cold feeling, pain, burning, numbness, and, in the final stages, a false sense of warmth. The skin first turns red, then pale. It may be bluish in color and then may appear frosty or waxy white. The skin may feel hard, may not be movable over the joints and bony prominences, or may be frozen. Identification of deep versus superficial frostbite is difficult to determine and often requires three to seven days after rewarming for medical personnel to diagnose. Blisters, swelling, and pain may occur after thawing.

(d) *Treatment.* Using the buddy system is one of the primary ways to prevent frostbite. Buddies must watch each other for symptoms of frostbite and provide mutual aid if frostbite occurs. Frostbite should be identified early with prompt first-aid care applied to prevent further damage.

- Treat early signs of frostbite by rewarming with skin-to-skin contact or by sheltering the body part under the clothing next to the body. *Do this immediately.* If tissues have frozen, evacuate the victim before they thaw. If the feet are involved, evacuate the victim as a litter patient.
- Thawing of a frostbitten victim is a hospital procedure. If the victim has frostbite with frozen extremities, protect the frozen parts and evacuate as a litter patient.
- If frostbite is not recognized before it thaws, do not let the area refreeze since this causes more damage. The most often-affected body parts are the hands, fingers, toes, feet, ears, chin, and nose. If evacuation of the victim as a litter case

is not possible and the body part has not yet thawed, have the victim walk out on his own. Walking out on frozen feet is better than having them thaw and refreeze. Self-evacuation may be tactically necessary. Walking on frozen feet does less harm than walking on thawed feet.

- If reheating is inevitable, do not overheat the affected body parts near flame; the warming temperature should not be greater than normal body temperature. Do not rub the parts—the crystallized tissues may break internally and cause more damage. Do not pop blisters; cover them with a dry, sterile dressing. Keep the victim warm (apply loose, bulky bandages to separate toes and fingers.)
- Once a part is rewarmed it will become painful. Pain may be managed with narcotic analgesics.
- Once the foot is rewarmed it will swell and putting the boot back on will not be possible.

(7) ***Constipation.*** Constipation is the infrequent or difficult passage of stools.

(a) *Contributing Factors.* Factors that contribute to constipation are a lack of fluids, improper nutrition, and not defecating when needed.

(b) *Symptoms.* Symptoms include headache, cramping, lack of bowel movement, painful bowel movement, and loss of appetite.

(c) *Treatment.* Constipation is prevented by consuming adequate amounts and varieties of food, drinking from four to six liters of liquid each day, and defecating regularly. If allowed to progress beyond self-care stages, victims will need medical aid.

(8) ***Carbon Monoxide Poisoning.*** This is the replacement of oxygen in the blood with carbon monoxide.

(a) *Contributing Factor.* A contributing factor is inhaling fumes from burning fuel, such as fires, stoves, heaters, and running engines, without proper ventilation.

(b) *Symptoms.* Symptoms are similar to other common illnesses and include headaches, fatigue, excessive yawning, nausea, dizziness, drowsiness, confusion, and unconsciousness. Death may occur. The one visible symptom is bright red lips, mouth, and inside of the eyelids.

(c) *Treatment.* Remove the victim from the source of contamination; administer oxygen, if available; and evacuate to a medical facility. Severe complications may develop even in casualties who appear to have recovered. If the victim is unconscious, administer rescue breathing and CPR as needed.

2-10. HEAT INJURIES

Heat injuries, although associated with hot weather, can occur in cold-weather environments. Most heat injuries can be avoided by planning, periodic inspections of personnel clothing (ventilation) and equipment, a balance of water and food intake, and rest.

a. **Heat Cramps.** Heat cramps are caused by an accumulation of lactic acid in the muscles and a loss of salt through perspiration.

(1) ***Contributing Factor.*** Strenuous exertion causes the body to heat up and to produce heavy perspiration.

(2) ***Symptoms.*** Symptoms of heat cramps include pain and cramping in the arms, legs, back, and stomach. The victim sweats profusely and cannot quench his thirst.

(3) ***Treatment.*** Have the victim rest in a cool, shady area, breath deeply, and stretch the cramped muscle as soon as possible to obtain relief. Loosen the victim's clothing and have him drink cool water. Monitor his condition and seek medical attention if pain and cramps continue.

b. **Heat Exhaustion.** Heat exhaustion may occur when a soldier exerts himself in any environment and he overheats. The blood vessels in the skin become so dilated that the blood flow to the brain and other organs is reduced.

(1) ***Contributing Factors.*** Factors that contribute to heat exhaustion are strenuous activity in hot areas, unacclimatized troops, inappropriate diet, and not enough water or rest.

(2) ***Symptoms.*** Symptoms of heat exhaustion may be similar to fainting but may also include weakness; dizziness; confusion; headache; cold, clammy skin; and nausea. The victim may also have a rapid but weak pulse.

(3) ***Treatment.*** Move the victim to a cool, shady area and loosen his clothes and boots. Have the victim drink water and, if possible, immerse him in water to aid in cooling. Elevate the victim's legs to help restore proper circulation. Monitor his condition and seek medical attention if the symptoms persist.

c. **Heat Stroke.** Heat stroke is a life-threatening situation caused by overexposure to the sun. The body is so depleted of liquids that its internal cooling mechanisms fail to function.

(1) ***Contributing Factors.*** Factors that contribute to heat stroke are prolonged exposure to direct sunlight, overexertion, dehydration, and depletion of electrolytes.

(2) ***Symptoms.*** Symptoms of heat stroke include hot, dry skin; dizziness; confusion and incoherency; headache; nausea; seizures; breathing difficulty; a slow pulse; and loss of consciousness.

(3) ***Treatment.*** Cool the victim at once, and restore breathing and circulation. If the victim is conscious, administer water. If possible, submerge the victim in water to reduce his temperature, treat for shock, and prepare for immediate evacuation.

2-11. ACUTE MOUNTAIN SICKNESS

Acute mountain sickness is a temporary illness that may affect both the beginner and experienced climber. Soldiers are subject to this sickness in altitudes as low as 5,000 feet. Incidence and severity increases with altitude, and when quickly transported to high altitudes. Disability and ineffectiveness can occur in 50 to 80 percent of the troops who are rapidly brought to altitudes above 10,000 feet. At lower altitudes, or where ascent to altitudes is gradual, most personnel can complete assignments with moderate effectiveness and little discomfort.

a. Personnel arriving at moderate elevations (5,000 to 8,000 feet) usually feel well for the first few hours; a feeling of exhilaration or well-being is not unusual. There may be an initial awareness of breathlessness upon exertion and a need for frequent pauses to rest. Irregular breathing can occur, mainly during sleep; these changes may cause apprehension. Severe symptoms may begin 4 to 12 hours after arrival at higher altitudes with symptoms of nausea, sluggishness, fatigue, headache, dizziness, insomnia, depression, uncaring attitude, rapid and labored breathing, weakness, and loss of appetite.

b. A headache is the most noticeable symptom and may be severe. Even when a headache is not present, some loss of appetite and a decrease in tolerance for food occurs. Nausea, even without food intake, occurs and leads to less food intake. Vomiting may occur and contribute to dehydration. Despite fatigue, personnel are unable to sleep. The symptoms usually develop and increase to a peak by the second day. They gradually subside over the next several days so that the total course of AMS may extend from five to seven days. In some instances, the headache may become incapacitating and the soldier should be evacuated to a lower elevation.

c. Treatment for AMS includes the following:

- Oral pain medications such as ibuprofen or aspirin.
- Rest.
- Frequent consumption of liquids and light foods in small amounts.
- Movement to lower altitudes (at least 1,000 feet) to alleviate symptoms, which provides for a more gradual acclimatization.
- Realization of physical limitations and slow progression.
- Practice of deep-breathing exercises.
- Use of acetazolamide in the first 24 hours for mild to moderate cases.

d. AMS is nonfatal, although if left untreated or further ascent is attempted, development of high-altitude pulmonary edema (HAPE) and or high-altitude cerebral edema (HACE) can be seen. A severe persistence of symptoms may identify soldiers who acclimatize poorly and, thus, are more prone to other types of mountain sickness.

2-12. CHRONIC MOUNTAIN SICKNESS

Although not commonly seen in mountaineers, chronic mountain sickness (CMS) (or Monge's disease) can been seen in people who live at sufficiently high altitudes (usually at or above 10,000 feet) over a period of several years. CMS is a right-sided heart failure characterized by chronic pulmonary edema that is caused by years of strain on the right ventricle.

2-13. UNDERSTANDING HIGH-ALTITUDE ILLNESSES

As altitude increases, the overall atmospheric pressure decreases. Decreased pressure is the underlying source of altitude illnesses. Whether at sea level or 20,000 feet the surrounding atmosphere has the same percentage of oxygen. As pressure decreases the body has a much more difficult time passing oxygen from the lungs to the red blood cells and thus to the tissues of the body. This lower pressure means lower oxygen levels in the blood and increased carbon dioxide levels. Increased carbon dioxide levels in the blood cause a systemic vasodilatation, or expansion of blood vessels. This increased vascular size stretches the vessel walls causing leakage of the fluid portions of the blood into the interstitial spaces, which leads to cerebral edema or HACE. Unless treated, HACE will continue to progress due to the decreased atmospheric pressure of oxygen. Further ascent will hasten the progression of HACE and could possibly cause death.

While the body has an overall systemic vasodilatation, the lungs initially experience pulmonary vasoconstriction. This constricting of the vessels in the lungs causes increased workload on the right ventricle, the chamber of the heart that receives de-oxygenated blood from the right atrium and pushes it to the lungs to be re-oxygenated. As the right ventricle works harder to force blood to the lungs, its overall output is decreased thus decreasing the overall pulmonary perfusion. Decreased pulmonary perfusion causes decreased cellular respiration—the transfer of oxygen from the alveoli to the red blood cells. The body is now experiencing increased carbon dioxide levels due to the decreased oxygen levels, which now causes pulmonary vasodilatation. Just as in HACE, this expanding of the vascular structure causes leakage into interstitial space resulting in pulmonary edema or HAPE. As the edema or fluid in the lungs increases, the capability to pass oxygen to the red blood cells decreases thus creating a vicious cycle, which can quickly become fatal if left untreated.

2-14. HIGH-ALTITUDE PULMONARY EDEMA

HAPE is a swelling and filling of the lungs with fluid, caused by rapid ascent. It occurs at high altitudes and limits the oxygen supply to the body.

a. HAPE occurs under conditions of low oxygen pressure, is encountered at high elevations (over 8,000 feet), and can occur in healthy soldiers. HAPE may be considered a form of, or manifestation of, AMS since it occurs during the period of susceptibility to this disorder.

b. HAPE can cause death. Incidence and severity increase with altitude. Except for acclimatization to altitude, no known factors indicate resistance or immunity. Few cases have been reported after 10 days at high altitudes. When remaining at the same altitude, the incidence of HAPE is less frequent than that of AMS. No common indicator dictates how a soldier will react from one exposure to another. Contributing factors are:

- A history of HAPE.
- A rapid or abrupt transition to high altitudes.
- Strenuous physical exertion.
- Exposure to cold.
- Anxiety.

c. Symptoms of AMS can mask early pulmonary difficulties. Symptoms of HAPE include:

- Progressive dry coughing with frothy white or pink sputum (this is usually a later sign) and then coughing up of blood.
- Cyanosis—a blue color to the face, hands, and feet.
- An increased ill feeling, labored breathing, dizziness, fainting, repeated clearing of the throat, and development of a cough.
- Respiratory difficulty, which may be sudden, accompanied by choking and rapid deterioration.
- Progressive shortness of breath, rapid heartbeat (pulse 120 to 160), and coughing (out of contrast to others who arrived at the same time to that altitude).
- Crackling, cellophane-like noises (rales) in the lungs caused by fluid buildup (a stethoscope is usually needed to hear them).
- Unconsciousness, if left untreated. Bubbles form in the nose and mouth, and death results.

d. HAPE is prevented by good nutrition, hydration, and gradual ascent to altitude (no more than 1,000 to 2,000 feet per day to an area of sleep). A rest day, with no gain in altitude or heavy physical exertion, is planned for every 3,000 feet of altitude gained. If a soldier develops symptoms despite precautions, immediate descent is mandatory where he receives prompt treatment, rest, warmth, and oxygen. He is quickly evacuated to lower altitudes as a litter patient. A descent of 300 meters may help; manual descent is not delayed to await air evacuation. If untreated, HAPE may become irreversible and cause death. Cases that are recognized early and treated promptly may expect to recover with no aftereffects. Soldiers who have had previous attacks of HAPE are prone to second attacks.

e. Treatment of HAPE includes:

- Immediate descent (2,000 to 3,000 feet minimum) if possible; if not, then treatment in a monoplace hyperbaric chamber.
- Rest (litter evacuation)
- Supplemental oxygen if available.
- Morphine for the systemic vasodilatation and reduction of preload. This should be carefully considered due to the respiratory depressive properties of the drug.
- Furosemide (Lasix), which is a diuretic, given orally can also be effective.
- The use of mannitol should not be considered due to the fact that it crystallizes at low temperatures. Since almost all high-altitude environments are cold, using mannitol could be fatal.
- Nifidipine (Procardia), which inhibits calcium ion flux across cardiac and smooth muscle cells, decreasing contractility and oxygen demand. It may also dilate coronary arteries and arterioles.
- Diphenhydramine (Benadryl), which can help alleviate the histamine response that increases mucosal secretions.

2-15. HIGH-ALTITUDE CEREBRAL EDEMA

HACE is the accumulation of fluid in the brain, which results in swelling and a depression of brain function that may result in death. It is caused by a rapid ascent to altitude without progressive acclimatization. Prevention of HACE is the same as for HAPE. HAPE and HACE may occur in experienced, well-acclimated mountaineers without warning or obvious predisposing conditions. They can be fatal; when the first symptoms occur, immediate descent is mandatory.

a. Contributing factors include rapid ascent to heights over 8,000 feet and aggravation by overexertion.

b. Symptoms of HACE include mild personality changes, paralysis, stupor, convulsions, coma, inability to concentrate, headaches, vomiting, decrease in urination, and lack of coordination. The main symptom of HACE is a severe headache. A headache combined with any other physical or psychological disturbances should be assumed to be manifestations of HACE. Headaches may be accompanied by a loss of coordination, confusion, hallucinations, and unconsciousness. These may be combined with symptoms of HAPE. The victim is often mistakenly left alone since others may think he is only irritable or temperamental; no one should ever be ignored. The symptoms may rapidly progress to death. Prompt descent to a lower altitude is vital.

c. Preventive measures include good eating habits, maintaining hydration, and using a gradual ascent to altitude. Rest, warmth, and oxygen at lower elevations enhance recovery. Left untreated, HACE can cause death.

d. Treatment for HACE includes:

- Dexamethasone injection immediately followed by oral dexamethasone.
- Supplemental oxygen.
- Rapid descent and medical attention.
- Use of a hyberbaric chamber if descent is delayed.

2-16. HYDRATION IN HAPE AND HACE

HAPE and HACE cause increased proteins in the plasma, or the fluid portion of the blood, which in turn increases blood viscosity. Increased viscosity increases vascular pressure. Vascular leakage caused by stretching of the vessel walls is made worse because of this increased vascular pressure. From this, edema, both cerebral and pulmonary, occurs. Hydration simply decreases viscosity.

CHAPTER 3
MOUNTAINEERING EQUIPMENT

Commanders at every level must understand the complexity of operations in a mountainous environment where every aspect of combat operations becomes more difficult. Leaders must understand that each individual has a different metabolism and, therefore, cools down and heats up differently, which requires soldiers to dress-up and dress-down at different intervals. Provided all tactical concerns are met, the concept of uniformity is outdated and only reduces the unit's ability to fight and function at an optimum level. The extreme cold weather clothing system (ECWCS) is specifically designed to allow for rapid moisture transfer and optimum heat retention while protecting the individual from the elements. Every leader is responsible for ensuring that the ECWCS is worn in accordance with the manufacturers' recommendations. Commanders at all levels must also understand that skills learned at an Army mountaineering school are perishable and soldiers need constant practice to remain proficient. The properly trained mountain soldier of today can live better, move faster, and fight harder in an environment that is every bit as hostile as the enemy.

Section I. EQUIPMENT DESCRIPTION AND MAINTENANCE

With mountainous terrain encompassing a large portion of the world's land mass, the proper use of mountaineering equipment will enhance a unit's combat capability and provide a combat multiplier. The equipment described in this chapter is produced by many different manufacturers; however, each item is produced and tested to extremely high standards to ensure safety when being used correctly. The weak link in the safety chain is the user. Great care in performing preventative maintenance checks and services and proper training in the use of the equipment is paramount to ensuring safe operations. The manufacturers of each and every piece of equipment provide recommendations on how to use and care for its product. It is imperative to follow these instructions explicitly.

3-1. FOOTWEAR

Currently, CTA 50-900 provides adequate footwear for most operations in mountainous terrain. In temperate climates a combination of footwear is most appropriate to accomplish all tasks.

a. The hot weather boot provides an excellent all-round platform for movement and climbing techniques and should be the boot of choice when the weather permits. The intermediate cold weather boot provides an acceptable platform for operations when the weather is less than ideal. These two types of boots issued together will provide the unit with the footwear necessary to accomplish the majority of basic mountain missions.

b. Mountain operations are encumbered by extreme cold, and the extreme cold weather boot (with vapor barrier) provides an adequate platform for many basic mountain missions. However, plastic mountaineering boots should be incorporated into training as soon as possible. These boots provide a more versatile platform for any condition that would be encountered in the mountains, while keeping the foot dryer and warmer.

c. Level 2 and level 3 mountaineers will need mission-specific footwear that is not currently available in the military supply system. The two types of footwear they will need are climbing shoes and plastic mountaineering boots.

(1) Climbing shoes are made specifically for climbing vertical or near vertical rock faces. These shoes are made with a soft leather upper, a lace-up configuration, and a smooth "sticky rubber" sole (Figure 3-1). The smooth "sticky rubber" sole is the key to the climbing shoe, providing greater friction on the surface of the rock, allowing the climber access to more difficult terrain.

(2) The plastic mountaineering boot is a double boot system (Figure 3-1). The inner boot provides support, as well as insulation against the cold. The inner boot may or may not come with a breathable membrane. The outer boot is a molded plastic (usually with a lace-up configuration) with a lug sole. The welt of the boot is molded in such a way that crampons, ski bindings, and snowshoes are easily attached and detached.

Note: Maintenance of all types of footwear must closely follow the manufacturers' recommendations.

Figure 3-1. Climbing shoes and plastic mountaineering boots.

3-2. CLOTHING

Clothing is perhaps the most underestimated and misunderstood equipment in the military inventory. The clothing system refers to every piece of clothing placed against the skin, the insulation layers, and the outer most garments, which protect the soldier from the elements. When clothing is worn properly, the soldier is better able to accomplish his tasks. When worn improperly, he is, at best, uncomfortable and, at worst, develops hypothermia or frostbite.

a. **Socks.** Socks are one of the most under-appreciated part of the entire clothing system. Socks are extremely valuable in many respects, if worn correctly. As a system, socks provide cushioning for the foot, remove excess moisture, and provide insulation from cold temperatures. Improper wear and excess moisture are the biggest causes of hot spots and blisters. Regardless of climatic conditions, socks should always be worn in layers.

(1) The first layer should be a hydrophobic material that moves moisture from the foot surface to the outer sock.

(2) The outer sock should also be made of hydrophobic materials, but should be complimented with materials that provide cushioning and abrasion resistance.

(3) A third layer can be added depending upon the climatic conditions.

(a) In severe wet conditions, a waterproof type sock can be added to reduce the amount of water that would saturate the foot. This layer would be worn over the first two layers if conditions were extremely wet.

(b) In extremely cold conditions a vapor barrier sock can be worn either over both of the original pairs of socks or between the hydrophobic layer and the insulating layer. If the user is wearing VB boots, the vapor barrier sock is not recommended.

b. **Underwear.** Underwear should also be made of materials that move moisture from the body. Many civilian companies manufacture this type of underwear. The primary material in this product is polyester, which moves moisture from the body to the outer layers keeping the user drier and more comfortable in all climatic conditions. In colder environments, several pairs of long underwear of different thickness should be made available. A lightweight set coupled with a heavyweight set will provide a multitude of layering combinations.

c. **Insulating Layers.** Insulating layers are those layers that are worn over the underwear and under the outer layers of clothing. Insulating layers provide additional warmth when the weather turns bad. For the most part, today's insulating layers will provide for easy moisture movement as well as trap air to increase the insulating factor. The insulating layers that are presently available are referred to as pile or fleece. The ECWCS (Figure 3-2, page 3-4) also incorporates the field jacket and field pants liner as additional insulating layers. However, these two components do not move moisture as effectively as the pile or fleece.

d. **Outer Layers.** The ECWCS provides a jacket and pants made of a durable waterproof fabric. Both are constructed with a nylon shell with a laminated breathable membrane attached. This membrane allows the garment to release moisture to the environment while the nylon shell provides a degree of water resistance during rain and snow. The nylon also acts as a barrier to wind, which helps the garment retain the warm air trapped by the insulating layers. Leaders at all levels must understand the importance of wearing the ECWCS correctly.

Note: Cotton layers must not be included in any layer during operations in a cold environment.

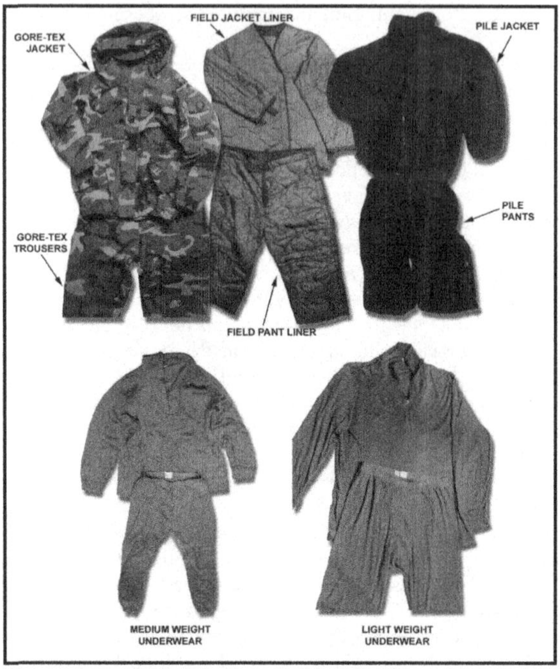

Figure 3-2. Extreme cold weather clothing system.

e. **Gaiters.** Gaiters are used to protect the lower leg from snow and ice, as well as mud, twigs, and stones. The use of waterproof fabrics or other breathable materials laminated to the nylon makes the gaiter an integral component of the cold weather clothing system. Gaiters are not presently fielded in the standard ECWCS and, in most cases, will need to be locally purchased. Gaiters are available in three styles (Figure 3-3).

(1) The most common style of gaiter is the open-toed variety, which is a nylon shell that may or may not have a breathable material laminated to it. The open front allows the boot to slip easily into it and is closed with a combination of zipper, hook-pile tape, and snaps. It will have an adjustable neoprene strap that goes under the boot to keep it snug to the boot. The length should reach to just below the knee and will be kept snug with a drawstring and cord lock.

(2) The second type of gaiter is referred to as a full or randed gaiter. This gaiter completely covers the boot down to the welt. It can be laminated with a breathable material and can also be insulated if necessary. This gaiter is used with plastic mountaineering boots and should be glued in place and not removed.

(3) The third type of gaiter is specific to high-altitude mountaineering or extremely cold temperatures and is referred to as an overboot. It is worn completely over the boot and must be worn with crampons because it has no traction sole.

Figure 3-3. Three types of gaiters.

f. **Hand Wear.** During operations in mountainous terrain the use of hand wear is extremely important. Even during the best climatic conditions, temperatures in the mountains will dip below the freezing point. While mittens are always warmer than gloves, the finger dexterity needed to do most tasks makes gloves the primary cold weather hand wear (Figure 3-4, page 3-6).

(1) The principals that apply to clothing also apply to gloves and mittens. They should provide moisture transfer from the skin to the outer layers—the insulating layer must insulate the hand from the cold and move moisture to the outer layer. The outer layer must be weather resistant and breathable. Both gloves and mittens should be required for all soldiers during mountain operations, as well as replacement liners for both. This will provide enough flexibility to accomplish all tasks and keep the users' hands warm and dry.

(2) Just as the clothing system is worn in layers, gloves and mittens work best using the same principle. Retention cords that loop over the wrist work extremely well when the wearer needs to remove the outer layer to accomplish a task that requires fine finger dexterity. Leaving the glove or mitten dangling from the wrist ensures the wearer knows where it is at all times.

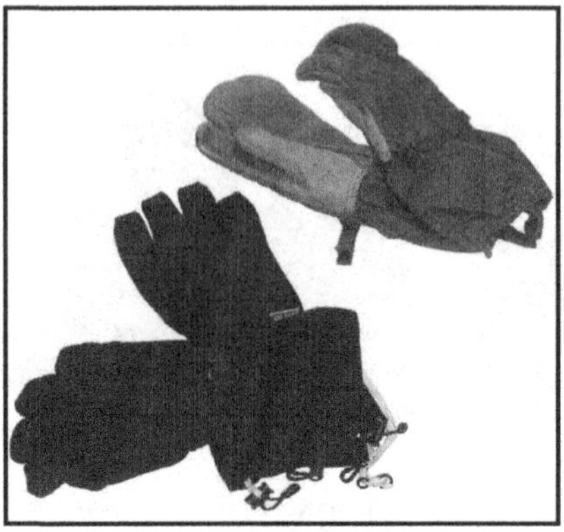

Figure 3-4. Hand wear.

g. **Headwear.** A large majority of heat loss (25 percent) occurs through the head and neck area. The most effective way to counter heat loss is to wear a hat. The best hat available to the individual soldier through the military supply system is the black watch cap. Natural fibers, predominately wool, are acceptable but can be bulky and difficult to fit under a helmet. As with clothes and hand wear, man-made fibers are preferred. For colder climates a neck gaiter can be added. The neck gaiter is a tube of man-made material that fits around the neck and can reach up over the ears and nose (Figure 3-5). For extreme cold, a balaclava can be added. This covers the head, neck, and face leaving only a slot for the eyes (Figure 3-5). Worn together the combination is warm and provides for moisture movement, keeping the wearer drier and warmer.

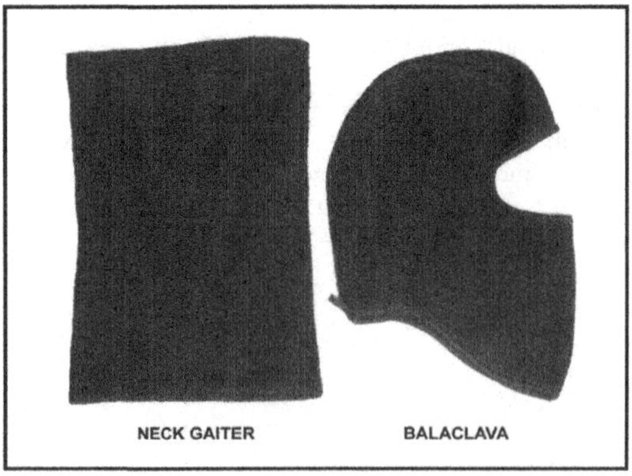

Figure 3-5. Neck gaiter and balaclava.

h. **Helmets**. The Kevlar ballistic helmet can be used for most basic mountaineering tasks. It must be fitted with parachute retention straps and the foam impact pad (Figure 3-6). The level 2 and 3 mountaineer will need a lighter weight helmet for specific climbing scenerios. Several civilian manufacturers produce an effective helmet. Whichever helmet is selected, it should be designed specifically for mountaineering and adjustable so the user can add a hat under it when needed.

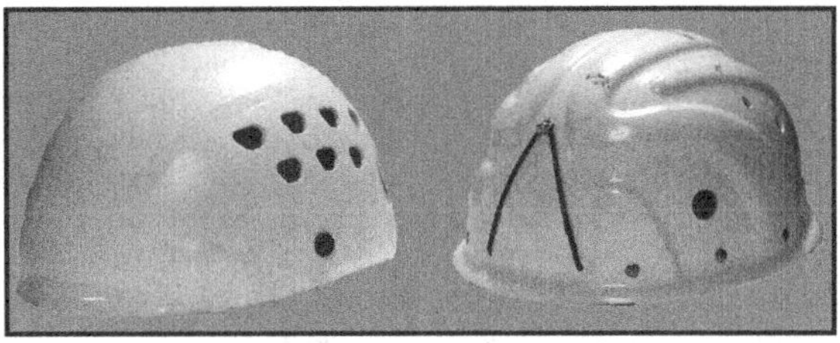

Figure 3-6. Helmets.

i. **Eyewear.** The military supply system does not currently provide adequate eyewear for mountaineering. Eyewear is divided into two catagories: glacier glasses and goggles (Figure 3-7). Glacier glasses are sunglasses that cover the entire eye socket. Many operations in the mountains occur above the tree line or on ice and snow surfaces where the harmful UV rays of the sun can bombard the eyes from every angle increasing the likelihood of snowblindness. Goggles for mountain operations should be antifogging. Double or triple lenses work best. UV rays penetrate clouds so the goggles should be UV protected. Both glacier glasses and goggles are required equipment in the mountains. The lack of either one can lead to severe eye injury or blindness.

Figure 3-7. Glacier glasses and goggles.

j. **Maintenance of Clothing.** Clothing and equipment manufacturers provide specific instructions for proper care. Following these instructions is necessary to ensure the equipment works as intended.

3-3. CLIMBING SOFTWARE

Climbing software refers to rope, cord, webbing, and harnesses. All mountaineering specific equipment, to include hardware (see paragraph 3-4), should only be used if it has the UIAA certificate of safety. UIAA is the organization that oversees the testing of mountaineering equipment. It is based in Paris, France, and comprises several commissions. The safety commission has established standards for mountaineering and climbing equipment that have become well recognized throughout the world. Their work continues as new equipment develops and is brought into common use. Community Europe (CE) recognizes UIAA testing standards and, as the broader-based testing facility for the combined European economy, meets or exceeds the UIAA standards for all climbing and mountaineering equipment produced in Europe. European norm (EN) and CE have been combined to make combined European norm (CEN). While the United States has no specific standards, American manufacturers have their equipment tested by UIAA to ensure safe operating tolerances.

a. **Ropes and Cord.** Ropes and cords are the most important pieces of mountaineering equipment and proper selection deserves careful thought. These items are your lifeline in the mountains, so selecting the right type and size is of the utmost importance. All ropes and cord used in mountaineering and climbing today are constructed with the same basic configuration. The construction technique is referred to as Kernmantle, which is, essentially, a core of nylon fibers protected by a woven sheath, similar to parachute or 550 cord (Figure 3-8).

Figure 3-8. Kernmantle construction.

(1) Ropes come in two types: static and dynamic. This refers to their ability to stretch under tension. A static rope has very little stretch, perhaps as little as one to two percent, and is best used in rope installations. A dynamic rope is most useful for climbing and general mountaineering. Its ability to stretch up to 1/3 of its overall length makes it the right choice any time the user might take a fall. Dynamic and static ropes come in various

diameters and lengths. For most military applications, a standard 10.5- or 11-millimeter by 50-meter dynamic rope and 11-millimeter by 45-meter static rope will be sufficient.

(2) When choosing dynamic rope, factors affecting rope selection include intended use, impact force, abrasion resistance, and elongation. Regardless of the rope chosen, it should be UIAA certified.

(3) Cord or small diameter rope is indispensable to the mountaineer. Its many uses make it a valuable piece of equipment. All cord is static and constructed in the same manner as larger rope. If used for Prusik knots, the cord's diameter should be 5 to 7 millimeters when used on an 11-mm rope.

b. **Webbing and Slings.** Loops of tubular webbing or cord, called slings or runners, are the simplest pieces of equipment and some of the most useful. The uses for these simple pieces are endless, and they are a critical link between the climber, the rope, carabiners, and anchors. Runners are predominately made from either 9/16-inch or 1-inch tubular webbing and are either tied or sewn by a manufacturer (Figure 3-9). Runners can also be made from a high-performance fiber known as spectra, which is stronger, more durable, and less susceptible to ultraviolet deterioration. Runners should be retired regularly following the same considerations used to retire a rope. For most military applications, a combination of different lengths of runners is adequate.

(1) Tied runners have certain advantages over sewn runners—they are inexpensive to make, can be untied and threaded around natural anchors, and can be untied and retied to other pieces of webbing to create extra long runners.

(2) Sewn runners have their own advantages—they tend to be stronger, are usually lighter, and have less bulk than the tied version. They also eliminate a major concern with the homemade knotted runner—the possibility of the knot untying. Sewn runners come in four standard lengths: 2 inches, 4 inches, 12 inches, and 24 inches. They also come in three standard widths: 9/16 inch, 11/16 inch, and 1 inch.

Figure 3-9. Tied or sewn runners.

c. **Harnesses.** Years ago climbers secured themselves to the rope by wrapping the rope around their bodies and tying a bowline-on-a-coil. While this technique is still a viable way of attaching to a rope, the practice is no longer encouraged because of the increased possibility of injury from a fall. The bowline-on-a-coil is best left for low-angle climbing or an emergency situation where harness material is unavailable. Climbers today can select from a wide range of manufactured harnesses. Fitted properly, the harness should ride high on the hips and have snug leg loops to better distribute the force of a fall to the entire pelvis. This type of harness, referred to as a seat harness, provides a comfortable seat for rappelling (Figure 3-10).

(1) Any harness selected should have one very important feature—a double-passed buckle. This is a safety standard that requires the waist belt to be passed over and back through the main buckle a second time. At least 2 inches of the strap should remain after double-passing the buckle.

(2) Another desirable feature on a harness is adjustable leg loops, which allows a snug fit regardless of the number of layers of clothing worn. Adjustable leg loops allow the soldier to make a latrine call without removing the harness or untying the rope.

(3) Equipment loops are desirable for carrying pieces of climbing equipment. For safety purposes always follow the manufacturer's directions for tying-in.

(4) A field-expedient version of the seat harness can be constructed by using 22 feet of either 1-inch or 2-inch (preferred) tubular webbing (Figure 3-10). Two double-overhand knots form the leg loops, leaving 4 to 5 feet of webbing coming from one of the leg loops. The leg loops should just fit over the clothing. Wrap the remaining webbing around the waist ensuring the first wrap is routed through the 6- to 10-inch long strap between the double-overhand knots. Finish the waist wrap with a water knot tied as tightly as possible. With the remaining webbing, tie a square knot without safeties over the water knot ensuring a minimum of 4 inches remains from each strand of webbing.

(5) The full body harness incorporates a chest harness with a seat harness (Figure 3-10). This type of harness has a higher tie-in point and greatly reduces the chance of flipping backward during a fall. This is the only type of harness that is approved by the UIAA. While these harnesses are safer, they do present several disadvantages—they are more expensive, are more restrictive, and increase the difficulty of adding or removing clothing. Most mountaineers prefer to incorporate a separate chest harness with their seat harness when warranted.

(6) A separate chest harness can be purchased from a manufacturer, or a field-expedient version can be made from either two runners or a long piece of webbing. Either chest harness is then attached to the seat harness with a carabiner and a length of webbing or cord.

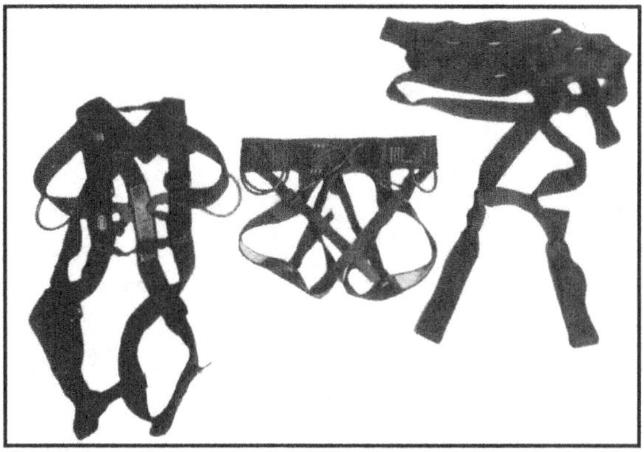

Figure 3-10. Seat harness, field-expedient harness, and full body harness.

3-4. CLIMBING HARDWARE

Climbing hardware refers to all the parts and pieces that allow the trained mountain soldier to accomplish many tasks in the mountains. The importance of this gear to the mountaineer is no less than that of the rifle to the infantryman.

a. **Carabiners.** One of the most versatile pieces of equipment available to the mountaineer is the carabiner. This simple piece of gear is the critical connection between the climber, his rope, and the protection attaching him to the mountain. Carabiners must be strong enough to hold hard falls, yet light enough for the climber to easily carry a quantity of them. Today's high tech metal alloys allow carabiners to meet both of these requirements. Steel is still widely used, but is not preferred for general mountaineering, given other options. Basic carabiner construction affords the user several different shapes. The oval, the D-shaped, and the pear-shaped carabiner are just some of the types currently available. Most models can be made with or without a locking mechanism for the gate opening (Figure 3-11, page 3-12). If the carabiner does have a locking mechanism, it is usually referred to as a locking carabiner. When using a carabiner, great care should be taken to avoid loading the carabiner on its minor axis and to avoid three-way loading (Figure 3-12, page 3-12).

Note: Great care should be used to ensure all carabiner gates are closed and locked during use.

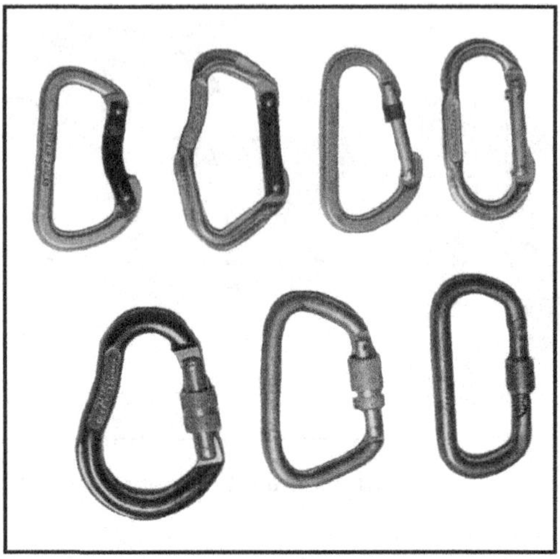

Figure 3-11. Nonlocking and locking carabiners.

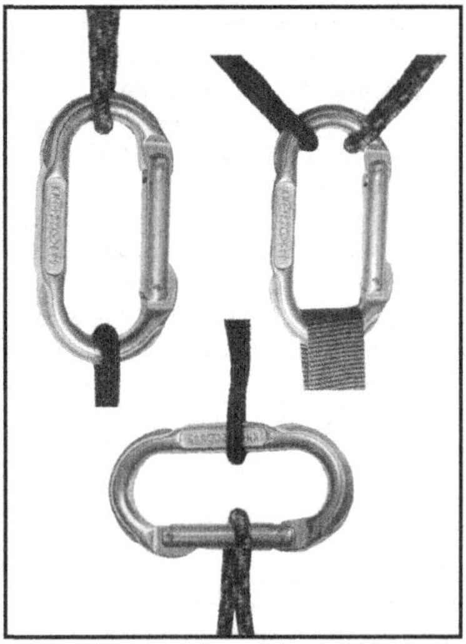

Figure 3-12. Major and minor axes and three-way loading.

(1) The major difference between the oval and the D-shaped carabiner is strength. Because of the design of the D-shaped carabiner, the load is angled onto the spine of the

carabiner thus keeping it off the gate. The down side is that racking any gear or protection on the D-shaped carabiner is difficult because the angle of the carabiner forces all the gear together making it impossible to separate quickly.

(2) The pear-shaped carabiner, specifically the locking version, is excellent for clipping a descender or belay device to the harness. They work well with the munter hitch belaying knot.

(3) Regardless of the type chosen, all carabiners should be UIAA tested. This testing is extensive and tests the carabiner in three ways—along its major axis, along its minor axis, and with the gate open.

b. **Pitons.** A piton is a metal pin that is hammered into a crack in the rock. They are described by their thickness, design, and length (Figure 3-13, page 3-14). Pitons provide a secure anchor for a rope attached by a carabiner. The many different kinds of pitons include: vertical, horizontal, wafer, and angle. They are made of malleable steel, hardened steel, or other alloys. The strength of the piton is determined by its placement rather than its rated tensile strength. The two most common types of pitons are: blades, which hold when wedged into tight-fitting cracks, and angles, which hold blade compression when wedged into a crack.

(1) ***Vertical Pitons***. On vertical pitons, the blade and eye are aligned. These pitons are used in flush, vertical cracks.

(2) ***Horizontal Pitons***. On horizontal pitons, the eye of the piton is at right angles to the blade. These pitons are used in flush, horizontal cracks and in offset or open-book type vertical or horizontal cracks. They are recommended for use in vertical cracks instead of vertical pitons because the torque on the eye tends to wedge the piton into place. This provides more holding power than the vertical piton under the same circumstances.

(3) ***Wafer Pitons***. These pitons are used in shallow, flush cracks. They have little holding power and their weakest points are in the rings provided for the carabiner.

(4) ***Knife Blade Pitons***. These are used in direct-aid climbing. They are small and fit into thin, shallow cracks. They have a tapered blade that is optimum for both strength and holding power.

(5) ***Realized Ultimate Reality Pitons***. Realized ultimate reality pitons (RURPs) are hatchet-shaped pitons about 1-inch square. They are designed to bite into thin, shallow cracks.

(6) ***Angle Pitons***. These are used in wide cracks that are flush or offset. Maximum strength is attained only when the legs of the piton are in contact with the opposite sides of the crack.

(7) ***Bong Pitons***. These are angle pitons that are more than 3.8 centimeters wide. Bongs are commonly made of steel or aluminum alloy and usually contain holes to reduce weight and accommodate carabiners. They have a high holding power and require less hammering than other pitons.

(8) ***Skyhook (Cliffhangers).*** These are small hooks that cling to tiny rock protrusions, ledges, or flakes. Skyhooks require constant tension and are used in a downward pull direction. The curved end will not straighten under body weight. The base is designed to prevent rotation and aid stability.

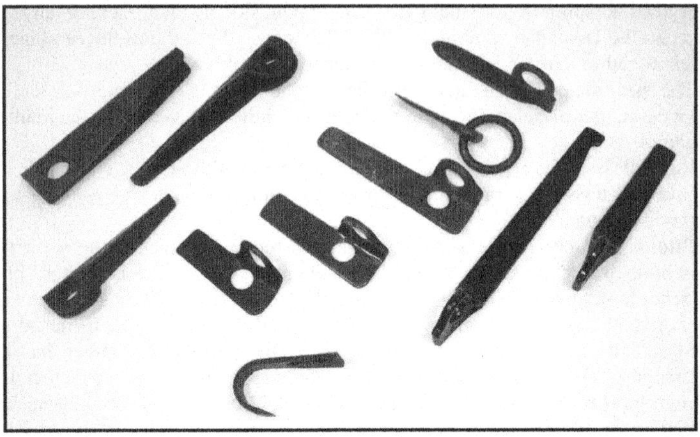

Figure 3-13. Various pitons.

c. **Piton Hammers.** A piton hammer has a flat metal head; a handle made of wood, metal, or fiberglass; and a blunt pick on the opposite side of the hammer (Figure 3-14). A safety lanyard of nylon cord, webbing, or leather is used to attach it to the climber The lanyard should be long enough to allow for full range of motion. Most hammers are approximately 25.5 centimeters long and weigh 12 to 25 ounces. The primary use for a piton hammer is to drive pitons, to be used as anchors, into the rock. The piton hammer can also be used to assist in removing pitons, and in cleaning cracks and rock surfaces to prepare for inserting the piton. The type selected should suit individual preference and the intended use.

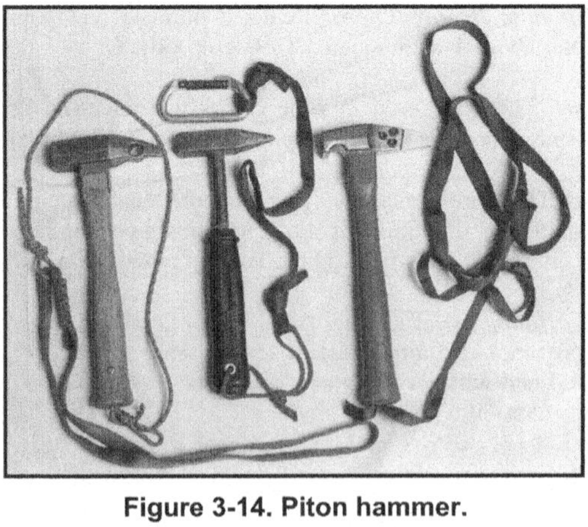

Figure 3-14. Piton hammer.

d. **Chocks.** "Chocks" is a generic term used to describe the various types of artificial protection other than bolts or pitons. Chocks are essentially a tapered metal wedge constructed in various sizes to fit different sized openings in the rock (Figure 3-15). The design of a chock will determine whether it fits into one of two categories—wedges or cams. A wedge holds by wedging into a constricting crack in the rock. A cam holds by slightly rotating in a crack, creating a camming action that lodges the chock in the crack or pocket. Some chocks are manufactured to perform either in the wedging mode or the camming mode. One of the chocks that falls into the category of both a wedge and cam is the hexagonal-shaped or "hex" chock. This type of chock is versatile and comes with either a cable loop or is tied with cord or webbing. All chocks come in different sizes to fit varying widths of cracks. Most chocks come with a wired loop that is stronger than cord and allows for easier placement. Bigger chocks can be threaded with cord or webbing if the user ties the chock himself. Care should be taken to place tubing in the chock before threading the cord. The cord used with chocks is designed to be stiffer and stronger than regular cord and is typically made of Kevlar. The advantage of using a chock rather than a piton is that a climber can carry many different sizes and use them repeatedly.

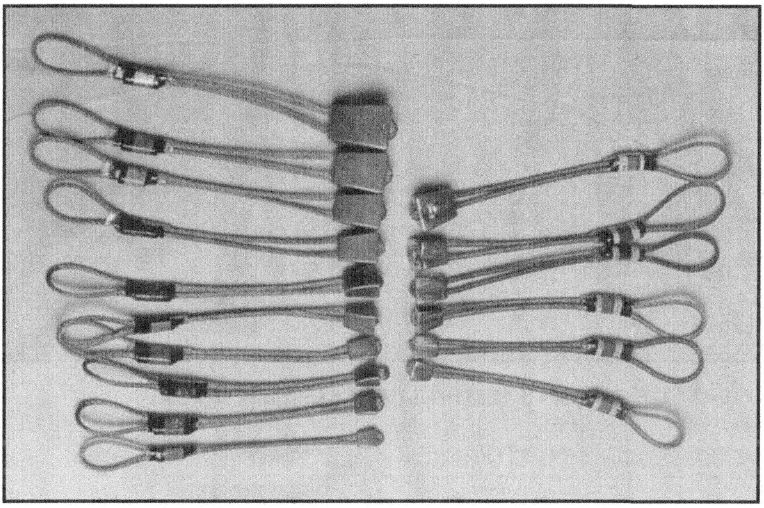

Figure 3-15. Chocks.

e. **Three-Point Camming Device.** The three-point camming device's unique design allows it to be used both as a camming piece and a wedging piece (Figure 3-16). Because of this design it is extremely versatile and, when used in the camming mode, will fit a wide range of cracks. The three-point camming device comes in several different sizes with the smaller sizes working in pockets that no other piece of gear would fit in.

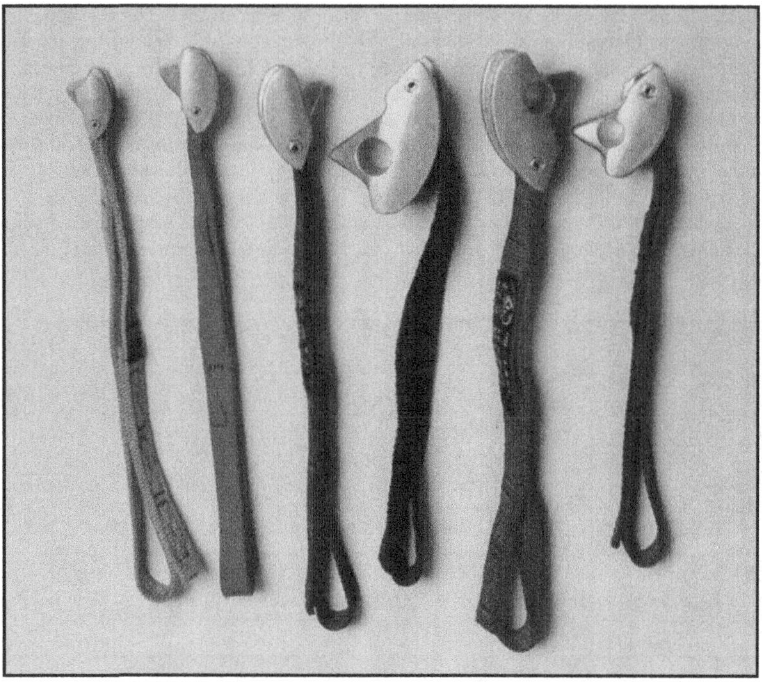

Figure 3-16. Three-point camming device.

f. **Spring-Loaded Camming Devices.** Spring-loaded camming devices (SLCDs) (Figure 3-17) provide convenient, reliable placement in cracks where standard chocks are not practical (parallel or flaring cracks or cracks under roofs). SLCDs have three or four cams rotating around a single or double axis with a rigid or semi-rigid point of attachment. These are placed quickly and easily, saving time and effort. SLCDs are available in many sizes to accommodate different size cracks. Each fits a wide range of crack widths due to the rotating cam heads. The shafts may be rigid metal or semi-rigid cable loops. The flexible cable reduces the risk of stem breakage over an edge in horizontal placements.

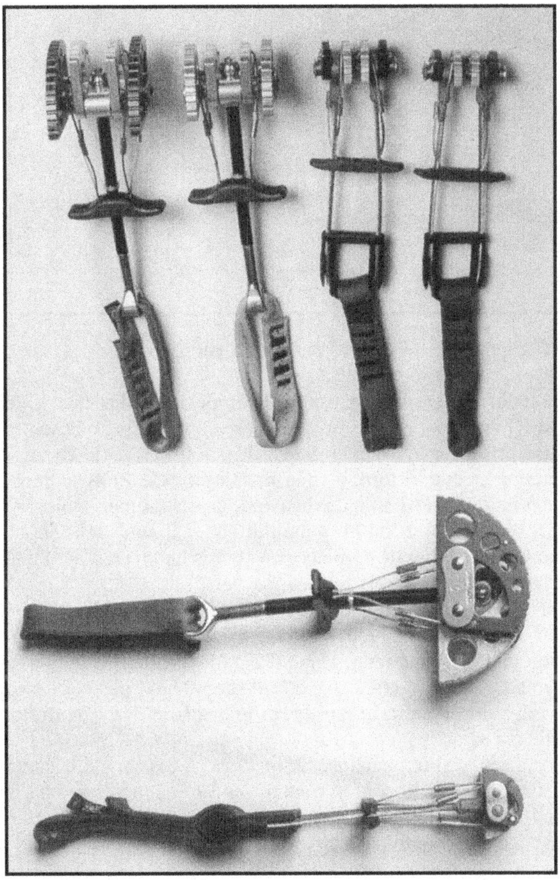

Figure 3-17. Spring-loaded camming devices.

g. **Chock Picks.** Chock picks are primarily used to extract chocks from rock when the they become severely wedged (Figure 3-18). They are also handy to clean cracks with. Made from thin metal, they can be purchased or homemade. When using a chock pick to extract a chock be sure no force is applied directly to the cable juncture. One end of the chock pick should have a hook to use on jammed SLCDs.

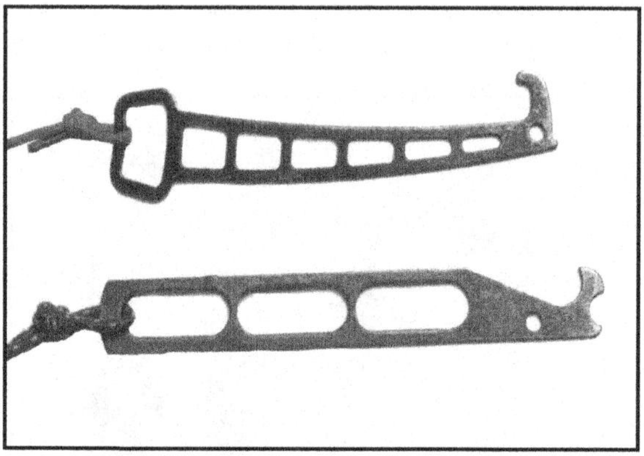

Figure18. Chock picks.

h. **Bolts.** Bolts are screw-like shafts made from metal that are drilled into rock to provide protection (Figure 3-19). The two types are contraction bolts and expansion bolts. Contraction bolts are squeezed together when driven into a rock. Expansion bolts press around a surrounding sleeve to form a snug fit into a rock. Bolts require drilling a hole into a rock, which is time-consuming, exhausting, and extremely noisy. Once emplaced, bolts are the most secure protection for a multidirectional pull. Bolts should be used only when chocks and pitons cannot be emplaced. A bolt is hammered only when it is the nail or self-driving type.

(1) A hanger (for carabiner attachment) and nut are placed on the bolt. The bolt is then inserted and driven into the hole. Because of this requirement, a hand drill must be carried in addition to a piton hammer. Hand drills (also called star drills) are available in different sizes, brands, and weights. A hand drill should have a lanyard to prevent loss.

(2) Self-driving bolts are quicker and easier to emplace. These require a hammer, bolt driver, and drilling anchor, which is driven into the rock. A bolt and carrier are then secured to the emplaced drilling anchor. All metal surfaces should be smooth and free of rust, corrosion, dirt, and moisture. Burrs, chips, and rough spots should be filed smooth and wire-brushed or rubbed clean with steel wool. Items that are cracked or warped indicate excessive wear and should be discarded.

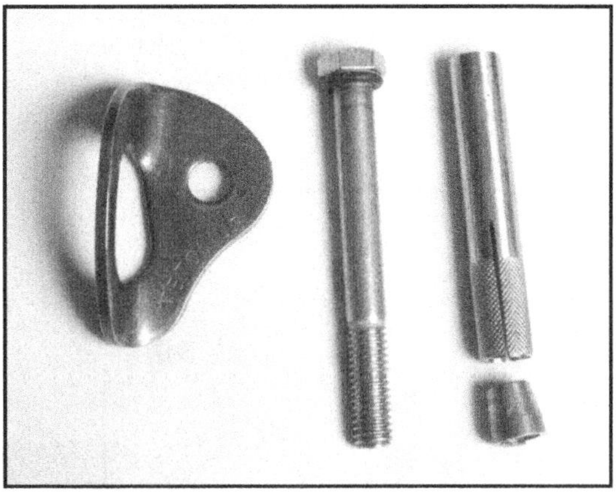

Figure 3-19. Bolts and hangers.

i. **Belay Devices.** Belay devices range from the least equipment intensive (the body belay) to high-tech metal alloy pieces of equipment. Regardless of the belay device choosen, the basic principal remains the same—friction around or through the belay device controls the ropes' movement. Belay devices are divided into three categories: the slot, the tuber, and the mechanical camming device (Figure 3-20).

(1) The slot is a piece of equipment that attaches to a locking carabiner in the harness; a bight of rope slides through the slot and into the carabiner for the belay. The most common slot type belay device is the Sticht plate.

(2) The tuber is used exactly like the slot but its shape is more like a cone or tube.

(3) The mechanical camming device is a manufactured piece of equipment that attaches to the harness with a locking carabiner. The rope is routed through this device so that when force is applied the rope is cammed into a highly frictioned position.

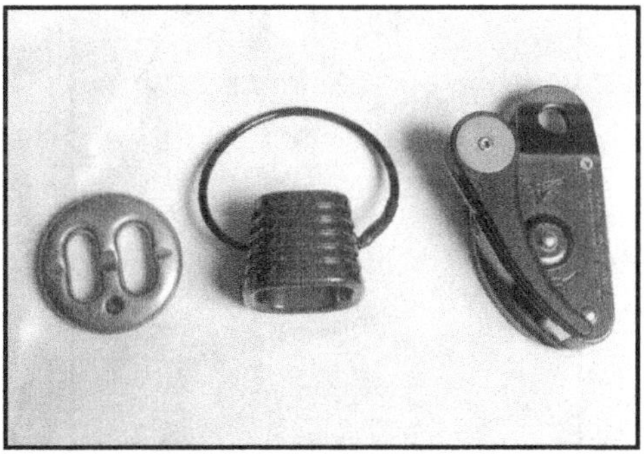

Figure 3-20. Slot, tuber, mechanical camming device.

j. **Descenders.** One piece of equipment used for generations as a descender is the carabiner. A figure-eight is another useful piece of equipment and can be used in conjunction with the carabiner for descending (Figure 3-21).

Note: All belay devices can also be used as descending devices.

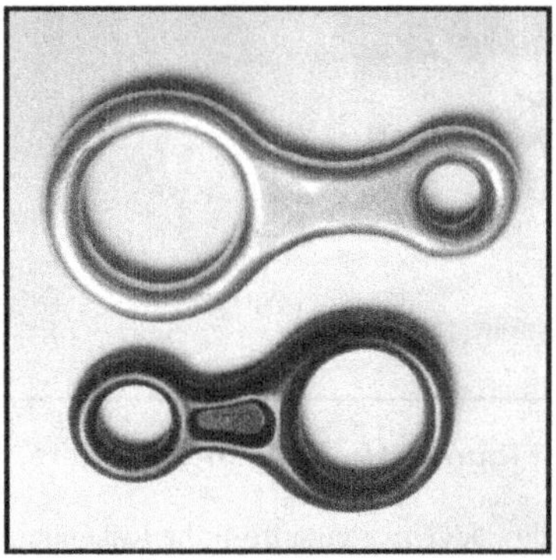

Figure 3-21. Figure-eights.

k. **Ascenders.** Ascenders may be used in other applications such as a personal safety or hauling line cam. All modern ascenders work on the principle of using a cam-like device to allow movement in one direction. Ascenders are primarily made of metal alloys and come in a variety of sizes (Figure 3-22). For difficult vertical terrain, two ascenders work best. For lower angle movement, one ascender is sufficient. Most manufacturers make ascenders as a right and left-handed pair.

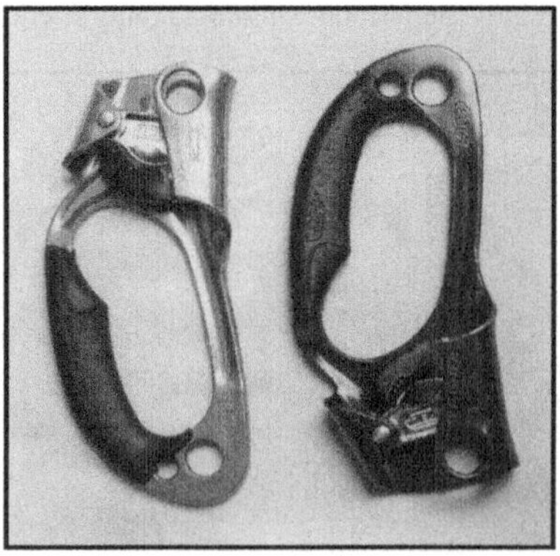

Figure 3-22. Ascenders.

l. **Pulleys.** Pulleys are used to change direction in rope systems and to create mechanical advantage in hauling systems. A pulley should be small, lightweight, and strong. They should accommodate the largest diameter of rope being used. Pulleys are made with several bearings, different-sized sheaves (wheel), and metal alloy sideplates (Figure 3-23). Plastic pulleys should always be avoided. The sideplate should rotate on the pulley axle to allow the pulley to be attached at any point along the rope. For best results, the sheave diameter must be at least four times larger than the rope's diameter to maintain high rope strength.

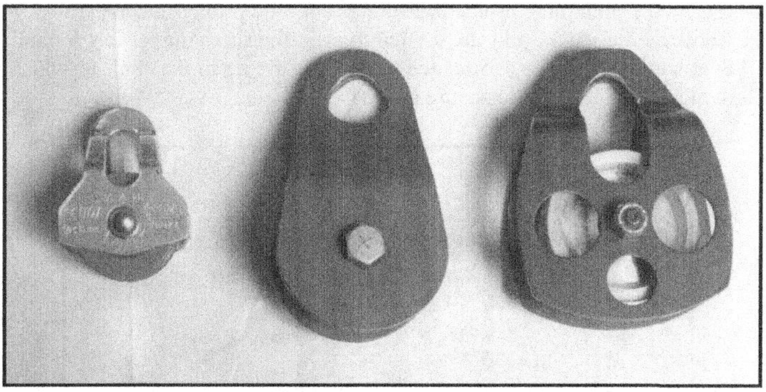

Figure 3-23. Pulley.

3-5. SNOW AND ICE CLIMBING HARDWARE

Snow and ice climbing hardware is the equipment that is particular to operations in some mountainous terrain. Specific training on this type of equipment is essential for safe use. Terrain that would otherwise be inaccessible—snowfields, glaciers, frozen waterfalls—can now be considered avenues of approach using the snow and ice climbing gear listed in this paragraph.

a. **Ice Ax.** The ice ax is one of the most important tools for the mountaineer operating on snow or ice. The climber must become proficient in its use and handling. The versatility of the ax lends itself to balance, step cutting, probing, self-arrest, belays, anchors, direct-aid climbing, and ascending and descending snow and ice covered routes.

(1) Several specific parts comprise an ice ax: the shaft, head (pick and adze), and spike (Figure 3-24, page 3-22).

(a) The shaft (handle) of the ax comes in varying lengths (the primary length of the standard mountaineering ax is 70 centimeters). It can be made of fiberglass, hollow aluminum, or wood; the first two are stronger, therefore safer for mountaineering.

(b) The head of the ax, which combines the pick and the adze, can have different configurations. The pick should be curved slightly and have teeth at least one-fourth of its length. The adze, used for chopping, is perpendicular to the shaft. It can be flat or curved along its length and straight or rounded from side to side. The head can be of one-piece construction or have replaceable picks and adzes. The head should have a hole directly above the shaft to allow for a leash to be attached.

(c) The spike at the bottom of the ax is made of the same material as the head and comes in a variety of shapes.

(2) As climbing becomes more technical, a shorter ax is much more appropriate, and adding a second tool is a must when the terrain becomes vertical. The shorter ax has all the attributes of the longer ax, but it is anywhere from 40 to 55 centimeters long and can have a straight or bent shaft depending on the preference of the user.

b. **Ice Hammer.** The ice hammer is as short or shorter than the technical ax (Figure 3-24). It is used for pounding protection into the ice or pitons into the rock. The only difference between the ice ax and the ice hammer is the ice hammer has a hammerhead instead of an adze. Most of the shorter ice tools have a hole in the shaft to which a leash is secured, which provides a more secure purchase in the ice.

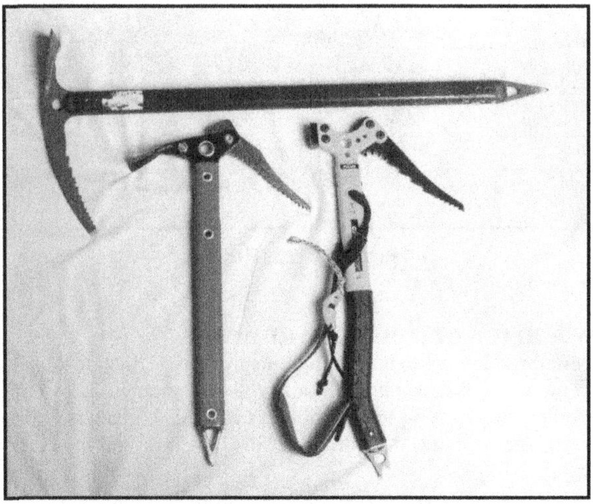

Figure 3-24. Ice ax and ice hammers.

c. **Crampons.** Crampons are used when the footing becomes treacherous. They have multiple spikes on the bottom and spikes protruding from the front (Figure 3-25). Two types of crampons are available: flexible and rigid. Regardless of the type of crampon chosen, fit is the most important factor associated with crampon wear. The crampon should fit snugly on the boot with a minimum of 1 inch of front point protruding. Straps should fit snugly around the foot and any long, loose ends should be trimmed. Both flexible and rigid crampons come in pairs, and any tools needed for adjustment will be provided by the manufacturer.

(1) The hinged or flexible crampon is best used when no technical ice climbing will be done. It is designed to be used with soft, flexible boots, but can be attached to plastic

mountaineering boots. The flexible crampon gets its name from the flexible hinge on the crampon itself. All flexible crampons are adjustable for length while some allow for width adjustment. Most flexible crampons will attach to the boot by means of a strap system. The flexible crampon can be worn with a variety of boot types.

(2) The rigid crampon, as its name implies, is rigid and does not flex. This type of crampon is designed for technical ice climbing, but can be used on less vertical terrain. The rigid crampon can only be worn with plastic mountaineering boots. Rigid crampons will have a toe and heel bail attachment with a strap that wraps around the ankle.

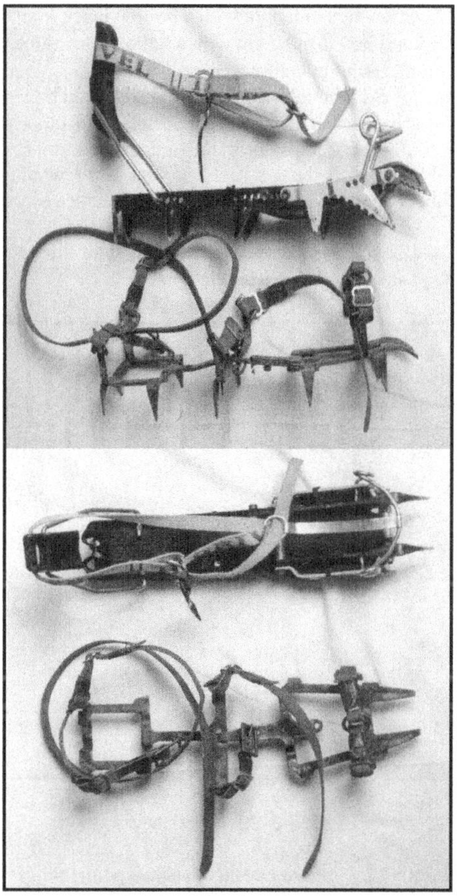

Figure 3-25. Crampons.

d. **Ice Screws.** Ice screws provide artificial protection for climbers and equipment for operations in icy terrain. They are screwed into ice formations. Ice screws are made of chrome-molybdenum steel and vary in lengths from 11 centimeters to 40 centimeters (Figure 3-26). The eye is permanently affixed to the top of the ice screw. The tip consists of milled or hand-ground teeth, which create sharp points to grab the ice when being emplaced. The ice screw has right-hand threads to penetrate the ice when turned clockwise.

(1) When selecting ice screws, choose a screw with a large thread count and large hollow opening. The close threads will allow for ease in turning and better strength. The large hollow opening will allow snow and ice to slide through when turning.

- Type I is 17 centimeters in length with a hollow inner tube.
- Type II is 22 centimeters in length with a hollow inner tube.
- Other variations are hollow alloy screws that have a tapered shank with external threads, which are driven into ice and removed by rotation.

(2) Ice screws should be inspected for cracks, bends, and other deformities that may impair strength or function. If any cracks or bends are noticed, the screw should be turned in. A file may be used to sharpen the ice screw points. Steel wool should be rubbed on rusted surfaces and a thin coat of oil applied when storing steel ice screws.

Note: Ice screws should always be kept clean and dry. The threads and teeth should be protected and kept sharp for ease of application.

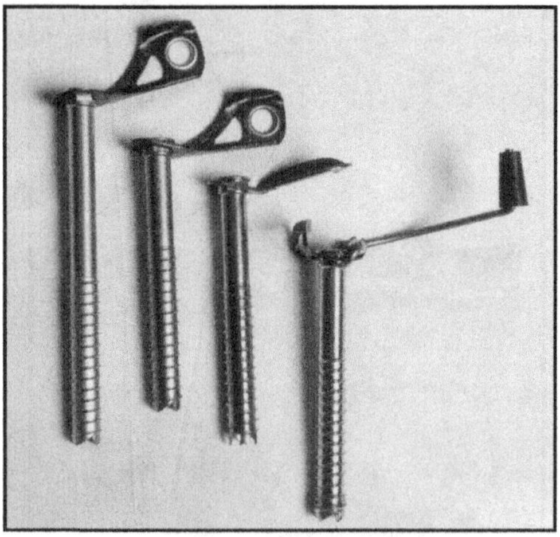

Figure 3-26. Ice screws.

e. **Ice Pitons.** Ice pitons are used to establish anchor points for climbers and equipment when conducting operations on ice. They are made of steel or steel alloys (chrome-molybdenum), and are available in various lengths and diameters (Figure 3-27). They are tubular with a hollow core and are hammered into ice with an ice hammer. The eye is permanently fixed to the top of the ice piton. The tip may be beveled to help grab the ice to facilitate insertion. Ice pitons are extremely strong when placed properly in hard ice. They can, however, pull out easily on warm days and require a considerable amount of effort to extract in cold temperatures.

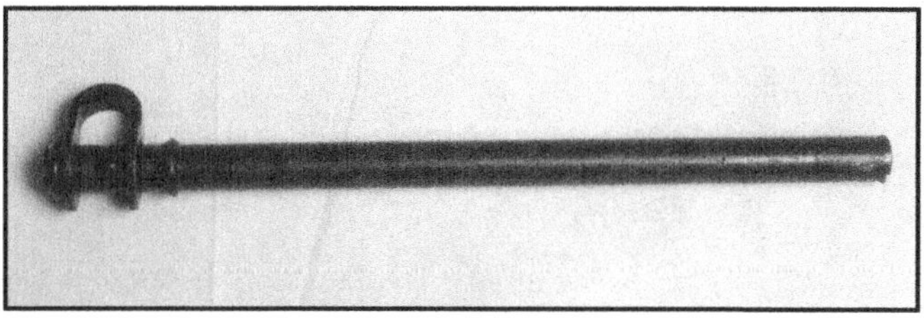

Figure 3-27. Ice piton.

f. **Wired Snow Anchors.** The wired snow anchor (or fluke) provides security for climbers and equipment in operations involving steep ascents by burying the snow anchor into deep snow (Figure 3-28, page 3-26). The fluted anchor portion of the snow anchor is made of aluminum. The wired portion is made of either galvanized steel or stainless steel. Fluke anchors are available in various sizes—their holding ability generally increases with size. They are available with bent faces, flanged sides, and fixed cables. Common types are:

- Type I is 22 by 14 centimeters. Minimum breaking strength of the swaged wire loop is 600 kilograms.
- Type II is 25 by 20 centimeters. Minimum breaking strength of the swaged wire loop is 1,000 kilograms.

The wired snow anchor should be inspected for cracks, broken wire strands, and slippage of the wire through the swage. If any cracks, broken wire strands, or slippage is noticed, the snow anchor should be turned in.

g. **Snow Picket.** The snow picket is used in constructing anchors in snow and ice (Figure 3-28, page 3-26). The snow picket is made of a strong aluminum alloy 3 millimeters thick by 4 centimeters wide, and 45 to 90 centimeters long. They can be angled or T–section stakes. The picket should be inspected for bends, chips, cracks, mushrooming ends, and other deformities. The ends should be filed smooth. If bent or cracked, the picket should be turned in for replacement.

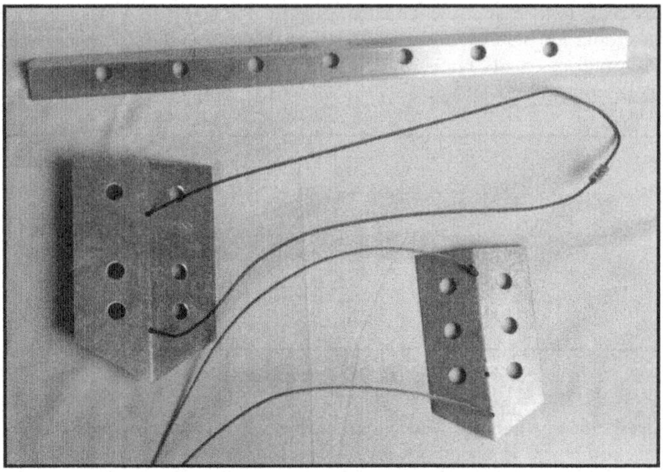

Figure 3-28. Snow anchors, flukes, and pickets.

3-6. SUSTAINABILITY EQUIPMENT

This paragraph describes all additional equipment not directly involved with climbing. This equipment is used for safety (avalanche equipment, wands), bivouacs, movement, and carrying gear. While not all of it will need to be carried on all missions, having the equipment available and knowing how to use it correctly will enhance the unit's capability in mountainous terrain.

a. **Snow Saw**. The snow saw is used to cut into ice and snow. It can be used in step cutting, in shelter construction, for removing frozen obstacles, and for cutting snow stability test pits. The special tooth design of the snow saw easily cuts into frozen snow and ice. The blade is a rigid aluminum alloy of high strength about 3 millimeters thick and 38 centimeters long with a pointed end to facilitate entry on the forward stroke. The handle is either wooden or plastic and is riveted to the blade for a length of about 50 centimeters. The blade should be inspected for rust, cracks, warping, burrs, and missing or dull teeth. A file can repair most defects, and steel wool can be rubbed on rusted areas. The handle should be inspected for cracks, bends, and stability. On folding models, the hinge and nuts should be secure. If the saw is beyond repair, it should not be used.

b. **Snow Shovel.** The snow shovel is used to cut and remove ice and snow. It can be used for avalanche rescue, shelter construction, step cutting, and removing obstacles. The snow shovel is made of a special, lightweight aluminum alloy. The handle should be telescopic, folding, or removable to be compact when not in use. The shovel should have a flat or rounded bottom and be of strong construction. The shovel should be inspected for cracks, bends, rust, and burrs. A file and steel wool can remove rust and put an edge on the blade of the shovel. The handle should be inspected for cracks, bends, and stability. If the shovel is beyond repair, it should be turned in.

c. **Wands.** Wands are used to identify routes, crevasses, snow-bridges, caches, and turns on snow and glaciers. Spacing of wands depends on the number of turns, number of hazards identified, weather conditions (and visibility), and number of teams in the climbing party. Carry too many wands is better than not having enough if they become lost. Wands are 1 to 1.25 meters long and made of lightweight bamboo or plastic shafts pointed on one end with a plastic or nylon flag (bright enough in color to see at a distance) attached to the other end. The shafts should be inspected for cracks, bends, and deformities. The flag should be inspected for tears, frays, security to the shaft, fading, and discoloration. If any defects are discovered, the wands should be replaced.

d. **Avalanche rescue equipment.** Avalanche rescue equipment (Figure 3-29) includes the following:

(1) ***Avalanche Probe.*** Although ski poles may be used as an emergency probe when searching for a victim in an avalanche, commercially manufactured probes are better for a thorough search. They are 9-millimeter thick shafts made of an aluminum alloy, which can be joined to probe up to 360 centimeters. The shafts must be strong enough to probe through avalanche debris. Some manufacturers of ski poles design poles that are telescopic and mate with other poles to create an avalanche probe.

(2) ***Avalanche Transceivers.*** These are small, compact radios used to identify avalanche burial sites. They transmit electromagnetic signals that are picked up by another transceiver on the receive mode.

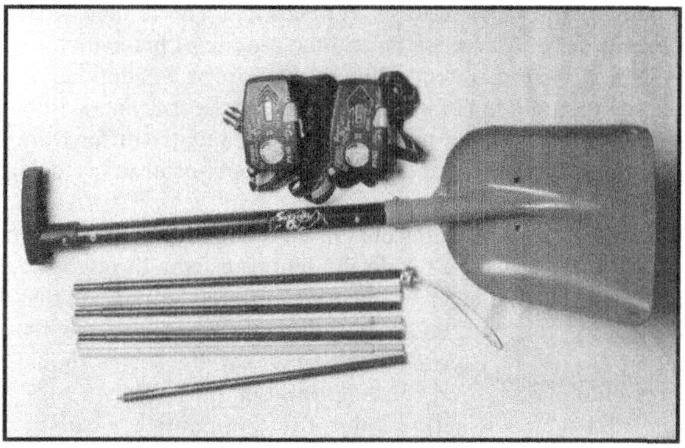

Figure 3-29. Avalanche rescue equipment.

e. **Packs.** Many types and brands of packs are used for mountaineering. The two most common types are internal and external framed packs.

(1) Internal framed packs have a rigid frame within the pack that help it maintain its shape and hug the back. This assists the climber in keeping their balance as they climb or ski. The weight in an internal framed pack is carried low on the body assisting with balance. The body-hugging nature of this type pack also makes it uncomfortable in warm weather.

(2) External framed packs suspend the load away from the back with a ladder-like frame. The frame helps transfer the weight to the hips and shoulders easier, but can be cumbersome when balance is needed for climbing and skiing.

(3) Packs come in many sizes and should be sized appropriately for the individual according to manufacturer's specifications. Packs often come with many unneeded features. A good rule of thumb is: The simpler the pack, the better it will be.

f. **Stoves.** When selecting a stove one must define its purpose—will the stove be used for heating, cooking or both? Stoves or heaters for large elements can be large and cumbersome. Stoves for smaller elements might just be used for cooking and making water, and are simple and lightweight. Stoves are a necessity in mountaineering for cooking and making water from snow and ice. When choosing a stove, factors that should be considered are weight, altitude and temperature where it will be used, fuel availability, and its reliability.

(1) There are many choices in stove design and in fuel types. White gas, kerosene, and butane are the common fuels used. All stoves require a means of pressurization to force the fuel to the burner. Stoves that burn white gas or kerosene have a hand pump to generate the pressurization and butane stoves have pressurized cartridges. All stoves need to vaporize the liquid fuel before it is burned. This can be accomplished by burning a small amount of fuel in the burner cup assembly, which will vaporize the fuel in the fuel line.

(2) Stoves should be tested and maintained prior to a mountaineering mission. They should be easy to clean and repair during an operation. The reliability of the stove has a huge impact on the success of the mission and the morale of personnel.

g. **Tents.** When selecting a tent, the mission must be defined to determine the number of people the tent will accommodate. The climate the tents will be used in is also of concern. A tent used for warmer temperatures will greatly differ from tents used in a colder, more harsh environment. Manufacturers of tents offer many designs of different sizes, weights, and materials.

(1) Mountaineering tents are made out of a breathable or weatherproof material. A single-wall tent allows for moisture inside the tent to escape through the tent's material. A double-wall tent has a second layer of material (referred to as a fly) that covers the tent. The fly protects against rain and snow and the space between the fly and tent helps moisture to escape from inside. Before using a new tent, the seams should be treated with seam sealer to prevent moisture from entering through the stitching.

(2) The frame of a tent is usually made of an aluminum or carbon fiber pole. The poles are connected with an elastic cord that allows them to extend, connect, and become long and rigid. When the tent poles are secured into the tent body, they create the shape of the tent.

(3) Tents are rated by a "relative strength factor," the speed of wind a tent can withstand before the frame deforms. Temperature and expected weather for the mission should be determined before choosing the tent.

h. **Skis.** Mountaineering skis are wide and short. They have a binding that pivots at the toe and allows for the heel to be free for uphill travel or locked for downhill. Synthetic skins with fibers on the bottom can be attached to the bottom of the ski and allow the ski to travel forward and prevent slipping backward. The skins aid in traveling uphill and slow down the rate of descents. Wax can be applied to the ski to aid in ascents instead of skins. Skis can decrease the time needed to reach an objective depending on the ability of the user. Skis can make crossing crevasses easier because of the load distribution, and they can become a makeshift stretcher for casualties. Ski techniques can be complicated and require thorough training for adequate proficiency.

i. **Snowshoes.** Snowshoes are the traditional aid to snow travel that attach to most footwear and have been updated into small, lightweight designs that are more efficient than older models. Snowshoes offer a large displacement area on top of soft snow preventing tiresome post-holing. Some snowshoes come equipped with a crampon like binding that helps in ascending steep snow and ice. Snowshoes are slower than skis, but are better suited for mixed terrain, especially if personnel are not experienced with the art of skiing. When carrying heavy packs, snowshoes can be easier to use than skis.

j. **Ski poles**. Ski poles were traditionally designed to assist in balance during skiing. They have become an important tool in mountaineering for aid in balance while hiking, snowshoeing, and carrying heavy packs. They can take some of the weight off of the lower body when carrying a heavy pack. Some ski poles are collapsible for ease of packing when not needed (Figure 3-30). The basket at the bottom prevents the pole from plunging deep into the snow and, on some models, can be detached so the pole becomes an avalanche or crevasse probe. Some ski poles come with a self-arrest grip, but should not be the only means of protection on technical terrain.

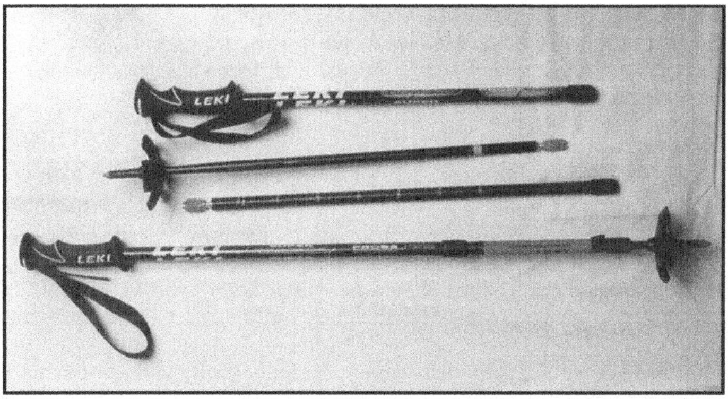

Figure 3-30. Collapsible ski poles.

k. **Sleds**. Sleds vary greatly in size, from the squad-size Ahkio, a component of the 10-man arctic tent system, to the one-person skow. Regardless of the size, sleds are an invaluable asset during mountainous operations when snow and ice is the primary surface on which to travel. Whichever sled is chosen, it must be attachable to the person or people that will be pulling it. Most sleds are constructed using fiberglass bottoms with or without exterior runners. Runners will aid the sleds ability to maintain a true track in the snow. The sled should also come with a cover of some sort—whether nylon or canvas, a cover is essential for keeping the components in the sled dry. Great care should be taken when packing the sled, especially when hauling fuel. Heavier items should be carried towards the rear of the sled and lighter items towards the front.

l. **Headlamps**. A headlamp is a small item that is not appreciated until it is needed. It is common to need a light source and the use of both hands during limited light conditions in mountaineering operations. A flashlight can provide light, but can be cumbersome when both hands are needed. Most headlamps attach to helmets by means of elastic bands.

(1) When choosing a headlamp, ensure it is waterproof and the battery apparatus is small. All components should be reliable in extreme weather conditions. When the light is being packed, care should be taken that the switch doesn't accidentally activate and use precious battery life.

(2) The battery source should compliment the resupply available. Most lights will accept alkaline, nickel-cadmium, or lithium batteries. Alkaline battery life diminishes quickly in cold temperatures, nickel-cadmium batteries last longer in cold but require a recharging unit, and lithium batteries have twice the voltage so modifications are required.

Section II. EQUIPMENT PACKING

Equipment brought on a mission is carried in the pack, worn on the body, or hauled in a sled (in winter). Obviously, the rucksack and sled (or Ahkio) can hold much more than a climber can carry. They would be used for major bivouac gear, food, water, first aid kits, climbing equipment, foul weather shells, stoves, fuel, ropes, and extra ammunition and demolition materials, if needed.

3-7. CHOICE OF EQUIPMENT

Mission requirements and unit SOP will influence the choice of gear carried but the following lists provide a sample of what should be considered during mission planning.

a. **Personal Gear.** Personal gear includes emergency survival kit containing signaling material, fire starting material, food procurement material, and water procurement material. Pocket items should include a knife, whistle, pressure bandage, notebook with pen or pencil, sunglasses, sunblock and lip protection, map, compass and or altimeter.

b. **Standard Gear.** Standard gear that can be individually worn or carried includes cushion sole socks; combat boots or mountain boots, if available; BDU and cap; LCE with canteens, magazine pouches, and first aid kit; individual weapon; a large rucksack containing waterproof coat and trousers, polypropylene top, sweater, or fleece top; helmet; poncho; and sleeping bag.

> **CAUTION**
> Cotton clothing, due to its poor insulating and moisture-wicking characteristics, is virtually useless in most mountain climates, the exception being hot, desert, or jungle mountain environments. Cotton clothing should be replaced with synthetic fabric clothing.

c. **Mountaineering Equipment and Specialized Gear.** This gear includes:

- Sling rope or climbing harness.
- Utility cord(s).
- Nonlocking carabiners.
- Locking carabiner(s).
- Rappelling gloves.
- Rappel/belay device.
- Ice ax.
- Crampons.
- Climbing rope, one per climbing team.
- Climbing rack, one per climbing team.

d. **Day Pack.** When the soldier plans to be away from the bivouac site for the day on a patrol or mountaineering mission, he carries a light day pack. This pack should contain the following items:

- Extra insulating layer: polypropylene, pile top, or sweater.
- Protective layer: waterproof jacket and pants, rain suit, or poncho.
- First aid kit.
- Flashlight or headlamp.
- Canteen.
- Cold weather hat or scarf.
- Rations for the time period away from the base camp.
- Survival kit.
- Sling rope or climbing harness.
- Carabiners.
- Gloves.
- Climbing rope, one per climbing team.
- Climbing rack, one per climbing team.

e. **Squad or Team Safety Pack.** When a squad-sized element leaves the bivouac site, squad safety gear should be carried in addition to individual day packs. This can either be loaded into one rucksack or cross-loaded among the squad members. In the event of an injury, casualty evacuation, or unplanned bivouac, these items may make the difference between success and failure of the mission.

- Sleeping bag.
- Sleeping mat.

- Squad stove.
- Fuel bottle.

f. **The Ten Essentials.** Regardless of what equipment is carried, the individual military mountaineer should always carry the "ten essentials" when moving through the mountains.

(1) ***Map.***

(2) ***Compass, Altimeter, and or GPS.***

(3) ***Sunglasses and Sunscreen.***

(a) In alpine or snow-covered sub-alpine terrain, sunglasses are a vital piece of equipment for preventing snow blindness. They should filter 95 to 100 percent of ultraviolet light. Side shields, which minimize the light entering from the side, should permit ventilation to help prevent lens fogging. At least one extra pair of sunglasses should be carried by each independent climbing team.

(b) Sunscreens should have an SPF factor of 15 or higher. For lip protection, a total UV blocking lip balm that resists sweating, washing, and licking is best. This lip protection should be carried in the chest pocket or around the neck to allow frequent reapplication.

(4) ***Extra Food.*** One day's worth extra of food should be carried in case of delay caused by bad weather, injury, or navigational error.

(5) ***Extra Clothing.*** The clothing used during the active part of a climb, and considered to be the basic climbing outfit, includes socks, boots, underwear, pants, blouse, sweater or fleece jacket, hat, gloves or mittens, and foul weather gear (waterproof, breathable outerwear or waterproof rain suit).

(a) Extra clothing includes additional layers needed to make it through the long, inactive hours of an unplanned bivouac. Keep in mind the season when selecting this gear.

- Extra underwear to switch out with sweat-soaked underwear.
- Extra hats or balaclavas.
- Extra pair of heavy socks.
- Extra pair of insulated mittens or gloves.
- In winter or severe mountain conditions, extra insulation for the upper body and the legs.

(b) To back up foul weather gear, bring a poncho or extra-large plastic trash bag. A reflective emergency space blanket can be used for hypothermia first aid and emergency shelter. Insulated foam pads prevent heat loss while sitting or lying on snow. Finally, a bivouac sack can help by protecting insulating layers from the weather, cutting the wind, and trapping essential body heat inside the sack.

(6) ***Headlamp and or Flashlight.*** Headlamps provide the climber a hands-free capability, which is important while climbing, working around the camp, and employing weapons systems. Miniature flashlights can be used, but commercially available headlamps are best. Red lens covers can be fabricated for tactical conditions. Spare batteries and spare bulbs should also be carried.

(7) ***First-aid Kit.*** Decentralized operations, the mountain environment—steep, slick terrain and loose rock combined with heavy packs, sharp tools, and fatigue—requires each climber to carry his own first-aid kit. Common mountaineering injuries that can be expected are punctures and abrasions with severe bleeding, a broken bone, serious sprain,

and blisters. Therefore, the kit should contain at least enough material to stabilize these conditions. Pressure dressings, gauze pads, elastic compression wrap, small adhesive bandages, butterfly bandages, moleskin, adhesive tape, scissors, cleanser, latex gloves and splint material (if above tree line) should all be part of the kit.

(8) ***Fire Starter.*** Fire starting material is key to igniting wet wood for emergency campfires. Candles, heat tabs, and canned heat all work. These can also be used for quick warming of water or soup in a canteen cup. In alpine zones above tree line with no available firewood, a stove works as an emergency heat source.

(9) ***Matches and Lighter.*** Lighters are handy for starting fires, but they should be backed up by matches stored in a waterproof container with a strip of sandpaper.

(10) ***Knife.*** A multipurpose pocket tool should be secured with cord to the belt, harness, or pack.

g. **Other Essential Gear.** Other essential gear may be carried depending on mission and environmental considerations.

(1) ***Water and Water Containers.*** These include wide-mouth water bottles for water collection; camel-back type water holders for hands-free hydration; and a small length of plastic tubing for water procurement at snow-melt seeps and rainwater puddles on bare rock.

(2) ***Ice Ax.*** The ice ax is essential for travel on snowfields and glaciers as well as snow-covered terrain in spring and early summer. It helps for movement on steep scree and on brush and heather covered slopes, as well as for stream crossings.

(3) ***Repair Kit.*** A repair kit should include:

- Stove tools and spare parts.
- Duct tape.
- Patches.
- Safety pins.
- Heavy-duty thread.
- Awl and or needles.
- Cord and or wire.
- Small pliers (if not carrying a multipurpose tool).
- Other repair items as needed.

(4) ***Insect Repellent.***

(5) ***Signaling Devices.***

(6) ***Snow Shovel.***

3-8. TIPS ON PACKING

When loading the internal frame pack the following points should be considered.

a. In most cases, speed and endurance are enhanced if the load is carried more by the hips (using the waist belt) and less by the shoulders and back. This is preferred for movement over trails or less difficult terrain. By packing the lighter, more compressible items (sleeping bag, clothing) in the bottom of the rucksack and the heavier gear (stove, food, water, rope, climbing hardware, extra ammunition) on top, nearer the shoulder blades, the load is held high and close to the back, thus placing the most weight on the hips.

b. In rougher terrain it pays to modify the pack plan. Heavy articles of gear are placed lower in the pack and close to the back, placing more weight on the shoulders and back. This lowers the climber's center of gravity and helps him to better keep his balance.

c. Equipment that may be needed during movement should be arranged for quick access using either external pockets or placing immediately underneath the top flap of the pack. As much as possible, this placement should be standardized across the team so that necessary items can be quickly reached without unnecessary unpacking of the pack in emergencies.

d. The pack and its contents should be soundly waterproofed. Clothing and sleeping bag are separately sealed and then placed in the larger wet weather bag that lines the rucksack. Zip-lock plastic bags can be used for small items, which are then organized into color-coded stuffsacks. A few extra-large plastic garbage bags should be carried for a variety of uses—spare waterproofing, emergency bivouac shelter, and water procurement, among others.

e. The ice ax, if not carried in hand, should be stowed on the outside of the pack with the spike up and the adze facing forward or to the outside, and be securely fastened. Mountaineering packs have ice ax loops and buckle fastening systems for this. If not, the ice ax is placed behind one of the side pockets, as stated above, and then tied in place.

f. Crampons should be secured to the outside rear of the pack with the points covered.

CHAPTER 4

ROPE MANAGEMENT AND KNOTS

The rope is a vital piece of equipment to the mountaineer. When climbing, rappelling, or building various installations, the mountaineer must know how to properly utilize and maintain this piece of equipment. If the rope is not managed or maintained properly, serious injury may occur. This chapter discusses common rope terminology, management techniques, care and maintenance procedures, and knots.

SECTION I. PREPARATION, CARE AND MAINTENANCE, INSPECTION, TERMINOLOGY

The service life of a rope depends on the frequency of use, applications (rappelling, climbing, rope installations), speed of descent, surface abrasion, terrain, climate, and quality of maintenance. Any rope may fail under extreme conditions (shock load, sharp edges, misuse).

4-1. PREPARATION

The mountaineer must select the proper rope for the task to be accomplished according to type, diameter, length, and tensile strength. It is important to prepare all ropes before departing on a mission. Avoid rope preparation in the field.

a. **Packaging.** New rope comes from the manufacturer in different configurations—boxed on a spool in various lengths, or coiled and bound in some manner. Precut ropes are usually packaged in a protective cover such as plastic or burlap. Do not remove the protective cover until the rope is ready for use.

b. **Securing the Ends of the Rope:** If still on a spool, the rope must be cut to the desired length. All ropes will fray at the ends unless they are bound or seared. Both static and dynamic rope ends are secured in the same manner. The ends must be heated to the melting point so as to attach the inner core strands to the outer sheath. By fusing the two together, the sheath cannot slide backward or forward. Ensure that this is only done to the ends of the rope. If the rope is exposed to extreme temperatures, the sheath could be weakened, along with the inner core, reducing overall tensile strength. The ends may also be dipped in enamel or lacquer for further protection.

4-2. CARE AND MAINTENANCE

The rope is a climber's lifeline. It must be cared for and used properly. These general guidelines should be used when handling ropes.

a. Do not step on or drag ropes on the ground unnecessarily. Small particles of dirt will be ground between the inner strands and will slowly cut them.

b. While in use, do not allow the rope to come into contact with sharp edges. Nylon rope is easily cut, particularly when under tension. If the rope must be used over a sharp edge, pad the edge for protection.

c. Always keep the rope as dry as possible. Should the rope become wet, hang it in large loops off the ground and allow it to dry. Never dry a rope with high heat or in direct sunlight.

d. Never leave a rope knotted or tightly stretched for longer than necessary. Over time it will reduce the strength and life of the rope.

e. Never allow one rope to continuously rub over or against another. Allowing rope-on-rope contact with nylon rope is extremely dangerous because the heat produced by the friction will cause the nylon to melt.

f. Inspect the rope before each use for frayed or cut spots, mildew or rot, or defects in construction (new rope).

g. The ends of the rope should be whipped or melted to prevent unraveling.

h. Do not splice ropes for use in mountaineering.

i. Do not mark ropes with paints or allow them to come in contact with oils or petroleum products. Some of these will weaken or deteriorate nylon.

j. Never use a mountaineering rope for any purpose except mountaineering.

k. Each rope should have a corresponding rope log (DA Form 5752-R, Rope History and Usage), which is also a safety record. It should annotate use, terrain, weather, application, number of falls, dates, and so on, and should be annotated each time the rope is used (Figure 4-1). DA Form 5752-R is authorized for local reproduction on 8 1/2- by 11-inch paper.

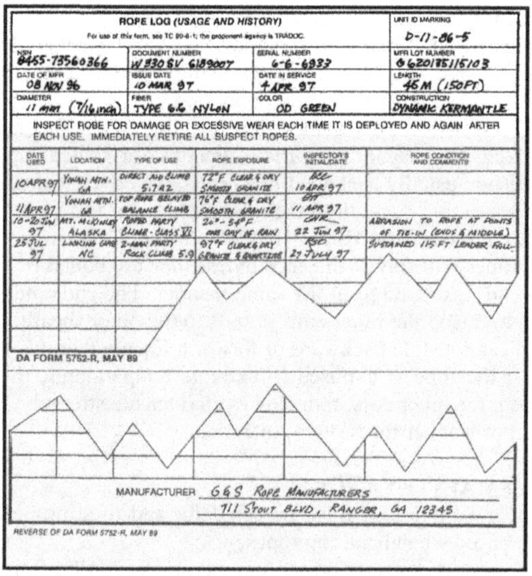

ROPE LOG (USAGE AND HISTORY)

For use of this form, see TC 90-6-1; the proponent agency is TRADOC.

UNIT ID MARKING: D-11-86-5

NSN	DOCUMENT NUMBER	SERIAL NUMBER	MFR LOT NUMBER
8455-73560366	W330SV 6189007	6-6-6933	0620175115103
DATE OF MFR	ISSUE DATE	DATE IN SERVICE	LENGTH
08 NOV 96	10 MAR 97	4 APR 97	45 M (150 FT)
DIAMETER	FIBER	COLOR	CONSTRUCTION
11 mm (7/16 inch)	TYPE 6.6 NYLON	OD GREEN	DYNAMIC KERMANTLE

INSPECT ROBE FOR DAMAGE OR EXCESSIVE WEAR EACH TIME IT IS DEPLOYED AND AGAIN AFTER EACH USE. IMMEDIATELY RETIRE ALL SUSPECT ROPES.

DATE USED	LOCATION	TYPE OF USE	ROPE EXPOSURE	INSPECTOR'S INITIAL/DATE	ROPE CONDITION AND COMMENTS
10 APR 97	YONAH MTN. GA	DIRECT AID CLIMB 5.7 A2	72°F CLEAR & DRY SMOOTH GRANITE	BCC 10 APR 97	
11 APR 97	YONAH MTN. GA	TOP ROPE BELAYED BALANCE CLIMB	76°F CLEAR & DRY SMOOTH GRANITE	[illegible] 11 APR 97	
10-20 JUN 97	MT. MCKINLEY ALASKA	ROPED PARTY CLIMB - CLASS VI	20°-54°F ONE DAY OF RAIN	CHR 22 JUN 97	ABRASION TO ROPE AT POINTS OF TIE-IN (ENDS & MIDDLE)
25 JUL 97	LOOKING GLASS NC	2-MAN PARTY ROCK CLIMB 5.9	97°F CLEAR & DRY GRANITE & QUARTZITE	RSD 27 JULY 97	SUSTAINED 115 FT LEADER FALL

DA FORM 5752-R, MAY 89

MANUFACTURER G & S ROPE MANUFACTURERS
711 STOUT BLVD, RANGER, GA 12345

REVERSE OF DA FORM 5752-R, MAY 89

Figure 4-1. Example of completed DA Form 5752-R.

l. Never subject the rope to high heat or flame. This will significantly weaken it.

m. All ropes should be washed periodically to remove dirt and grit, and rinsed thoroughly. Commercial rope washers are made from short pieces of modified pipe that connect to any faucet. Pinholes within the pipe force water to circulate around and scrub the rope as you slowly feed it through the washer. Another method is to machine wash, on a gentle cycle, in cold water with a nylon safe soap, never bleach or harsh cleansers. Ensure that only front loading washing machine are used to wash ropes.

n. Ultraviolet radiation (sunlight) tends to deteriorate nylon over long periods of time. This becomes important if rope installations are left in place over a number of months.

o. When not in use, ropes should be loosely coiled and hung on wooden pegs rather than nails or other metal objects. Storage areas should be relatively cool with low humidity levels to prevent mildew or rotting. Rope may also be loosely stacked and placed in a rope bag and stored on a shelf. Avoid storage in direct sunlight, as the ultraviolet radiation will deteriorate the nylon over long periods

4-3. INSPECTION

Ropes should be inspected before and after each use, especially when working around loose rock or sharp edges.

a. Although the core of the kernmantle rope cannot be seen, it is possible to damage the core without damaging the sheath. Check a kernmantle rope by carefully inspecting the sheath before and after use while the rope is being coiled. When coiling, be aware of how the rope feels as it runs through the hands. Immediately note and tie off any lumps or depressions felt.

b. Damage to the core of a kernmantle rope usually consists of filaments or yarn breakage that results in a slight retraction. If enough strands rupture, a localized reduction in the diameter of the rope results in a depression that can be felt or even seen.

c. Check any other suspected areas further by putting them under tension (the weight of one person standing on a Prusik tensioning system is about maximum). This procedure will emphasize the lump or depression by separating the broken strands and enlarging the dip. If a noticeable difference in diameter is obvious, retire the rope immediately.

d. Many dynamic kernmantle ropes are quite soft. They may retain an indention occasionally after an impact or under normal use without any trauma to the core. When damage is suspected, patiently inspect the sheath for abnormalities. Damage to the sheath does not always mean damage to the core. Inspect carefully.

4-4. TERMINOLOGY

When using ropes, understanding basic terminology is important. The terms explained in this section are the most commonly used in military mountaineering. (Figure 4-2, page 4-4, illustrates some of these terms.)

a. **Bight**. A bight of rope is a simple bend of rope in which the rope does not cross itself.

b. **Loop.** A loop is a bend of a rope in which the rope does cross itself.

c. **Half Hitch**. A half hitch is a loop that runs around an object in such a manner as to lock or secure itself.

d. **Turn.** A turn wraps around an object, providing 360-degree contact.

e. **Round Turn**. A round turn wraps around an object one and one-half times. A round turn is used to distribute the load over a small diameter anchor (3 inches or less). It may also be used around larger diameter anchors to reduce the tension on the knot, or provide added friction.

f. **Running End**. A running end is the loose or working end of the rope.

g. **Standing Part**. The standing part is the static, stationary, or nonworking end of the rope.

h. **Lay.** The lay is the direction of twist used in construction of the rope.

i. **Pigtail.** The pigtail (tail) is the portion of the running end of the rope between the safety knot and the end of the rope.

j. **Dress.** Dress is the proper arrangement of all the knot parts, removing unnecessary kinks, twists, and slack so that all rope parts of the knot make contact.

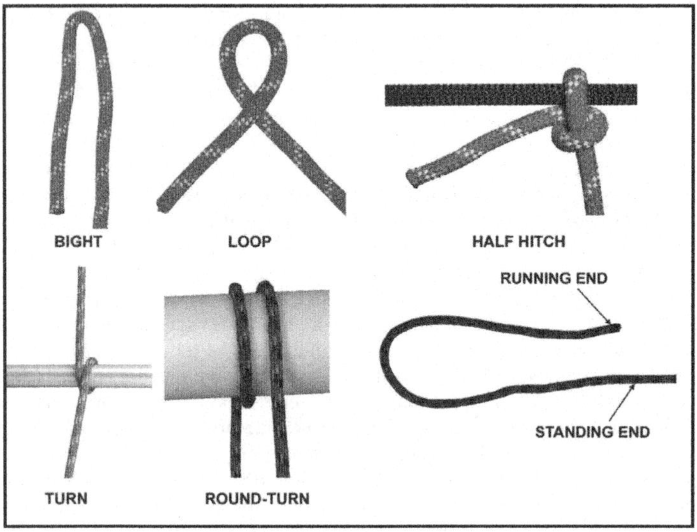

Figure 4-2. Examples of roping terminology.

Section II. COILING, CARRYING, THROWING

The ease and speed of rope deployment and recovery greatly depends upon technique and practice.

4-5. COILING AND CARRYING THE ROPE

Use the butterfly or mountain coil to coil and carry the rope. Each is easy to accomplish and results in a minimum amount of kinks, twists, and knots later during deployment.

a. **Mountain Coil.** To start a mountain coil, grasp the rope approximately 1 meter from the end with one hand. Run the other hand along the rope until both arms are outstretched. Grasping the rope firmly, bring the hands together forming a loop, which is laid in the hand closest to the end of the rope. This is repeated, forming uniform loops that run in a clockwise direction, until the rope is completely coiled. The rope may be given a 1/4 twist as each loop is formed to overcome any tendency for the rope to twist or form figure-eights.

(1) In finishing the mountain coil, form a bight approximately 30 centimeters long with the starting end of the rope and lay it along the top of the coil. Uncoil the last loop and, using this length of the rope, begin making wraps around the coil and the bight, wrapping toward the closed end of the bight and making the first wrap bind across itself so as to lock it into place. Make six to eight wraps to adequately secure the coil, and then route the end of the rope through the closed end of the bight. Pull the running end of the bight tight, securing the coil.

(2) The mountain coil may be carried either in the pack (by forming a figure eight), doubling it and placing it under the flap, or by placing it over the shoulder and under the opposite arm, slung across the chest. (Figure 4-3 shows how to coil a mountain coil.)

Figure 4-3. Mountain coil.

b. **Butterfly Coil.** The butterfly coil is the quickest and easiest technique for coiling (Figure 4-4).

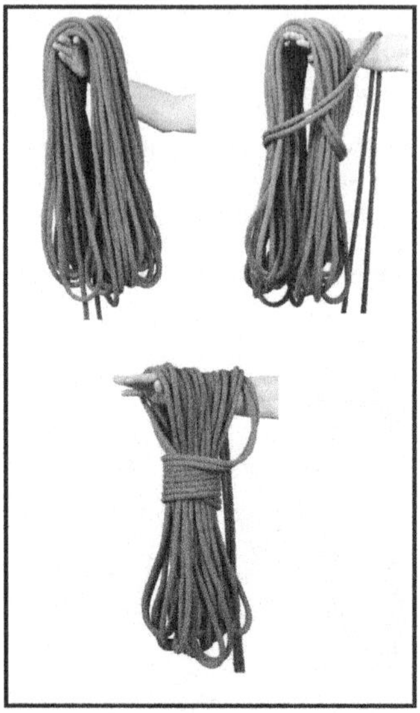

Figure 4-4. Butterfly coil.

(1) ***Coiling.*** To start the double butterfly, grasp both ends of the rope and begin back feeding. Find the center of the rope forming a bight. With the bight in the left hand, grasp both ropes and slide the right hand out until there is approximately one arms length of rope. Place the doubled rope over the head, draping it around the neck and on top of the shoulders. Ensure that it hangs no lower than the waist. With the rest of the doubled rope in front of you, make doubled bights placing them over the head in the same manner as the first bight. Coil alternating from side to side (left to right, right to left) while maintaining equal-length bights. Continue coiling until approximately two arm-lengths of rope remain. Remove the coils from the neck and shoulders carefully, and hold the center in one hand. Wrap the two ends around the coils a minimum of three doubled wraps, ensuring that the first wrap locks back on itself.

(2) ***Tie-off and Carrying.*** Take a doubled bight from the loose ends of rope and pass it through the apex of the coils. Pull the loose ends through the doubled bight and dress it down. Place an overhand knot in the loose ends, dressing it down to the apex of the bight securing coils. Ensure that the loose ends do not exceed the length of the coils. In this configuration the coiled rope is secure enough for hand carrying or carrying in a rucksack, or for storage. (Figure 4-5 shows a butterfly coil tie-off.)

Figure 4-5. Butterfly coil tie-off.

c. **Coiling Smaller Diameter Rope.** Ropes of smaller diameters may be coiled using the butterfly or mountain coil depending on the length of the rope. Pieces 25 feet and shorter (also known as cordage, sling rope, utility cord) may be coiled so that they can be hung from the harness. Bring the two ends of the rope together, ensuring no kinks are in the rope. Place the ends of the rope in the left hand with the two ends facing the body. Coil the doubled rope in a clockwise direction forming 6- to 8-inch coils (coils may be larger depending on the length of rope) until an approximate 12-inch bight is left. Wrap that bight around the coil, ensuring that the first wrap locks on itself. Make three or more wraps. Feed the bight up through the bights formed at the top of the coil. Dress it down tightly. Now the piece of rope may be hung from a carabiner on the harness.

e. **Uncoiling, Back-feeding, and Stacking.** When the rope is needed for use, it must be uncoiled and stacked on the ground properly to avoid kinks and snarls.

(1) Untie the tie-off and lay the coil on the ground. Back-feed the rope to minimize kinks and snarls. (This is also useful when the rope is to be moved a short distance and

coiling is not desired.) Take one end of the rope in the left hand and run the right hand along the rope until both arms are outstretched. Next, lay the end of the rope in the left hand on the ground. With the left hand, re-grasp the rope next to the right hand and continue laying the rope on the ground.

(2) The rope should be laid or stacked in a neat pile on the ground to prevent it from becoming tangled and knotted when throwing the rope, feeding it to a lead climber, and so on. This technique can also be started using the right hand.

4-6. THROWING THE ROPE

Before throwing the rope, it must be properly managed to prevent it from tangling during deployment. The rope should first be anchored to prevent complete loss of the rope over the edge when it is thrown. Several techniques can be used when throwing a rope. Personal preference and situational and environmental conditions should be taken into consideration when determining which technique is best.

a. Back feed and neatly stack the rope into coils beginning with the anchored end of the rope working toward the running end. Once stacked, make six to eight smaller coils in the left hand. Pick up the rest of the larger coils in the right hand. The arm should be generally straight when throwing. The rope may be thrown underhanded or overhanded depending on obstacles around the edge of the site. Make a few preliminary swings to ensure a smooth throw. Throw the large coils in the right hand first. Throw up and out. A slight twist of the wrist, so that the palm of the hand faces up as the rope is thrown, allows the coils to separate easily without tangling. A smooth follow through is essential. When a slight tug on the left hand is felt, toss the six to eight smaller coils out. This will prevent the ends of the rope from becoming entangled with the rest of the coils as they deploy. As soon as the rope leaves the hand, the thrower should sound off with a warning of "ROPE" to alert anyone below the site.

b. Another technique may also be used when throwing rope. Anchor, back feed, and stack the rope properly as described above. Take the end of the rope and make six to eight helmet-size coils in the right hand (more may be needed depending on the length of the rope). Assume a "quarterback" simulated stance. Aiming just above the horizon, vigorously throw the rope overhanded, up and out toward the horizon. The rope must be stacked properly to ensure smooth deployment.

c. When windy weather conditions prevail, adjustments must be made. In a strong cross wind, the rope should be thrown angled into the wind so that it will land on the desired target. The stronger the wind, the harder the rope must be thrown to compensate.

SECTION III. KNOTS

All knots used by a mountaineer are divided into four classes: Class I—joining knots, Class II—anchor knots, Class III—middle rope knots, and Class IV—special knots. The variety of knots, bends, bights, and hitches is almost endless. These classes of knots are intended only as a general guide since some of the knots discussed may be appropriate in more than one class. The skill of knot tying can perish if not used and practiced. With experience and practice, knot tying becomes instinctive and helps the mountaineer in many situations.

4-7. SQUARE KNOT

The square knot is used to tie the ends of two ropes of equal diameter (Figure 4-6). It is a joining knot.

a. **Tying the Knot.**

STEP 1. Holding one working end in each hand, place the working end in the right hand over the one in the left hand.

STEP 2. Pull it under and back over the top of the rope in the left hand.

STEP 3. Place the working end in the left hand over the one in the right hand and repeat STEP 2.

STEP 4. Dress the knot down and secure it with an overhand knot on each side of the square knot.

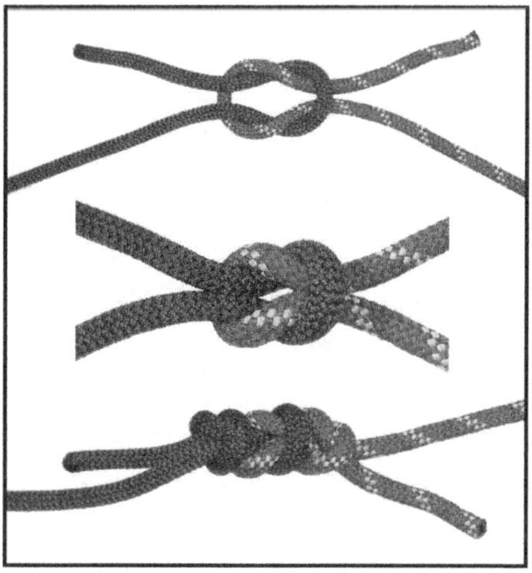

Figure 4-6. Square knot.

b. **Checkpoints.**

(1) There are two interlocking bights.

(2) The running end and standing part are on the same side of the bight formed by the other rope.

(3) The running ends are parallel to and on the same side of the standing ends with 4-inch minimum pig tails after the overhand safeties are tied.

4-8. FISHERMAN'S KNOT

The fisherman's knot is used to tie two ropes of the same or approximately the same diameter (Figure 4-7, page 4-10). It is a joining knot.

a. **Tying the Knot.**

STEP 1. Tie an overhand knot in one end of the rope.

STEP 2. Pass the working end of the other rope through the first overhand knot. Tie an overhand knot around the standing part of the first rope with the working end of the second rope.

STEP 3. Tightly dress down each overhand knot and tightly draw the knots together.

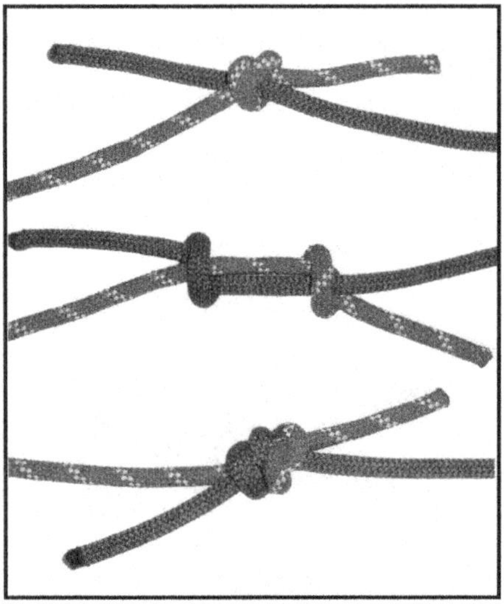

Figure 4-7. Fisherman's knot.

b. **Checkpoints.**

(1) The two separate overhand knots are tied tightly around the long, standing part of the opposing rope.

(2) The two overhand knots are drawn snug.

(3) Ends of rope exit knot opposite each other with 4-inch pigtails.

4-9. DOUBLE FISHERMAN'S KNOT

The double fisherman's knot (also called double English or grapevine) is used to tie two ropes of the same or approximately the same diameter (Figure 4-8). It is a joining knot.

a. **Tying the Knot.**

STEP 1. With the working end of one rope, tie two wraps around the standing part of another rope.

STEP 2. Insert the working end (STEP 1) back through the two wraps and draw it tight.

STEP 3. With the working end of the other rope, which contains the standing part (STEPS 1 and 2), tie two wraps around the standing part of the other rope (the working end in STEP 1). Insert the working end back through the two wraps and draw tight.

STEP 4. Pull on the opposing ends to bring the two knots together.

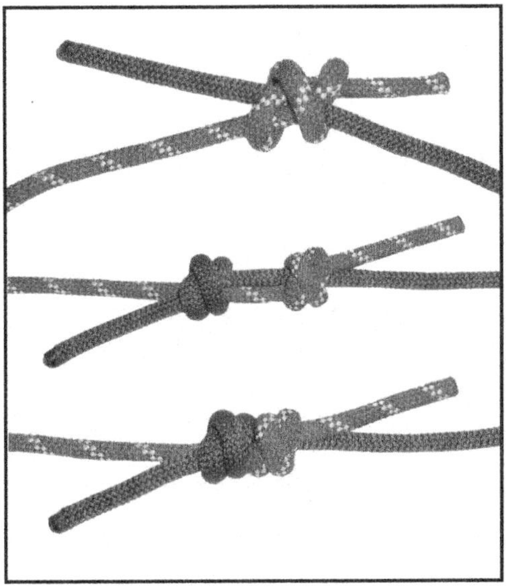

Figure 4-8. Double fisherman's knot.

b. **Checkpoints.**

(1) Two double overhand knots securing each other as the standing parts of the rope are pulled apart.

(2) Four rope parts on one side of the knot form two "x" patterns, four rope parts on the other side of the knot are parallel.

(3) Ends of rope exit knot opposite each other with 4-inch pigtails.

4-10. FIGURE-EIGHT BEND

The figure-eight bend is used to join the ends of two ropes of equal or unequal diameter within 5-mm difference (Figure 4-9, page 4-12).

a. **Tying the Knot.**

STEP 1. Grasp the top of a 2-foot bight.

STEP 2. With the other hand, grasp the running end (short end) and make a 360-degree turn around the standing end.

STEP 3. Place the running end through the loop just formed creating an in-line figure eight.

STEP 4. Route the running end of the other ripe back through the figure eight starting from the original rope's running end. Trace the original knot to the standing end.

STEP 5. Remove all unnecessary twists and crossovers. Dress the knot down.

Figure 4-9. Figure-eight bend.

b. **Checkpoints.**

(1) There is a figure eight with two ropes running side by side.

(2) The running ends are on opposite sides of the knot.

(3) There is a minimum 4-inch pigtail.

4-11. WATER KNOT

The water knot is used to attach two webbing ends (Figure 4-10). It is also called a ring bend, overhand retrace, or tape knot. It is used in runners and harnesses and is a joining knot.

a. **Tying the Knot.**

STEP 1. Tie an overhand knot in one of the ends.

STEP 2. Feed the other end back through the knot, following the path of the first rope in reverse.

STEP 3. Draw tight and pull all of the slack out of the knot. The remaining tails must extend at least 4 inches beyond the knot in both directions.

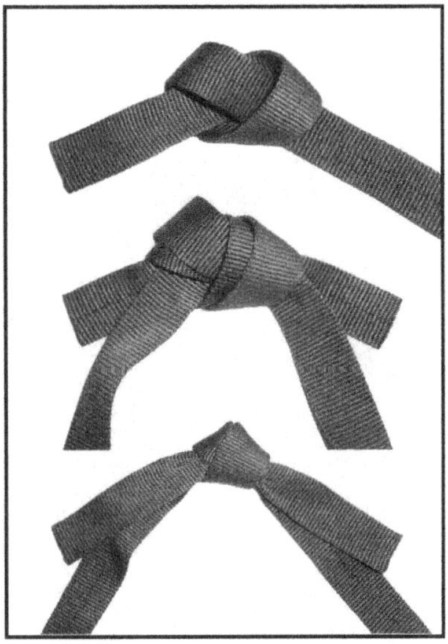

Figure 4-10. Water knot.

b. **Checkpoints.**

(1) There are two overhand knots, one retracing the other.

(2) There is no slack in the knot, and the working ends come out of the knot in opposite directions.

(3) There is a minimum 4-inch pigtail.

4-12. BOWLINE

The bowline is used to tie the end of a rope around an anchor. It may also be used to tie a single fixed loop in the end of a rope (Figure 4-11, page 4-14). It is an anchor knot.

a. **Tying the Knot.**

STEP 1. Bring the working end of the rope around the anchor, from right to left (as the climber faces the anchor).

STEP 2. Form an overhand loop in the standing part of the rope (on the climber's right) toward the anchor.

STEP 3. Reach through the loop and pull up a bight.

STEP 4. Place the working end of the rope (on the climber's left) through the bight, and bring it back onto itself. Now dress the knot down.

STEP 5. Form an overhand knot with the tail from the bight.

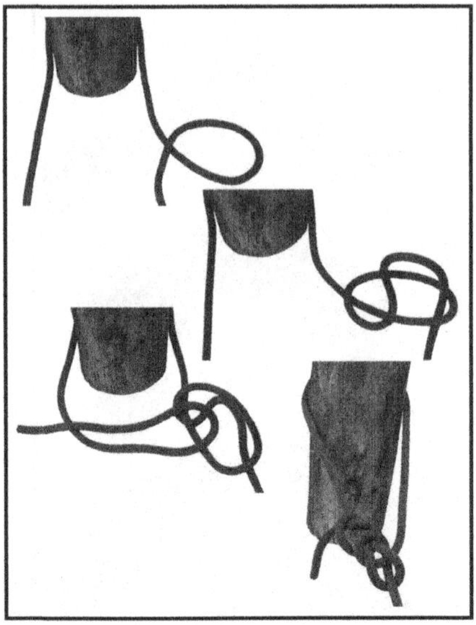

Figure 4-11. Bowline knot.

b. **Checkpoints.**

(1) The bight is locked into place by a loop.

(2) The short portion of the bight is on the inside and on the loop around the anchor (or inside the fixed loop).

(3) There is a minimum 4-inch pigtail after tying the overhand safety.

4-13. ROUND TURN AND TWO HALF HITCHES

This knot is used to tie the end of a rope to an anchor, and it must have constant tension (Figure 4-12). It is an anchor knot.

a. **Tying the Knot.**

STEP 1. Route the rope around the anchor from right to left and wrap down (must have two wraps in the rear of the anchor, and one in the front). Run the loop around the object to provide 360-degree contact, distributing the load over the anchor.

STEP 2. Bring the working end of the rope left to right and over the standing part, forming a half hitch (first half hitch).

STEP 3. Repeat STEP 2 (last half hitch has a 4 inch pigtail).

STEP 4. Dress the knot down.

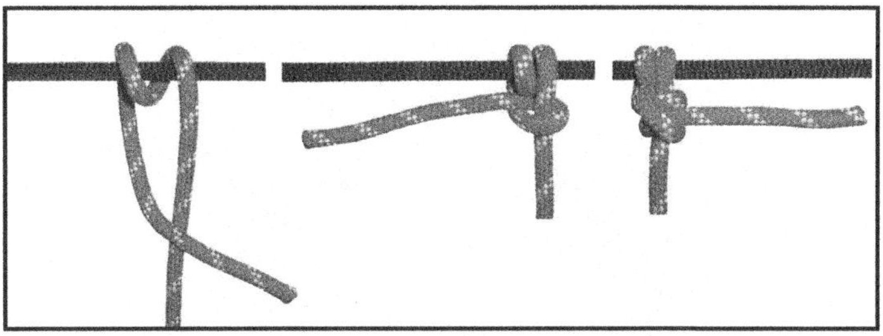

Figure 4-12. Round turn and two half hitches.

b. **Checkpoints.**

(1) A complete round turn should exist around the anchor with no crosses.

(2) Two half hitches should be held in place by a diagonal locking bar with no less than a 4-inch pigtail remaining.

4-14. FIGURE-EIGHT RETRACE (REROUTED FIGURE-EIGHT)

The figure-eight retrace knot produces the same result as a figure-eight loop. However, by tying the knot in a retrace, it can be used to fasten the rope to trees or to places where the loop cannot be used (Figure 4-13, page 4-16). It is also called a rerouted figure-eight and is an anchor knot.

a. **Tying the Knot.**

STEP 1. Use a length of rope long enough to go around the anchor, leaving enough rope to work with.

STEP 2. Tie a figure-eight knot in the standing part of the rope, leaving enough rope to go around the anchor. To tie a figure-eight knot form a loop in the rope, wrap the working end around the standing part, and route the working end through the loop. The finished knot is dressed loosely.

STEP 3. Take the working end around the anchor point.

STEP 4. With the working end, insert the rope back through the loop of the knot in reverse.

STEP 5. Keep the original figure eight as the outside rope and retrace the knot around the wrap and back to the long-standing part.

STEP 6. Remove all unnecessary twists and crossovers; dress the knot down.

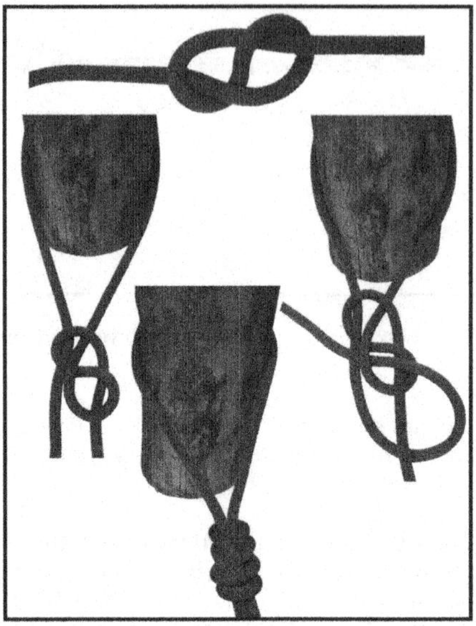

Figure 4-13. Figure-eight retrace.

b. **Checkpoints**

(1) A figure eight with a doubled rope running side by side, forming a fixed loop around a fixed object or harness.

(2) There is a minimum 4-inch pigtail.

4-15. CLOVE HITCH

The clove hitch is an anchor knot that can be used in the middle of the rope as well as at the end (Figure 4-14). The knot must have constant tension on it once tied to prevent slipping. It can be used as either an anchor or middle of the rope knot, depending on how it is tied.

a. **Tying the Knot.**

(1) ***Middle of the Rope.***

STEP 1. Hold rope in both hands, palms down with hands together. Slide the left hand to the left from 20 to 25 centimeters.

STEP 2. Form a loop away from and back toward the right.

STEP 3. Slide the right hand from 20 to 25 centimeters to the right. Form a loop inward and back to the left hand.

STEP 4. Place the left loop on top of the right loop. Place both loops over the anchor and pull both ends of the rope in opposite directions. The knot is tied.

(2) ***End of the Rope.***

Note: For instructional purposes, assume that the anchor is horizontal.

STEP 1. Place 76 centimeters of rope over the top of the anchor. Hold the standing end in the left hand. With the right hand, reach under the horizontal anchor, grasp the working end, and bring it inward.

STEP 2. Place the working end of the rope over the standing end (to form a loop). Hold the loop in the left hand. Place the working end over the anchor from 20 to 25 centimeters to the left of the loop.

STEP 3. With the right hand, reach down to the left hand side of the loop under the anchor. Grasp the working end of the rope. Bring the working end up and outward.

STEP 4. Dress down the knot.

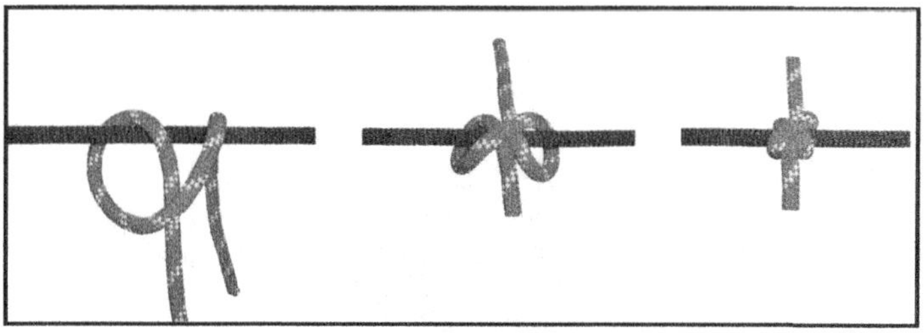

Figure 4-14. Clove hitch.

b. **Checkpoints.**

(1) The knot has two round turns around the anchor with a diagonal locking bar.
(2) The locking bar is facing 90 degrees from the direction of pull.
(3) The ends exit 180 degrees from each other.
(4) The knot has more than a 4-inch pigtail remaining.

4-16. WIREMAN'S KNOT

The wireman's knot forms a single, fixed loop in the middle of the rope (Figure 4-15, page 4-18). It is a middle rope knot.

a. **Tying the Knot.**

STEP 1. When tying this knot, face the anchor that the tie-off system will be tied to. Take up the slack from the anchor, and wrap two turns around the left hand (palm up) from left to right.

STEP 2. A loop of 30 centimeters is taken up in the second round turn to create the fixed loop of the knot.

STEP 3. Name the wraps from the palm to the fingertips: heel, palm, and fingertip.

STEP 4. Secure the palm wrap with the right thumb and forefinger, and place it over the heel wrap.

STEP 5. Secure the heel wrap and place it over the fingertip wrap.

STEP 6. Secure the fingertip wrap and place it over the palm wrap.

STEP 7. Secure the palm wrap and pull up to form a fixed loop.

STEP 8. Dress the knot down by pulling on the fixed loop and the two working ends.

STEP 9. Pull the working ends apart to finish the knot.

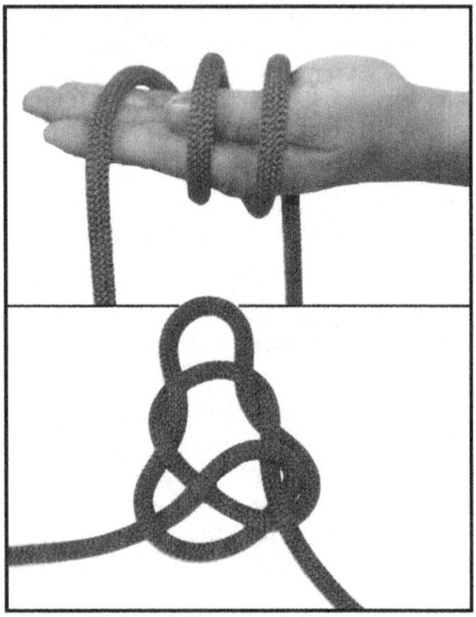

Figure 4-15. Wireman's knot.

b. **Checkpoints.**

(1) The completed knot should have four separate bights locking down on themselves with the fixed loop exiting from the top of the knot and laying toward the near side anchor point.

(2) Both ends should exit opposite each other without any bends.

4-17. DIRECTIONAL FIGURE-EIGHT

The directional figure-eight knot forms a single, fixed loop in the middle of the rope that lays back along the standing part of the rope (Figure 4-16). It is a middle rope knot.

a. **Tying the Knot.**

STEP 1. Face the far side anchor so that when the knot is tied, it lays inward.

STEP 2. Lay the rope from the far side anchor over the left palm. Make one wrap around the palm.

STEP 3. With the wrap thus formed, tie a figure-eight knot around the standing part that leads to the far side anchor.

STEP 4. When dressing the knot down, the tail and the bight must be together.

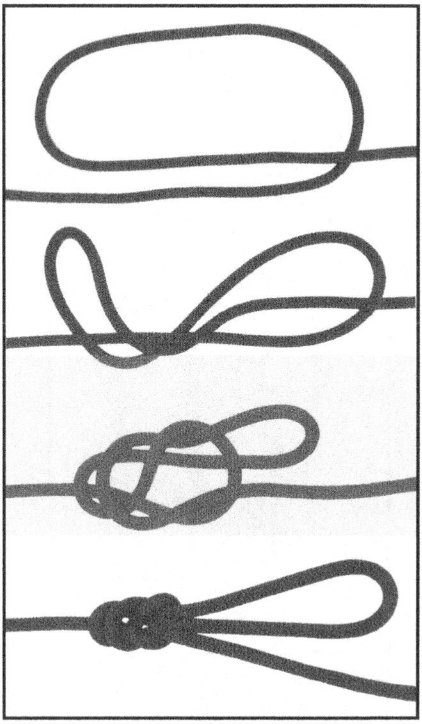

Figure 4-16. Directional figure-eight.

b. **Checkpoints.**

(1) The loop should be large enough to accept a carabiner but no larger than a helmet-size loop.

(2) The tail and bight must be together.

(3) The figure eight is tied tightly.

(4) The bight in the knot faces back toward the near side.

4-18. BOWLINE-ON-A-BIGHT (TWO-LOOP BOWLINE)

The bowline-on-a-bight is used to form two fixed loops in the middle of a rope (Figure 4-17, page 4-20). It is a middle rope knot.

a. **Tying the Knot.**

STEP 1. Form a bight in the rope about twice as long as the finished loops will be.

STEP 2. Tie an overhand knot on a bight.

STEP 3. Hold the overhand knot in the left hand so that the bight is running down and outward.

STEP 4. Grasp the bight with the right hand; fold it back over the overhand knot so that the overhand knot goes through the bight.

STEP 5. From the end (apex) of the bight, follow the bight back to where it forms the cross in the overhand knot. Grasp the two ropes that run down and outward and pull up, forming two loops.

STEP 6. Pull the two ropes out of the overhand knot and dress the knot down.

STEP 7. A final dress is required: grasp the ends of the two fixed loops and pull, spreading them apart to ensure the loops do not slip.

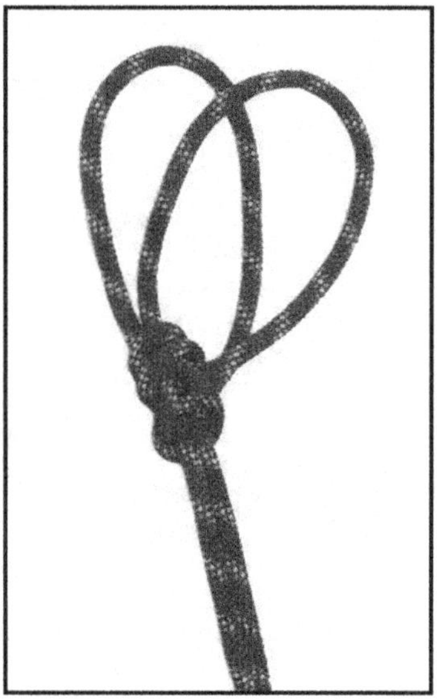

Figure 4-17. Bowline-on-a-bight.

b. **Checkpoints.**

(1) There are two fixed loops that will not slip.

(2) There are no twists in the knot.

(3) A double loop is held in place by a bight.

4-19. TWO-LOOP FIGURE-EIGHT

The two-loop figure-eight is used to form two fixed loops in the middle of a rope (Figure 4-18.) It is a middle rope knot.

a. **Tying the Knot.**

STEP 1. Using a doubled rope, form an 18-inch bight in the left hand with the running end facing to the left.

STEP 2. Grasp the bight with the right hand and make a 360-degree turn around the standing end in a counterclockwise direction.

STEP 3. With the working end, form another bight and place that bight through the loop just formed in the left hand.

STEP 4. Hold the bight with the left hand, and place the original bight (moving toward the left hand) over the knot.

STEP 5. Dress the knot down.

Figure 4-18. Two-loop figure-eight.

b. **Checkpoints.**

(1) There is a double figure-eight knot with two loops that share a common locking bar.

(2) The two loops must be adjustable by means of a common locking bar.

(3) The common locking bar is on the bottom of the double figure-eight knot.

4-20. FIGURE-EIGHT LOOP (FIGURE-EIGHT-ON-A-BIGHT)

The figure-eight loop, also called the figure-eight-on-a-bight, is used to form a fixed loop in a rope (Figure 4-19). It is a middle of the rope knot.

a. **Tying the Knot.**

STEP 1. Form a bight in the rope about as large as the diameter of the desired loop.

STEP 2. With the bight as the working end, form a loop in rope (standing part).

STEP 3. Wrap the working end around the standing part 360 degrees and feed the working end through the loop. Dress the knot tightly.

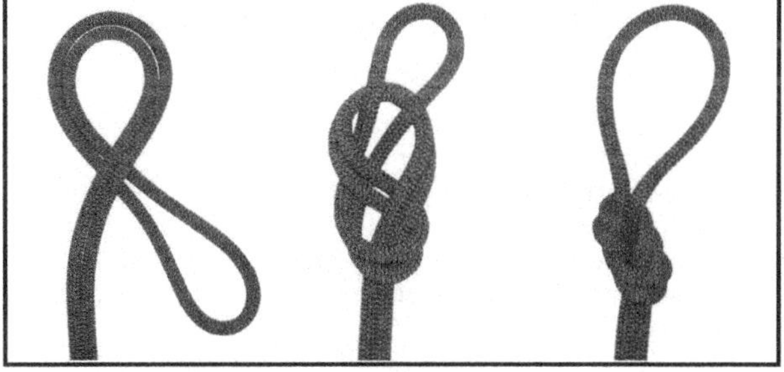

Figure 4-19. Figure-eight loop.

b. **Checkpoints.**

(1) The loop is the desired size.

(2) The ropes in the loop are parallel and do not cross over each other.

(3) The knot is tightly dressed.

4-21. PRUSIK KNOT

The Prusik knot is used to put a moveable rope on a fixed rope such as a Prusik ascent or a tightening system. This knot can be tied as a middle or end of the rope Prusik. It is a specialty knot.

a. **Tying the Knot.**

(1) ***Middle-of-the-Rope Prusik.*** The middle-of-the-rope Prusik knot can be tied with a short rope to a long rope as follows (Figure 4-20.):

STEP 1. Double the short rope, forming a bight, with the working ends even. Lay it over the long rope so that the closed end of the bight is 12 inches below the long rope and the remaining part of the rope (working ends) is the closest to the climber; spread the working end apart.

STEP 2. Reach down through the 12-inch bight. Pull up both of the working ends and lay them over the long rope. Repeat this process making sure that the working ends pass in the middle of the first two wraps. Now there are four wraps and a locking bar working across them on the long rope.

STEP 3. Dress the wraps and locking bar down to ensure they are tight and not twisted. Tying an overhand knot with both ropes will prevent the knot from slipping during periods of variable tension.

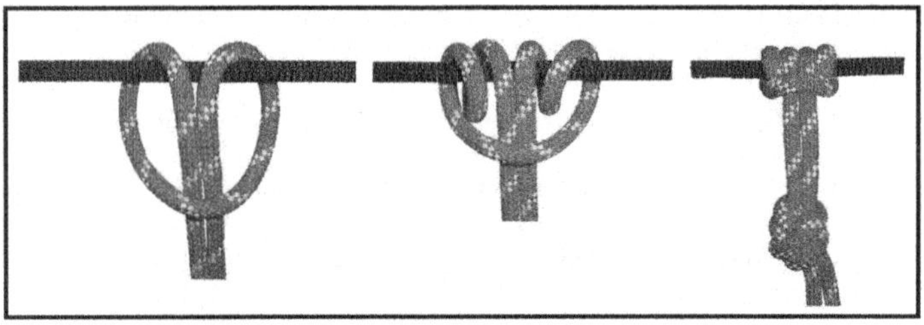

Figure 4-20. Middle-of-the-rope Prusik.

(2) ***End-of-the-Rope Prusik*** (Figure 4-21).

STEP 1. Using an arm's length of rope, and place it over the long rope.

STEP 2. Form a complete round turn in the rope.

STEP 3. Cross over the standing part of the short rope with the working end of the short rope.

STEP 4. Lay the working end under the long rope.

STEP 5. Form a complete round turn in the rope, working back toward the middle of the knot.

STEP 6. There are four wraps and a locking bar running across them on the long rope. Dress the wraps and locking bar down. Ensure they are tight, parallel, and not twisted.

STEP 7. Finish the knot with a bowline to ensure that the Prusik knot will not slip out during periods of varying tension.

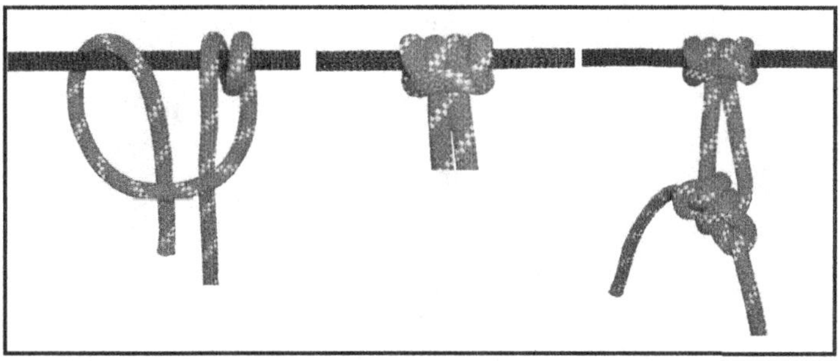

Figure 4-21. End-of-the-rope Prusik knot.

b. **Checkpoints.**

(1) Four wraps with a locking bar.

(2) The locking bar faces the climber.

(3) The knot is tight and dressed down with no ropes twisted or crossed.

(4) Other than a finger Prusik, the knot should contain an overhand or bowline to prevent slipping.

4-22. BACHMAN KNOT

The Bachman knot provides a means of using a makeshift mechanized ascender (Figure 4-22, page 4-24). It is a specialty knot.

a. **Tying the Knot.**

STEP 1. Find the middle of a utility rope and insert it into a carabiner.

STEP 2. Place the carabiner and utility rope next to a long climbing rope.

STEP 3. With the two ropes parallel from the carabiner, make two or more wraps around the climbing rope and through the inside portion of the carabiner.

Note: The rope can be tied into an etrier (stirrup) and used as a Prusik-friction principle ascender.

b. **Checkpoints.**

(1) The bight of the climbing rope is at the top of the carabiner.

(2) The two ropes run parallel without twisting or crossing.

(3) Two or more wraps are made around the long climbing rope and through the inside portion of the carabiner.

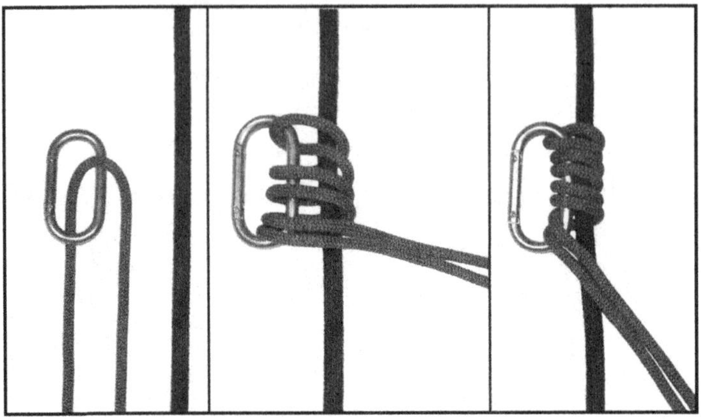

Figure 4-22. Bachman knot.

4-23. BOWLINE-ON-A-COIL

The bowline-on-a-coil is an expedient tie-in used by climbers when a climbing harness is not available (Figure 4-23). It is a specialty knot.

a. **Tying the Knot.**

STEP 1. With the running end, place 3 feet of rope over your right shoulder. The running end is to the back of the body.

STEP 2. Starting at the bottom of your rib cage, wrap the standing part of the rope around your body and down in a clockwise direction four to eight times.

STEP 3. With the standing portion of the rope in your left hand, make a clockwise loop toward the body. The standing portion is on the bottom.

STEP 4. Ensuring the loop does not come uncrossed, bring it up and under the coils between the rope and your body.

STEP 5. Using the standing part, bring a bight up through the loop. Grasp the running end of the rope with the right hand. Pass it through the bight from right to left and back on itself.

STEP 6. Holding the bight loosely, dress the knot down by pulling on the standing end.

STEP 7. Safety the bowline with an overhand around the top, single coil. Then, tie an overhand around all coils, leaving a minimum 4-inch pigtail.

b. **Checkpoints.**

(1) A minimum of four wraps, not crossed, with a bight held in place by a loop.

(2) The loop must be underneath all wraps.

(3) A minimum 4-inch pigtail after the second overhand safety is tied.

(4) Must be centered on the mid-line of the body.

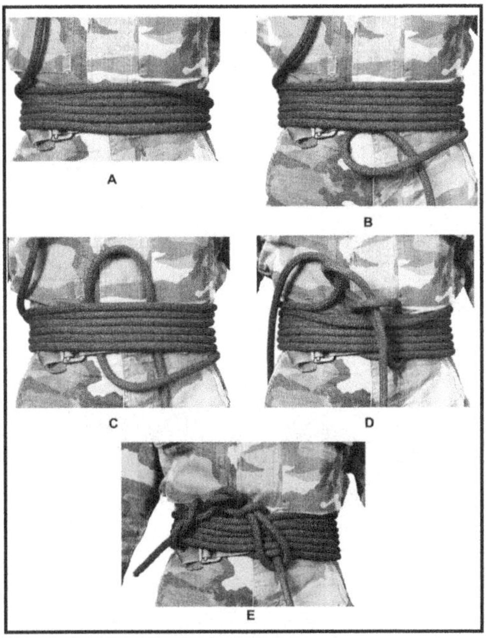

Figure 4-23. Bowline-on-a-coil.

4-24. THREE-LOOP BOWLINE

The three-loop bowline is used to form three fixed loops in the middle of a rope (Figure 4-24, page 4-26). It is used in a self-equalizing anchor system. It is a specialty knot.

a. **Tying the Knot.**

STEP 1. Form an approximate 24-inch bight.

STEP 2. With the right thumb facing toward the body, form a doubled loop in the standing part by turning the wrist clockwise. Lay the loops to the right.

STEP 3. With the right hand, reach down through the loops and pull up a doubled bight from the standing part of the rope.

STEP 4. Place the running end (bight) of the rope (on the left) through the doubled bight from left to right and bring it back on itself. Hold the running end loosely and dress the knot down by pulling on the standing parts.

STEP 5. Safety it off with a doubled overhand knot.

Figure 4-24. Three-loop bowline.

b. **Checkpoints.**

(1) There are two bights held in place by two loops.

(2) The bights form locking bars around the standing parts.

(3) The running end (bight) must be on the inside of the fixed loops.

(4) There is a minimum 4-inch pigtail after the double overhand safety knot is tied.

4-25. FIGURE-EIGHT SLIP KNOT

The figure eight slip knot forms an adjustable bight in a rope (Figure 4-25). It is a specialty knot.

a. **Tying the Knot.**

STEP 1. Form a 12-inch bight in the end of the rope.

STEP 2. Hold the center of the bight in the right hand. Hold the two parallel ropes from the bight in the left hand about 12 inches up the rope.

STEP 3. With the center of the bight in the right hand, twist two complete turns clockwise.

STEP 4. Reach through the bight and grasp the long, standing end of the rope. Pull another bight (from the long standing end) back through the original bight.

STEP 5. Pull down on the short working end of the rope and dress the knot down.

STEP 6. If the knot is to be used in a transport tightening system, take the working end of the rope and form a half hitch around the loop of the figure eight knot.

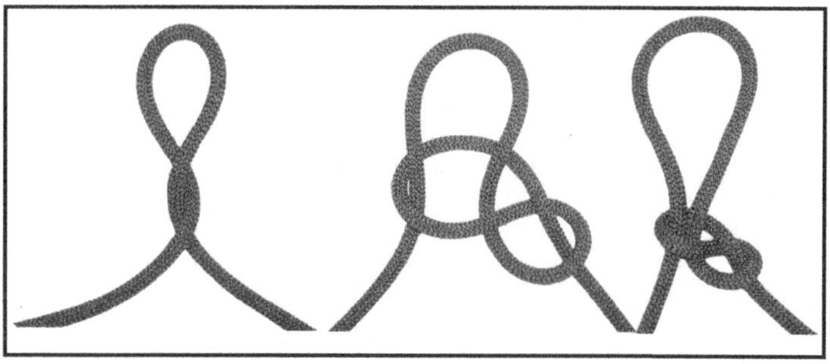

Figure 4-25. Figure-eight slip knot.

b. **Checkpoints.**

(1) The knot is in the shape of a figure eight.

(2) Both ropes of the bight pass through the same loop of the figure eight.

(3) The sliding portion of the rope is the long working end of the rope.

4-26. TRANSPORT KNOT (OVERHAND SLIP KNOT/MULE KNOT)

The transport knot is used to secure the transport tightening system (Figure 4-26, page 4-28). It is simply an overhand slip knot.

a. **Tying the Knot.**

STEP 1. Pass the running end of the rope around the anchor point passing it back under the standing portion (leading to the far side anchor) forming a loop.

STEP 2. Form a bight with the running end of the rope. Pass over the standing portion and down through the loop and dress it down toward the anchor point.

STEP 3. Secure the knot by tying a half hitch around the standing portion with the bight.

Figure 4-26. Transport knot.

b. **Check Points.**

(1) There is a single overhand slip knot.

(2) The knot is secured using a half hitch on a bight.
(3) The bight is a minimum of 12 inches long.

4-27. KLEIMHIEST KNOT

The Kleimhiest knot provides a moveable, easily adjustable, high-tension knot capable of holding extremely heavy loads while being pulled tight (Figure 4-27). It is a special-purpose knot.

a. **Tying the Knot.**

STEP 1. Using a utility rope or webbing offset the ends by 12 inches. With the ends offset, find the center of the rope and form a bight. Lay the bight over a horizontal rope.

STEP 2. Wrap the tails of the utility rope around the horizontal rope back toward the direction of pull. Wrap at least four complete turns.

STEP 3. With the remaining tails of the utility rope, pass them through the bight (see STEP 1).

STEP 4. Join the two ends of the tail with a joining knot.

STEP 5. Dress the knot down tightly so that all wraps are touching.

Note: Spectra should not be used for the Kleimhiest knot. It has a low melting point and tends to slip .

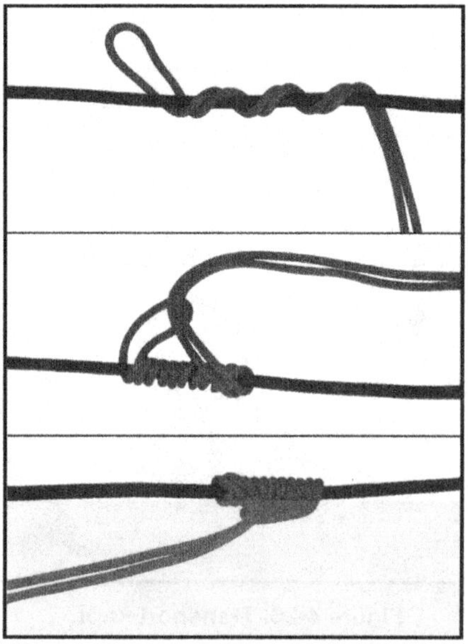

Figure 4-27. Kleimhiest knot.

b. **Checkpoints.**
(1) The bight is opposite the direction of pull.
(2) All wraps are tight and touching.
(3) The ends of the utility rope are properly secured with a joining knot.

4-28. FROST KNOT

The frost knot is used when working with webbing (Figure 4-28, page 4-30). It is used to create the top loop of an etrier. It is a special-purpose knot.

a. **Tying the Knot.**

STEP 1. Lap one end (a bight) of webbing over the other about 10 to 12 inches.

STEP 2. Tie an overhand knot with the newly formed triple-strand webbing; dress tightly.

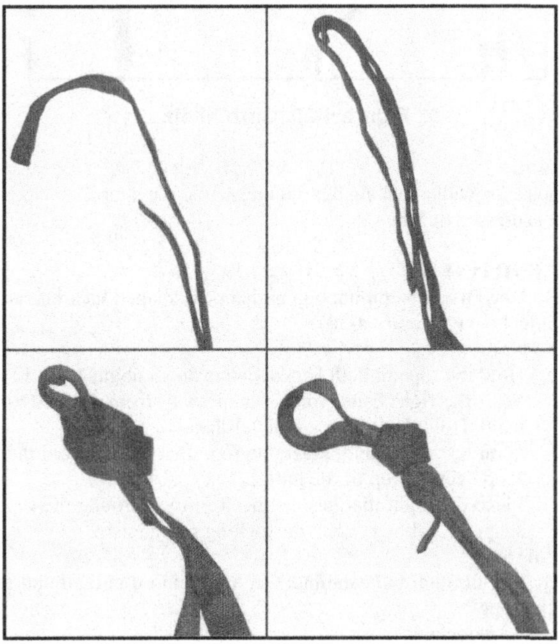

Figure 4-28. Frost knot.

b. **Checkpoints.**
(1) The tails of the webbing run in opposite directions.
(2) Three strands of webbing are formed into a tight overhand knot.
(3) There is a bight and tail exiting the top of the overhand knot.

4-29. GIRTH HITCH

The girth hitch is used to attach a runner to an anchor or piece of equipment (Figure 4-29). It is a special-purpose knot.

a. **Tying the Knot.**

STEP 1: Form a bight.
STEP 2: Bring the runner back through the bight.
STEP 3: Cinch the knot tightly.

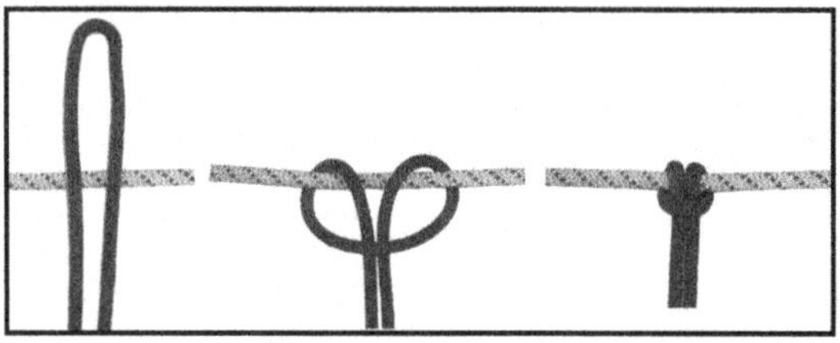

Figure 4-29. Girth hitch.

b. **Checkpoint.**

(1) Two wraps exist with a locking bar running across the wraps.
(2) The knot is dressed tightly.

4-30. MUNTER HITCH

The munter hitch, when used in conjunction with a pear-shaped locking carabiner, is used to form a mechanical belay (Figure 4-30).

a. **Tying the Knot.**

STEP 1. Hold the rope in both hands, palms down about 12 inches apart.
STEP 2. With the right hand, form a loop away from the body toward the left hand. Hold the loop with the left hand.
STEP 3. With the right hand, place the rope that comes from the bottom of the loop over the top of the loop.
STEP 4. Place the bight that has just been formed around the rope into the pear shaped carabiner. Lock the locking mechanism.

b. **Check Points.**

(1) A bight passes through the carabiner, with the closed end around the standing or running part of the rope.
(2) The carabiner is locked.

Figure 4-30. Munter hitch.

4-31. RAPPEL SEAT

The rappel seat is an improvised seat rappel harness made of rope (Figure 4-31, page 4-32). It usually requires a sling rope 14 feet or longer.

a. **Tying the Knot.**

STEP 1. Find the middle of the sling rope and make a bight.

STEP 2. Decide which hand will be used as the brake hand and place the bight on the opposite hip.

STEP 3. Reach around behind and grab a single strand of rope. Bring it around the waist to the front and tie two overhands on the other strand of rope, thus creating a loop around the waist.

STEP 4. Pass the two ends between the legs, ensuring they do not cross.

STEP 5. Pass the two ends up under the loop around the waist, bisecting the pocket flaps on the trousers. Pull up on the ropes, tightening the seat.

STEP 6. From rear to front, pass the two ends through the leg loops creating a half hitch on both hips.

STEP 7. Bring the longer of the two ends across the front to the nonbrake hand hip and secure the two ends with a square knot safetied with overhand knots. Tuck any excess rope in the pocket below the square knot.

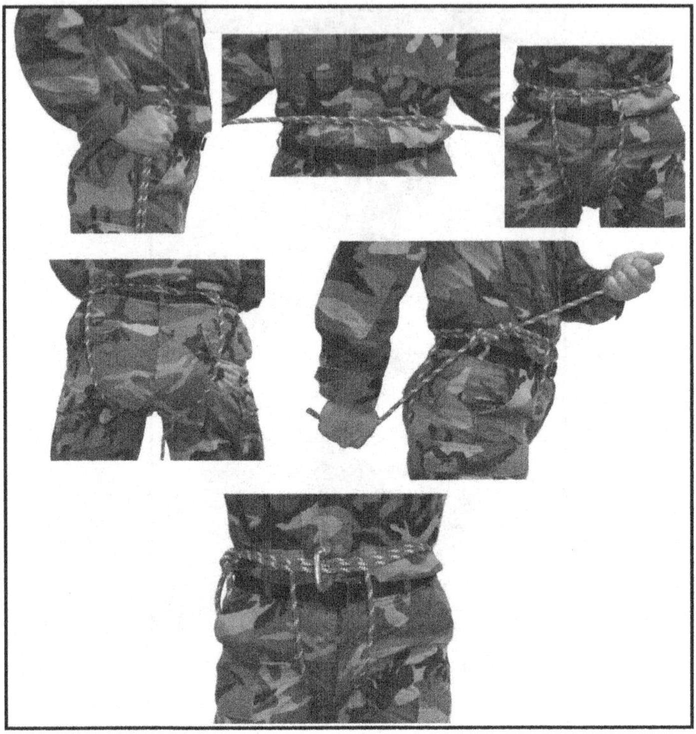

Figure 4-31. Rappel seat.

b. **Check Points.**

(1) There are two overhand knots in the front.

(2) The ropes are not crossed between the legs.

(3) A half hitch is formed on each hip.

(4) Seat is secured with a square knot with overhand safeties on the non-brake hand side.

(5) There is a minimum 4-inch pigtail after the overhand safeties are tied.

4-32. GUARDE KNOT

The guarde knot (ratchet knot, alpine clutch) is a special purpose knot primarily used for hauling systems or rescue (Figure 4-32). The knot works in only one direction and cannot be reversed while under load.

a. **Tying the Knot.**

STEP 1. Place a bight of rope into the two anchored carabiners (works best with two like carabiners, preferably ovals).

STEP 2. Take a loop of rope from the non-load side and place it down into the opposite cararabiner so that the rope comes out between the two carabiners.

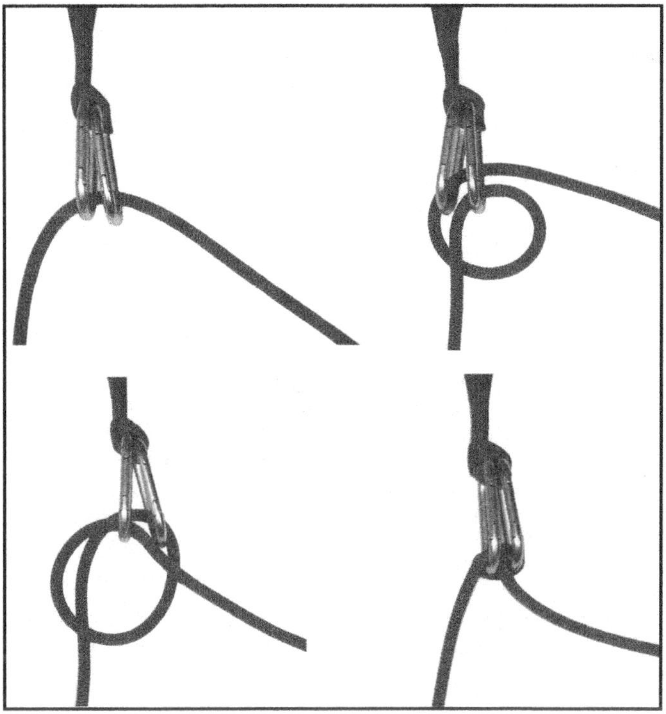

Figure 4-32. Guarde knot.

b. **Check Points.**

(1) When properly dressed, rope can only be pulled in one direction.

(2) The knot will not fail when placed under load.

CHAPTER 5
ANCHORS

This chapter discusses different types of anchors and their application in rope systems and climbing. Proper selection and placement of anchors is a critical skill that requires a great deal of practice. Failure of any system is most likely to occur at the anchor point. If the anchor is not strong enough to support the intended load, it will fail. Failure is usually the result of poor terrain features selected for the anchor point, or the equipment used in rigging the anchor was placed improperly or in insufficient amounts.

When selecting or constructing anchors, always try to make sure the anchor is "bombproof." A bombproof anchor is stronger than any possible load that could be placed on it. An anchor that has more strength than the climbing rope is considered bombproof.

Section I. NATURAL ANCHORS

Natural anchors should be considered for use first. They are usually strong and often simple to construct with minimal use of equipment. Trees, boulders, and other terrain irregularities are already in place and simply require a method of attaching the rope. However, natural anchors should be carefully studied and evaluated for stability and strength before use. Sometimes the climbing rope is tied directly to the anchor, but under most circumstances a sling is attached to the anchor and then the climbing rope is attached to the sling with a carabiner(s). (See paragraph 5-7 for slinging techniques.)

5-1. TREES

Trees are probably the most widely used of all natural anchors depending on the terrain and geographical region (Figure 5-1). However, trees must be carefully checked for suitability.

a. In rocky terrain, trees usually have a shallow root system. This can be checked by pushing or tugging on the tree to see how well it is rooted. Anchoring as low as possible to prevent excess leverage on the tree may be necessary.

b. Use padding on soft, sap producing trees to keep sap off ropes and slings.

Figure 5-1. Trees used as anchors.

5-2. BOULDERS

Boulders and rock nubbins make ideal anchors (Figure 5-2). The rock can be firmly tapped with a piton hammer to ensure it is solid. Sedimentary and other loose rock formations are not stable. Talus and scree fields are an indicator that the rock in the area is not solid. All areas around the rock formation that could cut the rope or sling should be padded.

Figure 5-2. Boulders used as anchors.

5-3. CHOCKSTONES

A chockstone is a rock that is wedged in a crack because the crack narrows downward (Figure 5-3). Chockstones should be checked for strength, security, and crumbling and should always be tested before use. All chockstones must be solid and strong enough to support the load. They must have maximum surface contact and be well tapered with the surrounding rock to remain in position.

a. Chockstones are often directional—they are secure when pulled in one direction but may pop out if pulled in another direction.

b. A creative climber can often make his own chockstone by wedging a rock into position, tying a rope to it, and clipping on a carabiner.

c. Slings should not be wedged between the chockstone and the rock wall since a fall could cut the webbing runner.

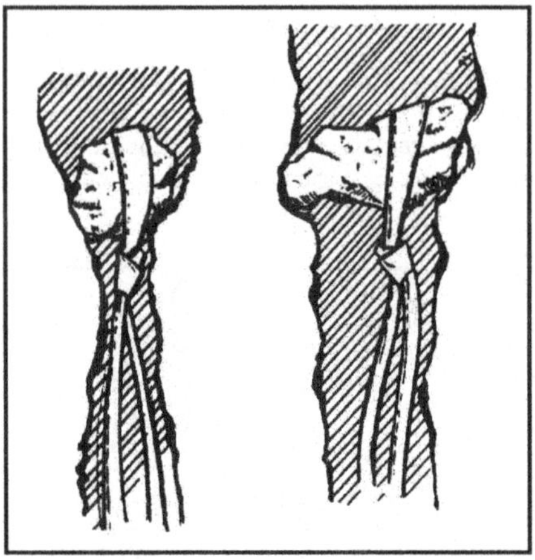

Figure 5-3. Chockstones.

5-4. ROCK PROJECTIONS

Rock projections (sometimes called nubbins) often provide suitable protection (Figure 5-4). These include blocks, flakes, horns, and spikes. If rock projections are used, their firmness is important. They should be checked for cracks or weathering that may impair their firmness. If any of these signs exist, the projection should be avoided.

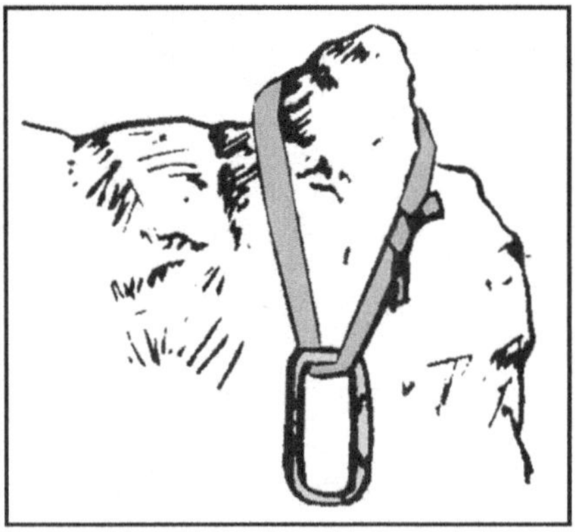

Figure 5-4. Rock projections.

5-5. TUNNELS AND ARCHES

Tunnels and arches are holes formed in solid rock and provide one of the more secure anchor points because they can be pulled in any direction. A sling is threaded through the opening hole and secured with a joining knot or girth hitch. The load-bearing hole must be strong and free of sharp edges (pad if necessary).

5-6. BUSHES AND SHRUBS

If no other suitable anchor is available, the roots of bushes can be used by routing a rope around the bases of several bushes (Figure 5-5). As with trees, the anchoring rope is placed as low as possible to reduce leverage on the anchor. All vegetation should be healthy and well rooted to the ground.

Figure 5-5. Bushes and shrubs.

5-7. SLINGING TECHNIQUES

Three methods are used to attach a sling to a natural anchor—drape, wrap, and girth. Whichever method is used, the knot is set off to the side where it will not interfere with normal carabiner movement. The carabiner gate should face away from the ground and open away from the anchor for easy insertion of the rope. When a locking carabiner cannot be used, two carabiners are used with gates opposed. Correctly opposed gates should open on opposite sides and form an "X" when opened (Figure 5-6).

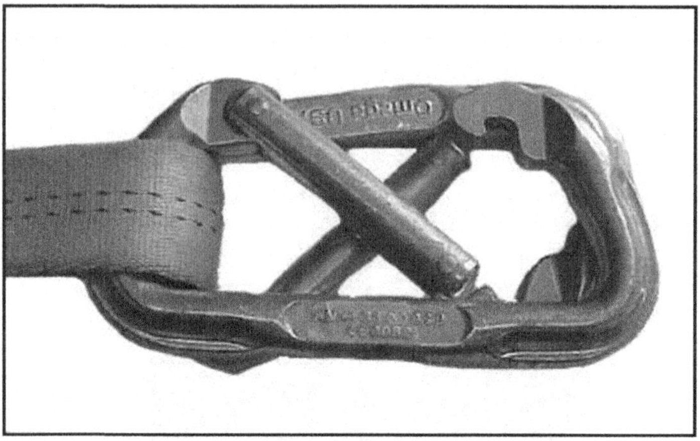

Figure 5-6. Correctly opposed carabiners.

a. **Drape.** Drape the sling over the anchor (Figure 5-7). Untying the sling and routing it around the anchor and then retying is still considered a drape.

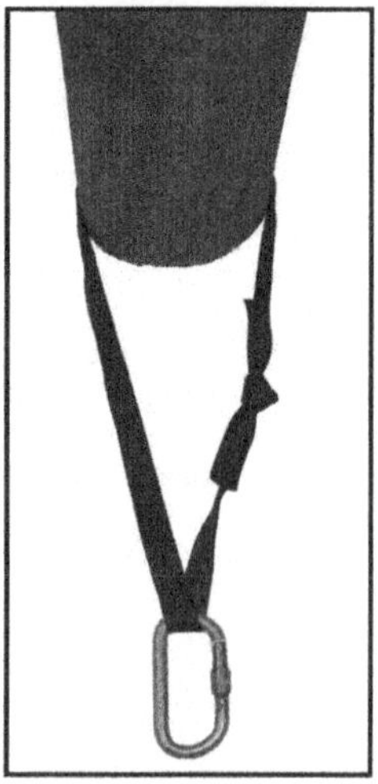

Figure 5-7. Drape.

b. **Wrap.** Wrap the sling around the anchor and connect the two ends together with a carabiner(s) or knot (Figure 5-8).

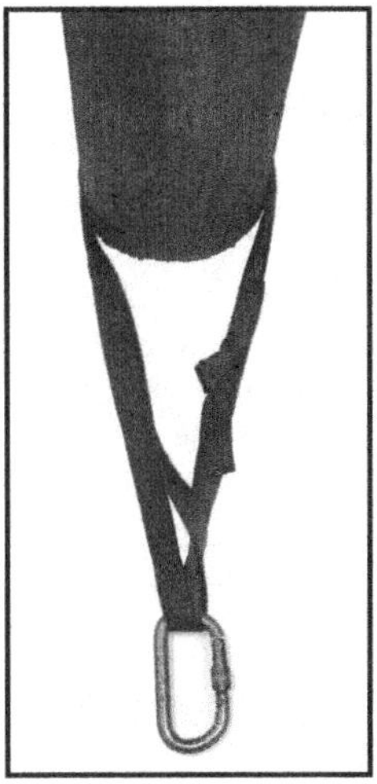

Figure 5-8. Wrap.

c. **Girth.** Tie the sling around the anchor with a girth hitch (Figure 5-9). Although a girth hitch reduces the strength of the sling, it allows the sling to remain in position and not slide on the anchor.

Figure 5-9. Girth.

Section II. ANCHORING WITH THE ROPE

The climbing or installation rope can be tied directly to the anchor using several different techniques. This requires less equipment, but also sacrifices some rope length to tie the anchor. The rope can be tied to the anchor using an appropriate anchor knot such as a bowline or a rerouted figure eight. Round turns can be used to help keep the rope in position on the anchor. A tensionless anchor can be used in high-load installations where tension on the attachment point and knot is undesirable.

5-8. ROPE ANCHOR

When tying the climbing or installation rope around an anchor, the knot should be placed approximately the same distance away from the anchor as the diameter of the anchor (Figure 5-10). The knot shouldn't be placed up against the anchor because this can stress and distort the knot under tension.

Figure 5-10. Rope tied to anchor with anchor knot.

5-9. TENSIONLESS ANCHOR

The tensionless anchor is used to anchor the rope on high-load installations such as bridging and traversing (Figure 5-11, page 5-8). The wraps of the rope around the anchor absorb the tension of the installation and keep the tension off the knot and carabiner. The anchor is usually tied with a minimum of four wraps, more if necessary, to absorb the tension. A smooth anchor may require several wraps, whereas a rough barked tree might only require a few. The rope is wrapped from top to bottom. A fixed loop is placed into the end of the rope and attached loosely back onto the rope with a carabiner.

Figure 5-11. Tensionless anchor.

Section III. ARTIFICIAL ANCHORS

Using artificial anchors becomes necessary when natural anchors are unavailable. The art of choosing and placing good anchors requires a great deal of practice and experience. Artificial anchors are available in many different types such as pitons, chocks, hexcentrics, and SLCDs. Anchor strength varies greatly; the type used depends on the terrain, equipment, and the load to be placed on it.

5-10. DEADMAN

A "deadman" anchor is any solid object buried in the ground and used as an anchor.

a. An object that has a large surface area and some length to it works best. (A hefty timber, such as a railroad tie, would be ideal.) Large boulders can be used, as well as a bundle of smaller tree limbs or poles. As with natural anchors, ensure timbers and tree limbs are not dead or rotting and that boulders are solid. Equipment, such as skis, ice axes, snowshoes, and ruck sacks, can also be used if necessary.

b. In extremely hard, rocky terrain (where digging a trench would be impractical, if not impossible) a variation of the deadman anchor can be constructed by building above the ground. The sling is attached to the anchor, which is set into the ground as deeply as possible. Boulders are then stacked on top of it until the anchor is strong enough for the load. Though normally not as strong as when buried, this method can work well for light-load installations as in anchoring a hand line for a stream crossing.

Note: Artificial anchors, such as pitons and bolts, are not widely accepted for use in all areas because of the scars they leave on the rock and the environment. Often they are left in place and become unnatural, unsightly fixtures in the natural environment. For training planning, local laws and courtesies should be taken into consideration for each area of operation.

5-11. PITONS

Pitons have been in use for over 100 years. Although still available, pitons are not used as often as other types of artificial anchors due primarily to their impact on the environment. Most climbers prefer to use chocks, SLCDs and other artificial anchors rather than pitons because they do not scar the rock and are easier to remove. Eye protection should always be worn when driving a piton into rock.

Note: The proper use and placement of pitons, as with any artificial anchor, should be studied, practiced, and tested while both feet are firmly on the ground and there is no danger of a fall.

a. **Advantages.** Some advantages in using pitons are:

- Depending on type and placement, pitons can support multiple directions of pull.
- Pitons are less complex than other types of artificial anchors.
- Pitons work well in thin cracks where other types of artificial anchors do not.

b. **Disadvantages.** Some disadvantages in using pitons are:

- During military operations, the distinct sound created when hammering pitons is a tactical disadvantage.
- Due to the expansion force of emplacing a piton, the rock could spread apart or break causing an unsafe condition.
- Pitons are more difficult to remove than other types of artificial anchors.
- Pitons leave noticeable scars on the rock.
- Pitons are easily dropped if not tied off when being used.

c. **Piton Placement.** The proper positioning or placement of pitons is critical. (Figure 5-12, page 5-10, shows examples of piton placement.) Usually a properly sized piton for a rock crack will fit one half to two thirds into the crack before being driven with the piton hammer. This helps ensure the depth of the crack is adequate for the size piton selected. As pitons are driven into the rock the pitch or sound that is made will change with each hammer blow, becoming higher pitched as the piton is driven in.

(1) Test the rock for soundness by tapping with the hammer. Driving pitons in soft or rotten rock is not recommended. When this type of rock must be used, clear the loose rock, dirt, and debris from the crack before driving the piton completely in.

(2) While it is being driven, attach the piton to a sling with a carabiner (an old carabiner should be used, if available) so that if the piton is knocked out of the crack, it will not be lost. The greater the resistance overcome while driving the piton, the firmer the anchor will be. The holding power depends on the climber placing the piton in a sound crack, and on the type of rock. The piton should not spread the rock, thereby loosening the emplacement.

Note: Pitons that have rings as attachment points might not display much change in sound as they are driven in as long as the ring moves freely.

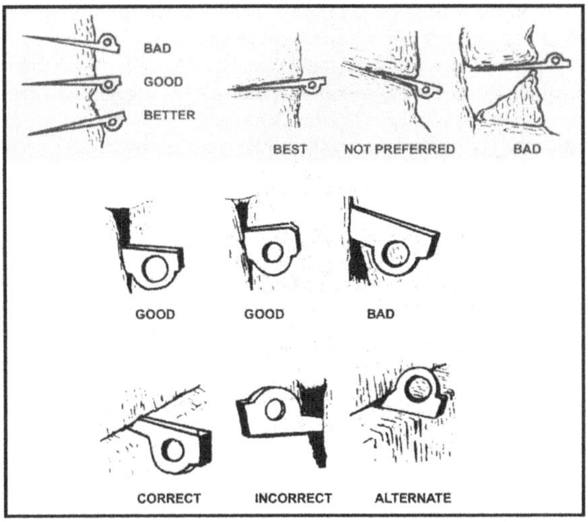

Figure 5-12. Examples of piton placements.

(3) Military mountaineers should practice emplacing pitons using either hand. Sometimes a piton cannot be driven completely into a crack, because the piton is too long. Therefore, it should be tied off using a hero-loop (an endless piece of webbing) (Figure 5-13). Attach this loop to the piton using a girth hitch at the point where the piton enters the rock so that the girth hitch is snug against the rock. Clip a carabiner into the loop.

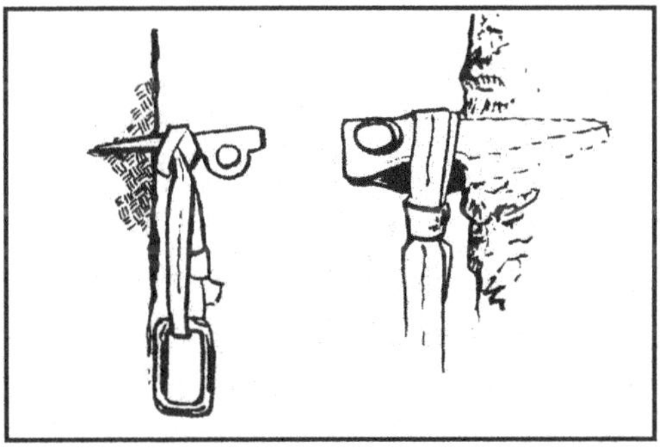

Figure 5-13. Hero-loop.

d. **Testing.** To test pitons pull up about 1 meter of slack in the climbing rope or use a sling. Insert this rope into a carabiner attached to the piton, then grasp the rope at least 1/2 meter from the carabiner. Jerk vigorously upward, downward, to each side, and then outward while observing the piton for movement. Repeat these actions as many times as necessary. Tap the piton to determine if the pitch has changed. If the pitch has changed greatly, drive the piton in as far as possible. If the sound regains its original pitch, the emplacement is probably safe. If the piton shows any sign of moving or if, upon driving it, there is any question of its soundness, drive it into another place. Try to be in a secure position before testing. This procedure is intended for use in testing an omni-directional anchor (one that withstands a pull in any direction). When a directional anchor (pull in one direction) is used, as in most free and direct-aid climbing situations, and when using chocks, concentrate the test in the direction that force will be applied to the anchor.

e. **Removing Pitons.** Attach a carabiner and sling to the piton before removal to eliminate the chance of dropping and losing it. Tap the piton firmly along the axis of the crack in which it is located. Alternate tapping from both sides while applying steady pressure. Pulling out on the attached carabiner eventually removes the piton (Figure 5-14).

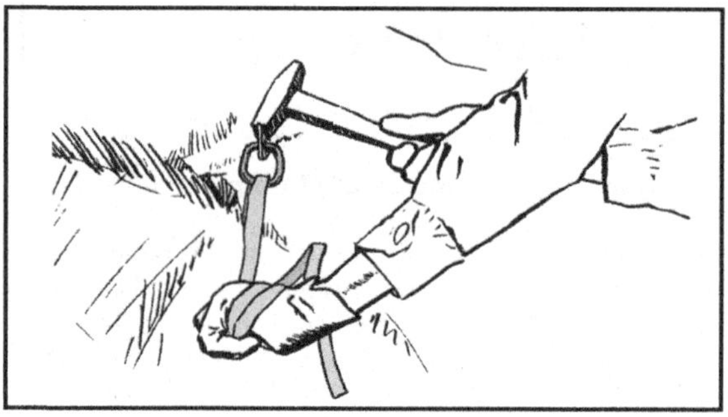

Figure 5-14. Piton removal.

f. **Reusing Pitons.** Soft iron pitons that have been used, removed, and straightened may be reused, but they must be checked for strength. In training areas, pitons already in place should not be trusted since weather loosens them in time. Also, they may have been driven poorly the first time. Before use, test them as described above and drive them again until certain of their soundness.

5-12. CHOCKS

Chock craft has been in use for many decades. A natural chockstone, having fallen and wedged in a crack, provides an excellent anchor point. Sometimes these chockstones are in unstable positions, but can be made into excellent anchors with little adjustment. Chock craft is an art that requires time and technique to master—simple in theory, but complex in practice. Imagination and resourcefulness are key principles to chock craft. The skilled climber must understand the application of mechanical advantage, vectors, and other forces that affect the belay chain in a fall.

a. **Advantages.** The advantages of using chocks are:

- Tactically quiet installation and recovery.
- Usually easy to retrieve and, unless severely damaged, are reusable.
- Light to carry.
- Easy to insert and remove.
- Minimal rock scarring as opposed to pitons.
- Sometimes can be placed where pitons cannot (expanding rock flakes where pitons would further weaken the rock).

b. **Disadvantages.** The disadvantages of using chocks are:

- May not fit in thin cracks, which may accept pitons.
- Often provide only one direction of pull.
- Practice and experience necessary to become proficient in proper placement.

c. **Placement.** The principles of placing chocks are to find a crack with a constriction at some point, place a chock of appropriate size above and behind the constriction, and set the chock by jerking down on the chock loop (Figure 5-15). Maximum surface contact with a tight fit is critical. Chocks are usually good for a single direction of pull.

(1) Avoid cracks that have crumbly (soft) or deteriorating rock, if possible. Some cracks may have loose rock, grass, and dirt, which should be removed before placing the chock. Look for a constriction point in the crack, then select a chock to fit it.

(2) When selecting a chock, choose one that has as much surface area as possible in contact with the rock. A chock resting on one small crystal or point of rock is likely to be unsafe. A chock that sticks partly out of the crack is avoided. Avoid poor protection. Ensure that the chock has a wire or runner long enough; extra ropes, cord, or webbing may be needed to extend the length of the runner.

(3) End weighting of the placement helps to keep the protection in position. A carabiner often provides enough weight

(4) Parallel-sided cracks without constrictions are a problem. Chocks designed to be used in this situation rely on camming principles to remain emplaced. Weighting the emplacement with extra hardware is often necessary to keep the chocks from dropping out.

(a) Emplace the wedge-shaped chock above and behind the constriction; seat it with a sharp downward tug.

(b) Place a camming chock with its narrow side into the crack, then rotate it to the attitude it will assume under load; seat it with a sharp downward tug.

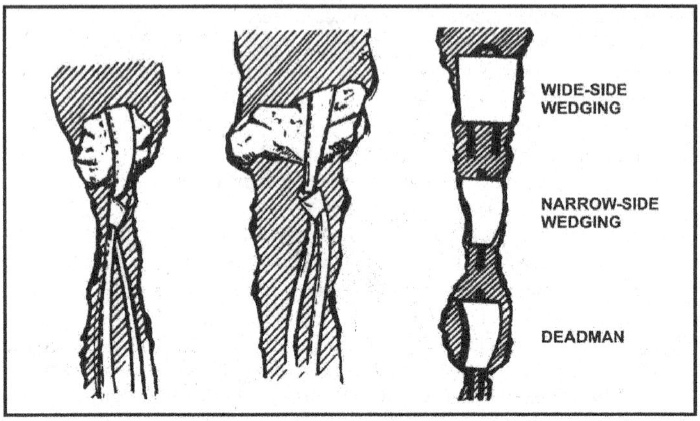

Figure 5-15. Chock placements.

d. **Testing.** After seating a chock, test it to ensure it remains in place. A chock that falls out when the climber moves past it is unsafe and offers no protection. To test it, firmly pull the chock in every anticipated direction of pull. Some chock placements fail in one or more directions; therefore, use pairs of chocks in opposition.

5-13. SPRING-LOADED CAMMING DEVICE

The SLCD offers quick and easy placement of artificial protection. It is well suited in awkward positions and difficult placements, since it can be emplaced with one hand. It can usually be placed quickly and retrieved easily (Figure 5-16, page 5-14).

a. To emplace an SLCD hold the device in either hand like a syringe, pull the retractor bar back, place the device into a crack, and release the retractor bar. The SLCD holds well in parallel-sided hand- and fist-sized cracks. Smaller variations are available for finger-sized cracks.

b. Careful study of the crack should be made before selecting the device for emplacement. It should be placed so that it is aligned in the direction of force applied to it. It should not be placed any deeper than is needed for secure placement, since it may be impossible to reach the extractor bar for removal. An SLCD should be extended with a runner and placed so that the direction of pull is parallel to the shaft; otherwise, it may rotate and pull out. The versions that have a semi-rigid wire cable shaft allow for greater flexibility and usage, without the danger of the shaft snapping off in a fall.

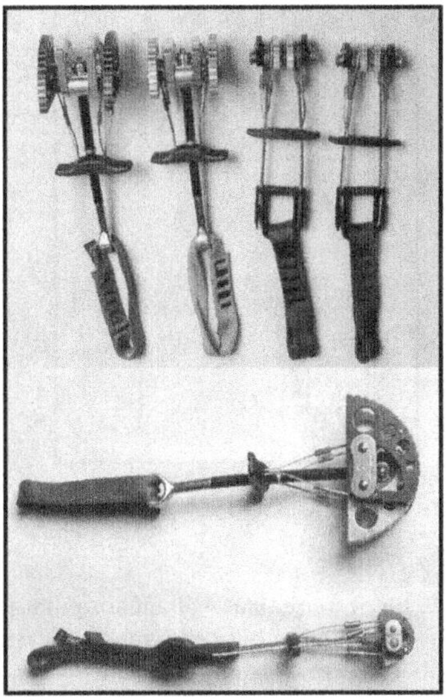

Figure 5-16. SLCD placements.

5-14. BOLTS

Bolts are often used in fixed-rope installations and in aid climbing where cracks are not available.

a. Bolts provide one of the most secure means of establishing protection. The rock should be inspected for evidence of crumbling, flaking, or cracking, and should be tested with a hammer. Emplacing a bolt with a hammer and a hand drill is a time-consuming and difficult process that requires drilling a hole in the rock deeper than the length of the bolt. This normally takes more than 20 minutes for one hole. Electric or even gas-powered drills can be used to greatly shorten drilling time. However, their size and weight can make them difficult to carry on the climbing route.

b. A hanger (carrier) and nut are placed on the bolt, and the bolt is inserted and then driven into the hole. A climber should never hammer on a bolt to test or "improve" it, since this permanently weakens it. Bolts should be used with carriers, carabiners, and runners.

c. When using bolts, the climber uses a piton hammer and hand drill with a masonry bit for drilling holes. Some versions are available in which the sleeve is hammered and turned into the rock (self-drilling), which bores the hole. Split bolts and expanding sleeves are common bolts used to secure hangers and carriers (Figure 5-17). Surgical tubing is useful in blowing dust out of the holes. Nail type bolts are emplaced by driving the nail with a hammer to expand the sleeve against the wall of the drilled hole. Safety glasses should always be worn when emplacing bolts.

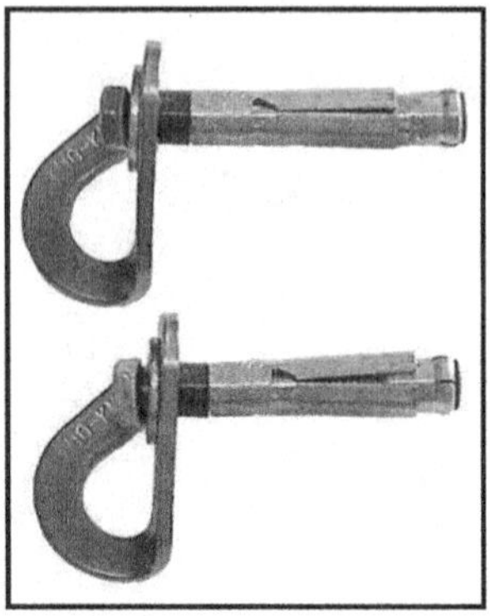

Figure 5-17. Bolt with expanding sleeve.

5-15. EQUALIZING ANCHORS

Equalizing anchors are made up of more than one anchor point joined together so that the intended load is shared equally. This not only provides greater anchor strength, but also adds redundancy or backup because of the multiple points.

a. **Self-equalizing Anchor.** A self-equalizing anchor will maintain an equal load on each individual point as the direction of pull changes (Figure 5-18). This is sometimes used in rappelling when the route must change left or right in the middle of the rappel. A self-equalizing anchor should only be used when necessary because if any one of the individual points fail, the anchor will extend and shock-load the remaining points or even cause complete anchor failure.

Figure 5-18. Self-equalizing anchors.

b. **Pre-equalized Anchor.** A pre-equalized anchor distributes the load equally to each individual point (Figure 5-19). It is aimed in the direction of the load. A pre-equalized anchor prevents extension and shock-loading of the anchor if an individual point fails. An anchor is pre-equalized by tying an overhand or figure-eight knot in the webbing or sling.

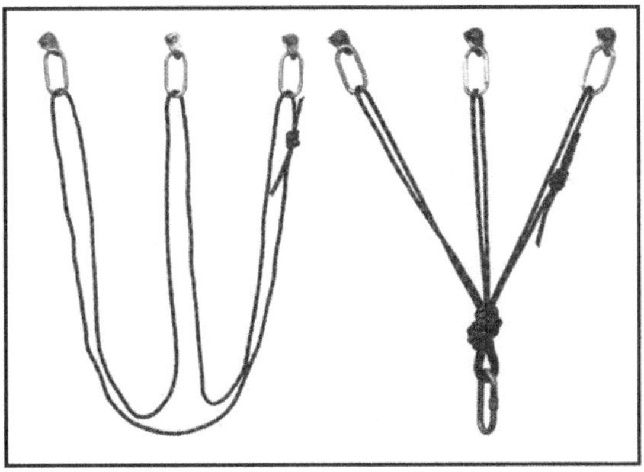

Figure 5-19. Pre-equalized anchor.

Note: When using webbing or slings, the angles of the webbing or slings directly affect the load placed on an anchor. An angle greater than 90 degrees can result in anchor failure (Figure 5-20).

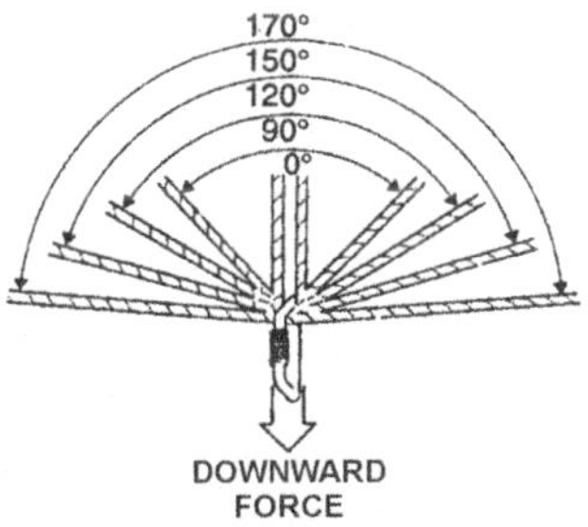

TENSION IN MULTIPLE ANCHOR RIGGING	
ANGLE	RESULTING LEG TENSION*
170°	1150%
150°	200%
120°	100%
90°	70%
0°	50%

*On each leg relative to downward force.

Figure 5-20. Effects of angles on an anchor.

CHAPTER 6
CLIMBING

A steep rock face is a terrain feature that can be avoided most of the time through prior planning and good route selection. Rock climbing can be time consuming, especially for a larger unit with a heavy combat load. It can leave the climbing party totally exposed to weather, terrain hazards, and the enemy for the length of the climb.

Sometimes steep rock cannot be avoided. Climbing relatively short sections of steep rock (one or two pitches) may prove quicker and safer than using alternate routes. A steep rock route would normally be considered an unlikely avenue of approach and, therefore, might be weakly defended or not defended at all.

All personnel in a unit preparing for deployment to mountainous terrain should be trained in the basics of climbing. Forward observers, reconnaissance personnel, and security teams are a few examples of small units who may require rock climbing skills to gain their vantage points in mountainous terrain. Select personnel demonstrating the highest degree of skill and experience should be trained in roped climbing techniques. These personnel will have the job of picking and "fixing" the route for the rest of the unit.

Rock climbing has evolved into a specialized "sport" with a wide range of varying techniques and styles. This chapter focuses on the basics most applicable to military operations.

Section I. CLIMBING FUNDAMENTALS

A variety of refined techniques are used to climb different types of rock formations. The foundation for all of these styles is the art of climbing. Climbing technique stresses climbing with the weight centered over the feet, using the hands primarily for balance. It can be thought of as a combination of the balanced movement required to walk a tightrope and the technique used to ascend a ladder. No mountaineering equipment is required; however, the climbing technique is also used in roped climbing.

6-1. ROUTE SELECTION

The experienced climber has learned to climb with the "eyes." Even before getting on the rock, the climber studies all possible routes, or "lines," to the top looking for cracks, ledges, nubbins, and other irregularities in the rock that will be used for footholds and handholds, taking note of any larger ledges or benches for resting places. When picking the line, he mentally climbs the route, rehearsing the step-by-step sequence of movements that will be required to do the climb, ensuring himself that the route has an adequate number of holds and the difficulty of the climb will be well within the limit of his ability.

6-2. TERRAIN SELECTION FOR TRAINING

Route selection for military climbing involves picking the easiest and quickest possible line for all personnel to follow. However, climbing skill and experience can only be developed by increasing the length and difficulty of routes as training progresses. In the training environment, beginning lessons in climbing should be performed CLOSE to the ground on lower-angled rock with plenty of holds for the hands and feet. Personnel not climbing can act as "spotters" for those climbing. In later lessons, a "top-rope" belay can be used for safety, allowing the individual to increase the length and difficulty of the climb under the protection of the climbing rope.

6-3. PREPARATION

In preparation for climbing, the boot soles should be dry and clean. A small stick can be used to clean out dirt and small rocks that might be caught between the lugs of the boot sole. If the soles are wet or damp, dry them off by stomping and rubbing the soles on clean, dry rock. All jewelry should be removed from the fingers. Watches and bracelets can interfere with hand placements and may become damaged if worn while climbing. Helmets should be worn to protect the head from injury if an object, such as a rock or climbing gear, falls from climbers above. Most climbing helmets are not designed to provide protection from impact to the head if the wearer falls, but will provide a minimal amount of protection if a climber comes in contact with the rock during climbing.

CAUTION

Rings can become caught on rock facial features and or lodged into cracks, which could cause injuries during a slip or fall.

6-4. SPOTTING

Spotting is a technique used to add a level of safety to climbing without a rope. A second man stands below and just outside of the climbers fall path and helps (spots) the climber to land safely if he should fall. Spotting is only applicable if the climber is not going above the spotters head on the rock. Beyond that height a roped climbing should be conducted. If an individual climbs beyond the effective range of the spotter(s), he has climbed TOO HIGH for his own safety. The duties of the spotter are to help prevent the falling climber from impacting the head and or spine, help the climber land feet first, and reduce the impact of a fall.

CAUTION

The spotter should not catch the climber against the rock because additional injuries could result. If the spotter pushes the falling climber into the rock, deep abrasions of the skin or knee may occur. Ankle joints could be twisted by the fall if the climber's foot remained high on the rock. The spotter might be required to fully support the weight of the climber causing injury to the spotter.

6-5. CLIMBING TECHNIQUE

Climbing involves linking together a series of movements based on foot and hand placement, weight shift, and movement. When this series of movements is combined correctly, a smooth climbing technique results. This technique reduces excess force on the limbs, helping to minimize fatigue. The basic principle is based on the five body parts described here.

a. **Five Body Parts.** The five body parts used for climbing are the right hand, left hand, right foot, left foot, and body (trunk). The basic principle to achieve smooth climbing is to move only one of the five body parts at a time. The trunk is not moved in conjunction with a foot or in conjunction with a hand, a hand is not moved in conjunction with a foot, and so on. Following this simple technique forces both legs to do all the lifting simultaneously.

b. **Stance or Body Position.** Body position is probably the single most important element to good technique. A relaxed, comfortable stance is essential. (Figure 6-1 shows a correct climbing stance, and Figure 6-2, page 6-4, shows an incorrect stance.) The body should be in a near vertical or erect stance with the weight centered over the feet. Leaning in towards the rock will cause the feet to push outward, away from the rock, resulting in a loss of friction between the boot sole and rock surface. The legs are straight and the heels are kept low to reduce fatigue. Bent legs and tense muscles tire quickly. If strained for too long, tense muscles may vibrate uncontrollably. This vibration, known as "Elvis-ing" or "sewing-machine leg" can be cured by straightening the leg, lowering the heel, or moving on to a more restful position. The hands are used to maintain balance. Keeping the hands between waist and shoulder level will reduce arm fatigue.

Figure 6-1. Correct climbing stance—balanced over both feet.

Figure 6-2. Incorrect stance—stretched out.

(1) Whenever possible, three points of contact are maintained with the rock. Proper positioning of the hips and shoulders is critical. When using two footholds and one handhold, the hips and shoulders should be centered over both feet. In most cases, as the climbing progresses, the body is resting on one foot with two handholds for balance. The hips and shoulders must be centered over the support foot to maintain balance, allowing the "free" foot to maneuver.

(2) The angle or steepness of the rock also determines how far away from the rock the hips and shoulders should be. On low-angle slopes, the hips are moved out away from the rock to keep the body in balance with the weight over the feet. The shoulders can be moved closer to the rock to reach handholds. On steep rock, the hips are pushed closer to the rock. The shoulders are moved away from the rock by arching the back. The body is still in balance over the feet and the eyes can see where the hands need to go. Sometimes, when footholds are small, the hips are moved back to increase friction between the foot and the rock. This is normally done on quick, intermediate holds. It should be avoided in the rest position as it places more weight on the arms and hands. When weight must be placed on handholds, the arms should be kept straight to reduce fatigue. Again, flexed muscles tire quickly.

c. **Climbing Sequence.** The steps defined below provide a complete sequence of events to move the entire body on the rock. These are the basic steps to follow for a smooth climbing technique. Performing these steps in this exact order will not always be necessary because the nature of the route will dictate the availability of hand and foot placements. The basic steps are weight, shift, and movement (movement being either the foot, hand, or body). (A typical climbing sequence is shown in Figure 6-3, pages 6-6 through 6-8.)

STEP ONE: Shift the weight from both feet to one foot. This will allow lifting of one foot with no effect on the stance.

STEP TWO:	Lift the unweighted foot and place it in a new location, within one to two feet of the starting position, with no effect on body position or balance (higher placement will result in a potentially higher lift for the legs to make, creating more stress, and is called a high step) The trunk does not move during foot movement.
STEP THREE:	Shift the weight onto both feet. (Repeat steps 1 through 3 for remaining foot.)
STEP FOUR:	Lift the body into a new stance with both legs.
STEP FIVE:	Move one hand to a new position between waist and head height. During this movement, the trunk should be completely balanced in position and the removed hand should have no effect on stability.
STEP SIX:	Move the remaining hand as in Step 5.

Now the entire body is in a new position and ready to start the process again. Following these steps will prevent lifting with the hands and arms, which are used to maintain stance and balance. If both legs are bent, leg extension can be performed as soon as one foot has been moved. Hand movements can be delayed until numerous foot movements have been made, which not only creates shorter lifts with the legs, but may allow a better choice for the next hand movements because the reach will have increased.

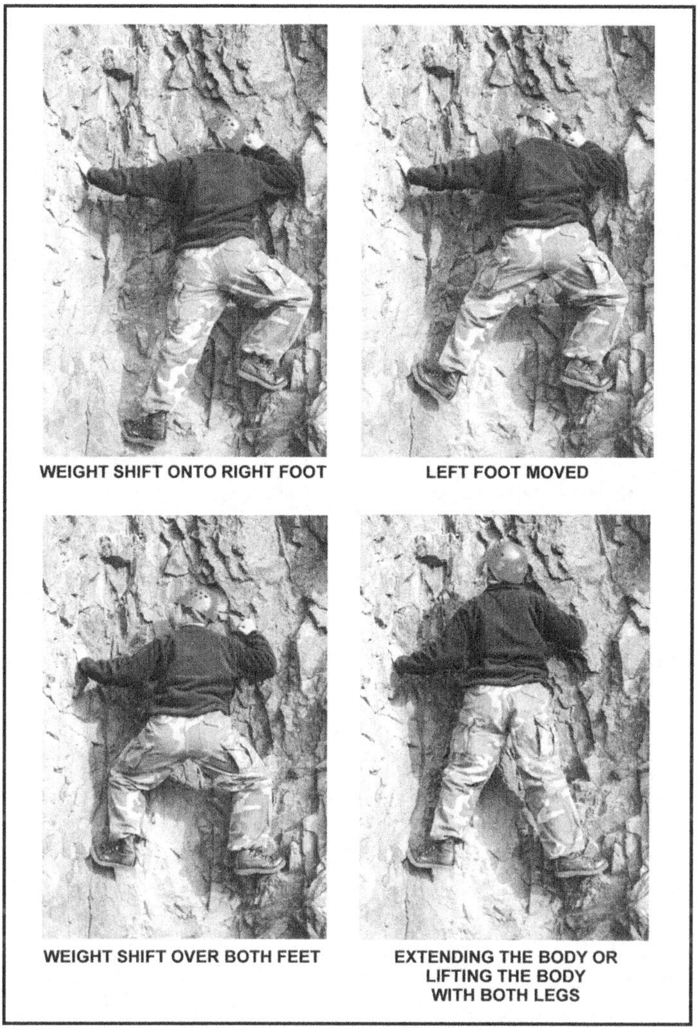

Figure 6-3. Typical climbing sequence.

Figure 6-3. Typical climbing sequence (continued).

Figure 6-3. Typical climbing sequence (continued).

(1) Many climbers will move more than one body part at a time, usually resulting in lifting the body with one leg or one leg and both arms. This type of lifting is inefficient, requiring one leg to perform the work of two or using the arms to lift the body. Proper climbing technique is lifting the body with the legs, not the arms, because the legs are much stronger.

(2) When the angle of the rock increases, these movements become more critical. Holding or pulling the body into the rock with the arms and hands may be necessary as the angle increases (this is still not lifting with the arms). Many climbing routes have angles greater than ninety degrees (overhanging) and the arms are used to support partial body weight. The same technique applies even at those angles.

(3) The climber should avoid moving on the knees and elbows. Other than being uncomfortable, even painful, to rest on, these bony portions of the limbs offer little friction and "feel" on the rock.

6-6. SAFETY PRECAUTIONS

The following safety precautions should be observed when rock climbing.

a. While ascending a seldom or never traveled route, you may encounter precariously perched rocks. If the rock will endanger your second, it may be possible to remove it from the route and trundle it, tossing it down. This is extremely dangerous to climbers below and should not be attempted unless you are absolutely sure no men are below. If you are not sure that the flight path is clear, do not do it. Never dislodge loose rocks carelessly. Should a rock become loose accidentally, immediately shout the warning "ROCK" to alert climbers below. Upon hearing the warning, personnel should seek immediate cover behind any rock bulges or overhangs available, or flatten themselves against the rock to minimize exposure.

b. Should a climber fall, he should do his utmost to maintain control and not panic. If on a low-angle climb, he may be able to arrest his own fall by staying in contact with the rock, grasping for any possible hold available. He should shout the warning "FALLING" to alert personnel below.

CAUTION

Grasping at the rock in a fall can result in serious injuries to the upper body. If conducting a roped climb, let the rope provide protection.

c. When climbing close to the ground and without a rope, a spotter can be used for safety. The duties of the spotter are to ensure the falling climber does not impact the head or spine, and to reduce the impact of a fall.

d. Avoid climbing directly above or below other climbers (with the exception of spotters). When personnel must climb at the same time, following the same line, a fixed rope should be installed.

e. Avoid climbing with gloves on because of the decreased "feel" for the rock. The use of gloves in the training environment is especially discouraged, while their use in the mountains is often mandatory when it is cold. A thin polypropylene or wool glove is best for rock climbing, although heavier cotton or leather work gloves are often used for belaying.

f. Be extremely careful when climbing on wet or moss-covered rock; friction on holds is greatly reduced.

g. Avoid grasping small vegetation for handholds; the root systems can be shallow and will usually not support much weight.

6-7. MARGIN OF SAFETY

Besides observing the standard safety precautions, the climber can avoid catastrophe by climbing with a wide margin of safety. The margin of safety is a protective buffer the climber places between himself and potential climbing hazards. Both subjective (personnel-related) and objective (environmental) hazards must be considered when applying the margin of safety. The leader must apply the margin of safety taking into account the strengths and weaknesses of the entire team or unit.

a. When climbing, the climber increases his margin of safety by selecting routes that are well within the limit of his ability. When leading a group of climbers, he selects a route well within the ability of the weakest member.

b. When the rock is wet, or when climbing in other adverse weather conditions, the climber's ability is reduced and routes are selected accordingly. When the climbing becomes difficult or exposed, the climber knows to use the protection of the climbing rope and belays. A lead climber increases his margin of safety by placing protection along the route to limit the length of a potential fall.

Section II. USE OF HOLDS

The climber should check each hold before use. This may simply be a quick, visual inspection if he knows the rock to be solid. When in doubt, he should grab and tug on the hold to test it for soundness BEFORE depending on it. Sometimes, a hold that appears weak can actually be solid as long as minimal force is applied to it, or the force is applied in a direction that strengthens it. A loose nubbin might not be strong enough to support the climber's weight, but it may serve as an adequate handhold. Be especially careful when climbing on weathered, sedimentary-type rock.

6-8. CLIMBING WITH THE FEET

"Climb with the feet and use the hands for balance" is extremely important to remember. In the early learning stages of climbing, most individuals will rely heavily on the arms, forgetting to use the feet properly. It is true that solid handholds and a firm grip are needed in some combination techniques; however, even the most strenuous techniques require good footwork and a quick return to a balanced position over one or both feet. Failure to climb any route, easy or difficult, is usually the result of poor footwork.

a. The beginning climber will have a natural tendency to look up for handholds. Try to keep the hands low and train your eyes to look down for footholds. Even the smallest irregularity in the rock can support the climber once the foot is positioned properly and weight is committed to it.

b. The foot remains on the rock as a result of friction. Maximum friction is obtained from a correct stance over a properly positioned foot. The following describes a few ways the foot can be positioned on the rock to maximize friction.

(1) ***Maximum Sole Contact.*** The principle of using full sole contact, as in mountain walking, also applies in climbing. Maximum friction is obtained by placing as much of the boot sole on the rock as possible. Also, the leg muscles can relax the most when the entire foot is placed on the rock. (Figure 6-4 shows examples of maximum and minimum sole contact.)

(a) Smooth, low-angled rock (slab) and rock containing large "bucket" holds and ledges are typical formations where the entire boot sole should be used.

(b) On some large holds, like bucket holds that extend deep into the rock, the entire foot cannot be used. The climber may not be able to achieve a balanced position if the foot is stuck too far underneath a bulge in the rock. In this case, placing only part of the foot on the hold may allow the climber to achieve a balanced stance. The key is to use as much of the boot sole as possible. Remember to keep the heels low to reduce strain on the lower leg muscles.

MINIMUM SOLE CONTACT (POOR)

MAXIMUM SOLE CONTACT (BEST)

MINUMUM SOLE CONTACT (POOR)

MAXIMUM SOLE CONTACT (BEST)

Figure 6-4. Examples of maximum and minimum sole contact.

(2) ***Edging.*** The edging technique is used where horizontal crack systems and other irregularities in the rock form small, well-defined ledges. The edge of the boot sole is placed on the ledge for the foothold. Usually, the inside edge of the boot or the edge area around the toes is used. Whenever possible, turn the foot sideways and use the entire inside edge of the boot. Again, more sole contact equals more friction and the legs can rest more when the heel is on the rock. (Figure 6-5, page 6-12, shows examples of the edging technique.)

(a) On smaller holds, edging with the front of the boot, or toe, may be used. Use of the toe is most tiring because the heel is off the rock and the toes support the climber's weight. Remember to keep the heel low to reduce fatigue. Curling and stiffening the toes in the boot increases support on the hold. A stronger position is usually obtained on small ledges by turning the foot at about a 45-degree angle, using the strength of the big toe and the ball of the foot.

(b) Effective edging on small ledges requires stiff-soled footwear. The stiffer the sole, the better the edging capability. Typical mountain boots worn by the US military have a relatively flexible lugged sole and, therefore, edging ability on smaller holds will be somewhat limited.

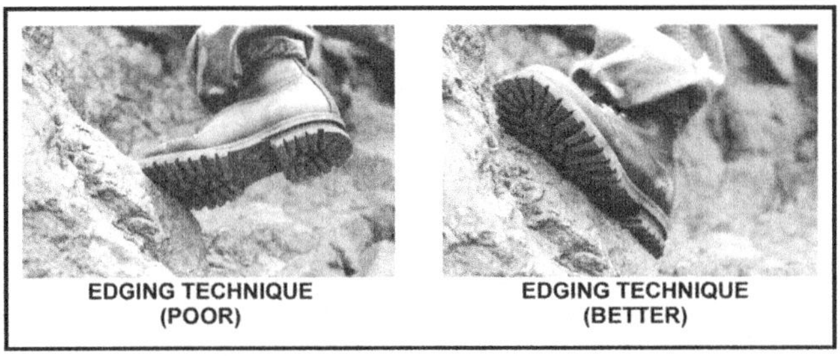

Figure 6-5. Examples of edging technique.

(3) ***Smearing.*** When footholds are too small to use a good edging technique, the ball of the foot can be "smeared" over the hold. The smearing technique requires the boot to adhere to the rock by deformation of the sole and by friction. Rock climbing shoes are specifically designed to maximize friction for smearing; some athletic shoes also work well. The Army mountain boot, with its softer sole, usually works better for smearing than for edging. Rounded, down-sloping ledges and low-angled slab rock often require good smearing technique. (Figure 6-6 shows examples of the smearing technique.)

(a) Effective smearing requires maximum friction between the foot and the rock. Cover as much of the hold as possible with the ball of the foot. Keeping the heel low will not only reduce muscle strain, but will increase the amount of surface contact between the foot and the rock.

(b) Sometimes flexing the ankles and knees slightly will place the climber's weight more directly over the ball of the foot and increase friction; however, this is more tiring and should only be used for quick, intermediate holds. The leg should be kept straight whenever possible.

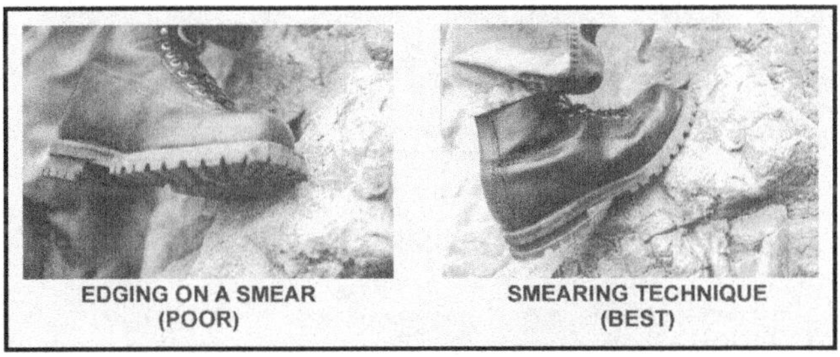

Figure 6-6. Examples of the smearing technique.

(4) ***Jamming.*** The jamming technique works on the same principal as chock placement. The foot is set into a crack in such a way that it "jams" into place, resisting a downward pull. The jamming technique is a specialized skill used to climb vertical or

near vertical cracks when no other holds are available on the rock face. The technique is not limited to just wedging the feet; fingers, hands, arms, even the entire leg or body are all used in the jamming technique, depending on the size of the crack. Jam holds are described in this text to broaden the range of climbing skills. Jamming holds can be used in a crack while other hand/foot holds are used on the face of the rock. Many cracks will have facial features, such as edges, pockets, and so on, inside and within reach. Always look or feel for easier to use features. (Figure 6-7 shows examples of jamming.)

(a) The foot can be jammed in a crack in different ways. It can be inserted above a constriction and set into the narrow portion, or it can be placed in the crack and turned, like a camming device, until it locks in place tight enough to support the climber's weight. Aside from these two basic ideas, the possibilities are endless. The toes, ball of the foot, or the entire foot can be used. Try to use as much of the foot as possible for maximum surface contact. Some positions are more tiring, and even more painful on the foot, than others. Practice jamming the foot in various ways to see what offers the most secure, restful position.

(b) Some foot jams may be difficult to remove once weight has been committed to them, especially if a stiffer sole boot is used. The foot is less likely to get stuck when it is twisted or "cammed" into position. When removing the boot from a crack, reverse the way it was placed to prevent further constriction.

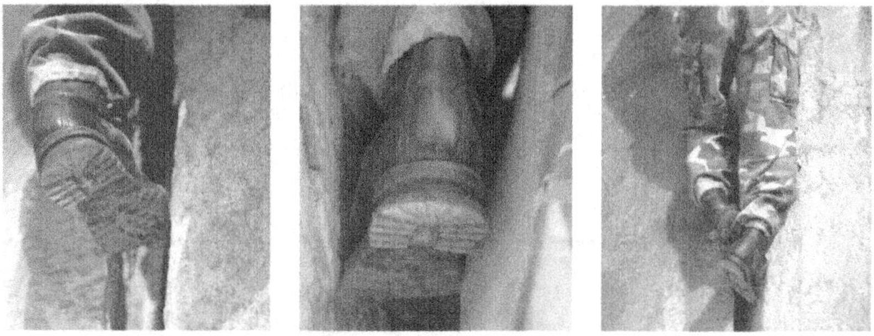

FOOT JAM

FOOT JAMMED INTO POSITION, HEEL IS PUSHED TO THE RIGHT

BOTH FEET JAMMED

Figure 6-7. Examples of jamming.

6-9. USING THE HANDS

The hands can be placed on the rock in many ways. Exactly how and where to position the hands and arms depends on what holds are available, and what configuration will best support the current stance as well as the movement to the next stance. Selecting handholds between waist and shoulder level helps in different ways. Circulation in the arms and hands is best when the arms are kept low. Secondly, the climber has less tendency to "hang" on his arms when the handholds are at shoulder level and below. Both of these contribute to a relaxed stance and reduce fatigue in the hands and arms.

a. As the individual climbs, he continually repositions his hands and arms to keep the body in balance, with the weight centered over the feet. On lower-angled rock, he may simply need to place the hands up against the rock and extend the arm to maintain balance; just like using an ice ax as a third point of contact in mountain walking. Sometimes, he will be able to push directly down on a large hold with the palm of the hand. More often though, he will need to "grip" the rock in some fashion and then push or pull against the hold to maintain balance.

b. As stated earlier, the beginner will undoubtedly place too much weight on the hands and arms. If we think of ourselves climbing a ladder, our body weight is on our legs. Our hands grip, and our arms pull on each rung only enough to maintain our balance and footing on the ladder. Ideally, this is the amount of grip and pull that should be used in climbing. Of course, as the size and availability of holds decreases, and the steepness of the rock approaches the vertical, the grip must be stronger and more weight might be placed on the arms and handholds for brief moments. The key is to move quickly from the smaller, intermediate holds to the larger holds where the weight can be placed back on the feet allowing the hands and arms to relax. The following describes some of the basic handholds and how the hand can be positioned to maximize grip on smaller holds.

(1) ***Push Holds.*** Push holds rely on the friction created when the hand is pushed against the rock. Most often a climber will use a push hold by applying "downward pressure" on a ledge or nubbin. This is fine, and works well; however, the climber should not limit his use of push holds to the application of down pressure. Pushing sideways, and on occasion, even upward on less obvious holds can prove quite secure. Push holds often work best when used in combination with other holds. Pushing in opposite directions and "push-pull" combinations are excellent techniques. (Figure 6-8 shows examples of push holds.)

(a) An effective push hold does not necessarily require the use of the entire hand. On smaller holds, the side of the palm, the fingers, or the thumb may be all that is needed to support the stance. Some holds may not feel secure when the hand is initially placed on them. The hold may improve or weaken during the movement. The key is to try and select a hold that will improve as the climber moves past it.

(b) Most push holds do not require much grip; however, friction might be increased by taking advantage of any rough surfaces or irregularities in the rock. Sometimes the strength of the hold can be increased by squeezing, or "pinching," the rock between the thumb and fingers (see paragraph on pinch holds).

Figure 6-8. Examples of push holds.

(2) ***Pull Holds.*** Pull holds, also called "cling holds," which are grasped and pulled upon, are probably the most widely used holds in climbing. Grip plays more of a role in a pull hold, and, therefore, it normally feels more secure to the climber than a push hold. Because of this increased feeling of security, pull holds are often overworked. These are the holds the climber has a tendency to hang from. Most pull holds do not require great strength, just good technique. Avoid the "death grip" syndrome by climbing with the feet. (Figure 6-9, page 6-16, shows examples of pull holds.)

(a) Like push holds, pressure on a pull hold can be applied straight down, sideways, or upward. Again, these are the holds the climber tends to stretch and reach for, creating an unbalanced stance. Remember to try and keep the hands between waist and shoulder level, making use of intermediate holds instead of reaching for those above the head.

(b) Pulling sideways on vertical cracks can be very secure. There is less tendency to hang from "side-clings" and the hands naturally remain lower. The thumb can often push against one side of the crack, in opposition to the pull by the fingers, creating a stronger hold. Both hands can also be placed in the same crack, with the hands pulling in opposite directions. The number of possible combinations is limited only by the imagination and experience of the climber.

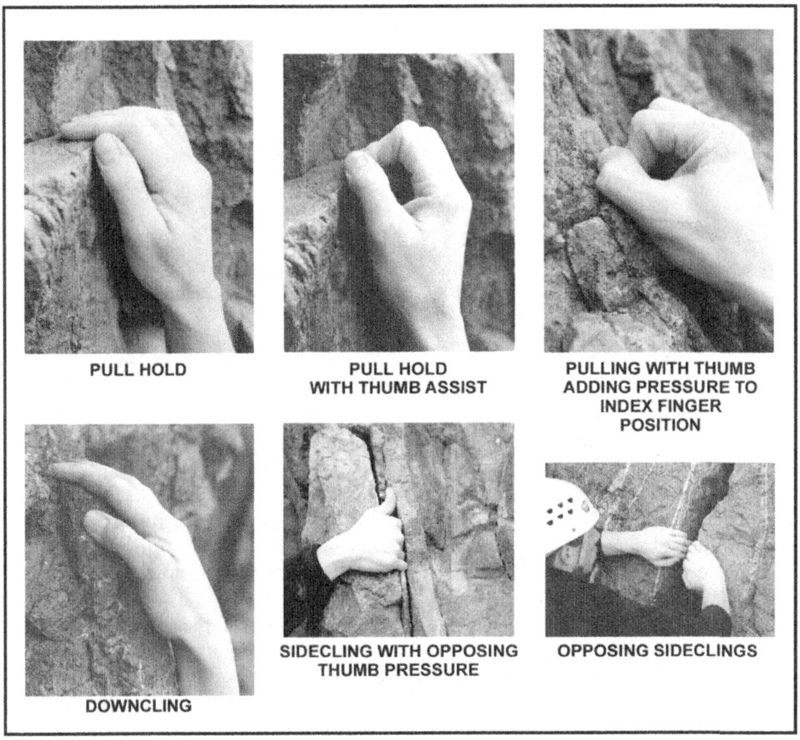

Figure 6-9. Examples of pull holds.

(c) Friction and strength of a pull hold can be increased by the way the hand grips the rock. Normally, the grip is stronger when the fingers are closed together; however, sometimes more friction is obtained by spreading the fingers apart and placing them between irregularities on the rock surface. On small holds, grip can often be improved by bending the fingers upward, forcing the palm of the hand to push against the rock. This helps to hold the finger tips in place and reduces muscle strain in the hand. Keeping the forearm up against the rock also allows the arm and hand muscles to relax more.

(d) Another technique that helps to strengthen a cling hold for a downward pull is to press the thumb against the side of the index finger, or place it on top of the index finger and press down. This hand configuration, known as a "ring grip," works well on smaller holds.

(3) ***Pinch Holds.*** Sometimes a small nubbin or protrusion in the rock can be "squeezed" between the thumb and fingers. This technique is called a pinch hold. Friction is applied by increasing the grip on the rock. Pinch holds are often overlooked by the

novice climber because they feel insecure at first and cannot be relied upon to support much body weight. If the climber has his weight over his feet properly, the pinch hold will work well in providing balance. The pinch hold can also be used as a gripping technique for push holds and pull holds. (Figure 6-10 shows examples of pinch holds.)

PINCH HOLD PINCH HOLD LARGE PINCH HOLD

Figure 6-10. Examples of pinch holds.

(4) ***Jam Holds.*** Like foot jams, the fingers and hands can be wedged or cammed into a crack so they resist a downward or outward pull. Jamming with the fingers and hands can be painful and may cause minor cuts and abrasions to tender skin. Cotton tape can be used to protect the fingertips, knuckles, and the back of the hand; however, prolonged jamming technique requiring hand taping should be avoided. Tape also adds friction to the hand in jammed position. (Figure 6-11, page 6-18, shows examples of jam holds.)

(a) The hand can be placed in a crack a number of ways. Sometimes an open hand can be inserted and wedged into a narrower portion of the crack. Other times a clenched fist will provide the necessary grip. Friction can be created by applying cross pressure between the fingers and the back of the hand. Another technique for vertical cracks is to place the hand in the crack with the thumb pointed either up or down. The hand is then clenched as much as possible. When the arm is straightened, it will twist the hand and tend to cam it into place. This combination of clenching and camming usually produces the most friction, and the most secure hand jam in vertical cracks.

(b) In smaller cracks, only the fingers will fit. Use as many fingers as the crack will allow. The fingers can sometimes be stacked in some configuration to increase friction. The thumb is usually kept outside the crack in finger jams and pressed against the rock to increase friction or create cross pressure. In vertical cracks it is best to insert the fingers with the thumb pointing down to make use of the natural camming action of the fingers that occurs when the arm is twisted towards a normal position.

(c) Jamming technique for large cracks, or "off widths," requiring the use of arm, leg, and body jams, is another technique. To jam or cam an arm, leg, or body into an off width, the principle is the same as for fingers, hands, or feet—you are making the jammed appendage "fatter" by folding or twisting it inside the crack. For off widths, you may place your entire arm inside the crack with the arm folded and the palm pointing outward. The leg can be used, from the calf to the thigh, and flexed to fit the crack. Routes requiring this type of climbing should be avoided as the equipment normally used for protection might not be large enough to protect larger cracks and openings. However, sometimes a narrower section may be deeper in the crack allowing the use of "normal" size protection.

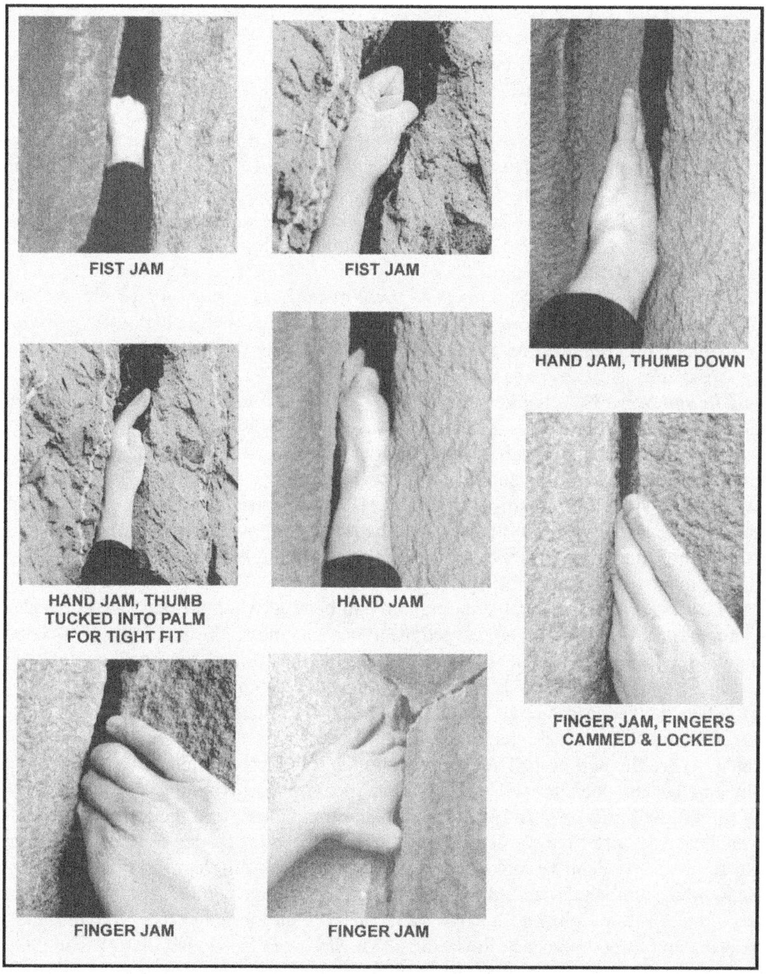

Figure 6-11. Examples of jam holds.

6-10. COMBINATION TECHNIQUES

The positions and holds previously discussed are the basics and the ones most common to climbing. From these fundamentals, numerous combination techniques are possible. As the climber gains experience, he will learn more ways to position the hands, feet, and body in relation to the holds available; however, he should always strive to climb with his weight on his feet from a balanced stance.

a. Sometimes, even on an easy route, the climber may come upon a section of the rock that defies the basic principles of climbing. Short of turning back, the only alternative is to figure out some combination technique that will work. Many of these type problems require the hands and feet to work in opposition to one another. Most will place more weight on the hands and arms than is desirable, and some will put the climber in an "out of balance" position. To make the move, the climber may have to "break the rules" momentarily. This is not a problem and is done quite frequently by experienced climbers. The key to using these type of combination techniques is to plan and execute them deliberately, without lunging or groping for holds, yet quickly, before the hands, arms, or other body parts tire. Still, most of these maneuvers require good technique more than great strength, though a certain degree of hand and arm strength certainly helps.

b. Combination possibilities are endless. The following is a brief description of some of the more common techniques.

(1) ***Change Step.*** The change step, or hop step, can be used when the climber needs to change position of the feet. It is commonly used when traversing to avoid crossing the feet, which might put the climber in an awkward position. To prevent an off balance situation, two solid handholds should be used. The climber simply places his weight on his handholds while he repositions the feet. He often does this with a quick "hop," replacing the lead foot with the trail foot on the same hold. Keeping the forearms against the rock during the maneuver takes some of the strain off the hands, while at the same time strengthening the grip on the holds.

(2) ***Mantling.*** Mantling is a technique that can be used when the distance between the holds increases and there are no immediate places to move the hands or feet. It does require a ledge (mantle) or projection in the rock that the climber can press straight down upon. (Figure 6-12, page 6-20, shows the mantling sequence.)

(a) When the ledge is above head height, mantling begins with pull holds, usually "hooking" both hands over the ledge. The climber pulls himself up until his head is above the hands, where the pull holds become push holds. He elevates himself until the arms are straight and he can lock the elbows to relax the muscles. Rotating the hands inward during the transition to push holds helps to place the palms more securely on the ledge. Once the arms are locked, a foot can be raised and placed on the ledge. The climber may have to remove one hand to make room for the foot. Mantling can be fairly strenuous; however, most individuals should be able to support their weight, momentarily, on one arm if they keep it straight and locked. With the foot on the ledge, weight can be taken off the arms and the climber can grasp the holds that were previously out of reach. Once balanced over the foot, he can stand up on the ledge and plan his next move.

(b) Pure mantling uses arm strength to raise the body; however, the climber can often smear the balls of the feet against the rock and "walk" the feet up during the maneuver to take some of the weight off the arms. Sometimes edges will be available for short steps in the process.

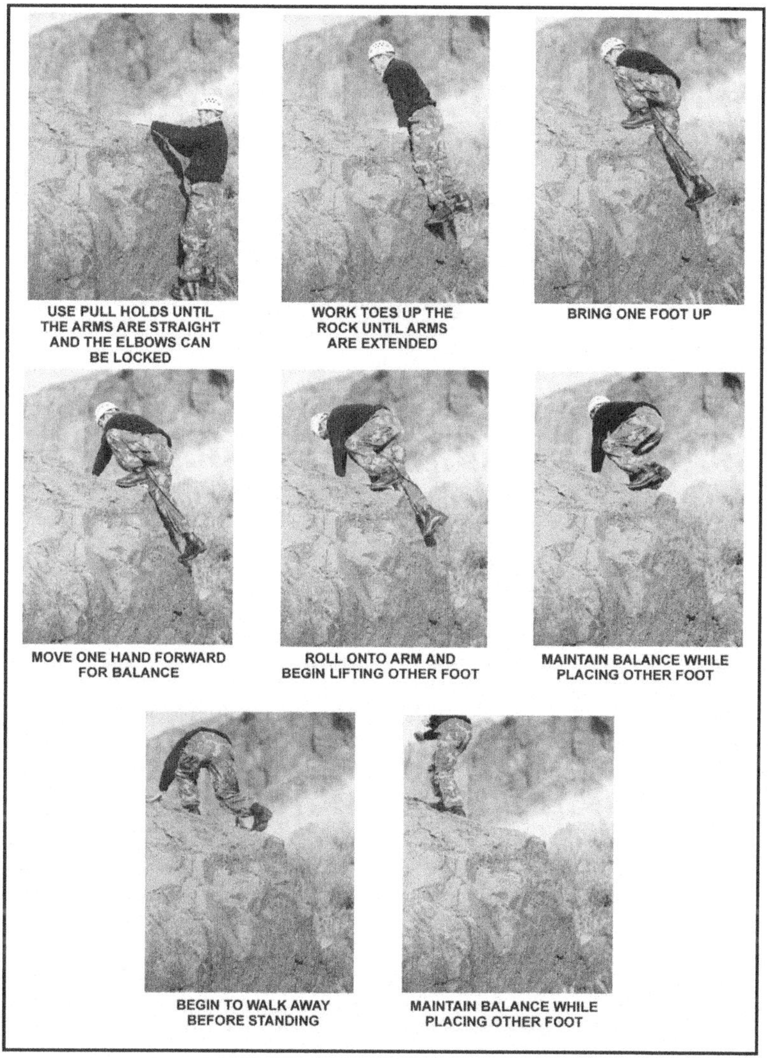

Figure 6-12. Mantling sequence.

(3) ***Undercling.*** An "undercling" is a classic example of handholds and footholds working in opposition (Figure 6-13). It is commonly used in places where the rock

projects outward, forming a bulge or small overhang. Underclings can be used in the tops of buckets, also. The hands are placed "palms-up" underneath the bulge, applying an upward pull. Increasing this upward pull creates a counterforce, or body tension, which applies more weight and friction to the footholds. The arms and legs should be kept as straight as possible to reduce fatigue. The climber can often lean back slightly in the undercling position, enabling him to see above the overhang better and search for the next hold.

Figure 6-13. Undercling.

(4) ***Lieback.*** The "lieback" is another good example of the hands working in opposition to the feet. The technique is often used in a vertical or diagonal crack separating two rock faces that come together at, more or less, a right angle (commonly referred to as a dihedral). The crack edge closest to the body is used for handholds while the feet are pressed against the other edge. The climber bends at the waist, putting the body into an L-shaped position. Leaning away from the crack on two pull holds, body tension creates friction between the feet and the hands. The feet must be kept relatively high to maintain weight, creating maximum friction between the sole and the rock surface. Either full sole contact or the smearing technique can be used, whichever seems to produce the most friction.

(a) The climber ascends a dihedral by alternately shuffling the hands and feet upward. The lieback technique can be extremely tiring, especially when the dihedral is near vertical. If the hands and arms tire out before completing the sequence, the climber will likely fall. The arms should be kept straight throughout the entire maneuver so the climber's weight is pulling against bones and ligaments, rather than muscle. The legs should be straightened whenever possible.

(b) Placing protection in a lieback is especially tiring. Look for edges or pockets for the feet in the crack or on the face for a better position to place protection from, or for a rest position. Often, a lieback can be avoided with closer examination of the available face features. The lieback can be used alternately with the jamming technique, or vice versa, for variation or to get past a section of a crack with difficult or nonexistent jam possibilities. The lieback can sometimes be used as a face maneuver (Figure 6-14).

Figure 6-14. Lieback on a face.

(5) ***Stemming.*** When the feet work in opposition from a relatively wide stance, the maneuver is known as stemming. The stemming technique can sometimes be used on faces, as well as in a dihedral in the absence of solid handholds for the lieback (Figure 6-15).

Figure 6-15. Stemming on a face.

(a) The classic example of stemming is when used in combination with two opposing push holds in wide, parallel cracks, known as chimneys. Chimneys are cracks in which the walls are at least 1 foot apart and just big enough to squeeze the body into. Friction is created by pushing outward with the hands and feet on each side of the crack. The climber ascends the chimney by alternately moving the hands and feet up the crack (Figure 6-16). Applying pressure with the back and bottom is usually necessary in wider chimneys. Usually, full sole contact of the shoes will provide the most friction, although smearing may work best in some instances. Chimneys that do not allow a full stemming position can be negotiated using the arms, legs, or body as an integral contact point. This technique will often feel more secure since there is more body to rock contact.

Figure 6-16. Chimney sequence.

Figure 6-16. Chimney sequence (continued).

(b) The climber can sometimes rest by placing both feet on the same side of the crack, forcing the body against the opposing wall. The feet must be kept relatively high up under the body so the force is directed sideways against the walls of the crack. The arms should be straightened with the elbows locked whenever possible to reduce muscle strain. The climber must ensure that the crack does not widen beyond the climbable width before committing to the maneuver. Remember to look for face features inside chimneys for more security in the climb.

(c) Routes requiring this type of climbing should be avoided as the equipment normally used for protection might not be large enough to protect chimneys. However, face features, or a much narrower crack in one or both corners, may sometimes be found deeper in the chimney allowing the use of normal size protection.

(6) ***Slab Technique.*** A slab is a relatively smooth, low-angled rock formation that requires a slightly modified climbing technique (Figure 6-17). Since slab rock normally contains few, if any holds, the technique requires maximum friction and perfect balance over the feet.

(a) On lower-angled slab, the climber can often stand erect and climb using full sole contact and other mountain walking techniques. On steeper slab, the climber will need to apply good smearing technique. Often, maximum friction cannot be attained on steeper slab from an erect stance. The climber will have to flex the ankles and knees so his weight is placed more directly over the balls of the feet. He may then have to bend at the waist to place the hands on the rock, while keeping the hips over his feet.

(b) The climber must pay attention to any changes in slope angle and adjust his body accordingly. Even the slightest change in the position of the hips over the feet can mean the difference between a good grip or a quick slip. The climber should also take advantage of any rough surfaces, or other irregularities in the rock he can place his hands or feet on, to increase friction.

Figure 6-17. Slab technique.

(7) ***Down Climbing.*** Descending steep rock is normally performed using a roped method; however, the climber may at some point be required to down climb a route. Even if climbing ropes and related equipment are on hand, down climbing easier terrain is often quicker than taking the time to rig a rappel point. Also, a climber might find himself confronted with difficulties part way up a route that exceed his climbing ability, or the abilities of others to follow. Whatever the case may be, down climbing is a skill well worth practicing.

CAUTIONS

1. Down climbing can inadvertently lead into an unforeseen dangerous position on a descent. When in doubt, use a roped descent.
2. Down climbing is accomplished at a difficulty level well below the ability of the climber. When in doubt, use a roped descent.

(a) On easier terrain, the climber can face outward, away from the rock, enabling him to see the route better and descend quickly. As the steepness and difficulty increase, he can often turn sideways, still having a good view of the descent route, but being better able to use the hands and feet on the holds available. On the steepest terrain, the climber will have to face the rock and down climb using good climbing techniques.

(b) Down climbing is usually more difficult than ascending a given route. Some holds will be less visible when down climbing, and slips are more likely to occur. The climber must often lean well away from the rock to look for holds and plan his movements. More weight is placed on the arms and handholds at times to accomplish this, as well as to help lower the climber to the next foothold. Hands should be moved to holds as low as waist level to give the climber more range of movement with each step. If the handholds are too high, he may have trouble reaching the next foothold. The climber must be careful not to overextend himself, forcing a release of his handholds before reaching the next foothold.

CAUTION

Do not drop from good handholds to a standing position. A bad landing could lead to injured ankles or a fall beyond the planned landing area.

(c) Descending slab formations can be especially tricky. The generally lower angle of slab rock may give the climber a false sense of security, and a tendency to move too quickly. Down climbing must be slow and deliberate, as in ascending, to maintain perfect balance and weight distribution over the feet. On lower-angle slab the climber may be able to stand more or less erect, facing outward or sideways, and descend using good flat foot technique. The climber should avoid the tendency to move faster, which can lead to uncontrollable speed.

(d) On steeper slab, the climber will normally face the rock and down climb, using the same smearing technique as for ascending. An alternate method for descending slab is to face away from the rock in a "crab" position (Figure 6-18). Weight is still concentrated over the feet, but may be shifted partly onto the hands to increase overall friction. The climber is able to maintain full sole contact with the rock and see the entire descent route. Allowing the buttocks to "drag behind" on the rock will decrease the actual weight on the footholds, reducing friction, and leading to the likelihood of a slip. Facing the rock, and down-climbing with good smearing technique, is usually best on steeper slab.

Figure 6-18. Descending slab in the crab position.

Section III. ROPED CLIMBING

When the angle, length, and difficulty of the proposed climbing route surpasses the ability of the climbers' safety margin (possibly on class 4 and usually on class 5 terrain), ropes must be used to proceed. Roped climbing is only safe if accomplished correctly. Reading this manual does not constitute skill with ropes—much training and practice is necessary. Many aspects of roped climbing take time to understand and learn. Ropes are normally not used in training until the basic principles of climbing are covered.

Note: A rope is completely useless for climbing unless the climber knows how to use it safely.

6-11. TYING-IN TO THE CLIMBING ROPE

Over the years, climbers have developed many different knots and procedures for tying-in to the climbing rope. Some of the older methods of tying directly into the rope require minimal equipment and are relatively easy to inspect; however, they offer little support to the climber, may induce further injuries, and may even lead to strangulation in a severe fall. A severe fall, where the climber might fall 20 feet or more and be left dangling on the end of the rope, is highly unlikely in most instances, especially for most personnel involved in military climbing. Tying directly into the rope is perfectly safe for many roped party climbs used in training on lower-angled rock. All climbers should know how to properly tie the rope around the waist in case a climbing harness is unavailable.

6-12. PRESEWN HARNESSES

Although improvised harnesses are made from readily available materials and take little space in the pack or pocket, presewn harnesses provide other aspects that should be considered. No assembly is required, which reduces preparation time for roped movement. All presewn harnesses provide a range of adjustability. These harnesses have a fixed buckle that, when used correctly, will not fail before the nylon materials connected to it. However, specialized equipment, such as a presewn harness, reduce the flexibility of gear. Presewn harness are bulky, also.

a. **Seat Harness.** Many presewn seat harnesses are available with many different qualities separating them, including cost.

(1) The most notable difference will be the amount and placement of padding. The more padding the higher the price and the more comfort. Gear loops sewn into the waist belt on the sides and in the back are a common feature and are usually strong enough to hold quite a few carabiners and or protection. The gear loops will vary in number from one model/manufacturer to another.

(2) Although most presewn seat harnesses have a permanently attached belay loop connecting the waist belt and the leg loops, the climbing rope should be run around the waist belt and leg loop connector. The presewn belay loop adds another link to the chain of possible failure points and only gives one point of security whereas running the rope through the waist belt and leg loop connector provides two points of contact.

(3) If more than two men will be on the rope, connect the middle position(s) to the rope with a carabiner routed the same as stated in the previous paragraph.

(4) Many manufactured seat harnesses will have a presewn loop of webbing on the rear. Although this loop is much stronger than the gear loops, it is not for a belay anchor. It is a quick attachment point to haul an additional rope.

b. **Chest Harness.** The chest harness will provide an additional connecting point for the rope, usually in the form of a carabiner loop to attach a carabiner and rope to. This type of additional connection will provide a comfortable hanging position on the rope, but otherwise provides no additional protection from injury during a fall (if the seat harness is fitted correctly).

(1) A chest harness will help the climber remain upright on the rope during rappelling or ascending a fixed rope, especially while wearing a heavy pack. (If rappelling or ascending long or multiple pitches, let the pack hang on a drop cord below the feet and attached to the harness tie-in point.)

(2) The presewn chest harnesses available commercially will invariably offer more comfort or performance features, such as padding, gear loops, or ease of adjustment, than an improvised chest harness.

c. **Full-Body Harness.** Full-body harnesses incorporate a chest and seat harness into one assembly. This is the safest harness to use as it relocates the tie-in point higher, at the chest, reducing the chance of an inverted position when hanging on the rope. This is especially helpful when moving on ropes with heavy packs. A full-body harness only affects the body position when hanging on the rope and will not prevent head injury in a fall.

CAUTION

This type of harness does not prevent the climber from falling head first. Body position during a fall is affected only by the forces that generated the fall, and this type of harness promotes an upright position only when hanging on the rope from the attachment point.

6-13. IMPROVISED HARNESSES

Without the use of a manufactured harness, many methods are still available for attaching oneself to a rope. Harnesses can be improvised using rope or webbing and knots.

a. **Swami Belt.** The swami belt is a simple, belt-only harness created by wrapping rope or webbing around the waistline and securing the ends. One-inch webbing will provide more comfort. Although an effective swami belt can be assembled with a minimum of one wrap, at least two wraps are recommended for comfort, usually with approximately ten feet of material. The ends are secured with an appropriate knot.

b. **Bowline-on-a-Coil.** Traditionally, the standard method for attaching oneself to the climbing rope was with a bowline-on-a-coil around the waist. The extra wraps distribute the force of a fall over a larger area of the torso than a single bowline would, and help prevent the rope from riding up over the rib cage and under the armpits. The knot must be tied snugly around the narrow part of the waist, just above the bony portions of the hips (pelvis). Avoid crossing the wraps by keeping them spread over the waist area. "Sucking in the gut" a bit when making the wraps will ensure a snug fit.

(1) The bowline-on-a-coil can be used to tie-in to the end of the rope (Figure 6-19). The end man should have a minimum of four wraps around the waist before completing the knot.

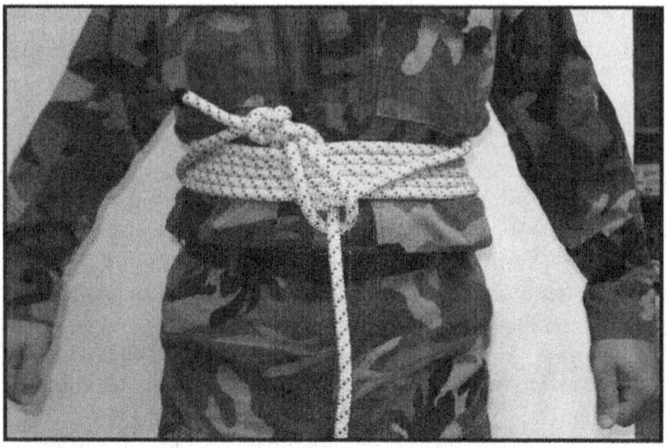

Figure 6-19. Tying-in with a bowline-on-a-coil.

(2) The bowline-on-a-coil is a safe and effective method for attaching to the rope when the terrain is low-angled, WITHOUT THE POSSIBILITY OF A SEVERE FALL. When the terrain becomes steeper, a fall will generate more force on the climber and this will be felt through the coils of this type of attachment. A hard fall will cause the coils to ride up against the ribs. In a severe fall, any tie-in around the waist only could place a "shock load" on the climber's lower back. Even in a relatively short fall, if the climber ends up suspended in mid-air and unable to regain footing on the rock, the rope around the waist can easily cut off circulation and breathing in a relatively short time.

(3) The climbing harness distributes the force of a fall over the entire pelvic region, like a parachute harness. Every climber should know how to tie some sort of improvised climbing harness from sling material. A safe, and comfortable, seat/chest combination harness can be tied from one-inch tubular nylon.

c. **Improvised Seat Harness.** A seat harness can be tied from a length of webbing approximately 25 feet long (Figure 6-20, page 6-30).

(1) Locate the center of the rope. Off to one side, tie two fixed loops approximately 6 inches apart (overhand loops). Adjust the size of the loops so they fit snugly around the thigh. The loops are tied into the sling "off center" so the remaining ends are different lengths. The short end should be approximately 4 feet long (4 to 5 feet for larger individuals).

(2) Slip the leg loops over the feet and up to the crotch, with the knots to the front. Make one complete wrap around the waist with the short end, wrapping to the outside, and hold it in place on the hip. Keep the webbing flat and free of twists when wrapping.

(3) Make two to three wraps around the waist with the long end in the opposite direction (wrapping to the outside), binding down on the short end to hold it in place. Grasping both ends, adjust the waist wraps to a snug fit. Connect the ends with the

appropriate knot between the front and one side so you will be able to see what you are doing.

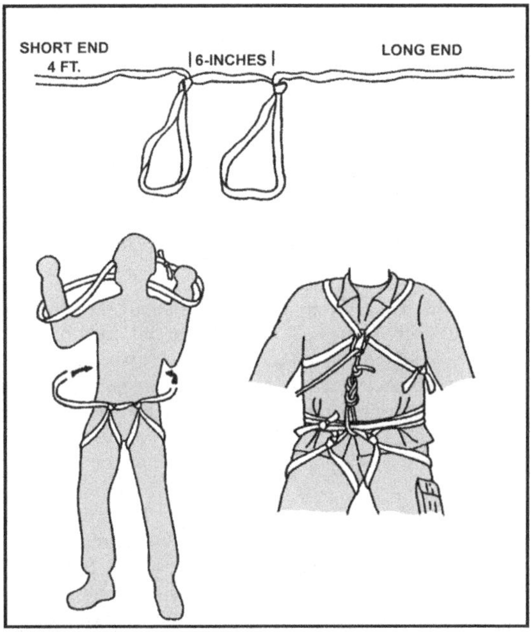

Figure 6-20. Improvised seat and chest harness.

d. **Improvised Chest Harness.** The chest harness can be tied from rope or webbing, but remember that with webbing, wider is better and will be more comfortable when you load this harness. Remember as you tie this harness that the remaining ends will need to be secured so choose the best length. Approximately 6 to 10 feet usually works.

(1) Tie the ends of the webbing together with the appropriate knot, making a sling 3 to 4 feet long.

(2) Put a single twist into the sling, forming two loops.

(3) Place an arm through each loop formed by the twist, just as you would put on a jacket, and drape the sling over the shoulders. The twist, or cross, in the sling should be in the middle of the back.

(4) Join the two loops at the chest with a carabiner. The water knot should be set off to either side for easy inspection (if a pack is to be worn, the knot will be uncomfortable if it gets between the body and the pack). The chest harness should fit just loose enough to allow necessary clothing and not to restrict breathing or circulation. Adjust the size of the sling if necessary.

e. **Improvised Full-Body Harness.** Full-body harnesses incorporate a chest and seat harness into one assembly.

(1) The full-body harness is the safest harness because it relocates the tie-in point higher, at the chest, reducing the chance of an inverted hanging position on the rope. This is especially helpful when moving on ropes with heavy packs. A full-body harness affects the body position only when hanging on the rope.

CAUTION

A full-body harness does not prevent falling head first; body position in a fall is caused by the forces that caused the fall.

(2) Although running the rope through the carabiner of the chest harness does, in effect, create a type of full-body harness, it is not a true full-body harness until the chest harness and the seat harness are connected as one piece. A true full-body harness can be improvised by connecting the chest harness to the seat harness, but not by just tying the rope into both—the two harnesses must be "fixed" as one harness. Fix them together with a short loop of webbing or rope so that the climbing rope can be connected directly to the chest harness and your weight is supported by the seat harness through the connecting material.

f. **Attaching the Rope to the Improvised Harness.** The attachment of the climbing rope to the harness is a CRITICAL LINK. The strength of the rope means nothing if it is attached poorly, or incorrectly, and comes off the harness in a fall. The climber ties the end of the climbing rope to the seat harness with an appropriate knot. If using a chest harness, the standing part of the rope is then clipped into the chest harness carabiner. The seat harness absorbs the main force of the fall, and the chest harness helps keep the body upright.

CAUTION

The knot must be tied around all the waist wraps and the 6-inch length of webbing between the leg loops.

(1) A middleman must create a fixed loop to tie in to. A rethreaded figure-eight loop tied on a doubled rope or the three loop bowline can be used. If using the three loop bowline, ensure the end, or third loop formed in the knot, is secured around the tie-in loops with an overhand knot. The standing part of the rope going to the lead climber is clipped into the chest harness carabiner.

Note: The climbing rope is not clipped into the chest harness when belaying.

(2) The choice of whether to tie-in with a bowline-on-a-coil or into a climbing harness depends entirely on the climber's judgment, and possibly the equipment

available. A good rule of thumb is: "Wear a climbing harness when the potential for severe falls exists and for all travel over snow-covered glaciers because of the crevasse fall hazard."

(3) Under certain conditions many climbers prefer to attach the rope to the seat harness with a locking carabiner, rather than tying the rope to it. This is a common practice for moderate snow and ice climbing, and especially for glacier travel where wet and frozen knots become difficult to untie.

CAUTION

Because the carabiner gate may be broken or opened by protruding rocks during a fall, tie the rope directly to the harness for maximum safety.

Section IV. BELAY TECHNIQUES

Tying-in to the climbing rope and moving as a member of a rope team increases the climber's margin of safety on difficult, exposed terrain. In some instances, such as when traveling over snow-covered glaciers, rope team members can often move at the same time, relying on the security of a tight rope and "team arrest" techniques to halt a fall by any one member. On steep terrain, however, simultaneous movement only helps to ensure that if one climber falls, he will jerk the other rope team members off the slope. For the climbing rope to be of any value on steep rock climbs, the rope team must incorporate "belays" into the movement.

Belaying is a method of managing the rope in such a way that, if one person falls, the fall can be halted or "arrested" by another rope team member (belayer). One person climbs at a time, while being belayed from above or below by another. The belayer manipulates the rope so that friction, or a "brake," can be applied to halt a fall. Belay techniques are also used to control the descent of personnel and equipment on fixed rope installations, and for additional safety on rappels and stream crossings.

Belaying is a skill that requires practice to develop proficiency. Setting up a belay may at first appear confusing to the beginner, but with practice, the procedure should become "second nature." If confronted with a peculiar problem during the setup of a belay, try to use common sense and apply the basic principles stressed throughout this text. Remember the following key points:

- Select the best possible terrain features for the position and use terrain to your advantage.
- Use a well braced, sitting position whenever possible.
- Aim and anchor the belay for all possible load directions.
- Follow the "minimum" rule for belay anchors—2 for a downward pull, 1 for an upward pull.
- Ensure anchor attachments are aligned, independent, and snug.
- Stack the rope properly.
- Choose a belay technique appropriate for the climbing.

- Use a guide carabiner for rope control in all body belays.
- Ensure anchor attachments, guide carabiner (if applicable), and rope running to the climber are all on the guidehand side.
- The brake hand remains on the rope when belaying.

CAUTION
Never remove the brake hand from the rope while belaying. If the brake hand is removed, there is no belay.

- Ensure you are satisfied with your position before giving the command "BELAY ON."
- The belay remains in place until the climber gives the command "OFF BELAY."

CAUTION
The belay remains in place the from the time the belayer commands *"BELAY ON"* until the climber commands *"OFF BELAY."*

6-14. PROCEDURE FOR MANAGING THE ROPE
A number of different belay techniques are used in modern climbing, ranging from the basic "body belays" to the various "mechanical belays," which incorporate some type of friction device.

a. Whether the rope is wrapped around the body, or run through a friction device, the rope management procedure is basically the same. The belayer must be able to perform three basic functions: manipulate the rope to give the climber slack during movement, take up rope to remove excess slack, and apply the brake to halt a fall.

b. The belayer must be able to perform all three functions while maintaining "total control" of the rope at all times. Total control means the brake hand is NEVER removed from the rope. When giving slack, the rope simply slides through the grasp of the brake hand, at times being fed to the climber with the other "feeling" or guide hand. Taking up rope, however, requires a certain technique to ensure the brake hand remains on the rope at all times. The following procedure describes how to take up excess rope and apply the brake in a basic body belay.

(1) Grasping the rope with both hands, place it behind the back and around the hips. The hand on the section of rope between the belayer and the climber would be the guide hand. The other hand is the brake hand.

(2) Take in rope with the brake hand until the arm is fully extended. The guide hand can also help to pull in the rope (Figure 6-21, step 1, page 6-34).

(3) Holding the rope in the brake hand, slide the guide hand out, extending the arm so the guide hand is father away from the body than the brake hand (Figure 6-21, step 2).

(4) Grasp both parts of the rope, to the front of the brake hand, with the guide hand (Figure 6-21, step 3).

(5) Slide the brake hand back towards the body (Figure 6-21, step 4).

(6) Repeat step 5 of Figure 6-21. The brake can be applied at any moment during the procedure. It is applied by wrapping the rope around the front of the hips while increasing grip with the brake hand (Figure 6-21, step 6).

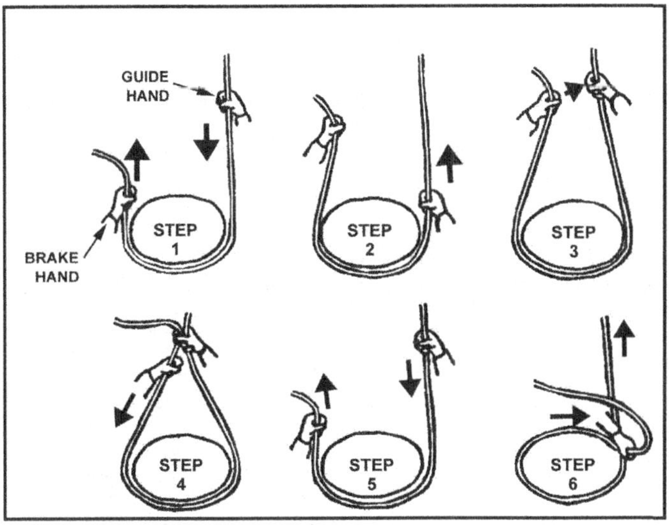

Figure 6-21. Managing the rope.

6-15. CHOOSING A BELAY TECHNIQUE

The climber may choose from a variety of belay techniques. A method that works well in one situation may not be the best choice in another. The choice between body belays and mechanical belays depends largely on equipment available, what the climber feels most comfortable with, and the amount of load, or fall force, the belay may have to absorb. The following describes a few of the more widely used techniques, and the ones most applicable to military mountaineering.

a. **Body Belay.** The basic body belay is the most widely used technique on moderate terrain. It uses friction between the rope and the clothed body as the rope is pressured across the clothing. It is the simplest belay, requiring no special equipment, and should be the first technique learned by all climbers. A body belay gives the belayer the greatest "feel" for the climber, letting him know when to give slack or take up rope. Rope management in a body belay is quick and easy, especially for beginners, and is effective in snow and ice climbing when ropes often become wet, stiff, and frozen. The body belay, in its various forms, will hold low to moderate impact falls well. It has been known to arrest some severe falls, although probably not without inflicting great pain on the belayer.

CAUTION

The belayer must ensure he is wearing adequate clothing to protect his body from rope burns when using a body belay. Heavy duty cotton or leather work gloves can also be worn to protect the hands.

(1) ***Sitting Body Belay.*** The sitting body belay is the preferred position and is usually the most secure (Figure 6-22). The belayer sits facing the direction where the force of a fall will likely come from, using terrain to his advantage, and attempts to brace both feet against the rock to support his position. It is best to sit in a slight depression, placing the buttocks lower than the feet, and straightening the legs for maximum support. When perfectly aligned, the rope running to the climber will pass between the belayer's feet, and both legs will equally absorb the force of a fall. Sometimes, the belayer may not be able to sit facing the direction he would like, or both feet cannot be braced well. The leg on the "guide hand" side should then point towards the load, bracing the foot on the rock when possible. The belayer can also "straddle" a large tree or rock nubbin for support, as long as the object is solid enough to sustain the possible load.

Figure 6-22. Sitting body belay.

(2) ***Standing Body Belay.*** The standing body belay is used on smaller ledges where there is no room for the belayer to sit (Figure 6-23). What appears at first to be a fairly unstable position can actually be quite secure when belay anchors are placed at or above shoulder height to support the stance when the force will be downward.

Figure 6-23. Standing body belay.

(a) For a body belay to work effectively, the belayer must ensure that the rope runs around the hips properly, and remains there under load when applying the brake. The rope should run around the narrow portion of the pelvic girdle, just below the bony high points of the hips. If the rope runs too high, the force of a fall could injure the belayer's midsection and lower rib cage. If the rope runs too low, the load may pull the rope below the buttocks, dumping the belayer out of position. It is also possible for a strong upward or downward pull to strip the rope away from the belayer, rendering the belay useless.

(b) To prevent any of these possibilities from happening, the belay rope is clipped into a carabiner attached to the guide hand side of the seat harness (or bowline-on-a-coil). This "guide carabiner" helps keep the rope in place around the hips and prevents loss of control in upward or downward loads (Figure 6-24).

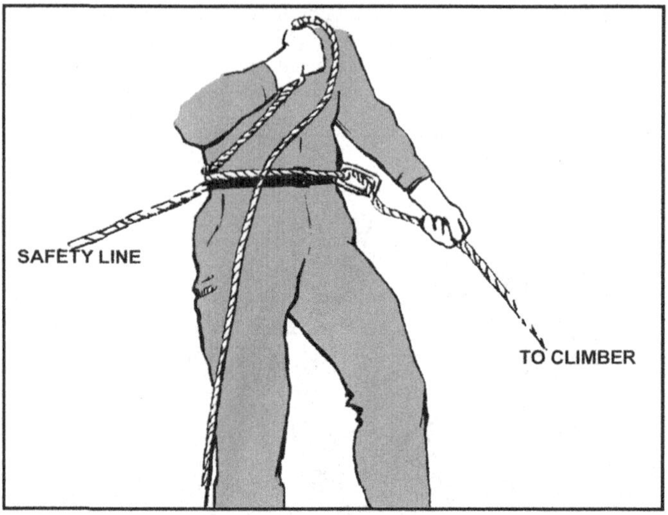

Figure 6-24. Guide carabiner for rope control in a body belay.

b. **Mechanical Belay.** A mechanical belay **must** be used whenever there is potential for the lead climber to take a severe fall. The holding power of a belay device is vastly superior to any body belay under high loads. However, rope management in a mechanical belay is more difficult to master and requires more practice. For the most part, the basic body belay should be totally adequate on a typical military route, as routes used during military operations should be the easiest to negotiate.

(1) ***Munter Hitch.*** The Munter hitch is an excellent mechanical belay technique and requires only a rope and a carabiner (Figure 6-25, page 6-38). The Munter is actually a two-way friction hitch. The Munter hitch will flip back and forth through the carabiner as the belayer switches from giving slack to taking up rope. The carabiner must be large enough, and of the proper design, to allow this function. The locking pear-shaped carabiner, or pearabiner, is designed for the Munter hitch.

(a) The Munter hitch works exceptionally well as a lowering belay off the anchor. As a climbing belay, the carabiner should be attached to the front of the belayer's seat harness. The hitch is tied by forming a loop and a bight in the rope, attaching both to the carabiner. It's fairly easy to place the bight on the carabiner backwards, which forms an obvious, useless hitch. Put some tension on the Munter to ensure it is formed correctly, as depicted in the following illustrations.

(b) The Munter hitch will automatically "lock-up" under load as the brake hand grips the rope. The brake is increased by pulling the slack rope away from the body, towards the load. The belayer must be aware that flipping the hitch DOES NOT change the function of the hands. The hand on the rope running to the climber, or load, is always the guide hand.

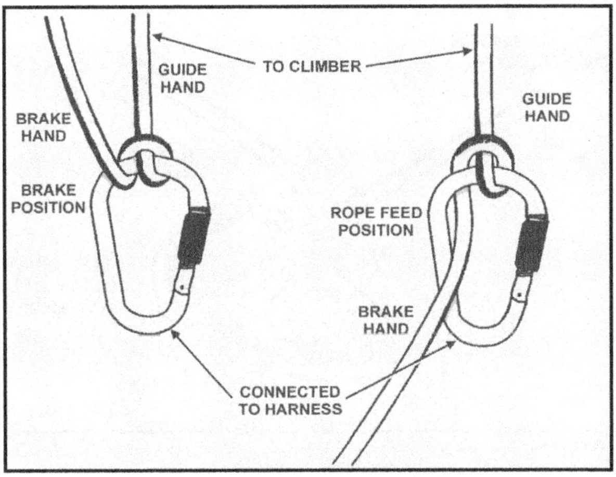

Figure 6-25. Munter hitch.

(2) ***Figure-Eight Device.*** The figure-eight device is a versatile piece of equipment and, though developed as a rappel device, has become widely accepted as an effective mechanical belay device (Figure 6-26). The advantage of any mechanical belay is friction required to halt a fall is applied on the rope through the device, rather than around the belayer's body. The device itself provides rope control for upward and downward pulls and excellent friction for halting severe falls. The main principle behind the figure-eight device in belay mode is the friction developing on the rope as it reaches and exceeds the 90-degree angle between the rope entering the device and leaving the device. As a belay device, the figure-eight works well for both belayed climbing and for lowering personnel and equipment on fixed-rope installations.

(a) As a climbing belay, a bight placed into the climbing rope is run through the "small eye" of the device and attached to a locking carabiner at the front of the belayer's seat harness. A short, small diameter safety rope is used to connect the "large eye" of the figure eight to the locking carabiner for control of the device. The guide hand is placed on the rope running to the climber. Rope management is performed as in a body belay. The brake is applied by pulling the slack rope in the brake hand towards the body, locking the rope between the device and the carabiner.

(b) As a lowering belay, the device is normally attached directly to the anchor with the rope routed as in rappelling.

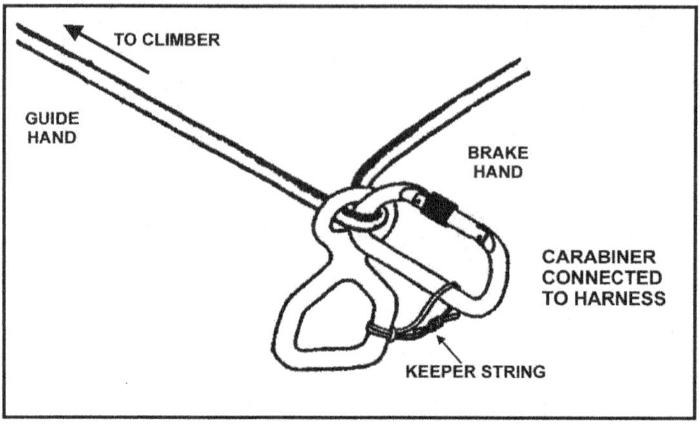

Figure 6-26. Figure-eight device.

Note: Some figure-eight descenders should not be used as belay devices due to their construction and design. Always refer to manufacturer's specifications and directions before use.

(3) ***Mechanical Camming Device.*** The mechanical camming device has an internal camming action that begins locking the rope in place as friction is increased. Unlike the other devices, the mechanical camming device can stop a falling climber without any input from the belayer. A few other devices perform similarly to this, but have no moving parts. Some limitations to these type devices are minimum and maximum rope diameters.

(4) ***Other Mechanical Belay Devices.*** There are many other commercially available mechanical belay devices. Most of these work with the same rope movement direction and the same braking principle. The air traffic controller (ATC), slotted plate, and other tube devices are made in many different shapes. These all work on the same principle as the figure-eight device—friction increases on the rope as it reaches and exceeds the 90-degree angle between the rope entering the device and leaving the device.

6-16. ESTABLISHING A BELAY

A belay can be established using either a direct or indirect connection. Each type has advantages and disadvantages. The choice will depend on the intended use of the belay.

a. **Direct Belay.** The direct belay removes any possible forces from the belayer and places this force completely on the anchor. Used often for rescue installations or to bring a second climber up to a new belay position in conjunction with the Munter hitch, the belay can be placed above the belayer's stance, creating a comfortable position and ease of applying the brake. Also, if the second falls or weights the rope, the belayer is not locked into a position. Direct belays provide no shock-absorbing properties from the belayer's attachment to the system as does the indirect belay; therefore, the belayer is apt to pay closer attention to the belaying process.

b. **Indirect Belay.** An indirect belay, the most commonly used, uses a belay device attached to the belayer's harness. This type of belay provides dynamic shock or weight absorption by the belayer if the climber falls or weights the rope, which reduces the direct force on the anchor and prevents a severe shock load to the anchor.

6-17. SETTING UP A BELAY

In rock climbing, climbers must sometimes make do with marginal protection placements along a route, but belay positions must be made as "bombproof" as possible. Additionally, the belayer must set up the belay in relation to where the fall force will come from and pay strict attention to proper rope management for the belay to be effective. All belay positions are established with the anchor connection to the front of the harness. If the belay is correctly established, the belayer will feel little or no force if the climber falls or has to rest on the rope. Regardless of the actual belay technique used, five basic steps are required to set up a sound belay.

a. **Select Position and Stance.** Once the climbing line is picked, the belayer selects his position. It's best if the position is off to the side of the actual line, putting the belayer out of the direct path of a potential fall or any rocks kicked loose by the climber. The position should allow the belayer to maintain a comfortable, relaxed stance, as he could be in the position for a fairly long time. Large ledges that allow a well braced, sitting stance are preferred. Look for belay positions close to bombproof natural anchors. The position must at least allow for solid artificial placements.

b. **Aim the Belay.** With the belay position selected, the belay must now be "aimed." The belayer determines where the rope leading to the climber will run and the direction the force of a fall will likely come from. When a lead climber begins placing protection, the fall force on the belayer will be in some upward direction, and in line with the first protection placement. If this placement fails under load, the force on the belay could be straight down again. The belayer must aim his belay for all possible load directions, adjusting his position or stance when necessary. The belay can be aimed through an anchor placement to immediately establish an upward pull; however, the belayer must always be prepared for the more severe downward fall force in the event intermediate protection placements fail.

c. **Anchor the Belay.** For a climbing belay to be considered bombproof, the belayer must be attached to a solid anchor capable of withstanding the highest possible fall force. A solid natural anchor would be ideal, but more often the belayer will have to place pitons or chocks. A single artificial placement should never be considered adequate for anchoring a belay (except at ground level). Multiple anchor points capable of supporting both upward and downward pulls should be placed. The rule of thumb is to place two anchors for a downward pull and one anchor for an upward pull as a MINIMUM. The following key points also apply to anchoring belays.

(1) Each anchor must be placed in line with the direction of pull it is intended to support.

(2) Each anchor attachment must be rigged "independently" so a failure of one will not shock load remaining placements or cause the belayer to be pulled out of position.

(3) The attachment between the anchor and the belayer must be snug to support the stance. Both belayer's stance and belay anchors should absorb the force of a fall.

(4) It is best for the anchors to be placed relatively close to the belayer with short attachments. If the climber has to be tied-off in an emergency, say after a severe fall, the belayer can attach a Prusik sling to the climbing rope, reach back, and connect the sling to one of the anchors. The load can be placed on the Prusik and the belayer can come out of the system to render help.

(5) The belayer can use either a portion of the climbing rope or slings of the appropriate length to connect himself to the anchors. It's best to use the climbing rope whenever possible, saving the slings for the climb. The rope is attached using either figure eight loops or clove hitches. Clove hitches have the advantage of being easily adjusted. If the belayer has to change his stance at some point, he can reach back with the guide hand and adjust the length of the attachment through the clove hitch as needed.

(6) The anchor attachments should also help prevent the force of a fall from "rotating" the belayer out of position. To accomplish this, the climbing rope must pass around the "guide-hand side" of the body to the anchors. Sling attachments are connected to the belayer's seat harness (or bowline-on-a-coil) on the guide-hand side.

(7) Arrangement of rope and sling attachments may vary according to the number and location of placements. Follow the guidelines set forth and remember the key points for belay anchors; "in line", "independent", and "snug". Figure 6-27 shows an example of a common arrangement, attaching the rope to the two "downward" anchors and a sling to the "upward" anchor. Note how the rope is connected from one of the anchors back to the belayer. This is not mandatory, but often helps "line-up" the second attachment.

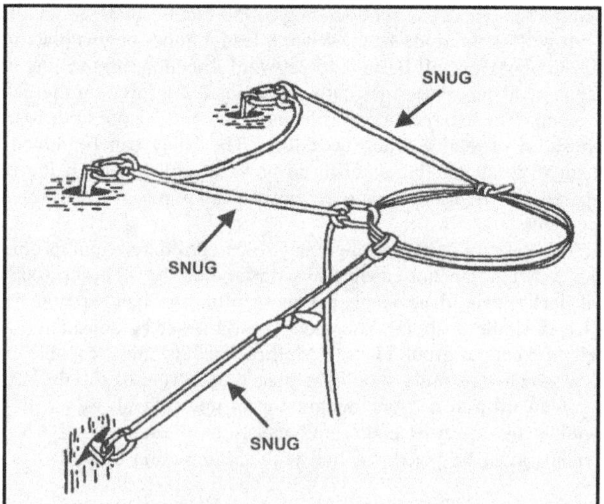

Figure 6-27. Anchoring a belay.

d. **Stack the Rope.** Once the belayer is anchored into position, he must stack the rope to ensure it is free of twists and tangles that might hinder rope management in the belay. The rope should be stacked on the ground, or on the ledge, where it will not get caught in cracks or nubbins as it is fed out to the climber.

(1) On small ledges, the rope can be stacked on top of the anchor attachments if there is no other place to lay it, but make sure to stack it carefully so it won't tangle with the anchored portion of the rope or other slings. The belayer must also ensure that the rope will not get tangled around his legs or other body parts as it "feeds" out.

(2) The rope should never be allowed to hang down over the ledge. If it gets caught in the rock below the position, the belayer may have to tie-off the climber and come out of the belay to free the rope; a time-consuming and unnecessary task. The final point to remember is the rope must be stacked "from the belayer's end" so the rope running to the climber comes off the "top" of the stacked pile.

e. **Attach the Belay.** The final step of the procedure is to attach the belay. With the rope properly stacked, the belayer takes the rope coming off the top of the pile, removes any slack between himself and the climber, and applies the actual belay technique. If using a body belay, ensure the rope is clipped into the guide carabiner.

(1) The belayer should make one quick, final inspection of his belay. If the belay is set up correctly, the anchor attachments, guide carabiner if applicable, and the rope running to the climber will all be on the "guide hand" side, which is normally closest to the rock (Figure 6-28). If the climber takes a fall, the force, if any, should not have any negative effect on the belayer's involvement in the system. The brake hand is out away from the slope where it won't be jammed between the body and the rock. The guide hand can be placed on the rock to help support the stance when applying the brake.

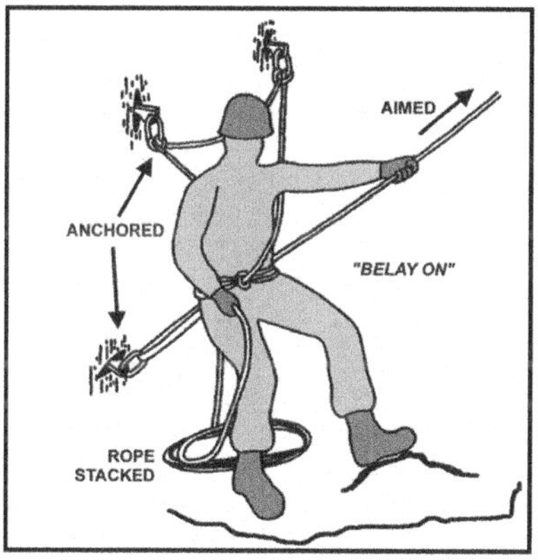

Figure 6-28. Belay setup.

(2) When the belayer is satisfied with his position, he gives the signal, "BELAY ON!". When belaying the "second", the same procedure is used to set up the belay. Unless the belay is aimed for an upward pull, the fall force is of course downward and the

belayer is usually facing away from the rock, the exception being a hanging belay on a vertical face. If the rope runs straight down to the climber and the anchors are directly behind the position, the belayer may choose to brake with the hand he feels most comfortable with. Anchor attachments, guide carabiner, and rope running to the climber through the guide hand must still be aligned on the same side to prevent the belayer from being rotated out of position, unless the belayer is using an improvised harness and the anchor attachment is at the rear.

6-18. TOP-ROPE BELAY

A "top-rope" is a belay setup used in training to protect a climber while climbing on longer, exposed routes. A solid, bombproof anchor is required at the top of the pitch. The belayer is positioned either on the ground with the rope running through the top anchor and back to the climber, or at the top at the anchor. The belayer takes in rope as the climber proceeds up the rock. If this is accomplished with the belayer at the bottom, the instructor can watch the belayer while he coaches the climber through the movements.

> **CAUTION**
> Do not use a body belay for top-rope climbing. The rope will burn the belayer if the climber has to be lowered.

Section V. CLIMBING COMMANDS

Communication is often difficult during a climb. As the distance between climber and belayer increases, it becomes harder to distinguish one word from another and the shortest sentence may be heard as nothing more than jumbled syllables. A series of standard voice commands were developed over the years to signal the essential rope management functions in a belayed climb. Each command is concise and sounds a bit different from another to reduce the risk of a misunderstanding between climber and belayer. They must be pronounced clearly and loudly so they can be heard and understood in the worst conditions.

6-19. VERBAL COMMANDS

Table 6-1, page 6-44, lists standard rope commands and their meanings in sequence as they would normally be used on a typical climb. (Note how the critical "BELAY" commands are reversed so they sound different and will not be confused.)

BELAYER	CLIMBER	MEANING/ACTION TAKEN
"BELAY ON"		The belay is on; you may climb when ready; the rope will be managed as needed.
	"CLIMBING" (as a courtesy)	I am ready to climb.
"CLIMB" (as a courtesy)		Proceed, and again, the rope will be managed as necessary.
"ROCK"	"ROCK"	**PROTECT YOURSELF FROM FALLING OBJECTS.** Signal will be echoed by all climbers in the area. If multipitch climbing, ensure climbers below hear.
	"TAKE ROPE"	Take in excess rope between us without pulling me off the route. Belayer takes in rope.
	"SLACK"	Release all braking/tension on the rope so I can have slack without pulling the rope. Belayer removes brake/tension.
	"TENSION"	Take all the slack, apply brake, and hold me. My weight will be on the rope. Belayer removes slack and applies brake.
	"FALLING"	I am falling. Belayer applies brake to arrest the fall.
"TWEN-TY-FIVE"		You have approximately 25 feet of rope left. Start looking for the next belay position. Climber selects a belay position.
"FIF-TEEN"		You have approximately 15 feet of rope left. Start looking for the next belay position. Climber selects a belay position within the next few feet.
"FIVE"	Set up the belay.	You have 5 feet of rope left. Set up the belay position. You have no more rope. Climber sets up the belay.
Removes the belay, remains anchored. Prepares to climb.	"OFF BELAY"	I have finished climbing and I am anchored. You may remove the belay. Belayer removes the belay and, remaining anchored, prepares to climb.

Table 6-1. Rope commands.

6-20. ROPE TUG COMMANDS

Sometimes the loudest scream cannot be heard when the climber and belayer are far apart. This is especially true in windy conditions, or when the climber is around a corner, above an overhang, or at the back of a ledge. It may be necessary to use a series of "tugs" on the rope in place of the standard voice commands. To avoid any possible confusion with interpretation of multiple rope tug commands, use only one.

a. While a lead climb is in progress, the most important command is "BELAY ON." This command is given only by the climber when the climber is anchored and is prepared for the second to begin climbing. With the issue of this command, the second knows the climber is anchored and the second prepares to climb.

b. For a rope tug command, the leader issues three distinct tugs on the rope AFTER anchoring and putting the second on belay. This is the signal for "BELAY ON" and signals the second to climb when ready. The new belayer keeps slack out of the rope.

Section VI. ROPED CLIMBING METHODS

In military mountaineering, the primary mission of a roped climbing team is to "fix" a route with some type of rope installation to assist movement of less trained personnel in the unit. This duty falls upon the most experienced climbers in the unit, usually working in two- or three-man groups or teams called assault climbing teams. Even if the climbing is for another purpose, roped climbing should be performed whenever the terrain becomes difficult and exposed.

6-21. TOP-ROPED CLIMBING

Top-roped climbing is used for training purposes only. This method of climbing is not used for movement due to the necessity of pre-placing anchors at the top of a climb. If you can easily access the top of a climb, you can easily avoid the climb itself.

a. For training, top-roped climbing is valuable because it allows climbers to attempt climbs above their skill level and or to hone present skills without the risk of a fall. Top-roped climbing may be used to increase the stamina of a climber training to climb longer routes as well as for a climber practicing protection placements.

b. The belayer is positioned either at the base of a climb with the rope running through the top anchor and back to the climber or at the top at the anchor. The belayer takes in rope as the climber moves up the rock, giving the climber the same protection as a belay from above. If this is accomplished with the belayer at the bottom, the instructor is able to keep an eye on the belayer while he coaches the climber through the movements.

6-22. LEAD CLIMBING

A lead climb consists of a belayer, a leader or climber, rope(s), and webbing or hardware used to establish anchors or protect the climb. As he climbs the route, the leader emplaces "intermediate" anchors, and the climbing rope is connected to these anchors with a carabiner. These "intermediate" anchors protect the climber against a fall-thus the term "protecting the climb."

Note: Intermediate anchors are commonly referred to as "protection," "pro," "pieces," "pieces of pro," "pro placements," and so on. For standardization within this publication, these specific anchors will be referred to as "protection;" anchors established for other purposes, such as rappel points, belays, or other rope installations, will be referred to as "anchors."

> **CAUTION**
> During all lead climbing, each climber in the team is either anchored or being belayed.

a. Lead climbing with two climbers is the preferred combination for movement on technically difficult terrain. Two climbers are at least twice as fast as three climbers, and are efficient for installing a "fixed rope," probably the most widely used rope installation in the mountains. A group of three climbers are typically used on moderate snow, ice, and snow-covered glaciers where the rope team can often move at the same time, stopping occasionally to set up belays on particularly difficult sections. A group or team of three climbers is sometimes used in rock climbing because of an odd number of personnel, a shortage of ropes (such as six climbers and only two ropes), or to protect and assist an individual who has little or no experience in climbing and belaying. Whichever technique is chosen, a standard roped climbing procedure is used for maximum speed and safety.

b. When the difficulty of the climbing is within the "leading ability" of both climbers, valuable time can be saved by "swinging leads." This is normally the most efficient method for climbing multipitch routes. The second finishes cleaning the first pitch and continues climbing, taking on the role of lead climber. Unless he requires equipment from the other rack or desires a break, he can climb past the belay and immediately begin leading. The belayer simply adjusts his position, re-aiming the belay once the new leader begins placing protection. Swinging leads, or "leap frogging," should be planned before starting the climb so the leader knows to anchor the upper belay for both upward and downward pulls during the setup.

c. The procedures for conducting a lead climb with a group of two are relatively simple. The most experienced individual is the "lead" climber or leader, and is responsible for selecting the route. The leader must ensure the route is well within his ability and the ability of the second. The lead climber carries most of the climbing equipment in order to place protection along the route and set up the next belay. The leader must also ensure that the second has the necessary equipment, such as a piton hammer, nut tool, etc., to remove any protection that the leader may place.

(1) The leader is responsible for emplacing protection frequently enough and in such a manner that, in the event that either the leader or the second should fall, the fall will be neither long enough nor hard enough to result in injury. The leader must also ensure that the rope is routed in a way that will allow it to run freely through the protection placements, thus minimizing friction, or "rope drag".

(2) The other member of the climbing team, the belayer (sometimes referred to as the "second"), is responsible for belaying the leader, removing the belay anchor, and retrieving the protection placed by the leader between belay positions (also called "cleaning the pitch").

(3) Before the climb starts, the second will normally set up the first belay while the leader is arranging his rack. When the belay is ready, the belayer signals, "BELAY ON", affirming that the belay is "on" and the rope will be managed as necessary. When the leader is ready, he double checks the belay. The leader can then signal, "CLIMBING", only as a courtesy, to let the belayer know he is ready to move. The belayer can reply with "CLIMB", again, only as a courtesy, reaffirming that the belay is "on" and the rope will be managed as necessary. The leader then begins climbing.

(4) While belaying, the second must pay close attention to the climber's every move, ensuring that the rope runs free and does not inhibit the climber's movements. If he cannot see the climber, he must "feel" the climber through the rope. Unless told otherwise by the climber, the belayer can slowly give slack on the rope as the climber proceeds on the route. The belayer should keep just enough slack in the rope so the climber does not have to pull it through the belay. If the climber wants a tighter rope, it can be called for. If the belayer notices too much slack developing in the rope, the excess

rope should be taken in quickly. It is the belayer's responsibility to manage the rope, whether by sight or feel, until the climber tells him otherwise.

(5) As the leader protects the climb, slack will sometimes be needed to place the rope through the carabiner (clipping), in a piece of protection above the tie-in point on the leaders harness. In this situation, the leader gives the command "SLACK" and the belayer gives slack, (if more slack is needed the command will be repeated). The leader is able to pull a bight of rope above the tie-in point and clip it into the carabiner in the protection above. When the leader has completed the connection, or the clip, the command "TAKE ROPE" is given by the leader and the belayer takes in the remaining slack.

(6) The leader continues on the route until either a designated belay location is reached or he is at the end of or near the end of the rope. At this position, the leader sets an anchor, connects to the anchor and signals "OFF BELAY". The belayer prepares to climb by removing all but at least one of his anchors and secures the remaining equipment. The belayer remains attached to at least one anchor until the command "BELAY ON" is given.

d. When the leader selects a particular route, he must also determine how much, and what types, of equipment might be required to safely negotiate the route. The selected equipment must be carried by the leader. The leader must carry enough equipment to safely protect the route, additional anchors for the next belay, and any other items to be carried individually such as rucksacks or individual weapons.

(1) The leader will assemble, or "rack," the necessary equipment onto his harness or onto slings around the head and shoulder. A typical leader "rack" consists of:

- Six to eight small wired stoppers on a carabiner.
- Four to six medium to large wired stoppers on a carabiner.
- Assorted hexentrics, each on a separate carabiner.
- SLCDs of required size, each on a separate carabiner.
- Five to ten standard length runners, with two carabiners on each.
- Two to three double length runners, with two carabiners on each.
- Extra carabiners.
- Nut tool.

Note: The route chosen will dictate, to some degree, the necessary equipment. Members of a climbing team may need to consolidate gear to climb a particular route.

(2) The belayer and the leader both should carry many duplicate items while climbing.

- Short Prusik sling.
- Long Prusik sling.
- Cordellette.
- 10 feet of 1-inch webbing.
- 20 feet of 1-inch webbing.
- Belay device (a combination belay/rappel device is multifunctional).
- Rappel device (a combination belay/rappel device is multifunctional).
- Large locking carabiner (pear shape carabiners are multifunctional).

- Extra carabiners.
- Nut tool (if stoppers are carried).

Note: If using an over the shoulder gear sling, place the items in order from smallest to the front and largest to the rear.

e. Leading a difficult pitch is the most hazardous task in roped climbing. The lead climber may be exposed to potentially long, hard falls and must exercise keen judgment in route selection, placement of protection, and routing of the climbing rope through the protection. The leader should try to keep the climbing line as direct as possible to the next belay to allow the rope to run smoothly through the protection with minimal friction. Protection should be placed whenever the leader feels he needs it, and BEFORE moving past a difficult section.

CAUTION
The climber must remember he will fall twice the distance from his last piece of protection before the rope can even begin to stop him.

(1) ***Placing Protection.*** Generally, protection is placed from one stable position to the next. The anchor should be placed as high as possible to reduce the potential fall distance between placements. If the climbing is difficult, protection should be placed more frequently. If the climbing becomes easier, protection can be placed farther apart, saving hardware for difficult sections. On some routes an extended diagonal or horizontal movement, known as a traverse, is required. As the leader begins this type of move, he must consider the second's safety as well as his own. The potential fall of the second will result in a pendulum swing if protection is not adequate to prevent this. The danger comes from any objects in the swinging path of the second.

CAUTION
Leader should place protection prior to, during, and upon completion of any traverse. This will minimize the potential swing, or pendulum, for both the leader and second if either should fall.

(2) ***Correct Clipping Technique.*** Once an anchor is placed, the climber "clips" the rope into the carabiner (Figure 6-29, page 6-50). As a carabiner hangs from the protection, the rope can be routed through the carabiner in two possible ways. One way will allow the rope to run smoothly as the climber moves past the placement; the other way will often create a dangerous situation in which the rope could become "unclipped" from the carabiner if the leader were to fall on this piece of protection. In addition, a

series of incorrectly clipped carabiners may contribute to rope drag. When placing protection, the leader must ensure the carabiner on the protection does not hang with the carabiner gate facing the rock; when placing protection in a crack ensure the carabiner gate is not facing into the crack.

- Grasp the rope with either hand with the thumb pointing down the rope towards the belayer
- Pull enough rope to reach the carabiner with a bight
- Note the direction the carabiner is hanging from the protection
- Place the bight into the carabiner so that, when released, the rope does not cause the carabiner to twist.

(a) If the route changes direction, clipping the carabiner will require a little more thought. Once leaving that piece of protection, the rope may force the carabiner to twist if not correctly clipped. If the clip is made correctly, a rotation of the clipped carabiner to ensure that the gate is not resting against the rock may be all that is necessary.

CAUTION

Ensure the carabiner gate is not resting against a protrusion or crack edge in the rock surface; the rock may cause the gate to open.

(b) Once the rope is clipped into the carabiner, the climber should check to see that it is routed correctly by pulling on the rope in the direction it will travel when the climber moves past that position.

(c) Another potential hazard peculiar to leading should be eliminated before the climber continues. The carabiner is attached to the anchor or runner with the gate facing away from the rock and opening down for easy insertion of the rope. However, in a leader fall, it is possible for the rope to run back over the carabiner as the climber falls below the placement. If the carabiner is left with the gate facing the direction of the route there is a chance that the rope will open the gate and unclip itself entirely from the placement. To prevent this possibility, the climber should ensure that after the clip has been made, the gate is facing away from the direction of the route. There are two ways to accomplish this: determine which direction the gate will face before the protection or runner is placed or once clipped, rotate the carabiner upwards 180 degrees. This problem is more apt to occur if bent gate carabiners are used. Straight gate ovals or "Ds" are less likely to have this problem and are stronger and are highly recommended. Bent gate carabiners are easier to clip the rope into and are used mostly on routes with bolts preplaced for protection. Bent gate carabiners are not recommended for many climbing situations.

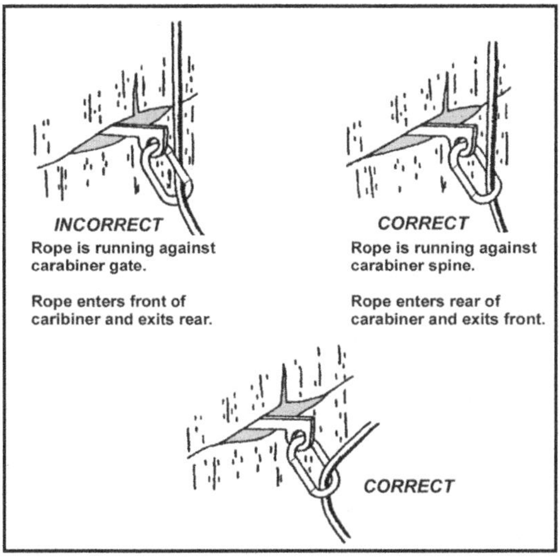

Figure 6-29. Clipping on to protection.

(3) ***Reducing Rope Drag; Using Runners***. No matter how direct the route, the climber will often encounter problems with "rope drag" through the protection positions. The friction created by rope drag will increase to some degree every time the rope passes through a carabiner, or anchor. It will increase dramatically if the rope begins to "zigzag" as it travels through the carabiners. To prevent this, the placements should be positioned so the rope creates a smooth, almost straight line as it passes through the carabiners (Figure 6-30). Minimal rope drag is an inconvenience; severe rope drag may actually pull the climber off balance, inducing a fall.

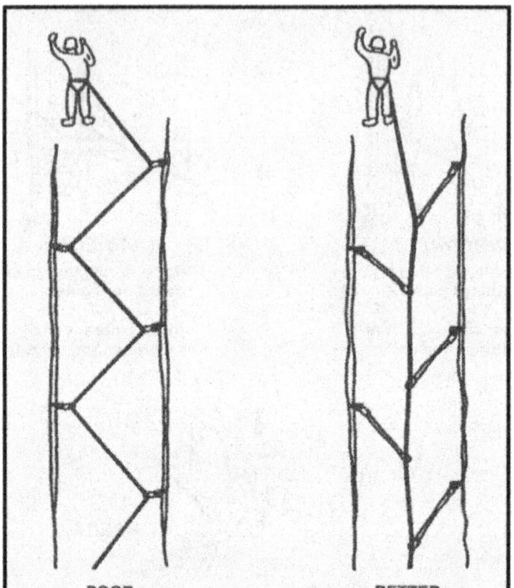

Figure 6-30. Use of slings on protection.

CAUTION

Rope drag can cause confusion when belaying the second or follower up to a new belay position. Rope drag can be mistaken for the climber, causing the belayer to not take in the necessary slack in the rope and possibly resulting in a serious fall.

(a) If it is not possible to place all the protection so the carabiners form a straight line as the rope moves through, you should "extend" the protection (Figure 6-31, page 6-52). Do this by attaching an appropriate length sling, or runner, to the protection to extend the rope connection in the necessary direction. The runner is attached to the protection's carabiner while the rope is clipped into a carabiner at the other end of the runner. Extending placements with runners will allow the climber to vary the route slightly while the rope continues to run in a relatively straight line.

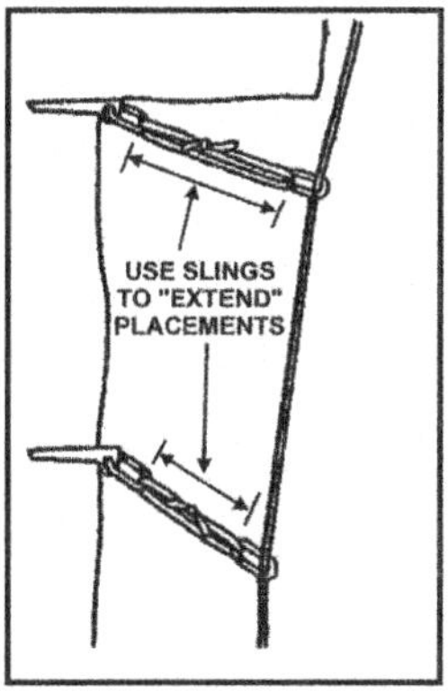

Figure 6-31. Use of slings to extend placement positions.

(b) Not only is rope drag a hindrance, it can cause undue movement of protection as the rope tightens between any "out of line" placements. Rope drag through chock placements can be dangerous. As the climber moves above the placements, an outward or upward pull from rope drag may cause correctly set chocks to pop out, even when used "actively". Most all chocks placed for leader protection should be extended with a runner, even if the line is direct to eliminate the possibility of movement.

(c) Wired chocks are especially prone to wiggling loose as the rope pulls on the stiff cable attachment. All wired chocks used for leader protection should be extended to reduce the chance of the rope pulling them out (Figure 6-32). Some of the larger chocks, such as roped Hexentrics and Tri-Cams, have longer slings pre-attached that will normally serve as an adequate runner for the placement. Chocks with smaller sling attachments must often be extended with a runner. Many of today's chocks are manufactured with pre-sewn webbing installed instead of cable.

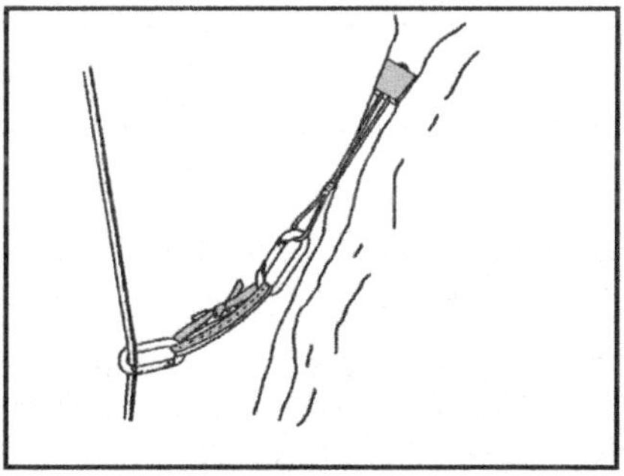

Figure 6-32. Use of sling on a wired stopper.

(d) When a correctly placed piton is used for protection, it will normally not be affected by rope drag. A correctly placed piton is generally a multi-directional anchor, therefore, rope drag through pitons will usually only affect the leader's movements but will continue to protect as expected.

(e) Rope drag will quite often move SLCDs out of position, or "walk" them deeper into the crack than initially placed, resulting in difficult removal or inability to remove them at all. Furthermore, most cases of SLCD movement result in the SLCD moving to a position that does not provide protection in the correct direction or no protection at all due to the lobes being at different angles from those at the original position.

Note: Any placement extended with a runner will increase the distance of a potential fall by the actual length of the sling. Try to use the shortest runners possible, ensuring they are long enough to function properly.

f. Belaying the follower is similar to belaying a top-roped climb in that the follower is not able to fall any farther than rope stretch will allow. This does not imply there is no danger in following. Sharp rocks, rock fall, and inadequately protected traverses can result in damage to equipment or injury to the second.

g. Following, or seconding, a leader has a variety of responsibilities. The second has to issue commands to the leader, as well as follow the leader's commands. Once the lead climber reaches a good belay position, he immediately establishes an anchor and connects to it. When this is completed he can signal "OFF BELAY" to the belayer. The second can now remove the leader's belay and prepare to climb. The second must remain attached to at least one of the original anchors while the leader is preparing the next belay position. The removed materials and hardware can be organized and secured on the second's rack in preparation to climb.

(1) When the leader has established the new belay position and is ready to belay the follower, the "new" belayer signals "BELAY ON." The second, now the climber, removes any remaining anchor hardware/materials and completes any final preparations. The belayer maintains tension on the rope, unless otherwise directed, while the final preparations are taking place, since removal of these remaining anchors can introduce slack into the rope. When the second is ready, he can, as a courtesy, signal "CLIMBING," and the leader can, again as a courtesy, reply with "CLIMB."

(2) Upon signaling "BELAY ON," the belayer must remove and keep all slack from the rope. (This is especially important as in many situations the belayer cannot see the follower. A long pitch induces weight and sometimes "drag" on the rope and the belayer above will have difficulty distinguishing these from a rope with no slack.)

h. When removing protection, the man cleaning the pitch should rack it properly to facilitate the exchange and or arrangement of equipment at the end of the pitch. When removing the protection, or "cleaning the pitch", SLCDs or chocks may be left attached to the rope to prevent loss if they are accidentally dropped during removal. If necessary, the hardware can remain on the rope until the second reaches a more secure stance. If removing a piton, the rope should be unclipped from the piton to avoid the possibility of damaging the rope with a hammer strike.

(1) The second may need to place full body weight on the rope to facilitate use of both hands for protection removal by giving the command "TENSION." The second

must also ensure that he does not climb faster than the rope is being taken in by the belayer. If too much slack develops, he should signal "TAKE ROPE" and wait until the excess is removed before continuing the climb. Once the second completes the pitch, he should immediately connect to the anchor. Once secured, he can signal "OFF BELAY." The leader removes the belay, while remaining attached to an anchor. The equipment is exchanged or organized in preparation for the next pitch or climb.

(2) When the difficulty of the climbing is within the "leading ability" of both climbers, valuable time can be saved by "swinging leads." This is normally the most efficient method for climbing multi-pitch routes. The second finishes cleaning the first pitch and continues climbing, taking on the role of lead climber. Unless he requires equipment from the belayer or desires a break, he can climb past the belay and immediately begin leading. The belayer simply adjusts his position, re-aiming the belay once the new leader begins placing protection. Swinging leads, or "leap frogging," should be planned before starting the climb so the leader knows to anchor the upper belay for both upward and downward pulls during the setup.

6-23. AID CLIMBING

When a route is too difficult to free climb and is unavoidable, if the correct equipment is available you might aid climb the route. Aid climbing consists of placing protection and putting full body weight on the piece. This allows you to hang solely on the protection you place, giving you the ability to ascend more difficult routes than you can free climb. Clean aid consists of using SLCDs and chocks, and is the simplest form of aid climbing.

a. **Equipment**. Aid climbing can be accomplished with various types of protection. Regardless of the type of protection used, the method of aid climbing is the same. In addition to the equipment for free climbing, other specialized equipment will be needed.

(1) ***Pitons***. Pitons are used the same as for free climbing. Most piton placements will require the use of both hands. Piton usage will usually leave a scar in the rock just by virtue of the hardness of the piton and the force required to set it with a hammer. Swinging a hammer to place pitons will lead to climber fatigue sooner than clean aid. Since pitons are multidirectional, the strength of a well-placed piton is more secure than most clean aid protection. Consider other forms of protection when noise could be hazardous to tactics.

(2) ***Bolts***. Bolts are used when no other protection will work. They are a more permanent form of protection and more time is needed to place them. Placing bolts creates more noise whether drilled by hand or by motorized drill. Bolts used in climbing are a multi-part expanding system pounded into predrilled holes and then tightened to the desired torque with a wrench or other tool. Bolts are used in many ways in climbing today. The most common use is with a hanger attached and placed for anchors in face climbing. However, bolts can be used for aid climbing, with or without the hanger.

(a) Placing bolts for aid climbing takes much more time than using pitons or clean aid. Bolting for aid climbing consists of consecutive bolts about 2 feet apart. Drilling a deep enough hole takes approximately thirty minutes with a hand drill and up to two minutes with a powered hammer drill. A lot of time and work is expended in a short distance no matter how the hole is drilled. (The weight of a powered hammer drill becomes an issue in itself.) Noise will also be a factor in both applications. A constant pounding with a hammer on the hand drill or the motorized pounding of the powered drill may alert the enemy to the position. The typical climbing bolt/hanger combination normally is left in the hole where it was placed.

(b) Other items that can be used instead of the bolt/hanger combination are the removable and reusable "spring-loaded removable bolts" such as rivets (hex head threaded bolts sized to fit tightly into the hole and pounded in with a hammer), split-shaft rivets, and some piton sizes that can be pounded into the holes. When using rivets or bolts without a hanger, place a loop of cable over the head and onto the shaft of the rivet or bolt and attach a carabiner to the other end of the loop (a stopper with the chock slid back will suffice). Rivet hangers are available that slide onto the rivet or bolt after it is placed and are easily removed for reuse. Easy removal means a slight loss of security while in use.

(3) ***SLCDs.*** SLCDs are used the same as for free climbing, although in aid climbing, full body weight is applied to the SLCD as soon as it is placed.

(4) ***Chocks.*** Chocks are used the same as for free climbing, although in aid climbing, weight is applied to the chock as soon as it is placed.

(5) ***Daisy Chains.*** Daisy chains are tied or presewn loops of webbing with small tied or presewn loops approximately every two inches. The small loops are just large enough for two or three carabiners. Two daisy chains should be girth-hitched to the tie-in point in the harness.

(6) ***Etriers (or Aiders).*** Etriers (aiders) are tied or presewn webbing loops with four to six tied or presewn internal loops, or steps, approximately every 12 inches. The internal loops are large enough to easily place one booted foot into. At least two etriers (aiders) should be connected by carabiner to the free ends of the daisy chains.

(7) ***Fifi Hook.*** A fifi hook is a small, smooth-surfaced hook strong enough for body weight. The fifi hook should be girth-hitched to the tie-in point in the harness and is used in the small loops of the daisy chain. A carabiner can be used in place of the fifi hook, although the fifi hook is simpler and adequate.

(8) ***Ascenders.*** Ascenders are mechanical devices that will move easily in one direction on the rope, but will lock in place if pushed or pulled the other direction. (Prusiks can be used but are more difficult than ascenders.)

b. **Technique.** The belay will be the same as in normal lead climbing and the rope will be routed through the protection the same way also. The big difference is the movement up the rock. With the daisy chains, aiders, and fifi hook attached to the rope tie-in point of the harness as stated above, and secured temporarily to a gear loop or gear sling, the climb continues as follows:

(1) The leader places the first piece of protection as high as can safely be reached and attaches the appropriate sling/carabiner

(2) Attach one daisy chain/aider group to the newly placed protection

(3) Clip the rope into the protection, (the same as for normal lead climbing)

(4) Insure the protection is sound by weighting it gradually; place both feet, one at a time, into the steps in the aider, secure your balance by grasping the top of the aider with your hands.

(5) When both feet are in the aider, move up the steps until your waist is no higher than the top of the aider.

(6) Place the fifi hook (or substituted carabiner) into the loop of the daisy chain closest to the daisy chain/aider carabiner, this effectively shortens the daisy chain;

maintain tension on the daisy chain as the hook can fall out of the daisy chain loop if it is unweighted.

Note: Moving the waist higher than the top of the aider is possible, but this creates a potential for a fall to occur even though you are on the aider and "hooked" close to the protection with the daisy chain. As the daisy chain tie-in point on the harness moves above the top of the aider, you are no longer supported from above by the daisy chain, you are now standing above your support. From this height, the fifi hook can easily fall out of the daisy chain loop if it is unweighted. If this happens, you could fall the full length of the daisy chain resulting in a static fall on the last piece of protection placed.

(7) Release one hand from the aider and place the next piece of protection, again, as high as you can comfortably reach; if using pitons or bolts you may need both hands free-"lean" backwards slowly, and rest your upper body on the daisy chain that you have "shortened" with the fifi hook

(8) Clip the rope into the protection

(9) Attach the other daisy chain/aider group to the next piece of protection

(10) Repeat entire process until climb is finished

c. **Seconding.** When the pitch is completed, the belayer will need to ascend the route. To ascend the route, use ascenders instead of Prusiks, ascenders are much faster and safer than Prusiks. Attach each ascender to a daisy chain/aider group with carabiners. To adjust the maximum reach/height of the ascenders on the rope, adjust the effective length of the daisy chains with a carabiner the same as with the fifi hook; the typical height will be enough to hold the attached ascender in the hand at nose level. When adjusted to the correct height, the arms need not support much body weight. If the ascender is too high, you will have difficulty reaching and maintaining a grip on the handle.

(1) Unlike lead climbing, there will be a continuous load on the rope during the cleaning of the route, this would normally increase the difficulty of removing protection. To make this easier, as you approach the protection on the ascenders, move the ascenders, one at a time, above the piece. When your weight is on the rope above the piece, you can easily unclip and remove the protection.

CAUTION

If both ascenders should fail while ascending the pitch, a serious fall could result. To prevent this possibility, *tie-in short* on the rope every 10-20 feet by tying a figure eight loop and clipping it into the harness with a separate locking carabiner as soon as the ascent is started. After ascending another 20 feet, repeat this procedure. Do not unclip the previous figure eight until the new knot is attached to another locking carabiner. Clear each knot as you unclip it.

Notes: 1. Ensure the loops formed by the short tie-ins do not catch on anything below as you ascend.

2. If the nature of the rock will cause the "hanging loop" of rope, formed by tying in at the end of the rope, to get caught as you move upward, do not tie into the end of the rope.

(2) Seconding an aid pitch can be done in a similar fashion as seconding free-climbed pitches. The second can be belayed from above as the second "climbs" the protection. However, the rope is unclipped from the protection before the aider/daisy chain is attached.

d. **Seconding Through a Traverse.** While leading an aid traverse, the climber is hanging on the protection placed in front of the current position. If the second were to clean the section by hanging on the rope while cleaning, the protection will be pulled in more than one direction, possibly resulting in the protection failing. To make this safer and easier, the second should hang on the protection just as the leader did. As the second moves to the beginning of the traverse, one ascender/daisy chain/aider group is removed from the rope and clipped to the protection with a carabiner, (keep the ascenders attached to the daisy chain/aider group for convenience when the traverse ends). The second will negotiate the traverse by leapfrogging the daisy chain/aider groups on the next protection just as the leader did. Cleaning is accomplished by removing the protection as it is passed when all weight is removed from it. This is in effect a self-belay. The second maintains a shorter safety tie-in on the rope than for vertical movement to reduce the possibility of a lengthy pendulum if the protection should pull before intended.

e. **Clean Aid Climbing**. Clean aid climbing consists of using protection placed without a hammer or drill involvement: chocks, SLCDs, hooks, and other protection placed easily by hand. This type of aid climbing will normally leave no trace of the climb when completed. When climbing the aiders on clean aid protection, ensure the protection does not "move" from it's original position.

(1) Hooks are any device that rests on the rock surface without a camming or gripping action. Hooks are just what the name implies, a curved piece of hard steel with a hole in one end for webbing attachment. The hook blade shape will vary from one model to another, some have curved or notched "blades" to better fit a certain crystal shape on a face placement. These types of devices due to their passive application, are only secure while weighted by the climber.

(2) Some featureless sections of rock can be negotiated with hook use, although bolts can be used. Hook usage is faster and quieter but the margin of safety is not there unless hooks are alternated with more active forms of protection. If the last twenty foot section of a route is negotiated with hooks, a forty foot fall could result.

6-24. THREE-MAN CLIMBING TEAM

Often times a movement on steep terrain will require a team of more than two climbers, which involves more difficulties. A four-man team (or more) more than doubles the difficulty found in three men climbing together. A four-man team should be broken down into two groups of two unless prevented by a severe lack of gear.

a. Given one rope, a three-man team is at a disadvantage on a steep, belayed climb. It takes at least twice as long to climb an average length pitch because of the third climber and the extra belaying required. The distance between belay positions will be halved if only one rope is used because one climber must tie in at the middle of the rope. Two ropes are recommended for a team of three climbers.

Note: Time and complications will increase when a three-man team uses only one rope. For example: a 100-foot climb with a 150-foot rope would normally require two belays for two climbers; a 100-foot climb with a 150-foot rope would require six belays for three climbers.

b. At times a three-man climb may be unavoidable and personnel should be familiar with the procedure. Although a team of three may choose from many different methods, only two are described below. If the climb is only one pitch, the methods will vary.

CAUTION

When climbing with a team of three, protected traverses will require additional time. The equipment used to protect the traverse must be left in place to protect both the second and third climbers.

(1) The first method can be used when the belay positions are not large enough for three men. If using one rope, two climbers tie in at each end and the other at the mid point. When using two ropes, the second will tie in at one end of both ropes, and the other two climbers will each tie in to the other ends. The most experienced individual is the leader, or number 1 climber. The second, or number 2 climber, is the stronger of the remaining two and will be the belayer for both number 1 and number 3. Number 3 will be the last to climb. Although the number 3 climber does no belaying in this method, each climber should be skilled in the belay techniques required. The sequence for this method (in one pitch increments) is as follows (repeated until the climb is complete):

(a) Number 1 ascends belayed by number 2. Number 2 belays the leader up the first pitch while number 3 is simply anchored to the rock for security (unless starting off at ground level) and manages the rope between himself and number 2. When the leader completes the pitch, he sets up the next belay and belays number 2 up.

(b) Number 2 ascends belayed by number 1, and cleans the route (except for traverses). Number 2 returns the hardware to the leader and belays him up the next pitch. When the leader completes this pitch, he again sets up a new belay. When number 2 receives "OFF BELAY" from the leader, he changes ropes and puts number 3 on belay. He should not have to change anchor attachments because the position was already aimed for a downward as well as an upward pull when he belayed the leader.

(c) Number 3 ascends belayed by number 2. When number 3 receives "BELAY ON," he removes his anchor and climbs to number 2's position. When the pitch is completed he secures himself to one of number 2's belay anchors. When number 1's belay is ready, he brings up number 2 while number 3 remains anchored for security. Number 2 again cleans the pitch and the procedure is continued until the climb is completed.

(d) In this method, number 3 performs no belay function. He climbs when told to do so by number 2. When number 3 is not climbing, he remains anchored to the rock for security. The standard rope commands are used; however, the number 2 climber may include the trailing climber's name or number in the commands to avoid confusion as to who should be climbing.

(d) Normally, only one climber would be climbing at a time; however, the number 3 climber could ascend a fixed rope to number 2's belay position using proper ascending technique, with no effect on the other two members of the team. This would save time for a team of three, since number 2 would not have to belay number 3 and could be either belaying number 1 to the next belay or climbing to number 1. If number 3 is to ascend a fixed rope to the next belay position, the rope will be loaded with number 3's weight, and positioned directly off the anchors established for the belay. The rope should be located so it does not contact any sharp edges. The rope to the ascending number 3 could be secured to a separate anchor, but this would require additional time and gear.

(2) The second method uses either two ropes or a doubled rope, and number 2 and number 3 climb simultaneously. This requires either a special belay device that accepts two ropes, such as the tuber type, or with two Munter hitches. The ropes must travel through the belay device(s) without affecting each other.

(a) As the leader climbs the pitch, he will trail a second rope or will be tied in with a figure eight in the middle of a doubled rope. The leader reaches the next belay position and establishes the anchor and then places both remaining climbers on belay. One remaining climber will start the ascent toward the leader and the other will start when a gap of at least 10 feet is created between the two climbers. The belayer will have to remain alert for differences in rope movement and the climbers will have to climb at the same speed. One of the "second" climbers also cleans the pitch.

(b) Having at least two experienced climbers in this team will also save time. The belayer will have additional requirements to meet as opposed to having just one second. The possible force on the anchor will be twice that of one second. The second that is not cleaning the pitch can climb off route, but staying on route will usually prevent a possible swing if stance is not maintained.

CHAPTER 7

ROPE INSTALLATIONS

Obstacles on the battlefield today are inevitable. They can limit the battlefield and, even worse, prevent a unit from accomplishing its mission. However, with highly skilled personnel trained on rope installations, leaders can be assured that even a unit with limited mountain skills and experience will be able to successfully move and operate in terrain that would otherwise have been impassable.

Section I. FIXED ROPE

A fixed rope is a rope anchored in place to assist soldiers in movement over difficult terrain. Its simplest form is a rope tied off on the top of steep terrain. As terrain becomes steeper or more difficult, fixed rope systems may require intermediate anchors along the route. Moving on a fixed rope requires minimal equipment. The use of harnesses, ascenders, and other technical gear makes fixed rope movement easier, faster, and safer, but adds to total mission weight.

7-1. INSTALLATION

To install a fixed rope, two experienced climbers rope up for a roped climb. The leader must have the necessary equipment to rig the anchor at the top of the pitch. Although leader protection is usually not needed on a typical slope, additional hardware can be brought along and placed at the leader's discretion. The second will establish a belay if protection is being placed. Otherwise, he will stack and manage the rope. He ensures the rope runs smoothly up the slope and does not get tangled as the climber ascends. Upon reaching the end of the pitch, the leader will establish the top anchor. Once the anchor is rigged, the leader will take up any remaining slack between himself and the second. He will anchor the installation rope and remain tied into the rope. The second unties from his end of the rope and begins to climb. If the leader placed protection, the second will clean the pitch on his way up.

7-2. UTILIZATION

All personnel using the fixed rope grasp the rope with the palm downward and use it for assistance as they ascend the slope (Figure 7-1, page 7-2). An individual can easily prevent a long fall by attaching himself to the rope with a sling using a friction knot (for example, Prusik, autoblock). The knot is slid along the rope as the individual ascends. If the climber slips and loses control of the rope, the friction knot will grab the rope and arrest the fall. The friction knot used in this manner is referred to as a self-belay (Figure 7-2, page 7-2).

Figure 7-1. Using a fixed rope.

Figure 7-2. Using a self-belay.

7-3. RETRIEVAL

If the fixed rope is to be used on the descent, it can be left in place and recovered after the last rappel. If not, the last climber will tie into the rope and be belayed from above. The climber now can easily free the rope if it gets caught on anything as it is taken up from the belayer.

7-4. FIXED ROPE WITH INTERMEDIATE ANCHORS

Whenever the route varies from the fall line of the slope, the fixed rope must be anchored at intermediate anchor points (Figure 7-3). Intermediate anchor points should also be used on any long routes that exceed the length of a single rope. The use of intermediate anchor points creates independent sections and allows for changes in direction from one section to the next. The independent sections allow for more personnel to move on the fixed rope. This type of fixed rope is commonly used along exposed ridges and narrow mountain passes.

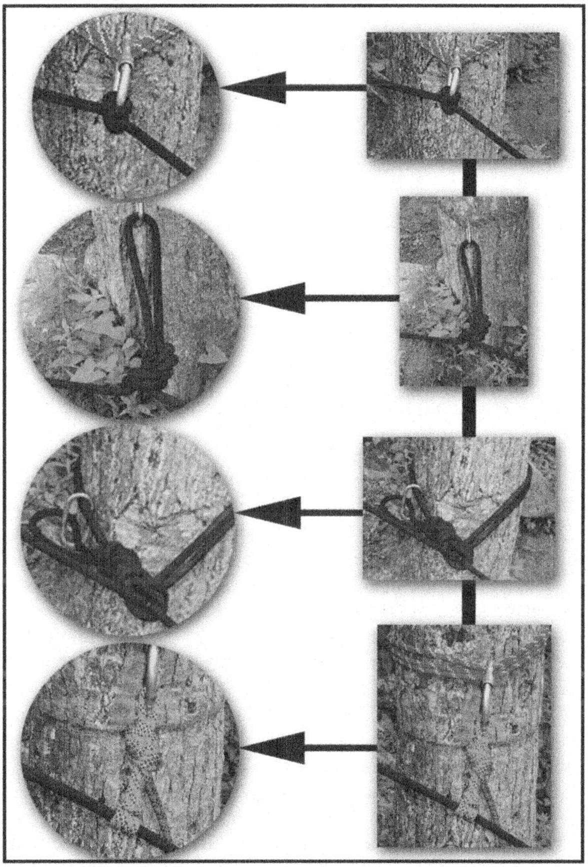

Figure 7-3. Fixed rope with intermediate anchors.

a. **Installation.** Two experienced climbers prepare for a roped climb. The leader will carry a typical rack with enough hardware to place an adequate number of intermediate anchor points. The second sets up a standard belay. The route they select must have the following characteristics:

- Most suitable location, ease of negotiation, avoids obstacles.
- Availability of anchors (natural and artificial).
- Area is safe from falling rock or ice.
- Tactical considerations are met.
- A rope routed between knee and chest height (waist high preferred).
- Rope crossovers should be avoided.

As the leader climbs the route, he will place the anchors and route the climbing rope as in a typical roped climb. The leader makes use of any available natural anchors.

Note: Sling attachments should be kept as short and snug as possible to ensure that a load on the fixed rope below the anchor is placed only on that anchor. This will prevent one section from affecting another section.

(1) The leader places an anchor at all points where a change of direction occurs. He also makes every attempt to route the rope so personnel will not have to cross back and forth over the rope between sections.

(2) When the leader reaches the end of the pitch, he temporarily anchors the rope. He should use a sling to anchor himself if there is any chance of slipping and falling. He then takes up any excess slack, and attaches the rope to the anchor.

Note: Enough slack must be left in the rope so the second can tie the knots necessary to fix the rope.

(3) The second unties from the rope and anchors it at the bottom. He attaches himself to the rope with a sling using a friction knot to create a self-belay. The self-belay will protect the second as he climbs and fixes the rope to the intermediate anchor points. When he reaches an anchor point, he unclips the climbing rope so he can advance the self-belay beyond the anchor point. He then takes the slack out of the section below the anchor point. He ensures that the fixed rope will be approximately knee to chest level as climbers negotiate the installation. He then attaches the rope using an anchor knot (for example, middle-of-the-rope clove hitch, double figure-eight). The second then moves to next anchor point and repeats the process.

(4) If a long runner is to be used at any anchor point, the second should adjust the section below it so the runner is oriented in the direction where the load or pull on the anchor will come from. This will help isolate the section.

(5) The sections are normally adjusted fairly snug between anchor points. A slack section may be necessary to move around obstacles in the route or large bulges in the terrain. If clove hitches are used, adjusting the clove hitches at each end of the section can leave any amount of slack.

(6) A middle-of-the-rope Prusik safetied with a figure eight may be used when utility ropes are available. These are used to adjust the rope height (either higher or lower).

(7) In addition to the fixed rope, the second could anchor etriers to be used as footholds.

(8) When the second reaches the end of the pitch, the rope is removed from the top anchor and the remaining slack is removed from the last section. The rope is reattached to the anchor. If additional fixed rope is required the procedure is repeated using another rope. The second will tie the ropes together before anchoring the next section, creating one continuous fixed rope.

b. **Utilization.** Personnel should be attached to the fixed rope during movement for safety reasons.

(1) If a self-belay is desired, a harness should be worn. A friction knot will be tied to the installation rope using a short sling. The sling will then be attached to the harness. Another short sling will be used as a safety line. One end of the sling will be attached to the harness and the other will have a carabiner inserted. This safety line is also attached to the fixed rope during movement. Once the climber reaches an anchor point, he removes his safety line and attaches it to the anchor or attaches it to the next section of rope. He will then untie the friction knot and tie another friction knot beyond the anchor point. The use of a mechanical ascender in the place of the friction knot could greatly speed up movement.

(2) There will be many situations where a self-belay may not be required. In these situations an individual may attach himself to the fixed rope using only a safety line. The individual will tie into the middle of a sling rope approximately 12 feet long. Fixed loops are tied into the running ends and a carabiner is attached into each of the fixed loops. The individual now has two points of attachments to the fixed rope. Upon reaching an anchor point, one safety line is removed and advanced beyond the anchor point onto the next section. Then the next safety is removed and placed on the next section. This way the individual is always secured to the fixed rope at all times.

(3) Personnel will move one at a time per section during the entire movement. Once an individual changes over to the next section he signals the next man to climb. When descending on the fixed rope, personnel can down climb using the installation for assistance. Another option would be to descend using a hasty rappel.

c. **Retrieval.** When the installation is retrieved, the next to last man on the system will untie the knots at the intermediate anchor points and reclips the rope as he ascends. He will be attached to rope using a self-belay. Once he reaches the top of the pitch, the rope should be running the same as when the leader initially placed it. The last man will untie the rope from the bottom anchor and tie into the rope. He will the clean the pitch as he climbs while being belayed from above.

Section II. RAPPELLING

When an individual or group must descend a vertical surface quickly, a rappel may be performed. Rappelling is a quick method of descent but it is extremely dangerous. These dangers include anchor failure, equipment failure, and individual error. Anchors in a mountainous environment should be selected carefully. Great care must be taken to load the anchor slowly and to ensure that no excessive stress is placed on the anchor. To ensure this, bounding rappels should be prohibited, and only walk down rappels used. Constant vigilance to every detail will guarantee a safe descent every time.

7-5. SELECTION OF A RAPPEL POINT

The selection of the rappel point depends on factors such as mission, cover, route, anchor points, and edge composition (loose or jagged rocks). There must be good anchors (primary and secondary). The anchor point should be above the rappeller's departure point. Suitable loading and off-loading platforms should be available.

7-6. INSTALLATION OF THE RAPPEL POINT

A rappel lane should have equal tension between all anchor points by establishing primary and secondary anchor points. The rappel rope should not extend if one anchor point fails. The following methods of establishing an anchor can be performed with a single or double rope. A double rope application should be used when possible for safety purposes.

a. If a rappel lane is less than half the rope length, the climber may apply one of the following techniques:

(1) Double the rope and tie a three-loop bowline around the primary anchor to include the primary anchor inside two loops and enough rope in the third loop to run to the secondary anchor (another three-loop bowline secured with an overhand knot).

(2) Bowline secured with an overhand knot (or any appropriate anchor knot).

(3) Double the rope and establish a self-equalizing anchor system with a three-loop bowline or any other appropriate anchor knot

b. If a rappel lane is greater than half the rope length, the climber may apply one of the following techniques:

(1) Use two ropes. With both ropes, tie a round turn anchor bowline around a primary anchor point. Take the remaining rope (the tail from the primary anchor bowline) and tie another round turn anchor bowline to a secondary anchor point. The secondary anchor point should be in a direct line behind the primary anchor point. The anchor can be either natural or artificial.

(2) Use two ropes. Establish a multi-point anchor system using a bowline on a bight or any other appropriate anchor knot.

c. Situations may arise where, due to the length of the rappel, the rappel rope cannot be tied to the anchor (if the rope is used to tie the knots, it will be too short to accomplish the rappel). The following techniques can be used:

(1) When using a natural anchor, tie a sling rope, piece of webbing, or another rope around the anchor using proper techniques for slinging natural anchors. The rappel rope will have a fixed loop tied in one end, which is attached to the anchor created.

(2) When using an artificial anchor, tie off a sling rope, piece of webbing, runner, or another rope to form a loop. Use this loop to create an equalizing or pre-equalized anchor, to which the rappel rope will be attached.

7-7. OPERATION OF THE RAPPEL POINT

Due to the inherent dangers of rappelling, special care must be taken to ensure a safe and successful descent.

a. **Communication.** Climbers at the top of a rappel point must be able to communicate with those at the bottom. During a tactical rappel, radios, hand signals, and rope signals are considered. For training situations use the commands shown in Table 7-1.

COMMAND	GIVEN BY	MEANING
LANE NUMBER ___, ON RAPPEL	Rappeller	I am ready to begin rappelling.
LANE NUMBER ___, ON BELAY	Belayer	I am on belay and you may begin your rappel.
LANE NUMBER ___, OFF RAPPEL	Rappeller	I have completed the rappel, cleared the rappel lane, and am off the rope.
LANE NUMBER ___, OFF BELAY	Belayer	I am off belay.

Table 7-1. Rappel commands.

Notes: 1. In a training environment, the lane number must be understood.
2. In a tactical situation, a series of tugs on the rope may be substituted for the oral commands to maintain noise discipline. The number of tugs used to indicate each command is IAW the unit SOP.

b. **Duties and Responsibilities.**

(1) Duties of the rappel point commander are as follows:

- Ensures that the anchors are sound and the knots are properly tied.
- Ensures that loose rock and debris are cleared from the loading platform.
- Allows only one man on the loading platform at a time and ensures that the rappel point is run orderly.
- Ensures that each man is properly prepared for the particular rappel: gloves on, sleeves down, helmet with chin strap fastened, gear prepared properly, and rappel seat and knots correct (if required). He also ensures that the rappeller is hooked up to the rope correctly and is aware of the proper braking position.
- Ensures that the proper signals or commands are used.
- Dispatches each man down the rope.
- Is the last man down the rope.

(2) Duties of the first rappeller down are as follows:

- Selects a smooth route, for the rope, that is clear of sharp rocks.
- Conducts a self-belay.
- Clears the route, placing loose rocks far enough back on ledges to be out of the way, which the rope may dislodge.
- Ensures the rope reaches the bottom or is at a place from which additional rappels can be made.
- Ensures that the rope will run freely around the rappel point when pulled from below.
- Clears the rappel lane by straightening all twists and tangles from the ropes.
- Belays subsequent rappellers down the rope or monitors subsequent belayers
- Takes charge of personnel as they arrive at the bottom (off-loading platform).

Note: A rappeller is always belayed from the bottom, except for the first man down. The first man belays himself down the rope using a self-belay attached to his rappel seat, which is hooked to the rappel rope with a friction knot. As the first man rappels down the rope, he "walks" the friction knot down with him.

(3) Each rappeller down clears the ropes, and shouts, "Off rappel," (if the tactical situation permits). After the rope is cleared and the rappeller is off rappel, he acts as the belayer for next rappeller.

(4) Soldiers wear gloves for all types of rappels to protect their hands from rope burns.

(5) Rappellers descend in a smooth, controlled manner.

(6) The body forms an L-shape with the feet shoulder-width apart, legs straight, and buttocks parallel to the ground. When carrying equipment or additional weight, a modified L-shape is used with the legs slightly lower than the buttocks to compensate for the additional weight. The rappeller's back is straight. He looks over the brake shoulder. The guide hand is extended on the rope with the elbow extended and locked. The rope slides freely through the guide hand. The guide hand is used to adjust equipment and assist balance during descent. The rappeller grasps the rope firmly with the brake hand and places it in the brake position. Releasing tension on the rope and moving the brake hand regulates the rate of descent. The rappeller never lets go of the ropes with his brake hand until the rappel is complete.

c. **Tying Off During the Rappel.** It may be necessary to stop during descent. This can be accomplished by passing the rope around the body and placing three or more wraps around the guide-hand-side leg, or by tying off using the appropriate knot for the rappel device.

7-8. RECOVERY OF THE RAPPEL POINT

After almost all personnel have descended, only two personnel will remain at the top of the rappel point. They will be responsible for establishing a retrievable rappel.

a. **Establishing the Retrievable Rappel.** To set up a retrievable rappel point, a climber must apply one of the following methods:

(1) Double the rope when the rappel is less than half the total length of the rope. Place the rope, with the bight formed by the midpoint, around the primary anchor. Join the tails of the rappel rope and throw the rope over the cliff. Tie a clove hitch around a carabiner, just below the anchor point, with the locking bar outside the carabiner away from the gate opening end and facing uphill. Snap the opposite standing portion into the carabiner. When the rappeller reaches the bottom, he pulls on that portion of the rope to which the carabiner is secured to allow the rope to slide around the anchor point.

(2) When the length of the rappel is greater than half the length of the rope used, join two ropes around the anchor point with an appropriate joining knot (except the square knot). Adjust the joining knot so that it is away from the anchor. Tie a clove hitch around a carabiner just below the anchor point with the locking bar outside the carabiner away from the gate opening end and facing uphill. Snap the opposite standing portion into the carabiner. Upon completion of the rappel, pull the rope to which the carabiner is secured to allow the rope to slide around the anchor point.

Notes: 1. When setting up a retrievable rappel, use only a primary point; care is taken in selecting the point.
2. Ensure the soldiers have a safety line when approaching the rappel point, with only the rappeller going near the edge.

b. **Retrieving the Rappel Rope.** The next to last rappeller will descend the lane, removing any twists, and routes the rope for easiest retrieval. Once he reaches the end of the rappel, he tests the rope for retrieval. If the rappel is retrievable, the last man will rappel down. Once he is off rappel, he pulls the lane down.

7-9. TYPES OF RAPPELS

During military mountaineering operations, many types of rappels may be used. The following paragraphs describe some these rappels.

a. **Hasty Rappel** (Figure 7-4). The hasty rappel is used only on moderate pitches. Its main advantage is that it is easier and faster than other methods. Gloves are worn to prevent rope burns.

(1) Facing slightly sideways to the anchor, the rappeller places the ropes horizontally across his back. The hand nearest to the anchor is his guide hand, and the other is the brake hand.

(2) To stop, the rappeller brings his brake hand across in front of his body locking the rope. At the same time, he turns to face up toward the anchor point.

Figure 7-4. Hasty rappel.

b. **Body Rappel** (Figure 7-5). The rappeller faces the anchor point and straddles the rope. He then pulls the rope from behind, and runs it around either hip, diagonally across the chest, and back over the opposite shoulder. From there, the rope runs to the brake hand, which is on the same side of the hip that the rope crosses (for example, the right hip to the left shoulder to the right hand). The rappeller leads with the brake hand down and faces slightly sideways. The foot corresponding to the brake hand precedes the guide hand at all times. The rappeller keeps the guide hand on the rope above him to guide himself--not to brake himself. He must lean out at a sharp angle to the rock. He keeps his legs spread well apart and relatively straight for lateral stability, and his back straight to reduce friction. The BDU collar is turned up to prevent rope burns on the neck. Gloves are worn, and other clothing may be used to pad the shoulders and buttocks. To brake, the rappeller leans back and faces directly toward the rock area so his feet are horizontal to the ground.

Figure 7-5. Body rappel.

Notes: 1. Hasty rappels and body rappels are not used on pitches that have overhangs; feet must maintain surface contact.

2. Hasty rappels and body rappels are not belayed from below.

c. **Seat-Hip Rappel** (Figure 7-6). The seat rappel differs from the body rappel in that the friction is absorbed by a carabiner that is inserted in a sling rope seat and fastened to the rappeller. This method provides a faster and more frictional descent than other methods. Gloves can be worn to prevent rope burns.

Figure 7-6. Seat-hip rappel.

(1) An alternate technique is to insert two carabiners opposite and opposed. Then insert a locking carabiner into the two carabiners with opening gate on brake hand side. Then run the rope through the single carabiner. This helps to keep the rappel rope away from the harness.

(2) To hook up for the seat-hip method, stand to one side of the rope. If using a right-hand brake, stand to the left of the rappel rope facing the anchor; if using a left-hand brake, stand to the right of the rappel rope. Place the rappel rope(s) into the locking carabiner; slack is taken between the locking carabiner and anchor point and wrapped around the shaft of the locking carabiner and placed into the gate so that a round turn is made around the shaft of the locking carabiner (Figure 7-7, page 7-12). Any remaining slack is pulled toward the uphill anchor point. If a single rope is used, repeat this process to place two round turns around the shaft of the locking carabiner. Face the anchor point and descend using the upper hand as the guide and the lower hand as the brake. This method has minimal friction, and is fast and safe. However, care is taken that the rope is hooked correctly into the carabiner to avoid the gate being opened by the rope. Loose clothing or equipment around the waist may be accidentally pulled into the locking carabiner and lock (stop) the rappel. For this reason, the rappeller must tuck in his shirt and keep his equipment out of the way during his descent.

Figure 7-7. Proper hookup using carabiner wrap.

d. **Figure-Eight Descender**. The figure-eight descender puts less kinks in the rope, and it can be used with one or two ropes (Figure 7-8).

(1) To use the figure-eight descender, pass a bight through the large eye and then over the small eye onto the neck. Place the small eye into a locking carabiner. To reduce the amount of friction on the figure-eight, place the original bight into the carabiner and not around the neck of the descender. (Less friction requires more braking force from the rappeller.)

(2) The guide hand goes on the rope that is running from the anchor. The brake hand goes on the slack rope. The brake is applied by moving the brake hand to the rear or downward.

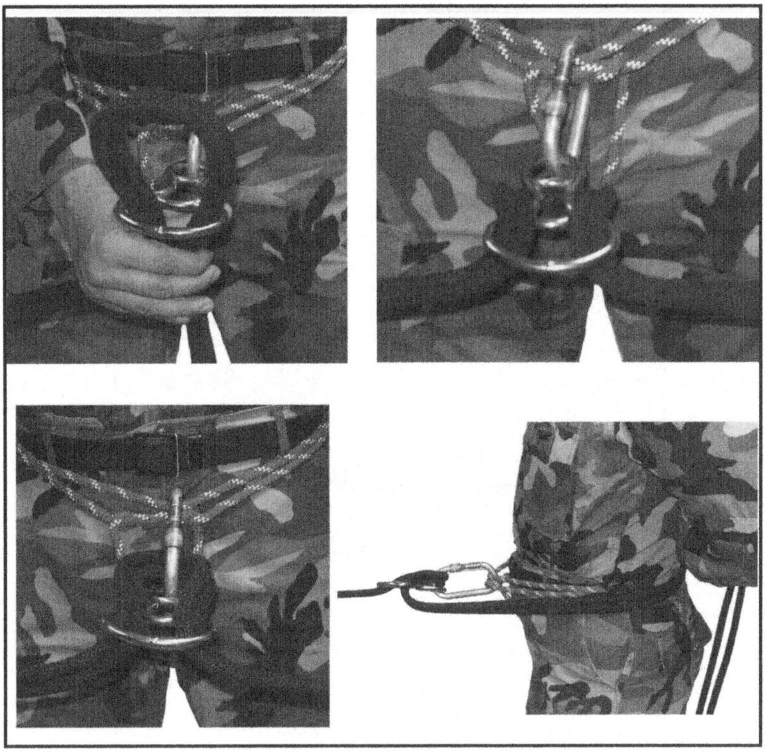

Figure 7-8. Figure-eight descender.

d. **Other Devices.** Many different types of devices are similar in design and operation to the basic plate. These include slots or plates and tubers. Most of these devices can accommodate two ropes not greater than 7/16 of an inch in size. Follow manufacturer's directions for using these devices for rappelling.

e. **Extending the Rappel Device.** The rappel device can be extended using either a piece of webbing or cordage to move the device away from the body and the harness, preventing accidental damage (Figure 7-9, page 7-14). It also allows for easier self-belay.

f. **Self-Belay Techniques.** A friction knot can be used as a belay for a rappeller (Figure 7-9, page 7-14). The knot acts as the brake hand when the rappeller must work or negotiate an obstacle requiring the use of both hands. The knot acts as a belay if the rappeller loses control of the rope.

Figure 7-9. Extended hookup with self-belay.

Section III. ONE-ROPE BRIDGE

The one-rope bridge is constructed using a static rope. The rope is anchored with an anchor knot on the far side of the obstacle and is tied off at the near end with a tightening system. A one-rope bridge may be built many ways, depending upon the tactical situation and area to be crossed (crossing a gorge above the tree line may require constructing artificial anchors). However, they all share common elements to safely construct and use the bridge: two suitable anchors; good loading and unloading platforms; a rope about 1-meter (waist) high for loading and unloading; a tightening system; and a rope tight enough for ease of crossing. Which side the tightening system is utilized, or whether an anchor knot or retrievable bowline is used, depends on the technique.

7-10. SITE SELECTION

A suitable crossing site must have "bombproof" anchors on both the near side and far side. These anchors must be extremely strong due to the amount of tension that will be placed upon them. Natural anchors, such as large trees and solid rock formations, are always preferred. The site must also have suitable loading and off-loading platforms to facilitate safe personnel movement.

7-11. INSTALLATION USING TRANSPORT TIGHTENING SYSTEM

The transport tightening system provides a mechanical advantage without requiring additional equipment.

a. The rope must first be anchored on the far side of the obstacle. If crossing a stream, the swimmer must be belayed across. If crossing a ravine or gorge, crossing may involve rappelling and a roped climb. Once across, the swimmer/climber will temporarily anchor the installation rope.

b. One man on the near side ties a fixed-loop knot (for example, wireman's, figure-eight slip knot) approximately 3 feet from the near side anchor and places the carabiner into the loop of the knot. The opening gate must be up and away from the loop. If two carabiners are used, the gates will be opposing. At that time, soldiers route the remainder of the rope around the near side anchor point and hook the rope into the carabiner. This system is known as a transport-tightening system (Figure 7-10). The man on the far side pulls the knot out four to six feet from the near anchor.

c. Once the knot has been pulled out, the far side man anchors the rope using a tensionless anchor. The anchor should be waist high.

Figure 7-10. Transport tightening system.

d. A three-man pull team on the near side pulls the slack out of the installation rope. The knot should be close enough to the near side anchor to allow personnel to easily load the installation.

Note: No more than three personnel should be used to tighten the rope. Using more personnel can over-tighten the rope and bring the rope critically close to failure.

e. The rope the can be secured using one of three methods: transport knot (Figure 7-11), round turn around anchor and two half hitches on a bight (Figure 7-12), or a tensionless anchor knot (Figure 7-13).

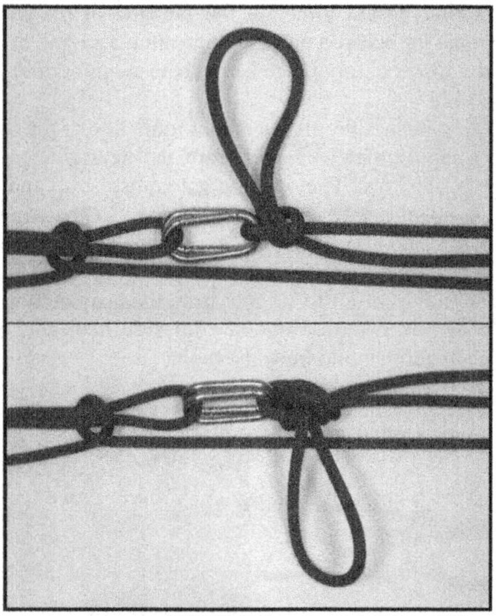

Figure 7-11. Transport knot.

Figure 7-12. Round turn around anchor and two half hitches on a bight.

Figure 7-13. Tensionless anchor knot.

Note: During training, a second static rope may be installed under less tension and alongside the tight rope to increase safety. An individual would clip into both ropes when crossing, thus having a backup in case of failure of the tighter rope.

7-12. INSTALLATION USING Z-PULLEY TIGHTENING SYSTEM

The Z-pulley tightening system (Figure 7-14) is another method for gaining a mechanical advantage.

a. The rope is brought across the obstacle the same way as discussed in paragraph 7-10.

b. Once across, the far side man anchors the rope.

c. One soldier ties a friction knot (autoblock, web wrap, Kleimheist) with a sling rope onto the bridging rope on the near side bank. Two steel carabiners are inserted with opposing gates into the friction knot.

d. The rope is routed around the near side anchor and through the carabiners, from inside to outside, and is run back to the near side anchor.

e. A second sling rope is tied to the bridge rope and then anchored to the near side anchor. This knot will be used as a progress capture device.

Figure 7-14. Z-pulley tightening system.

f. The three-man pull team on the near side then pulls on the rope, creating a pulley effect that tightens the system. As the rope is pulled tight, one man pushes the friction knot back toward the far side.

g. When the rope is tight, it is tied off with a tensionless anchor knot, transport knot, or round turn around anchor and two half hitches on a bight.

7-13. UTILIZATION

The rope bridge can be used to move personnel and equipment over obstacles. There are several methods of accomplishing this.

a. **Method of Crossing**. If dry crossing is impossible, soldiers will use the rope bridge as a hand line. Preferably, all soldiers will tie a safety line and attach it to the rope installation as they cross. If the soldier must cross with his rucksack, he may wear it over both shoulders, although the preferred method is to place another carabiner into the top of the rucksack frame, attach it to the bridge, and pull the rucksack across. Soldiers will always cross on the downstream side of the installation. If a dry crossing is possible soldiers will use one of three methods: commando crawl, monkey crawl, and Tyrolean traverse.

(1) ***Commando Crawl*** (Figure 7-15). The soldier lies on top of the rope with the upstream foot hooked on the rope and the knee bent close to the buttocks; the downstream leg hangs straight to maintain balance. He progresses by pulling with his hands and arms. To recover if he falls over, the soldier hooks one leg and the opposite arm over the rope, and then pushes down with the other hand to regain position.

Figure 7-15. Commando crawl.

Note: Only one man at a time is allowed on the bridge while conducting a commando crawl.

(2) ***Monkey Crawl*** (Figure 7-16, page 7-20). The soldier hangs below the rope suspended by his hands with both heels crossed over the rope. He pulls with his hands and arms, and pushes with his feet to make progress.

Figure 7-16. Monkey crawl.

(3) ***Rappel Seat Method*** (Figure 7-17). The soldier ties a rappel seat (or dons a seat harness) with the carabiner facing up and away from his body. He then faces the rope and clips into the rope bridge. He rotates under the rope and pulls with his hands and arms to make progress. The rappel seat method is the preferred method. If crossing with rucksacks, a carabiner is inserted into the frame and attached to the rope bridge. The soldier the places one or both legs through the shoulder carrying straps and pulls the rucksack across.

Figure 7-17. Rappel seat method.

b. **Rigging Special Equipment**. Any special equipment, such as crew-served weapons, ammunition, or supplies, must be rigged for movement across the rope bridge. A unit SOP may dictate the rigging of these items, but many expedient methods exist. The rigging should use various items that would be readily available to a deployed unit. Some of these items include tubular nylon webbing, cordage (various sizes), and carabiners.

(1) ***Machine Guns***. To rig machine guns, use a sling rope and tie a rerouted figure-eight around the spine of the front sight post. Then tie two evenly spaced fixed loops. Finally, anchor the sling rope to the buttstock of the machine gun. Additional tie downs may be necessary to prevent accidental disassembly of the weapon.

(2) ***ALICE Packs.*** ALICE packs can be joined together with a sling to facilitate moving more than one rucksack at one time.

7-14. HAULING LINE

A hauling line may be used to move rucksacks or casualties across the rope bridge (Figure 7-18).

a. **Construction**. An additional rope is brought across the rope bridge and anchored to the far side. The other end is anchored on the near side. All the slack is pulled to the near side, and a figure-eight slip knot is tied at the loading platform. A carabiner is inserted into the loop and clipped onto the rope bridge.

Figure 7-18. Hauling line.

b. **Moving Rucksacks**. Use carabiners to attach the rucksack frames to the rope bridge. Then clip the carabiner of the hauling line into the carabiner of the rucksack closest to the far side. Personnel on the far side pull the rucksacks across using the hauling line while personnel on the near side manages the slack at all times.

c. **Moving Litters**. The carabiner of the hauling line will remain on the rope bridge. On each side of this carabiner, using the hauling line tie a middle-of-the-rope clove hitch around both of the horizontal lift straps of the litter. Remove the slack between the carabiners. Then place the carabiners in each of the lift straps onto the rope bridge. The same technique used for the rucksacks is used to pull the litter across.

7-15. RETRIEVAL

Once all except two troops have crossed the rope bridge, the bridge team commander (BTC) chooses either the wet or dry method to dismantle the rope bridge.

a. If the BTC chooses the dry method, he should have anchored his tightening system with the transport knot.

(1) The BTC back-stacks all of the slack coming out of the transport knot, then ties a fixed loop and places a carabiner into the fixed loop.

(2) The next to last man to cross attaches the carabiner to his rappel seat or harness, and then moves across the bridge using the Tyrolean traverse method.

(3) The BTC then removes all knots from the system. The far side remains anchored. The rope should now only pass around the near side anchor.

(4) A three-man pull team, assembled on the far side, takes the end brought across by the next to last man and pulls the rope tight again and holds it.

(5) The BTC then attaches himself to the rope bridge and moves across.

(6) Once across, the BTC breaks down the far side anchor, removes the knots, and then pulls the rope across.

(b) If the BTC chooses a wet crossing, any method can be used to anchor the tightening system.

(1) All personnel cross except the BTC or the strongest swimmer.

(2) The BTC then removes all knots from the system.

(3) The BTC ties a fixed loop, inserts a carabiner, and attaches it to his rappel seat or harness. He then manages the rope as the slack is pulled to the far side.

(4) The BTC then moves across the obstacle while being belayed from the far side.

Section IV. SUSPENSION TRAVERSE

The suspension traverse is used to move personnel and equipment over rivers, ravines, chasms, and up or down a vertical rock face (Figure 7-19). The system may be established on a plane, varying from horizontal to near vertical.

Figure 7-19. Suspension traverse.

7-16. SITE SELECTION

The crossing site must have bombproof anchors at the near side and the far side, and suitable loading and off-loading platforms. If the anchors do not provide sufficient height to allow clearance, an A-frame must be used.

7-17. INSTALLATION

Installation of a suspension traverse can be time-consuming and equipment-intensive. All personnel must be well trained and well rehearsed in the procedures.

a. **A-frames**. Even in wooded mountainous terrain constructing an A-frame may be necessary due to the lack of height where the installation is needed. Site selection determines whether more height is needed; mission requirements determine site selection. The two main installations that use A-frames are the suspension traverse and vertical hauling line.

b. **Equipment**. Two sturdy poles are needed. The exact size of the poles depends on the type of load and location of the installation. The average size A-frame pole should be at least 3 inches in diameter and 9 to 12 feet long. Three to five 14-foot sling ropes are needed, depending on the size of the poles used for the A-frame.

c. **Construction**. Place two poles with the butt ends flush, and mark the apex on both poles.

(1) Ensure that proper height is attained and that the installation runs in a straight line between the two anchors. An A-frame placed out of proper alignment can cause the system to collapse. Try to find natural pockets in which to place the base of the A-frame poles.

(2) With a sling rope, tie a clove hitch around the left pole (standing at the base of the poles and facing the top) 3 inches above the apex marking, leaving about 18 inches of the sling rope free on top of the clove hitch. Place the locking bar on the outside edge of the pole. Make sure the rope end is pointing down as it is tied. (See Figure 7-20A, page 7-24.)

(3) Place the poles side by side and wrap the sling rope horizontally around both poles six to eight times, wrapping down from the clove hitch (Figure 7-20B, page 7-24). It may be necessary to join another sling rope to the first by using a square knot secured with overhand knots. Position this knot on the outside of one of the poles so as not to interfere with the vertical wraps. Make at least two additional wraps below the joining square knot. (See Figure 7-20C, page 7-24.)

(4) On the last horizontal wrap (ensure there are at least two wraps below the joining knot) to which the clove hitch is not tied, pass the rope between the poles below the wraps, and make four to six tight vertical wraps around the horizontal wraps (Figure 7-20D, page 7-24). Make the wraps as tight as possible. The vertical wraps must be as flat as possible next to each other. When starting the first vertical wrap, ensure it is in the same direction as the 18-inch tail on the top of the clove hitch. Insert a carabiner into the last two vertical wraps (Figure 7-20E, page 7-24).

(5) On the last vertical wrap, pass the rope between the poles above the horizontal wraps. Tie it off with a square knot in the section of rope coming from the clove hitch. Secure with overhand knots tied in the tails. (See Figure 7-20F, page 7-24.)

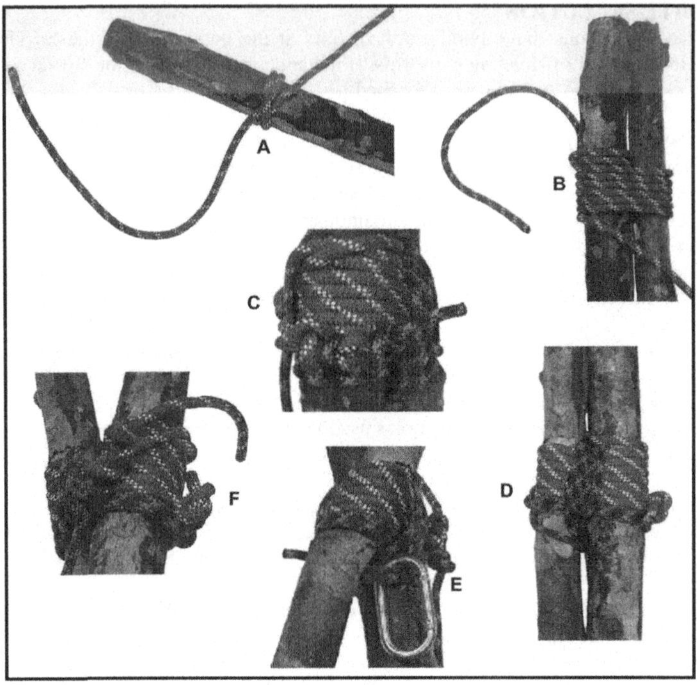

Figure 7-20. A-frame horizontal and vertical wraps.

(6) Use a spreader rope to prevent the A-frame from collapsing from pressure applied at the apex (Figure 7-21). If the ground is soft, dig the legs in about 6 inches. Tie a sling rope between the legs with a round turn with two half hitches around each leg. Remove all slack in the rope between the legs.

(7) If the ground is a hard surface, tie end-of-the-rope clove hitches with the locking portions facing to the rear, the direction of kick. Tie the tails off at a 45-degree angle with a round turn and two half hitches to a secondary anchor point. The spreader rope should be no more than 6 inches above ground level. The use of clove hitches and half hitches permits easy adjustment of the spreader rope. If more than one sling rope is needed, tie the two ropes together with a square knot and secure with half hitches or overhand knots.

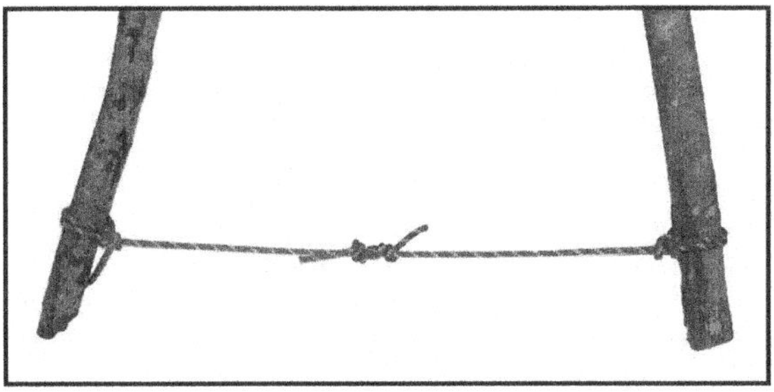

Figure 7-21. A-frame spreader.

d. **Installation Construction**. One man rappels down the pitch and secures two installation (traverse) ropes to the far anchor with an anchor knot. Place a transport tightening system in each installation rope at the near (upper) anchor. Run the installation ropes through or around the anchor in opposite directions and tie off. Anchor the traverse ropes as close together as possible so that the ropes do not cross.

(1) Place the A-frame (if needed) so that both traverse ropes run over the apex and the A-frame splits the angle formed between the near (upper) and far (lower) anchors, with the legs firmly emplaced or anchored with pitons. Ensure that the A-frame is in line with the anchors. Adjust the A-frame under the traverse ropes after tightening to firmly implant the A-frame.

(2) Tighten the installation ropes using either the transport tightening system (paragraph 7-11) or the z-pulley tightening system (paragraph 7-12).

(3) Anchor the A-frame to the traverse rope. Tie a clove hitch at the center of a sling rope. Place it over one of the poles above the apex and move down to the apex so that the locking bar of the clove hitch is to the inside of the A-frame. Secure each end of the sling rope to one of the tightened static lines with two Prusik knots-one forward and one to the rear of the A-frame on the same static line rope (Figure 7-22).

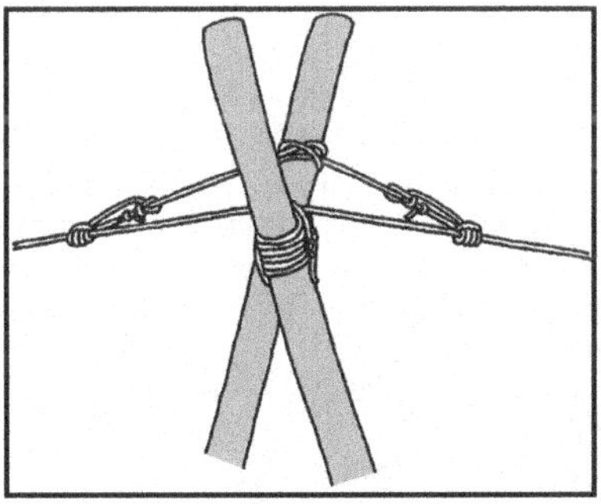

Figure 7-22. Anchoring the A-frame to the traverse rope.

Note: The A-frame should be positioned so that the angles created by the A-frame bisecting the installation rope are approximately equal on both sides. This creates downward pressure holding the A-frame in position, not forcing it in a lateral direction. It must also be placed in a straight line between the upper and lower anchor points.

(4) Use a carrying rope to attach loads to the traverse ropes (Figure 7-23). Join the ends of a 14-foot sling rope with a square knot and two overhand knots. Displace the knot one-third of the distance down the loop and tie an overhand knot both above and below the square knot. This forms two small loops and one large loop that is longer than the two small loops combined.

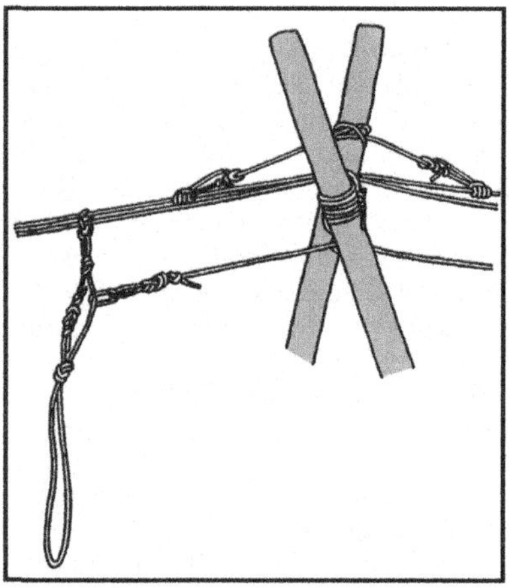

Figure 7-23. Carrying rope for use on a traverse.

(5) Attach the carrying rope to the traverse ropes with carabiners (or a pulley) that have the gates reversed and opening in opposite directions. Attach a belay rope to the center loop of the carrying rope using a fixed loop or locking carabiner on the side opposite the joining knot (Figure 7-23). When the suspension traverse is near horizontal, a second rope may be needed to pull the load across and should be attached to the carrying rope the same as the first.

(6) Insert second carabiner into the one placed into the wraps of the A-frame. This is where a belay rope will be attached

(7) With a sling rope, tie a six wrap middle-of-the-rope Prusik knot to both static ropes near the far side off-loading point. This acts as a stopper knot for the man descending, preventing him from hitting the lower anchor.

(8) Attach the load by running the long loop of the carrying rope through the load or through the soldier's harness and attaching the bottom loop to the traverse rope carabiner.

Descent must be belayed slowly and be controlled. Soldiers descending should hold onto the carrying rope and keep their feet high to avoid contact with the ground. Due to the constant tension maintained on the belayer, use a mechanical belay. If the belayer cannot view the entire descent route, use a relay man.

7-18. RETRIEVAL

The suspension traverse is not as readily retrievable as the one-rope bridge. Therefore, the installing unit should dismantle it after it is no longer needed.

Section V. VERTICAL HAULING LINE

The vertical hauling line is an installation used to move men and equipment up vertical or near-vertical slopes (Figure 7-24). It is often used with a fixed rope for personnel movement. The hauling line is used to move equipment, such as mortars or other crew-served weapons, rucksacks, or supplies

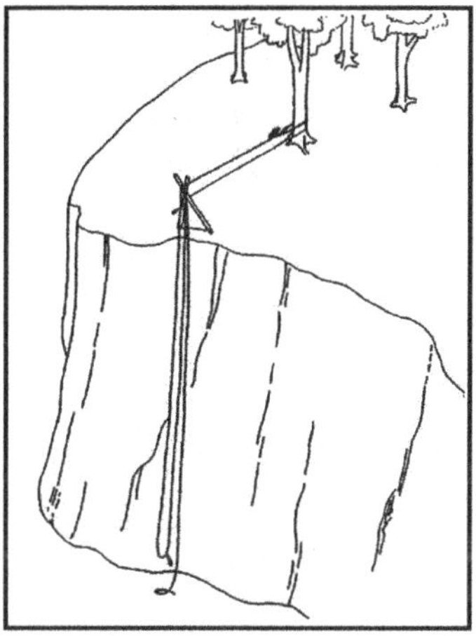

Figure 7-24. Vertical hauling line.

7-19. SITE SELECTION

The first and most important task is to determine where to construct the vertical hauling line. The site must have an appropriate top anchor that is secure enough to hold the system and load. Loading and unloading platforms should be easily accessible natural platforms that provide a safe working area. The ideal platform at the top allows construction of the vertical hauling line without the use of an A-frame. The site should also have sufficient clearance to allow for space between the slope and pulley rope for easy hauling of troops or equipment.

7-20. INSTALLATION

Construct an A-frame, if necessary, and anchor it. Double one installation rope, find the middle, and lay the middle of the installation rope over the apex of the A-frame; a 30-centimeter (12-inch) bight should hang below the apex.

a. To maintain the 12-inch bight, tie clove hitches above the A-frame lashing on each side of the apex with the installation rope, ensuring that the locking bars of the clove hitches are on the inside. Ensure that the portion of the rope that forms the bight comes out of the bottom of the clove hitch. (See Figure 7-25.)

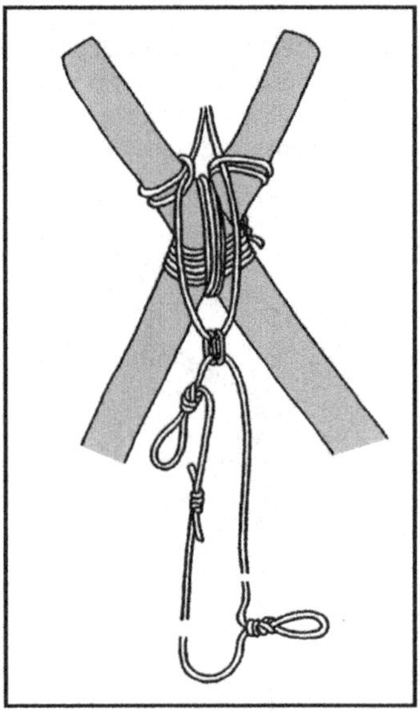

Figure 7-25. Attaching the anchor rope to the A-frame.

b. To anchor the A-frame, use a transport tightening system with the doubled rope, which is tied to the A-frame. Tie this off at an anchor point to the rear of the A-frame installation and adjust the angle of the A-frame so it leans out over the cliff edge. The angle should be 15 to 25 degrees unloaded. The A-frame should not lean outward more than 45 degrees once loaded since the legs can lose their position.

c. Tie the ends of another installation rope together with a joining knot to form the hauling line. Attach the rope to the system by two carabiners with gates up and opposed or one mountain rescue pulley with a locking steel carabiner in the 12-inch bight hanging from the apex of the A-frame. Tie fixed loops (wireman's, directional figure-eight, or single butterfly) on opposite sides of the endless rope at the loading and unloading platforms.

d. Attach equipment to the hauling line 12 inches above the joining knot by a carabiner in the fixed loop.

e. Additional fixed loops may be tied in the hauling line for more control over the object when moving large loads. Attach personnel to the hauling line by use of a rappel seat or seat harness.

Note: Mortar tubes and similar objects are attached to the line by two knots so that the tube stays parallel and as close to the hauling line as possible.

f. When personnel are moved using a vertical hauling line, make a knotted hand line; anchor it in line with, or to, the primary anchor (round turn with a bowline); and place it over the spreader on the legs of the A-frame. Space the overhand knots in the knotted hand line 12 inches apart, with about 20 feet of rope without knots at one end for the anchor. Throw the knotted hand line over the A-frame spreader rope and down the side of the cliff. Personnel ascending the vertical hauling line use this as a simple fixed rope.

g. Use as many men as needed to pull the load to the top by pulling on the rope opposite the load. If equipment and personnel are only being lowered, belay from the top using the hauling line. Station two climbers at the unloading platform to retrieve loads.

h. If only equipment is being hauled up, it is not necessary to use the knotted hand line rope, but it may be necessary to use a belay rope. To move materials or troops up on one side of the hauling line, pull the other side from below.

Note: Personnel using the hauling line for movement must apply all related principles of climbing. Always station two operators at the top of the vertical hauling line to aid men or to retrieve loads when they reach the top. They will always be safetied while working near the edge. When in use, the A-frame should lean slightly over the edge of the cliff to prevent excessive wear on the ropes that pass over sharp rocks. Reduce excessive friction on the system. Remove all obstacles and any loose objects that could be dislodged by personnel and equipment.

7-21. RETRIEVAL

The vertical hauling line is used along a main supply route. When it is no longer needed, the installing unit will return and dismantle the system.

Section VI. SIMPLE RAISING SYSTEMS

Moving heavy objects with limited manpower may be necessary in mountainous terrain. To reduce fatigue of those personnel moving the load, simple rigging techniques can be used to increase the mechanical advantage of the hauling system.

7-22. Z-PULLEY SYSTEM

The Z-pulley system is a simple, easily constructed hauling system (Figure 7-26, page 7-30).

a. **Considerations.** Anchors must be sturdy and able to support the weight of the load. Site selection is governed by different factors: tactical situation, weather, terrain, equipment, load weight, and availability of anchors.

b. **Theory.** Use carabiners as a substitute if pulleys are not available. The mechanical advantage obtained in theory is 3:1. The less friction involved the greater the mechanical advantage. Friction is caused by the rope running through carabiners, the load rubbing against the rock wall, and the rope condition.

c. **Construction**. Use the following procedures to construct a Z-pulley system.

(1) Establish an anchor (anchor pulley system [APS]). Place a carabiner on the runner at the anchor point, place a pulley into the carabiner, and run the hauling rope through the pulley.

(2) With a sling rope (preferably 7 millimeter), tie a middle-of-rope Prusik knot secured with a figure-eight knot on the load side of the pulley. This will be used as a progress capture device (PCD). A mechanical descender may be used in place of the Prusik knot. Take the tails exiting the figure-eight and tie a Munter hitch secured by a mule knot. Ensure the Munter hitch is loaded properly before tying the mule knot.

(3) At an angle away from the APS, establish a moveable pulley system (MPS) to create a "Z" in the hauling rope. Tie another Prusik knot on the load side of the hauling rope. Secure it with a figure-eight knot. Using the tails tie a double-double figure-eight knot. Insert a locking carabiner into the two loops formed, then place the working end into the carabiner. Mechanical ascenders should not be used as an MPS. Move the working end back on a parallel axis with the APS. Provide a pulling team on the working end with extra personnel to monitor the Prusik knots.

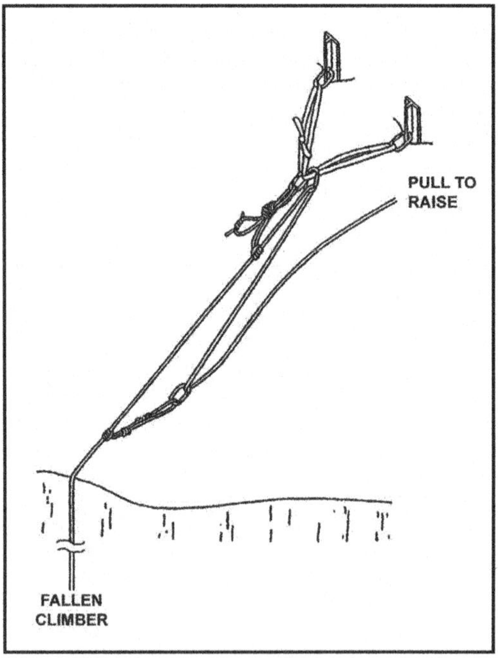

Figure 7-26. Z-pulley system.

d. **Other Factors**. If the two pulleys touch, the "Z" is lost along with the mechanical advantage. For greater efficiency, the main anchor should be well back from the edge and all ropes should pull parallel to the load.

Note: Avoid the possibility of overstressing the anchors. Be aware of reduced sensitivity to the load due to the mechanical advantage. Use belays and backup safeties. Protect the rope from edges and other abrasive parts of the rock.

7-23. U-PULLEY SYSTEM

The U-pulley system is another simple, easily-constructed hauling system (Figure 7-27, page 7-32).

a. **Considerations**. Anchors must be sturdy and able to support the weight of the load. Site selection is governed by different factors: tactical situation, weather, terrain, equipment, load weight, and availability of anchors.

b. **Theory.** Use carabiners as a substitute if pulleys are not available. The mechanical advantage obtained in theory is 2:1. The less friction involved the greater the mechanical advantage. Friction is caused by the rope running through carabiners, the load rubbing against the rock wall, and the rope condition.

c. **Construction.** Use the following procedures construct a U-pulley system.

(1) Anchor the hauling rope.

(2) Prepare the load or casualty for hauling. Place a locking carabiner the on to the harness or the rigged load.

(3) Lower a bight to the casualty or the load.

(4) Place the bight into the carabiner; or place the bight on to a pulley and then place pulley into the carabiner.

(5) Construct a second anchor. Attach a locking carabiner to the anchor.

(6) Tie a middle of the rope Prusik onto the haul rope exiting the pulley. Secure the Prusik with a double-double figure eight. This is the PCD. Place the fixed loops into the locking carabiner of the second anchor.

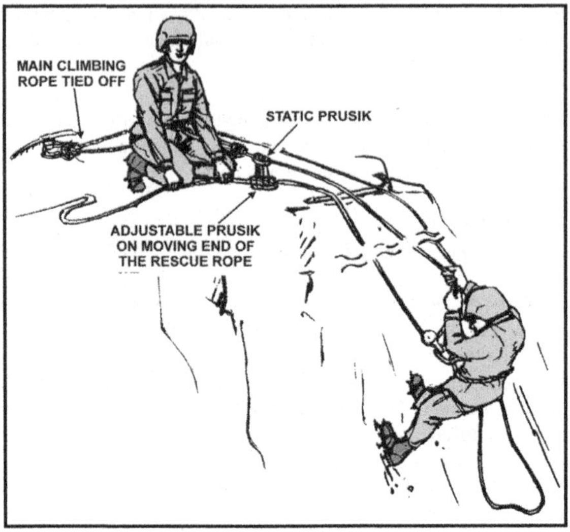

Figure 7-27. U-pulley system.

CHAPTER 8
MOUNTAIN WALKING TECHNIQUES

Mountain travel encompasses the full spectrum of techniques used to negotiate steep, rugged terrain. Mountain walking on rock and snow, technical rock and ice climbing, skiing or snow shoeing, rappelling, and stream crossing are the key travel skills a military mountaineer must possess.

8-1. BASIC PRINCIPLES

Up scree or talus, through boulder fields or steep wooded mountainsides, over snow or grass-covered slopes, the basic principles of mountain walking remain the same.

a. The soldier's weight is centered directly over the feet at all times. He places his foot flat on the ground to obtain as much (boot) sole-ground contact as possible. Then, he places his foot on the uphill side of grass tussocks, small talus and other level spots to avoid twisting the ankle and straining the Achilles tendon. He straightens the knee after each step to allow for rest between steps, and takes moderate steps at a steady pace. An angle of ascent or descent that is too steep is avoided, and any indentations in the slope are used to advantage.

b. In addition to proper technique, pace is adapted to conditions. The mountaineer sets a tempo, or number of steps per minute, according to the pace of the unit in which he is moving. (Physical differences mean that the tempos of two people moving at the same speed will not always be the same.) The soldier maintains tempo and compensates for changes of slope or terrain by adjusting the length of his stride. Tempo, pace, and rhythm are enhanced when an interval of three to five paces is kept between individuals. This interval helps lessen the "accordion" effect of people at the end of the file who must constantly stop and start.

c. The terrain, weather, and light conditions affect the rate of climb. The more adverse the conditions, the slower the pace. Moving too fast, even under ideal conditions, produces early fatigue, requires more rest halts, and results in loss of climbing time. A soldier can only move as fast as his lungs and legs will allow. The trained, conditioned and acclimatized soldier has greater endurance and moves more efficiently. Rest, good nutrition and hydration, conditioning, acclimatization, proper training, and the will to climb are key to successful mountain operations.

d. Breaks are kept to a minimum. When a moderate pace is set, the need for rest halts decreases, the chance of personnel overheating is lessened, and a unit can cover a given distance in a minimal time. If possible, rests should be taken on level ground avoiding steeper inclines.

(1) During the first half-hour of movement an adjustment halt should be taken. Soldiers will loosen or tighten bootlaces as needed, adjust packs and add or remove appropriate layers of clothing.

(2) Following the first halt, a well-conditioned party may take a short rest every 1 to 1.5 hours. If possible, soldiers lean against a tree, rock, or hillside to relieve the shoulders of pack weight, breathe deeply, hydrate, and snack on trail food. These halts are kept short enough to avoid muscles stiffening (one to two minutes).

(3) Later in the march longer halts may be necessary due to fatigue or mission requirements. At these halts soldiers should immediately put on additional clothing to avoid chilling—it is much easier to keep a warm body warm than to warm up a cold one.

(4) After a climb, a good rest is needed to revive tired muscles.

e. The rest step is used for steep slopes, snowfields, and higher elevations. It controls pace and limits fatigue by giving the lungs and legs a moment to recuperate between steps. Pace is kept slow and rhythmic.

(1) After each step forward, the soldier pauses briefly, relaxing the muscles of the forward leg while resting his entire bodyweight on the rear leg. The rear leg is kept straight with the knee locked so that bone, not muscle, supports the weight.

(2) Breathing is synchronized with the rest step. The number of breaths per step will change depending on the difficulty of the climb. Steeper slopes or higher elevations may require several breaths per step. When the air thins at altitude it is especially important to breathe deeply, using the "pressure breathing" technique. The soldier exhales strongly, enabling an easier, deeper inhale.

(3) This slow, steady, halting rest step is more efficient than spurts of speed, which are rapidly exhausting and require longer recovery.

f. Downhill walking uses less energy than uphill but is much harder on the body. Stepping down can hammer the full bodyweight onto the feet and legs. Blisters and blackened toenails, knee damage, and back pain may follow. To avoid these problems the soldier should start by tightening bootlaces to ensure a snug fit (also keep toenails trimmed). A ski pole, ice ax, or walking stick will help take some of the load and give additional stability. (Refer to Chapter 11 for techniques and use of the ice ax.) Keep a moderate pace and walk with knees flexed to absorb shock.

g. Side hill travel on any surface should be avoided whenever possible. Weighted down with a rucksack, the soldier is vulnerable to twisted ankles, back injury, and loss of balance. If side hill travel is necessary, try to switchback periodically, and use any lower angle flat areas such as rocks, animal trails, and the ground above grass or brush clumps to level off the route.

8-2. TECHNIQUES

Mountain walking techniques can be divided according to the general formation, surface, and ground cover such as walking on hard ground, on snow slopes and grassy slopes, through thick brush, and on scree and talus slopes.

a. **Hard Ground.** Hard ground is firmly compacted, rocky soil that does not give way under the weight of a soldier's step. It is most commonly found under mature forest canopy, in low brush or heather, and areas where animals have beaten out multiple trails.

(1) When ascending, employ the rest step to rest the leg muscles. Steep slopes can be traversed rather than climbed straight up. To turn at the end of each traverse, the soldier should step off in the new direction with the uphill foot. This prevents crossing the feet and possible loss of balance. While traversing, the full sole-to-ground principle is accomplished by rolling the ankle downhill on each step. For small stretches the herringbone step may be used—ascending straight up a slope with toes pointed out. A normal progression, as the slope steepens, would be from walking straight up, to a herringbone step, and then to a traverse on the steeper areas.

(2) Descending is best done by walking straight down the slope without traversing. The soldier keeps his back straight and bends at the knees to absorb the shock of each step. Body weight is kept directly over the feet and the full boot sole is placed on the ground with each step. Walking with a slight forward lean and with the feet in a normal position make the descent easier.

b. **Snow Slopes.** Snow-covered terrain can be encountered throughout the year above 1,500 meters in many mountainous areas. Talus and brush may be covered by hardened snowfields, streams made crossable with snowbridges. The techniques for ascending and descending moderate snow slopes are similar to walking on hard ground with some exceptions.

(1) ***Diagonal Traverse Technique.*** The diagonal traverse is the most efficient means to ascend snow. In conjunction with the ice ax it provides balance and safety for the soldier. This technique is a two-step sequence. The soldier performs a basic rest step, placing the leading (uphill) foot above and in front of the trailing (downhill) foot, and weighting the trail leg. This is the in-balance position. The ice ax, held in the uphill hand, is placed in the snow above and to the front. The soldier shifts his weight to the leading (uphill) leg and brings the unweighted trail (downhill) foot ahead of the uphill foot. He shifts weight to the forward (downhill) leg and then moves the uphill foot up and places it out ahead of the trail foot, returning to the in-balance position. At this point the ax is moved forward in preparation for the next step.

(2) ***Step Kicking.*** Step kicking is a basic technique used when crampons are not worn. It is best used on moderate slopes when the snow is soft enough to leave clear footprints. On softer snow the soldier swings his foot into the snow, allowing the leg's weight and momentum to carve the step. Fully laden soldiers will need to kick steps, which take half of the boot. The steps should be angled slightly into the slope for added security. Succeeding climbers will follow directly in the steps of the trailbreaker, each one improving the step as he ascends. Harder snow requires more effort to kick steps, and they will not be as secure. The soldier may need to slice the step with the side of his boot and use the diagonal technique to ascend.

(3) ***Descending Snow.*** If the snow is soft and the slope gentle, simply walk straight down. Harder snow or steeper slopes call for the plunge step, which must be done in a positive, aggressive manner. The soldier faces out, steps off, and plants his foot solidly, driving the heel into the snow while keeping his leg straight. He shifts his weight to the new foot plant and continues down with the other foot. On steeper terrain it may be necessary to squat on the weighted leg when setting the plunge step. The upper body should be kept erect or canted slightly forward.

(4) ***Tips on Snow Travel.*** The following are tips for travelling on snow.

(a) Often the best descent is on a different route than the ascent. When looking for a firmer travel surface, watch for dirty snow—this absorbs more heat and thus hardens faster than clean snow.

(b) In the Northern Hemisphere, slopes with southern and western exposures set up earlier in the season and quicker after storms, but are more prone to avalanches in the spring. These slopes generally provide firm surfaces while northern and eastern exposures remain unconsolidated.

(c) Travel late at night or early in the morning is best if daytime temperatures are above freezing and the sun heats the slopes. The night's cold hardens the snow surface.

(d) Avoid walking on snow next to logs, trees, and rocks as the subsurface snow has melted away creating hidden traps.

c. **Grassy Slopes.** Grassy slopes are usually composed of small tussocks of growth rather than one continuous field.

(1) When ascending, step on the upper side of each hummock or tussock, where the ground is more level.

(2) When descending a grassy slope, the traverse technique should be used because of the uneven nature of the ground. A climber can easily build up too much speed and fall if a direct descent is tried. The hop-skip step can be useful on this type of slope. In this technique, the lower leg takes all of the weight, and the upper leg is used only for balance. When traversing, the climber's uphill foot points in the direction of travel. The downhill foot points about 45 degrees off the direction of travel (downhill). This maintains maximum sole contact and prevents possible downhill ankle roll-out.

Note: Wet grass can be extremely slippery; the soldier must be aware of ground cover conditions.

d. **Thick Brush.** For the military mountaineer, brush is both a help and a hindrance. Brush-filled gullies can provide routes and rally points concealed from observation; on the other hand steep brushy terrain is hazardous to negotiate. Cliffs and steep ravines are hidden traps, and blow downs and thickets can obstruct travel as much as manmade obstacles. When brush must be negotiated take the most direct route across the obstacle; look for downed timber to use as raised paths through the obstacle; or create a tunnel through the obstacle by prying the brush apart, standing on lower branches and using upper limbs for support.

e. **Scree Slopes.** Slopes composed of the smallest rocks are called scree slopes. Scree varies in size from the smallest gravel to about the size of a man's fist.

(1) Ascending scree slopes is difficult and tiring and should be avoided, if possible. All principles of ascending hard ground and snow apply, but each step is carefully chosen so that the foot does not slide down when weighted. This is done by kicking in with the toe of the upper foot (similar to step-kicking in snow) so that a step is formed in the loose scree. After determining that the step is stable, weight is transferred to the upper leg, the soldier then steps up and repeats the process with the lower foot.

(2) The best method for descending scree slopes is to come straight down the slope using a short shuffling step with the knees bent, back straight, feet pointed downhill, and heels dug in. When several climbers descend a scree slope together, they should be as close together as possible (one behind the other at single arm interval) to prevent injury from dislodged rocks. Avoid running down scree as this can cause a loss of control. When the bottom of the slope (or run out zone) cannot be seen, use caution because drop-offs may be encountered.

(3) Scree slopes can be traversed using the ice ax as a third point of contact. Always keep the ax on the uphill side. When the herringbone or diagonal method is used to ascend scree, the ax can be used placing both hands on the top and driving the spike into the scree slope above the climber. The climber uses the ax for balance as he moves up to it, and then repeats the process.

f. **Talus Slopes.** Talus slopes are composed of rocks larger than a man's fist. When walking in talus, ascending or descending, climbers should always step on the uphill side of rocks and stay alert for movement underfoot. Disturbing unstable talus can cause rockslides. Climbers must stay in close columns while walking through talus so that dislodged rocks do not reach dangerous speeds before reaching lower soldiers. To prevent rock fall injuries, avoid traversing below other climbers. All other basics of mountain walking apply.

8-3. SAFETY CONSIDERATIONS

The mountain walking techniques presented here are designed to reduce the hazards of rock fall and loss of control leading to a fall. Carelessness can cause the failure of the best-planned missions.

a. Whenever a rock is kicked loose, the warning, "Rock!" is shouted immediately. Personnel near the bottom of the cliff immediately lean into the cliff to reduce their exposure, and do not look up. Personnel more than 3 meters away from the bottom of the cliff may look up to determine where the rock is heading and seek cover behind an obstacle. Lacking cover, personnel should anticipate which way the rock is falling and move out of its path to the left or right.

b. If a soldier slips or stumbles on sloping terrain (hard ground, grass, snow, or scree) he must immediately self-arrest, digging into the slope with hands, elbows, knees and toes. If he falls backwards and rolls over he must immediately try to turn over onto his stomach with his legs downhill and self-arrest with hands and toes.

c. When traveling through steep terrain, soldiers should be trained in the use of the ice ax for self-arrest. The ax can be used to arrest a fall on solid ground, grass and scree as well as snow. It may also be used as a third point of contact on difficult terrain. If not in use the ice ax is carried in or on the rucksack with its head down and secured.

8-4. NAVIGATION

Navigation is the process of determining one's present position, the location of a target objective, and selecting and following a route between these two points. Navigation consists of three distinct stages: orientation, navigation, and route finding.

- Orientation is simply figuring out exactly where one is. The use of the map, compass and identifiable terrain features, assisted by an altimeter and GPS, is the foundation of good navigation.
- Navigation includes the determination of the objective's location and the direction from the soldier's starting point to the objective. The same skills and equipment used in orientation are essential for good navigation.
- Route finding is picking the best line of travel that matches the equipment and capabilities of the team. Good route finding incorporates a comprehensive awareness of terrain, a solid base of mountaineering experience, good judgement and sound tactical instincts.

a. **Compasses.** The magnetic compass is the simplest and most widely used instrument for measuring directions and angles in the mountains. The lensatic compass is most commonly used in the military and can be employed in a variety of ways for either day or night navigation.

b. **Altimeters.** The altimeter is a vital piece of navigational equipment that can save valuable time in determining position through elevation.

(1) The standard altimeter is a modified barometer. A barometer is an instrument that measures the weight of a column of air above itself and displays the result on a scale marked in units of pressure, usually inches of mercury, millimeters of mercury, or millibars. Since air pressure drops uniformly as elevation is gained, it can be used to read altitude by means of the altimeter's scale, marked in feet or meters of elevation above sea level. By measuring air pressure, the altimeter/barometer gives the navigator new techniques for position finding, route planning, checking progress and terrain identification. It also gives the navigator valuable weather information specific to his immediate location.

(2) Changes in the weather are usually accompanied by air pressure changes, which are reflected in the altimeter. As the air pressure drops due to the approach of inclement weather for instance, the displayed elevation will rise by a corresponding amount. This means that a barometric pressure change of one inch of mercury equals roughly 1,000 feet of elevation. If the altimeter displays an elevation gain of 300 feet, a loss of barometric pressure of .3 inches has occurred, and bad weather should be expected.

(3) Altimeters come in two types: wrist-mounted digital altimeters and analog altimeters, usually attached to a cord.

(4) Because the altimeter is sensitive to changes in air pressure it must be recalibrated whenever a point of known elevation (summits, saddles, stream-trail intersections, survey monuments, and so forth) is reached. This is especially important when weather fronts are moving rapidly through the area.

(5) The altimeter may expand or contract because of changes in temperature. This can result in faulty elevation readings. Although some altimeters are temperature-compensated, rapid ascents or descents sometime overcome the adjustment, causing them to give poor readings.

(6) Keep the altimeter at a constant temperature. This is best accomplished by storing the altimeter (analog) in a pocket or on a cord around the neck, or on the wrist under the parka and hand gear (digital).

(7) Even though altimeters can be precise they are affected by both pressure and temperature changes and should be monitored carefully. The soldier should become familiar with the specific altimeter he employs and understand its capabilities and limitations.

c. **Global Positioning System.** The GPS is a space-based, global, all-weather, continuously available radio positioning navigation system. It is highly accurate in determining position location derived from a satellite constellation system. It can determine the latitude, longitude and elevation of the individual user. Location information is also displayed in military grid coordinates.

(1) The GPS provides precise steering information as well as position locations. The receiver can accept many checkpoints entered in any coordinate system by the user and convert them to the desired coordinate system. The user then calls up the desired checkpoint and the receiver will display direction and distance to the checkpoint. It can also compute travel time to the next checkpoint.

(2) Because the GPS does not need visible landmarks to operate, it can provide position (accurate up to 16 meters) in whiteouts or on featureless terrain. It also does not compound navigational errors as compass use can.

(3) During route planning, after choosing critical checkpoints, start point and objective, enter their coordinates as way points. The best use of the GPS is to verify these as they are reached, as a backup to terrain association and compass navigation.

(4) Since the 21-satellite constellation is not yet complete, coverage may be limited to specific hours of the day in certain areas of the world. The GPS navigational signals are similar to light rays, so anything that blocks light will reduce or block the effectiveness of the signals. The more unobstructed the view of the sky, the better the system performs. Although the GPS can be used in any terrain, it is performs best in more open areas such as the desert.

(5) Because the GPS requires horizon to horizon views for good satellite reception its use can be limited in the mountains. Canyons, deep valleys, saddles, and steep mountainsides are all problematic spots to use for shots. Ridgelines, spurs, summits, open valleys, or plateaus are better.

(6) When using GPS in regions with questionable surveying and mapping products, operational datum of the local maps must be reconciled with the datum used in navigational and targeting systems. Identify the spheroid and datum information on the pertinent map sheets and then check that the GPS receiver has the compatible datum loaded. If not then you must contact the S2 for updated datum or maps. Otherwise, the GPS will show different locations than those on the map.

(7) Extremely cold temperatures (-4 degrees F and below) and high elevations will adversely affect the operation of the GPS, due to the freezing of the batteries and the LCD screen. Battery life and overall performance can be improved by placing the GPS inside the parka or coat.

d. **Navigation Techniques.** The choice of movement technique often determines the route and navigational technique. For navigation, three techniques can be used: dead reckoning, terrain association, or altimeter navigation. The three are not mutually exclusive and are normally used together, with one chosen as the primary technique. The GPS can be used to supplement these techniques, but due to the problems associated with the restricted line of sight in the mountains, it should not be used as the main technique.

(1) ***Dead Reckoning.*** Because of the complex nature of mountainous terrain, dead reckoning is usually of limited value on most movements. The compass is generally employed more to support terrain association and to orient the map, than as a primary navigational aid. The main exception is during periods of limited visibility on featureless terrain. Heavy fog, snowy or whiteout conditions on a snowfield, glacier, large plateau or valley floor all would call for dead reckoning as a primary navigational technique.

(2) ***Terrain Association.*** The standard terrain association techniques all apply. Handrails, checkpoints, catching features, navigational corridors, boxing-in areas, and attack points are all used. When a small objective lies near or on an easily identifiable feature, that feature becomes an expanded objective. This simplifies the navigational problem by giving a large feature to navigate to first. The altimeter may finalize the search for the objective by identification through elevation. Rough compass headings are used to establish a general direction to the next checkpoint; used when the checkpoint headed toward is a linear feature, and not a precise point. The shape, orientation, size,

elevation, slope (SOSES) strategy is especially valuable in mountain terrain association and should be practiced extensively (FM 3-25.26).

(a) After extensive study of the map and all available sources of information it helps to create a mental image of the route. This will enable the navigator to make the terrain work in his favor. Avoid brush for speed and ease of movement; the military crest of spurs and ridgelines generally provides the best route while providing terrain masking effects. When clear cut, burned-over, or large avalanche slide areas are encountered, it may be necessary to box or contour around them as they may be full of slash or brushy second-growth small trees. Old-growth forest provides the easiest travel.

(b) The following situations will result in objects appearing closer than they actually are:

- When most of the object is visible and offers a clear outline.
- When you are looking across a partially cleared depression.
- When looking down a straight, open road or track.
- When looking over a smooth, uniform surface, such as snow, water, or desert.
- When the light is bright and the sun is shining from behind the observer.
- When the object is in sharp contrast to the background.
- When seen in the clear air of high altitude.
- When looking down from high ground to low ground.

(c) The following situations will result in objects appearing farther away than they actually are:

- When only part of the object is seen or it is small in relation to its surroundings.
- When you are looking across an exposed depression.
- When looking up from low ground to high ground.
- When your vision is narrowly confined.
- When the light is poor, such as dawn, dusk, or low visibility weather; or when the sun is in your eyes, but not behind the object being viewed.
- When the object blends into the background.

(2) ***Altimeter Navigation.*** Altimeters provide assistance to the navigator in several ways. They aid in orientation, in computing rates of ascent or descent, in resection, and in weather prediction.

(a) When moving along any linear feature such as a ridge, watercourse, or trail which is shown on the map, check the altimeter. The point where the indicated elevation contour crosses that feature is your location.

(b) The navigator frequently finds it necessary to determine his position through the use of resection. A modified resection can be performed by shooting an azimuth to a known, clearly visible summit or similar feature and then plotting the back azimuth on the map. By determining your present elevation and finding where that particular contour crosses the back azimuth you should locate your position. This can be difficult when in low ground, as mountain summits can rarely be clearly seen from valley floors. In addition, most mountaintops are so large that there is usually no specific point to shoot at. In this case, the soldier should take multiple azimuths to known features. If he is located on a good linear feature he will have a decent idea of where he is. The altimeter can be used to verify elevation and establish a notional linear feature—a contour line. The point where the resecting back azimuths cross the contour line is the navigator's location.

(c) Using the altimeter to calculate rates of ascent can help in sound decision-making. Rates of travel, along with weather conditions, light conditions (time of day), and the physical condition of the team, are all key variables that can influence the success or failure of the mission.

(d) Altimeters can be used as barometers to assist in weather prediction.

e. **Approach Observations.** Watch the mountain during the approach march, studying it for climbing routes. Distant views can reveal large-scale patterns of ridges, cliffs, snowfields and glaciers. General angles of the large rock masses can be seen from afar.

(1) Closer viewing displays these patterns and angles on a smaller scale. Fault lines, gross bedding planes of rock, cliff bands, and crevasse zones become visible. Snowy or vegetated ledge systems appear. Weaknesses in the mountain walls, such as couloirs or gullies, may present themselves.

(2) Most of these features repeat themselves at increasingly finer levels, as they are generally derived from the overall structure of the particular mountain group. A basic knowledge of mountain geology, combined with the specific geological background of the operational area, pays off in more efficient travel.

f. **Natural Indicators of Direction in the Northern Hemisphere.** Southern slopes are sunnier and drier than northern slopes, with sparser or different types of vegetation. Northern slopes can be snowier and, because of more intense glaciation in past ages, are often steeper.

Note: Opposite rules apply in the Southern Hemisphere.

g. **Winter Route Selection.** The following must be considered when selecting a route in the winter.

(1) Conduct a thorough map reconnaissance considering the weather, individual ski abilities, avalanche danger, vegetation, water features, terrain relief, and the size of the unit.

(2) Weather conditions will affect the chosen route. During calm weather, your rate of movement will be significantly faster than during periods of inclement weather.

(3) Individual ski abilities will affect your rate of movement, constrain your choice of terrain, and impact on your route choices.

(4) Avalanche danger zones must be identified by map review and data gathered during route planning. During movement, snow pits, shovel tests, and ski shear tests must be conducted prior to crossing an avalanche danger zone. Bottom line: avoid avalanche danger areas. If you must cross one, cross above the starting zone or below the run-out zone.

(5) Vegetation can work for you or against you. Thickly forested areas usually have a deep snow pack. For weaker skiers, forested areas are full of potentially dangerous obstacles. On slopes with an angle of 30 to 45 degrees that are sparsely vegetated an avalanche danger is still present. If the weather turns bad, forested areas provide welcome relief from wind and blowing snow.

(6) Water features provide valuable navigation aids. Under deep snow pack small creeks and ponds may be hard to locate. Large frozen lakes and rivers can provide excellent means of increasing your rate of march.

(7) During ski movements, efficient use of terrain will greatly improve morale and reduce fatigue. While traveling in mountainous terrain, do not needlessly give up elevation gained. Maintain a steady climb rate and avoid over exertion. Avoid north, east, and south facing slopes when the avalanche danger is high. Avoid cornices and be aware of their probable and improbable fracture lines. Weather and tactical situation permitting, travel on the windward side of ridgelines. If weak skiers are in the group, stay away from restrictive terrain with sheer drop-offs. When touring use climbing skins to maintain control and lessen lost time per hour due to individuals falling.

(8) The following are additional hints for navigation in snowy conditions:

- Keep the compass warm.
- If no terrain features exist for steering marks, use your back azimuth and tracks to maintain course.
- Limit steering marks to shorter distances since visibility can change quickly.
- Never take azimuths near metallic objects. Hold the compass far enough from your weapon, ice ax, and so on to get accurate readings.
- Make frequent compass checks.
- Preset azimuths on your compass.
- Use a steady, unshifting wind to aid you in maintaining course.

h. **Problems.** The following conditions and characteristics of cold weather and mountainous regions make accurate navigation difficult.

(1) In winter, short hours of daylight, fog, snowfall, blizzards, whiteouts, and drifting snow, especially above tree line, drastically limit visibility. At times, an overcast sky and snow-covered terrain create a phenomenon called flat light, which makes recognition of irregularities in the terrain extremely difficult.

(2) Heavy snow may completely cover existing tracks, trails, outlines of small lakes and similar landmarks. Because the appearance of the terrain is quite different in winter from that in summer, particular attention must be paid to identifying landmarks, both on the ground and from aerial photographs.

(3) Magnetic disturbances, caused by large ore deposits, are frequently encountered and make magnetic compass readings difficult and sometimes unreliable.

(4) Handling maps, compasses, and other navigation instruments in low temperatures with bare hands is difficult. Removing hand wear may only be possible for short periods.

(5) Keeping count of pace is extremely difficult in winter and mountain environments. Thick vegetation and rough, steep slopes hamper attempts at accurate pace counts. The most reliable method is the use of a 50-meter long piece of field wire or rope.

8-5. ROUTE PLANNING

Proper route planning can make the difference between success and failure on long mountain movements. Careful map reconnaissance, knowledge of the enemy situation, terrain analysis of the operational area, and an accurate assessment of the unit's capabilities are all key parts of the planning process.

a. **Map Reconnaissance.** Topographic maps provide the primary source of information concerning the area of operations. A 1:25,000 map depicts greater detail than a 1:50,000 map and should be used whenever possible. Because examination of the micro-terrain is so important for mountain operations, even larger scale maps are desirable. Civilian 1:12,000 maps can be used if available. Aerial, oblique angle, photographs give details not always shown on maps (crags and overhangs). Sketch maps supplement other sources of information but should not be relied on for accuracy since they are seldom drawn to scale. Along with sketch maps, verbal descriptions, documented information gathered from units previously in the area, or published sources such as alpine journals or climbing guides may help. Forest service and logging and mining company maps provide additional information, often showing the most recent changes to logging trails and mining access roads. Standard military topographic maps are generally accurate graphic depictions of the operational area.

(1) When conducting a map reconnaissance, pay close attention to the marginal information. Mountain-specific terrain features may be directly addressed in the legend. In addition, such facilities as ski lifts, cable and tramways are often found. Check the datum descriptor (for foreign maps) to ensure compatibility with entered datum in GPS units. Along with the standard topographic map color scheme, there are some commonly seen applications for mountainous terrain. White with blue contours indicates glaciers or permanent snowfields. The outline of the snow or ice is shown by dashed blue lines while their contour lines are solid blue. High ice cliffs which are equal to or exceed the contour interval will be shown. Low ice cliffs and ice caves may be indicated if they provide local landmarks. Brown contour lines on white mean dry areas without significant forest cover. Areas above tree line, clear cuts, rock or avalanche slide paths and meadows are all possible. Study the surrounding terrain and the legend for other clues. An important point to remember is that thick brush in small gullies and streambeds may not be depicted by green, but should still be expected.

(2) Obstacles, such as rivers and gorges, will require technical equipment to cross if bridges are not present. Fords and river crossing sites should be identified. Due to the potential for hazardous weather conditions, potential bivouac sites are noted on the map. Ruins, barns, sheds and terrain-protected hollows are all possible bivouac sites. Danger areas in the mountains; isolated farms and hamlets, bridges, roads, trails, and large open areas above tree line, are factored in, and plans made to avoid them. Use of terrain-masking becomes essential because of the extended visibility offered by enemy observation points on the dominant high ground.

(3) Helicopter support, weather permitting, requires identification of tentative landing zones for insertions, extractions, resupply and medevac. The confined nature of mountain travel means that crucial passes become significant chokepoints and planners should designate overwatches/surveillance positions beforehand. Alternate routes should be chosen with weather imposed obstacles in mind: spring flood or afternoon snowmelt turns small streams into turbulent, impassable torrents. Avalanche danger prohibits travel on certain slopes or valley floors.

b. **Enemy Situation.** Route selection should only be done after reviewing all available information about the friendly and enemy situation.—Is the enemy force on his own ground? Are they accustomed to the terrain and the weather? Are they trained mountain troops with specialized equipment?—Only after answering these and other questions can an effective route plan be completed. If the enemy force is better prepared to maneuver in the mountains, they have a marked advantage, and route selection must be scrutinized.

c. **Analysis of the Operational Area.** Not all mountainous terrain is created equal and not all movement plans have the same expectation of success. Planners must undertake a thorough analysis of the general terrain to be crossed, including the geology, mountain structure and forms, and ground cover.

(1) Heavily glaciated granite mountains pose different problems than does river-carved terrain. The U-shaped valley bulldozed out by a glacier forces maneuver elements down to the valley floor or up to the ridge tops, while the water-cut V-shape of river valleys allows movement throughout the compartment.

(2) Routes through granite rock (long cracks, good friction; use of pitons, chocks and camming units) will call for different equipment and technique than that used for steep limestone (pockets, smooth rock; bolts, camming units).

(3) Operations above tree line in temperate climates or in the brushy zone of arid mountains means that material for suspension traverse A-frames must be packed. The thick brush and krummholtz mats of subalpine zones and temperate forested mountains can create obstacles that must be bypassed.

(4) Heavy spruce/fir tangles slow progress to a crawl, therefore planners should ensure routes do not blindly traverse these zones.

d. **Unit Assessment.** When assessing unit movement capabilities the key indices are training and conditioning levels. Soldiers who have received basic military mountaineer training, who know how to move through rough terrain, and who have been hardened with training hikes through the mountains, will perform better than troops without this background.

e. **Time-Distance Formulas.** Computing march rates in the mountains is extremely difficult, especially when there is snow cover. The following rates are listed as a guide (Table 8-1). Rates are given for movement over flat or gently rolling terrain for individuals carrying a rifle and loaded rucksack.

	UNBROKEN TRAIL	BROKEN TRAIL
On foot (no snow cover)	2 to 3 kph (cross-country)	3 to 4 kph (trail walking)
On foot (less than 1 foot of snow)	1.6 to 3.2 kph	2 to 3.2 kph
On foot (more than 1 foot of snow)	.4 to 1.2 kph	2 to 3.2 kph
Snowshoeing	1.6 to 3.2 kph	3.2 to 4 kph
Skiing	1. to 5.6 kph	4.8 to 5.6 kph
Skijoring	N/A	3 to 24 kph

Table 8-1. Time-distance formulas.

(1) March distances in mountainous terrain are often measured in time rather than distance units. In order to do this, first measure the map distance. This distance plus 1/3 is a good estimate of actual ground distance. Add one hour for each 1,000 feet of ascent or 2,000 feet of descent to the time required for marching a map distance.

(2) As Table 8-1 indicates, snow cover will significantly affect rates of march. Since snow can be expected in the mountains most months of the year, units should have some experience at basic snow travel.

(3) Individual loads also affect march rates. Combined soldier loads that exceed 50 pounds per man can be expected to slow movement significantly in mountainous terrain. Given the increased weight of extra ammunition for crew-served weapons, basic mountaineering gear, and clothing for mountain travel, it becomes obvious that soldiers will be carrying weights well in excess of that 50-pound limit. Units should conduct cross-country movements in the mountains with the expected rucksack and LCE weights in order to obtain accurate, realistic rates of march.

(4) In the harsh environment of the mountains, helicopter support cannot be relied on. The process of transporting extra equipment and sustainment supplies will result in vastly increased movement times. The heavier loads will exhaust soldiers mentally and physically. Tactical movements, such as patrolling or deliberate assaults, should take this into account.

8-6. ROUTE SELECTION

Many variables affect the selection of the proper route. The following guidelines apply to all situations.

a. **Select a Current Map.** Check the date of the map for an indication of the reliability of the map in depicting vegetation, clearings, roads, and trails accurately. The leader should use all the latest topographic data he can find.

b. **Gather Intelligence Information.** The most important consideration in every leader's mind when plotting a movement is "where is the enemy?" The latest intelligence reports are essential. Additionally, weather reports, snow condition reports, avalanche probability, aerial photos, and any past or recent history of operations in the area may be of help.

c. **Select a Route.** Identify the starting point and determine the movement objective. Plot start and end points. Carefully scrutinize the area in between and begin to select the route. Consider the following:

(1) ***Trafficability.*** This includes degree of slopes, vegetation, width of trails, snow depth, avalanche probability, and the likelihood of crevasses.

(2) ***Time-Distance Formula.*** Time allotted and distance to be covered must be considered.

(3) ***Required Equipment.*** Carry enough equipment to move along the route and to survive if an extended stay becomes necessary. Do not plan a route beyond the means of your equipment.

(4) ***Location of Enemy.*** Plan a route that allows maximum use of the masking effect of the terrain. Avoid danger areas or areas of recent enemy activity unless required by the mission. Use vegetation to mask your movement if possible (especially coniferous forests). Avoid silhouetting on ridgelines.

(5) ***Communications.*** Communications will be severely limited in the mountains. Dead spaces or communications holes are common. Use all available information and plan accordingly.

(6) ***Conditions/Capabilities of Unit.*** The unit must be able to negotiate the route chosen. Take into consideration their present health, as well as their training level when selecting your intended route.

(7) ***Checkpoints/Control Points.*** When plotting a route on the map, utilize prominent terrain features on either side of the route as checkpoints. Ensure that when you select

your checkpoints they are visually significant (elevation) and that they are easily identifiable. Avoid the use of manmade features as checkpoints due to their unreliability and lack of permanence. Select features that are unique to the area.

CHAPTER 9
MOUNTAIN STREAM CROSSINGS

Operations conducted in mountainous terrain may often require the crossing of swift flowing rivers or streams. Such crossings should not be taken lightly. The force of the flowing water may be extremely great and is most often underestimated. All rivers and streams are obstacles to movement. They should be treated as danger areas and avoided whenever possible. When rivers or streams must be crossed, there are a variety of techniques the small-unit leader may choose from, depending upon the type of stream, its width, speed of the current, and depth of the water.

There are limits on the safe use of these techniques. Not all mountain rivers or streams will be fordable with these techniques. If a water obstacle is too wide, swift, or deep, an alternate route should be used, or the crossing will require major bridging by engineers. It may require the use of rafts or boats. Reconnaissance of questionable crossing sites is essential. This chapter covers the techniques for crossing mountain streams that have a depth generally not exceeding waist deep.

9-1. RECONNAISSANCE

Reconnaissance of the route (map, photo, and or aerial) may not always reveal that a water obstacle exists. In a swamp, for example, unfordable sloughs may not show on the map, and they may be concealed from aerial observation by a canopy of vegetation. Whenever it is possible that a unit will be required to cross a water obstacle, its commander must plan some type of crossing capability.

a. Site selection is extremely important once you determine that you must make a crossing (Figure 9-1). Look for a high place from which you can get a good view of the obstacle and possible crossing sites. A distant view, perhaps from a ridge, is sometimes better than a hundred close views from a riverbank. Site selection must be made before the arrival of the main body.

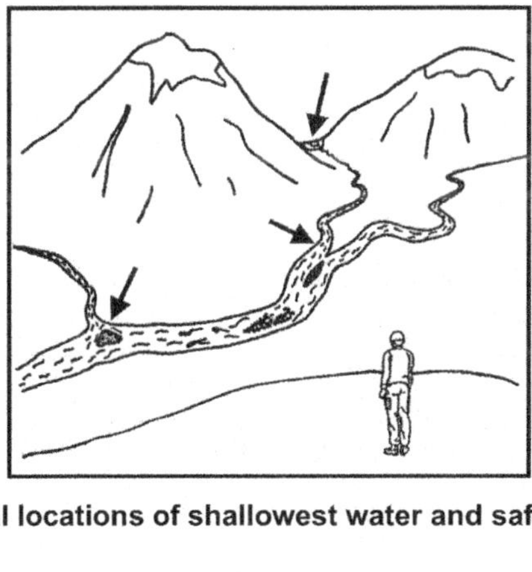

Figure 9-1. Normal locations of shallowest water and safest crossing sites.

b. A dry crossing on fallen timber or log jams is preferable to attempting a wet crossing. Depending upon the time of year, the closer you are to the source, or headwaters, the better your chances are of finding a natural snow or ice bridge for crossing. If a dry crossing is unavailable, the following considerations should be made:

(1) The time of day the crossing can be an important factor. Although early morning is generally best because the water level is normally lower during this period, recent weather is a big factor; there may have been heavy rain in the last eight hours. As glaciers, snow, or ice melt during the day, the rivers rise, reaching their maximum height between mid afternoon and late evening, depending on the distance from the source. Crossings, if made during the early morning, will also allow clothing to dry more quickly during the heat of the day.

(2) A crossing point should normally be chosen at the widest, and thus shallowest, point of the river or stream. Sharp bends in the river should be avoided since the water is likely to be deep and have a strong current on the outside of the bend. Crossings will be easiest on a smooth, firm bottom. Large rocks and boulders provide poor footing and cause a great deal of turbulence in the water.

(3) Many mountain streams, especially those which are fed by glacier run-off, contain sections with numerous channels. It is often easier to select a route through these braided sections rather than trying to cross one main channel. A drawback to crossing these braided channels, however, is the greater distance to the far bank may increase exposure time and often the sand and gravel bars between the channels will offer little cover or concealment, if any.

(4) The crossing site should have low enough banks on the near and far side to allow a man carrying equipment to enter and exit the stream with relative ease. If a handline or rope bridge is to be constructed, the crossing site should have suitable anchors on the near and far bank, along with safe loading and unloading areas. Natural anchors are not a necessity, however the time required to find a site with solid natural anchors will probably be less than the time required to construct artificial anchors. In some areas, above the tree line for example, artificial anchors may have to be constructed. Deadman anchors buried in the ground, or under a large pile of boulders work well.

(5) Log jams and other large obstructions present their own hazards. Logs floating downstream will generally get hung up in shallower sections creating the jam. Once a log jam is formed, however, the water forced to flow around it will erode the stream bottom. Eventually deep drop-offs or holes may develop, especially around the sides and off the downstream end of the log jam. A log jam that totally bridges a section of the stream may be the best way to cross. A wet crossing in the vicinity of a log jam should be performed a good distance below or above it. Some things to consider when crossing near log jams are:

- Cross well to the downstream side when possible.
- Keep a sharp lookout for floating timber that could knock you off your feet.
- If you must cross on the upstream side, stay well upstream from the log jam. If a person is swept off his feet and caught in the debris of the jam, he could easily drown. A handline will greatly increase safety here.

(6) When possible, select a crossing site that has enough natural protection on the near and far banks so that security teams may be placed out and enough cover and concealment is available for the size of the element making the crossing. When cover and

concealment is minimal, as in the higher alpine zones, the crossings must be conducted as efficiently as possible to minimize exposure to enemy observation.

9-2. PREPARATION OF TROOPS AND EQUIPMENT

Prepare men and equipment for a crossing as far in advance as feasible. Final preparation should be completed in a security perimeter on the near side just before crossing. Preparation includes the following.

a. Waterproof water-sensitive items. Wrap radios, binoculars, SOI, papers, maps and any extra clothing in waterproof bags (trash bags also work well), if available. These bags also provide additional buoyancy in case of a fall.

b. Trousers are unbloused and shirts are pulled out of the trousers. All pockets are buttoned. This allows water to escape through the clothing. Otherwise the clothing would fill up and retain water, which would weigh the body down. This is especially critical if an individual must swim to shore. Depending on the circumstances of the crossing (for example, tactical situation, temperature of the air and water), the crossing can be made in minimal clothing so that dry clothing is available after the crossing. Boots should be worn to protect feet from rocks; however, socks and inner soles should be removed. On the far side, the boots can be drained and dry socks replaced.

c. Load-carrying equipment harness and load-bearing vest (LBV) is unbuckled and worn loosely. It is extremely difficult to remove a buckled harness in the water in an emergency.

d. Helmets are normally removed and placed in the rucksack in slow moving streams with sandy or gravel bottoms. If you have to resort to swimming it is easier done without the helmet. However, when crossing swift flowing streams, especially those with large rocks and other debris, the risk of head injury if a person slips is high. In this case the helmet should be worn with the chinstrap fastened.

e. The rucksack should be worn well up on the shoulders and snug enough so it does not flop around and cause you to lose your balance. The waist strap MUST be unbuckled so you can get rid of the pack quickly if you are swept off your feet and have to resort to swimming. If a pack has a chest strap it must also be unbuckled. Secure everything well within your pack. It is easier to find one large pack than to find several smaller items.

f. Individual weapons should be attached to the pack or slung over the shoulder.

9-3. INDIVIDUAL CROSSINGS

Whenever possible, and when the degree of experience permits, streams should be forded individually for a speedier crossing. The average soldier should be able to cross most streams with mild to moderate currents and water depths of not much more than knee deep using proper techniques.

a. The individual should generally face upstream and slightly sideways, leaning slightly into the current to help maintain balance. At times, he may choose to face more sideways as this will reduce the surface area of the body against the current, thus reducing the current's overall force on the individual.

b. The feet should be shuffled along the bottom rather than lifted, with the downstream foot normally in the lead. He should take short, deliberate steps. Lunging steps and crossing the feet result in a momentary loss of balance and greatly increase the chance of a slip.

c. The individual should normally cross at a slight downstream angle so as not to fight the current. There is normally less chance of a slip when stepping off with the current as opposed to stepping off against the current.

d. The individual must constantly feel for obstacles, holes and drop-offs with the lead foot and adjust his route accordingly. If an obstacle is encountered, the feet should be placed on the upstream side of it where the turbulence is less severe and the water normally shallower.

e. To increase balance, and if available, a long ice ax, sturdy tree limb, or other staff can be used to give the individual a third point of contact (Figure 9-2). The staff should be used on the upstream side of the individual and slightly leaned upon for support. The staff should be moved first, then the feet shuffled forward to it. This allows two points of contact to be maintained with the streambed at all times. The individual still moves at a downstream angle with the downstream foot in the lead.

Figure 9-2. Individual crossing with staff.

9-4. TEAM CROSSING

When the water level begins to reach thigh deep, or anytime the current is too swift for personnel to safely perform an individual crossing, a team crossing may be used. For chain crossing, two or more individuals cross arms with each other and lock their hands in front of themselves (Figure 9-3). The line formed faces the far bank. The largest individual should be on the upstream end of the line to break the current for the group. The line formed will then move across the stream using the same principles as for individual crossings, but with the added support of each other. The line should cross parallel to the direction of the current. The team still moves at a slight downstream angle, stepping off with the downstream foot in the lead.

Figure 9-3. Chain method for stream crossing.

9-5. ROPE INSTALLATIONS

When the water level begins to reach waist deep or the current is too swift for even a team crossing, the chosen site must be closely examined. The stream at this point may be impassable. Many times though, a crossing site which may be unsafe for individual or team crossings can be made safe with the installation of a handline or rope bridge. Crossing on a handline will still require each individual to enter the water and get wet. If a one-rope bridge can be constructed, it may require only a couple of individuals to enter the water. Deciding whether to install a handline or a rope bridge will depend on the anchors available, height of the anchors above the water, and the distance from the near and far anchors. The maximum distance a one-rope bridge is capable of spanning is approximately 1/2 to 2/3 the length of the rope in use.

a. **Establishing the Far Anchor.** Whether a handline or rope bridge is to be installed, someone must cross the stream with one end of the rope and anchor it on the far side. This duty should be performed by the most capable and strongest swimmer in the party. The swimmer should be belayed across for his own safety. The belay position should be placed as far above the crossing as possible. In the event that the current is too strong for the individual, he will pendulum back to the near bank. Rescuers should be poised on the near bank at points where the individual will pendulum back, should he fail to reach the far bank. The initial crossing site should be free of obstacles that would snag the rope and prevent the pendulum back to the bank for an easy recovery.

(1) The individual may attach the belay rope to his seat harness or a swami belt with a carabiner. He should NEVER tie directly into the rope when being belayed for a stream crossing. If the swimmer should be swept away and become tangled, he must be able to release himself quickly from the rope and swim to shore as best he can. The individual may also choose to tie a fixed loop into the end of the belay rope and hang on to it, where he can immediately release it in an emergency.

(2) Anytime a crossing site must be used where the swimmer may encounter problems getting to the far bank, he should have on a life vest or other personal flotation

device (PFD). If the swimmer must release the rope at any time, he will have to rely on his own water survival skills and swimming ability to get to shore. A PFD will greatly increase his own personal safety. A PFD may also be used by the last man across, as he will release the rope from the anchor and be belayed across as the first man

b. **Installation of a Handline.** If it is possible to use a rope high enough above the water to enable soldiers to perform a dry crossing, then a rope bridge should be installed as such. If this is impossible, and the rope must be installed to assist in a wet crossing, then it should be installed as a handline (Figure 9-4).

(1) The far anchor should be downstream from the near anchor so that the rope will run at an angle downstream from the near anchor, approximately thirty to forty-five degrees, rather than straight across the stream. Here again, it is easier to move with the current as opposed to directly across or against it.

(2) The rope may be anchored immediately on the far bank, pulled tight, and anchored on the near bank, or it may be installed with a transport tightening system if a tighter rope is required.

(3) Crossing will always be performed on the downstream side of the handline, shuffling the feet with the downstream foot in the lead.

(4) A second climbing rope is used as a belay (Figure 9-5). One end of the belay rope will be on the near bank and the other end on the far bank. It should be sent across with the strong-swimmer. An appropriate knot is tied into the middle of the belay rope to form two fixed loops with each loop being approximately 6 inches long. One loop is connected to the handline with carabiner(s) and the individual crossing connects one loop to himself. The loops are short enough so the individual is always within arms reach of the handline should he slip and let go. The individuals are belayed from both the near and far banks. If a mishap should occur the individual can be retrieved from either shore, whichever appears easiest. The belay on the opposite shore can be released allowing the individual to pendulum to the bank. It is important that the belay rope NOT be anchored or tied to the belayer so that it may be quickly released if necessary.

Figure 9-4. Stream crossing using a handline.

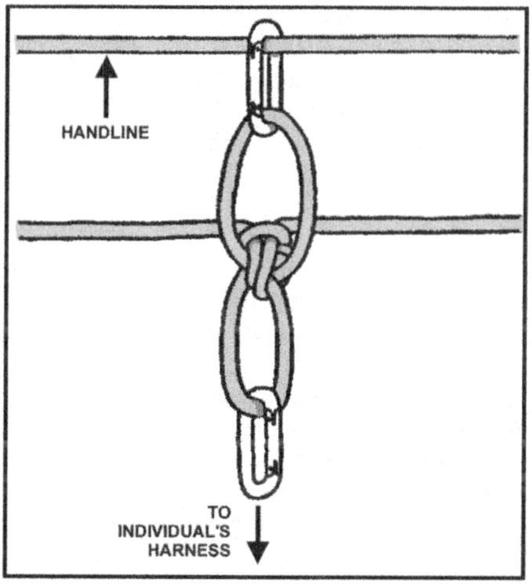

Figure 9-5. Belay rope for crossing using a handline.

(5) Under most circumstances, the handline should be crossed one person at a time. This keeps rope stretch and load on the anchors to a minimum.

(6) Rucksacks can be either carried on the back the same way as for individual crossings, or they can be attached to the handline and pulled along behind the individual.

(7) If a large amount of equipment must be moved across the stream, especially heavier weapons, such as mortars, recoilless rifles, and so on, then a site should be selected to install a rope bridge.

9-6. SAFETY

River and stream crossings present one of the most hazardous situations faced by the military mountaineer. The following safety procedures are minimum guidelines that should be followed when conducting a river or stream crossing.

a. All weak and nonswimmers should be identified before a crossing so that stronger swimmers may give assistance in crossing.

b. Not every river or stream can be crossed safely. It is always possible to cross at a different time or place, use a different technique, or choose another route.

c. The technique used is directly dependent upon water depth, speed of the current, stream bottom configuration, width of the stream, and individual experience.

d. The safest methods of crossing are always with the use of a handline or one-rope bridge.

e. If the installation of a handline or rope bridge becomes too difficult at a given crossing site, then that site should be considered too hazardous and another site selected.

f. A lookout should be posted 50 to 100 meters upstream to watch for any obstacles that may be carried downstream and interfere with the crossing.

g. When conducting individual crossings (those without a handline or rope bridge), lifeguards should be posted downstream with poles or ropes prepared to throw, for assistance or rescue.

h. When the unit knows a rope installation will be required for crossing, at least two life vests or other PFDs should be on hand to provide additional safety for the strong swimmer who must establish the far anchor, and the last man across who retrieves the system.

9-7. SWIMMING

There are times when you might be alone and have no choice but to swim across, or there may be a time that you find yourself suddenly plunged into a swift river or rapids. In either case, the following techniques could save your life.

a. Immediately jettison any equipment or clothing that restricts movement.

b. Do not try to fight the current. Maneuver towards shore in a position with the feet downstream, facing downstream, and fanning the hands alongside the body to add buoyancy and to fend off submerged rocks. Use the feet to protect the rest of the body and to fend off submerged rocks.

c. Keep the head above water to observe for obstacles and attempt to maneuver away from them.

d. Try to avoid backwater eddies and converging currents as they often contain dangerous swirls. Avoid bubbly water under falls as it has little buoyancy. Breath between the wave troughs.

e. If the shore is too difficult to reach, seek out the closest and safest spot, such as a sandbar, to get yourself out of the water as quickly as possible. Hypothermia will set in quickly in colder waters.

CHAPTER 10
MOVEMENT OVER SNOW AND ICE

Movement over snow- and ice-covered slopes presents its own unique problems. Movement on steeper slopes requires an ice ax, crampons, and the necessary training for this equipment. Personnel will also have to learn how to place solid anchors in snow and ice to protect themselves during these movements if roped. Snow-covered glaciers present crevasse fall hazards even when the slope is relatively flat, requiring personnel to learn unique glacier travel and crevasse rescue techniques.

All the principles of rock climbing, anchor placement, belays, and rope usage discussed throughout the previous chapters apply to snow and ice climbing as well. This chapter will focus on the additional skills and techniques required to move safely through snow-covered mountains and over glaciated terrain.

10-1. MOVEMENT OVER SNOW

The military mountaineer must be equally adept on both snow and ice due to route necessity and rapidly changing conditions. On steep slopes in deep snow, the climber may climb straight up facing the slope. The ice ax shaft, driven directly into the snow, provides a quick and effective self-belay in case of a slip—the deeper the shaft penetrates the snow, the better the anchor (Figure 10-1). It is usually best, however, to climb snow-covered slopes in a traversing fashion in order to conserve energy, unless there is significant avalanche danger.

Figure 10-1. Self-belay on snow.

a. The progression from walking on flat terrain to moving on steep terrain is the same as for moving over snow-free terrain. If the snow is packed the sole of the boot will generally hold by kicking steps, even on steep slopes. Where it is difficult to make an effective step with the boot, a cut made with the adze of the ice ax creates an effective step. In these situations crampons should be used for faster and easier movement.

b. When descending on snow, one can usually come straight downhill, even on steep terrain. Movement downhill should be slow and deliberate with the climber using an even pace. The heels should be kicked vigorously into the snow. The body may be kept erect with the aid of an ice ax, which may be jammed into the snow at each step for additional safety. Here again, crampons or step cutting may be necessary. A technique known as glissading may also be used as an easy method of descent and is covered in detail later in this chapter.

10-2. MOVEMENT OVER ICE

Ice is found in many areas of mountains when snow is present, and during the summer months also where perennial snowpack exists. Many times an ice area will be downslope of a snowfield and sometimes the ice pack itself will be lightly covered with snow. Even if using an ice ax and or crampons, movement will still be difficult without proper training.

10-3. USE OF ICE AX AND CRAMPONS

Movement over snow and ice is almost impossible without an ice ax and or crampons.

a. **Ice Ax.** When walking on snow or ice, the ice ax can be used as a third point of contact. When the terrain steepens, there are a number of ways to use the ice ax for snow or ice climbing. Some positions are more effective than others, depending on the intended result. You may find other ways to hold and use the ax, as long the security remains in effect.

(1) ***Cane Position.*** The ice ax can be used on gentle slopes as a walking stick or cane (Figure 10-2). The ax is held by the head with the spike down and the pick facing to the rear in preparation for self-arrest. When moving up or down gentle slopes the ice ax is placed in front as the third point of contact, and the climber moves toward it. When traversing, the ax is held on the uphill side, in preparation for a self-arrest.

Figure 10-2. Using the ice ax in the cane position.

(2) ***Cross Body Position or Port Arms Position.*** On steeper slopes the ax can be used in the port arms position, or cross body position (Figure 10-3). It is carried across the chest, upslope hand on the shaft, spike towards the slope. The head of the ax is held away from the slope with the pick to the rear in preparation for self-arrest. Ensure the leash is connected to the upslope hand, which allows the ax to be used in the hammer position on the upslope side of the climber. The spike, in this case, is used as an aid for maintaining balance.

Figure 10-3. Ice ax in the cross body or port arms position.

(3) ***Anchor Position***. As the slope continues to steepen, the ax may be used in the anchor position (Figure 10-4). The head is held in the upslope hand and the pick is driven into the slope. The spike is held in the downhill hand and pulled slightly away from the slope to increase the "bite" of the pick into the ice. If the climber is wearing a harness, the pick can be deeply inserted in the ice or hard snow and the ax leash could be connected to the tie-in point on the harness for an anchor (ensure the ax is placed for the intended direction of pull).

Figure 10-4. Ice ax in the anchor position.

(4) ***Push-Hold Position.*** Another variation on steep slopes is the push-hold position (Figure 10-5). The hand is placed on the shaft of the ax just below the head with the pick forward. The pick is driven into the slope at shoulder height. The hand is then placed on the top of the ax head for use as a handhold.

Figure 10-5. Ice ax in the push-hold position.

(5) ***Dagger Position.*** The dagger position is used on steep slopes to place a handhold above shoulder height (Figure 10-6). The hand grasps the head of the ax with the pick forward and the shaft hanging down. The ax is driven into the surface in a stabbing action. The hand is then placed on the ax head for use as a handhold.

Figure 10-6. Ice ax in the dagger position.

(6) ***Hammer Position.*** The hammer position will set the pick deepest in any snow or ice condition (Figure 10-7, page 10-6). The ax is used like a hammer with the pick being driven into the slope. On vertical or near-vertical sections, two axes used in the hammer position will often be required.

Figure 10-7. Ice ax in the hammer position.

b. **Crampons.** Walking in crampons is not complicated but it does present difficulties. When walking in crampons, the same principles are used as in mountain walking, except that when a leg is advanced it is swung in a slight arc around the fixed foot to avoid locking the crampons or catching them in clothing or flesh. The trousers should be bloused to prevent catching on crampons. All straps should be secured to prevent stepping on them and, potentially, causing a fall. The buckles should be located on the outside of each foot when the crampons are secured to prevent snagging. Remember, when the crampon snags on the pants or boots, a tear or cut usually results, and sometimes involves the skin on your leg and or a serious fall.

(1) Two methods of ascent are used on slopes: traversing and straight up.

(a) A traverse on ice or snow looks much like any mountain walking traverse, except that the ankles are rolled so that the crampons are placed flat on the surface (Figure 10-8). On snow the points penetrate easily; on ice the foot must be pressed or stamped firmly to obtain maximum penetration. At the turning points of a traverse, direction is changed with the uphill foot as in mountain walking.

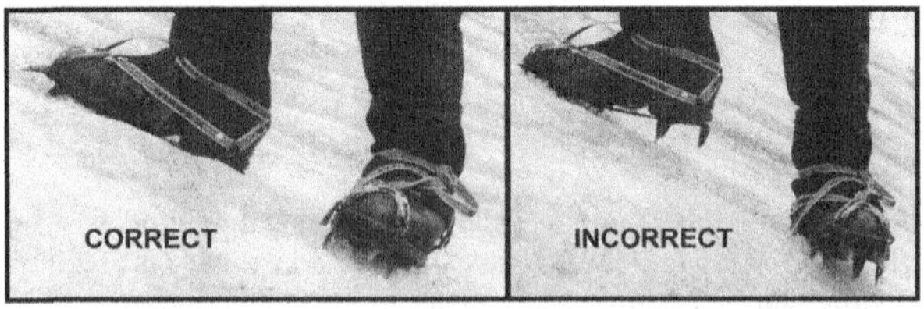

Figure 10-8. Correct and incorrect crampon technique.

(b) A straight up method is for relatively short pitches, since it is more tiring than a traverse. The climber faces directly up the slope and walks straight uphill. As the slope steepens, the herringbone step is used to maintain the flatfoot technique. For short steep pitches, the climber may also face downslope, squatting so the legs almost form a

90-degree angle at the knees, driving the spike of the ice ax into the slope at hip level, and then moving the feet up to the ax. By repeating these steps, the ax and crampon combination can be used to climb short, steep pitches without resorting to step cutting. This method can be tiring. The technique is similar to the crab position used for climbing on slab rock and can also be used for short descents.

(2) A technique known as "front-pointing" may be used for moving straight uphill (Figure 10-9). It is especially useful on steep terrain, in combination with the ice ax in the push-hold, dagger, or hammer position. Front-pointing is easiest with the use of more rigid mountain boots and rigid crampons. The technique is similar to doing calf raises on the tips of the toes and is much more tiring than flat-footing.

(a) The technique starts with the feet approximately shoulder width apart. When a step is taken the climber places the front points of the crampons into the ice with the toe of the boot pointing straight into the slope.

(b) When the front points have bitten into the ice the heel of the boot is lowered slightly so that the first set of vertical points can also bite. The body is kept erect, with the weight centered over the feet as in climbing on rock.

Figure 10-9. Front-pointing with crampons.

c. **Vertical Ice.** When a climb on ice reaches the 60- to 70-degree angle, two ice axes may be helpful, and will become necessary as the angle approaches 90 degrees. The same basic climbing techniques described in Chapter 6 should be applied. If leashes of the correct length and fit are attached to both axes, it may be possible to hang completely from the axes while moving the feet.

d. **Descending with Crampons and Ice Ax.** Whenever possible, descend straight down the fall line. As the slope steepens, gradually turn sideways; on steeper slopes, bend at the waist and knees as if sitting, keeping the feet flat to engage all vertical crampon points and keep the weight over the feet as in descending rock slab (Figures 10-10 and 10-11, page 10-8). On steep terrain, assume a cross body or port arms position with the ax, and traverse. The crab position or front-pointing may also be used for descending. Regardless of the technique used, always ensure the points of the crampons are inserted in the snow or ice and take short, deliberate steps to minimize the chance of tripping and falling down the slope.

Figure 10-10. Flat-footing in the crab position.

Figure 10-11. Use of ice ax in descent.

e. **Normal Progression.** The use of the ice ax and crampons follows a simple, logical progression. The techniques can be used in any combination, dictated by the terrain and skill of the individual. A typical progression could be as follows:

(1) ***Crampons.*** Use crampons in the following situations:

- Walking as on flat ground.

- Herringbone step straight up the slope.
- Traverse with feet flat.
- Backing up the slope (crab position).
- Front-pointing.

(2) ***Ice Ax.*** Use the ice ax in these situations:

- Cane position on flat ground.
- Cane position on uphill side as slope steepens.
- Port arms position with spike on uphill side.
- Anchor position with pick on uphill side.
- Push-hold position using front-pointing technique.
- Dagger position using front-pointing technique.
- Hammer position using front-pointing technique.

e. **Climbing Sequence.** Using most of these positions, a single ax can be "climbed" in steps to move upslope on low-angle to near vertical terrain (Figure 10-12). Begin by positioning the feet in a secure stance and placing the ax in the hammer position as high as possible. Slowly and carefully move the feet to higher positions alternately, and move the hand up the ax shaft. Repeat this until the hand is on top of the head of the ax. Remove the ax and place it at a higher position and begin again.

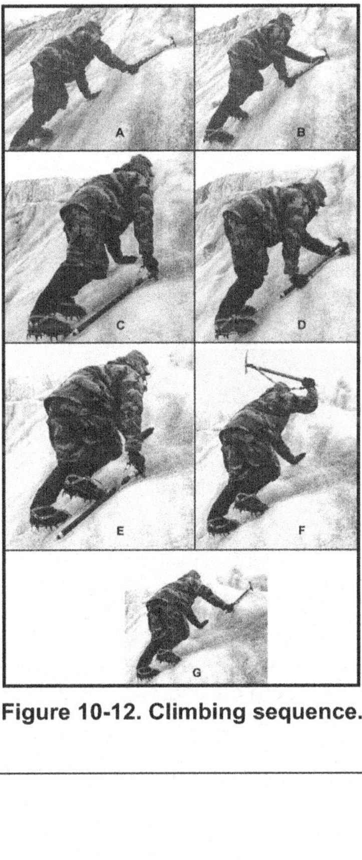

Figure 10-12. Climbing sequence.

f. **Step Cutting.** Step cutting is an extremely valuable technique that is a required skill for any military mountaineer (Figure 10-13). Using cut steps can save valuable time that would be spent in donning crampons for short stretches of ice and can, in some cases, save the weight of the crampons altogether. Steps may also have to be cut by the lead team to enable a unit without proper equipment to negotiate snow- or ice-covered terrain. As units continue to move up areas where steps have been cut they should continue to improve each step. In ascending, steps may be cut straight up the slope, although a traverse will normally be adopted. In descending, a traverse is also the preferred method. When changing direction, a step large enough for both feet and crampons must be made. Once the step is formed, the adze is best used to further shape and clean the step.

(1) ***Snow.*** On slopes of firm snow and soft ice, steps may be cut by swinging the ax in a near-vertical plane, using the inside corner of the adze for cutting. The step should be fashioned so that it slopes slightly inward and is big enough to admit the entire foot. Steps used for resting or for turning must be larger.

(2) ***Ice.*** Hard ice requires that the pick of the ax be used. Begin by directing a line of blows at right angles to the slope to make a fracture line along the base of the intended step. This technique will reduce the chance of an unwanted fracture in the ice breaking out the entire step. Next, chop above the fracture line to fashion the step. When using the pick it should be given an outward jerk as it is placed to prevent it from sticking in the ice.

(3) ***Step Cutting in a Traverse.*** When cutting steps in a traverse, the preferred cutting sequence is to cut one step at an arm's length from the highest step already cut, then cut one between those two. Cutting ahead one step then cutting an intermediate step keeps all of the steps relatively close to one another and maintains a suitable interval that all personnel can use.

(4) ***Handholds.*** If handholds are cut, they should be smaller than footholds, and angled more.

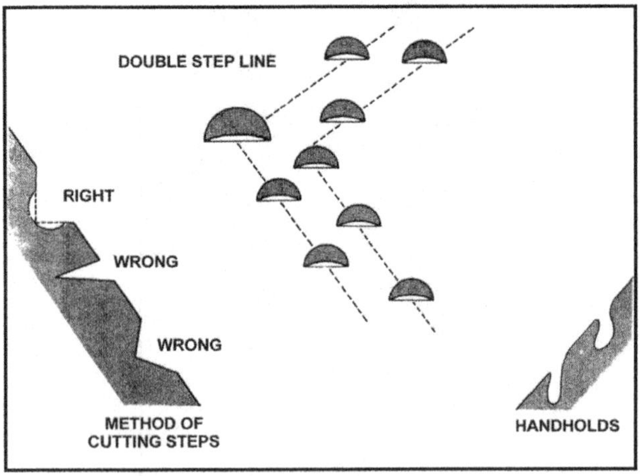

Figure 10-13. Step cutting and handhold cutting.

g. **Self-Arrest.** The large number of climbers injured or killed while climbing on snow and ice can be attributed to two major failings on the part of the climber: climbing unroped, and a lack of knowledge and experience in the techniques necessary to stop, or arrest, a fall (Figure 10-14). A climber should always carry an ice ax when climbing on steep snow or ice; if a fall occurs, he must retain possession and control of his ice ax if he is to successfully arrest the fall. During movement on steep ice, the ax pick will be in the ice solidly before the body is moved, which should prevent a fall of any significance (this is a self belay not a self-arrest).

CAUTION

Self-arrest requires the ax pick to gradually dig in to slow the descent. Self-arrest is difficult on steep ice because the ice ax pick instantly "bites" into the ice, possibly resulting in either arm or shoulder injury, or the ax is deflected immediately upon contact.

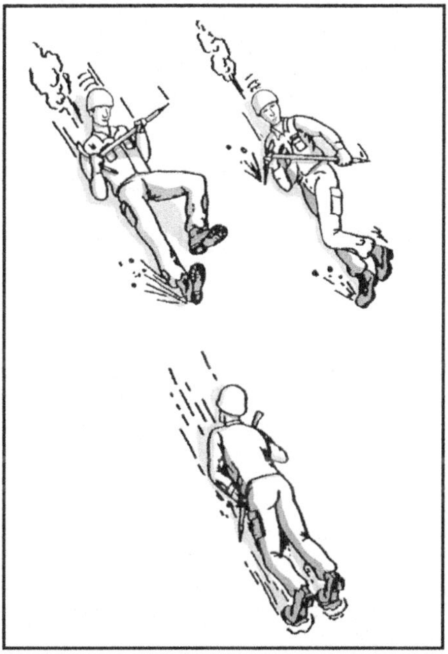

Figure 10-14. Self-arrest technique.

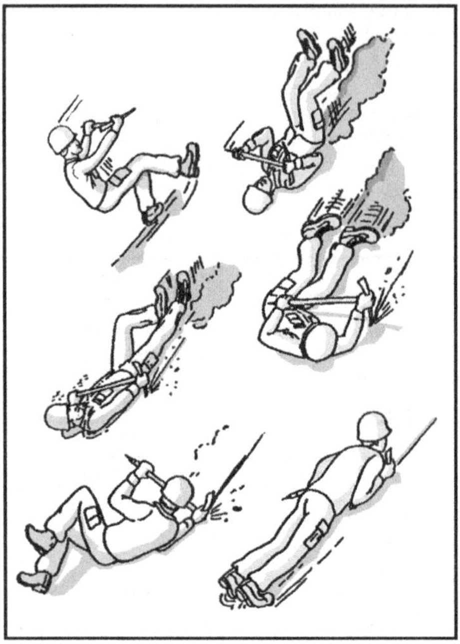

Figure 10-14. Self-arrest technique (continued).

(1) A climber who has fallen may roll or spin; if this happens, the climber must first gain control of his body, whether it is with his ice ax or simply by brute force. Once the roll or spin has been controlled, the climber will find himself in one of four positions.

- Head upslope, stomach on the slope, and feet pointed downslope.
- Head upslope, back to the slope, and feet pointed downslope.
- Head downslope, stomach on the slope, and feet pointed upslope.
- Head downslope, back to the slope, and feet pointed upslope.

(2) To place the body in position to arrest from the four basic fall positions the following must be accomplished.

(a) In the first position, the body is in proper relation to the slope for an arrest.

(b) In the second position, the body must first be rotated from face up to face down on the slope. This is accomplished by rolling the body toward the head of the ax.

(c) In the third position, the pick of the ice ax is placed upslope and used as a pivot to bring the body into proper position.

(d) In the fourth position, the head of the ax must be driven into the snow to the climber's side. This will cause the body to rotate into a head up, stomach down position.

(3) The final position when the arrest of the fall is completed should be with the head upslope, stomach on the slope, with the feet pointed downslope. If crampons are not

worn, the toe of the boots may be dug into the slope to help arrest the fall. The ax is held diagonally across the chest, with the head of the ax by one shoulder and the spike near the opposite hip. One hand grasps the head of the ax, with the pick pointed into the slope, while the other hand is on the shaft near the spike, lifting up on it to prevent the spike from digging into the slope.

Note: If crampons are worn, the feet must be raised to prevent the crampons from digging into the snow or ice too quickly. This could cause the climber to tumble and also, could severely injure his ankles and legs.

(4) When a fall occurs, the climber should immediately grasp the ax with both hands and hold it firmly as described above. Once sufficient control of the body is attained, the climber drives the pick of the ice ax into the slope, increasing the pressure until the fall is arrested. Raising the spike end of the shaft increases the biting action of the pick. It is critical that control of the ice ax be maintained at all times.

10-4. GLISSADING

Glissading is the intentional, controlled, rapid descent, or slide of a mountaineer down a steep slope covered with snow (Figure 10-15, page 10-14). Glissading is similar to skiing, except skis are not used. The same balance and control are necessary, but instead of skis the soles of the feet or the buttocks are used. The only piece of equipment required is the standard ice ax, which serves as the rudder, brake, and guide for the glissade. The two basic methods of glissading are:

a. **Squatting Glissade.** The squatting glissade is accomplished by placing the body in a semi-crouched position with both knees bent and the body weight directly over the feet. The ice ax is grasped with one hand on the head, pick, and adze outboard (away from the body), and the other hand on the shaft. The hand on the shaft grips it firmly in a position that allows control as well as the application of downward pressure on the spike of the ax.

b. **Sitting Glissade.** Using this method the glissader sits on the snow with the legs flat, and the heels and feet raised and pointed downslope. The ice ax is firmly grasped in the same manner as the squatting glissade, with the exception that the hand on the shaft must be locked against the hip for control. The sitting glissade is slower but easier to control than the squatting glissade.

c. **Safety.** A glissade should never be attempted on a slope where the bottom cannot be seen, since drop-offs may exist out of view. Also, a sitting glissade should not be used if the snow cover is thin, as painful injury could result.

Figure 10-15. Glissading techniques.

10-5. SNOW AND ICE ANCHORS

Ice and snow anchors consist of snow pickets, flukes, deadman-type anchors, ice screws, and ice pitons. Deadman anchors can be constructed from snowshoes, skis, backpacks, sleds, or any large items.

a. **Ice Pitons.** The ice piton is used to establish anchor points. The ice piton is not seen in modern ice climbing but may still be available to the military. The standard ice piton is made of tubular steel and is 10 inches in length. Ice pitons installed in pairs are a bombproof anchor; however, ice pitons have no threads for friction to hold them in the ice once placed and are removed easily. Safe use of ice pitons requires placement in pairs. Used singularly, ice pitons are a strong anchor but are easily removed, decreasing the perceived security of the anchor. Follow the instructions below for placing ice pitons in pairs.

(1) Cut a horizontal recess into the ice, and also create a vertical surface (two clean surfaces at right angles to each other).

(2) Drive one piton into the horizontal surface and another into the vertical surface so that the two pitons intersect at the necessary point (Figure 10-16).

(3) Connect the two rings with a single carabiner, ensuring the carabiner is not cross-loaded. Webbing or rope can be used if the rings are turned to the inside of the intersection.

(4) Test the piton pair to ensure it is secure. If it pulls out or appears weak, move to another spot and replace it. The pair of pitons, when placed correctly, are multidirectional.

Figure 10-16. Ice piton pair.

(5) The effective time and or strength for an ice piton placement is limited. The piton will heat from solar radiation, or the ice may crack or soften. Solar radiation can be nearly eliminated by covering the pitons with ice chips once they have been placed. If repeated use is necessary for one installation, such as top roping, the pitons should be inspected frequently and relocated when necessary. When an ice piton is removed, the ice that has accumulated in the tube must be removed before it freezes in position, making further use difficult.

c. **Ice Screws**. The ice screw is the most common type of ice protection and has replaced the ice piton for the most part (Figure 10-17). Some screws have longer "hangers" or handles, which allow them to be easily twisted into position by hand. Place ice screws as follows:

(1) Clear away all rotten ice from the surface and make a small hole with the ax pick to start the ice screw in.

(2) Force the ice screw in until the threads catch.

Figure 10-17. Placement of ice screw using the pick.

(3) Turn the screw until the eye or the hanger of the ice screw is flush with the ice and pointing down. The screw should be placed at an angle 90 to 100 degrees from the

lower surface. Use either your hand or the pick of the ice ax to screw it in. If you have a short ax (70 centimeters or less), you may be able to use the spike in the eye or hanger to ease the turning. (Remember that you may only have use of one hand at this point depending on your stance and the angle of the terrain.)

(4) As with ice pitons, melting of the ice around a screw over a period of time must be considered. The effective time and or strength of an ice screw placement is limited. The screw will heat from solar radiation, or the ice may crack or soften. Solar radiation can be nearly eliminated by covering the screw with ice chips once it has been emplaced. If repeated use is necessary for one installation, such as top roping, the screws should be inspected frequently and relocated when necessary. When an ice screw is removed, the ice that has accumulated in the tube, must be removed before further use.

d. **Horseshoe or Bollard Anchor.** This is an artificial anchor shaped generally like a horseshoe (Figure 10-18). It is formed from either ice or snow and constructed by either cutting with the ice ax or stamping with the boots. When constructed of snow, the width should not be less than 10 feet. In ice, this width may be narrowed to 2 feet, depending on the strength of the ice. The length of the bollard should be at least twice the width. The trench around the horseshoe should be stamped as deeply as possible in the snow and should be cut not less than 6 inches into the ice after all rotten ice is removed. The backside of the anchor must always be undercut to prevent the rope from sliding off and over the anchor.

(1) This type of anchor is usually available and may be used for fixed ropes or rappels. It must be inspected frequently to ensure that the rope is not cutting through the snow or ice more than one-third the length of the anchor; if it is, a new anchor must be constructed in a different location.

(2) A horseshoe anchor constructed in snow is always precarious, its strength depending upon the prevailing texture of the snow. For dry or wind-packed snow, the reliability of the anchor should always be suspect. The backside of the bollard can be reinforced with ice axes, pickets, or other equipment for added strength.

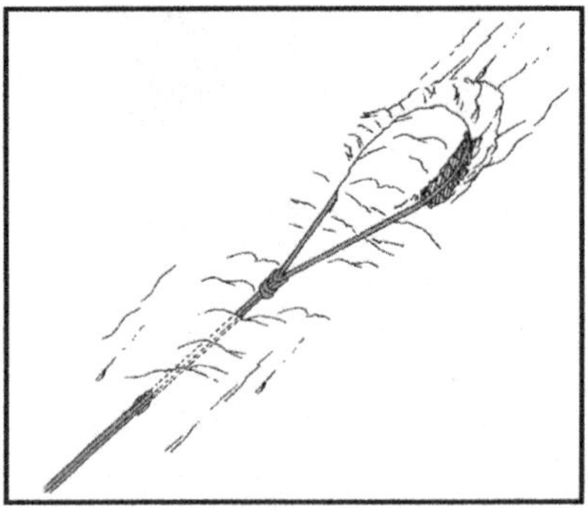

Figure 10-18. Horseshoe or bollard anchor.

e. **Pickets and Ice Axes.** Pickets and ice axes may be used as snow anchors as follows.

(1) The picket should be driven into the snow at 5 to 15 degrees off perpendicular from the lower surface. If the picket cannot be driven in all the way to the top hole, the carabiner should be placed in the hole closest to the snow surface to reduce leverage. The picket may also be tied off with a short loop of webbing or rope as in tying off pitons.

(2) An ice ax can be used in place of a picket. When using an ice ax as a snow anchor, it should be inserted with the widest portion of the ax shaft facing the direction of pull. The simplest connection to the ax is to use a sling or rope directly around the shaft just under the head. If using the leash ensure it is not worn, frayed, or cut from general use; is strong enough; and does not twist the ax when loaded. A carabiner can be clipped through the hole in the head, also.

(3) Whenever the strength of the snow anchor is suspect, especially when a picket or ax cannot be driven in all the way, the anchor may be buried in the snow and used as a "dead man" anchor. Other items suitable for dead man anchor construction are backpacks, skis, snowshoes, ski poles, or any other item large enough or shaped correctly to achieve the design. A similar anchor, sometimes referred to as a "dead guy," can be made with a large sack either stuffed with noncompressible items or filled with snow and buried. Ensure the attaching point is accessible before burying. The direction of pull on long items, such as a picket or ax, should be at a right angle to its length. The construction is identical to that of the dead man anchor used in earth.

f. **Equalized Anchors.** Snow and ice anchors must be constantly checked due to melting and changing snow or ice conditions.

(1) Whenever possible, two or more anchors should be used. While this is not always practical for intermediate anchor points on lead climbs or fixed ropes, it should be mandatory for main anchors at all belay positions, rappel points, or other fixed rope installations. (Figure 10-19, page 10-18, shows an example of three snow pickets configured to an equalized anchor.)

(2) As with multipoint anchors on rock, two or more snow or ice anchors can be joined together with a sling rope or webbing to construct one solid, equalized anchor. A bowline on a bight tied into the climbing rope can also be used instead of sling ropes or webbing.

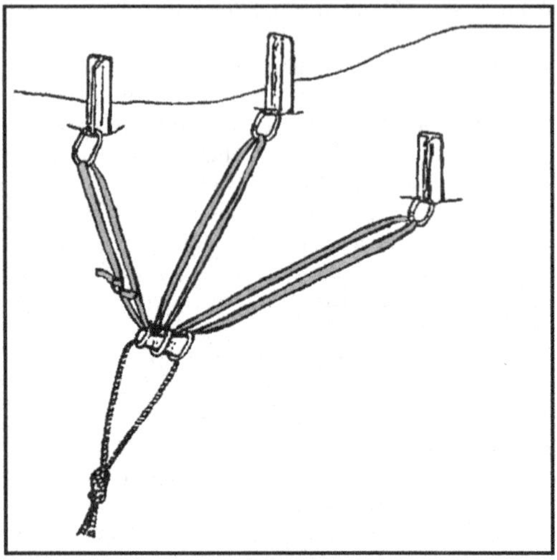

Figure 10-19. Equalized anchor using pickets.

10-6. ROPED CLIMBING ON ICE AND SNOW

When climbing on ice or snow team members tie into a climbing rope the same as when they climb on rock. When crevasses are expected, a three-man rope team is recommended.

a. **Tie-In Method.** For climbing on snow and ice, the tie-in procedure is normally the same as for rock climbing; however, when moving over snow-covered glaciers, the tie-in is modified slightly. (See paragraph 10-7, Movement on Glaciers, for more information).

b. **Movement.** For movement on gentle or moderate slopes where there is little chance of a serious fall, all climbers move simultaneously. Normally the climbers move in single file using the steps created by the lead climber and improving them when necessary. The rope between the climbers should be fully extended and kept reasonably tight. Should any member fall, he immediately yells "FALLING." The other rope team members immediately drop into a self-arrest position. The fallen climber also applies the self-arrest procedure. By using this method, called the "team arrest," the entire team as a whole arrests the fall of one member. On steeper slopes, and when crossing snow-covered crevasses where the snow bridges appear weak, the climbers use belayed climbing techniques as in rock climbing.

c. **Belaying on Snow and Ice.** The principles of belaying on ice and snow are the same as on rock. Generally, the high-force falls found in rock climbing are not present on snow and ice unless the pitch being climbed is extremely steep.

(1) ***Boot-Ax Belay.*** This belay can be useful in areas where the full length of the ice ax can penetrate the snow. The holding strength of the boot-ax belay is directly related to the firmness of the snow and to the strength of the ice ax shaft. The shaft of the ax is

tilted slightly uphill and jammed into the snow. The belayer places his uphill foot against the downhill side of the ax for support. A bight formed in the rope is placed over the boot and around the shaft of the ice ax. The brake is applied by wrapping the rope around the heel of the boot (Figure 10-20).

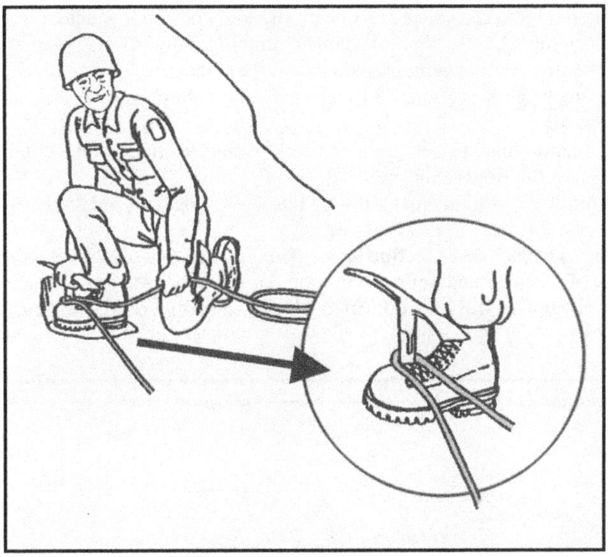

Figure 10-20. Boot-ax belay.

(2) ***Body Belay.*** The body belay can be used on snow and ice, also. The principles are the same as for belays on rock—solid anchors must be used and a well-braced position assumed. The position can be improved by digging depressions into the snow or ice for a seat and footholds. A strong platform should be constructed for the standing body belay.

(3) ***Munter Hitch.*** This belay technique is also used on snow and ice. When using the hitch off of the anchor, a two-point equalized anchor should be constructed as a minimum.

d. **Fixed Ropes.** The use of fixed ropes on ice is recommended for moving units through icefall areas on glaciers or other steep ice conditions. The procedures for emplacing fixed ropes on ice are basically the same as on rock with the exception that anchors need more attention, both in initial placement and in subsequent inspection, and steps may have to be cut to assist personnel.

10-7. MOVEMENT ON GLACIERS

Movement in mountainous terrain may require travel on glaciers. An understanding of glacier formation and characteristics is necessary to plan safe routes. A glacier is formed by the perennial accumulation of snow and other precipitation in a valley or draw. The accumulated snow eventually turns to ice due to metamorphosis. The "flow" or movement of glaciers is caused by gravity. There are a few different types of glaciers identifiable primarily by their location or activity.

- Valley glacier—resides and flows in a valley.
- Cirque glacier—forms and resides in a bowl.
- Hanging glacier—these are a result of valley or cirque glaciers flowing and or deteriorating. As the movement continues, portions separate and are sometimes left hanging on mountains, ridgelines, or cliffs.
- Piedmont glacier—formed by one or more valley glaciers; spreads out into a large area.
- Retreation glacier—a deteriorating glacier; annual melt of entire glacier exceeds the flow of the ice.
- Surging glacier—annual flow of the ice exceeds the melt; the movement is measurable over a period of time.

a. **Characteristics and Definitions.** This paragraph describes the common characteristics of glaciers, and defines common terminology used in reference to glaciers. (Figure 10-21 shows a cross section of a glacier, and Figure 10-22 depicts common glacier features.)

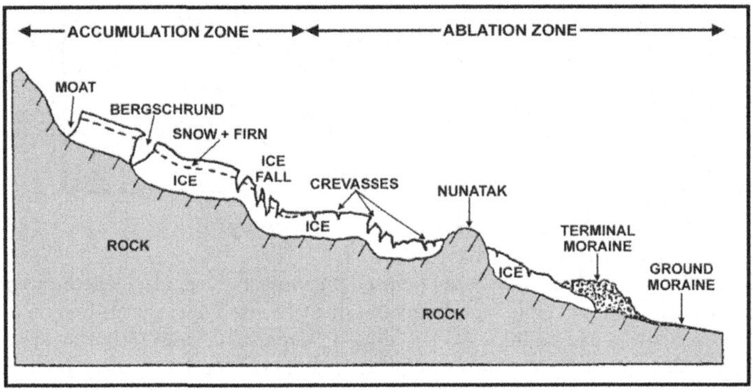

Figure 10-21. Glacier cross section.

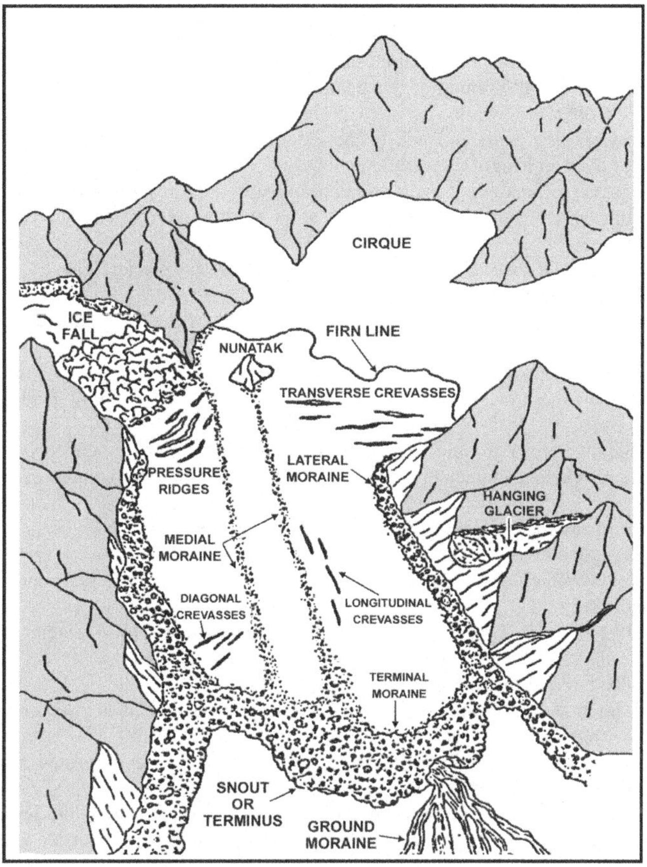

Figure 10-22. Glacier features.

(1) Firn is compacted granular snow that has been on the glacier at least one year. Firn is the building blocks of the ice that makes the glacier.

(2) The accumulation zone is the area that remains snow-covered throughout the year because of year-round snowfall. The snowfall exceeds melt.

(3) The ablation zone is the area where the snow melts off the ice in summer. Melt equals or exceeds snowfall.

(4) The firn line separates the accumulation and ablation zones. As you approach this area, you may see "strips" of snow in the ice. Be cautious, as these could be *snow bridges* remaining over crevasses. Remember that snow bridges will be weakest lower on the glacier as you enter the accumulation zone. The firn line can change annually.

(5) A bergschrund is a large crevasse at the head of a glacier caused by separation of active (flowing) and inactive (stationary) ice. These will usually be seen at the base of a major incline and can make an ascent on that area difficult.

(6) A moat is a wall formed at the head (start) of the glacier. These are formed by heat reflected from valley wall.

(7) A crevasse is a split or crack in the glacier surface. These are formed when the glacier moves over an irregularity in the bed surface.

(8) A transverse crevasse forms perpendicular to the flow of a glacier. These are normally found where a glacier flows over a slope with a gradient change of 30 degrees or more.

(9) Longitudinal crevasses form parallel to the flow of a glacier. These are normally found where a glacier widens.

(10) Diagonal crevasses form at an angle to the flow of a glacier. These are normally found along the edges where a glacier makes a bend.

(11) A snow bridge is a somewhat supportive structure of snow that covers a crevasse. Most of these are formed by the wind. The strength of a snow bridge depends on the snow itself.

(12) Icefalls are a jumble of crisscross crevasses and large ice towers that are normally found where a glacier flows over a slope with a gradient change of 25 degrees or more.

(13) Seracs are large pinnacles or columns of ice that are normally found in icefalls or on hanging glaciers.

(14) Ice avalanches are falling chunks of ice normally occurring near icefalls or hanging glaciers.

(15) The moraine is an accumulation of rock or debris on a glacier caused by rockfall or avalanche of valley walls.

(16) The lateral moraine is formed on sides of glacier.

(17) The medial moraine is in the middle of the glacier. This is also formed as two glaciers come together or as a glacier moves around a central peak.

(18) The terminal moraine is at the base of a glacier and is formed as moraines meet at the snout or terminus of a glacier.

(19) The ground moraine is the rocky debris extending out from the terminus of a glacier. This is formed by the scraping of earth as the glacier grew or surged and exposed as the glacier retreats.

(20) A Nunatak is a rock projection protruding through the glacier as the glacier flows around it.

(21) An ice mill is a hole in the glacier formed by swirling water on the surface. These can be large enough for a human to slip into.

(22) Pressure ridges are wavelike ridges that form on glacier normally after a glacier has flowed over icefalls.

(23) A glacier window is an opening at the snout of the glacier where water runs out of the glacier.

b. **Dangers and Obstacles.** The principle dangers and obstacles to movement in glacial areas are crevasses, icefalls, and ice avalanches. Snow-covered crevasses make movement on a glacier extremely treacherous. In winter, when visibility is poor, the difficulty of recognizing them is increased. Toward the end of the summer, crevasses are widest and covered by the least snow. Crossing snow bridges constitutes the greatest potential danger in movement over glaciers in the summer. On the steep pitch of a glacier, ice flowing over irregularities and cliffs in the underlying valley floor cause the ice to break up into ice blocks and towers, criss-crossed with crevasses. This jumbled cliff of ice is known as an icefall. Icefalls present a major obstacle to safe movement of troops on glaciers.

(1) Moving on glaciers brings about the hazard of falling into a crevasse. Although the crevasses are visible in the ablation zone in the summer (Figure 10-23), the accumulation zone will still have hidden crevasses. The risk of traveling in the accumulation zone can be managed to an acceptable level when ropes are used for connecting the team members (Figure 10-24, page 10-24). Crampons and an ice ax are all that is required to safely travel in the ablation zone in the summer.

Figure 10-23. Ablation zone of glacier in summer.

Figure 10-24. Rope teams moving in the accumulation zone of a glacier.

(2) When conditions warrant, three to four people will tie in to one rope at equal distances from each other. To locate the positions, if three people are on a team, double the rope and one ties into the middle and the other two at the ends. If four people are on a team, form a "z" with the rope and expand the "z" fully, keeping the end and the bight on each "side" of the "z" even. Tie in to the bights and the ends.

(3) Connect to the rope with the appropriate method and attach the Prusik as required. The rope should be kept relatively tight either by Prusik belay or positioning of each person. If the team members need to assemble in one area, use the Prusik to belay each other in.

(4) If a team member falls into a crevasse, the remaining members go into team arrest, assess the situation, and use the necessary technique to remove the person from the crevasse. The simplest and most common method for getting someone out of a crevasse is for the person to climb out while being belayed.

(5) All items should be secured to either the climber or the rope/harness to prevent inadvertent release and loss of necessary items or equipment. Packs should be secured to the rope/harness with webbing or rope. If traveling with a sled in tow, secure it not only to a climber to pull it, but connect it to the rope with webbing or rope also.

(6) If marking the route on the glacier is necessary for backtracking or to prevent disorientation in storms or flat-light conditions, use markers that will be noticeable against the white conditions. The first team member can place a new marker when the last team member reaches the previous marker.

c. **Roped Movement.** The first rule for movement on glaciers is to rope up (Figure 10-25). A roped team of two, while ideal for rock climbing, is at a disadvantage on a snow-covered glacier. The best combination is a three-man rope team. Generally, the rope team members will move at the same time with the rope fully extended and reasonably tight between individuals, their security being the team arrest. If an individual should break through a snow bridge and fall into a crevasse, the other members immediately perform self-arrest, halting the fall. At points of obvious weakness in the snow bridges, the members may decide to belay each other across the crevasse using one of the established belay techniques.

Figure 10-25. Preparation for roped movement.

(1) Even with proper training in crevasse rescue techniques, the probability exists that an individual may remain suspended in a crevasse for a fairly lengthy amount of time while trying to get himself out or while awaiting help from his rope team members. Because of this, it is strongly recommended that all personnel wear a seat/chest combination harness, whether improvised or premanufactured.

(2) Rope team members must be able to quickly remove the climbing rope from the harness(es) during a crevasse rescue. The standard practice for connecting to the rope for glacier travel is with a locking carabiner on a figure-eight loop to the harness. This allows quick detachment of the rope for rescue purposes. The appropriate standing part of the rope is then clipped to the chest harness carabiner.

(3) If a rope team consists of only two people, the rope should be divided into thirds, as for a four-person team. The team members tie into the middle positions on the rope, leaving a third of the rope between each team member and a third on each end of the rope. The remaining "thirds" of the rope should be coiled and either carried in the rucksack, attached to the rucksack, or carried over the head and shoulder. This gives each

climber an additional length of rope that can be used for crevasse rescue, should one of the men fall through and require another rope. If necessary, this excess end rope can be used to connect to another rope team for safer travel.

Note: The self-arrest technique used by one individual will work to halt the fall of his partner on a two-man rope team; however, the chance of it failing is much greater. Crevasse rescue procedures performed by a two-man rope team, by itself, may be extremely difficult. For safety reasons, movement over a snow-covered glacier by a single two-man team should be avoided wherever possible.

d. **Use of Prusik Knots.** Prusik knots are attached to the climbing rope for all glacier travel. The Prusiks are used as a self-belay technique to maintain a tight rope between individuals, to anchor the climbing rope for crevasse rescue, and for self-rescue in a crevasse fall. The Prusik slings are made from the 7-millimeter by 6-foot and 7-millimeter by 12-foot ropes. The ends of the ropes are tied together, forming endless loops or slings, with double fisherman's knots. Form the Prusik knot on the rope in front of the climber. An overhand knot can be tied into the sling just below the Prusik to keep equal tension on all the Prusik wraps. Attach this sling to the locking carabiner at the tie in point on the harness.

Note: An ascender can replace a Prusik sling in most situations. However, the weight of an ascender hanging on the rope during movement will become annoying, and it could be stepped on during movement and or climbing.

e. **Securing the Backpack/Rucksack.** If an individual should fall into a crevasse, it is essential that he be able to rid himself of his backpack. The weight of the average pack will be enough to hinder the climber during crevasse rescue, or possibly force him into an upside down position while suspended in the crevasse. Before movement, the pack should be attached to the climbing rope with a sling rope or webbing and a carabiner. A fallen climber can immediately drop the pack without losing it. The drop cord length should be minimal to allow the fallen individual to reach the pack after releasing it, if warm clothing is needed. When hanging from the drop cord, the pack should be oriented just as when wearing it (ensure the cord pulls from the top of the pack).

f. **Routes.** An individual operating in the mountains must appreciate certain limitations in glacier movement imposed by nature.

(1) Additional obstacles in getting onto a glacier may be swift glacier streams, steep terminal or lateral moraines, and difficult mountain terrain bordering the glacier ice. The same obstacles may also have to be overcome in getting on and off a valley glacier at any place along its course.

(2) Further considerations to movement on a glacier are steep sections, heavily crevassed portions, and icefalls, which may be major obstacles to progress. The use of current aerial photographs in conjunction with aerial reconnaissance is a valuable means of gathering advance information about a particular glacier. However, they only supplement, and do not take the place of, on-the-ground reconnaissance conducted from available vantage points.

g. **Crossing Crevasses.** Open crevasses are obvious, and their presence is an inconvenience rather than a danger to movement. Narrow cracks can be jumped, provided the take off and landing spots are firm and offer good footing. Wider cracks will have to be circumvented unless a solid piece of ice joins into an ice bridge strong enough to support at least the weight of one member of the team. Such ice bridges are often formed in the lower portion of a crevasse, connecting both sides of it.

(1) In the area of the firn line, the zone that divides seasonal melting from permanent falls of snow, large crevasses remain open, though their depths may be clogged with masses of snow. Narrow cracks may be covered. In this zone, the snow, which covers glacier ice, melts more rapidly than that which covers crevasses. The difference between glacier ice and narrow snow-covered cracks is immediately apparent; the covering snow is white, whereas the glacier ice is gray.

(2) Usually the upper part of a glacier is permanently snow covered. The snow surface here will vary in consistency from dry powder to consolidated snow. Below this surface cover are found other snow layers that become more crystalline in texture with depth, and gradually turn into glacier ice. It is in this snow-covered upper part of a glacier that crevasses are most difficult to detect, for even wide crevasses may be completely concealed by snow bridges.

h. **Snow Bridges.** Snow bridges are formed by windblown snow that builds a cornice over the empty interior of the crevasse. As the cornice grows from the windward side, a counter drift is formed on the leeward side. The growth of the leeward portion will be slower than that to the windward so that the juncture of the cornices occurs over the middle of the crevasse only when the contributing winds blow equally from each side. Bridges can also be formed without wind, especially during heavy falls of dry snow. Since cohesion of dry snow depends only on an interlocking of the branches of delicate crystals, such bridges are particularly dangerous during the winter. When warmer weather prevails the snow becomes settled and more compacted, and may form firmer bridges.

(1) Once a crevasse has been completely bridged, its detection is difficult. Bridges are generally slightly concave because of the settling of the snow. This concavity is perceptible in sunshine, but difficult to detect in flat light. If the presence of hidden crevasses is suspected, the leader of a roped team must probe the snow in front of him with the shaft of his ice ax. As long as a firm foundation is encountered, the team may proceed, but should the shaft meet no opposition from an underlying layer of snow, a crevasse is probably present. In such a situation, the prober should probe closer to his position to make sure that he is not standing on the bridge itself. If he is, he should retreat gently from the bridge and determine the width and direction of the crevasse. He should then follow and probe the margin until a more resistant portion of the bridge is reached. When moving parallel to a crevasse, all members of the team should keep well back from the edge and follow parallel but offset courses.

(2) A crevasse should be crossed at right angles to its length. When crossing a bridge that seems sufficiently strong enough to hold a member of the team, the team will generally move at the same time on a tight rope, with each individual prepared to go into self-arrest. If the stability of the snow bridge is under question, they should proceed as follows for a team of three glacier travelers:

(a) The leader and second take up a position at least 10 feet back from the edge. The third goes into a self-belay behind the second and remains on a tight rope.

(b) The second belays the leader across using one of the established belay techniques. The boot-ax belay should be used only if the snow is deep enough for the ax to be inserted up to the head and firm enough to support the possible load. A quick ice ax anchor should be placed for the other belays. Deadman or equalizing anchors should be used when necessary.

(c) The leader should move forward, carefully probing the snow and evaluating the strength of the bridge, until he reaches firm snow on the far side of the crevasse. He then continues as far across as possible so number two will have room to get across without number one having to move.

(d) The third assumes the middle person's belay position. The middle can be belayed across by both the first and last. Once the second is across, he assumes the belay position. Number one moves out on a tight rope and anchors in to a self-belay. Number two belays number three across.

(3) In crossing crevasses, distribute the weight over as wide an area as possible. Do not stamp the snow. Many fragile bridges can be crossed by lying down and crawling to the other side. Skis or snowshoes help distribute the weight nicely.

i. **Arresting and Securing a Fallen Climber.** The simplest and most common method for getting someone out of a crevasse is for the person to climb out while being belayed. Most crevasse falls will be no more than body height into the opening if the rope is kept snug between each person.

(1) To provide a quick means of holding an unexpected breakthrough, the rope is always kept taut. When the leader unexpectedly breaks through, the second and third immediately go into a self-arrest position to arrest the fall. A fall through a snow bridge results either in the person becoming jammed in the surface hole, or in being suspended in the crevasse by the rope. If the leader has fallen only partially through the snow bridge, he is supported by the snow forming the bridge and should not thrash about as this will only enlarge the hole and result in deeper suspension. All movements should be slow and aimed at rolling out of the hole and distributing the weight over the remainder of the bridge. The rope should remain tight at all times and the team arrest positions adjusted to do so. It generally is safer to retain the rucksack, as its bulk often prevents a deeper fall. Should a team member other than the leader experience a partial fall, the rescue procedure will be same as for the leader, only complicated slightly by the position on the rope.

(2) When the person falls into a crevasse, the length of the fall depends upon how quickly the fall is arrested and where in the bridge the break takes place. If the fall occurs close to the near edge of the crevasse, it usually can be checked before the climber has fallen more than 6 feet. However, if the person was almost across, the fall will cause the rope to cut through the bridge, and then even an instantaneous check by the other members will not prevent a deeper fall. The following scenario is an example of the sequence of events that take place after a fall by the leader in a three-person team. (This scenario is for a team of three, each person referred to by position; the leader is number 1.)

(a) Once the fall has been halted by the team arrest, the entire load must be placed on number 2 to allow number 3 to move forward and anchor the rope. Number 3 slowly releases his portion of the load onto number 2, always being prepared to go back into self-arrest should number 2's position begin to fail.

(b) Once number 2 is confident that he can hold the load, number 3 will proceed to number 2's position, using the Prusik as a self belay, to anchor the rope. In this way the rope remains reasonably tight between number 2 and number 3. Number 3 must always be prepared to go back into self-arrest should number 2's position begin to fail.

(c) When number 3 reaches number 2's position he will establish a bombproof anchor 3 to 10 feet in front of number 2 (on the load side), depending on how close number 2 is to the lip of the crevasse. This could be either a deadman or a two-point equalized anchor, as a minimum.

(d) Number 3 connects the rope to the anchor by tying a Prusik with his long Prusik sling onto the rope leading to number 1. An overhand knot should be tied into the long Prusik sling to shorten the distance to the anchor, and attached to the anchor with a carabiner. The Prusik knot is adjusted toward the load.

(e) Number 2 can then release the load of number 1 onto the anchor. Number 2 remains connected to the anchor and monitors the anchor.

(f) A fixed loop can be tied into the slack part of the rope, close to number 2, and attached to the anchor (to back up the Prusik knot).

(g) Number 3 remains tied in, but continues forward using a short Prusik as a self-belay. He must now quickly check on the condition of number 1 and decide which rescue technique will be required to retrieve him.

(3) These preliminary procedures must be performed before retrieving the fallen climber. If number 3 should fall through a crevasse, the procedure is the same except that number 1 assumes the role of number 3. Normally, if the middle person should fall through, number 1 would anchor the rope by himself. Number 3 would place the load on number 1's anchor, then anchor his rope and move forward with a Prusik self-belay to determine the condition of number 2.

j. **Crevasse Rescue Techniques.** Snow bridges are usually strongest at the edge of the crevasse, and a fall is most likely to occur some distance away from the edge. In some situations, a crevasse fall will occur at the edge of the snow bridge, on the edge of the ice. If a fall occurs away from the edge, the rope usually cuts deeply into the snow, thus greatly increasing friction for those pulling from above. In order to reduce friction, place padding, such as an ice ax, ski, ski pole, or backpack/rucksack, under the rope and at right angles to the stress. Push the padding forward as far as possible toward the edge of the crevasse, thus relieving the strain on the snow. Ensure the padding is anchored from falling into the crevasse for safety of the fallen climber.

(1) ***Use of Additional Rope Teams.*** Another rope team can move forward and assist in pulling the victim out of a crevasse. The assisting rope team should move to a point between the fallen climber and the remaining rope team members. The assisting team can attach to the arresting team's rope with a Prusik or ascender and both rope teams' members can all pull simultaneously. If necessary, a belay can be initiated by the fallen climber's team while the assisting team pulls. The arresting team member closest to the fallen climber should attach the long Prusik to themselves and the rope leading to the fallen climber, and the assisting team can attach their Prusik or ascender between this long Prusik and the arresting team member. As the assisting team pulls, the Prusik belay will be managed by the arresting team member at the long Prusik.

Note: Safety in numbers is obvious for efficient crevasse rescue techniques. Additional rope teams have the necessary equipment to improve the main anchor or establish new ones and the strength to pull a person out even if he is deep in the crevasse. Strength of other rope teams should always be used before establishing more time-consuming and elaborate rescue techniques.

(2) ***Fixed Rope.*** If the fallen climber is not injured, he may be able to climb out on a fixed rope. Number 1 clips number 3's rope to himself. He then climbs out using number 3's rope as a simple fixed line while number 2 takes up the slack in number 1's rope through the anchor Prusik for a belay.

(3) ***Prusik Ascending Technique.*** There may be times when the remaining members of a rope team can render little assistance to the person in the crevasse. If poor snow conditions make it impossible to construct a strong anchor, the rope team members on top may have to remain in self-arrest. Other times, it may just be easier for the fallen climber to perform a self-rescue. (Figure 10-26 shows the proper rope configuration.) The technique is performed as follows:

(a) The fallen climber removes his pack and lets it hang below from the drop cord.

(b) The individual slides their short Prusik up the climbing rope as far as possible.

(c) The long Prusik is attached to the rope just below the short Prusik. The double fisherman's knot is spread apart to create a loop large enough for one or both feet. The fallen climber inserts his foot/feet into the loop formed allowing the knot to cinch itself down.

(d) The individual stands in the foot loop, or "stirrup," of the long sling.

(e) With his weight removed from the short Prusik, it is slid up the rope as far as it will go. The individual then hangs from the short Prusik while he moves the long Prusik up underneath the short Prusik again.

(f) The procedure is repeated, alternately moving the Prusiks up the rope, to ascend the rope. Once the crevasse lip is reached, the individual can simply grasp the rope and pull himself over the edge and out of the hole.

(g) Besides being one of the simplest rope ascending techniques, the short Prusik acts as a self-belay and allows the climber to take as long a rest as he wants when sitting in the harness. The rope should be detached from the chest harness carabiner to make the movements less cumbersome. However, it is sometimes desirable to keep the chest harness connected to the rope for additional support. In this case the Prusik knots must be "on top" of the chest harness carabiner so they can be easily slid up the rope without interference from the carabiner. The long Prusik sling can be routed through the chest harness carabiner for additional support when standing up in the stirrup.

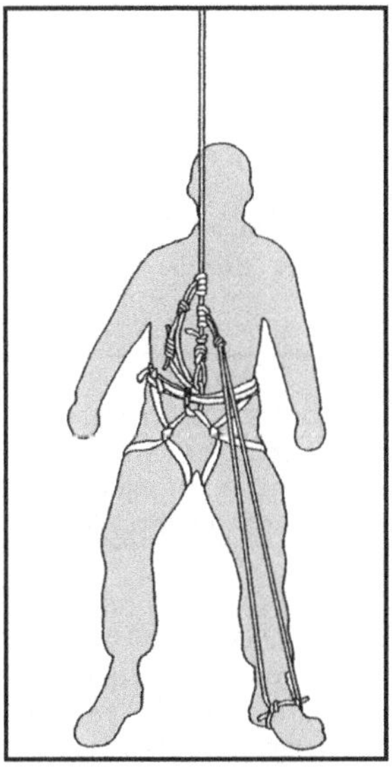

Figure 10-26. Prusik ascending technique.

(4) ***Z-Pulley Hauling System.*** If a fallen climber is injured or unconscious, he will not be able to offer any assistance in the rescue. If additional rope teams are not immediately available, a simple raising system can be rigged to haul the victim out of the crevasse. The Z-pulley hauling system is one of the simplest methods and the one most commonly used in crevasse rescue (Figure 10-27, page 10-32). The basic Z rig is a "3-to-1" system, providing mechanical advantage to reduce the workload on the individuals operating the haul line. In theory, it would only take about 33 pounds of pull on the haul rope to raise a 100-pound load with this system. In actual field use, some of this mechanical advantage is lost to friction as the rope bends sharply around carabiners and over the crevasse lip. The use of mechanical rescue pulleys can help reduce this friction in the system. The following describes rigging of the system. (This scenario is for a team of three, each person referred to by position; the leader is number 1.)

(a) After the rope team members have arrested and secured number 1 to the anchor, and they have decided to install the Z rig, number 2 will attach himself to the anchor without using the rope and clear the connecting knot used. Number 3 remains connected to the rope.

(b) The slack rope exiting the anchor Prusik is clipped into a separate carabiner attached to the anchor. A pulley can be used here if available.

(c) Number 3 will use number 2's short Prusik to rig the haul Prusik. He moves toward the crevasse lip (still on his own self-belay) and ties number 2's short Prusik onto number 1's rope (load rope) as close to the edge as possible.

(d) Another carabiner (and pulley if available) is clipped into the loop of the haul Prusik and the rope between number 3's belay Prusik and the anchor is clipped (or attached through the pulley). Number 3's rope becomes the haul rope.

(e) Number 3 then moves towards the anchor and number 2. Number 2 could help pull if necessary but first would connect to the haul rope with a Prusik just as number 3. If the haul Prusik reaches the anchor before the victim reaches the top, the load is simply placed back on the anchor Prusik and number 3 moves the haul Prusik back toward the edge. The system is now ready for another haul.

CAUTION

The force applied to the fallen climber through use of the Z-pulley system can be enough to destroy the harness-to-rope connection or injure the fallen climber if excess force is applied to the pulling rope.

Notes: 1. The Z-pulley adds more load on the anchor due to the mechanical advantage. The anchor should be monitored for the duration of the rescue.

2. With the "3-to-1" system, the load (fallen climber) will be raised 1 foot for every 3 feet of rope taken up during the haul.

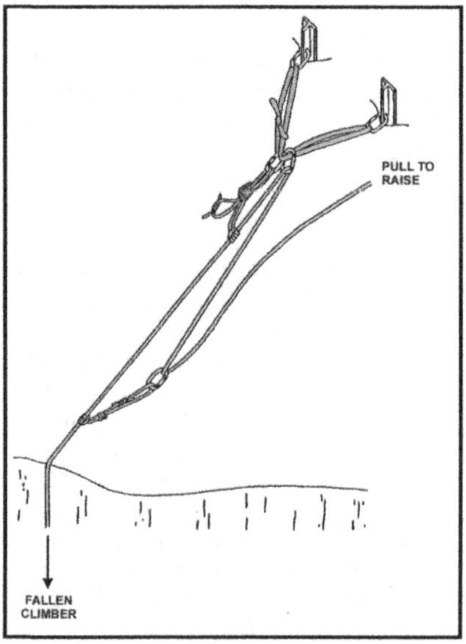

Figure 10-27. Z-pulley hauling system.

10-8. GLACIER BIVOUAC PROCEDURES

When locating a bivouac site or a gathering area where the team might need or want to unrope, at least one person will need to "probe" the area for hidden crevasses. The best type of probe will be the manufactured collapsing probe pole, at least eight feet in length. Other items could be used but the length and strength of the probe is most important. Other rope team members will belay the probers. The prober is "feeling" for a solid platform to place the tent by pushing the probe as hard and deep as possible into the surface. Probing should be in 2-foot intervals in all directions within the site.

a. If the probe suddenly has no resistance while pushing down, a crevasse is present. Attempts to outline the crevasse can be futile if the crevasse is large. Normally, the best decision is to relocate the proposed bivouac area far enough away to avoid that crevasse. (Sometimes only a few feet one way or the other is all that's needed to reach a good platform.) Probe the tent site again after digging to the desired surface. Mark boundaries with wands or other items such as skis, poles, and so on.

b. Occasionally while probing, increased pressure will be noticed without reaching a solid platform. The amount of snowfall may be such that even after digging into the snow, the probe still doesn't contact a hard surface. Try to find a solid platform.

c. There should be no unroped movement outside the probed/marked areas. If a latrine area is needed, probe a route away from the bivouac site and probe the latrine area also. If a dugout latrine is necessary, probe again after digging.

d. Multiple tent sites can be connected, which keeps tents closer together. Probe all areas between the tents if you plan to move in those areas. Closer tents will make communicating between tent groups and rope teams easier.

e. If there is a chance for severe storms with high winds, snow walls may be constructed to protect the tent site from wind. The walls can be constructed from loose snow piled on the perimeter, or blocks can be cut from consolidated snow layers. In deep soft snow, digging three or four feet to find a consolidated layer will result in enough snow moved to build up decent walls around the tent site.

(1) For block construction, move the soft snow from the surface into the wall foundation areas (down to a consolidated layer of snow).

(2) Cut blocks approximately 1 by 1 by 2 feet, and construct the walls by interlocking the blocks with overlapping placements. The walls should be slightly higher than the tent. At a minimum, build walls on the windward side of the tent site.

(3) Snow walls can also provide shelter from wind for food preparation.

CHAPTER 11
MOUNTAIN RESCUE AND EVACUATION

Steep terrain and adverse weather are common in mountainous environments. Under these conditions, relatively minor injuries may require evacuation. The evacuation technique chosen is determined by the type of injury, distance to be moved, terrain, and existing installations. Air evacuation is preferred; however, the weather, tactical situation, or operational ceiling of the aircraft may make this impossible. It is, therefore, imperative that all personnel are trained in mountain evacuation techniques and are self-sufficient. Casualties should be triaged before evacuation. Triage is performed by the most experienced medical personnel available (physician, physician's assistant, medic).

Performing a rescue operation can be a significant emotional event. Rescue scenarios must be practiced and rehearsed until rescue party members are proficient in the many tasks required to execute a rescue. To perform most of the high-angle rescues, Level I and Level II mountaineers are required with a Level III supervising.

11-1. CONSIDERATIONS

The techniques of evacuation are proven techniques. They are, however, all subject to improvement and should be discarded or modified as better methods of handling victims are developed.

a. When evacuating a victim from mountainous areas keep in mind that the purpose of a rescue operation is to save a life, and physical risk to the rescuers must be weighed against this purpose. However, there is no excuse for failing to make the maximum effort within this limitation. Work and expense should be no deterrent when a life is at stake.

b. Rescues will be unplanned (improvised) or planned rescue operations. For a planned rescue, equipment that is especially suited and designed for rescue should be used. For training missions always have a medical plan developed before an emergency arises (plan for the worst and hope for the best). Ensure that the MEDEVAC plan is a comprehensive plan and must be thought out and understood by all that may be involved in a potential rescue.

c. The following actions will be done immediately at the rescue scene.

(1) Assume command. One person, and one person only, is overall in charge at all times.

(2) Prevent further injuries to the victim and to others. Use reasonable care in reaching the victim.

(3) Immediately ensure the victim has an open airway, resume victim's breathing, control serious bleeding, and maintain moderate body warmth. If the victim is unconscious, continually monitor pulse. Protect the patient from environmental hazards.

(4) Do not move the victim until you have ascertained the extent of injuries, unless it is necessary to prevent further injuries or the victim is located in a dangerous location (for example, avalanche run-out zone, hanging glacier, possibility of falling rocks).

(5) Do nothing more until you have thoroughly considered the situation. Resist the urge for action. Speed is less important than correct action.

(6) Decide whether to evacuate with available facilities or to send for help. Speed in getting to a hospital must be balanced against the probability of further injury if working with inexperienced people, lack of equipment or wrong equipment, and terrain at hand.

(7) When the evacuation route is long and arduous, a series of litter relay points or stations should be established. These stations must be staffed with the minimum medical personnel to provide proper emergency treatment. When a victim develops signs of shock or worsens while being evacuated, he should be treated and retained at one of these stations until his condition allows evacuation.

(8) Helicopters or heated vehicles, if available, should be used for evacuation. While the use of aircraft or vehicles is preferred and can expedite a rescue operation, evacuation of a seriously wounded soldier should never be delayed to await aircraft, vehicle, or a change in weather.

11-2. PLANNING RESCUE OPERATIONS

Every commander should have a medical evacuation plan before undertaking an operation. This plan should have contingencies included so as not to rely on a single asset.

a. When rescuing a casualty (victim) threatened by hostile action, environmental hazard, or any other immediate hazard, the rescuer should not take action without first determining the extent of the hazard and his ability to handle the situation. THE RESCUER MUST NOT BECOME A CASUALTY.

b. The rescue team leader must evaluate the situation and analyze the factors involved. This evaluation can be divided into three major steps:

- Identify the task.
- Evaluate the circumstances of the rescue.
- Plan the action.

c. The task must be identified. In planning a rescue, the rescuer tries to obtain the following information:

- Who, what, where, when, why, and how the situation happened.
- Number of casualties by precedence (urgent, priority, routine, tactical immediate),
- number of casualties by type (litter or ambulatory), and the nature of their injuries.
- Terrain features and location of the casualties.
- Tactical situation.
- If adequate assistance is available to aid in security, rescue, treatment, and evacuation.
- If treatment can be provided at the scene; if the victims require movement to a safer location.
- Equipment required for the rescue operation.

d. Circumstances of the rescue are as follows:

(1) After identifying the task, relate it to the circumstances of the situation.

- Are additional personnel, security, medical, or special rescue equipment needed?
- Are there circumstances, such as aircraft accidents (mass casualties), that may require specialized skills?

- What is the weather condition?
- Is the terrain hazardous?
- How much time is available?

(2) The time element may cause a rescuer to compromise planning stages or treatment (beyond first aid). Make a realistic estimate of time available as quickly as possible to determine the action time remaining. The key elements are the casualty's condition and environment.

(3) Mass casualties are to be expected on the modern battlefield. All problems or complexities of rescue are now multiplied by the number of casualties. Time becomes the critical element.

(4) Considerations for the main rescue group for a planned rescue are as follows:

(a) Carry all needed equipment, hot food and drinks, stove, sleeping bags, tents, bivouac sacks, warm clothes, ropes, and stretchers.

(b) Prepare the evacuation route (ground transport to hospital, walking trails, fixed lines, lowering lines, anchor points, and rescue belay points). If the victim is airlifted out, attach a paper with the medical actions that were performed on the ground (for example, blood pressure, pulse rate, drugs started, and so on).

(c) When performing all rescues, the rescuers are always tied in for safety. With all rescue techniques, remember to think things through logically for safety and to prevent the rescuer from accidentally untying himself or the fallen climber.

(d) Constantly inform the casualty (if they are conscious) as to what you are doing and what he must do.

e. The rescue plan should proceed as follows:

(1) In estimating time available, the casualties' ability to endure is of primary importance. Age and physical condition may vary. Time available is a balance of the endurance time of the casualty, the situation, and the personnel and equipment available.

(2) Consider altitude and visibility. Maximum use of secure, reliable trails or roads is essential.

(3) Ensure that blankets and rain gear are available. Even a mild rain can complicate a normally simple rescue. In high altitudes, extreme cold, or gusting winds, available time is drastically reduced.

(4) High altitudes and gusting winds reduce the ability of fixed-wing or rotary-wing aircraft to assist in operations. Rotary-wing aircraft may be available to remove casualties from cliffs or inaccessible sites, and to quickly transport casualties to a medical treatment facility. Relying on aircraft or specialized equipment is a poor substitute for careful planning.

11-3. MASS CASUALTIES

When there are mass casualties, an orderly rescue may involve further planning.

a. To manage a mass casualty rescue or evacuation, separate stages are taken.

- FIRST STAGE: Remove personnel who are not trapped among debris or who can be easily evacuated.
- SECOND STAGE: Remove personnel who may be trapped by debris, but whose extraction only requires the equipment on hand and little time.

- THIRD STAGE: Remove the remaining personnel who are trapped in extremely difficult or time-consuming situations, such as moving large amounts of debris or cutting through a wall.
- FOURTH STAGE: Remove dead personnel.

b. Evacuation of wounded personnel is based on the victim's condition and is prioritized as follows:

- PRIORITY ONE: Personnel with life-threatening injuries that require immediate emergency care to survive; first aid and stabilization are accomplished before evacuation.
- PRIORITY TWO: Personnel with injuries that require medical care but speed of evacuation is not essential.
- PRIORITY THREE: Injured personnel who can evacuate themselves with minimal assistance.
- PRIORITY FOUR: The logistics removal of dead personnel.

11-4. SPECIAL TRAINING

Before receiving training in basic mountain evacuation, litter teams should receive instruction in military mountaineering and basic first aid. Litter bearers and medics must know the use and care of rope as an item of equipment. The members of litter teams must be proficient in the techniques of belaying and choosing belay points. Proper support and protection must be given to victims and litter bearers when evacuating over steep, difficult terrain.

11-5. PREPARATION FOR EVACUATION

Although the wounded soldier's life may have been saved by applying first aid, it can be lost through carelessness, rough handling, or inadequate protection from the elements. Therefore, before trying to move the wounded soldier, the type and extent of his injury must be evaluated. Dressings over wounds must be reinforced, and fractured bones must be properly immobilized and supported. Based upon the evaluation of the type and extent of the soldier's injury, the best method of manual transportation is selected.

11-6. MANUAL CARRIES

Personnel who are not seriously injured but cannot evacuate themselves may be assisted by fellow soldiers. Personnel who are injured and require prompt evacuation should not be forced to wait for mobile evacuation or special equipment.

a. **One-Man Carries.** The basic carries taught in the Soldier's Manual of Common Tasks (fireman's carry, two-hand, four-hand, saddleback, piggyback, pistol belt, and poncho litter) are viable means of transporting injured personnel; however, the mountainous terrain lends itself to several other techniques. One-man carries include the sling-rope carry and the rope coil carry.

(1) ***Sling-Rope Carry.*** The sling-rope carry (Figure 11-1) requires a 4.5-meter sling rope and two men—one as the bearer and the other as an assistant to help secure the casualty to the bearer's back. Conscious or unconscious casualties may be transported this way.

(a) The bearer kneels on all fours.

(b) The assistant places the casualty face down on the bearer's back ensuring the casualty's armpits are even with the bearer's shoulders.

(c) The assistant then finds the middle of the sling rope and places it between the casualty's shoulders.

(d) The assistant runs the ends of the sling rope under the casualty's armpits, crosses the ends, and runs the ends over the bearer's shoulders and back under the bearer's arms.

(e) The assistant runs the ends of the rope between the casualty's legs, around the casualty's thighs, and back around to the front of the bearer. The rope is tied with a square knot with two overhand knots just above the bearer's belt buckle.

(f) The rope must be tight. Padding, when available, should be placed where the rope passes over the bearer's shoulders and under the casualty's thighs.

Figure 11-1. Sling-rope carry.

(2) ***Rope Coil Carry.*** The rope coil carry requires a bearer and a 36 1/2-meter coiled rope. It can be used to transport a conscious or unconscious victim.

(a) Place the casualty on his back.

(b) Separate the loops on one end of the coil, forming two almost equal groups.

(c) Slide one group of loops over the casualty's left leg and the other group over the right leg. The wraps holding the coil should be in the casualty's crotch with the loops on the other end of the coil extending upward toward the armpits.

(d) The bearer lies on his back between the casualty's legs and slides his arms through the loops. He then moves forward until the coil is extended.

(e) Grasping the casualty's arm, the bearer rolls over (toward the casualty's uninjured side), pulling the casualty onto his back.

(f) Holding the casualty's wrists, the bearer carefully stands, using his legs to lift up and keeping his back as straight as possible.

(g) A sling rope around both the casualty and bearer, tied with a joining knot at chest level, aids in keeping an unconscious victim upright. This also prevents the coils from slipping off the carrier's chest.

Note: The length of the coils on the rope coil and the height of the bearer must be considered. If the coils are too long and the bearer is shorter, the rope must be uncoiled and recoiled with smaller coils. If this is not done, the casualty will hang too low on the bearer's back and make it a cumbersome evacuation. A sling-rope harness can be used around the victim's back and bearer's chest, which frees the bearer's hands.

b. **Buddy Rappel.** The carrier can also conduct a seat-hip rappel with a victim secured to his back. In this case, the rappeller faces the cliff and assumes a modified L-shape body position to compensate for the weight of the victim on his back. The victim is top-rope belayed from above, which provides the victim with a point of attachment to a secured rope. The methods for securing a victim to a rappeller's back are described below.

(1) To secure the victim to the carrier's back with a rope, the carrier ties a standard rappel seat (brake hand of choice, depending on the injury) and rests his hands on his knees while the victim straddles his back.

(2) A 4.2-meter sling rope is used. A 45-centimeter tail of the sling is placed on the victim's left hip. (This method describes the procedure for a seat-hip rappel with right-hand brake.)

(3) The remaining long end of the sling rope is routed under the victim's buttocks, and passed over the victim's and carrier's right hip. The rope is run diagonally, from right to left, across the carrier's chest, over his left shoulder, and back under the victim's left armpit.

(4) The rope is then run horizontally, from left to right, across the victim's back. The rope is passed under the victim's right armpit and over the carrier's right shoulder.

(5) The rope is run diagonally, from right to left, across the carrier's chest and back across the carrier's and victim's left hip.

(6) The two rope ends should now meet. The two ends are tied together with a square knot and overhand knots.

(7) The knot is positioned on the victim's left hip. The carrier's shoulders may need to be padded to prevent cutting by the rope.

(8) An alternate method is to use two pistol belts hooked together and draped over the carrier's shoulders. The victim straddles the carrier, and the belay man secures the loose ends of the pistol belts under the victim's buttocks. Slack in the pistol belt sling should be avoided, since the carrier is most comfortable when the victim rests high on his back (see FM 8-35).

(9) A large rucksack can be slit on the sides near the bottom so that the victim can step into it. The victim is belayed from the top with the carrier conducting a standard rappel. The carrier wears the rucksack with the victim inside.

(10) A casualty secured to a carrier, as described above, can be rappelled down a steep cliff using a seat-shoulder or seat-hip rappel. The casualty's and rappeller's shoulders should be padded where the sling rope and rappel lines cross if a seat-shoulder

rappel is used. The buddy team should be belayed from above with a bowline tied around the victim's chest under his armpits. The belay rope must run over the rappeller's guide hand shoulder.

11-7. LITTERS

Many types of litters are available for evacuating casualties in rough mountain terrain. Casualties may be secured to litters in many different ways, depending on the terrain, nature of injuries, and equipment available. **All casualties must be secured.** This should be done under medical supervision after stabilization. It is also important to render psychological support to any victim awaiting evacuation.

If the litter must be carried, belayed, and then carried again, a sling rope should be wound around the litter end and tied off in a l-meter-long loop. This enables the carriers to hook and unhook the litter from the belay. Slings are available to aid the soldiers with litter carrying. Utility rope or webbing 6 meters long may be used. The rope is folded in half, and the loose ends are tied together with an overhand knot. These slings are attached to the litter rails (two or three to a side, depending on the number of litter bearers) by a girth hitch, and then routed up along the handling arm, over the shoulder, behind the neck, and then down along the other arm. The knot can be adjusted to help the outside arm grip the webbing. These slings help distribute the load more evenly, which is important if a great distance must be traveled.

a. **Manufactured Litters.** The following litters are readily available to mountaineering units.

(1) The poleless, nonrigid litter (NSN 6530-00-783-7510) is best issued for company medics since it is lightweight, easy to carry, and readily available. Casualties should be secured with the chest strap and pelvic straps, which are sewn on one side. This litter may be used when rappelling, on traverse lines, and on hauling lines in the vertical or horizontal position. It can be improvised with poles.

(2) The poleless semi-rigid litter (NSN 6530-00-783-7600) may be used the same as the nonrigid litter. It offers more victim protection and back support because of the wooden slats sewn into it.

(3) The mountain basket-type rigid litter (NSN 6530-00-181-7767) is best suited for areas where several casualties are to be transported. All other litters may be placed inside this litter basket and transported across traverse lines. This litter is rectangular and has no vertical leg divider so that it will accommodate other litters. It is also known as a modified Stokes litter.

(4) The Stokes metal litter (NSN 6530-00-042-8131) is suited for situations as above; however, the casualty must be moved in and out of the litter since no other litter will fit inside it. Some Stokes litter frames have a central weld on the frame end, which is a potential breaking point. Winding the rope around the frame end will distribute the force over a wider area and stabilize the system. (See FM 8-10-6 or USAF TO 00-75-5 for additional information on the Stokes litter.)

(5) The standard collapsible litter (NSN 6530-00-783-7905) (rigid pole folding litter) is most readily available in all units and, although heavy and unsuited to forward deployment, may be rigged for movement over rough or mountainous terrain. The folding aluminum litter (NSN 6530-00-783-7205) is a compact version of the pole litter and is better suited for forward deployment.

(6) The UT 2000 is manufactured in Austria and is specifically designed for mountaineering operations. The litter consists of two parts that join together to form a rigid litter. Each part has shoulder and waist straps that can be used to man-pack the litter making it extremely light and portable. When joined together the shoulder and waist straps are used to secure the casualty to the litter. Strapping is also provided to make a secure hoist point for aircraft extraction and high-angle rescues. Wheel sets are another accessory to the UT 2000 litter (either two wheels or one); they attach to the litter for use during a low-angle rescue.

(7) The patient rescue and recovery system (NSN 6530-01-260-1222) provides excellent patient support and protection (Figure 11-2). However, it is not a spinal immobilization device. A backboard must be used with this system for patients who have injuries to the shoulder area. This system will accommodate long and short backboards, scoop stretchers, and most other immobilization equipment.

Figure 11-2. Rescue and recovery system (NSN 6530-01-260-1222).

b. **Field-Expedient Litters.** A litter can be improvised from many different things. Most flat-surface objects of suitable size can be used as litters. Such objects include boards, doors, window shutters, benches, ladders, cots, and poles. Some may need to be tied together to obtain the required size. If possible, these objects should be padded.

(1) Litters can also be made by securing poles inside blankets, ponchos, shelter halves, tarpaulin, jackets, shirts, sacks, bags, or mattress covers. Poles can be improvised from strong branches, tent supports, skis, and other similar items.

(2) If poles cannot be found, a large item, such as a blanket, can be rolled from both sides toward the center. Then the rolls can be used to obtain a firm grip to carry the victim. If a poncho is used, the hood must be up and under the victim, not dragging on the ground.

(3) A rope litter is prepared using one rope (Figure 11-3). It requires 20 to 30 minutes to prepare and should be used only when other materials are not available. Four to six bearers are required to carry the litter. The rope litter is the most commonly used field-expedient litter.

Note: Above the tree line, little material exists to construct litters.

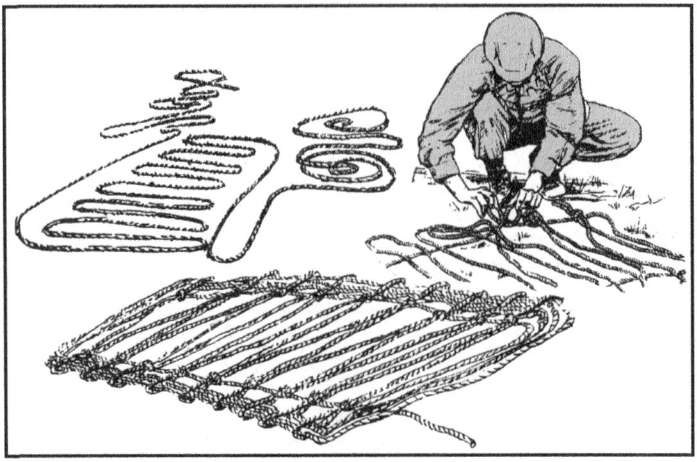

Figure 11-3. Rope litter.

(a) Make 24 bights about 45 to 61 centimeters long, starting in the middle of the rope so that two people can work on the litter at one time.

(b) With the remainder of the rope, make a clove hitch over each bight. Each clove hitch should be about 15 centimeters from the closed end of the bight when the litter is complete.

(c) Pass the remainder of the rope through the bights outside of the clove hitches. Dress the clove hitches down toward the closed end of the bight to secure the litter and tie off the ends of the rope with clove hitches.

(d) Line the litter with padding such as clothing, sleeping bags, empty boxes.

(e) Make the rope litter more stable by making it about 6 inches wider. After placing the clove hitches over the bights, slide them in (away from the closed end) about 15 centimeters. Take two 3- to 4-meter poles, 8 centimeters in diameter at the butt ends, and slide each pole down through the bights on each side. Dress down the clove hitches against the poles. Take two 1-meter poles, and tie them off across the head and foot of the litter with the remaining tails of the climbing rope.

Note: The above measurements may have to be altered to suit the overall length of rope available.

11-8. RESCUE SYSTEMS

Rescue systems are indispensable when conducting rescue operations. A large number of soldiers will not always be available to help with a rescue. Using a mechanical advantage rescue system allows a minimal amount of rescuers to perform tasks that would take a larger number of people without it.

a. **Belay Assist.** This system is used to bring a climber over a section that he is unable to climb, but will continue climbing once he is past the difficult section.

(1) First, tie off the following climber at the belay with a mule knot.

(2) Tie a Prusik knot with short Prusik cord about 12 inches below the mule knot, and place a carabiner into the loop. Place the tail from the mule knot into the carabiner in the Prusik cord.

(3) Untie the mule knot without letting the following climber descend any more than necessary. Do not remove the belay.

(4) Maintain control of the brake side of the rope and pull all of the slack through the carabiner in the Prusik cord.

(5) Pull up on the rope. The rope will automatically feed through the belay.

(6) If the leader has to pull for a distance he can tie another mule knot at the belay to secure the second climber before sliding the Prusik down to get more pulling distance.

(7) After the climber can continue climbing, the leader secures the belay with a mule knot.

(8) Remove the Prusik cord and carabiner, then untie the mule knot and continue belaying normally.

Notes: With all rescue techniques make sure that you always think everything through, and double check all systems to ensure that you don't accidentally untie the fallen climber or find yourself without back-up safety. Do not compound the problem! When doing any rescue work the rescuers will always be tied in for safety.

b. **Belay Escape.** The belay escape is used when a climber has taken a serious fall and cannot continue. The belayer is anchored and is performing an indirect belay, and must assist the injured climber or go for assistance. To accomplish this he must escape the belay system. The belayer will remain secured to an anchor at all times.

(1) After a climber has been injured, tie off the belay device on your body using a mule knot. To improve this system, clip a nonlocking carabiner through the loop in the overhand knot and clip it over the rope.

(2) Attach a short Prusik cord to the load rope and secure it to the anchor with a releasable knot.

(3) Using a guard knot or Munter mule, attach the climbing rope from the belay device.

(4) Untie the mule knot in the belay device attached to the harness and slowly lower the climber, transferring the weight of the climber onto the Prusik.

(5) Remove the climbing rope from the belay device attached to the harness.

(6) Release the mule knot securing the Prusik, transferring the weight to the anchor.

(7) At this point the climber is secured by the rope to the anchor system and the belayer can now assist the injured climber.

11-9. LOW-ANGLE EVACUATION

Cliffs and ridges, which must be surmounted, are often encountered along the evacuation path. Raising operations place a greater load on all elements of the system than do lowering operations. Since all means of raising a victim (pulley systems, hand winches, and power winches) depend on mechanical advantage, it becomes easy to overstress and break anchors and hand ropes. Using mechanical raising systems tends to reduce the soldier's sensitivity to the size of the load. It becomes important to monitor the system and to understand the forces involved.

a. **Raising Systems, Belays, and Backup Safeties.** Raising systems, belays, and backup safeties are of special importance in any raising operation. The primary raising system used is the Z-pulley system, which theoretically gives three pounds of lift for each pound of force expended. In practice, these numbers decrease due to rope-pulley friction, rope-edge friction, and other variables. A separate belay rope is attached to the litter and belayed from a separate anchor. Backup Prusik safeties should be installed in case any part of the pulley system fails.

(1) ***Raising System***. When considering a raising system for evacuations, the Z-pulley system is the most adaptable. It can be rigged with the equipment on hand, and can be modified and augmented to handle heavier loads. Although the vertical or horizontal hauling lines can also be used, the Z-pulley system offers a mechanical advantage that requires less exertion by the transport team.

(2) ***Belays***. Whenever ropes are used for an evacuation operation, the overriding safety concern is damage to the ropes. This is the main reason for two-rope raising systems (raising rope and belay rope).

(3) ***Backup Safeties.*** Because the stresses generated by the Z-pulley system can cause anchors to fail, backup safety Prusiks are used to safeguard the system. These should be attached to alternate anchor points, if possible.

b. **Raising the Litter.** The litter is prepared as already described.

(1) The raising ropes and belay ropes are secured to top anchors and are thrown down to the litter crew.

(2) Padding is placed at the cliff edge and over any protrusions to protect the ropes from abrasion.

(3) The litter attendants secure the ropes to the litter.

(4) The raising crew sets up the Z-pulley system.

(5) One member of the crew secures himself to an anchor and moves to the edge of the cliff to transmit signals and directions. (This is the signalman or relay man.)

Note: If the load is too heavy at this time, another pulley is added to the system to increase the mechanical advantage.

(6) Attendants guide the litter around obstacles while the crew continues to raise the system.

(7) As the litter nears the cliff edge, the signalman assists the attendant in moving the litter over the edge and onto the loading platform, taking care not to jar the casualty.

c. **Descending Slopes.** When descending a moderately steep slope that can be down-climbed, the litter and victim are prepared as described earlier (Figure 11-4, page 11-12).

(1) One man serves as the belay man and another takes his position on the rope in front of the belay man, assisting him in lowering the litter. The litter bearers take their positions and move the litter down with the speed of descent controlled by the belay man.

Figure 11-4. Low-angle evacuation—descending.

(2) The extra man may assist with the litter or precede the team to select a trail, clearing away shrubs and vines. He reconnoiters so that the team need not retrace its steps if a cliff is encountered.

(3) The most direct, practical passage should be taken utilizing available natural anchors as belay positions.

11-10. HIGH-ANGLE EVACUATION

Evacuation down cliffs should be used only when absolutely necessary and only by experienced personnel. The cliffs with the smoothest faces are chosen for the route. Site selection should have the following features: suitable anchor points, good loading and unloading platforms, clearance for the casualty along the route, and anchor points for the A-frame, if used. There are many ways to lower a casualty down a steep slope. As long as safety principals are followed, many different techniques can be used. One of the easiest and safest techniques is as follows (Figure 11-5):

a. Use multiple anchors for the litter and litter tenders.

b. Secure the litter to the lowering rope with a minimum of four tie-in points (one at each corner of the litter). Lengths of sling rope or 7-millimeter cordage work best. Make the attached ropes adjustable with Prussik knots so that each corner of the litter can be raised or lowered to keep the litter stable during descent. Tie the top of the ropes with loops and attach to the lowering rope with a pear shaped locking carabiner.

c. Two litter tenders will descend with the litter to control the descent and to monitor the casualty. They can be attached to separate anchors and either self-belay themselves or be lowered by belayers.

d. Once the steep slope has been negotiated, continue the rescue with a low-angle evacuation.

Figure 11-5. Cliff evacuation descent.

APPENDIX A

LEVELS OF MILITARY MOUNTAINEERING

Military mountaineering training provides tactical mobility in mountainous terrain that would otherwise be inaccessible. Soldiers with specialized training who are skilled in using mountain climbing equipment and techniques can overcome the difficulties of obstructing terrain. Highly motivated soldiers who are in superior physical condition should be selected for move advanced military mountaineering training (Levels 2 and 3) conducted at appropriate facilities. Soldiers who have completed advanced mountaineering training should be used as trainers, guides, and lead climbers during collective training. They may also serve as supervisors of installation teams and evacuation teams. Properly used these soldiers can drastically improve mobility and have a positive impact disproportionate to their numbers. Units anticipating mountain operations should strive to have approximately ten percent of their force achieve advanced mountaineering skills.

A-1. LEVEL 1: BASIC MOUNTAINEER

The basic mountaineer should be a graduate of a basic mountaineering course and have the fundamental travel and climbing skills necessary to move safely and efficiently in mountainous terrain. These soldiers should be comfortable functioning in this environment and, under the supervision of qualified mountain leaders or assault climbers, can assist in the rigging and use of all basic rope installations.

a. On technically difficult terrain, the basic mountaineer should be capable of performing duties as the "follower" or "second" on a roped climbing team, and should be well trained in using all basic rope systems. These soldiers may provide limited assistance to soldiers unskilled in mountaineering techniques.

b. Particularly adept soldiers may be selected as members of special purpose teams led and supervised by advanced mountaineers. Figure A-1 lists the minimum knowledge and skills required of basic mountaineers.

c. In a unit training program, Level 1 qualified soldiers should be identified and prepared to serve as assistant instructors to train unqualified soldiers in basic mountaineering skills. All high-risk training, however, must be conducted under the supervision of qualified Level 2 or 3 personnel.

• Characteristics of the mountain environment (summer and winter). • Mountaineering safety. • Use, care, and packing of individual cold weather clothing and equipment. • Care and use of basic mountaineering equipment. • Mountain bivouac techniques. • Mountain communications. • Mountain travel and walking techniques. • Hazard recognition and route selection. • Mountain navigation. • Basic medical evacuation.	• Rope management and knots. • Natural anchors. • Familiarization with artificial anchors. • Belay and rappel techniques. • Use of fixed ropes (lines). • Rock climbing fundamentals. • Rope bridges and lowering systems. • Individual movement on snow and ice. • Mountain stream crossings (to include water survival techniques). • First aid for mountain illnesses and injuries.

Figure A-1. Level 1, basic mountaineering tasks.

A-2. LEVEL 2: ASSAULT CLIMBER

Assault climbers are responsible for the rigging, inspection, use, and operation of all basic rope systems. They are trained in additional rope management skills, knot tying, and belay and rappel techniques, as well as using specialized mountaineering equipment. Assault climbers are capable of rigging complex, multipoint anchors, and high-angle raising/lowering systems. Level 2 qualification is required to supervise all high-risk training associated with Level 1. At a minimum, assault climbers should possess the additional knowledge and skills shown in Figure A-2.

• Use specialized mountaineering equipment.
• Perform multipitch climbing:
  – Free climbing and aid climbing.
  – Leading on class 4 and 5 terrain.
• Conduct multipitch rappelling.
• Establish and operate hauling systems.
• Establish fixed ropes with intermediate anchors.
	• Move on moderate angle snow and ice.
	• Establish evacuation systems and perform high-angle rescue.
	• Perform avalanche hazard evaluation and rescue techniques.
	• Be familiar with movement on glaciers.

Figure A-2. Level 2, assault climber tasks.

A-3. LEVEL 3: MOUNTAIN LEADER

Mountain leaders possess all the skills of the assault climber and have extensive practical experience in a variety of mountain environments in both winter and summer conditions. Level 3 mountaineers should have well-developed hazard evaluation and safe route finding skills over all types of mountainous terrain. Mountain leaders are best qualified to advise commanders on all aspects of mountain operations, particularly the preparation and leadership required to move units over technically difficult, hazardous, or exposed terrain. The mountain leader is the highest level of qualification and is the principle trainer for conducting mountain operations. Instructor experience at a military mountaineering center or as a member of a special operations forces (SOF) mountain team is critical to acquiring Level 3 qualification. Figure A-3 outlines the additional knowledge and skills expected of mountain leaders. Depending on the specific AO, mountain leaders may need additional skills such as snowshoeing and all-terrain skiing.

• Recognizing and evaluating peculiar terrain, weather, and hazards. • Preparing route, movement, bivouac, and risk management plans for all conditions and elevation levels. • Using roped movement techniques on steep snow and ice. • Performing multipitch climbing on mixed terrain (rock, snow, and ice).	• Performing glacier travel and crevice rescue. • Establishing and operating technical high-angle, multipitch rescue and evacuation systems. • Usiing winter shelters and survival techniques. • Leading units over technically difficult, hazardous, or exposed terrain in both winter and summer conditions.

Figure A-3. Level 3, mountaineer leader tasks.

APPENDIX B
MEASUREMENT CONVERSION FACTORS

MULTIPLY	BY	TO OBTAIN
Millimeters	.03937	Inches
Centimeters	.3937	Inches
Centimeters	.03281	Feet
Meters	39.37	Inches
Meters	3.281	Feet
Meters	1.0936	Yards
Kilometers	.62137	Miles
Knots	1.1516	MPH

APPENDIX C

AVALANCHE SEARCH AND RESCUE TECHNIQUES

The effect of an avalanche can be disastrous. Chances of survival after burial by an avalanche are approximately 90 percent if the victim is located within the first 15 minutes. Probability of survival drops rapidly and, after two hours, chances of survival are remote. Suffocation accounts for 65 percent of avalanche fatalities, collision with obstacles such as rocks and trees accounts for 25 percent, and hypothermia and shock accounts for 10 percent.

The best chance of survival in snow country is to avoid an avalanche; but, if a member of your group ***is*** *in an avalanche, they are depending on you for rescue!*

C-1. IMMEDIATE ACTION

Survivors at the avalanche site organize into the first rescue team and immediately start rescue operations. If any indication of the location of the victim is found, random probing starts in that vicinity. The tip and edges of the slide are also likely areas to search. A human body is bulky and is apt to be thrown toward the surface or the sides.

C-2. GENERAL PROCEDURES

Establish from witnesses where the victim was located just before the avalanche to determine the point where the victim disappeared—the "last seen" point. Using this and any other information, establish a probable victim trajectory line leading to high priority search areas. Make a quick but systematic check of the slide area and the deposition area, and mark all clues. Look for skis, poles, ice axes, packs, gloves, hats, goggles, boots, or any other article the person may have been carrying—it might still be attached to the victim.

a. Organize initial searchers and probers. If using avalanche beacons, immediately select personnel to begin a beacon search. Ensure all other beacons are turned off or to receive to eliminate erroneous signals. All personnel should have a shovel or other tool for digging or, if enough personnel are available, a digger can be standing by to assist when needed. If the initial search reveals items from the victim, make an initial probe search in that area. This probing should take only a few seconds.

b. Make a coarse probe of all likely areas of burial, and repeat it as long as a live rescue remains possible. Resort to the fine probe only when the possibility of a live rescue is highly improbable. Unless otherwise indicated, start the coarse probe at the deposition area.

C-3. ESTABLISHING THE VICTIM'S MOST PROBABLE LOCATION

In many respects, a moving avalanche resembles a liquid. A human body, with a higher density than the flowing snow, would be expected to sink deeper and deeper into the avalanche; however, several factors influence the body's location. Turbulence, terrain, and the victim's own efforts to extricate himself all interact to determine the final burial position. Study of a large number of case histories leads to the following conclusions.

- The majority of buried victims are carried to the place of greatest deposition, usually the toe of the slide.
- If two points of the victim's trajectory can be established, a high probability exists that the victim will be near the downhill flow line passing through these two points.
- Any terrain features that catch and hold avalanche debris are also apt to catch a victim.
- If an avalanche follows a wandering gully, all debris deposit areas are likely burial spots. The likelihood of a victim being buried in a particular bend is proportional to the amount of debris deposited there.
- Vegetation, rocks, and other obstacles act as snares. The victim tends to be retained above the obstacle. An obstacle may simply delay the victim's motion, leading to final burial down flow from the obstacle.
- Maximum speed of the flowing snow occurs at the avalanche center. Friction reduces flow velocity along the edges. The closer the victim's trajectory is to the center of the slide, the greater will be his burial depth.
- Efforts of the victim to extricate himself by vigorous motion and "swimming" definitely minimize burial depth. Conversely, the limp body of an unconscious victim is likely to be buried deeply.
- An occasional exception to the above is emphasized. The victim may not be buried but may have been hurled away from the avalanche by wind blast. In the case of large and violent avalanches, a search of the surrounding terrain is advisable. Victims have been located in tree tops outside the slide area.

Use of avalanche transceivers is the most efficient method of searching for an avalanche victim, but only if the victim is wearing an active transceiver. Many models of transceivers are available, each with its own manufacturer's instructions for proper use and care. All currently available transceivers are compatible, although they may operate differently in the search mode.

C-4. PROBING FOR AVALANCHE VICTIMS

Probing offers the advantage of requiring simple equipment that can be operated by personnel with no previous training. Although the probers do not need previous training the search leader must be familiar with the technique to ensure proper execution of the probe line.

a. **Probe Poles.** Rigid steel tubing approximately 3/4-inch in diameter and approximately 10 feet long is recommended for the primary probe pole. Longer poles are difficult to manage, especially in a high wind. Although this type of pole performs best, it is difficult to transport to the avalanche site because of its length and weight.

(1) Each person operating in avalanche areas should carry folding sectional poles. These poles are similar to folding tent poles, but are stronger and are connected with cable instead of bungee cord. These poles should be carried on the outside of the pack for immediate access.

(2) If no probing poles are available, initial probing attempts can be started using ski poles in one of two ways: the ski pole can be reversed, probing with the wrist strap down; or the basket can be removed so that the point is down (the preferred method), which allows the ski pole to penetrate the snow more easily.

b. **Probing Lines.** For the probing operation to be effective, probing lines must be orderly and properly spaced. To ensure systematic and orderly probing, the number of personnel per line should be limited. Twenty per line is satisfactory, while thirty is normally the upper limit. The number of probers in the line will be dictated by not only the width of the area to be probed but the number of personnel available. A string may be used to keep the probe lines aligned, but will require added time to maintain.

(1) The probe line maintains a steady advance upslope. Advancing uphill automatically helps set the pace and permits easy probing to the full length of the probe. Probing does not come to a halt when a possible contact is made. The probe is left in contact and the line continues. A shovel crew follows up on the strike by digging down along the pole. Extra probes are carried by the shovel crew to replace those left in contact. Such a plan of operation is especially important when more than one victim is buried.

(2) Striking a body gives a distinct feel to the probe, which is easily recognizable in soft snow but less recognizable in hard compacted snow. A common problem is encountering debris within the snow that can be mistaken for the victim. The only sure check is by digging.

c. **Probing Techniques.** Two distinct probing methods are recognized: coarse probe and fine probe. As evidenced by their names, coarse probing implies a wider spacing of probe pole insertions with emphasis on speed. Fine probing involves close-spaced probing with emphasis on thoroughness. Coarse probing is used during initial phases of the search when live recovery is anticipated. Fine probing is the concluding measure, which almost guarantees finding the body. The coarse probe technique has a 70-percent chance of locating the victim on a given pass, while the fine probe has, essentially, a 100-percent chance of locating the body.

(1) The coarse probe functions as follows:

(a) Probers are spaced along a line 30 inches center to center, with feet about 15 inches apart.

(b) A single probe pole insertion is made at the center of the straddle span.

(c) On command of the probe line commander, the group advances 20 inches and repeats the single probe.

(d) Three commands are used for the complete sequence:

- "DOWN PROBE."
- "UP PROBE."
- "STEP FORWARD."

By using these commands, the leader can maintain closer control of the advancing probe line. It is important that the commands be adjusted to a rhythm that enforces the maximum reasonable pace. A string may also be used along the probe line to keep the probers dressed, although this requires the use of two soldiers to control the string. Strict discipline and firm, clear commands are essential for efficient probing. The probers themselves work silently.

(2) The fine probe functions as follows:

(a) Probers are spaced the same as for the coarse probe. Each man probes in front of his left foot, then in the center of his straddled position, and finally in front of his right foot.

(b) On command, the line advances 1 foot and repeats the probing sequence. Each probe is made 10 inches from the adjacent one.

(c) The commands for the fine probe are:

- "LEFT PROBE."
- "UP PROBE."
- "CENTER PROBE."
- "UP PROBE."
- "RIGHT PROBE."
- "UP PROBE."
- "STEP FORWARD."

(d) Good discipline and coordinated probing is even more important in fine probing than with the coarse probe. Careless or irregular probing can negate the advantages of fine probing. Use of a string to align the probers is especially important with the fine probe. The three insertions are made along the line established by the string, which is then moved ahead 1 foot.

GLOSSARY

AI	alpine ice
ALICE	all-purpose, lightweight, individual carrying equipment
AMS	acute mountain sickness
APS	anchor pulley system
ATC	air traffic controller
BDU	battle dress uniform
BTC	bridge team commander
C	Celsius
CE	community Europe
CEN	combined European norm
cm	centimeter
CMS	chronic mountain sickness
CPR	cardio-pulmonary resuscitation
CTA	common table of allowance
DA	Department of the Army
ECWCS	extreme cold weather clothing system
EN	European norm
F	Fahrenheit
FM	field manual
ft	foot; feet
GPS	Global Positioning System
GTA	graphic training aid
HACE	high-altitude cerebral edema
HAPE	high-altitude pulmonary edema
IAW	in accordance with
LBV	load-bearing vest
LCD	liquid crystal diode
LCE	load-carrying equipment
MEDEVAC	medical evacuation
MPH	miles per hour
MPS	moveable pulley system
MRE	meal ready-to-eat

NSN	national stock number
OP	observation post
PABA	para-amino benzoic acid
PCD	progress capture device
PFD	personal flotation device
RURP	realized ultimate reality piton
SE	southeast
SLCD	spring-loaded camming device
SOF	special operations forces
SOP	standard operating procedure
SOSES	shape, orientation, size, elevation, slope
SPF	sun protection factor
TOE	table of organization and equipment
UIAA	Union des International Alpine Association
U.S.	United States
USAF	United States Air Force
USN	United States Navy
UV	ultraviolet
UVA	ultraviolet A (radiation wavelengths between 320 and 400 nanometers)
UVB	ultraviolet B (radiation wavelengths between 295 and 320 nanometers)
WI	waterfall ice
YDS	Yosemite Decimal System

INDEX

ROPE LOG *(USAGE AND HISTORY)*

For use of this form, seeFM 3-97.61; the proponent agency is TRADOC

			UNIT ID MARKING
NSN	DOCUMENT NUMBER	SERIAL NUMBER	MFR LOT NUMBER
DATE OF MFR	ISSUE DATE	DATE IN SERVICE	LENGTH
DIAMETER	FIBER	COLOR	CONSTRUCTION

INSPECT ROPE FOR DAMAGE OR EXCESSIVE WEAR EACH TIME IT IS DEPLOYED AND AGAIN AFTER EACH USE. IMMEDIATELY RETIRE ALL SUSPECT ROPES.

DATE USED	LOCATION	TYPE OF USE	ROPE EXPOSURE	INSPECTOR'S INITIAL/DATE	ROPE CONDITION AND COMMENTS

DA FORM 5752-R, MAY 89

DATE USED	LOCATION	TYPE OF USE	ROPE EXPOSURE	INSPECTOR'S INITIAL/DATE	ROPE CONDITION AND COMMENTS

MANUFACTURER ______________________

____

REVERSE OF DA FORM 5752-R, MAY 89